A VOYAGE OF
DISCOVERY

SIR GEORGE DEACON, F.R.S.

A VOYAGE OF DISCOVERY

edited by

MARTIN ANGEL

Institute of Oceanographic Sciences,
Wormley Godalming, Surrey

PERGAMON PRESS

OXFORD · NEW YORK · TORONTO · SYDNEY · PARIS · FRANKFURT

U.K.	Pergamon Press Ltd., Headington Hill Hall, Oxford OX3 0BW, England
U.S.A.	Pergamon Press Inc., Maxwell House, Fairview Park, Elmsford, New York 10523, U.S.A.
CANADA	Pergamon of Canada Ltd., 75 The East Mall, Toronto, Ontario, Canada
AUSTRALIA	Pergamon Press (Aust.) Pty. Ltd., 19a Boundary Street, Rushcutters Bay, N.S.W. 2011, Australia
FRANCE	Pergamon Press SARL, 24 rue des Ecoles, 75240 Paris, Cedex 05, France
WEST GERMANY	Pergamon Press GmbH, 6242 Kronberg-Taunus, Pferdstrasse 1, West Germany

First edition 1977

Library of Congress Cataloging in Publication Data

Main entry under title:

George Deacon: 70th anniversary volume.

Bibliography: p.
1. Oceanography—Addresses, essays, lectures.
2. Marine biology—Addresses, essays, lectures.
3. Earth sciences—Addresses, essays, lectures.
4. Deacon, George Edward Raven, Sir, 1906-
GC26.G46   1977   551.4'6   76-57958
ISBN 0-08-021380-4

Printed in Great Britain by Thomson Litho Ltd, East Kilbride, Scotland

Preface

Sir George Deacon's 70th birthday was on 21 March 1976. This anniversary came as so much of a surprise to many of us that no one had given a thought as to how to celebrate his three score and ten years. Sir George is the doyen of British oceanography, having nursed the science in this country through the years of its development on an international scale. A volume to celebrate his birthday seemed the most appropriate way of paying tribute to him. The 'choice' of editor left me feeling a little overwhelmed since it might be said that "he doth bestride the narrow world like a Colossus and we petty men walk under his huge legs and peep about…" (Shakespeare, *Julius Caesar*). Oceanography is both multidisciplinary and international, so it takes men of exceptional calibre to make an effective impact on it.

Sir George began his oceanography on the early Discovery Investigations Expeditions to the Antarctic. He participated in the second commission in R.R.S. *William Scoresby* in 1927–28 and three of the first four commissions in R.R.S. *Discovery II* 1929–31, 1931–33 and 1935–37. On the 1935–37 commission he was principal scientist. These commissions were not the relatively short cruises of today, but lengthy marathons. Sir Alister Hardy's book *Great Waters* gives a sense of the spirit of these early expeditions; any participant must have developed an almost innate feeling from such extensive practical experience of the processes at work in the sea. The aims of these early Discovery Investigations Expeditions were to establish the baselines for the Antarctic whaling industry; this was over 40 years before 'baseline' became acceptable in ecological jargon. Sir George's careful chemical analyses laid the foundation of our understanding of the oceanic water masses and current systems in the Southern Ocean. It is salutary to see that modern oceanography with all its sophisticated technology is mainly filling in the details within the broad patterns derived from the early unsophisticated but well-planned observations.

From 1939 to 1944, Sir George was involved with anti-submarine warfare initially with H.M.S. *Osprey* and then with His Majesty's Anti-submarine Experimental Establishment at Fairlie. In 1944 he took over Group W at the Admiralty Research laboratory which investigated waves breaking on beaches as a vital prelude to what was to be expected during a potentially long island-hopping war in the Pacific. He remained at A.R.L. until 1949.

During the war the British Government decided to set up a multi-discipline oceanographic institute combining physical and biological sciences. The National Institute of Oceanography was created in 1949 with its main core formed by combining the A.R.L. Group W and the biologists of Discovery Investigations, under Sir George's direction. From its inception it included designers and engineers who translated the scientist's requirements into practical functional equipment. The new N.I.O. soon began to grow in size, scope and reputation. It was run as a team with the Director maintaining personal contact with every one of the staff. As the institute grew in size he never lost contact with his staff, even as the burden of committee work and administration took up more and more of his time. Here is a list of a few of his appointments, none of them sinecures, that he held during his time as Director, as an indication of the immense work load he carried:

Vice-President of the Royal Geographical Society (1965–70), President of the Institute of Navigation (1961–64), President of the International Association for Physical Oceanography (1960–63), a founder member and a Vice-President of the Scientific Committee on Oceanic Research (1958–64), a member of the Council of the Marine Biological Association of the United Kingdom (1946–73) and the NATO Subcommittee on Oceanographic Research. He is still a member of the Scientific Committee on Antarctic Research, Chairman of the National Committee for Oceanic Research, on the National Committee on Ocean Engineering, the National Committee for the History of Science, Medicine and Technology, and the Committee of the Scott Polar Research Institute.

His contributions both as a first-class scientist and as a scientific administrator have been recognized by medals, honorary doctorates and fellowships from universities and learned societies from all over the world. It was a great delight to us, when he was awarded a knighthood in 1971 just prior to his retirement as Director of the N.I.O.

After such a full career many men on retirement would relax and grow roses. To Sir George, retirement was the longed for opportunity to shed administrative work and to return to the unresolved scientific problems of Southern Ocean hydrography. He has plunged back into the science with a vigour that both belies his years and seems lacking in many younger men. He has returned to the Antarctic and is beginning to produce a fresh flow of original papers. He found the time and energy to organize and run a seven-week course at Trieste on physical oceanography.

Sir George is an unassuming man. Success and reputation have not come to him through his striving after fame, but from his unswerving pursuit of science. His philosophy has always been to encourage and guide rather than dictate. This volume is the result of the desire of many to acknowledge the debt that both they and oceanography owe to him. So many people were bitterly disappointed not to have anything suitable to contribute within the deadlines set for publication, including many of Sir George's old colleagues and shipmates from the early *Discovery II* days. Many have contributed anonymously as referees or in helping and advising the editor in other ways. Even the immediate willingness of the publishers to produce such a volume acknowledges Sir George's long-continued help and guidance with *Deep-Sea Research*. The scope of the papers from history, physical oceanography, and biology to engineering and geophysics reflects both the breadth of the science and of Sir George's own interests. Any Festschrift is inevitably uneven in the standard of its contents, but let no would-be critic of this one fail to recognize that it is the product of our gratitude both to the scientist and to the man. With it comes the wish that Sir George will be helping, guiding and leading us for many more years to come.

M. V. ANGEL

Contents

Contents

Selected Bibliography of Papers

G. E. R. DEACON

(compiled by D. W. PRIVETT)

1927 Studies of equilibria in the systems sodium chloride–lead chloride–water, lithium chloride–lead chloride–water. *Journal of the Chemical Society*, 1927, 2063–5.

1931 Velocity of deep currents in the South Atlantic. *Nature, London* (3224), **128**, 267.

1933 An examination of 'eau der mer normale P 13, 28.6.1929' (with an appendix by D. J. Matthews). *Journal du Conseil*, **8** (1), 59–63.

1933 A general account of the hydrology of the South Atlantic Ocean. *'Discovery' Reports*, **7**, 171–238.

1934 Die Nordgrenzen antarktischen und subantarktischen Wassers im Weltmeer. *Annalen der Hydrographie u maritime Meteorologie*, **62**, 129–136.

1934 Nochmals: Wie entsteht die Antarktische Konvergenz? *Annalen der Hydrographie u maritime Meteorologie*, **62**, 475–478.

1935 [with CLOWES, A.J.] The deep water circulation of the Indian Ocean. *Nature, London*, **136**, 936–938.

1937 The hydrology of the Southern Ocean. *'Discovery' Reports*, **15**, 1–124.

1937 Note on the dynamics of the Southern Ocean. *'Discovery' Reports*, **15**, 125–152.

1938 The work of the R.R.S. 'Discovery II' (1935–1937). *Cape Naturalist*, **1**(5), 163–165.

1939 The Antarctic voyages of R.R.S. 'Discovery II' and R.R.S. 'William Scoresby' 1935–37. *Geographical Journal*, **93**(3), 185–209.

1940 Carbon dioxide in Arctic and Antarctic seas. *Nature, London*, **145**, 250–252.

1941 New work on the Gulf Stream. *Geographical Journal*, **98**(2), 84–90.

1942 The Sargasso Sea. *Geographical Journal*, **99**(1), 16–28.

1945 Water circulation and surface boundaries in the oceans. [Symons Memorial Lecture.] *Quarterly Journal of the Royal Meteorological Society*, **71**, 11–25.

1947 Relations between sea waves and microseisms. *Nature, London*, **160**, 419–421.

1949 Waves and swell. [Symons Memorial Lecture.] *Quarterly Journal of the Royal Meteorological Society*, **75**, 227–238.

1949 Recent studies of waves and swell. *Annals of the New York Academy of Sciences*, **51**(3), 475–482.

1951 The National Institute of Oceanography. *Polar Record*, **6**, 88–90.

1951 Applications of oceanographical research to navigation. *Journal of the Institute of Navigation*, **4**, 276–287.

1952 Progress in oceanographical research. *Journal of the Institute of Navigation*, **5**, 168–173.

1952 The 'Manihine' expedition to the Gulf of Aqaba, 1948–1949, II. Preliminary hydrological report. *Bulletin of the British Museum (Natural History)*, Zoology, **1**(8), 159–162.

1952 Analysis of sea waves. Pp. 209–214 in *Gravity waves* [Proceedings of N.B.S. Symposium on Gravity Waves, held June 1951]. Washington, D.C.: Government Printing Office [National Bureau of Standards, Circular 521].

1954 The National Institute of Oceanography. *Journal of the Institute of Navigation*, **7**(3), 252–261.

1954 Exploration of the deep sea. *Journal of the Institute of Navigation*, **7**(2), 165–174.

1955 Information from electric currents in the sea. *Journal of the Institute of Navigation*, **8**(2), 117–120.

1955 The Discovery Investigations in the Southern Ocean. *Transactions of the American Geophysical Union*, **36**(5), 877–880.

1955 Applications of oceanography to navigation. *Journal of Marine Research*, **14**(4), 333–336.

1955 [with SVERDRUP, H. U., *et al.*] Discussions on the relationships between meteorology and oceanography. *Journal of Marine Research*, **14**(4), 499–501.

1956 Marine physics. [James Forrest Lecture.] *Proceedings of the Institution of Civil Engineers*, **5**, 661–676.

1957 Origin and effects of long period waves in ports. Pt. I. Long waves. *19th International Navigation Congress, London 1957*. Paper SII–C1, pp. 1–4.

1957 Recent advances in science: physical oceanography. *Science Progress*, **45**(177), 75–86.

1957 International cooperation in marine research. *Nature, London*, **180**(4592), 894–895.

1957 Marine research—the work of the National Institute of Oceanography. *Proceedings of the Royal Society of Edinburgh*, A, **44**(23), 350–368.

1958   The use of oceanography. *Impact of Science on Society*, **9**, 79–92.
1959   Recent approaches to problems in marine research. Pp. 19–27 in *Journées des 24 et 25 Février 1958*. Bruxelles: Centre Belge d'Océanographie et de recherches sousmarines, 222 pp.
1959   The Antarctic Ocean. *Science Progress*, **47**, 647–660.
1960   International cooperation in marine science. *Science Progress*, **48**(192), 667–672.
1960   The southern cold temperate zone. *Proceedings of the Royal Society*, B, **152**(949), 441–446.
1962   Navigation and the science of the sea. [2nd Duke of Edinburgh's Lecture.] *Journal of the Institute of Navigation*, **15**, 1–13.
1962   The development and present status of oceanography as a scientific discipline. *I.C.S.U. Review*, **4**(2), 82–91.
1962   *Oceans*, [Editor] London: Paul Hamlyn, 297 pp.
1963   The Southern Ocean. Pp. 281–296 in *The Sea*, edited by M. N. HILL, Vol. 2. London: Interscience Publishers.
1963   Present status of wave research. Pp. 3–6 of *Ocean wave spectra*. New Jersey: Prentice-Hall Inc. 357 pp. (Paper presented at Conference at Easton Maryland, 1–4 May 1961.)
1963   The value of research to navigation. [Presidential Address.] *Journal of the Institute of Navigation*, **16**, 1–8.
1964   Sea current measurements. *The Royal Society I.G.Y. Expedition, Halley Bay, 1955–59*, **4**, 348–352.
1964   Review of recent advances in physical oceanography. *Transactions of the Institute of Naval Architects, London*, **106**, 27–38.
1964   The Southern Ocean. Pp. 292–307 of *Antarctic research*, edited by R. PRIESTLEY, *et al*. London: Butterworths.
1964   Antarctic oceanography: the physical environment. Pp. 81–86 of *Antarctic biology*, edited by R. CARRICK, *et al*. Paris: Hermann. [Proceedings SCAR Symposium, Paris, 2–8 Sept. 1962.]
1965   The Southern Ocean and the Convergence. *Anais de Academia Brasileira de Ciencias*, **37**, Supplemento, 23–29.
1965   Waves and ships. [Presidential address.] *Journal of the Institute of Navigation*, **18**, 1–9.
1966   Hans Pettersson, 1888–1966. Elected For. Mem. R.S. 1956. *Biographical Memoirs of Fellows of the Royal Society*, **12**, 405–421.
1968   Progress in oceanographic research and technology. [The 40th Thomas Lowe Gray Lecture.] *Proceedings of the Institution of Mechanical Engineers*, **182**, Part I, 846–857.
1968   Early scientific studies of the Antarctic Ocean. Pp. 269–279 in *Premier Congrès International d'Histoire de l'Oceanographie, Monaco, 1966*. Monaco: Musée Océanographique. [Bulletin de l'Institut Océanographique, No. spécial 2.]
1969   Oceanography and navigation. *Journal of the Institute of Navigation*, **22**(1), 77–91.
1969   [with DEACON M.] Captain Cook as a navigator. *Notes and Records of the Royal Society of London*, **24**(1), 33–42.
1971   [*et al*.] A discussion on ocean currents and their dynamics. *Philosophical Transactions of the Royal Society*, A, **270**(1206), 349–465.
1971   Problems of the pack ice zone. Pp. 96–98 in *Symposium on Antarctic ice and water masses* [held Tokyo, 19 September 1970], edited by G. E. R. DEACON. Cambridge: Scientific Committee on Antarctic Research, 113 pp.
1972   [with DEACON M.] The 'Challenger' Expedition: the first oceanographic expedition. *Geographical Magazine*, **44**(12), 863–866.
1974   Water exchanges near the Antarctic continent. Pp. 23–25 in *Processus de formation des eaux oceaniques profondes*...[*Paris 4–7 Oct. 1972*]. Paris: C.N.R.S. 278 pp. [Colloques Internationaux du C.N.R.S. No. 215.]
1974   A Hyde Park memorial. *Geographical Journal*, **140**(1), 167–9.
1975   Southern Ocean exploration. *Oceanus*, **18**(4), 2–7.
1975   The oceanographical observations of Scott's last expedition. *Polar Record*, **17**(109), 391–396.
1975   [with MOOREY J. A.] The boundary region between currents from the Weddell Sea and Drake Passage. *Deep-Sea Research*, **22**(4), 265–268.
1975   Bicentenary of Captain Cook's landing on South Georgia. *Polar Record*, **17**(111), 692–694.
1976   The cyclonic circulation in the Weddell Sea. *Deep-Sea Research*, **23**(1), 125–126.

Staff-Commander Tizard's journal and the voyages of H.M. Ships *Knight Errant* and *Triton* to the Wyville Thomson Ridge in 1880 and 1882

Margaret Deacon

National Maritime Museum, Greenwich, London, U.K.

In 1958 Dr. Daniel Merriman and the late Mrs. Mary Merriman published some letters written between 1877 and 1881 by Sir Charles Wyville Thomson (1830–1882) to Staff-Commander Thomas Henry Tizard.[1] Thomson, Regius Professor of Natural History at Edinburgh University, had been leader of the team of civilian scientists during the voyage of H.M.S. *Challenger*, 1872–1876; Tizard, one of the ship's officers. In 1880 and 1882 as Captain of the *Knight Errant* and of H.M.S. *Triton* he led further expeditions to the Faeroe–Shetland Channel in order to throw new light on observations made by Thomson and W. B. Carpenter (1813–1885) in H.M.SS. *Lightning* and *Porcupine* in the 1860s. Some of Tizard's journals were presented to the National Maritime Museum by his son, the late Sir Henry Tizard, F.R.S. They include his journals kept in the *Knight Errant* and *Triton* between 1880 and 1890 and these contain Tizard's diary of events on the two scientific voyages.[2]

Thomas Henry Tizard was born in 1839, a younger son of Joseph Tizard, a Weymouth ship-owner and coal merchant.[3] He was educated at the Royal Naval Hospital School, Greenwich, and entered the Navy in 1854 as a master's assistant. He first served in the paddle steamer H.M.S. *Dragon*, in the Baltic during the Crimean War and then in the Mediterranean.[4] In 1857 he was appointed to H.M.S. *Indus* on the North American and West Indian Station.[5] He was promoted to second master in 1860 and then spent rather more than a year attached to the training ship H.M.S. *Britannia*. His future commanding officer, G. S. Nares (1831–1915), was one of the *Britannia's* lieutenants.

Tizard's career as a surveyor began in earnest in 1861 when he was appointed to the survey ship H.M.S. *Rifleman*, working in the Far East and China seas.[6] During the latter part of his 6 years there he was often working independently, in command of the tender *Saracen*, having been promoted to master in 1864. He was invalided home in 1867.[7] In 1868 he was appointed to H.M.S. *Newport*, Commander G. S. Nares, on survey work in the Mediterranean. The *Newport* was principally occupied in the survey of Sicily and the North African coast. She was present at the opening of the Suez Canal and later surveyed it and began work in the Red Sea. In 1871 she was replaced by H.M.S. *Shearwater*.[8]

It was on the voyage out to Egypt that the *Shearwater* spent a week working in the Strait of Gibraltar helping W. B. Carpenter to repeat and extend the current observations and salinity and temperature measurements made the year before in H.M.S. *Porcupine* and

Editorial note. The usual *Deep-Sea Research* style for references has been replaced in this paper by the numbering style usually employed in historical papers, since many of the references are to original manuscripts.

1

designed by him to show the presence of an undercurrent in the Strait as an example of ocean circulation in miniature.[9]

Carpenter's dogged work on behalf of marine exploration, and in particular his defence of his favourite theory of ocean circulation sustained by temperature differences, led to the voyage of H.M.S. *Challenger*. Nares was appointed Captain and Tizard Navigating Lieutenant. Nares was recalled from Hong Kong at the end of 1874 to act as leader of the Arctic Expedition of 1875. He was replaced by Captain F. T. Thomson (d. 1884) but it was Tizard who was put in charge of the surveying work and navigation for the remainder of the voyage.[10] He had been promoted to staff-commander in July 1874.

From the return of the *Challenger* in 1876 until 1879 Tizard worked on the charts and narrative of the expedition at the Hydrographic Office. In 1879 he returned to the sea in the survey ship *Porcupine*, Staff-Captain J. Parsons. He was appointed to command the *Knight Errant*, surveying on the west coast of Britain and Ireland in 1880. In 1882 he commissioned H.M.S. *Triton*, surveying on the east coast, and continued this work until 1890, being promoted to staff-captain in 1889. He was appointed Assistant Hydrographer and elected a Fellow of the Royal Society in 1891. He retired in 1907 and died in 1924. Admiral Day has described him as "one of the most distinguished surveyors of his day".[7]

In the *Challenger* Tizard had been responsible for the serial temperature measurements. Though personally less interested in this side of the work than in the biological discoveries, Wyville Thomson was anxious that the results should be fully worked out. This involved making new corrections for the effect of pressure upon thermometers since it had been discovered that the original ones were defective. This was done by P. G. Tait (1831–1901), Professor of Natural Philosophy at Edinburgh. It would also involve interpretation of the results, once finalized, but meanwhile, Thomson was anxious to throw new light on some general aspects of the *Challenger's* work on temperatures and the earlier results obtained with Carpenter.

The original purpose of the *Lightning* and *Porcupine* voyages in 1868–1870 was the extension of dredging to depths in the sea hitherto generally supposed to be empty of life.[11] Carpenter became interested in the temperature records being made in the *Lightning*, which had been issued with thermometers of various makes in order to test their correctness under pressure at considerable depths. Measurements revealed the existence in the depths of the Faeroe–Shetland Channel of two bodies of water, one cold, the other about 17°F warmer, in close proximity to each other. At the time the explanation which seemed the most likely was that a current of warm water from the south was banking up a flow of cold water from the Arctic Ocean. Contrary to what has since been written [12] the idea that there might be a submarine ridge dividing the two did occur but was dismissed without much discussion because it seemed as though they had explored the area sufficiently thoroughly for it to have shown up if it existed.[13]

Carpenter argued that the warm water was derived from a general northerly movement of the surface layers from tropical regions, of which the Gulf Stream was a part. This movement was to compensate for a contrary flow of cold dense water from Arctic regions, through channels other than the Faeroe–Shetland Channel, into the depths of the Atlantic. This circulation, which he believed would be found to exist in all the oceans, was generated by the difference in specific gravity between the water of polar seas and the warmer and therefore lighter water of equatorial regions.[14]

Carpenter's theories met much opposition, particularly from those who for various reasons favoured the then generally accepted view that winds were the sole cause of ocean currents.[15] During the *Porcupine* voyages, during his second visit to the Strait of Gibraltar in the *Shearwater* and in organizing the voyage of the *Challenger*, his principal aim was to provide fresh evidence for his ideas.

By the time the *Challenger* had been at sea a year few would have denied, however much they might dispute Carpenter's views of the mechanism of ocean circulation either in detail or *in toto*, that the deep water of the Atlantic was of polar origin. It also soon became clear that submarine topography must play a part in regulating the distribution of the deep water masses. As early as September 1873 Nares was writing to G. H. Richards (1819–1896), the Hydrographer, that the temperature of the bottom water on the western side of the Atlantic was lower than on the eastern side, suggesting the presence of "if not a continuous bank...a chain of banks" down the centre of the ocean.[16] The existence of a mid-ocean ridge, though this term was not yet used, was demonstrated by soundings made in the South Atlantic by the *Challenger* on her return in 1876.

The most striking example of the effect of topography on vertical temperature distribution was found in the East Indies. Tizard wrote:

> The temperatures obtained in the seas partially enclosed by the Indian Archipelago, prove that they have each of them deep basins cut off from the general oceanic circulation by ridges connecting the islands which surround them; for although in each sea, soundings of over 2,000 fathoms were obtained, in no case did the temperature decrease in a regular curve from the surface to the bottom, as is usual in the open ocean; in every case, after attaining a certain depth the temperature below that depth remained the same; thus in the Banda and China seas the temperature remained the same from 900 fathoms to the bottom, in the Celebes sea from 700 fathoms to the bottom, and in the Sulu sea from 400 fathoms to the bottom.[17]

In fact, the cause was here inferred. Tizard later explained:

> As the voyage of the *Challenger* was devoted to general oceanic research, it was found impracticable to spend much time over particular localities without lengthening the voyage considerably, and consequently there was no opportunity of testing by actual soundings the correctness or otherwise of this theory.[18]

Wyville Thomson related how, during the voyage, he and Tizard discussed these findings in relation to the work of the *Lightning* and *Porcupine*.[19] Both came to believe that the temperature distribution in the Faeroe–Shetland Channel must be governed by a submarine ridge linking the Faeroes with Shetland which prevented the cold Arctic water from flowing through the depths of the Channel into the Atlantic. On their return Thomson was anxious to reinvestigate the area, principally because he felt that if the existence of the ridge and its influence could be clearly shown there, then it would be legitimate to infer the presence of similar submarine barriers when discussing the *Challenger* temperature observations.[20]

Thomson's letters to Tizard show that he obtained the consent of the Hydrographer, F. J. O. Evans (1805–1885), to a voyage in 1879 but had to postpone it due to ill-health.[21] In 1880 he renewed his application. Evans replied that he could make no special provision for such a voyage but that he could send Tizard who was already at work on the west coast to the area in the *Knight Errant*.[22] Thomson and his assistants would be able to sail with him to search for the ridge and to carry out dredging, provided they made their own equipment available and did not interfere with the survey work. Thomson already knew that he would not be able to go himself for reasons of health but he arranged for John Murray (1841–1914), who had stayed on after the *Challenger* as his deputy at the *Challenger* Office, to go instead. The drawbacks to the expedition were to be not the rather stringent conditions imposed by Evans but the unsuitability of the *Knight Errant* for the purpose.

Thomson arranged to borrow a steam winch for dredging from the firm of David and

William Henderson[23] at Partick.[24] The *Knight Errant* was moored at their Meadowside works from 19 to 21 July 1880 while the winch was fitted and supplies of rope, bottles, spirits, etc., were taken on board. Thomson and his colleagues went to and fro between Edinburgh and Glasgow making the final arrangements and Murray and Frederick Pearcy, formerly Thomson's laboratory assistant in the *Challenger*, joined the *Knight Errant* before she sailed. They arrived at Oban the following day and early on the 23rd picked up Thomson and 'party'. In fact the only companion named was his son Frank. The second day's voyage took them through the Sound of Sleat, where they hailed J. Y. Buchanan's[25] yacht, to Gairloch. The next day they crossed the Minch to Stornoway. No work, either oceanographic or surveying, was done during this trip save for a trial sounding with Buchanan's sounding tube and depth indicator.

At Stornoway Tizard's first concern was to see about coal supplies, for the *Knight Errant* used large quantities but had little storage capacity. He was relieved to find that there was plenty to be had and, rather ungratefully, recorded that "Stornoway has an ancient and fish-like odour prevailing through it".

They sailed on Monday, 26 July, coaling completed, at 5 p.m. Thomson must after all have been hoping to make the voyage but Tizard wrote: "A fresh easterly breeze with a slightly falling barometer induced the Professor to remain behind in which I think he acts wisely." At 8 p.m. abreast of Tolsta Head, he recorded: "A nasty swell outside. All of us more or less feeling the motion of this little beast." Tizard had no illusions about his craft.

Early next morning they sighted Sulisker (Sula Sgeir) and North Rona and began sounding north-northwestwards towards the Faeroe Bank (see Fig. 1). The weather was not bad and Tizard grumbled: "The sea moderate and the wind a nice ENE breeze but this little beast of a ship is like a ⌊half tide?⌋ rock and takes the water in on all sides, besides having a most unpleasant motion. In fact she is not fit for this work at all. Murray much better this morning but F. Thomson still seedy and unable to eat his breakfast." They sounded all day "on the warm side of the supposed ridge", that is, on its southern side, taking sights when the sun broke through, and making hauls with the tow-net. At night the dredge was put over. By this time nearly half of the coal taken on at Stornoway had been used.

The following day they completed the line of soundings to the edge of the Faeroe Bank and began a new line parallel to and northeastwards from the first one, obtaining three soundings in the cold area before evening. Next morning they were only able to obtain three further soundings before the weather became too bad to continue. Tizard decided to head for port, but having run the distance to Sulisker and not sighted it and weather and visibility having worsened, he kept the *Knight Errant* steaming head to sea all night. In the early morning the weather cleared a little. Tizard wrote: "At 8 a.m. having run the distance to the Butt of Lewis I steered in for the land (Sco^d.) and had the satisfaction about half past 8 of sighting the Butt itself in a clear between the rain squalls and ran down to Stornoway where we arrived at about 1 p.m."

After a week-end rest and coaling on the Monday, they sailed again on Tuesday, 3 August, with the weather appearing more settled. On the way north they stopped to shoot some sea birds and take hauls with the tow-net "in order to trace the food supply of both birds and herrings". A line of soundings was made between the Butt of Lewis and Sulisker. They dredged in the evening and again overnight. The weather had not lived up to its promise and they experienced rain and then fog and the barometer fell continuously. On the 4th they steered north ½ west from Rona to continue the line of soundings which had been abandoned on the previous trip. "At 5.15 a.m. we took our first sounding and proceeded to search for the ridge", Tizard wrote. "We continued sounding out to the NNW^d from Rona island and succeeded in obtaining all our casts but one on the ridge though probably not on the shoalest part." At 5.30 p.m. "the weather still looking nasty", he decided to head for Stornoway "not wishing to be caught outside in another gale". When they arrived in port, Wyville Thomson came on board "and I arranged to take him to Gairloch tomorrow as he considered the essential part of the work was now done as I do myself". In the event the weather was too bad and Thomson took the *Clansman* to Oban on the Monday. The *Knight Errant* sailed too, after ferrying his luggage to the steamer, and took a line of soundings north-north west from the Butt "in the direction of what the Professor calls the *Holtenia* ground".[26] During the night the wind rose to gale force. It moderated in the morning barely long enough for a trawl to be made. Sights showed them to be east of the *Holtenia* ground and it was not until the following afternoon that they finally succeeded in trawling in the right area. The

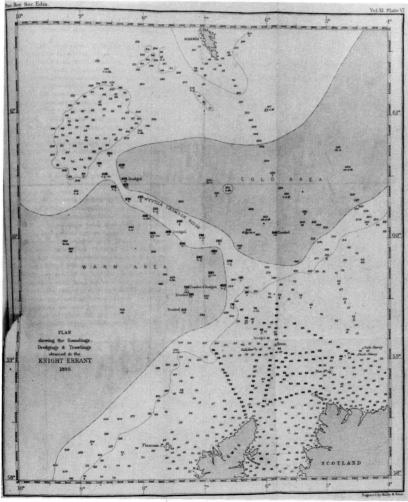

Fig. 1.   Plan showing the soundings, dredging and trawlings obtained in *Knight Errant*, 1880. From *Proceedings of the Royal Society of Edinburgh* (1880–1882) Vol. II, pl. 6. Reproduced by permission of the Royal Society of Edinburgh.

result was disappointing. Thomson had particularly wanted further specimens of sponges to add to the collection for his memoir for the *Challenger Report* [27] but none was obtained, save a fragment in the dredge the following morning. Tizard wrote:

> I now considered that I had obtained as many hauls of the trawl and dredge over the warm area as could be expected of this ship and decided to return to Stornoway, not having enough coals left to proceed to the cold area. Besides which this is a most unsuitable vessel to carry on work up here. She knocks about so much that sleep (even to those accustomed to serve in winter in her) is almost impossible and she is a source of constant anxiety from her small coal carrying capacity and want of facility for battening down, distilling water etc. and utter helplessness should the engines break down.
>
> For real work here it is necessary a ship should be used that could keep the sea for 3 weeks or a month at a time.

The weather was fine as they headed back to Stornoway and he

> regretted our coals did not allow of our proceeding to the cold area without risk. It is true I had enough to go there and steam back provided the weather continued as today but only just enough, that would have left us without any reserves of fuel and I did not think the object to be gained was worth the risk.

Back at Stornoway he received a telegram from Thomson reading:

> You know the circumstances best weather looks wonderfully fine settled now another trip of two or three days might do much if you think there would be no objection Early still.

Tizard replied:

> Arrived 8 p.m. yesterday [12 August] having obtained three trawlings and one dredging in warm area. Could not proceed to cold area being short of coals. Think as we have now obtained the more important soundings and dredgings it would be inadvisable to pursue this work further in this ship.

Later in the day, however, he received a letter from Evans instructing him to carry out further soundings to the northwest of Scotland if practicable. Accordingly the *Knight Errant* left on her final voyage to the north on 16 August.

Frank Thomson had elected to go home but Murray and Pearcey were still on board. When the ship reached Rona,[28] after making a line of soundings from the Butt, they landed with Tizard and explored the island. On the morning of the 17th they sounded in the cold area, took serial temperatures and hauls of the Hearder trawl and the tow-net. This was the last of the oceanographical work. In the afternoon the *Knight Errant* sounded between Rona and Sule and Stack skerries and next day to the Nun Rock and Cape Wrath. She reached Stornoway on 19 August and 2 days later left for Glasgow where the winch was dismantled. She sailed for Ireland on the 28th. Murray sailed with them to Lough Larne and finally left the ship on 4 September for Edinburgh.

The enterprise had been successful in that the existence of the ridge had been demonstrated. However, it was still only imperfectly known and the subsidiary aims of dredging and exploring the nature of the bottom had only been moderately successful. It was not only the deficiencies of the *Knight Errant* that were to blame. They had been unlucky too. Tizard and Murray wrote: "We were unfortunate in having bad weather during all the cruises. This circumstance together with the fact that the *Knight Errant* could not carry a large supply of coal, prevented our doing so much work as was originally intended."[29]

Wyville Thomson's preliminary report on the results of the voyage was published in *Nature* on 2 September 1880.[19] Though its existence had been confirmed, he pointed out, "The highest line of the ridge has probably not been found, and the details of the temperature have yet to be traced out more accurately along the line and for a short distance on either side."[30] His hope of obtaining supplementary specimens for the *Challenger* work had not been very successful "owing to the boisterous weather and the insufficiency of the vessel". He felt the area would repay further investigation by a full-scale dredging expedition.

Two months later Thomson applied to the Council of the Royal Society of London for help in arranging another expedition to the Faeroe–Shetland Channel.[31] Echoing a point made in *Nature*, he wrote: "The vessel would require to be of some strength, to have stowage for coal for at least a fortnight's steaming, and to carry sail enough to enable her to lie to in a breeze." The ship must be fully equipped for dredging and trawling and for making temperature observations. He proposed that the management of the work be entrusted to a committee, consisting as in *Porcupine* days, of himself, Carpenter and J. Gwyn Jeffreys (1809–1885).

Thomson's letter was carefully constructed. He spelled out his immediate aims in the necessary fullness:

> My special object in my present proposal is to determine in detail the behaviour of two ocean currents at different temperatures separated to a certain height by a continuous barrier; to work out more fully the relations between the organisms living on the surface of the sea and those forming the bottom sediments; to acquire some definite knowledge of the nature and extent of the fauna of a limited area of the abyssal region; and to compare and contrast at a point where they appear in close contact, the abyssal fauna of a region fed by an Arctic indraught with one where the under current appears to be subject to southern influences.
>
> My reason for urging the Council to press the Admiralty for the use of a vessel next summer, is that with our additional knowledge many points might now be settled by the careful examination of an accessible locality, which would throw much light upon questions which fall to be discussed in the concluding volume of the *Challenger Report*. This volume ought to be written during the year 1882.

In fact this programme could no more be carried out in a single cruise than the *Challenger Report* was to be completed in 1882, nor did Thomson propose that it should be. In another paragraph he again harked back to the fact that the *Challenger* expedition had only begun the exploration of the sea: "many important questions in Physical Geography were roused and finally settled, but many were left undecided and many were little more than suggested. I have already more than once dwelt upon the importance of following up some of these lines of research." He stressed the importance to the British Museum of having a representative collection of British deep-water fauna. Most significantly, he suggested that the work might not necessarily be completed in a year but that sections of the area might be examined in turn.

It seems clear that Thomson saw this as the way in which he hoped British oceanography would develop. He did not know that he had not long to live and perhaps the untimely nature of his death and the muddle into which the affairs of the *Challenger* Commission descended at that time have drawn attention since, away from what his hopes for the future might have been. He was frustrated in the end both by his own failing health and by political restraint on public expenditure which meant that for many years the development of marine science in Britain was more limited in scope than he envisaged. The voyages of the *Knight Errant* and *Triton* have been seen as appendages to the *Challenger* expedition, a mere tidying-up of loose ends. It seems clear, however, that Thomson intended them not only as the closing of an old chapter but also as the beginning of a new one.

For some time it even looked as though there would be no second voyage. In May 1881 Thomson wrote to T. H. Huxley (1825–1895), President of the Royal Society, to ask the Society to discover from the Admiralty if a voyage could be made that year.[32] He said he had heard unofficially that the request would be granted but the official letter, when it came, dashed his hopes, announcing that their Lordships regretted "their inability to detach a vessel on the Service referred to as all the suitable vessels on the Home Station are at present urgently required on other services".[33]

Fig. 2. H.M.S. *Triton*, about 1911. Reproduced by permission of the Trustees, National Maritime Museum, Greenwich.

Thomson died the following March. Murray, who succeeded him as head of the *Challenger* Office, made a fresh application and this was granted. Again it was Tizard who was to command the expedition.

On 2 May 1882 Tizard commissioned the *Triton* (see Fig. 2) at Sheerness. She was newly built by Samuda Brothers at Poplar and had a displacement of 410 tons. She was what was known as a composite paddle steamer, which meant that she had sails as well as steam power (350 h.p., opposed to the *Knight Errant's* 80 h.p.). Her normal complement was forty men, as opposed to twenty-four in the *Knight Errant*, and for this voyage Tizard was allowed to take on four extra seamen and three extra stokers. The preliminary preparations were made at Sheerness Dockyard, where a deck engine was fitted, and the *Triton* sailed on 23 June, arriving at Granton Harbour on the Firth of Forth 2 days later.

The Triton stayed at Edinburgh for 2 days while the scientists' gear was loaded. Tizard went into town to meet the scientists involved with the work, Murray, Tait, Alexander Buchan (1829–1907), the Secretary of the Scottish Meteorological Society, and George Chrystal (1851–1911), Professor of Mathematics at Edinburgh University. They in turn visited the ship, as did Frank Thomson. On the afternoon of the 26th Tizard went to Linlithgow to call on Lady Thomson, returning to Edinburgh in the evening to dine with Murray at his club. He also had a talk with the Shetland Island pilot.

The *Triton* sailed in the evening on 27 June. She spent a fortnight surveying off the east coast of Scotland, accompanied by Dr. Day.[34] Dredging and tow-netting were carried out. Tizard then proceeded north to obtain soundings round Fair Isle. This done, the *Triton* headed for Stornoway which she reached on 25 July. Murray and Pearcey arrived by steamer the following day. Tizard checked the fitting of a boom for sounding and dredging and of hatchway covers for battening down. He heard from Evans that a surgeon had been appointed to the *Triton* and they had to wait until 3 August for him to arrive.

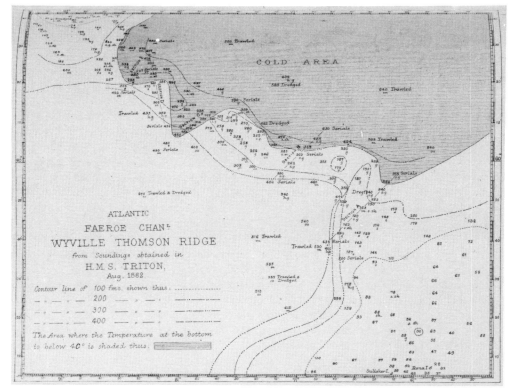

Fig. 3.   The Wyville Thomson Ridge from soundings obtained in H.M.S. *Triton*, August 1882. From *Proceedings of the Royal Society of London* (1883) Vol. 35, pl. 4. Reproduced by permission of the Royal Society of London.

In the larger and more sea-worthy *Triton*, which could carry nearly 60 tons of coal, they were able to stay at sea for much longer. During the month spent working in the Faeroe–Shetland Channel they returned to port only twice, spending periods of 9, 11, and 8 days at sea. The work consisted mainly of establishing the line of the ridge by sounding and making temperature measurements during transverse sections across the ridge and by a more detailed exploration of the ridge itself (see Fig. 3). They also made excursions to work on either side, the sequence of activities still being partly governed by the weather.

On the morning of the 4th Tizard wrote: "At 10½ a.m. commenced running the first sectional line across the Wyville Thomson ridge obtaining 200 fms on the warm side at the SE part of the ridge and then steering across it getting a sounding at about every 3 miles— Took serial temperatures at every 20 fms at first cast."[35] In the afternoon he wrote: "A confused sea. This ship however much better than the *Knight Errant*. Murray worked the tow net very successfully finding a large amount of life at the surface. The Doctor rather sea sick."

On the morning of the 5th they were back in the warm area but it was too misty to get the sights necessary to run a sectional line so after two soundings and a trawling, Tizard headed for the Faeroe Bank. He wrote: "I much regretted being unable to proceed with the sectional lines but it was useless doing so whilst the weather remained so foggy and the sea high—was only too glad we were able to get the trawling." Next day, which was Sunday, began foggy, then the wind freshened and a considerable sea arose. Work was resumed the following morning but sights soon showed that they were to the west of their expected position and they steamed east to pick up soundings of 1880. These turned out to be in the cold area so they retreated westwards again and in the evening located the summit of the ridge at 205 fathoms in 62°23′N, 8°41′W. They proceeded under sails and easy steam towards the Faeroe Bank overnight and next morning sounded and trawled on the Bank and then ran a line of soundings south towards the 205-fathom point of the day before and beyond. In the afternoon they suddenly found themselves in deep water on the southwest, warm, side of the ridge and turned northeastwards to run another sectional line across it and pick up the crest again. Next day they recrossed the ridge southwards. That night the trawl got caught up at the bottom and was eventually retrieved much torn. On sounding they found themselves in the cold area, having drifted to the northeast during the night. After a further sounding they headed back but the weather deteriorated and they were forced to lie to. Next day Tizard headed slowly for Stornoway through 15-foot waves and they arrived early on 12 August.

The *Triton* sailed again on 15 August and headed back to the southeastern part of the ridge. On the 16th they obtained a line of soundings across it, though fog prevented them from obtaining any sights. On the 17th they took a long line westwards into the warm area and then a more direct line back into the cold area. The next 2 days were spent in a detailed investigation of the central part of the ridge. Tizard wrote on the 18th: "obtaining soundings in NW central part of ridge all day, finding it very narrow in some parts. Some little difficulty in working owing to the strong wind and considerable sea. The sky clearer than we have yet had it.... In the afternoon we came across a gap in the ridge and traced the cold water running through it [over?] 305 fms down the warm side to 370 fathoms when it being 8 p.m. we were obliged to stop for the night."

Next morning they got on the same line farther westwards and went back to the place where they had stopped, sounding as they went. They then continued towards the gap, the bottom temperature falling as they approached, and sounded through it into the cold area.

Next day, Sunday, the brief spell of fine weather came to an end with a gale. On the 21st they sounded in the cold area and worked the tow-nets. In the afternoon conditions improved enough to allow dredging—but without much success. Next morning they sounded in 640 fathoms, some miles to the southeast, and trawled. At the second attempt they obtained a reasonable catch. On the 23rd the weather turned fine and they made two sectional lines across the southeast part of the ridge. Tizard wrote: "Quite a treat to be able to dry our sails and gear and see the decks somewhat dry between the operations of sounding and trawling, these however always keep a certain amount of dampness about." Then, having completed this part of the ridge, they headed for the *Holtenia* ground. There, next day, at the second attempt they obtained "a magnificent haul of Fish, Starfish, Echini, Holothurians, Shrimps, Sea Spiders and amongst other things a piece of wood...sodden with water and completely eaten through with Teredos, etc. Altogether this has been one of the most successful hauls of the trawl I have ever seen—strange to say however we got no sponges." They then headed for Stornoway where they arrived on 25 August.

The *Triton* left for the last voyage north on the evening of Sunday, 27 August. They went first to the *Holtenia* ground where they made two casts with a dredge designed by Murray, but without the success of the previous attempt. The 29th was spent exploring the area of the gap in the ridge. Tizard noted:[36] "Every now and then on the ridge we came across patches of *smooth* like those in a tide way over uneven ground evidently showing a movement of the water."

The following day they made a final visit to the cold area. They sounded with a Baillie tube to get a sample of the bottom and trawled twice without catching anything more than a few shrimps. Tizard wrote: "That the trawl had been on the ground both in the forenoon and afternoon I feel as certain as one can be of anything without actually seeing it and can only conclude that here we were unfortunate enough to hit on a patch where there were no animals at the bottom." Next day they repeated the exercise in the warm area but both the Baillie rod and the trawl came up empty, though a tow-net attached to the trawl contained some specimens. In the afternoon the dredge brought up a large quantity of globigerina ooze and various forms of life.

It was a cold day, though calm. Tizard recorded: "Found I had a chilblain this morning." Surely a record even for Scottish summers! In the afternoon there were signs of bad weather approaching. The wind freshened and the barometer began to fall. After the last dredging Tizard had set a course for the Butt of Lewis but during the night the weather got worse and he was obliged to reduce speed and, in the morning, to tack. He wrote: "This wind a great nuisance as we are but 60 miles to the Butt of Lewis and can't steam against it without flooding the deck. We were in hopes we should be in Stornoway this evening." In the event it took them another 2 days, finally arriving at midnight on 3 September, having been forced by the wind and high seas to lie to for most of the previous day.

It was presumably at this point that someone composed the verse, obviously meant to be sung to the tune of the policemen's song from Act II of the *Pirates of Penzance* by Gilbert and Sullivan:[37]

> When the Wyville Thomson ridge is quite completed
> When the trawlings and the dredgings are all done
> Then the Tritons, by the elements ill treated,
> Southwards turn in hopes they yet may see the sun.
> The wind and rain have been a dreadful bother
> The fog and choppy seas wet everyone
> Taking one consideration with another
> The Faeroe Channel trip's not been a pleasant one.

On reaching port they were met by Alexander Buchan. From this most reliable source Tizard learnt to his surprise "that but little wind had been experienced at Stornoway during the last two days".

The expedition was not quite over for they were due to pick up Chrystal at Oban and sail to deeper water off the continental shelf[38] to make observations on the effects of pressure in connection with Tait's work on the *Challenger* deep-sea temperatures. They left Stornoway on 5 September and after squalls on the first day arrived at Oban on the 6th. There they also found J. Y. Buchanan.

The *Triton* sailed from Oban on Monday, 11 September, with Murray and Chrystal on board. Further squalls obliged them to take shelter off Colonsay but next day they headed for a point between Rockall and the northwest of Ireland. There, on the 13th, they began by lowering the pressure gauges to 500 fathoms. This had little effect on them so they lowered two batches to 800 fathoms and seven gauges to 1360 fathoms. One of the boxes came up empty, apparently because the pressure had caused the fastening of the lid to give way and the gauges had fallen out. In the afternoon they repeated the work in a depth of 1345 fathoms, 2 miles to the east.

Chrystal being satisfied with the work, Tizard steered eastwards, intending to trawl on the edge of the continental shelf next day. However, the weather proved unsuitable so they headed for the Clyde but ended by taking shelter in Lough Swilly. They stayed there for a day until the weather looked more settled. On the 16th they reached the Clyde, spending the night off Arran in Lamlash harbour, and finally tied up at Greenock on Sunday, 17 September. Next day Murray and Chrystal landed with their equipment and the *Triton* prepared to return to her normal duties.[39]

This was unfortunately to be the end for the time being of the fruitful co-operation in the field between oceanographers and the Navy which had begun in the *Lightning* 14 years before. Naval surveyors, particularly former *Challenger* officers, continued to provide data and specimens from all round the world but civilian scientists were rarely taken to sea and there were no official expeditions. The new potential source of state aid for marine research, without which little could be accomplished, was the provision of money for fisheries research, as, for example, under the Act of 1882 which reconstituted the Scottish Fisheries Board. It was this source which Murray hoped to tap when he set up his marine station at Granton in 1884 but, for a variety of reasons, not least internecine strife in the scientific community, he failed to establish the station on a permanent footing. His subsequent work with H. R. Mill and others on the sea lochs and fresh-water lochs of Scotland, though important, was undoubtedly for him a second best. There is no doubt that given the chance he would, like Thomson, have preferred to continue the exploration of the deep oceans and felt that this was where the important results were to be won. This field, however, was to remain almost wholly closed to British scientists until the establishment of the *Discovery* Investigations 10 years after Murray's death.

Acknowledgements—The author is extremely grateful to Dr. J. C. Swallow, F.R.S. and Mrs. J. C. Swallow of the Institute of Oceanographic Sciences for help with the preparation of this paper. The quotations from Tizard's journal are reproduced by kind permission of the Trustees of the National Maritime Museum, Greenwich, and from Wyville Thomson's letter of 2 November 1880, etc., by permission of the Royal Society. Thanks are also due to Mr. A. W. H. Pearsall, Historian, Mrs. A. M. Shirley, Custodian of Manuscripts, and Mr. George Osbon, National Maritime Museum, and to the Librarian of the Royal Society and his staff.

NOTES

1. Merriman, D. and M. (1958) Sir C. Wyville Thomson's letters to Staff-Commander Thomas H. Tizard, 1877–1881. *Journal of Marine Research*, **17**, 347–374.
2. National Maritime Museum MSS (TIZ/5, 6 and 7) Journals of T. H. Tizard in H.M. survey ships *Knight Errant* and *Triton*, 1880–1890. For the other items see (4) to (6) below. His personal logbooks or journals for the *Newport*, *Shearwater* and *Challenger*, if any, were not in the family's possession.
3. Some details of T. H. Tizard and his background are given in the biography of his son, Sir Henry Tizard: Clark R. W. (1965) *Tizard*, Methuen, London.
4. National Maritime Museum MSS (TIZ/1 and 2). Journal and logbook of H.M.S. *Dragon*, 1854–1857, kept by T. H. Tizard.
5. National Maritime Museum MSS (TIZ/2 and 3). Logbooks of H.M.S. *Indus*, 1857–1860, kept by T. H. Tizard.
6. National Maritime Museum MS (TIZ/4). Abstract logbook of H.M.S. *Rifleman*, 1861–1867, kept by T. H. Tizard.
7. Day A. (1967) *The Admiralty Hydrographic Service, 1795–1919*. H.M.S.O., London, pp. 156–157.
8. Public Record Office MSS, ADM. 53/9971, 9972 and 9973. Ship's logbooks of H.M.S. *Newport*, 1868–1871; and ADM. 53/10,387. Ship's logbook of H.M.S. *Shearwater*, 1871–1872.
9. Carpenter W. B. (1872) Report on scientific researches carried on during the months of August, September, and October, 1871, in H.M. surveying-ship *Shearwater*. *Proceedings of the Royal Society of London*, **20**, 535–644.
10. Ritchie G. S. (1967) *The Admiralty Chart, British Naval hydrography in the nineteenth century*, Hollis & Carter, London, p. 330.

11. THOMSON C. W. (1873) *The Depths of the Sea*, Macmillan, London.
12. Cf. TIZARD T. H. and JOHN MURRAY (1882) Exploration of the Faeroe Channel, during the summer of 1880, in H.M.'s hired ship *Knight Errant*. *Proceedings of Royal Society of Edinburgh*, **11**, 638–724; pp. 639–640.
13. THOMSON C. W., *op. cit.*, pp. 313–314: "There is no great difference in depth between the two series of soundings; and there is no indication of a ridge separating them."
14. CARPENTER W. B. (1868) Preliminary report of dredging operations in the seas to the north of the British Islands, carried on in H.M. Steam-vessel *Lightning*. *Proceedings of the Royal Society of London*, **17**, 168–200. The idea of circulation due to density differences was not, of course, new. Carpenter largely ignored the possible effect of salinity differences. As these sometimes reinforce and sometimes counteract the density effects of temperature a much more complicated circulatory pattern than the one he suggested is produced and this has been, and still is being, pieced together by oceanographers of the twentieth century.
15. DEACON M. B. (1971) *Scientists and the sea, 1650–1900. A study of marine science*, Academic Press, London and New York, pp. 318–328.
16. NARES G. S. (1873) Letter to G. H. Richards. In: H.M.S. *Challenger. Reports of Captain G. S. Nares, R.N. with abstract of soundings and diagrams of ocean temperatures in North and South Atlantic oceans*, p. 10. This was one of a series of seven preliminary reports of the *Challenger* work. No bibliographic details are given but they were presumably printed for the Admiralty by the Stationery Office.
17. TIZARD T. H. (1875) Remarks on the temperatures of the China, Sulu, Celebes, and Banda seas, H.M.S. *Challenger* No. 4. *Report on ocean soundings and temperatures, Pacific ocean, China and adjacent seas*, p. 4. This is one of the series described in (16).
18. TIZARD T. H. (1883) Remarks on the soundings and temperatures obtained in the Faeroe Channel during the summer of 1882, *Proceedings of the Royal Society of London*, **35**, 202–226; p. 203.
19. THOMSON C. W. (1880) The cruise of the *Knight Errant. Nature*, **22**, 405–407.
20. Thomson in fact did not wait on his own account for further confirmation. See his discussion of the effect of continuous barriers on the vertical distribution of ocean temperature: THOMSON C. W. (1877) *The Voyage of the* Challenger. *The Atlantic. A preliminary account of the general results of the exploring voyage of H.M.S.* Challenger *during the year 1873 and the early part of the year 1876*, Macmillan, London, **2**, 323 ff.
21. MERRIMAN D. and M., *op. cit.*, pp. 362–367.
22. The *Knight Errant* was hired by the Admiralty and therefore appears in the *Merchant Navy List* of the period. She was built at Hull in 1862 and was registered at Liverpool in 1869. Her owner was George Percival of that city. She was a paddle steamer, just under 127 feet long and with a displacement of 180 tons (gross).
23. This was the firm which built Murray's yacht the *Medusa* in 1884. He later married into the family.
24. The description of the voyages of the *Knight Errant* and *Triton* to the Faeroe-Shetland Channel in 1880 and 1882 is taken from Tizard's surveying diary: National Maritime Museum MS (TIZ/5). All unnumbered quotations are from this source. Tizard and Murray published an account of the first voyage (1882), *op. cit.*, which included reports on the zoological results, on sea water and on the rocks of North Rona, by various authors.
25. John Young Buchanan (1844–1925) was the chemist in the *Challenger*. He continued working independently on marine research after the return of the expedition.
26. THOMSON C. W. (1873) *Depths of the Sea*, *op. cit.*, p. 104. This was the submarine slope northwest of the islet of North Rona. There in 1868 they had obtained (p. 70) "an extraordinary number of silicious sponges of most remarkable and novel forms". One of these was *Holtenia carpenteri* and Thomson gave its name to the area.
27. Owing to Thomson's death this section of the report was written by Franz Eilhard Schulze (1840–   ) who was to have collaborated with him on it. SCHULZE F. E. (1887) Report on the Hexactinellida collected by H.M.S. *Challenger* during the years 1873–1876. *Report on the scientific results of the voyage of H.M.S.* Challenger *during the years 1872 to 1876*, H.M.S.O., London. *Zoology*, **21**, No. 1, 513 pp.
28. This tiny island was perhaps more notable for events in the vicinity than for any history of its own. The route taken by the nineteenth-century expeditions setting out to search for the North-West Passage was through the North Sea to the Orkneys and then westwards through the Pentland Firth. It was off North Rona on 4 June 1845 that the *Erebus* and *Terror* had parted from the steamers which had towed them from the south and set out on their fateful voyage from which no one returned.
29. TIZARD T. H. and JOHN MURRAY (1882), *op. cit.*, p. 648.
30. THOMSON C. W. (1882), *op. cit.*, p. 407.
31. Royal Society MSS, Miscellaneous Correspondence, **12**, No. 108. Thomson to the Council of the Royal Society, 2 November 1880.
32. *Ibid.*, **12**, No. 166. Thomson to T. H. Huxley, 29 May 1881.
33. *Ibid.*, **12**, No. 178, Admiralty to Royal Society, 14 June 1881.
34. This was presumably Francis Day (1829–1889), ichthyologist and author of *Fishes of Great Britain and Ireland*, Williams & Norgate, London and Edinburgh, 2 vols. (1880–1884). He left the ship on 12 July "having obtained as many specimens and as much information as to the Fishery question as he required for the time".
35. The Merrimans, *op. cit.*, pp. 367–369, show that Thomson wished to call the ridge after the Norse god Ymir

but after his death Tizard and Murray (1882), *op. cit.*, p. 649, named it the Wyville Thomson Ridge in his memory.

36. He had previously noted this phenomenon on crossing the ridge on 8 August.

37. First produced in 1879 and brought to London in 1880.

38. This term was not used by Tizard who referred to it as "the edge of the bank" but is employed to make the meaning clearer.

39. TIZARD T. H. (1883) Remarks on the soundings and temperatures obtained in the Faeroe Channel during the summer of 1882. *Proceedings of the Royal Society of London*, **35**, 202–226. This is a general account of the voyage, concentrating on the survey work. Papers on the zoological results by various authors were published in volume 12 of the *Proceedings of the Royal Society of Edinburgh* (1882–1884). The *Triton*, after her surveying career was over, was used as a training ship between the wars and then moored in the Thames until sold for breaking up in the early 1960s.

Water characteristics of the Southern Ocean
south of the Polar Front

Eddy C. Carmack

Canada Centre for Inland Waters, 4160 Marine Drive, West Vancouver, B.C., Canada

Abstract—A volumetric census of the Southern Ocean south of the Polar Front is computed from 945 selected hydrographic stations occupied around the Antarctic continent. Water masses of the oceanic domain ($137.6 \times 10^6 \, \text{km}^3$) and continental shelf domain ($1.3 \times 10^6 \, \text{km}^3$) are considered separately, and attention is drawn to quantitative variations among the Atlantic, Indian and Pacific Ocean sectors of the Southern Ocean. Near-bottom values of potential temperature, salinity, dissolved oxygen and a density parameter (σ^4) are mapped to examine the abyssal circulation of Antarctic Bottom Water varieties in relation to their source regions.

Warm Deep Water ($\theta > 0°C$, $S \geq 34.65\%_{00}$) is by far the most voluminous water mass of the oceanic domain comprising nearly 58% of the total volume; followed by Antarctic Bottom Water ($\theta \leq 0°C$, $S \geq 34.62\%_{00}$) with about 28%. The remaining 14% of oceanic water is composed of surface and transitional water masses. Water on the continental shelves surrounding Antarctica displays dominant univariate modes in the class intervals $\theta = -2.0$ to $-1.6°C$, $S = 34.5$ to $34.7\%_{00}$. High Salinity Shelf Water ($S \geq 34.6\%_{00}$), a principal mixing component of Antarctic Bottom Water, forms about 25% of the total summer-time reservoir of continental shelf water. The Weddell Sea is shown to produce the bulk of Antarctic Bottom Water, although the Ross Sea is also cited as a major source; only minor contributions appear to be derived from other regions. A rough calculation based on estimated shelf residence times yields a total bottom water production rate of 5 to $10 \times 10^6 \, \text{m}^3 \, \text{s}^{-1}$.

1. INTRODUCTION

THE ocean surrounding Antarctica has long received attention as an important region of water mass assembly, and as the major connection for heat and mass exchange in the world ocean.

The general oceanography of the Southern Ocean is well known from the early work of Wüst (1933), Mosby (1934) and Deacon (1973). Recent studies by Jacobs, Amos and Bruchhausen (1970), Seabrooke, Elder and Hufford (1971), Reid and Lynn (1971), Reid and Nowlin (1971), Callahan (1972), Carmack and Foster (1975 a, b), Foster and Carmack (1976 a, b), and especially Gordon (1966, 1971 a, b, c, 1972, 1974, 1975) have greatly expanded our descriptive knowledge of antarctic water mass structure and circulation. However, a unified *quantitative* account of the water masses surrounding Antarctica has not been previously reported. The purpose of the present study is, therefore, to identify and describe the spatial distribution of water masses in the Southern Ocean (Section 2), infer the spreading of bottom water varieties from their source regions (Section 3), and describe the frequency distribution of potential temperature–salinity (θ–S) characteristics of antarctic water masses by volumetric analysis (Section 4).

Briefly, volumetric analysis entails computing the volume of water within specified bivariate class intervals on a θ–S correlation diagram, and relating this information to the

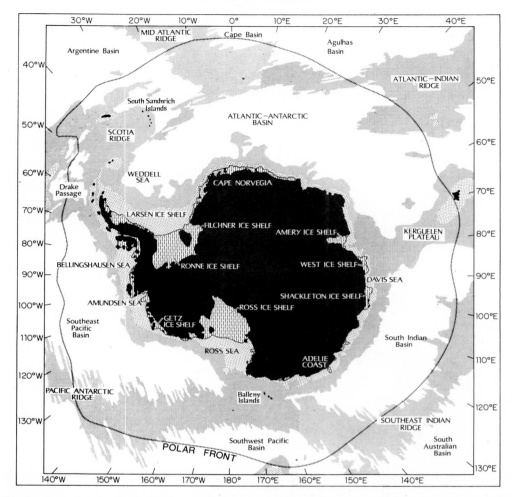

Fig. 1.   Area of study. The polar front is depicted by the dotted line surrounding the Antarctic continent. Depths less than 1000 m are indicated by coarse stippling; depths 1000–4000 m are indicated by fine stippling. Continental ice shelves are distinguished by broken hatch lines.

basic water mass structure of an oceanic region (cf. MONTGOMERY, 1958). Knowledge of the amount of water with given θ–S characteristics provides a useful starting-point for more detailed water mass, circulation and property budget studies. In the present census (Fig. 1) water masses of the continental shelf domain are considered separately from those of the open ocean domain, and the Southern Ocean is further divided into Atlantic (60°W to 20°E), Indian (20°E to 170°E) and Pacific (170°E to 60°W) sectors. The Polar Front was chosen as a logical northern boundary for the census since it is generally recognized as the transition region between Antarctic and subantarctic waters (DEACON, 1937).

The volumetric census, θ–S scatter diagrams and bottom property charts were all constructed from the same set of over 900 hydrographic stations (Fig. 2, Appendix I) selected from historical and recent unpublished data reports. An attempt was made to

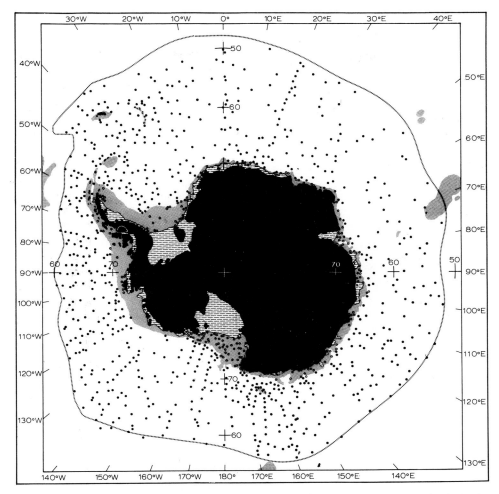

Fig. 2.   Location of hydrographic stations. Open circles denote continental shelf stations; solid circles denote oceanic stations.

keep the geographical coverage as uniform as possible. Greater emphasis was placed on selection of stations from recent expeditions because they generally reached closer to the bottom, used a closer vertical sampling interval and are of greater accuracy.

2. WATER MASS CHARACTERISTICS

The vertical structure of the Southern Ocean is composed of several recognizable layers of water, each originally derived from surface water in a particular area of the ocean. Conditioning of water at the surface, sinking to equilibrium depth and spreading along appropriate density surfaces is a continuous or repetitive process that eventually leads to large volumes of water having relatively uniform values of temperatures and salinity. Following the original suggestion of HELLAND-HANSEN (1916), a *water type* is represented by a single point on a θ–S diagram, while a *water mass* is defined by an area or segment of a curve.

Water masses are traditionally identified by a core property, i.e. that part of a layer where some oceanographic characteristic reaches an extreme value. This method assumes that the core represents both the main axis of spreading and the maximum percentage of a particular water type within a layer. Water masses may also stand out as densely patterned areas or modes on a θ–S correlation diagram. Boundaries between water masses, on the other hand, are more difficult to define. Divisions may be based on slope changes in the θ–S curve, volumetric considerations, or simply taken as the mid-point between two cores. The recent work by REID, NOWLIN and PATZERT (1977) demonstrates that stability maxima surfaces, extending over wide geographical areas, often interlay core properties, and thus may represent true water mass boundaries between layers from different sources.

A limitation of volumetric analysis is that water mass identification must be based solely on absolute position in θ–S space, rather than the primary identification characteristics discussed above. The aim of this section is thus to review the θ–S characteristics of Antarctic water masses so that a comparison to the volumetric census can later be made. A water mass classification scheme, based on the discussion below, is schematically illustrated in Fig. 3. This classification is intended to be as simple as possible, for com-

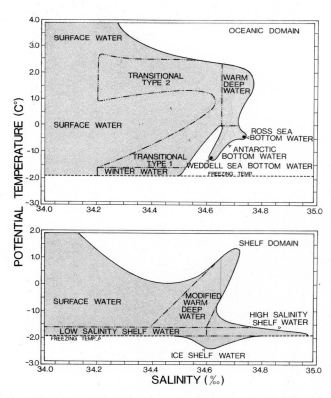

Fig. 3.   Simplified schematic representation of water masses of the oceanic (top) and continental (bottom) domains. Abbreviations are explained in text.

parison to the volumetric diagrams, and is not intended to reveal the complexities of Antarctic water mass structure.

Oceanic domain

Following DEACON (1937), the oceanic water column south of the Polar Front can be conveniently separated into three major strata: Surface Water, Warm Deep Water (or Circumpolar Deep Water) and Antarctic Bottom Water, as illustrated by θ–S scatter diagrams for the Atlantic (Fig. 4), Indian (Fig. 5) and Pacific (Fig. 6) Ocean sectors of the Southern Ocean.

The surface layer includes water within and above a shallow (50–300 m) temperature minimum, and with salinities below 34.5‰. The temperature minimum layer is believed to be a remnant of wintertime convection, and is designated Winter Water. Near the continent, and in the central Weddell and Ross Seas, Winter Water has temperatures near the freezing-point, and salinities approaching the critical value for the onset of the cabbeling instability (FOFONOFF, 1956; FOSTER, 1972).

Warm Deep Water is generally characterized by an intermediate depth temperature maximum, and a somewhat deeper salinity maximum; these two core layers have been used by GORDON (1971) to identify upper and lower deep water. Although the principal source of the warm and saline water is thought to be North Atlantic Deep Water, the isopycnal charts of CALLAHAN (1972) clearly show a complex history of exchange and

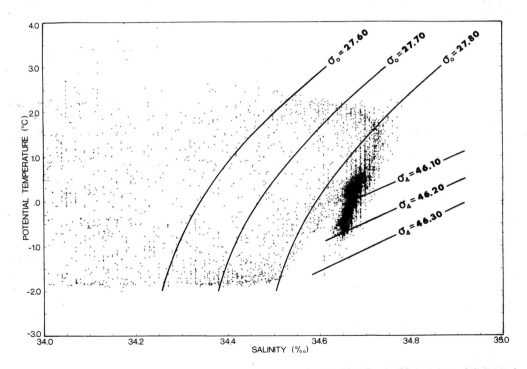

Fig. 4.   Potential temperature–salinity scatter diagram from stations in the Atlantic Ocean sector of the oceanic domain.

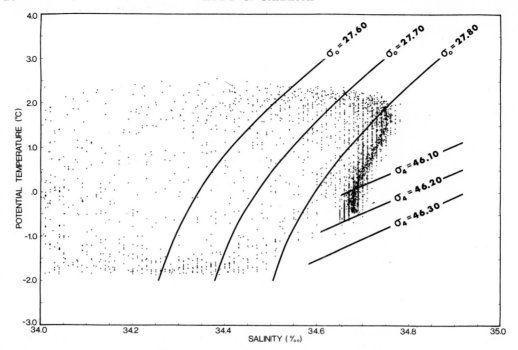

Fig. 5. Potential temperature–salinity scatter diagram from stations in the Indian Ocean sector of the oceanic domain.

modification within this layer. The detailed structure of Warm Deep Water has been further emphasized by GORDON (1975) and REID, NOWLIN and PATZERT (1977). For example, θ–S patterns on the Atlantic sector diagram suggest at least three stages of Warm Deep Water; a small number of points, from near the Polar Front, show high salinities indicative of North Atlantic Deep Water; an intermediate number of observations, which form the curve between $\theta = 2.0$ to $5.0°C$, characterize Circumpolar Deep Water entering the Atlantic sector via the Drake Passage; and finally the densely populated area below $0.5°C$ which includes the deep water of the Weddell Sea. In the volumetric census that follows, however, Warm Deep Water is treated as a single water mass, and no attempt is made to examine its detailed structure.

Antarctic Bottom Water occupies the lower portion of the water column and is generally, although somewhat arbitrarily, recognized by temperatures below about $0°C$ (GORDON, 1971). As noted by GORDON (1974), two major bottom water varieties are found: the cold, or low salinity Weddell Sea Bottom Water, and the warmer, more saline Ross Sea Bottom Water. CARMACK and FOSTER (1975b) distinguished between classical Antarctic Bottom Water and Weddell Sea Bottom Water, which is characterized by potential temperatures below $-0.7°C$ suggesting recent formation in the southwestern and western Weddell Sea. In the present report, the bottom water of the Weddell Sea and Ross Sea will be considered as particular subclasses of Antarctic Bottom Water.

GORDON and TCHERNIA (1972) have also shown a low salinity bottom water variety is also derived from the Adélie Coast. Additional bottom water may be produced near the Davis Sea (TRESHNIKOV, GIRS, BARANOV and YEFIMOV, 1973), in the eastern portion

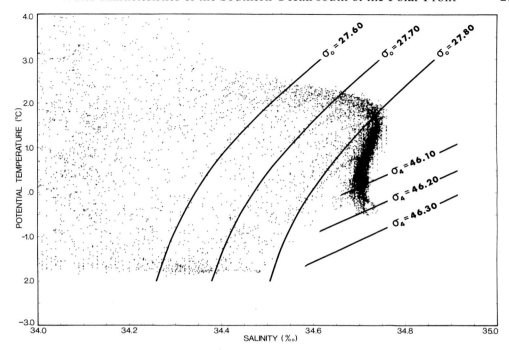

Fig. 6.   Potential temperature–salinity scatter diagram from stations in the Pacific Ocean sector of the oceanic domain.

of the Atlantic–Antarctic Basin (JACOBS and GEORGI [this volume pp. 43–84]) and in the Bransfield Strait (CLOWES, 1934).

The Weddell Sea variety of bottom water is predominant in the Atlantic sector (Fig. 4) as shown by the tail of monotonically decreasing θ and S values. Ross Sea Bottom Water, identified in Fig. 6 by the spur of high salinity water branching off from the base of Warm Deep Water at about $S = 34.7\%_{o}$, is clearly the densest bottom water in the Pacific sector. The Indian sector (Fig. 5) reveals an inverted 'Y' at its base, indicating bottom water contributions from both major sources.

Two major transitional water masses are suggested by patterns on the θ–S scatter diagrams. The first is indicated by the relatively high density of points stretching between the cores of Winter Water and Warm Deep Water. The vertical stability within this layer is weak, and likely influenced by the cabbeling instability (FOSTER and CARMACK, 1976a). The second transitional water mass is shown by the tail of points extending horizontally from upper Warm Deep Water towards about 2°C and 34.2‰. This water apparently contains mixtures of Warm Deep Water with Antarctic Intermediate Water, which is believed to form near the Polar Front.

Continental shelf domain

The water mass structure of the continental shelf of the Weddell Sea has been described by SEABROOKE, HUFFORD and ELDER (1971), CARMACK (1974) and CARMACK and FOSTER

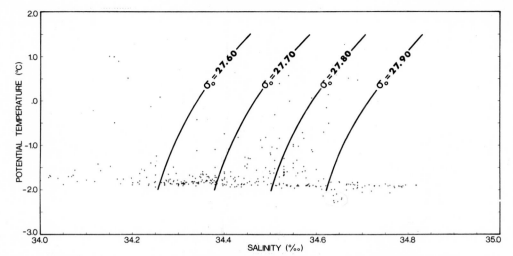

Fig. 7. Potential temperature–salinity scatter diagram from stations in the Atlantic Ocean sector of the continental shelf domain.

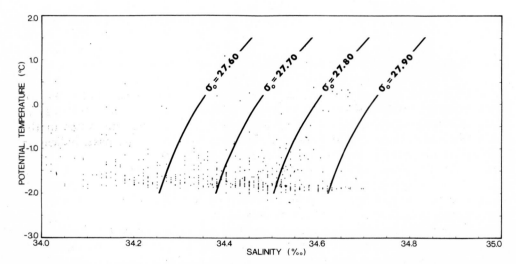

Fig. 8. Potential temperature–salinity scatter diagram from stations in the Indian Ocean sector of the continental shelf domain.

(1975); of the ROSS Sea by JACOBS, AMOS and BRUCHHAUSEN (1970); and of the Adélie Coast by GORDON and TCHERNIA (1972). The basic characteristics of shelf water are illustrated by θ–S scatter diagrams for the Atlantic sector in Fig. 7; for the Indian sector in Fig. 8; and for the Pacific sector in Fig. 9.

In general, during summer, the surface water on the continental shelf is somewhat warmed, and diluted by ice melting. Below the surface layer, water with temperatures near the freezing-point covers most of the shelf floor. In relation to bottom water production, it is convenient to distinguish between Low Salinity Shelf Water and High Salinity Shelf Water by the 34.6‰ isohaline, since this value represents the minimum

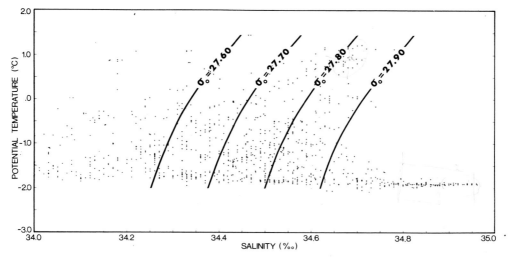

Fig. 9.   Potential temperature–salinity scatter diagram from stations in the Pacific Ocean sector of the continental shelf domain.

salinity that shelf water must attain to participate in bottom water formation (cf. Mosby, 1934).

GORDON (1974) examined shelf salinity distributions around Antarctica and noted a close relationship between shelf width and bottom salinity, with the most saline water lying above wide continental shelves. Correspondingly, the θ–S scatter diagrams show the Pacific sector (Ross Sea) to have the highest shelf water salinities, up to 35.0‰, followed closely by the Atlantic sector (Weddell Sea) with shelf salinities up to 34.8‰. The lowest shelf water salinities are associated with the narrow continental shelves in the Indian Ocean, although some shelf water with salinities above 34.6‰ has been observed in the Adélie Coast and Amery Ice Shelf regions.

Modified Warm Deep Water is represented on the θ–S scatter diagrams by the warm limb of water extending upwards from about $-1.8°C$ and 34.5‰. This water has θ–S characteristics similar to the transition water between Winter Water and Warm Deep Water in the Oceanic Domain. Modified Warm Deep Water exists as an intermediate layer interleaving between the surface layer and High Salinity Shelf Water, particularly near the shelf break in Weddell and Ross Seas. The southward penetration of Modified Warm Deep Water onto the continental shelf is believed to partially compensate the offshore flow of newly formed bottom water, and also contribute to bottom water production by mixing with High Salinity Shelf Water (GILL, 1973; FOSTER and CARMACK, 1976b).

The final shelf water mass, unique to antarctic waters, is the body of water colder than the freezing temperature of sea water at one atmosphere pressure, observed primarily near the Filchner, Ronne and Ross ice shelves. LUSQUIÑOS (1963) suggested that because the freezing-point of water is depressed by increasing hydrostatic pressure, the extreme temperatures are likely effected below the sea surface by cooling along the undersides of floating glacial ice shelves at depths of 200 m or more; thus this water is designated Ice Shelf Water, and potential temperatures as low as $-2.4°C$ have been observed (HUFFORD and SEABROOKE, 1970).

3. ABYSSAL CIRCULATION

A volumetric census becomes more meaningful when its relationship to basic circulation features and water mass source regions is established. Bottom property distributions have been used in the past (cf. Wüst, 1939; Deacon, 1937; Klepikov, 1958; Gordon, 1966, 1972; and Mantyla, 1975) to infer the flow of Antarctic Bottom Water. Owing to the number of reliable observations now available, a more detailed and unified examination of bottom water spreading is possible.

The use of bottom property maps to infer circulation features relies on the assumption that the main pathways of flow are revealed by patterns or tongues in bottom property distributions. However, bottom property maps are not necessarily a sound index of near-bottom *velocity*. For example, depth variations may also contribute to bottom property gradients regardless of the slope of associated density surfaces. Thus, the maps presented

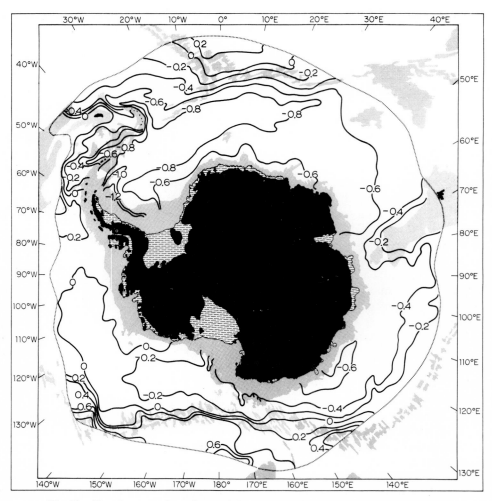

Fig. 10.   Near-bottom values of potential temperature for depths greater than 2000 m.

in this section serve only to indicate the location of the bottom water varieties tabulated in the volumetric diagrams, and, from pattern interpretation, to show their most likely paths of spreading from source regions.

Because more than one region of bottom water production exists around Antarctica, care must be taken when inferring flow from a single bottom property map. For example, Weddell Sea Bottom Water is recognized by its low salinity, whereas Ross Sea Bottom Water is indicated by its high salinity. Dissolved oxygen is usually a qualitative indicator of the 'freshness' of a bottom water variety. Finally, when water from different sources meet, the denser one would be expected to underflow the other, so that a density parameter also proves useful. The mean depth of the oceanic domain is about 4000 m. Hence, the density parameter, σ_4, has been chosen for use here, since it represents the density that a parcel of water would have if it were moved adiabatically to a pressure of 4000 decibars. It has units of $\sigma_4 = (\rho_4 - 1)10^3$. In this section bottom charts of potential

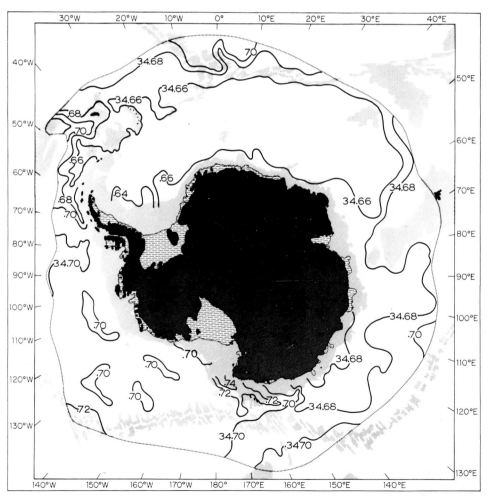

Fig. 11 Near-bottom values of salinity for depths greater than 2000m.

temperature (Fig. 10), salinity (Fig. 11), oxygen (Fig. 12), and potential density at 4000 dbars (Fig. 13), for depths below 2000 m, are presented to examine the abyssal circulation of the Southern Ocean. The 3000-m isobath is included on the charts to demonstrate the importance of bottom topography in shaping pathways of flow.

The extreme characteristics of bottom water derived from the Weddell Sea are clearly defined in all the bottom property maps. This water has the lowest values of potential temperature and salinity, and the highest values of oxygen and σ_4 observed in the abyssal Southern Ocean. The newly formed Weddell Sea Bottom Water begins to leave the continental shelf in the southwestern Weddell Sea between 35–40°W, and flows clockwise as a contour-following current adjacent to the continental margin in the western Weddell Sea. As this water moves eastward along the southern Scotia Ridge, small amounts appear to penetrate northwards into the Scotia Basin through gaps between 30–40°W (cf. GORDON, 1966). There is, however, no indication of this water moving farther westward into the southeast Pacific Basin.

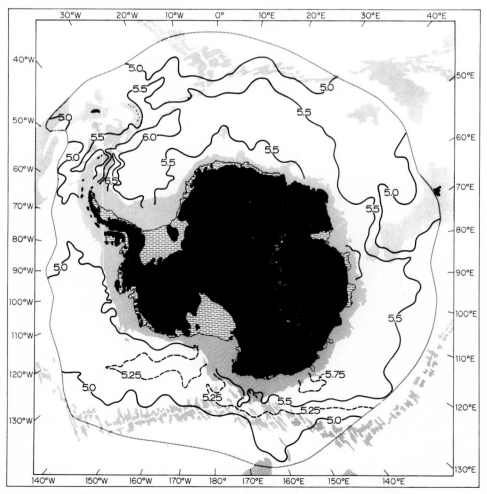

Fig. 12.   Near-bottom values of dissolved oxygen for depths greater than 2000 m.

East of the Scotia Arc Weddell Sea Bottom Water appears to split into two branches, as indicated by the $-0.8°C$, 34.66‰, 5.5 ml l^{-1} and 46.20 contours. The narrow tongue extending northwards along the Scotia Arc is largely the result of bathymetry as it represents Weddell Sea Bottom Water within the deep (>8000 m) South Sandwich Trench. There is no evidence of the extreme form of Weddell Sea Bottom Water north of the trench and, indeed, it seems unlikely that this dense water would flow up and out of the depression. Instead, since the water north of the trench has the θ–S characteristics of water at 200–800 m above the bottom in the northwestern Weddell Sea, it most likely represents a stratum of water from deep within the Weddell gyre which slides over the denser water filling the trench and flows northwards into the Argentine Basin. The most intense flow continuing into the Atlantic Ocean follows along the western boundary of the Argentine Basin. The major portion of Weddell Sea Bottom Water remains within the Atlantic–Antarctic Basin, and spreads eastward along the south flank of the Atlantic–Indian Ridge. Strong gradients along this ridge mainly reflect the northward shoaling

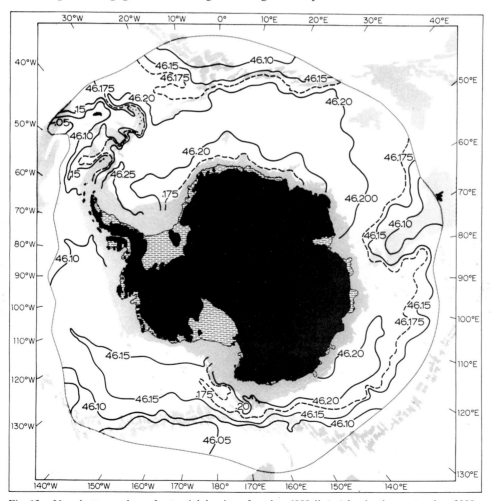

Fig. 13.   Near-bottom values of potential density referred to 4000 db (σ_4) for depths greater than 2000 m.

of the Bottom. Weddell Sea Bottom Water does not appear to enter the Cape or Agulhas Basins south of Africa.

Between 40–70°E water from the Weddell Sea, as indicated by the -0.6°C, 34.66‰, 5.5 ml l^{-1} and 46.20 contours, is turned southwards along the shoaling topography of the Kerguelen Plateau. There is some indication that part of bottom water turns westwards along the base of the continental margin, to form an elongate cyclonic gyre within the Atlantic–Antarctic Basin. Upon re-entering the southwestern Weddell Sea, this 'older' bottom water would, by virtue of its lower density, override the newly formed Weddell Sea Bottom Water; with the net effect being a gradual spiraling upwards of water mass layers.

Water entering the South Indian Basin south of the Kerguelen Plateau has temperatures above -0.6°C, salinities above 34.66‰, and oxygen values above 5.5 ml l^{-1}, and is thus similar to water at the base of the continental margin in the Atlantic–Antarctic Basin between 30°E and 30°W. East of the Plateau, the bottom flow again divides into two branches; some of the water moves north into the Indian Ocean along the eastern bank of the plateau, while the remainder spreads evenly across the floor of the South Indian Basin.

Bottom water derived from the Ross Sea is distinguished mainly by high salinity values. This water, as indicated by the 34.72‰ isohaline, leaves the continental shelf in the southwestern Ross Sea and initially flows west as a contour-following current along the continental margin. Near the Balleny Islands two branches of high salinity bottom water are observed. One branch continues westward into the South Indian Basin, while the other turns north towards the Pacific Antarctic Ridge, where it presumably bends eastward into the southeast Pacific Basin.

The contribution of bottom water from the Adélie Coast (Gordon and Tschernia, 1972) is clearly visible as a tongue of cold ($\theta \leq -0.6$°C) and well-oxygenated ($O_2 > 5.75$ ml l^{-1}) water extending westward from the coast of Antarctica at about 135–140°E. This water has θ–S characteristics similar to bottom water derived from the Weddell Sea and, therefore, it is difficult to trace any separate influence beyond 110–120°E.

The bottom water of the southeast Pacific Basin has θ–S characteristics common to the juncture of the high and low salinity spurs on the θ–S diagram and thus its origin is difficult to uniquely determine from the bottom property charts. Since the $\sigma_4 = 46.20$ contour does not extend continuously from the Atlantic into the Indian and Pacific sectors, its presence implies a local source. The $\sigma_4 = 46.15$ contour, on the other hand, does extend from the Atlantic sector into the southeast Pacific Basin, and transfer of water of Weddell Sea origin into the Pacific sector is, therefore, possible. It is suggested here that when eastward-flowing bottom water from the Weddell Sea enters the southeastern Indian Ocean, and encounters the denser Ross Sea and Adélie Coast bottom waters, that an overriding of the Weddell Sea water occurs. This circulation is supported by the presence of a deep salinity minimum layer immediately above Ross Sea Bottom Water in the southeastern Indian and southwestern Pacific Oceans (cf. Gordon, 1975). Salinity values within the deep salinity minimum layer tend to increase toward the east consistent with the hypothesis of water from the Weddell Sea interflowing and mixing with the more saline Warm Deep Water above and Ross Sea Bottom Water below. Farther to the east, salinity values once again reach a relative minimum at the bottom. The above can be taken as evidence that the abyssal water of the southeast Pacific Basin likely represents a blend of bottom waters from both the Weddell and Ross Seas with Warm Deep Water.

4. VOLUMETRIC ANALYSIS

Method

Volumetric diagrams for the Southern Ocean were prepared from the hydrographic data listed in Appendix. I. Each station was visually assigned an area of influence to represent that portion of the sea surface plane which lies closer to it than any other station. This method, although somewhat subjective in terms of geographical representativeness, maintains the actual water mass structure of each station. The alternative procedure, based on the analysis of area-averaged data (e.g. 5- or 10-degree squares), was not chosen as the resultant mean profiles may or may not represent actual water column structure. Potential temperature–salinity correlation curves for each hydrographic station were examined next to determine the thickness of water occupying specified θ–S class intervals on the correlation diagram. The volume of water in each θ–S class interval for individual stations was then obtained by multiplying the water thickness in each interval by the station's assigned surface area. The final diagrams were obtained by summing the partial volumes within each class for all hydrographic stations. The above computations were performed by computer analysis.

Bivariate class intervals $\Delta\theta = 0.5°C$, $\Delta S = 0.05\%_{oo}$ for the oceanic domain; and $\Delta\theta = 0.2°C$, $\Delta S = 0.05\%_{oo}$ for the continental shelf domain were chosen to make the volumetric diagrams as simple as possible, and yet maintain the integrity of water mass identification (Fig. 3). A more detailed description of Antarctic Bottom Water is also given using fine-scale class intervals $\Delta\theta = 0.1°C$ and $\Delta S = 0.02\%_{oo}$.

Oceanic domain

The volumetric distribution of potential temperature and salinity in the oceanic domain (volume $= 137.6 \times 10^6 \text{ km}^3$) of the Southern Ocean is shown in Fig. 14. Volumetric characteristics by ocean sector and water mass are summarized in Table 1. In describing the diagram, we will begin at the 'bottom' of the water column and work upwards.

The volume of Antarctic Bottom Water existing south of the Polar Front is approximately $38.0 \times 10^6 \text{ km}^3$ (Table 1b), or nearly 28% of the total Southern Ocean volume. This yields a mean thickness above the ocean floor of about 1100 m. For comparison, the volumetric diagram of MONTGOMERY (1958) shows approximately $45 \times 10^6 \text{ km}^3$ of this water mass in the entire world ocean. Hence, nearly 85% of all classical Antarctic Bottom Water lies south of the Polar Front. The apparent confinement of Antarctic Bottom Water properties to the Southern Ocean may initially appear to contradict the well-known effects of Antarctic waters on the abyssal properties of the world ocean. It should be noted, however, that the term Antarctic Bottom Water mass is used only subjectively when discussing abyssal water masses of the world ocean, which may contain less than 20% of the original water type (cf. WÜST, 1933). Furthermore, REID, NOWLIN and PATZERT (in press) have shown that the Antarctic water which flows north into the Atlantic Ocean is, indeed, drawn from lower Circumpolar Deep Water and the deep water of the Weddell Sea.

Water in the class interval $\theta = 0$ to $-0.5°C$, $S = 34.65$ to $34.70\%_{oo}$ comprises the dominant volumetric mode of Antarctic water. This Antarctic mode water, nearly $24 \times 10^6 \text{ km}^3$, represents over 50% of Antarctic Bottom Water in the World Ocean.

The Atlantic sector of the Southern Ocean contains the greatest volume ($18.8 \times 10^6\,\mathrm{km}^3$) of Antarctic Bottom Water, followed by the Indian sector ($15.8 \times 10^6\,\mathrm{km}^3$). The Pacific sector contains less than $3.3 \times 10^6\,\mathrm{km}^3$ of classical Antarctic Bottom Water despite its proximity to the Ross Sea.

Warm Deep Water is by far the most voluminous Antarctic water mass with $81 \times 10^6\,\mathrm{km}^3$ or nearly 58% of the total Southern Ocean volume. The mean thickness of this water is thus 2300 m. The general volumetric trend within this stratum of water is for increasing volume with decreasing temperature and salinity values; the volumetric mode of Warm Deep Water (volume $= 18.9 \times 10^6\,\mathrm{km}^3$) occurs in the class interval $\theta = 0.5$ to $0°\mathrm{C}$, $S = 34.65$ to $34.70‰$, immediately above the Antarctic Bottom Water.

The Pacific sector contains the largest reservoir of Warm Deep Water, about $34.7 \times 10^6\,\mathrm{km}^3$. This reflects the relative scarcity of cold bottom water in the southeast Pacific Basin, and the absence of Antarctic Bottom Water north of the Pacific–Antarctic Ridge. The second largest reservoir of Warm Deep Water, about $28.8 \times 10^6\,\mathrm{km}^3$, exists in the Indian sector. Inspection of individual diagrams for each sector reveals that the most saline Warm Deep Water, nearly 72% of all Warm Deep Water with salinities above $34.75‰$, is found in the Indian sector, although the major source of high salinity water is North Atlantic Deep Water derived from the Atlantic Ocean. This substantiates, in a quantitative manner, the isopycnal analysis of REID and LYNN (1971) and CALLAHAN (1972)

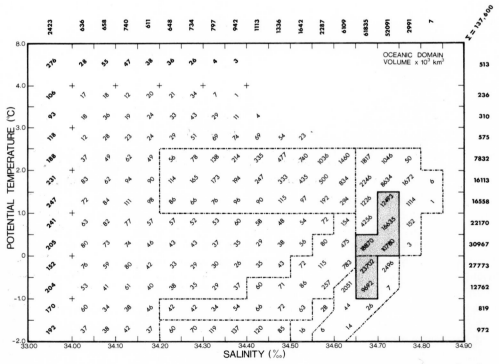

Fig. 14. Volumetric (θ–S) diagram for the oceanic domain of the Southern Ocean. Volumes ($\times 10^3\,\mathrm{km}^3$) are shown in bivariate class intervals, $0.5°\mathrm{C} \times 0.05‰$. Heavy numerals inside the frame (top and left) are the sums of points beyond the $0.5°\mathrm{C} \times 0.05‰$ grid. Univariate potential temperature and salinity distributions are exterior to the diagram at the top and right, respectively. Dashed lines delimit the water masses outlined in Fig. 3. Shading indicates the 67% frequency limits.

Water characteristics of the Southern Ocean sout

Table 1
(a) Volumetric characteristics

Sector	Surface area ($10^3 \times km^2$)	Volume ($10^3 \times km^3$)	Mean (m
Oceanic Domain			
Atlantic	10,302	41,261	40
Indian	13,770	51,561	37
Pacific	11,205	44,778	39
Total	35,277	137,600	39
Shelf Domain			
Atlantic	789	409	5
Indian	730	394	5
Pacific	945	489	5
Total	2,473	1,292	518

Total ... -1.43 ... 34.460

(b) Water Volumes ($10^3 \times km^3$)

Water mass	Atlantic	%	Indian	%	Pacific	%	Total	%
Oceanic Domain								
Surface Water	1,483	3.59	2,393	4.64	2,883	6.44	6,759	4.91
Winter Water	275	0.67	172	0.34	160	0.36	607	0.44
Transition I	1,443	3.50	801	1.55	310	0.69	2,554	1.86
Transition II	1,717	4.16	3,626	7.03	3,298	7.37	8,641	6.28
Warm Deep Water	17,494	42.40	28,754	55.77	34,753	77.61	81,001	58.87
Antarctic Bottom Water	18,849	45.68	15,815	30.67	3,374	7.53	38,038	27.64
Total	41,261		51,561		44,778		137,600	
Shelf Domain								
Surface Water	33.9	8.28	60.0	15.25	101.8	20.81	190.7	14.76
Modified Warm Deep	44.7	10.92	66.9	17.01	178.2	36.43	294.8	22.82
Low Salinity Shelf	154.8	37.81	224.0	56.94	99.6	20.36	478.4	37.02
High Salinity Shelf	152.6	37.27	38.6	9.81	105.9	21.65	297.1	23.00
Ice Shelf Water	23.4	5.72	3.9	0.99	3.7	0.76	31.0	2.40
Total	409.4		393.4		489.2		1292.0	

who noted the North Atlantic Deep Water to first move eastward around South Africa before crossing the Polar Front. It should also be noted that the Pacific sector contains the lowest percentage, about 10%, of Warm Deep Water with salinities above 34.75‰. This probably reflects the inflow and mixing of low salinity water from the southwestern Pacific Ocean suggested by CALLAHAN (1972).

As noted earlier on the θ–S scatter diagrams, two major transitional water masses overlie Warm Deep Water in different areas of the Southern Ocean. Figure 14 shows that these transitional water masses represent larger volumes of water than might first be expected, making up over 8% of the oceanic water column. Transition I between Winter Water and Warm Deep Water is the smallest of the two transitional waters. It is, however, more voluminous than its overlying parent water mass. The volumetric trend of Transition I is heavily weighted towards the high salinity end of the mixing curve, showing that this water is much richer in deep water than in surface water.

Transition II represents the third largest water mass in the oceanic domain. Unlike Transition I, which clearly connects the modes of Winter Water and Warm Deep Water, the volume of water in Transition II decreases uniformly with decreasing salinity, approximately along the 2°C isotherm, and exhibits no well-defined overlying mixing component. Transition II is especially predominant in the Pacific and Indian sectors.

Winter Water forms a small surface water mode between $\theta = -1.5$ to $-2.0°C$ and $S = 34.2$ to 34.5‰. Although easily identified by its core properties, Winter Water comprises less than 0.5% of the total volume of water in the Southern Ocean.

Shelf domain

The total volume of water occupying the continental shelf domain (Fig. 15) is $1292 \times 10^3 \, km^3$, or less than 1% of the volume of oceanic water.

A general characteristic of the shelf domain is the pronounced bivariate mode formed by Low Salinity Shelf Water and High Salinity Shelf Water; about $685 \times 10^3 \, km^3$ or over 50% of all shelf water occurs in the interval $\theta = -1.6$ to $-2.0°C$, and $S = 34.3$ to 34.9‰. Correspondingly, the univariate temperature mode for Antarctic shelf waters is $\theta = -1.8$ to $-2.0°C$; the univariate salinity mode is $S = 34.5$ to 34.6‰.

Low Salinity Shelf Water represents the most voluminous shelf water mass; about 37% of the total volume. The $\theta-S$ characteristics of this water mass are nearly identical to those of Winter Water in the oceanic domain. Since its density is less than that of Warm Deep Water or Antarctic Bottom Water, Low Salinity Shelf Water cannot directly cause deep convection and bottom water production.

High Salinity Shelf Water is by far the densest Antractic water mass with σ_0 values approaching 28.20 in the Ross Sea and 28.00 in the Weddell Sea. This water is easily dense enough to drain from the continental shelf and displace any offshore water mass. Indeed, most explanations of bottom water formation (cf. MOSBY, 1934; GILL, 1973; FOSTER and CARMACK, 1976) assume High Salinity Shelf Water to be an important mixing component in bottom water production. It is, therefore, important to note that even during summer, the volume of this water mass is $297 \times 10^3 \, km^3$ or 23% of the water on the continental shelf. Since most High Salinity Shelf Water is not trapped within topographic depressions, it must be held on the continental shelves by geostrophic circulation as discussed by GILL (1973). The largest pool of High Salinity Shelf Water occurs in the southwestern Weddell Sea, with about $153 \times 10^6 \, km^3$, followed by the Ross Sea with $106 \times 10^6 \, km^3$; only $39 \times 10^3 \, km^3$ is contained in the Indian sector, affecting in turn, the low production rate of bottom water in the Indian Ocean.

Modified Warm Deep Water, which forms the warm limb of water in Fig. 15, has been identified as an important mixing component in the bottom water formation models of GILL (1973) and FOSTER and CARMACK (1976). Its volume, $295 \times 10^3 \, km^3$, is nearly equal to that of High Salinity Shelf Water.

Ice Shelf Water, despite its interesting thermodynamic properties, appears to be the least voluminous shelf water mass. (In this analysis, Ice Shelf Water has been conveniently defined as water colder than $-2.0°C$, rather than the actual freezing temperature which is about $-1.9°C$, so the volume of water colder than the freezing temperature

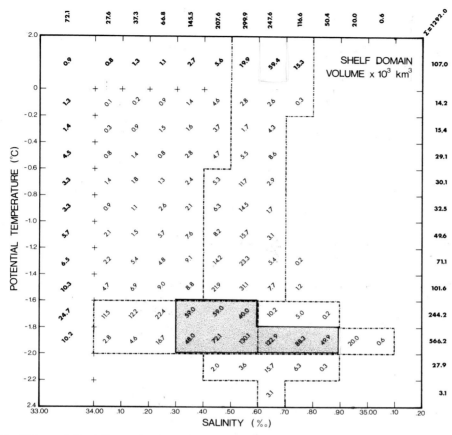

Fig. 15.   Volumetric $(\theta–S)$ diagram for the continental shelf domain of the Southern Ocean. Volumes ($\times 10^3$ km^3) are shown in bivariate class intervals, $0.2°C \times 0.05‰$. Heavy numerals inside the frame (top and left) are the sums of points beyond the $0.2°C \times 0.05‰$ grid. Univariate potential temperature and salinity distributions are exterior to the diagram at the top and right, respectively. Dashed lines delimit the water masses outline in Fig. 3. Shading indicates the 50% frequency limits.

at one atmosphere pressure is actually somewhat higher.) The largest volume of Ice Shelf Water (23×10^3 km^3) is observed in the Weddell Sea near the Filchner and Ronne Ice Shelves.

Bottom water characteristics

The fine-scale volumetric diagram shown in Fig. 16 allows a comparison of bottom water varieties and their quantitative distributions around Antarctica. Numbers interior to this diagram represent the class interval volumes for the Atlantic, Indian and Pacific sectors, and the total volume, respectively.

The 75% frequency limits in Fig. 16 show, as expected, that the bottom waters of the Atlantic sector are typically cold and low in salinity, and hence dominated by mixtures containing Weddell Sea Bottom Water. The 75% frequency boundary of the Pacific sector encloses water warmer than $-0.3°C$, and with salinities up to $34.74‰$

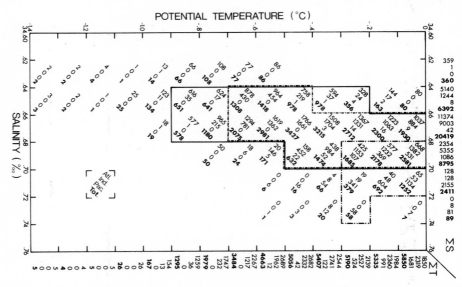

Fig. 16.　Fine-scale volumetric (θ–S) diagram covering the properties of Antarctic Bottom Water. Volumes ($\times 10^3\,\text{km}^3$) are shown in bivariate class intervals, $0.1°\text{C} \times 0.02‰$. Within each class interval, the first value refers to the volume occurring in the Atlantic Ocean sector; the second to that in the Indian Ocean sector; the third to that in the Pacific Ocean sector; and the final (dark) value refers to the total volume (see inset). The 75% frequency limits are also shown for the Atlantic (———), Indian (—·—) and Pacific (—··—) Ocean sectors.

thus indicating the influence of Ross Sea Bottom Water. The Indian sector, which receives bottom water inputs from both sources, mainly trends toward the low salinity bottom water variety, but is also relatively rich in bottom water that is intermediate between Weddell Sea Bottom Water and Ross Sea Bottom Water.

In compiling a volumetric census, a large number of hydrographic observations are integrated onto one θ–S diagram, with the result that water mass identification must be based solely on incremental values of potential temperature and salinity. Thus, water masses with overlapping θ–S characteristics, such as the low salinity bottom waters of the Weddell Sea and of the Adélie Coast cannot, in general, be uniquely distinguished. However, the total volume of Adélie Coast Bottom Water (calculated from *individual* hydrographic stations) appears to be about $15 \times 10^3\,\text{km}^3$ or less than 0.5% of the total Antarctic Bottom Water volume; contributions from other source regions are likely even smaller. It, therefore, seems reasonable, as a first approximation, to refer to bottom water in the low salinity tail on the θ–S diagram (cf. Fig. 9) as being derived by mixing with Weddell Sea Bottom Water, and in the high salinity tail as being derived by the admixture of Ross Sea Bottom Water. More specifically, Antarctic Bottom Water with $S \leqslant 34.68‰$ is considered to be a Weddell Sea Bottom Water mixture, with $S > 34.70‰$ to be a Ross Sea Bottom Water mixture, and with $34.68 < S‰ \leqslant 34.70$ to be a blend of both. On this basis, the total volume of Antarctic Bottom Water showing the Weddell Sea influence is $27.2 \times 10^6\,\text{km}^3$, of which 62% occurs in the Atlantic sector. The volume of Antarctic Bottom Water influenced by the Ross Sea is only $2.5 \times 10^6\,\text{km}^3$, while the common variety accounts for $8.8 \times 10^6\,\text{km}^3$.

5. APPLICATIONS

An advantage of the *quantitative θ–S* approach to water mass analysis is that it allows a rapid and concise summation of a large amount of hydrographic data in a given region of the ocean. Because of this, volumetric analysis provides useful baseline information for subsequent water mass structure, formation and circulation studies. In addition, a volumetric census can be used to gain insight into general heat and mass budgets, as the examples below illustrate.

Winter heat budget

As noted earlier, the oceanic water mass designated Winter Water is believed to be produced by wintertime convection. That is, the temperature of surface water is first cooled to the freezing temperature, then the salinity of the water is raised by brine rejection during sea-ice formation. In the volumetric census presented here, nearly all of the hydrographic stations were occupied during the austral summer, with the result being that the surface layer was generally warm and diluted by ice melting. Winter Water, therefore, existed only as a remnant water mass at depths of 50–300 m. It, therefore, seems reasonable to assume that the wintertime *volumetric* conditions can be approximated by assuming that during winter, the entire surface layer is transformed into Winter Water. Indeed, the relatively few winter hydrographic stations suggest this may be the case. Subsequently, it should be possible to approximate the winter heat budget of the surface layer by calculating the loss of heat required to convert all summertime surface water into water with Winter Water characteristics. This is accomplished by first determining the amount of cooling required to cool the surface water to the freezing point ($-1.9°C$); calculating the amount of sea-ice formation required to increase the salinity of this water up to that of Winter Water; and then computing the latent heat loss required to form this volume of ice. In order to cool all surface water south of the Polar Front down to the freezing-point requires the removal of approximately 20×10^{18} kcal or about 55 kcal cm^{-2}. Subsequently, to increase the salinity of surface water up to the minimum salinity of the Winter Water mode requires 870×10^{12} g salts, which in turn requires the freezing of about 30×10^{3} km^{3} seawater. (This would yield a mean thickness of 0.9 m, which is quite close to the generally accepted value of 1 m yr^{-1}.) The above freezing requires the additional removal of 2.5×10^{18} kcal or about 7 kcal cm^{-2}. The total *winter* heat budget of the oceanic domain of the Southern Ocean, assuming no significant advective exchanges or mixing with the warmer and more saline water below, is thus estimated at about 62 kcal cm^{-2}.

Bottom water budget

What is the relative production rate of bottom water derived from the Weddell Sea compared to that of the Ross Sea? Although the existing volume of a water mass cannot be taken as evidence of a formation rate (any more than standing stock of planktonic algae relates to primary production) the volumetric comparison at least allows an estimate of the relative effects of the two bottom water source regions on water column structure. The major difficulty in applying volumetric analysis to this question lies in defining unique and meaningful water mass boundaries. For example, CARMACK and FOSTER (1976b) argued that a subtle θ–S slope change at about $-0.7°C$ in the outflow of water from the Weddell Sea was effected by the introduction of newly formed bottom water. Furthermore, REID, NOWLIN and PATZERT (1977) showed evidence that such features may, indeed, represent true water mass boundaries. However, to critically examine individual hydrographic stations in this manner, and then apply a volumetric tabulation would indeed be a time consuming task. Instead, as a first approximation, the comparison can be based on the simple division of water masses suggested in Section 4. That is, all bottom water with $S \leqslant 34.68‰$ represents a mixture with Weddell Sea Bottom Water and all bottom water with $S > 34.70‰$ represents a mixture with Ross Sea Bottom Water. Then the total volume of classical Antarctic Bottom Water exhibiting Weddell Sea derived characteristics is $27.2 \times 10^6 \, km^3$, whereas the volume with Ross Sea characteristics is only $2.5 \times 10^6 \, km^3$. For comparison, a three-component mixing nomogram for Warm Deep Water, Weddell Sea Bottom Water and Ross Sea Bottom Water was constructed and applied to the volumetric diagram. The results of this calculation yield a somewhat higher percentage of Ross Sea Bottom Water, about 20%, comprising the total volume of Antarctic Bottom Water. It would, therefore, appear from this rough comparison that the Weddell Sea influence on the abyssal waters ($\theta < 0°C$) of the Southern Ocean is about 5 to 10 times greater than that of the Ross Sea.

Finally, a crude estimation of bottom water production can be made based on the residence time of High Salinity Shelf Water. GILL (1973) estimated the residence time of water on the continental shelf of the Weddell Sea to be about 3.5 to 7 years. Assuming this is representative of shelf water residence times, and taking the total volume of High Salinity Shelf Water to be $297 \times 10^3 \, km^3$, then the annual flux of High Salinity Shelf Water off the continental shelf is calculated at 42–$84 \, km^3 \, yr^{-1}$, or 1.3–$2.7 \times 10^6 \, m^3 \, s^{-1}$. From the bottom water formation model of FOSTER and CARMACK (1976b), classical Antarctic Bottom Water is one-fourth part High Salinity Shelf Water. So, on the basis of this rough budget estimate, the annual production of new bottom water is expected to be 5–$10 \times 10^6 \, m^3 \, s^{-1}$.

Acknowledgements—The author is grateful to T. D. FOSTER for suggesting that a volumetric census of the Southern Ocean should be prepared, and to J. L. REID, K. AAGAARD, S. S. JACOBS and R. J. DALEY for their critical reading of the original manuscript. It is a pleasure to acknowledge the assistance of S. LOWE for data reduction and computer programming, and J. MORLEY and S. W. FLYNN for preparing the figures used in the text. This work was supported by the office of Polar Programs, National Science Foundation under Grant GV 41578.

REFERENCES

CALLAHAN J. E. (1972) The structure and circulation of Deep Water in the Antarctic. *Deep-Sea Research*, **19**, 563–575.

CARMACK E. C. (1974) A quantitative characterization of water masses in the Weddell Sea during summer. *Deep-Sea Research*, **21**, 431–443.

CARMACK E. C. and T. D. FOSTER (1975a) Circulation and distribution of oceanographic properties near the Filchner Ice Shelf. *Deep-Sea Research*, **22**, 77–90.

CARMACK E. C. and T. D. FOSTER (1975b) On the flow of water out of the Weddell Sea. *Deep-Sea Research*, **22**, 711–724.

DEACON G. E. R. (1973) The hydrography of the Southern Ocean. '*Discovery*' *Reports*, **15**, 1–124.

FOFONOFF N. P. (1956) Some properties of sea water influencing the formation of Antarctic bottom water. *Deep-Sea Research*, **4**, 32–35.

FOSTER T. D. (1972) An analysis of the cabbeling instability in sea water. *Journal of Physical Oceanography*, **2**, 294–301.

FOSTER T. D. and E. C. CARMACK (1976a) Temperature and salinity structure in the Weddell Sea. *Journal of Physical Oceanography*, **6**, 36–44.

FOSTER T. D. and E. C. CARMACK (1976b) Frontal zone mixing and Antarctic Bottom Water formation in the southern Weddell Sea. *Deep-Sea Research*, **23**, 301–317.

GILL A. E. (1973) Circulation and bottom water formation in the Weddell Sea. *Deep-Sea Research*, **20**, 111–140.

GORDON A. L. (1966) Potential temperature, oxygen and circulation of bottom water in the Southern Ocean. *Deep-Sea Research*, **13**, 1125–1138.

GORDON A. L. (1971a) Oceanography of Antarctic Water. In: *Antarctic Oceanography I, Antarctic Research Series*, J. L. REID, editor, American Geophysical Union, **15**, pp. 169–203.

GORDON A. L. (1971b) Antarctic polar front zone. In: *Antarctic Oceanography I, Antarctic Research Series*, J. L. REID, editor, American Geophysical Union, **15**, pp. 205–221.

GORDON A. A. (1971c) Recent physical oceanographic studies of Antarctic waters. In: *Antarctic Research*, L. QUAM and H. PORTER, editors, American Association for the Advancement of Science, pp. 609–629.

GORDON A. L. (1972) Spreading of Antarctic bottom water II. In: *Studies in physical oceanography—a tribute to George Wüst on his 80th birthday*, A. L. GORDON, editor, Gordon & Breach, **2**, pp. 1–17.

GORDON A. L. (1974) Varieties and variability of Antarctic Bottom Water. *Colloques International du Centre national de la recherche scientifique No. 215, Processus de Formation des Eaux Océaniques Profondes*, pp. 33–47.

GORDON A. L. (1975) An Antarctic oceanographic section along 170°E. *Deep-Sea Research*, **22**, 357–377.

GORDON A. L. and P. TCHERNIA (1972) Waters of the continental margin off Adélie Coast, Antarctica. In: *Antarctic Oceanography II: The Australian–New Zealand Sector, Antarctic Research Series*, D. E. HAYES, editor, American Geophysical Union, **9**, pp. 59–69.

HELLAND-HANSEN (1916) Nogen hydrografiske medoder. *Scand. Naturforsker Möte*, pp. 357–359.

HUFFORD G. L. and J. M. SEABROOKE (1970) Oceanography of the Weddell Sea in 1969 (IWSOE). *United States Coast Guard Oceanographic Report CG 373-13*, 33 pp.

JACOBS S. S., A. F. MOSS and P. M. BRUCHHAUSEN (1970) Ross Sea Oceanography and Antarctic Bottom Water formation. *Deep-Sea Research*, **17**, 935–962.

JACOBS S. S. and D. T. GEORGI Observations on the Southwest Indian/Antarctic Ocean. In: *Voyage of Discovery*, M. V. ANGEL, editor, *Deep-Sea Research* supplement to Vol. **24**, 43–84.

KLEPIKOV V. V. (1958) The origin and diffusion of Antarctic ocean-bed water. *Problems of the North*, **1**, National Research Council of Canada, Ottawa, pp. 321–333.

LUSQUIÑOS A. J. (1963) Extreme temperatures in the Weddell Sea. *Årbok for Universitetet i Bergen, Matematisk-naturvitenskapelig serie*, **23**, 1–19.

MANTYLEA A. W. (1975) On the potential temperature in the Abyssal Pacific Ocean. *Journal of Marine Research*, **33**, 341–354.

MONTGOMERY R. B. (1958) Water characteristics of the Atlantic Ocean and of World ocean. *Deep-Sea Research*, **5**, 134–148.

MOSBY H. (1934) The waters of the Atlantic Antarctic Ocean. *Scientific Results of the Norwegian Antarctic Expeditions 1927–1928*, **1**(11), 131 pp.

REID J. L. and R. J. LYNN (1971) On the influence of the Norwegian–Greenland and Weddell Seas upon the bottom waters of the Indian and Pacific Oceans. *Deep-Sea Research*, **18**, 1063–1088.

REID J. L. and W. D. NOWLIN (1971) Transport of water through the Drake Passage. *Deep-Sea Research*, **18**, 51–64.

REID J. L., W. D. NOWLIN and W. C. PATZERT On the characteristics and circulation of the Southwestern Atlantic Ocean. *Journal of Physical Oceanography*, **7**, 62–91.

SEABROOKE J. M., G. L. HUFFORD and R. B. ELDER (1971) Formation of Antarctic Bottom Water in the Weddell Sea. *Journal of Geophysical Research*, **76**, 2164–2178.

TRESHNIKOV A. F., A. A. GIRS, G. I. BARANOV and V. A. YEFIMOV (1973) *Preliminary programme of the polar experiment for the south polar region*, Arctic and Antarctic Research Institute, Leningrad, 55 pp.

WÜST G. (1933) Das Bodenwasser und die Gliederung der atlantischen Tiefsee. *Wissenschaftliche Ergebnisse der Deutschen Atlantischen Expedition auf dem Vermessungs und Forchungsschiff "METEOR", 1925–57*, **6**, 107 pp.

WÜST G. (1939) Bodentemperatur and Bodenstrom in der atlantischen, indischen und pazifichen Tiefsee. *Beitrage zur Geophysik*, **54**, 1–8.

Appendix I
Hydrographic station list

Ship	Year	NODC ID number	Station number
Shelf Domain—Atlantic			(38 stations)
Glacier	1968	318037	1 3 9 14 15 17 19 24
Glacier	1969	318085	1 5 10 13 14 15 18 23
Glacier	1970	318154	2 8 21 25 26 38 40
San Martin	1955	080695	5
San Martin	1961	080002	1
San Martin	1965	080029	4
San Martin	1966	080035	4
San Martin	1969	080061	9
Staten Island	1957	310561	22
Discovery	1932–33	740038	242 323
Discovery	1936–39	740040	164
Deutschland	1912	060044	63
Ob	1963	099150	19
Ob	1965	099148	25 28 45
Ob	1966	099149	24
Shelf Domain—Indian			(59 stations)
Eltanin	1969	37*	987 996 999 1012 1062 1073 1083
Staten Island	1959	310613	14 15 16
Staten Island	1950	350712	11
Burton Island	1948	310592	10
TH	1934	580812	5
Ob	1956	900830	18 26 29 40 42 43
Ob	1957	900004	1 3 5 33 35 36 38 39 42 49 51 76 80 129
Ob	1958	900005	1 15 31 32
Ob	1960	900029	33
Ob	1961	900057	19 20 23 27 30 34 53 88
Ob	1963	900150	9
Ob	1965	900148	10 16 49 53 57 63 65 67 70 72
Ob	1966	900149	35 38
Shelf Domain—Pacific			(76 stations)
Eltanin	1967	318035	38
Eltanin	1967	311207	99 100 101 102 103 104 105 107
Eltanin	1968	311709	22 25 36 40 56 59
Glacier	1956	310563	3 5 6
Glacier	1960	310652	10
Glacier	1962	310868	4
Glacier	1967	310880	32
Atka	1958	310590	8 17
Atka	1964	310194	14 17 47
Staten Island	1958–59	310613	5 10
Staten Island	1960–61	310672	3 4 6 20 23 25 26 28 61 66 69 70 74 77
Staten Island	1965	310409	9
Edisto	1956	310514	7
Edisto	1963	310150	40 64 94
Burton Island	1962	310867	1 3 10 11 17 20 23 24 25
Eastwind	1962	310951	4 12
San Martin	1956	080695	28 30 33
San Martin	1965	080029	12
Ob	1956	90005	33
T. Washington	1971	*	66
W. Scoresby	1930	740037	440 441

Appendix I (cont.)

Ship	Year	NODC ID number	Station number
Shelf Domain—Pacific (cont.)			
Discovery	1930–31	740037	251 256 282 283
Discovery	1932–33	740038	237
Discovery	1936–39	740040	26 32 33 37 39
Oceanic Domain—Atlantic			(229 stations)
Glacier	1968	318037	25 27 28 29 30 31 33 35 37 39 41 47 51 54 55 56 58 61 64 49 72 80
Glacier	1969	318085	25 26 28 29
Glacier	1970	318154	28 32
Eltanin	1962	310187	88
Eltanin	1964	310707	1 2 4 6 10 13 14 15 16 17 18 20 21 22 27 37 39 41 42 43 48 49 51 52 54 62 75 84 85 86 88 91 93 94 96 97 98 99 109 168 169 170 171 172 173 174 175 176 179 183 185 186 188 191 192 193 194
Eltanin	1967	311207	4 6 11 12 13 14 17 18 19 20 21 22 23 24 25 26 27 28 29
Discovery	1930	740035	81
Discovery	1930–31	740037	135 161 169 192 237 335 343
Discovery	1932–33	740038	44 54 101 104 247 272 347 348 349 350 352 353 355 359
Discovery	1934–35	740039	11 101 107 111 115 159 168 169 218 219 220 221 223 225
Discovery	1936–39	740040	105 109 110 128 130 132 134 136 234 254 256 257 258 259 261 264 380 381 382 384 387 389 390 394 396 446 482 483 484 485 498 500 501 502 503 504 505 506 507 508 511 526 527 528 534 535 538 539 541
Deutschland	1912	060044	54 59
Meteor	1926	060004	113 115 117 118 124 125 126 129 130 131
San Martin	1961	080061	6 10 11 15 16 17 18 21
San Martin	1959	080697	4
San Martin	1968	080056	2
Ob	1958	900005	172
Ob	1961	900057	62 63
Ob	1962	900058	9
Ob	1965	900148	32
Knorr	1973	*	79 80 81 82 84 85 86 87 88 89 90 91
Oceanic Domain—Indian			(200 stations)
Eltanin	1968	311697	16 17 18 19 20 21 22 24
Eltanin	1968	311698	8 10 11 26 28
Eltanin	1968	311699	4
Eltanin	1970	312091	6 8 9 10 11 12 14 16 20 27
Eltanin	1968	34*	850
Eltanin	1969	37*	989 991 1067 1085 1030 1036
Eltanin	1969	38*	4 8
Eltanin	1969	39*	1161 1162
Eltanin	1970	41*	7 8 9 10 11 12 13
Eltanin	1970	44*	1230 1231 1233 1234 1236 1238
Eltanin	1970	45*	1251 1252 1254 1257 1258 1261
Eltanin	1971	47*	1281 1283 1291 1298 1299 1303 1309
Eltanin	1971	49*	1351 1353 1355 1357 1371 1377 1378
Eltanin	1971	50*	1407 1428 1430 1431 1435 1445 1448 1464 1582 1496 1508 1512

Appendix I (*cont.*)

Ship	Year	NODC ID number	Station number
Oceanic Domain—Indian (*cont.*)			
Eltanin	1972	54*	1564
Discovery	1930–31	740037	1
Discovery	1932–33	740038	118 119 120 121 122 123 125 126 128 129 130 131 132 150 152 153
Discovery	1934–35	740039	116 117 118 119 121 122 123 124 226 227 228 229 230 231 232 233 234 235 236 238 246
Discovery	1936–39	740040	20 23 25 61 62 64 66 68 69 70 71 301 302 304 305 306 307 308 309 310 311 312 313 314 329 331 332 334 337 391 392 412 464 465
Discovery	1930–31	740227	12 26 28 29 34
Discovery	1951	740702	156 158 159
Ob	1956	900830	16 19 46 105 109 110 111 113 114
Ob	1957	900004	8 17 31 43 45 53 58 74 78 82 121 123 132 136
Ob	1960	900029	34
Ob	1961	900057	17 24 25 89
Ob	1963	900150	25
Ob	1965	900148	55
Staten Island	1959	310613	17 18
Burton Island	1958	310592	8 9 12
Argo	1962	310184	118
Vema	1960	310834	32
Oceanic Domain—Pacific			
Eltanin	1963	310187	36
Eltanin	1964	310707	63 64 71 107 108 109 110 111 113 114 115 116 117 118 123 124 127 128 139 142 144 146 147 148 151 152 153 154 155 156 157 158 159 160 161 162 163 164 165 166 211 212 213 214 215 217 218 219 220 221 222 225 226 227 230 233 240 245 246 247 248 249 253 254 257 258 259 260 261 263 264 271 278 279 284 292 293 294 295 298 303 305
Eltanin	1965	310790	5 7 8 10 12 13 14 15 16 17 18 19 20 21 22 23 24 26 28 30 32 46 47 48 51 58 59 61 64 66 67 68 69 76 77 80 83 84 86 87 110 111
Eltanin	1967	311207	33 34 37 38 45 46 47 86 95 96 97 98 108 109 110 111 112 113 114 116
Eltanin	1968	311696	1 2 4 7 9 10 11 12 14 15 17
Eltanin	1968	311697	7
Eltanin	1968	311699	7
Eltanin	1969	37*	972 974
Eltanin	1969	38*	1
Eltanin	1968	311709	63 88
Eltanin	1970	42*	1170 1171 1172 1173 1174 1175 1176 1178 1180
Eltanin	1970	43*	1181 1182
Eltanin	1970	44*	1222 1223 1224 1226 1227 1228 1229
Discovery	1930–31	740037	285
Discovery	1932–33	740038	36 201 205 233 236
Discovery	1934–35	740039	42 43 44 51 52 54 56 57 58 59 60 61 62 70 74 75 77 81 82 179

Appendix I (cont.)

Ship	Year	NODC ID number	Station number
Oceanic Domain—Pacific (cont.)			
Discovery	1936–39	740040	341 343 355 357 358 362 363 366 367
Discovery	1951	740702	118 119 120 121
Staten Island	1960–61	310672	16 19 27 37 41 51 55 58
Glacier	1960	310652	12
Glacier	1962	310868	5
Atka	1964	310194	8 10 19 24 27
Eastwind	1960	310651	1 2 3
Northwind	1972		1 2 3 4 5 6 8 9 10 11 12 13 14 15 16 17 18 19 21 22
Argo	1961	310181	181
Hudson	1973	*	227
Ob	1958	900005	71 83 91
H.H.	1969	498001	58 59 60 61 62 63 64 65 66 67 69
Knorr	1973	*	77 78
Melville	1974	*	280 281 282 283 284 285 287 288 289 290
Meteor	1926	060004	108 109 110
T. Washington	1971	*	1 2 3 6 9 11 12 13 16 17 18 19 20 21 22 23 24 25 26 27 28 29 30 32 33 34 35 36 37 38 39 40 41 42 43 44 45 46 47 48 50 51 52 53 54 55 57 59 61 62 63
T. Washington	1970	*	10 12

(Pac. Domain)

Note

On EL-27 (1967) There were
≈ 20 shelf stns (# 629-649)
in Ross

On EL-32 (1968) There were >20
shelf stns in Ross (# 695-750

Representative stns only used by E.C.
(see p.29) (no area averages)

Class interval on shelf: θ= 0.2, S=.05

No shelf data more recent than 1971

364 shelf stns.

(but given
as 0.1 in Fig)

Observations on the southwest Indian/Antarctic Ocean*

S. S. JACOBS and D. T. GEORGI

Lamont-Doherty Geological Observatory of Columbia University Palisades, New York 10964

Abstract—Stratification and circulation at intermediate and deeper levels of the southwest Indian/Antarctic Ocean are discussed on the basis of *R. D. Conrad* STD stations taken between South Africa, the Crozet Plateau, Kerguelen Islands and Antarctica from January to April, 1974. Vertical sections of temperature, salinity, oxygen, silicate, phosphate and geostrophic velocity are presented. Medium-scale (10–100 m) density-compensated thermohaline interleaving near the Antarctic Inter-mediate Water (AAIW) salinity minimum (S_{min}) is taken as evidence of erosion through mixing with warmer, saltier water from the northern Indian Ocean. A direct surface origin for AAIW during the austral summer in this region is doubtful, but sufficient production of a precursor water mass appears likely. AAIW in the southwest Indian Ocean resembles its counterpart in the Tasman Sea, where there are some similar topographic features, more than it does the southwest Atlantic S_{min}, which extends much further north.

North Atlantic Deep Water (NADW) influence is strong throughout the southwest Indian Ocean and reaches the Antarctic coast east of the 50° meridian. The deep-water temperature maximum which occurs near or below the oxygen minimum can be traced from Antarctica to north of the Subtropical Convergence. Lateral processes seem dominant in the mixing processes at deep-water levels. θ/S relationships between the deep and bottom water on individual STD stations can clearly be separated into a few linear segments. Combining all stations, the slopes ($d\theta/dS$) of these lines fall into several discrete sets, each retaining a separate imprint of the past history of that water mass. The temperature and salinity values at discontinuities between segments can be correlated only over limited regions.

Water near the Antarctic continental slope plays an important role in the modification of deep water and is also related to the formation of Antarctic Bottom Water (AABW). AABW production is inferred near the eastern end of the Weddell–Enderby Basin. The low silicate characteristic of newly formed AABW is transformed into a bottom silicate maximum within a short distance of the Antarctic continental slope. The major passage for AABW into the northwest Indian Ocean lies between the Crozet and Kerguelen Islands. A variety of near-bottom boundary-layer phenomena are illustrated by the STD and the nephelometer data.

Baroclinic geostrophic transport relative to the deepest observations is approximately 122×10^6 $m^3 s^{-1}$ on each of two meridional sections south of the mid-ocean ridge. Two-thirds of the Antarctic Circumpolar Current passes south of Kerguelen. The Agulhas Current has a baroclinic transport of $136 \times 10^6 m^3 s^{-1}$, and the effects of this current system appear to extend to 4000 m depth. North-easterly currents are inferred below 4000 m on the southeast flank of the Agulhas Plateau and at shallower depths on the southeast flank of the Mozambique Plateau.

1. DATA

DEEP and shallow casts were made with Plessey 9006 and 9040 salinity-temperature-depth (STD) sensors, with data recorded in continuous analog form, and at $4c \cdot s^{-1}$ on a Plessey 8114A Digital Data Logger. Sensor lowering rates were near $0.5 m s^{-1}$ in shallow regions of high thermal gradients and $1 m s^{-1}$ in the more homogeneous deeper water. Only digital data recorded during sensor descent were subsequently processed. All digital data in the upper water column were utilized; only one of every four scans was used at deeper levels.

* Lamont-Doherty Geological Observatory Contribution Number 2504.

Shallow salinity data were corrected for dynamic errors introduced by conductivity and temperature probe time constant mismatch (e.g. Scarlett, 1975) and then smoothed with a 15-point running mean. A 7-point running mean was applied to temperature and depth data to remove digitization noise. A latch filter next passed only levels where depth increased by > 0.2 m. Salinity, temperature and depth data were reconciled to serial observations obtained with an Innerspace Technology surface-actuated-multiple sampler and with individual sample bottles attached to the STD wire. Polynomials up to the third order were necessary to correct the salinity data, which required adjustment for cell constant (salinity offset), temperature and pressure. Resulting accuracies are estimated to be ± 10 m, .01°C and .005‰ for D, T and S.

Dissolved oxygen was determined by the method of Carpenter (1965). Improper reagents caused the loss of data on stas. 214 to 237. Silicate and phosphate analyses were completed on a Beckman DU spectrophotometer on loan from the Scripps Institution of Oceanography and the Naval Undersea Center, San Diego. Methods closely followed those

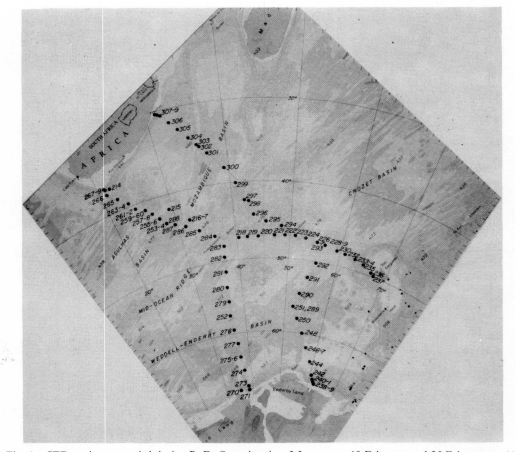

Fig. 1.   STD stations occupied during R. D. *Conrad* cruises 5 January to 19 February and 25 February to 11 April 1974. Multiple numbers at one location indicate deep and shallow casts or reoccupied stations. Topography after Heezen, Tharp and Bentley (1972).

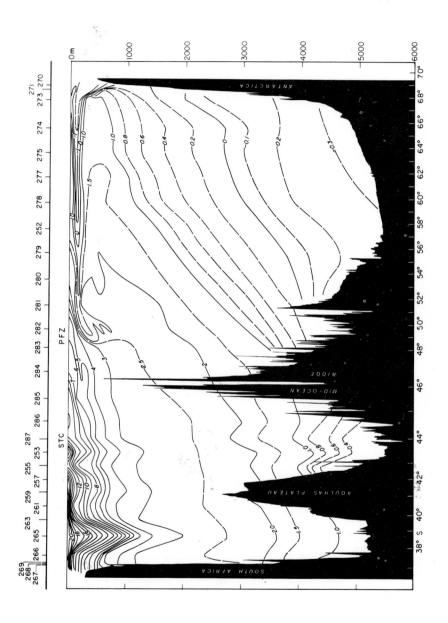

Fig. 2a. Temperature section between Cape Agulhas, Marion Island, and western Enderby Land (see Fig. 1). Isotherms derived from STD data. Vertical exaggeration 400/1 on all sections. Polar Front Zone (PFZ) and Subtropical Convergence (STC) positions approximate.

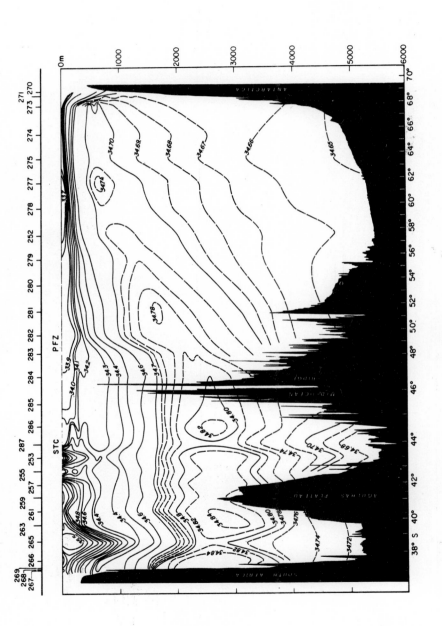

Fig. 3a. Salinity section between Cape Agulhas, Marion Island, and western Enderby Land. Isohalines derived from STD data.

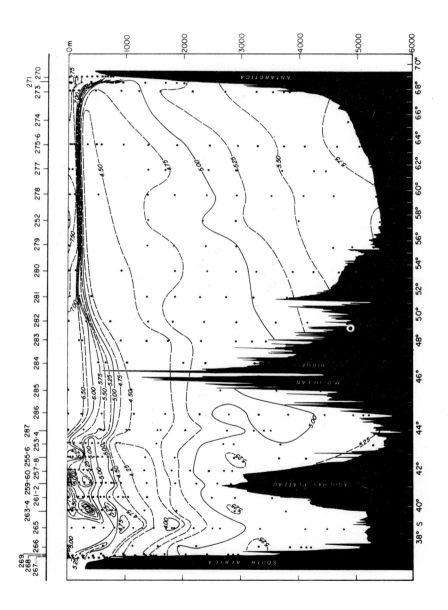

Fig. 4a.   Oxygen section between Cape Agulhas, Marion Island, and western Enderby Land. Dots indicate sample depths.

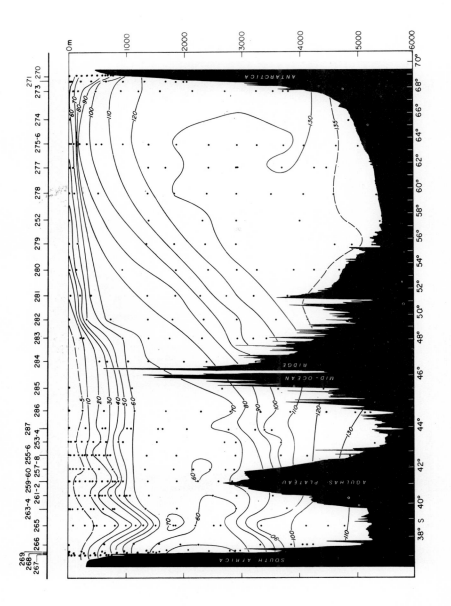

Fig. 5a.   Silicate section between Cape Agulhas, Marion Island, and western Enderby Land.

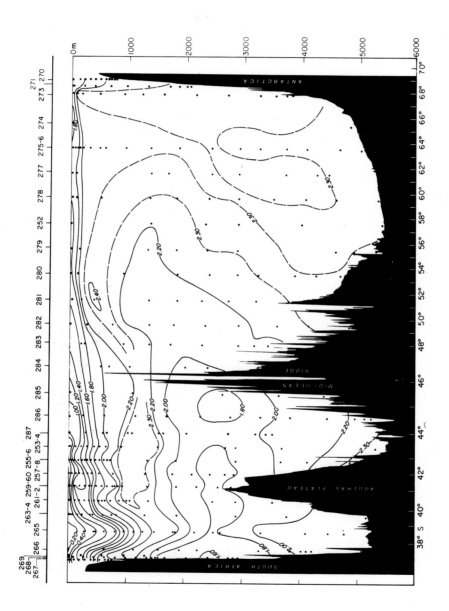

Fig. 6a. Phosphate section between Cape Agulhas, Marion Island, and western Enderby Land.

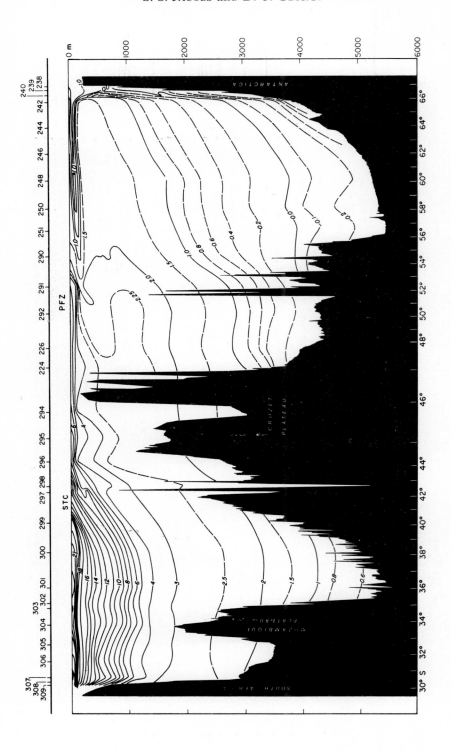

Fig. 2b. Temperature section between Durban, the Crozet Islands, and eastern Enderby Land (see Fig. 1).

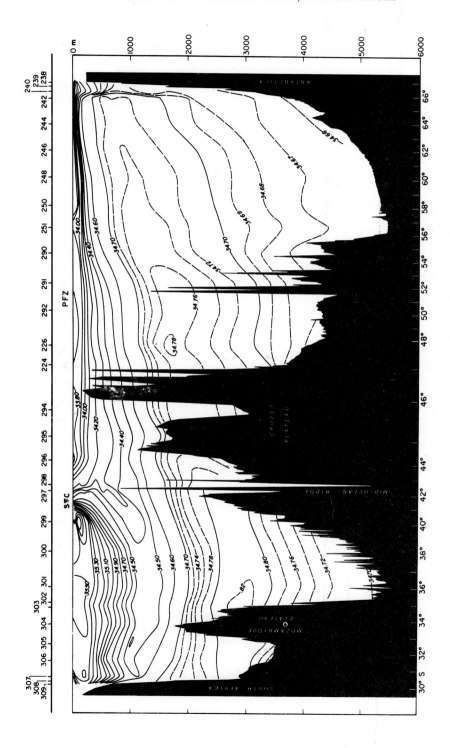

Fig. 3b. Salinity section between Durban, the Crozet Islands, and eastern Enderby Land.

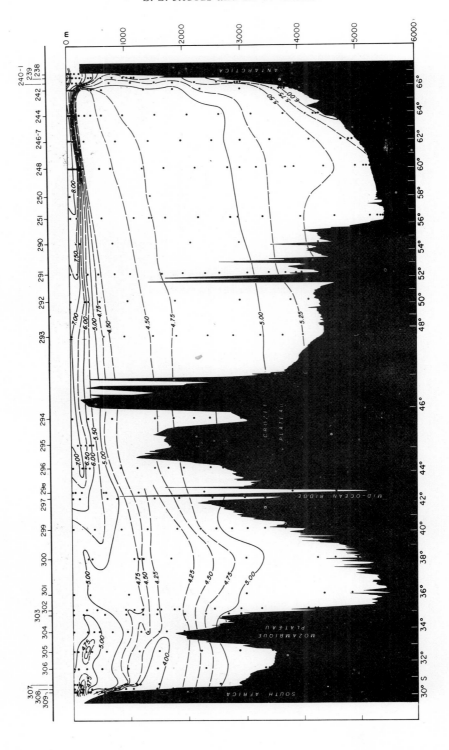

Fig. 4b. Oxygen section between Durban, the Crozet Islands, and eastern Enderby Land.

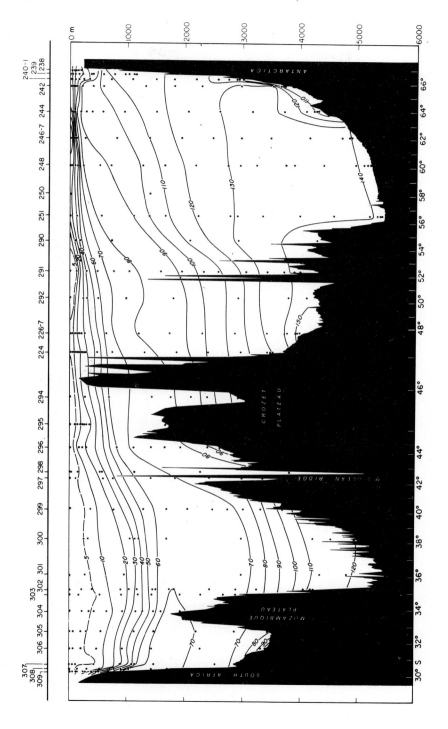

Fig. 5b.   Silicate section between Durban, the Crozet Islands, and eastern Enderby Land.

Note PO4 contoured incorrectly

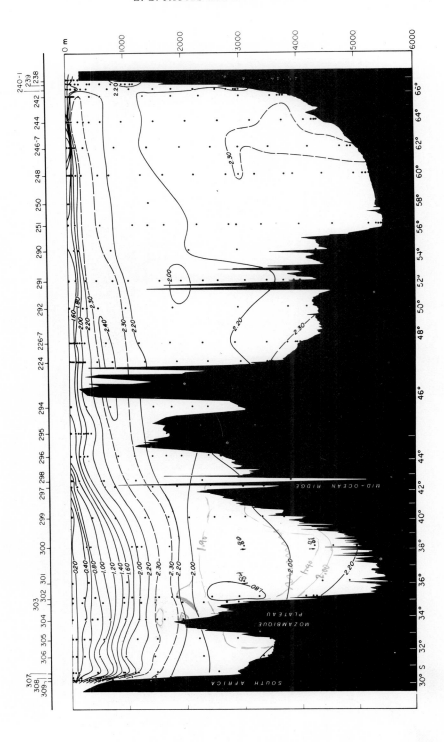

Fig. 6b. Phosphate section between Durban, the Crozet Islands, and eastern Enderby Land.

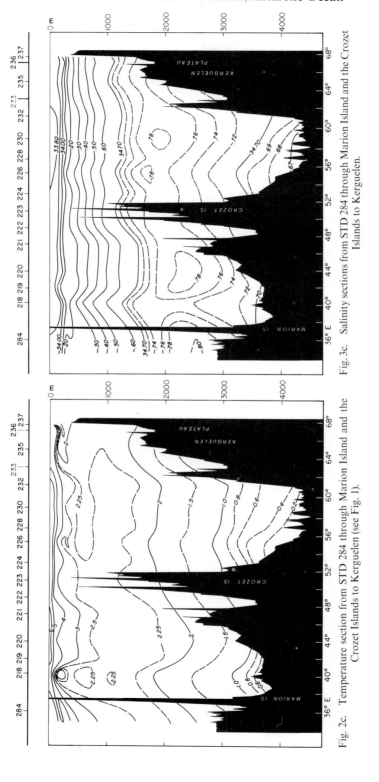

Fig. 3c.   Salinity sections from STD 284 through Marion Island and the Crozet Islands to Kerguelen.

Fig. 2c.   Temperature section from STD 284 through Marion Island and the Crozet Islands to Kerguelen (see Fig. 1).

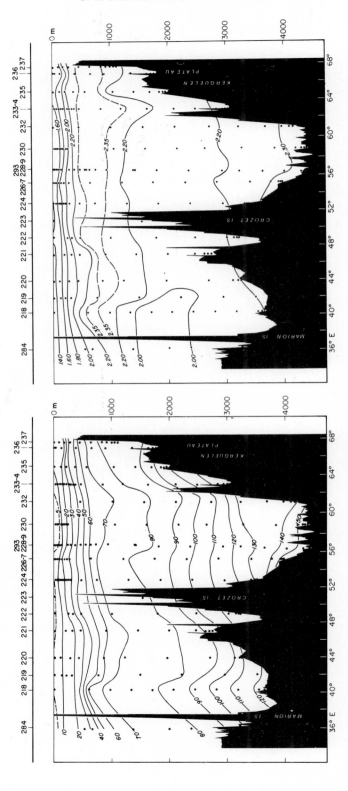

Fig. 6c.   Phosphate sections from STD 284 through Marion Island and the
Crozet Islands to Kerguelen.

Fig. 5c.   Silicate section from STD 284 through Marion Island and the Crozet
Islands to Kerguelen.

described in the Scripps Institution's *Marine Technicians' Handbook* (SIO, 1971) which include slight modifications of procedures outlined in STRICKLAND and PARSONS (1968).

The *Oceanographic Atlas of the International Indian Ocean Expedition* (WYRTKI, 1971) contains numerous horizontal charts of relevant oceanographic parameters in this region. We have thus chosen to present vertical sections of the *Conrad* hydrographic data (Figs. 2–6) and to make frequent reference to the Indian Ocean *Atlas*.

Major transects are (a) from Cape Agulhas via the Prince Edward (Marion) Islands to Antarctica, (b) from Durban through the Crozet Islands to Antarctica, and (c) from Marion Island through the Crozet Islands to Kerguelen. In the vertical sections south of Marion Island, STD 270 is plotted south of STD 271 because 270 is shallower; the correct station positions are shown in Fig. 1. Compromises involving available ship time, equipment malfunctions and competing geophysical interests, at times resulted in poor bottle spacing and in non-consecutive stations along transects. Figs. 2, 3 and 15 were prepared from the STD observations. Sample bottle depths are indicated by dots on the vertical chemistry sections (Figs. 4–6). Observations to within 20 m (av. 7.4 m) of the sea floor at sixty-five of the seventy-one station locations were facilitated by a Benthos 2216 Bottom Pinger.

2. INTERMEDIATE WATER

(a) *Characteristics and distribution*

Antarctic Intermediate Water (AAIW), also referred to in the literature as Subantarctic Intermediate Water, appears on Figs. 3a and 3b as a salinity minimum (S_{min}) between 500 and 1300 m north of the Subtropical Convergence. The S_{min} is above 600 m near the South African continental shelf break and deepest beneath the thick anticyclonic subtropical gyre and Agulhas Current system. North of 35°S, the S_{min} rises to 700 m near its northern limit around 10°S (WYRTKI, 1971, p. 288).

Salinity and temperature at the salinity minimum ranged from 34.153‰, 3.31°C (STD 283) to 34.559‰, 6.15°C (STD 309). Salinities were often significantly lower than reported by earlier expeditions in the same area. For example, DEACON (1937, p. 67) and WYRTKI (1971, p. 289) show the southwest Indian Ocean 34.5 isohaline near 37°S; these *Conrad* data, however, reveal 34.5‰ water to nearly 30°S. Further, a narrow tongue of < 34.5‰ water apparently extends north of 26°S above the west side of the Mozambique Basin (WARREN, 1974, fig. 2; ORREN, 1966, p. 15). This northward relocation of the 34.5 isohaline may result from the denser vertical salinity sampling achieved with *Conrad* STD equipment and *Atlantis* (WARREN, 1974) water bottles rather than from real changes at the S_{min}. The S_{min} layer is relatively thin beneath the Agulhas Current system (Fig. 3a, STD 265; Fig. 7b, STD 308) and could easily be missed by water bottles. Nonetheless, in this region the 34.5 to 34.7 isohalines remain 10° farther south than in the central and eastern Indian Ocean.

Dissolved oxygen varied from > 6.0 to near 4.0 ml · l^{-1} at the salinity minimum, decreasing toward the north. An oxygen maximum is sometimes associated with and slightly above the AAIW salinity minimum in this region. This is a common feature elsewhere (DEACON, 1933; REID, 1965) and is distinctly separate here from a shallower, more prominent oxygen maximum (e.g. STD 304, Fig. 7a).

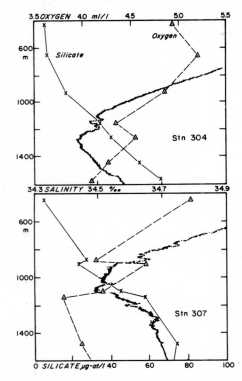

Fig. 7a.   Vertical profiles of salinity, oxygen and silicate at intermediate water levels on two stations southeast of Durban, South Africa.

Silicate varied from $< 20\,\mu$g-at$\cdot$l^{-1} to $> 40\mu$g-at$\cdot$l^{-1} near the salinity minimum, increasing toward the north. Similar values (25–45 μg-at$\cdot$l^{-1} were reported by IVANENKOV and GUBIN (1960, p. 77) for AAIW in the southeast Indian Ocean. The common silicate increase with depth is less rapid across the AAIW, but an actual silicate minimum at this level (Fig. 7a, STD 307) is uncommon. Phosphate ranged from 1.60 to 2.23 μg-at$\cdot$l^{-1} at the salinity minimum, but rarely reached a minimum at this level. The well-developed phosphate maximum (Fig. 6) $> 2.30\,\mu$g-at$\cdot$l^{-1} is about 500 m below the AAIW.

(b) *T/S relationships*

A *T/S* diagram (Fig. 7b) of representative stations north of the Polar Front Zone (PFZ) illustrates some of the AAIW variability. There are several points of particular interest.

(i) The salinity minimum is well defined between $\sigma_1 = 31.7$ and 32.0 near to and north of the Subtropical Convergence (STD 297, 301, 306, 308). It appears on a few stations south of the convergence (e.g. STD 284, but not 294 and 296) where the overlying water column is even lower in salinity. No subsurface salinity minima were observed southeast of a line parallel to the mid-ocean ridge between STD 283 and 298. Near to and southeast of the ridge a shallower subsurface S_{min} and associated T_{min} occur between $\sigma_1 = 31.3$ and 31.6 (STD 297, 284). This is a feature completely separate from the AAIW.

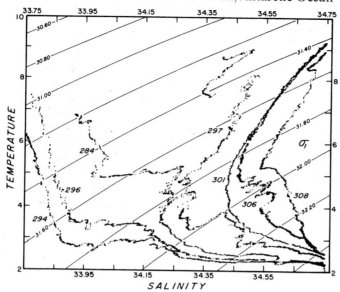

Fig. 7b. AAIW temperature/salinity diagram for representative *Conrad* stations. Intermediate water lies between 31.75 and 32.05.

(ii) The northerly progression to higher salinity and temperature at the S_{min} is not monotonic (see also Fig. 8). STD 296 is 2 degrees north of STD 284, for example, yet is colder and fresher with no well-developed subsurface S_{min}. These stations are near the ridge and Crozet Plateau which are of sufficient elevation to influence circulation and mixing at intermediate levels.

(iii) The smooth S_{min} encountered on STD 301 was an exception. More often the AAIW resembled STD 297, with several S_{min} reaching nearly the same salinity over a temperature range of about 1.5°C and depth range of 250 m. The most distorted core (STD 265, not plotted) was relatively far south and deep beneath the Agulhas Current system.

(iv) The T/S relationship for STD 306 is transitional between those for STD 301 and 308. It portrays active AAIW erosion via the interleaving along density surfaces of cold, fresh layers and warm, salty filaments. Similar structure has been reported in the Antarctic PFZ (GORDON, 1975, p. 363; GORDON, TAYLOR and GEORGI, 1974) and elsewhere in the ocean (KATZ, 1970; PINGREE, 1972). The major density-compensated interleaving T/S structure on STD 306 extends from 1029 to 1240 m, with individual layers from 10 to 100 m thick, temperature variations of .18 to .69°C, and salinity changes of .027 to .086‰. This is the same vertical scale as the thermohaline structure observed by HAMON (1967, fig. 8) near the northern extent of AAIW in the Indian Ocean. Our station spacing is inadequate to evaluate horizontal scales, which HAMON reported to be 5–10 km and KUKSA (1972a) described as lenses with 40–100-km dimensions. The observed medium-scale structure is likely to be an important component of the mixing processes involved in AAIW evolution. "The sharpest and most consistent medium-scale variations of the oceanographic characteristics...occur in the ranges of (the) intermediate layers" of the Indian Ocean (KUKSA, 1972, p. 27). Thermohaline inversions of the type illustrated by STD 306 were not predominantly associated with topographic highs, as may be the case for the core of Mediterranean water in the North Atlantic (KATZ, 1970).

(v) Red Sea water penetrates farthest south along the western margin of the Indian Ocean. WYRTKI (1971, p. 221, pp. 282–287) places its southern extent near AAIW-level density surfaces at about 25°S, with a salinity just below 34.8‰ and oxygen below $3.0 \, \mathrm{ml} \cdot \mathrm{l}^{-1}$. It mixes strongly with AAIW and provides the high-salinity component for the structure observed on STD 306. The dissolved oxygen data support this interpretation with relatively high-oxygen AAIW intruding into water with oxygen below $4.0 \, \mathrm{ml} \cdot \mathrm{l}^{-1}$ (Fig. 7a). The highly modified northern Indian Ocean water extends, as a tongue of low-oxygen, high-salinity water, south along the South African coast to about 33°S and then southeast across the Mozambique Basin (WYRTKI. 1971, pp. 251, 291). The presence of this northern Indian Ocean water mass in the southwest Indian Ocean was shown earlier by CLOWES and DEACON (1935).

(vi) Temperature minima (T_{min}) are often associated with the AAIW, from the S_{min} depth to a few hundred meters deeper (DEACON, 1933, p. 221). Small temperature inversions have also been reported below the AAIW (WÜST, 1936; DEACON, 1937, p. 68). These phenomena are illustrated by the portion of STD 297 near the $\sigma_1 = 32.0$ surface. They may derive from the T_{min} at the base of the Antarctic Surface Water or be induced as a result of filaments of warmer water mixing into the AAIW. The position of the T_{min} below the S_{min} could also result from the greater temperature/depth gradient above than below the S_{min} core, relative to the corresponding salinity/depth gradient, or from a horizontal displacement in the origin of these features and subsequent sinking along density surfaces. These data are inadequate to resolve the source (s) of the observed temperature minima.

(c) *Circumpolar comparisons*

In the southwest Indian Ocean the AAIW salinity-minimum appears as much as 200 m deeper than at other locations in the world ocean (WÜST, 1936, Beilage IX; REID, 1965, p. 51, 61; ROCHFORD, 1960, fig. 3). Large AAIW depth changes occur below the Agulhas Current system, over the South African continental slope and at the southern edge of the anticyclonic gyre. In the South Pacific, the steepest slopes (of the $80 \, \mathrm{cl} \cdot \mathrm{ton}^{-1}$ surface) occur in the western boundary currents and in the Antarctic Circumpolar Current (REID, 1965, p. 30). In the South Atlantic, the major depth changes occur in a zonal band between 49°S and 37°S (WÜST, 1936).

The S_{min} core is not coincident with a single isopycnal surface in all the southern oceans nor at all stations on a single meridional section. WÜST (1936) followed the Altantic salinity minimum from $\sigma_t = 27.05$ (100% AAIW) at the surface in the Polar Front Zone to $\sigma_t = 27.45$ (0% AAIW) at its northern limit. On the 1970 South Atlantic section taken by the Canadian research vessel *Hudson* along 30°W, the salinity minimum is between the 85 and $98 \, \mathrm{cl} \cdot \mathrm{ton}^{-1}$ surfaces. On these *Conrad* stations the S_{min} occurs between the 85 and $100 \, \mathrm{cl} \cdot \mathrm{ton}^{-1}$ surfaces (ave = $92 \, \mathrm{cl} \cdot \mathrm{ton}^{-1}$, $\sigma_t \approx 27.24$). GORDON, TAYLOR and GEORGI, (1974) found the salinity minimum near the $90 \, \mathrm{cl} \cdot \mathrm{ton}^{-1}$ surface on three north-south Antarctic sections in the Indian Ocean and near the $100 \, \mathrm{cl} \cdot \mathrm{ton}^{-1}$ surface on three Pacific sections. JOHNSON (1973) was able to model the South Pacific AAIW distribution on the $27.10 \, \sigma_t$ surface. AAIW in the South Pacific thus appears to be lighter than in the South Atlantic and South Indian Oceans. REID, (1965) utilized the $80 \, \mathrm{cl} \cdot \mathrm{ton}^{-1}$ surface in the South Pacific but based his choice primarily upon intermediate water characteristics in the tropical region. REID's figs. 3 and 6 show that a shallower specific volume surface would have been closer to the salinity minimum core south of 15°S.

There are striking differences between S_{min} values in the southwest regions of the three southern hemisphere oceans (Fig. 8; TAFT, 1963). In the Indian Ocean, the 34.4‰ line is near 40°S (WYRTKI, 1971, p. 289) and in the Tasman Sea it ranges from 35°–45°S REID, 1965, fig. 24). By contrast, the 34.4 isohaline extends as far as 5°S in the Atlantic (WÜST, 1936, Beilage Xa). These differences cannot be accounted for by T/S characteristics at the surface in the Polar Front Zone (SVERDRUP, 1940, p. 99), nor are there wide salinity variations in the 200 m circumpolar S_{min} band (OSTAPOFF, 1962, fig. 4). There must then be strong differences in AAIW circulation or production rates, or both, in various regions of the three southern hemisphere oceans.

Several authors (TAFT, 1963; KIRWAN, 1963; REID, 1965; BUSCAGLIA, 1971; WYRTKI, 1973; JOHNSON, 1973) have deduced an AAIW circulation more related to the wind-driven southern hemisphere anticyclonic gyres and large current systems than to a thermohaline-driven, primarily meridional flow toward the north (WÜST, 1936). AAIW enters the Tasman Sea primarily along its eastern margin and not adjacent to Tasmania and Australia (ROCHFORD, 1960). Similarly, direct northward penetration of AAIW along the South African margin is impeded by the Agulhas Current and subtropical gyre, which appear to entrain intermediate water from both the equatorial and Antarctic regions (TAFT, 1963, p. 133; ORREN, 1966; WYRTKI, 1973, p. 31). Supporting the interpretation of the circulation derived from salinity data, the oxygen concentrations at AAIW levels in the southwest Indian Ocean and the Tasman Sea are lower than in the adjacent South Indian and South Pacific Oceans, while the southwest Atlantic values are high relative to the remaining South Atlantic (TAFT, 1963). But why should northward AAIW transport be so much greater in the South Atlantic than in the South Indian or South Pacific? One possibility might be a secondary source region, which MARTINEAU (1953) has hypothesized for AAIW along the east coast of South America between 25°S and 45°S as a result of mixing between the Falkland and Brazil Currents. Further, both the Mozambique and Tasman Basins are restricted to the north and east at intermediate and deeper levels by ridges, plateaus and islands that have no South Atlantic analogs. Topographic control over oceanic intermediate water circulation thus seems feasible.

(d) Formation

Most authors state that AAIW is formed as a result of convergence and sinking of surface waters in or near the Polar Front Zone. While direct measurements of sinking have yet to be made, the surface origin has been inferred from the nearly straight line on a T/S diagram between AAIW core values and surface characteristics in the Polar Front Zone (WÜST, 1936, Abb. 17, 18). However, there are few stations with subsurface S_{min} within the 100% to 65% range of WÜST's AAIW spectrum (see also Fig. 7b, STD 294–296; SVERDRUP, 1940, fig. 13). WYRTKI (1971, p. 262) noted that "an S_{min} is usually not developed in depths less than 500 m; therefore the surfacing of the core cannot be easily determined and the position of the Polar Front is given as the line of origin of AAIW". Alternatively, a surface origin in the Polar Front Zone may be chosen because this is where representative AAIW specific volume surfaces often outcrop. On six Indian and Pacific Ocean meridional sections (GORDON, TAYLOR and GEORGI, 1974) AAIW-related specific volume levels reach the sea surface near the Polar Front Zone. On these *Conrad* stations the 90 cl · ton^{-1} level surfaces only at STD 273 (67.7°S), but it does shoal to < 200 m near the Polar Front Zone and south of there usually remains between the surface mixed layer and T_{min}. In summer, the Indian Ocean 27.2 isopycnal reaches to within 20–50 m of the sea surface; in winter it outcrops between 55° and 60°S (Wyrtki, 1971, p. 221). The core and isentropic methods are more useful for tracing water mass properties, however, than for elucidating the processes of water mass formation.

It is important to determine whether AAIW components originate directly at the sea surface or only in the subsurface layers. If the shallowest inputs derive from the T_{min} at the base of the Antarctic Surface Water, then direct AAIW access to the sea surface would occur mainly during the winter months, if at all. AAIW that varied seasonally in its properties or production rate could result. Reid (1965, p. 37) inferred that "some part of the intermediate water of the South Pacific derives its characteristic properties beneath the surface in high latitudes from diffusion through the pycnocline, rather than from direct contact with the atmosphere as surface water". He found that an AAIW-associated oxygen maximum slightly above the S_{min} supported this interpretation. In the Southwest Indian Ocean, a shallow well-developed oxygen maximum derives from the transition area south of the subtropical convergence (Wyrtki, 1973, p. 31). Its oxygen values are much higher than those encountered in the AAIW (Fig. 7a), which supposedly originates further south where surface oxygen is higher. This circumstance suggests either a predominantly sub-surface origin for AAIW or a relatively small component of surface water. Correspondingly, AAIW nutrients are rather high for a surface origin. Reid (1973, p. 43) has noted in this respect that surface nutrients are unusually high in the Antarctic. Surface silicate, however, decreases sharply to $< 10\ \mu g\text{-at} \cdot l^{-1}$ on a northward crossing of the Polar Front Zone in this region, possibly implying a small Subantarctic Surface Water contribution to the AAIW. The non-conservative nature of the oxygen and nutrients and the unknown residence time of AAIW within the transition region between Polar Front Zone and Subtropical Convergence may also be significant factors.

Deacon (1937, p. 66) described "a region extending 100–200 miles north of the Antarctic Convergence where a large volume of mixed water exists...the colder, heavier part sinks to form the intermediate current". This description might well be applied to Fig. 3a, where the extensive 34.2–34.3‰ region between STD 283 and 287 looks as if it were maintained by inflow near the Polar Front Zone and drained by AAIW outflow near the Subtropical Convergence.

Summer surface mixed layers near the Polar Front Zone in this region (Figs. 2a and 3a) are too warm and fresh to be the sole source of even the southernmost AAIW S_{min} observed north of the Polar Front Zone (Fig. 7a, STD 284). The few winter stations available indicate little surface salinity increase, and mixing to depths over 500 m would thus be needed to raise the salinity by $\approx .3$‰. Winter cooling of about 3°C could raise the surface density to AAIW levels, but an additional high-salinity, high-silicate, low-oxygen component would be required. A probable source of the latter may be from several hundred meter depths near the T_{max} in the Polar Front Zone, again necessitating deep mixing.

Particular regions are often given as preferred locations of AAIW formation or northward flow: southeast in the Pacific (Reid, 1965); southwest in the Atlantic (Wüst, 1936); or southeast and south-central in the Atlantic (Taft, 1963; Kirwan, 1963). Koopmann (1953) computed the circumpolar meridional Ekman convergence from the geostrophic wind field, finding maximum values along the Polar Front Zone from 10° to 50°E and in the south-central Pacific. As noted above, the southwest Indian Ocean is not a region where AAIW extends far north. The extensive mass of 34.1 to 34.3‰ water beneath the Subantarctic

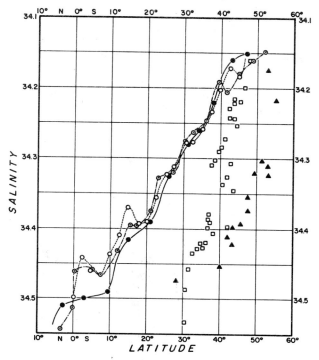

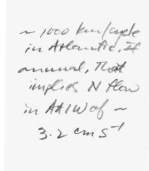

Fig. 8. AAIW salinity minimum vs. latitude. □ = *Conrad* 17, ▲ = *Eltanin* stations near 160°E (cruises 26, 28, 29, 34, 37 and 44); ● = *Discovery* 30°W, after DEACON (1933); ○ = *Hudson* 30°W; = 'GEOSECS' southwest Atlantic. Only Atlantic values, each a set from a single meridional profile, are connected.

Surface Water may, however, indicate local or upstream production of a precursor water mass. Possibly supporting KOOPMAN's calculations, the meridional extent of low salinity water at 200 m in the southwest and south-central Indian Ocean is about double that observed elsewhere around the continent (OSTAPOFF, 1962, fig. 4; GORDON and GOLDBERG, 1970, plate 4; WYRTKI, 1971, p. 77). The Southwest Indian Ocean is also the only major region where Ostapoff shows 200 m salinities to be less than 34.0‰.

DEACON (1933, Fig. 22) plotted AAIW salinity against latitude for a series of stations along 30°W in the South Atlantic. He ascribed variations from a linear relationship to seasonal differences in surface water in the region of AAIW origin. Other authors found corroborative evidence in the temperature and chemistry data. WÜST (1936) took issue with these analyses on the grounds that the irregularities were small compared to the accuracy of the measurements, that they could also be caused by sampling away from the AAIW core, that variations at the source would be damped by lateral diffusion well before they reached the equator, and that observed wavelengths were of the same spatial scale as the station spacing. Recent higher precision measurements are at closer intervals in the vertical and horizontal, and the *Discovery* 30°W section has been repeated by the research vessel *Hudson*. We have thus replotted DEACON's S_{min} vs latitude data along with that for several recent cruises in the southwest region of each ocean (Fig. 8). In the western South Atlantic, there are striking similarities between the *Hudson*, *Discovery* and 'GEOSECS' data.

The wavelike structure at intervals of 8–10° of latitude persists in the more recent data, with occasional inversions, but with no marked change in the wavelength. The large S_{min} fluctuations in the northern South Atlantic are not likely to derive solely from the lower-amplitude signal further south. Since the signals are nearly coherent even though the data sets range from October–May, the signal may result more from large permanent gyres and local boundary conditions than from seasonal variations in the formation region.

Deacon's model assumed continuous AAIW production and source characteristics that varied with the season. Seasonal peaks in AAIW production are also feasible, and perhaps related to deeper winter mixing in the formation region or to variations in the position of maximum westerlies and/or transport of the circumpolar current (Gordon, Taylor and Georgi, 1974). Near 161°E (Fig. 8), for example, the salinity minimum was observed south of 50°S and was fresher than 34.31‰ during winter (*Eltanin* cruises 34 and 44) but not during summer (*Eltanin* cruises 16, 36, and 37).

3. DEEP WATER

(a) *Characteristics at the major core layers*

An oxygen minimum occurs near the upper levels of the deep-water mass in Subantarctic and Antarctic regions. The depth and structure of this minimum is not well defined by the *Conrad* data, but its gross features are evident in Fig. 4. The lowest oxygen value at the minimum was $3.58 \, ml \cdot l^{-1}$ on STD 308, with observations below $4.0 \, ml \cdot l^{-1}$ on STD 265 midway between Cape Agulhas and the Agulhas Plateau, and on STD 305–309 between the Mozambique Plateau and Durban. The minimum reached as deep as 1832 m on STD 265 and shoaled to less than 200 m over the Antarctic continental rise. The oxygen minimum has generally been located below the temperature maximum (T_{max}) but is near or above it in some regions (Gordon, 1967, Figs. 1–3). It usually appears above the T_{max} on these *Conrad* stations, but this is somewhat uncertain because of the large vertical spacing between oxygen measurements. The oxygen minimum roughly coincides with a phosphate maximum observed throughout the region (Fig. 6) and with a silicate maximum near South Africa (Fig. 5).

The vertical relationship and structure of the T_{max} and salinity maximum (S_{max}) are illustrated in Fig. 9. Deacon (1937, pp. 94, 95) associated the T_{max} with the southerly-

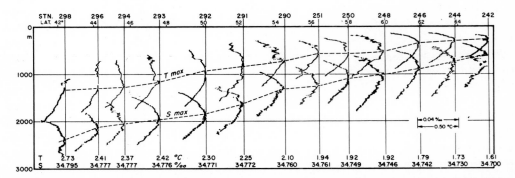

Fig. 9. Vertical structure and relationship of the T_{max} and S_{max} for the *Conrad* section (b) between the subtropical convergence and Antarctica. The region adjacent to the Antarctic continental slope is shown in Fig. 12a.

flowing deep water; GORDON (1967, p. 4) referred to the T_{max} as a feature induced by the T_{min} above. Like the T_{min} at the base of the Antarctic Surface Water, the T_{max} is often not traced north of the Polar Front Zone (GORDON, 1967, figs. 1, 2). In Fig. 9, the T_{max} and S_{max} are separated by several hundred meters; separations exceeding 1500 m were observed on STD 255 and 285. Throughout the region, the T_{max} is closer to the oxygen minimum than to the S_{max}. Both maxima shoal and converge toward Antarctica; in the narrow region over the Antarctic continental slope, these maxima deepen and decrease in value, merging together near bottom at the continental shelf break (see also SVERDRUP, 1940, p. 121). With these *Conrad* STD data, the T_{max} can be traced to depths >1500 m and north of the Subtropical Convergence, where its temperature rises above 3.0°C. Significantly, this indicates that the shallower T_{min} can also be observed in the same latitudes. North of the Polar Front Zone, the temperature increase at the T_{max} is typically less than 0.1°C. There are several temperature inversions near the T_{max} on some stations (e.g. STD 287).

The salinity maximum is a more prominent deep-water feature than the oxygen minimum or temperature maximum and will be discussed in greater detail. Circumpolar Deep Water (CDW) enters the Drake Passage with a relatively uniform S_{max} between 34.72 and 34.74‰ (GORDON, 1971, fig. 3). It interacts with the large cyclonic gyre of less saline Weddell Deep Water (DEACON, 1933; GORDON, 1967, p. 4; DEACON and MOOREY, 1975) and on transiting the South Atlantic/southwest Indian region mixes and merges with high-salinity North Atlantic Deep Water. The net result is a strong meridional salinity contrast. Salinity at the S_{max} on these *Conrad* stations (Fig. 3) ranged from 34.862‰ on STD 269 near the South African continental slope to values below 34.680‰ at its minimum depths of less than 600 m over the Antarctic continental rise. The S_{max} reached depths near 3000 m on STD 263 between South Africa and the Agulhas Plateau, and on STD 300–302 in the central Mozambique Basin.

North Atlantic Deep Water enters the Antarctic region with a relatively high temperature and oxygen and low nutrients (Figs. 2, 4, 5, 6). At the level of the salinity maximum between South Africa and Antarctica, silicate increased from 48 to 110 μg-at·l^{-1}, phosphate from 1.62 to 2.34 μg-at·l^{-1}, and potential temperature decreased from 2.54° to less than 1°C. Oxygen at the S_{max} was highest at STD 269 (5.23 ml·l^{-1}) and reached minimum values (4.55–4.60 ml·l^{-1}) near 60°S—well north of Antarctica.

(b) *Distribution*

In Fig. 10a, the θ/S region below 3°C and above 34.55‰ is plotted for most STD stations on the Cape Agulhas–Antarctica transect (Figs. 2a, 3a). The superimposed envelope 'A' is derived from Nansen bottle data of an *Eltanin* cruise 10 section along 75°W near the Drake Passage. θ/S properties at the S_{max} for *Conrad* stas. 252 and 242–244 (Fig. 1) are within this envelope and thus do not differ appreciably in thermohaline characteristics from the Drake Passage CDW. Oxygen and silicate at these *Conrad* stations are also similar to the *Eltanin* 10 S_{max} data. Stations to the north of STD 252 and 244 show the influence of warmer and saltier NADW, while stations south of STD 277 and 242 have lost heat and salt at the S_{max} relative to the envelope stations. The division between NADW and 'Weddell' influence is not sharp—STD 277 and 278 (Fig. 11) are ambiguous and probably represent the net effect of property interchange across the zone. The trend of the division across these two Antarctic sections parallels the isotherms and isohalines at the S_{max} and the lines of mass transport

(WYRTKI, 1971, pp. 297–8, 394). The envelope 'A' zone nears the Antarctic coast on the eastern section, indicating that all deep water east of the 50° meridian will have been modified to some extent by mixing with NADW.

Earlier literature includes frequent references to the influence of deep water from the North Indian Ocean in the far eastern Weddell–Enderby Basin, possibly even penetrating into the Weddell Gyre near Antarctica. The *Conrad* data do not support that thesis (see also CLOWES, 1933, p. 191). There is no significant difference between average oxygen at the

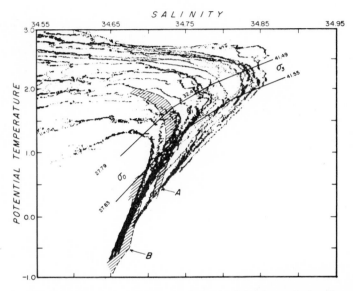

Fig. 10a.   Deep water θ/S diagram for *Conrad* Stas. 269–274 (Fig. 3a). *Eltanin* Cruise 10 stations along 75°W near the Drake Passage define shaded envelope 'A'; 1973 stations across the Weddell Sea (CARMACK and FOSTER, 1975) define envelope 'B'. Isopycnal bands referenced to different pressure surfaces link probable mixing paths at the $S_{\max}$.

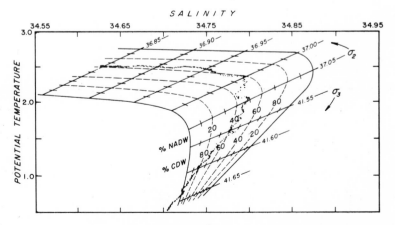

Fig. 10b.   Nomogram showing % NADW and % CDW applicable to Fig. 10a, with STD 287 superimposed.

S_{max} on the western and eastern Antarctic sections (STD 271–284 vs. 239–293), and a slight (4μg-at·1^{-1}) decrease occurs in average silicate at the S_{max} on the eastern section. Since North Indian Ocean deep water is high in silicate (WYRTKI, 1971, p. 295), whereas NADW is low, this decrease in silicate is consistent rather with an increased NADW component on the eastern section.

The deep water of the Agulhas and Mozambique Basins is primarily of North Atlantic origin (DEACON, 1937, p. 90; LE PICHON, 1960, p. 4061; WYRTKI, 1971, p. 262). Deep-water salinity is higher on the Cape Agulhas–Marion Island section (Fig. 3a) than on the Durban–Crozet Island section (Fig. 3b). The S_{max} of 34.862‰ at STD 269 agrees well with nearby *Meteor* stations and is consistent with DEACON's (1937, p. 90) observation that the core of NADW here lies "close round the Cape of Good Hope". Salinity and oxygen are relatively low and nutrients high at STD 265. These result from the deep-reaching Agulhas Current system and are consistent with a southwest flow extending down to near 4000 m (see also WYRTKI, 1973, p. 30, and fig. 6). It is possible, though less likely, that these anomalies near the STD 265 S_{max} could also result from an admixture of circumpolar deep water entrained by the Agulhas Return Current. The Agulhas Current lies primarily between STD 265 and 266, with the Return or Eastward Current (DARBYSHIRE, 1972) between STD 263 and 265. Within the limits of our station spacing, these positions agree with locations reported for this highly variable system.

Between the Agulhas Plateau and Marion Island (Fig. 3a), a marked waviness in the isohalines extends throughout the deep and bottom water, with station-to-station salinity differences exceeding .02‰ at the S_{max}. The low salinity, the θ/S relationship and the relatively low oxygen and high nutrients within and below the S_{max} at STD 287 dictate a southern origin for this water. There is a horizontal (salinity) separation below the S_{max} between groups of θ/S curves in Fig. 10a. Stations south of the mid-ocean ridge fall within the lower-salinity grouping; most stations north of the ridge are in the higher-salinity set. STD 287, however, lies within the low-salinity array, and STD 286 is intermediate between 0.25° and 1.0°C; both these stations are north of the mid-ocean ridge but appear to have readier access to deep water from the Weddell–Enderby Basin.

Southeast of Durban (Fig. 3b), the deep water is more homogeneous than it is southeast of Cape Agulhas. The highest salinites are near 3000 m, midway up the southeast side of the Mozambique Ridge. Oxygens are correspondingly high (Fig. 4b) and nutrients low (Figs. 5b, 6b), indicative of a northeast flow along the ridge in this vicinity (see also LE PICHON, 1960).

Salinity at the S_{max} also varies by $> .02‰$ between Marion and Kerguelen Islands (Fig. 3c). This section is near the Polar Front Zone along most of its length. The depth of the salinity maximum changes markedly beneath the Polar Front Zone; it varies along this section by nearly 1000 m.

(c) *Isopycnal mixing*

The salinity maximum for most stations in Fig. 10a may be enclosed by a succession of connected isopycnal bands. As the S_{max} shoals from north to south, it moves from the σ_3 range 41.49–41.55 to the σ_1 range 32.47–32.52 and then to the σ_0 range 27.79–27.83. Transitional regions from one band to the next occur near the Polar Front Zone and envelope 'A'. Whether or not the salinity maxima represent meridional components of flow, it is clear they can be linked by relatively simple "mixing paths, referred to appropriate pressure surfaces" (LYNN and REID, 1968, p. 596). CALLAHAN (1972) utilized the $30 \, \mathrm{cl \cdot ton^{-1}}$ surface to delineate spreading of the CDW, noting that this surface was representative of the S_{max} except near the Antarctic continent. Because of the strong NADW influence in the southwest Indian Ocean, however, the S_{max} in this region is near the $30 \, \mathrm{cl \cdot ton^{-1}}$ surface only near Antarctica. Its mean is $38 \, \mathrm{cl \cdot ton^{-1}}$ (± 5), with the highest value of $45 \, \mathrm{cl \cdot ton^{-1}}$ on station 300. On several stations in Fig. 10a (269, 284, 287), the NADW influence is further evidenced by a double S_{max}, the lighter of which occurs above the main σ_3 band.

The north-to-south decrease of salinity at the S_{max} is not monotonic (Fig. 3). Relative highs and lows in salinity have often been observed in horizontal and vertical sections constructed from serial bottle data, but there has always remained the possibility that particular extrema were missed by wide bottle spacing. The *Conrad* STD observations firmly establish the authenticity of these features.

A temporal exchange of properties at S_{max} levels is suggested by some of the thermohaline structures in Fig. 11. Both high- and low-salinity intrusions are impressed upon mean or 'smooth' θ/S curves. In Fig. 11a the complicated θ/S relationship for STD 248 lies intermediate between the relatively smooth STD 250 and 246. Interleaving aligns well with isopycnals, and the STD 248 (Fig. 11a) excursions just intersect θ/S curves of the adjoining stations. The apparent thickness of disturbances giving rise to the interleaving on STD 248 is less than 75 m. Ignoring the fact that information on property gradients perpendicular to

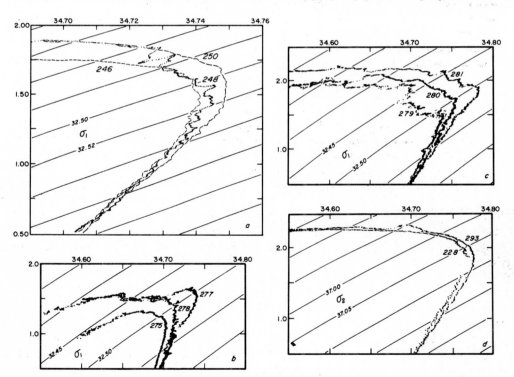

Fig. 11. Deep Water potential temperature/salinity structure for selected stations. Note the expanded scale in (a). A 7-point running mean in addition to the regular processing (1) has been applied to STD 248; Stas. 246 and 250 are represented by dashed lines. Stations 228 and 293(d) are at the same location 72 days apart.

the section is missing, it might be inferred that the S_{max} water on STD 248 derives from no farther than the adjacent stations. On STD 226, 277 and 281 (Figs. 3a, 3b and 11b), isolated high-salinity features are discernible while the S_{max} salinities are relatively low on STD 279. Unlike the interleaving structure of STD 248, these features have vertical scales of 400–500 m and appear to have horizontal dimensions on the order of the station spacing (≈ 250 km). In Fig. 11c, the stations reveal a variety of thermohaline scales.

From Fig. 10a it is apparent that the θ/S relationship between the deep and bottom water on individual stations is only slightly bowed from a straight line. However, the stations along this section (Fig. 3a) cannot be represented by a narrow envelope as in the Drake Passage, nor by a single CDW/AABW baseline, as along 170°E (GORDON, 1975). Rather, the individual θ/S curves are arrayed in a fan-like distribution between the extreme northern and southern stations. The array of individual curves is similar to diagrams that have been constructed for bottle data elsewhere in the oceans on the assumption that the intermediate types are formed by isopycnal mixing processes (SVERDRUP and FLEMING, 1941; MAMAYEV, 1975, section 53).

A similar analysis may be applied to these *Conrad* data. In Fig. 10b a nomogram is bounded on the left by a mean of the Drake Passage (CDW) envelope 'A' and on the right by STD 266, assumed to be 100% NADW. Dashed lines indicate % NADW and CDW along σ surfaces. The Weddell Deep Water θ/S region, colder and fresher than CDW, is not defined on the figure. The nearly linear θ/S relationship below the S_{max} on individual Antarctic stations is frequently interpreted as evidence for vertical mixing between CDW and AABW end points. It is evident from Fig. 10b, however, that isopycnal mixing can equally well produce the observed θ/S curve. Further, NADW influence (%) decreases with depth in the water column while the Weddell influence increases. The resulting θ/S curve on an individual station can remain nearly linear over major segments when the exchange of properties is accomplished laterally. Vertical mixing will further smooth segments of the curve [4].

(d) Variability

There are numerous references to variability in deep-water characteristics and movement in the Antarctic and adjacent regions (LUTJEHARMS, 1972; WARREN, 1973, p. 14). To account for an anomalously low S_{max} on *Discovery* sta. 850, section 8, DEACON (1937, p. 92) suggested that the southward flow of deep water might be "interrupted by eddies of colder water from the west". The apparently isolated high-salinity cores on the same *Discovery* section are equally striking (although not all were observed the same year) in that they are nearly duplicated by our nearby section (Fig. 3a). The three high-salinity cores centered at *Conrad* stas. 286, 281 and 277 in Fig. 3a may represent various stages in the decay of large, deep eddies detached from the NADW and moving downstream within the circumpolar current. The large seasonal variations of the Agulhas Current system (DARBYSHIRE, 1972), which reaches down to NADW levels, and the sharp southeastward turn of the circumpolar current as it crosses the mid-ocean ridge southeast of Marion Island [8] might both contribute to eddy formation.

Alternatively, IVANENKOV and GUBIN (1960, p. 46) have hypothesized that these high-salinity features represent continuous streams of NADW imbedded in the circumpolar current. Two reoccupied *Conrad* deep stations (STD 293 and STD 289) revealed significant changes of salinity, temperature and θ/S structure in the deep water over a period of about 2 months (Fig. 11d). S_{max} variations of up to .03‰ have been reported at SCOR-UNESCO reference stations in the southeast Indian Ocean (ROCHFORD, 1965). Temporal variations at a single position do not support continuous steady-state streams of high- or low-salinity deep water, but neither do they rule out meandering deep currents.

There is a marked decrease in the salinity and in the vertical dimension of the deep-water mass from the eastern section (Fig. 3b, STD 239–293) to the western section (Fig. 3a, STD 271–283). The western section is in or near the eastern end of the Weddell gyre, consistent with DEACON's (1937, p. 92) interpretation of the southern end of *Discovery* section 8. The historical data (WYRTKI 1971, p. 97), however, would place the eastern edge of the Weddell Gyre further west, between the 20° and 30° meridians. KOOPMANN (1953) found indications of a bimodal distribution of southwest Indian Ocean salinity, temperature and oxygen data and interpreted this as evidence for a variable eastward extension of the Weddell Gyre. If so, the relatively low salinities at the S_{max} in Fig. 3a would suggest we encountered the eastward extension mode. The Antarctic circumpolar current turns south and crosses the mid-ocean ridge southeast of Marion Island [8] and may effectively control the position of the eastern end of the Weddell Gyre. Temporal changes in the strength of the circumpolar current (McKEE, 1971) may also be related to a variable position of the eastern end of the gyre. A waxing and waning of the gyre, or an (intermittent?) input of higher salinity water from west of the Kerguelen Plateau (WÜST, 1933, p. 66; REID and LYNN, 1971, fig. 7) seem as likely as the steady-state interpretation (WYRTKI, 1971, p. 297).

4. BOTTOM WATER

Antarctic Bottom Water (AABW) is distinguished on deep stations throughout this region by a bottom potential temperature minimum, salinity minimum and oxygen maximum. The lowest *Conrad*-17 temperature ($\theta = -0.75$) and highest oxygen ($6.10 \, \text{ml} \cdot \text{l}^{-1}$) were observed at the base of the Antarctic continental slope near 60°E (Fig. 2b). Salinities there were below 34.660 and associated with a bottom silicate minimum ($104 \, \mu\text{g-at} \cdot \text{l}^{-1}$). These characteristics cannot have derived from a continuous easterly flow from the Weddell Sea, since all parameters, with the exception of salinity, do not indicate this AABW type at the continental slope base at 37°E (Fig. 2a). Similarly, there is no continuous westerly flow of this bottom water from the southeast Indian Basin (see, e.g., *Eltanin* cruise 47 stations near 80°E in JACOBS, BAUER, BRUCHHAUSEN, GORDON, ROOT and ROSSELOT, 1974). Bottom water is thus being formed along the Enderby Land/Prydz Bay coast as suggested earlier by SVERDRUP (1940) and GORDON (1974).

Recent near-bottom data over the Antarctic continental rise and slope have proven that significant bottom water sources exist outside the Weddell Sea (GORDON, 1974). Indeed, new source regions now seem to accumulate in rough proportion to the number of expeditions that reach the continental slope region. SVERDRUP (1931) hypothesized that bottom water formation takes place around the Antarctic continent, though he was later (1940) led to a retraction of that proposal. The importance of wind-driven motion as opposed to buoyancy forcing, and the necessity of utilizing the entire coastline of Antarctica as a bottom water source has also been emphasized in recent theoretical studies (KILLWORTH, 1973). Of particular significance here is that this source is located within the same basin that receives the major input of AABW from the Weddell Sea (DEACON, 1937).

The extent to which this locally produced bottom water contributes to the large mass of AABW in the Weddell–Enderby Basin is not clear. Of our two sections extending north from Antarctica, the one (Figs. 2a, 3a) nearer the eastern end of the Weddell Gyre contains a significantly greater quantity of bottom water and a proportionately diminished deep-water mass.

Envelope 'B' in Fig. 10a encloses the θ/S data below the S_{max} for a recent transect of stations across the Weddell Sea (CARMACK and FOSTER, 1975, fig. 1, stations 2–24). The Weddell Sea near-bottom salinities ($-0.1°$ to $-0.6°C$, 2750 to 3250 m) are at least .01‰ too high to have been derived from the southwest Indian sector. Further, salinities at these levels would be expected to decrease from the Indian to Atlantic region (WYRTKI, 1971, p. 297) due to the mixing of deep water with slope and bottom water. This suggests either production west of 37° in the Weddell–Enderby Basin of a higher-salinity (> 34.66) water mass in the stratum near $\theta = -0.5°C$, intermittent deep exchange with the Fig. 3b region of the far eastern Weddell–Enderby Basin, or the existence of more than one gyre in this Basin. DEACON (1976) has discussed the cyclonic circulation of the Weddell Sea in relation to the position of a nearly permanent low-pressure cell that appears on monthly charts of the atmospheric pressure field (TALJAARD, VAN LOON, CRUTCHER and JENNE, 1969). In 5 months of the year there are in fact two separate lows on the charts of this area, perhaps consistent with the presence of more than one gyre. TRESHNIKOV (1964, fig. 1) reports several gyres in the area. Indications of a subsurface constriction in the Weddell Gyre also appear west of the Greenwich meridian on several horizontal charts of temperature and oxygen (GORDON and GOLDBERG, 1970).

EDMOND (1970) and CARMACK (1973) have noted the characteristic low silicate of newly formed bottom water in the Weddell Sea region. It is apparent that low silicate (and phosphate) is also present in other regions adjacent to the Antarctic continent (Figs. 5b, 6b; JACOBS and AMOS, 1967, p. 221). Silicate is particularly non-conservative near bottom and cannot be traced very far north of the Antarctic continental margin. By the time bottom water exits the Weddell–Enderby Basin, its bottom silicate minimum has become a strong maximum. Of particular note are the high bottom silicates between the Crozet and Kerguelen Islands (Fig. 5c). (Deep *Conrad* silicate values are 20–40 μg-at·1^{-1} higher than nearby *Discovery* observations.) The rapid change from low to high silicate may result from several factors: the acquisition of silicate from the highly siliceous bottom sediments (EDMOND, 1970; HEATH, 1974); increased amounts of siliceous organisms and particulate matter carried down in the newly formed bottom water; higher turbidity levels near bottom over the continental slope (JACOBS, AMOS and BRUCHHAUSEN, 1970, Fig. 13), and in the bottom water (Fig. 14); and the continual input of detritus from higher up in the water column.

AABW movement out of Antarctic regions can be inferred from the distribution of bottom potential temperature. KOLLA, SULLIVAN, STREETER and LANGSETH (1976, fig. 2) have utilized archived data and these *Conrad* observations to produce such a map for the Indian Ocean. The major passage for bottom water into the Crozet Basin and northwest Indian Ocean is located on the western side of the Crozet–Kerguelen passage (Figs. 2c, 3c; DEACON, 1937) with a minimum temperature ($\theta = -.31$) and salinity (34.658) observed on the deepest station (230) midway across the passage. Lesser northward flow is probable between STD 218 and Marion Island.

West of the mid-ocean ridge the coldest bottom water between Africa and Marion Island was encountered in a layer between 4000 and 5000 m depth on the southeast flank of the Agulhas Plateau. The steep isolines and values of temperature, salinity, oxygen and phosphate (Figs. 2a, 3a, 4a and 6a) are indicative of a strong current with southern origins. The current is apparently weaker on the southeast flank of the Mozambique Plateau (Figs. 2b, 3b, 4b and 6b). Salinity and temperature between the Agulhas Plateau and mid-ocean ridge suggest at least one gyre in the bottom water, with alternating northeasterly and southwesterly flow. The mound of low salinity water centered at the bottom on STD 287 (Fig. 3a) is very similar to the feature at *Discovery* Sta. 848 (Deacon, 1937, Section 8) about 275 km to the southwest. There may be a permanent near-bottom gyre here, with a long dimension parallel to the topographic contours. Little AABW passes between South Africa and the Agulhas Plateau, where the bottom salinities remain above 34.71‰. Bottom water with a potential temperature below 0.4°C penetrates to the northern reaches of the Mozambique Basin. Further movement to the north and east is limited by the relatively shallow Mozambique Channel and Madagascar Ridge (Warren, 1974; Kolla, Sullivan, Streeter and Langseth, 1976). In the lower 1250 m of the 5126 m rift valley station 298, salinity changes by $<.01‰$, and temperature increases at three-fourths the adiabatic lapse rate.

5. THE ANTARCTIC CONTINENTAL SLOPE RÉGIME

Meridional hydrographic sections that include closely spaced stations over the Antarctic continental slope usually show a water mass there that is sharply separated from the deep water by a narrow frontal zone of great vertical extent (Sverdrup, 1940; Jacobs, Amos and Bruchhausen, 1970; Gordon and Tchernia 1972; Foster and Carmack, 1976). This feature is well illustrated by the *Conrad* data (Figs. 2b, 3b and 4b), where the frontal zone covers less than a degree of latitude and encompasses nearly 4000 m of the water column. The feature has a small horizontal dimension and does not often appear in averaged historical data (e.g. Gordon and Goldberg, 1970, plate 17, profile 4). The frequent observations of this frontal region attest to continuous or periodic renewal of the water mass over the Antarctic slope. Its characteristics vary at least spatially, as evidenced by the contrasts between parameters on STD 242–238 and 273–270 (Figs. 2a, b–6a, b).

Water over the continental slope is intimately related to the processes of AABW formation, either as a precursor or as a remnant insufficiently dense to reach the deep ocean, or both. Its low temperature, low salinity and high oxygen are tracers of its near-surface origin. The isolines of all parameters are bowed downward in the top several hundred meters over the continental slope (Figs. 2–6; see also Gill, 1973), an indication of the coastal current. The characteristics of this water mass over the continental slope ($< 0.4C$, $< 34.68‰ > 5 \text{ ml O}_2 \cdot \text{l}^{-1}$, 100–130 μg-at $\text{Si} \cdot \text{l}^{-1}$) are similar to those of AABW within the Weddell-Enderby Basin.

Thermohaline structure over the Antarctic continental slope is complex, with much interleaving. At depths between the T_{min} and T_{max}, where property gradients on density surfaces are largest, there are 50 m, 0.3°C inversions in temperature vs. depth (Fig. 12a). The deeper isopycnal surfaces slope sharply upward over the continental rise and slope, connecting the offshore region between deep and bottom water with shallower levels in the slope water. The predominant flow is likely to lie perpendicular to the plane of the section in Fig. 12a. Isopycnal motion would mix water from the upper continental shelf into the deep and bottom water, much as illustrated by Pingree (1972, p. 559) for the mixing of Mediterranean water into the North Atlantic.

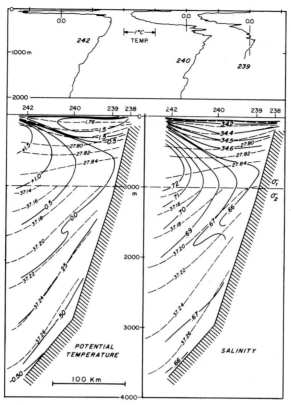

Fig. 12a.   Vertical temperature and salinity structure on STD 238–242 adjacent to Antarctica near 60°E. The isolines are smoothed above 1000 m. Maximum gradients on σ_t surfaces occur at the depths of maximum interleaving. Different σ surfaces are applied above and below 1000 m.

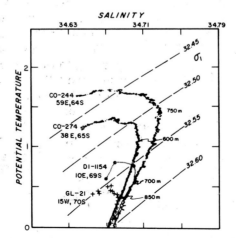

Fig. 12b.   Potential temperature/salinity curves for *Conrad, Discovery* and *Glacier* stations near the southern margin of the Weddell–Enderby basin. S_{max} depths are indicated.

It is apparent that this fresh water over the continental slope is also important in the modification of deep water via lateral mixing. Evidence is available from both the interleaving and from the marked changes that occur in deep water as it moves along the southern limb of the Weddell Gyre (Fig. 12b; DEACON, 1963, p. 285; DEACON, 1974). Weddell Deep Water is lower in salinity at the S_{max} than is CDW/NADW and must be modified by mixing along the periphery of its gyre by interaction with CDW, with water above the continental slope and with water of the Weddell–Scotia Confluence. The numerous cyclonic circulations in the Antarctic coastal zone surface water (TRESHNIKOV, 1964) may extend down into the slope water and cause mixing between the deep water and slope water. These processes are important in the circumpolar exchange of heat, salt and other properties. REID and LYNN (1971, p. 1080) have indicated that the lateral admixture of extreme Antarctic water back into the CDW is very limited. Property distributions on lateral surfaces only reveal net effects, however. It is possible that the slope water influence may be balanced by a meridional NADW flux that is larger than previously assumed.

The general movement of slope water has often been estimated and sometimes measured to be westerly (DEACON, 1963; HOLLISTER and ELDER, 1969; JACOBS, AMOS and BRUCHHAUSEN, 1970). This trend is consistent west of 20–30°E with the direction inferred for the southern limb of the Weddell cyclonic circulation (DEACON, 1976). Net westerly transport everywhere within the slope water in the Weddell–Enderby Basin, however, would require a significant barotropic component to counterbalance the easterly baroclinic transport across some meridians ([8], Fig. 15b; GORDON, 1975, p. 372). Westerly transport along the southern margin of this basin would supply potential AABW to the Weddell Sea régime. A net down-slope component of only 1° in the contour current could transport water 3000 m deeper in only 5° of longitude.

6. STRATIFICATION

Considerable thermohaline structure exists between the deep and bottom water (Fig. 13; WÜST, 1933, p. 93). On all *Conrad* deep STD stations, θ/S points between the bottom boundary layers [7] and $\theta = 1.0°C$ (an arbitrary upper limit) can be approximated by a series of straight line segments. Of forty-one stations analyzed, at least twenty-four have more than one segment (layer). The slopes $(d\theta/dS)$ of these segments fall into several discrete sets, with the highest slopes corresponding to the freshest bottom water and the lowest slopes near the NADW. The highest values of $d\theta/dS$, 83 and 50 $(10 \times °C/\text{‰})$ occurred on STD 240 and 273, respectively, near the base of the Antarctic continental slope. Eleven stations (e.g. the four stations in Fig. 13a) have segments with a mean slope of 36 (± 3), always near bottom and almost exclusively south of 53°S. Twenty-nine segments, all south and east of the mid-ocean ridge, have mean slopes of 24 (± 2), evenly distributed near bottom (STD 287, Fig. 13b) or above a deeper segment (all stations, Fig. 13a). Another category of eighteen stations shows a mean of 17 (± 1); these occur in all regions, but predominantly above bottom (e.g. the shallowest segments on STD 228 and 293, Fig. 13c). Three stations in the central Mozambique Basin (STD 300, 301 and 302) consist of single long linear segments with a slope of 15 (Fig. 13b). The only lower slopes (14 and 12) occur below the NADW salinity maximum northeast of Marion Island (e.g. the shallower segment on STD 287, Fig. 13b).

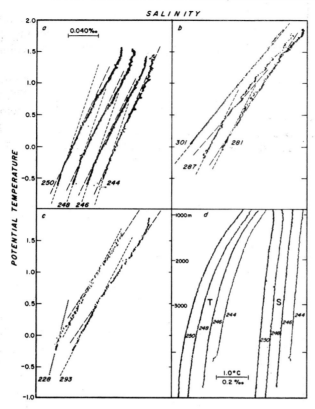

Fig. 13. (a–c) illustrate potential temperature/salinity relationships between the deep and bottom water. Salinity scales are offset laterally. Dotted and dashed lines indicate segments of the θ/S curves that can be approximated by straight lines and separated by distinct discontinuities. In (d), the temperature/depth and salinity/depth records are shown for the four stations in (a).

On most stations with multiple layers, the segments decrease successively in $d\theta/dS$ value with distance above bottom. Over the entire region, there is little correlation of the θ and S values at the points of slope change, though relationships are apparent over limited areas. For example, the four adjacent stations in Fig. 13a have similar slopes in the near-bottom segments (36, 37, 36, 40) and next-higher segments (25, 24, 24, 26); salinity varies little at the discontinuity points (34.668, .669, .674, .669), but θ ranges over wider limits (−.20, −.28, −.16, −.30). CARMACK and FOSTER (1975) note a discontinuity at $\theta = -0.5°C$ for their station 10, and this appears to be characteristic for that entire Weddell Sea transect (Fig. 10a, envelope 'B').

The slopes represent stratified layers between the deep and bottom water. These slopes are of greater importance as an imprint of the past water mass histories than are the temperature and salinity values at the discontinuities between segments. Vertical mixing and diffusion will alter the temperature and salinity end points more rapidly than it will alter the segment slopes. Deep discontinuities in hydrographic parameters have been traced over wide oceanic regions, but with continually varying temperature and chemical characteristics (CRAIG, CHUNG and FIADEIRO, 1972). Lateral processes must be important in maintaining the observed sharp transition points between segments. It would be difficult if not impossible to produce the multiple segments by simple vertical mixing between a single bottom water source and a single characteristic deep-water type. It is probable that each group of θ/S segments derives from a separate formation region, where local deep and bottom water characteristics define the distinctive slope.

We have discussed these features in θ/S space. They appear quite differently and frequently cannot be resolved in parameter/depth plots. In Fig. 13d, temperature/depth and salinity/depth STD traces are plotted for the four stations in Fig. 13a. The scales are different, but these data do not reveal the structural changes apparent in Fig. 13a. Alternatively, parameter/depth gradient changes, which are numerous throughout the region, sometimes appear on θ/S diagrams only as changes in the volumetric density of observed values.

A water layer will be modified during recirculation from a primary source region. The slope (31) of the Weddell Sea bottom water (Carmack and Foster, 1975, fig. 7, Sta. 10) falls between the slopes of two of the *Conrad* sets, and its very low temperature also indicates that this water mass does not persist unchanged into the southwest Indian Ocean. It is probable that modifications occur more through lateral than vertical exchange of properties. In areas remote from source regions and topographic influences, however, vertical mixing may have time to establish the long straight segments, as on Sta. 301 in the central Mozambique Basin (Fig. 13b).

The deep structure also varies in time, as evidenced by STD 228 and 293 (Fig. 13c) taken at the same location but 71 days apart. The deepest segment on STD 228 was missing at the time of reoccupation (STD 293); it was apparently a transient feature advected through the area. The deep changes at this location indicate that synoptic regional correlations of θ/S slopes may be difficult. The definition of fixed boundaries between water masses has always been arbitrary. Stability analyses (Shcerbinin, 1973) or volumetric θ/S slopes over an even grid of deep STD stations would better reveal the divisions.

7. BOTTOM BOUNDARY LAYERS

About one-third of the deep *Conrad* STD stations exhibited irregularities in near-bottom thermohaline gradients. These phenomena were more common near the Crozet and Agulhas Plateaus than south of 50°S, where they were observed only on stas. 244, 279 and 291. Several examples of the potential temperature/depth and salinity/depth structures are given in Fig. 14. All those shown are not within the AABW—STD 259, for example, is on the southeast Agulhas Plateau where the water near bottom is primarily NADW.

The near-bottom decrease in temperature generally involves more than one layer and gradient change, and at times terminates in a homogeneous layer of variable thickness. Accompanying salinity decreases are minor or non-existent; in only one case (STD 244, verified by bottles) did salinity increase near bottom. STD 306 revealed a step-like structure, rather deeper than this phenomenon is usually reported, and above the bottom boundary layer. The steps appear only in the temperature/depth plots and are typically .05 to .10°C and 15 to 40 m thick. Some rounding of the steps has occurred as a result of the data reduction methods [1]. STD 306 is about 200 km from the South African continental slope. It has been hypothesized that breaking internal waves in this type of boundary region might generate a "distorted, book-like structure" (Cooper, 1967, p. 88).

Light scattering (nephelometer, THORNDIKE, 1975) observations, made in conjunction with three stations (224, 282 and 306, Fig. 14), are plotted on the STD depth scale. STD bottom depths do not always correspond to nephelometer bottom depths as the casts were made separately and sometimes over rough topography. Nephelometer sta. N-42, on the southeast slope of the Crozet Plateau, displays two maxima, each about 70 m above the depth of the major temperature steps. Within the limits of the nephelometer depth resolution, the nephelometer and STD bottom depths are offset by a corresponding amount. Minor stability maxima, sufficient perhaps for preferential accumulation of light scattering particles, can be inferred at the depths of the sharp temperature decreases. Nephelometer 62, located at the STD 306 position, reveals a bottom nepheloid layer that complements the potential temperature structure, and a maximum in light scattering about 225 m above bottom that correlates with a temperature step.

Nephelometer Sta. 52, southeast of the Prince Edward Islands fracture zone, shows the strongest near-bottom nepheloid layer recorded on these cruises. Bottom photographs at this location indicate moderate to strong currents. The accompanying STD 282 reached within 10 m of the bottom but is nearly devoid of significant thermohaline layering. The nephelometer observations extend 200 m deeper and suggest strong near-bottom activity in a topographic depression missed by the STD cast.

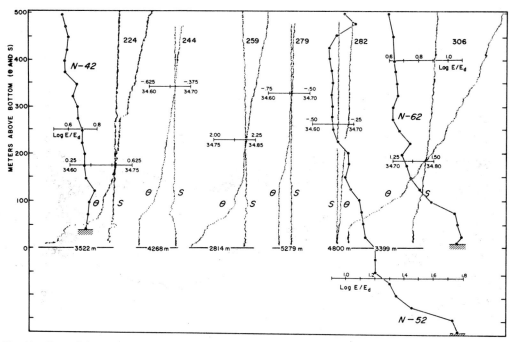

Fig. 14.   Potential temperature/depth and salinity/depth in the lower several hundred meters for selected stations with different types of bottom boundary layers. Note the step-like structure in the lower 400 m of STD 306. Nephelometer data at 25-m intervals for three stations at the same locations as STD 224, 282 and 306 are plotted on the STD depth scale. E = scattered light, E_d = attenuated direct light.

8. GEOSTROPHY

Velocity and transport were computed relative to the deepest common observed level for station pairs on the three transects, and transport was summed northward from Antarctica (Figs. 15 a, b, c). Positive values are to the east and north. This is a baroclinic transport and totals 129×10^6 m·s^{-1} on the Cape Agulhas to Antarctica transect, 111×10^6 m^3·s^{-1} on the Durban to Antarctica line, and 30×10^6 m^3·s^{-1} on the section from sta. 284 to Kerguelen. South of the mid-ocean ridge, on Figs. 15a and 15b, the transports are 123 and 121×10^6 m^3·s^{-1}. These figures are of the same order as results obtained across other Antarctic Ocean meridians (Gordon, 1975, p. 373).

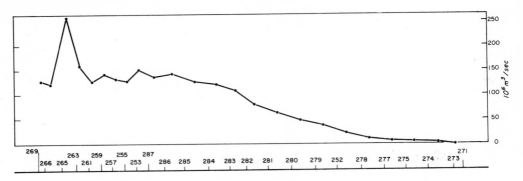

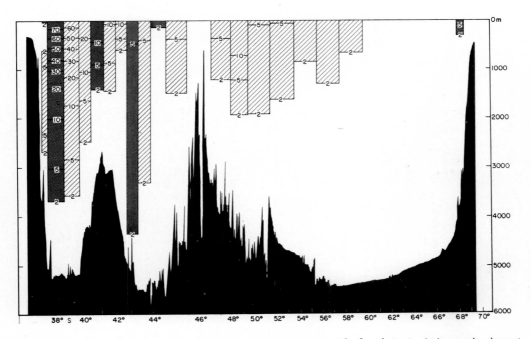

Fig. 15a.   Baroclinic velocity in cm·s^{-1} (bottom) and transport in 10^6 m^3·s^{-1} (top) relative to the deepest common observed level between station pairs. Section between Cape Agulhas and western Enderby Land. Lightly shaded regions indicate flow to the east or northeast (positive transport); heavy shading indicates flow to the west or southwest.

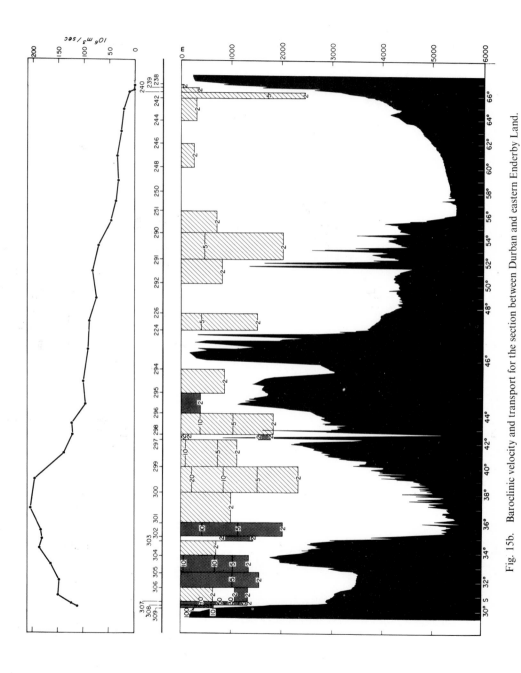

Fig. 15b.   Baroclinic velocity and transport for the section between Durban and eastern Enderby Land.

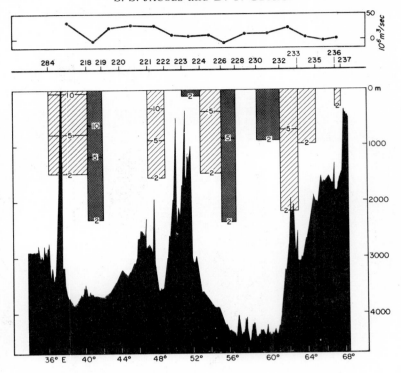

Fig. 15c.   Baroclinic velocity and transport for the section between STD 284 and the Kerguelen Islands.

On the Cape Agulhas–Antarctica section, most net flow occurs south of Marion Island and the mid-ocean ridge. It is apparent that the major portion of the circumpolar current crosses over the ridge between approximately 25°E and Marion Island (Wyrtki, 1971, pp. 388–394), possibly through one or more gaps in the fracture zone southwest of the Prince Edward Islands. The largest flux south of the ridge is between STD 282 and 283, just north of the Polar Front Zone. A slight westerly component over the Antarctic continental slope may reflect the coastal current that is found around much of the continent.

Northwest of the ridge, the largest transport is in the Agulhas Current between STD 265 and 266, $136 \times 10^6 \, \mathrm{m}^3 \cdot \mathrm{s}^{-1}$ to the southwest. A major part ($98 \times 10^6 \, \mathrm{m}^3 \cdot \mathrm{s}^{-1}$) of this flow is balanced by the Agulhas Return Current between stations 263 and 265. The Agulhas Current is strongly baroclinic, highly variable and is the strongest in the southern hemisphere (Wyrtki, 1973, p. 18). Earlier estimates for total transport of the Agulhas Current have usually been referenced to 1500 m or 2000 m levels and have ranged up to $80 \times 10^6 \, \mathrm{m}^3 \cdot \mathrm{s}^{-1}$. These data show an $80 \times 10^6 \, \mathrm{m}^3 \cdot \mathrm{s}^{-1}$ transport above 900 m and strong indications from the salinity and nutrient sections (Figs. 3a, 5a and 6a) that the Agulhas Current influence extends to 4000 m depths. Separating the Agulhas Current from the continental slope is a wedge of relatively cold and fresh water maintained by flow from the south and west. The baroclinic transport of this water mass is $7 \times 10^6 \, \mathrm{m}^3 \cdot \mathrm{s}^{-1}$ centered at 1700 m, and with maximum velocities over $6 \, \mathrm{cm} \cdot \mathrm{s}^{-1}$.

Southeast of the Agulhas Current system are a series of alternating southerly and northerly components that decrease in intensity toward Marion Island. These may represent smaller eddies east of the Agulhas system on the southern edge of the large anticyclonic gyre (WYRTKI, 1971, pp. 390–394). Some of these flows would be altered in direction or strength if absolute reference velocities could be applied. For example, the apparent deep southwesterly flow between STD 253 and 255 extends down into a region where the isolines and values of most parameters are indicative of flow to the northeast. The net transport below 3000 m between Cape Agulhas and the mid-ocean ridge must be near zero, as the Mozambique Basin is closed off to the north below this depth (WARREN, 1974, p. 3; KOLLA, SULLIVAN, STREETER and LANGSETH, 1976, fig. 2). A reference level in the 3800–4000 m vicinity between this station pair and near bottom elsewhere along the section would adjust the net flux below 3000 m from $6 \times 10^6 \, \text{m}^3 \cdot \text{s}^{-1}$ to near zero and result in bottom velocity components near $4 \, \text{cm} \cdot \text{s}^{-1}$ to the northeast.

Similarly, the net transport must be zero below 3000 m southeast of Durban (Fig. 15b). A northward current on the southeast flank of the Mozambique Plateau can be inferred from the temperature and salinity sections (Figs. 2b and 3b), although it is considerably weaker than southeast of the Agulhas Plateau. A reference level between 1800 and 1900 m, midway between the intermediate and deep water on station pair 301–302, would result in a transport balance below 3000 m, with a northward component of $1–2 \, \text{cm} \cdot \text{s}^{-1}$ adjacent to the Mozambique Plateau. A dense manganese nodule field, often indicative of bottom currents, was photographed at the STD 302 location, and STD 301 bottom pictures showed indications of at least a moderate bottom current. The salinity and chemistry are consistent with a northeast flow centred slightly above 3000 m [3]. Whether or not the estimated northeasterly flows below 3000 m on the southeast flanks of the Agulhas and Mozambique Plateaus constitute deep western boundary currents may be a moot point without a northern exit to the Mozambique Basin.

On the Durban–Antarctica transect (Fig. 15b) the Agulhas Current is over the continental slope; a transport of $37 \times 10^6 \, \text{m}^3 \cdot \text{s}^{-1}$ is observed between STD 307 and 309. This is only a quarter of the Agulhas transport on the southern section and about half the northeast component between Stas. 297 and 300 north of the ridge. Much of this latter flow adjacent to the ridge may pass into the Madagascar Basin over the Malagasy Fracture Zone between the Madagascar and Crozet Plateaus. Most of the net baroclinic transport across this section lies south of the Crozet Islands. In contrast to Fig. 15a, easterly baroclinic transport occurs adjacent to Antarctica. This is consistent with the convergence of circumpolar current streamlines south of the Kerguelen Plateau, shown in a numerical study by BOYER (1975, p. 336), and with a separation between easterly and westerly flows near Enderby Land (SVERDRUP, 1940, p. 120). Counterbalancing could result from significant westerly bottom flow in this active AABW formation region [6].

The Marion–Kerguelen Island section (Fig. 15c) contains only low velocity, alternating northerly and southerly flows, many of which would be reversed by relatively weak bottom currents. The net baroclinic flow between the Crozet and Kerguelen Islands is only $6 \times 10^6 \, \text{m}^3 \cdot \text{s}^{-1}$ leaving about $80 \times 10^6 \, \text{m}^3 \cdot \text{s}^{-1}$, or two-thirds of the net baroclinic transport south of the Crozet Islands to pass south of Kerguelen. With 85–90% of the baroclinic components above 2000 m and 60% above 1000 m on these sections, the Kerguelen Plateau may not present that much of an obstacle. The presence of the Kerguelen Plateau, however, and the sharp northeast trend of the mid-ocean ridge between Marion Island and 65°E may result in a broader or branched circumpolar current across these meridians.

Acknowledgements—Hydrographic data collection was carried out by S. Jacobs, D. Georgi, S. Patla, E. Bauer, T. Root, P. Bruchhausen, A. Amos, F. Rosselot, and by the officers, crew and technical support party of the R.D. *Conrad*. D. Muus and G. Anderson, Scripps Institution of Oceanography, facilitated the loan of a spectrophotometer and the reduction of nutrient data. C. Gaglio, L. Sullivan and T. Aitken assisted with portions of the data reduction. Helpful comments on the manuscript were made by A. Gordon, H. Taylor, E. Carmack, M. Rodman, R. Markl, V. Kolla and an anonymous reviewer. The interest and encouragement of A. Gordon have been critical to the progress of this study: in particular we have benefited from discussions of mixing processes. Lutjeharms (1972) has compiled a useful bibliography of the papers and data in the south-west Indian Ocean.

Typing was done by C. Teitch and A. Roy; drafting by A. Sotiropoulos. The 'GEOSECS', *Glacier* and *Hudson* data were obtained from the respective investigators on those expeditions, prior to formal publication. This study was supported by National Science Foundation grant 74-01213 from the Division of Environmental Sciences and by Energy Research and Development Administration grant AT (11-1)2185. The loan of equipment from U.S.N.S. *Eltanin* (now *Ara Islas Orcadas*) was arranged through the National Science Foundation's Division of Polar Programs.

REFERENCES

Boyer D. L. (1975) Numerical analysis of laboratory experiments on topographically controlled flow. *Numerical Models of Ocean Circulation*, Proceedings of Symposium, Durham, N.H., 17–20 October 1972. N.A.S., Washington, D.C.

Buscaglia J. C. (1971) On the circulation of the intermediate water in the Southwest Atlantic Ocean. *Journal of Marine Research*, **29**, 245–255.

Callahan J. E. (1972) Structure and circulation of deep water in the Antarctic. *Deep Sea Research*, **19**, 563–575.

Carmack E. C. (1973) Silicate and potential temperature in the deep and bottom waters of the western Weddell Sea. *Deep-Sea Research*, **20**, 927–932.

Carmack E. C. and T. D. Foster (1975) On the flow of water out of the Weddell Sea. *Deep-Sea Research*, **22**, 711–724.

Carpenter J. H. (1965) The Chesapeake Bay Institute technique for the Winkler dissolved oxygen method. *Limnology and Oceanography* **10**, 141–143.

Clowes A. J. (1933) Influence of the Pacific on the circulation in the southwest Atlantic Ocean. *Nature* **131**, 189–191.

Clowes A. J. and G. E. R. Deacon (1935) The deep water circulation of the Indian Ocean. *Nature*, **136**, 936–938.

Cooper L. H. N. (1967) Stratification in the deep ocean. *Science Progress, Oxford*, **55**, 73–90.

Craig H., Y. Chung and M. Fiadeiro (1972) A benthic front in the South Pacific. *Earth and Planetary Science Letters*, **16**, 50–65.

Darbyshire J. (1972) The effect of bottom topography on the Agulhas Current. *Pure and Applied Geophysics*, **101**, 208–220.

Deacon G. E. R. (1933) A general account of the hydrology of the South Atlantic Ocean. *Discovery Reports*, **7**, 171–238.

Deacon G. E. R. (1937) The hydrology of the Southern Ocean *Discovery Reports*, **15**, 1–124.

Deacon G. E. R. (1963) The Southern Ocean. *The Sea*, M. N. Hill, editor, vol. 2, pp. 281–296.

Deacon G. E. R. (1974) Water exchanges near the Antarctic continent. *Colloque international sur les processus de formation des eaux oceaniques profondes, Paris, 1972, Colloques Internationaux*, no. 215, pp. 23–25, C.N.R.S.

Deacon G. E. R. (1976) The cyclonic circulation in the Weddell Sea. *Deep-Sea Research*, **23**, 125–126.

Deacon G. E. R. and J. A. Moorey (1975) Boundary region between currents from the Weddell Sea and Drake Passage. *Deep-Sea Research*, **22**, 265–268.

Edmond J. M. (1970) Comments on the paper by T. L. Ku, Y. H. Li, G. G. Mathieu, and H. K. Wong, Radium in the Indian–Antarctic Ocean south of Australia, *Journal of Geophysical Research*, **75**(33), 6878–6883.

Foster T. D. and E. C. Carmack (1976) Frontal zone mixing and Antarctic Bottom Water formation in the southern Weddell Sea. *Deep-Sea Research*, **23**, 301–317.

Gill A. E. (1973) Circulation and bottom water production in the Weddell Sea. *Deep-Sea Research*, **20**, 111–140.

Gordon A. L. (1967) Structure of Antarctic waters between 20°W and 170°W *Antarctic Map Folio Series*, **6**, American Geographical Society, New York, N.Y.

Gordon A. L. (1971) Recent physical oceanographic studies of Antarctic waters, *Research in the Antarctic*, L. Quam, editor, A.A.A.S., Washington D.C. pp. 609–629.

Gordon A. L. (1974) Varieties and variability of Antarctic Bottom Water, *Colloques Internationaux C.N.R.S.* no. 215, *Processus de Formation des Eaux oceaniques profondes*, pp. 33–47.

Gordon A. L. (1975) An Antarctic oceanographic section along 170°E. *Deep-Sea Research*, **22**, 357–377.

GORDON A. L. and R. D. GOLDBERG (1970) Circumpolar characteristics of Antarctic waters. *Antarctic Map Folio Series*, **13**, American Geographical Society, New York, N.Y.

GORDON A. L., H. W. TAYLOR and D. T. GEORGI (1974) Antarctic oceanographic zonation. SCOR/SCAR Polar Oceans Conference, Montreal, Canada, May 1974 (Proceedings to be published).

GORDON A. L. and P. TCHERNIA (1972) Waters of the continental margin off Adelie Coast, Antartica. *Antarctic Oceanology II: The Australian–New Zealand Sector*, D. E. HAYES, editor Antarctic Research Series, **19**, 59–69, American Geophysical Union, Washington, D.C.

HAMON B. V. (1967) Medium-scale temperature and salinity structure in the upper 1500 m in the Indian Ocean. *Deep-Sea Research*, **14**, 169–181.

HEATH G. R. (1974) Dissolved silica and deep-sea sediments. *Studies in Paleo-oceanography*, W. W. HAY, editor, Soc. Econ. Paleo, and Mineral, Spec. Publ., no. 20.

HEEZEN B. C., M. THARP and C. BENTLEY (1972) Morphology of the earth in the Antarctic and Subantarctic. *Antarctic Map Folio Series*, **16**, American Geographical Society, New York, N.Y.

HOLLISTER C. D. and R. B. ELDER (1969) Contour currents in the Weddell Sea. *Deep-Sea Research*, **16**, 99–101.

IVANENKOV V. N. and F. A. GUBIN (1960) Water masses and hydrochemistry of the western and southern parts of the Indian Ocean. *Transactions of the Marine Hydrophysical Institute, Academy of Sciences U.S.S.R., Physics of the Sea, Hydrology*, **22**, (AGU translation, Jan. 1963, pp. 27–99).

JACOBS S. S. and A. F. AMOS (1967) Physical and chemical oceanographic observations in the southern oceans, USNS ELTANIN Cruises 22–27, 1966–67, T. R. no. ICU-1-67, 287 pp. Unpublished manuscript.

JACOBS S. S., A. F. AMOS and P. M. BRUCHHAUSEN (1970) Ross Sea oceanography and Antarctic Bottom Water formation. *Deep-Sea Research*, **17**, 935–962.

JACOBS S. S., E. B. BAUER, P. M. BRUCHHAUSEN, A. L. GORDON, T. F. ROOT and F. L. ROSSELOT (1974) *Eltanin* Reports, Cruises 47–50, 1971; 52–55, 1972, Hydrographic stations, Bottom photographs, Current measurements, Nephelometer profiles, TR CU-2-74, 502 pp. Unpublished manuscript.

JOHNSON R. E. (1973) Antarctic Intermediate Water in the South Pacific Ocean. International symposium on *Oceanography of the South Pacific*, Wellington, N.Z., Feb. 1972, *Proceedings*, pp. 55–69, UNESCO.

KATZ E. J. (1970) Diffusion of the core of Mediterranean Water above the Mid-Atlantic Ridge Crest. *Deep-Sea Research*, **17**, 611–625.

KILLWORTH P. D. (1973) A two-dimensional model for the formation of Antarctic Bottom Water. *Deep-Sea Research*, **20**, 941–971.

KIRWAN A. D. (1963) *Circulation of Antarctic Intermediate Water deduced through isentropic analysis*, Texas A & M Publ. 63–34F, 34 pp.

KOLLA V., L. SULLIVAN, S. S. STREETER and M. G. LANGSETH, Jr. (1976) Spreading of Antarctic Bottom Water and its effects on the floor of the Indian Ocean inferred from bottom-water potential temperature, turbidity and sea floor topography. *Marine Geology*, **21**, 171–189.

KOOPMAN G. (1953) Entstehung und Verbreitung von Divergenzen in der oberflächennahen Wasserbewegung der Antarktischen Gewässer. *Deutsche Hydrographische Zeitscrift* A, 2.

KUKSA V. I. (1972) Some peculiarities of the formation and distribution of intermediate layers in the Indian Ocean. *Oceanology*, **12**, 21–30.

LE PICHON X. (1960) The deep water circulation in the southwest Indian Ocean. *Journal of Geographical Research*, **65**, 4061–4074.

LUTJEHARMS J. E. (1972) A quantitative assessment of year-to-year variability in water movement in the south-west Indian Ocean. *Nature Physical Science*, **239**, 59–60.

LUTJEHARMSJ. R. E. (1972) A guide to research done concerning ocean currents and water masses in the south west Indian Ocean, Dept. of Oceanography, Univ. of Capetown, 577 pp.

LYNN R. J. and J. L. REID (1968) Characteristics and circulation of deep and abyssal waters. *Deep-Sea Research*, **15**, 577–598.

MAMAYEV O. (1975) *Temperature–salinity analysis of world ocean waters*, Elsevier Oceanographical Series, II Amsterdam.

MARTINEAU D. P. (1953) The influence of the current systems and lateral mixing upon Antarctic Intermediate Water in the South Atlantic. Woods Hole Oceanographical Institute Ref. no. 53–72. Unpublished manuscript.

McKEE W. D. (1971) A note on the sea level oscillation in the neighbourhood of the Drake Passage. *Deep-Sea Research*, **18**, 547–550.

ORREN M. J. (1966) *Hydrology of the SW Indian Ocean*. Report of the South African Department of Commerce and Industries, Division of Sea Fisheries Investigation Report, 55 pp.

OSTAPOFF F. (1962) The salinity distribution at 200 meters and the Antarctic frontal zones. *Deutsche Hydrographische Zeitschrift* **15**, 4, Sondendruck.

PINGREE R. D. (1972) Mixing in the deep stratified ocean. *Deep-Sea Research*, **19**, 549–561.

REID J. L. (1965) Intermediate waters of the Pacific Ocean. *Johns Hopkins Oceanographic Studies*, **2**, 85 pp.

REID J. L. and R. J. LYNN (1971) On the influence of the Norwegian–Greenland and Weddell Seas upon the bottom waters of the Indian and Pacific Oceans. *Deep-Sea Research*, **18**, 1063–1088.

Reid J. L. (1973) Transpacific hydrographic sections at lats 43°S and 28°S: the *Scorpio* Expedition-III. Upper water and a note on southward flow at mid-depth. *Deep-Sea Research*, **20**, 39–50.

Rochford D. J. (1960) The intermediate depth waters of the Tasman and Coral Seas I. The 27.20 σ_t Surface. *Australian Journal of Marine and Freshwater Research*, **II**, 127–147.

Rochford D. J. (1965) Rapid changes in the characteristics of the deep salinity maximum of the SE Indian Ocean. *Australian Journal of Marine and Freshwater Research*, **16**, 129–149.

Scarlett R. I. (1975) A data processing method for salinity, temperature, depth profiles. *Deep-Sea Research*, **22**, 509–515.

Scherbinin A. D. (1973) Geostrophic water circulation in the Indian Ocean. *Oceanology*, **13**, 649–651.

Scripps Institution of Oceanography (1971) Phosphate analysis. *Marine Technicians Handbook*, SIO Ref. 71–10, Sea Grant Publ. II, Institute of Marine Resources, La Jolla, California.

Strickland J. D. H. and T. R. Parsons (1968) *A practical handbook of seawater analysis*, Fisheries Research Board, Canada, Bulletin 167, 311 pp.

Sverdrup H. U. (1931) The origin of the deep water of the Pacific Ocean as indicated by the work of the *Carnegie*. *Gerlands Beiträge zur Geophysics*, **29**, Leipzig.

Sverdrup H. U. (1940) *Hydrology*, Section 2, discussion. *Report of the B.A.N.Z. Antarctic Research Expedition, 1921–31*, Series A, **3**, *Oceanography*, Pt. 2, sect. 2, pp. 88–126.

Sverdrup H. U. and R. H. Fleming (1941) The waters off the coast of Southern California, March to July, 1937. *S.I.O. Bulletin*, **4**, 261–378.

Taft B. A. (1963) Distribution of salinity and dissolved oxygen on surfaces of uniform potential specific volume in the South Atlantic, South Pacific and Indian Oceans. *Journal of Marine Research*, **21**, 129–146.

Taljaard J. J., H. van Loon, H. L. Crutcher and R. J. Jenne (1969) *Climate of the upper air, southern hemisphere*, Vol. I, USDC, ESSA.

Thorndike E. M. (1975) A deep-sea photographic nephelometer. *Ocean Engineering*, **3**, 1–15.

Treshnikov, A. F. (1964) Surface water circulation in the Antarctic Ocean. *Information Bulletin of the Soviet Antarctic Expedition*, **45**, 81–83 (translation).

Warren B. A. (1974) Transpacific hydrographic sections at Lats 43°S and 28°S: the *Scorpio* Expedition II. Deep Water. *Deep-Sea Research*, **20**, 9–38.

Warren B. A. (1974) Deep flow in the Madascar and Mascarene Basins. *Deep-Sea Research*, **21**, 1–22.

Wüst G. (1933) Das Bodenwasser und die Gliederung der Atlantischen Tiefsee. *Wissenschaftliche Ergebnisse der Deutschen Atlantischen Expedition auf dem Vermessungs und Forschungsschiff 'Meteor'*, 1925–27, **VI**, 1.

Wüst G. (1936) Schichtung und Zirkulation des Atlantischen Ozeans. Die Stratosphäre, 1. Ausbreitung und Vermischung der stratosphärischen Wasserarten in den Kernschichten. 2. Das subantarktische Zwischenwasser. *Wissenschaftliche Ergebnisse der Deutschen Atlantischen Expedition auf dem Vermessungs- und Forschungsschiff 'Meteor'*, 1925–27, **VI**.

Wyrtki K. (1971) *Oceanographic atlas of the International Indian Ocean Expedition*, NSF, U.S. Government Printing Office, Washington, D.C., 531 pp.

Wyrtki K. (1973) Physical oceanography of the Indian Ocean. *Biology of the Indian Ocean*. B. Zeitschel, and S. A. Gerlach, editors, *Ecological Studies* **3**, 18–36, Springer Verlag, Berlin.

Southern ocean temperature gradient near 2°C*

Arnold L. Gordon and Michael R. Rodman

Lamont-Doherty Geological Observatory of Columbia University, Palisades, N.Y. 10964

Abstract—U.S.N.S. *Eltanin* hydrographic STD stations south of Australia reveal an abrupt 50% increase in vertical temperature gradient below the 2°C isotherm (at depths of over 2000 m). The gradient discontinuity is confined to a 1200 km zone north of the polar front.

The vertical temperature gradient change occurs near a major junction of the temperature–salinity curve, separating the segment, extending from the circumpolar deep-water salinity maximum to the Antarctic bottom water, from the segment spanning the T/S region between the Antarctic Intermediate water and circumpolar deep water. The reduced temperature gradient above the 2°C discontinuity extends to 2.7°C and may be viewed as a weak thermostad. Silicate concentration versus depth reveals a reduction in silicate gradient accompanying the thermostad.

Two hypotheses are presented, which may explain the origin of the thermostad above 2°C.

The Advection Hypothesis suggests the thermostad is the product of advection from the southwest Atlantic Ocean. Its formation in the Atlantic results from intrusion of relatively cold low oxygen Pacific water into the layer immediately above the North Atlantic deep-water stratum.

The Isopycnal Stirring Hypothesis suggests the thermostad layer above 2°C is a product of cold water flux from the south. This cooling influence is derived from the temperature-minimum core layer.

1. INTRODUCTION

THE purpose of this paper is twofold. First, to describe the stratification observed near 2°C north of the polar front in the Australian sector of the Southern Ocean, specifically the reduced vertical temperature gradient immediately above 2°C. Second, to present two hypotheses which may explain the stratification.

2. THE FEATURE

The STD hydrographic stations obtained in the southeastern Indian Ocean reveal a change in temperature gradient near the 2°C isotherm (Fig. 1; GORDON, 1975). The region in which this discontinuity is observed extends northward from the polar front for a distance of approximately 1200 km, deepening from 1900 m at the polar front to 2800 m at its northern limits (Fig. 2). The 2°C isotherm north of the discontinuity reaches the Australian continental margins, but is not associated with a temperature gradient change (Fig. 3).† South of the polar front the 2°C isotherm occurs near the base of a relatively broad T_{max} layer (Fig. 4a).

---

* Lamont-Doherty Geological Observatory Contribution No. 2499.

† Note on Fig. 3, the reduced temperature and salinity gradients near 500 m (8–9°C and 34.35–34.50‰). McCartney (pp. 103–119) suggests that this 'stad' results from spreading of deep winter period convective mixed layer found north of the polar front.

North of the polar front the increase of the vertical temperature gradient near 2°C occurs slightly above the salinity maximum. Above the 2°C isotherm a nearly linear gradient extends to approximately the 2.7°C isotherm, located at the base of the thermocline associated with the salinity minimum layer of Antarctic Intermediate Water. The salinity gradient possesses an inflexion point near the 2.7°C isotherm. The nearly linear temperature gradient layer below the discontinuity reaches the 1°C isotherm, below which the water column is relatively homogeneous.

The average vertical potential temperature gradient (Fig. 5) above the discontinuity is 6.40×10^{-4}°C/m^{-1} (standard deviation of 1.11×10^{-4}), and below it averages 9.74×10^{-4}°C/m^{-1} (standard deviation of 1.39×10^{-4}). The average percentage increase is approximately 50%. The difference in gradient (Fig. 5) suggests slow attenuation of the contrast towards the east, from 5×10^{-4} to 2×10^{-4}°C/m^{-1}. Because of relatively low vertical gradient between 2°C to 2.7°C this stratum appears as a weak thermostad (SEITZ, 1967).

The T/S points at the 2°C feature (insert Fig. 6) indicate some variability. The average potential temperature and salinity are $1.91° \pm 0.09$°C, and $34.738\text{‰} \pm 0.013\text{‰}$, respectively. The salinity is about 0.02‰ lower than the salinity at the salinity maximum and for the most part occurs 100 to 300 m above the salinity maximum depth.

An expanded scale plot of the water column segment containing the 2°C temperature gradient discontinuity north of the polar front (Fig. 7) emphasizes the abruptness of the change. Though other changes in temperature gradient are observed, they are either altered or compensated within a short distance and therefore appear as fine-scale anomalies, close to the noise level of the STD. The gradient change near 2°C, while it occurs in a very limited depth interval, divides two macro-layers of different vertical gradients and can be thought of as a discontinuity.

As noted earlier, the relation of the salinity maximum to the temperature gradient discontinuity is somewhat variable: at Sta. 1391 they are coincident but at Sta. 1143 there is a separation of 375 m. Station 969 shows an increase in salinity gradient immediately

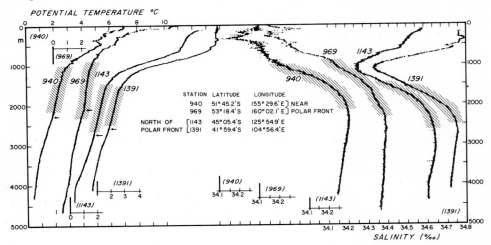

Fig. 1. Representative *Eltanin* STD stations demonstrating the 2°C temperature gradient discontinuity (shown by arrow) south of Australia. The pattern delineates the $\sigma_2 = 36.69$ to $\sigma_2 = 37.02$ range, which is the average density range of the thermostad (see text) for all STD temperature traces showing the discontinuity (see Fig. 2); note that the appropriate σt values which extend these σ_2 isopycnal surfaces above 1000 m are 27.50 and 27.76 respectively.

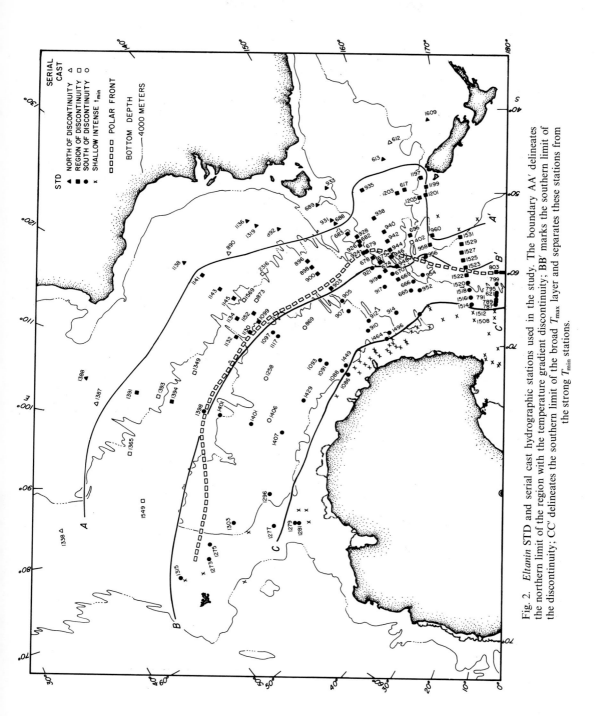

Fig. 2. *Eltanin* STD and serial cast hydrographic stations used in the study. The boundary AA′ delineates the northern limit of the region with the temperature gradient discontinuity; BB′ marks the southern limit of the discontinuity; CC′ delineates the southern limit of the broad T_{max} layer and separates these stations from the strong T_{min} stations.

below the temperature gradient discontinuity, similar to the transition zone associated with the benthic front of the southwest Pacific Ocean (CRAIG, CHUNG and FIADEIRO, 1972).

The θ/S relationship (θ is potential temperature and S is salinity) for standard level data (Fig. 6) indicates the water at the spatial gradient change occurs at a major gradient change in θ/S space: the interface between two basic segments of θ/S curve from the salinity maximum to the seafloor, and from 2.1°C to 2.7°C. Above the 2.7°C the θ/S curve becomes steeper as Antarctic Intermediate Water is approached.

Serial cast hydrographic stations, which indicate structure similar to the 2°C temperature gradient discontinuity, were chosen to provide oxygen and silicate data. The θ/O_2 plot (Fig. 8) indicates the oxygen concentration from 1°C to 2.3°C are somewhat higher than expected from the larger scale trend, determined from a straight line fit between the oxygen minimum (4.15 ml/l^{-1} to the local bottom water concentration (5.0 ml/l^{-1}). The θ/O_2 relation for stations north of the region possessing the 2°C temperature gradient discontinuity (Fig. 8) reveals decreased amplitude of the high oxygen anomaly. The θ/O_2 relation for stations south of the polar front reveal the oxygen minimum falls within the high oxygen anomaly north of the polar front.

The θ/Si distribution (Fig. 9) reveals a linear relation from the silicate concentration of 40 μM/l at the Antarctic Intermediate Water (above which the θ/Si gradient increases significantly) to bottom water silicate concentrations of 130 μM/l^{-1}. The temperature gradient discontinuity has silicate concentrations of approximately 80 to 85 μM/l^{-1}.

Curves of oxygen and silicate versus depth south of Australia (stas. 1343 and 1569 in Fig. 10a) reveal some structure near the 2°C temperature gradient discontinuity. The silicate versus depth gradient is strong above the 70 μM/l^{-1} silicate concentration and below the 90 μM/l^{-1} level, but weak between 70 to 90 μM/l^{-1}, which is the stratum coincidental with the thermostad; perhaps one can view this layer of weaker silicate gradient as a silicastad. The oxygen versus depth curve suggests that the 2°C temperature gradient

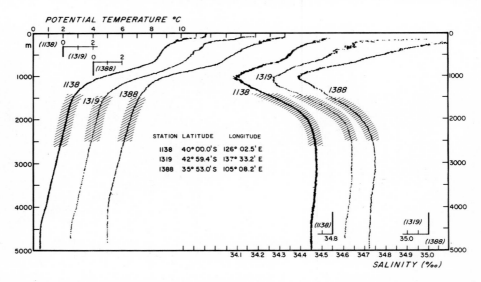

Fig. 3.   Representative *Eltanin* STD stations north of the 2°C temperature gradient discontinuity. The pattern shows the range $\sigma_2 = 36.69$ to $\sigma_2 = 37.02$, see Fig. 1 caption.

discontinuity is associated with a weak high oxygen anomaly (relative to a smoother, lower-order polynomial oxygen versus depth curve). This oxygen/depth high oxygen anomaly, as well as the oxygen/temperature high oxygen anomaly discussed above, apparently occurs both above and below the temperature gradient discontinuity.

Six STD hydrographic stations extending below 2000 m were obtained during U.S.N.S. *Eltanin* cruise 25 in the central Pacific sector of the Southern Ocean. Three of these (near 120°W) show evidence of a weak discontinuity in temperature gradient near 2°C

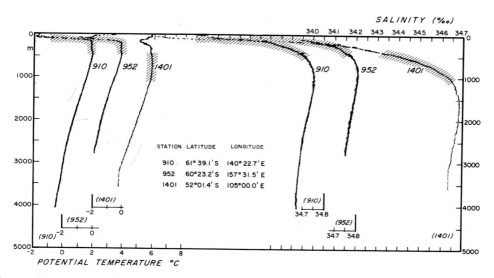

Fig. 4a.   Representative *Eltanin* STD stations south of the 2°C discontinuity, displaying the broad isothermal $T_{\max}$ layer. The pattern is the range from $\sigma_t = 27.50$ to $\sigma_t = 27.76$, see Fig. 1 caption.

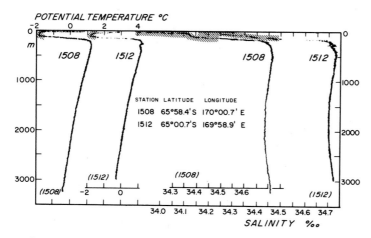

Fig. 4b.   Representative *Eltanin* STD stations showing the strong $T_{\min}$ south of the isothermal zone. Again the pattern delineates the $\sigma_t = 27.50$ to $\sigma_t = 27.76$ range.

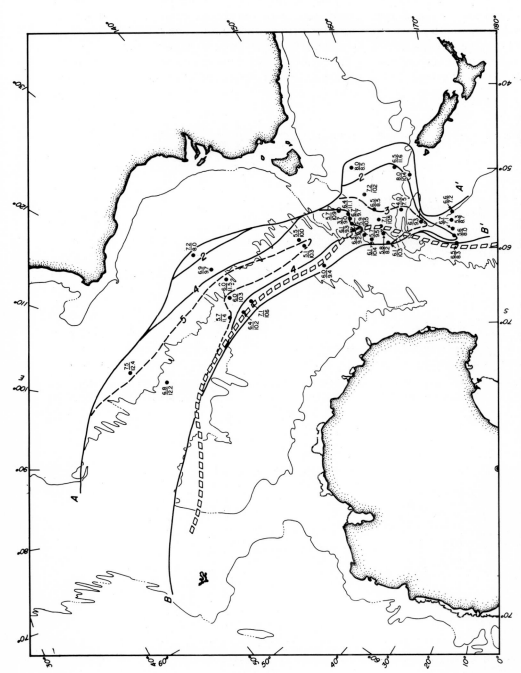

Fig. 5. Change of vertical temperature gradient (in $10^{-4}\,°C/m^{-1}$) across the discontinuity (upper gradient–lower gradient). The lines AA' and BB' are the northern and southern boundaries as in Fig. 2. The upper number given at each station is the gradient above the discontinuity, the lower number is the gradient below the discontinuity.

(*Eltanin* Stas. 598, 600, and 601); all are north of the polar front. DEACON (1937, p. 70) notes a weak temperature gradient in the South Pacific below the salinity minimum layer: "... between 2.5 and 2.25°C isotherm the temperature depth gradient was only very small, whilst there was a large increase of salinity, and it is very likely that the isotherms mark the boundary region between the intermediate layer and the highly saline deep current."

Numerous STD stations were occupied southeast of Africa during R/V *Conrad* cruise 17 (JACOBS, 1974; JACOBS and GEORGI, this volume). A few stations along 50°S, from the southern flank of the Crozet Basin to the northwestern end of the Kerguelen Plateau, reveal a temperature gradient discontinuity near 2°C (Stas. 219, 220, and 221 are particularly good examples).

3. VERTICAL PROCESS CONSIDERATIONS

CRAIG, CHUNG and FIADEIRO (1972) describe a feature in the southwest Pacific they name the 'Benthic Front'. They observed gradient changes in potential temperature, salinity, silicate, radium-226, oxygen, phosphates and nitrates. Below the Antarctic Intermediate Water silicate is shown to be conservative south of about 10°S, and on the Benthic Front silicate is at a maximum with a relative minimum at the deeper salinity maximum. Between the silicate maximum and the silicate minimum is an approximately 500-m-thick transition layer. Because the salinity gradient and the potential temperature gradient increase through the transition layer, it is also marked by a benthic pycnocline. They propose the transition layer between the benthic front and the salinity maximum marks the boundary between the southward moving deep waters of Pacific origin and northward moving bottom waters of Antarctic origin. Along the frontal surface the

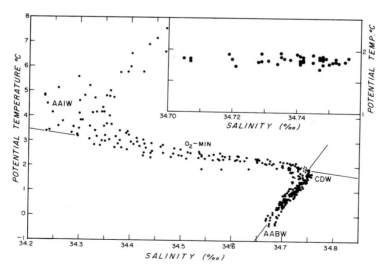

Fig. 6.   Potential temperature/salinity at standard levels for *Eltanin* STD stations showing the 2°C temperature gradient discontinuity (the θ/S point, shown by an X, is at the discontinuity). The insert is the θ/S of the discontinuity points. Added are linear regression best-fit lines for the θ/S regions of 2.7°C to the θ/S position of the discontinuity and from the S_{max} to the bottom water.

conservative properties are maintained by horizontal diffusion with a scale length (K_H/V), of 10^4 km for a simple one-dimensional model.

Chung (1975) determined the 'scale height' ($Z^* = k/w$) associated with the deep waters and bottom waters above and below the benthic front, respectively, by using the one-dimensional advection–diffusion equation solved by Munk (1966). Across the front the scale height changes roughly from 0.50 km for the bottom water to 1.0 km for the deep water. We did a similar analysis and found a Z^* between 1.33 to 2.0 km, but not changing significantly across the temperature gradient discontinuity.

The lack of a silicate maximum, a benthic pycnocline, and "scale height" change across the temperature gradient discontinuity south of Australia indicate the Antarctic discontinuity is not the same feature as the southwest Pacific benthic front. This is also

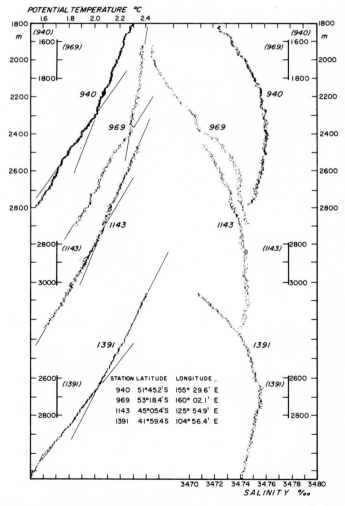

Fig. 7. Expanded scale of the select *Eltanin* STD stations from Fig. 1, displaying the 2°C temperature gradient discontinuity.

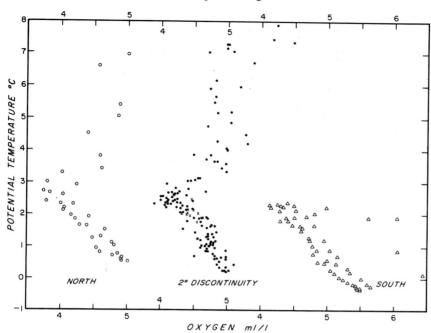

Fig. 8. Potential temperature/oxygen scatter for *Eltanin* serial cast hydrographic stations shown in Fig. 2. The solid circles are stations within the region of the discontinuity having an × marked at the discontinuity. The open circles are stations to the north, and the triangles are stations to the south of the discontinuity.

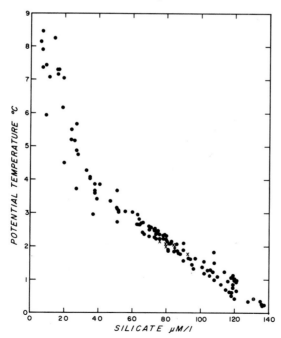

Fig. 9. Potential temperature/silicate of *Eltanin* serial cast hydrographic stations lying in the region of the discontinuity (Fig. 2). The position of the discontinuity is marked with an ×.

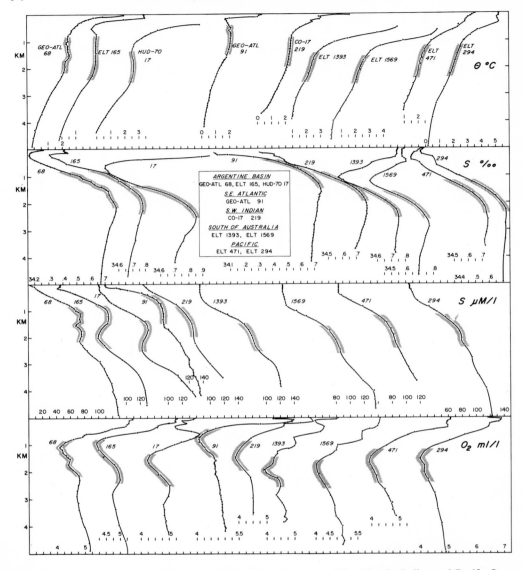

Fig. 10a.   Serial cast hydrographic stations in the Antarctic sectors of the Atlantic, Indian and Pacific Oceans, showing the circumpolar characteristic of the thermostad in temperature, salinity, silicate and oxygen. The density stratum $\sigma_t = 27.50$ (sigma-t is used since the upper part of the thermostad on some stations is above 1000 m) to $\sigma_2 = 37.02$ is shown by the patterned section of the trace.

GEOSECS–*Atlantic*	68	48 39.0S	46 01.0W
Eltanin	165	49 59.0S	39 59.0W
Hudson-70	17	39 55.0S	29 53.0W
GEOSECS–*Atlantic*	91	49 34.3S	11 28.0E
Conrad-17	219	47 03.4S	42 04.3E
Eltanin	1393	43 59.4S	105 00.6E
Eltanin	1569	48 04.7S	126 13.3E
Eltanin	471	52 58.0S	145 05.0W
Eltanin	294	57 01.0S	89 47.0W

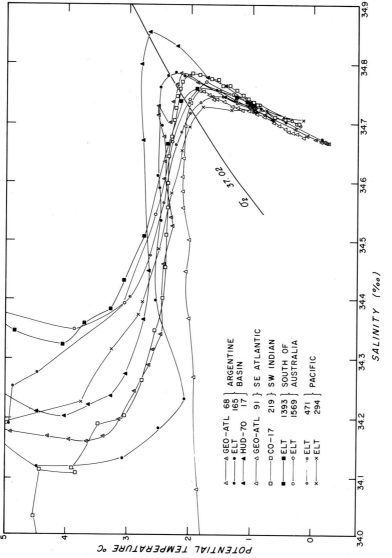

Fig. 10b.   Group θ/S diagram of stations used in Fig. 10a.

concluded by CRAIG, CHUNG and FIADEIRO (1972), who then go on to suggest the 2°C feature south of Australia is a remnant of North Atlantic Deep Water.

4. THE HYPOTHESES

Two hypotheses are now presented which may explain at least some aspects of the observed stratification north of the polar front.

1. *Advection hypothesis*

The vertical temperature profile may be a product of advection from the west with the circumpolar current. Inspection of the stratification in the South Atlantic Ocean reveals a similar feature. *Hudson* Sta. 17, *Eltanin* Sta. 165, and GEOSECS Sta. 68, within the Argentine Basin (Fig. 10), show (1) a thermostad in mid-depth between the 2° and 3°C isotherms; (2) with an accompanied silicastad; (3) an oxygen minimum in the upper part of the thermostad and a maximum at the base; (4) with a stabilizing halocline across the thermostad. Closer inspection of this feature shows the potential temperature attains a local minimum near the oxygen minimum, while the silicate attains a local minimum near the oxygen maximum.

The density range (Fig. 10) of the southwest Atlantic thermostad is the same as that of the thermostad to the east. On progressing eastward the temperature gradient within the thermostad goes from a slight reverse gradient to positive gradients. There is a slow reduction in the intensity of the gradient change at the base of the thermostad, as shown on Fig. 5 for the Australian segment of the southern ocean.

The local temperature minimum in the South Atlantic is noted by DEACON (1933, 1937), who attributes it to outflow from the shallow T_{min} layer (winter water) south of the polar front: "The low temperature suggests that the stratum is a sub-Antarctic mixture which contains a large proportion of the water which sinks from the cold stratum of the Antarctic layer" (p. 49, 1937). MANN, COOTE and GARNER (1973) concur on suggesting the silicate maximum–oxygen minimum layer in the western South Atlantic, lying below the Antarctic intermediate Water (see *Hudson* station in Fig. 10) and associated with a weak minimum in potential temperature, are Antarctic in origin since "all three extreme trace back to the Antarctic Convergence, where they run toward the surface". WÜST (1936) proposes the temperature minimum–oxygen minimum core is induced by more active renewal of high oxygen water above (Antarctic Intermediate Water) and below (North Atlantic Deep Water).

While the temperature minimum characteristic bespeaks an Antarctic origin the oxygen and silicate extreme do not; yet the temperature minimum characteristics must be renewed from a cold water source. Inspection of regional trends in the oxygen minimum layer suggests a possible source of the temperature minimum layer.

The oxygen minimum core layer maps (GORDON, 1967, plate 6; CALLAHAN, 1972; GORDON and MOLINELLI, 1975, plate 6B) suggest spreading of the Pacific $O_{2\,min}$ water of less than $4\,ml/l^{-1}$ between 1200 to 1500 m towards the east through the northern Drake Passage, and turning northward to enter the Argentine Basin. The temperature of this $O_{2\,min}$ varies between 2.2 to 2.4°C potential temperature (see $\theta/O_{2\,min}$ plate 6-S of GORDON and MOLINELLI, 1975), which corresponds to the characteristics at the temperature minimum (Fig. 10, 11). Below the temperature minimum layer in the South Atlantic the water

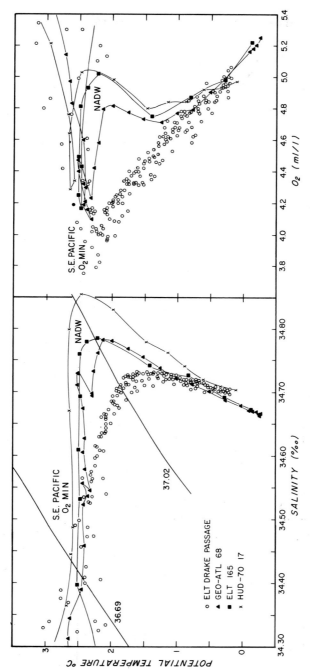

Fig. 11. Potential temperature/salinity and potential temperature/oxygen of *Eltanin* serial cast hydrographic stations within the Drake Passage (open circles). Three stations within the Argentine Basin are GEOSECS–Atlantic 68, *Eltanin* 165 and *Hudson-70* 17.

column is responding to the relatively warm saline–high oxygen–low silicate North Atlantic influence. The multi-inversion in all parameters revealed at GEOSECS Sta. 68 (Fig. 10, 11) near the oxygen minimum suggests the presence of a vigorous stirring action of Pacific and Atlantic waters.*

The influence of the Pacific oxygen minimum layer on the Atlantic water column is further indicated in the comparison of the Argentine Basin and Drake Passage θ/S and θ/O_2 curves (Fig. 11). Deviation of the Argentine Basin curves from those of the Drake Passage curves must represent the Atlantic Ocean influence. This is confined to the high saline–high oxygen excursion of the North Atlantic Deep Water and the extension of the curves towards more concentrated Antarctic Bottom Water, resulting in a steeper (lower stability) and less linear θ/S curve below the S_{max}. The oxygen minimum layer (a narrow temperature range intrusion in the Atlantic, i.e. the thermostad) has similar characteristics to the Drake Passage oxygen minimum, which supports the Pacific origin (assuming the average northern Drake Passage flow at the 1500-m level is towards the east, as indicated by the geostrophic field, REID and NOWLIN, 1971). Hypothesis One, in summary, is as follows: the colder-lower salinity, oxygen-poor, silicate-rich Pacific water is advected into the Atlantic and cools the water column immediately above NADW, inducing the observed thermostad; it is then advected by the Antarctic Circumpolar Current, with some alteration, into the South Atlantic and eastward across the Indian Ocean into the Pacific, as shown by the sequence of curves in Fig. 10. The thermostad remains within the same density stratum and the basic stratification of oxygen and silicate is preserved.

The continuous rotation of the temperature curve towards more positive slope (weakening of the thermostad) on progression from the Atlantic sector may be a result of mixing with waters of more positive vertical temperature gradients to the north and south (Figs. 1, 3, 4), possibly by an isopycnal process. The family of θ/S curves (Fig. 10b) revealing the upper part of the thermostad (with the $O_{2 \, min}$) in the Australian sector is similar to the northern segment of the South Atlantic thermostad, while the lower part of the Australian sector thermostad is closer to the southern segment of the South Atlantic thermostad. The significance of this is not clear, perhaps relative meridional motion occurs across the thermostad.

The advection hypothesis may explain most of the features observed south of Australia, but it does not explain the abrupt change in temperature gradient near 2°C, unless the maximum advective–diffusive lateral flux of the North Atlantic Deep Water component is confined to a narrow stratum near 2°C, or vertical mixing is weak, thus preserving the discontinuity observed in the Atlantic. The former is unlikely, since neither the salinity, oxygen, nor silicate data portray a sheet-like intrusion of the Deep Water at the 2°C, or any other layer. Rather, the Deep Water influences a broad stratum of water, as evident by the high oxygen anomaly observed in θ/O_2 space (Fig. 8), which may be attributed to NADW high-oxygen influence (Fig. 11).

2. *Isopycnal stirring hypothesis*

The thermostad between 2° to 2.7°C may be viewed as an anomalously cold stratum, induced by northward spreading of cold Antarctic characteristics.

---

* REID, NOWLIN and PATZERT (1977), also attribute the $O_{2 \, min}$ layer between the AAIW and NADW as Pacific in origin.

The density surface associated with the temperature gradient discontinuity is 37.02 (± 0.016) σ_2 units (the density of the water if moved adiabatically to 2000 m) (Fig. 1). The southward extension of this density surface (converting to the matching σ_t value of 27.76 at the 1000-m level) passes close to the base of the temperature maximum core layer south of the polar front (Fig. 4a) and to a position just below the temperature minimum closer to Antarctica (Fig. 4b). The density surface associated with the 2.7°C level is 36.69 (± 0.010) σ_2 units and 27.50 σ_t above 1000 m (Fig. 1), which, on extension southward of the polar front, is observed to pass through the base of the Antarctic Surface Water, i.e. the T_{min} layer (Fig. 4a). Further south the T_{min} becomes more saline (GORDON, TAYLOR and GEORGI, 1974) and the 27.5 σ_t surface passes to a position above the T_{min} (Fig. 4b). Therefore the density stratum 'containing' the anomalously cold thermostad north of the polar front extends into the T_{min} winter water far to the south.

The concept of isopycnal spreading of water characteristics stems directly from the isentropic concept applied to the atmosphere by ROSSBY (1936, 1937). Isentropic spreading represents adiabatic movement with no entropy change of particles (small volumes of water). For the ocean MONTGOMERY (1938, 1940) suggests that surfaces of constant potential density represent isentropes (surfaces of constant entropy), and the study of such surfaces represents 'isentropic analysis'. PARR (1938) suggests calling such an approach an 'isopycnic analysis'. MONTGOMERY continues to point out that potential density may be replaced by σ_t for the upper kilometer. MONTGOMERY (1938, p. 13) and LYNN and REID (1968) point out that potential density calculated in relation to a deeper more appropriate pressure may be a better approximation to entropes for application of isentropic analysis below 1000 m.

Naturally non-isentropic processes must occur in the ocean in order to re-establish

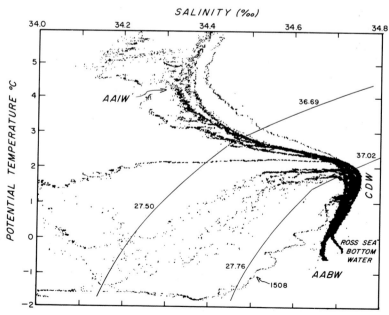

Fig. 12.   The potential temperature/salinity of the *Eltanin* STD stations of Figs. 1, 3, 4a and 4b. The σ_2 surfaces of 36.69 and 37.02 are converted at a 1000 m to a σ_t of 27.50 and 27.76, respectively.

the thermohaline gradients on isopycnal surfaces, which would otherwise be attenuated by isentropic processes. However, over much of the density surface area, removed from the sea surface, the isentropic processes may dominate the non-isentropic processes of vertical mixing in regard to transfer of water mass characteristics.

The use of 'stirring', rather than 'mixing', is recommended by ECKART (1948) and reiterated by MCLELLAN (1957), since mixing implies change in entropy.

The water below the 2°C level north of the polar front would not have access via a density surface to the Antarctic winter water and hence could not be cooled by isopycnal stirring from that layer. This may explain the abruptness of the gradient change. The density surface deeper than the $37.02\,\sigma_2$ extend to the Antarctic continental slope and may therefore have access to convected Antarctic waters over the continental slope, but this cooling influence would be expected to be small in comparison with the more extensive and colder open ocean T_{min} core layer.

The group θ/S diagram for stations north and south of the polar front with the σ_2 and σ_t isopycnals bounding the 2°C to 2.7°C thermostad (Fig. 12) show the density similarity of the thermostad with the cold, relatively saline, T_{min} layer of the southern areas (Sta. 1508 in Fig. 12).

As pointed out by DEACON (1933, 1937) the general upwelling of deep water into the near surface layers in the Southern Ocean induced by Ekman drift implies a net advection of the deep-water component towards the south and towards the surface. Assuming balance by the northward diffusive flux of Antarctic characteristics, determination of the ratio v to K_y from the meridional advection–diffusion equation has been made on three levels within the density stratum 36.69 to $37.02\,\sigma_2$ for three separate meridional sets of data. A value of about $5 \times 10^{-4}\,\mathrm{km}^{-1}$ seems most appropriate. Hence a K_y of $10^7\,\mathrm{cm}^2/\mathrm{s}^{-1}$ (KAUFMAN, TRIER, BROECKER and FEELY, 1973 suggest a K_y of 10^7 to 10^8 for the surface water) yields a southward meridional velocity of $0.05\,\mathrm{cm/s}^{-1}$.

How does the oxygen and silicate distribution relate to Hypothesis Two? The silicastad coincides with the thermostad and may be explained in similar fashion: the silicate concentration between the 36.69 and $37.02\,\sigma_2$ surfaces are lowered somewhat by northward stirring of the relatively low silicate T_{min} water in the southern extent of the winter water. Inspection of silicate values within the southern extent of the T_{min} (*Eltanin* cruises 27, 37 and 47), measured during the summer period, reveal values of 60–$80\,\mu\mathrm{M}/1$, which is below the silicastad value of 80–$85\,\mu\mathrm{M}/\mathrm{l}^{-1}$, and so the T_{min} can act as a low silicate source. The lack of an expression of the T_{min}/Si characteristic in θ/Si space (above $40\,\mu\mathrm{M}/\mathrm{l}^{-1}$, Fig. 9), does not support a T_{min} influence in the thermostad.

The oxygen minimum falls within the upper part of the thermostad (Fig. 10), while at the base of the thermostad, at the temperature gradient discontinuity, the oxygen levels are somewhat higher than expected. The source of the oxygen minimum core layer is believed to be at latitudes well north of the polar front (WYRTKI, 1962; CALLAHAN, 1972), where the oxygen consumption and renewal by advection and diffusion combine to yield a minimum in oxygen. Perhaps, as suggested in Hypothesis One, the oxygen minimum north of the polar front has a significant Pacific influence. The attenuation of the oxygen minimum core layer (Fig. 8) on approach to Antarctica would, by Hypothesis Two, be due to the high oxygen T_{min} water (values between 6–$7.5\,\mathrm{ml}/\mathrm{l}^{-1}$). The high oxygen near 2°C, as mentioned above, is most likely a result of the North Atlantic Deep Water influx, since it spans the temperature gradient discontinuity and is not a product of the stirring process of Hypothesis Two.

The basic weakness of Hypothesis Two is revealed on inspection of the thermostad characteristics around Antarctica (Fig. 10). The T_{min} is circumpolar, yet the thermostad, and its associated salinity and chemistry stratification, are most intense in the Atlantic sector. In addition, the Atlantic characteristics of the thermostad can be attributed to the influx of low oxygen Pacific water, without reference to the T_{min} core layer. One would not expect a separate formation process for the thermostad in the Pacific sector.

5. DISCUSSION

Hypothesis One appears the most likely candidate to explain the temperature gradient structure between 2°C to 2.7°C observed south of Australia. However, the process discussed in Hypothesis Two may have some relevance in explaining the temperature gradient discontinuity near 2°C. The water column above 37.02 σ_2, the average density at the discontinuity, can be influenced through isopycnal stirring by the waters of the southern T_{min} core, while the water below 37.02 σ_2 cannot. This retards attenuation of the thermostad and may account for the extensive spreading of the thermostad into the Pacific Ocean.

Acknowledgements—The Grant GA 41284 from the Division of Environmental Sciences of the National Science Foundation supported the initial stages of this research.

Grant OCE 74 15062 (a component of the International Studies of the Southern Ocean) from the Office of the International Decade of Ocean Exploration of the National Science Foundation supported the final preparation of the paper.

We are grateful to HOYT TAYLOR and STAN JACOBS for critically reviewing the manuscript, ANASTASIA SOTIROPOULOS for drafting the figures, and ANNE ROY for typing the manuscript.

REFERENCES

CALLAHAN J. E. (1972) The structure and circulation of deep water in the Antarctic. *Deep-Sea Research*, **19**, 563–575.

CHUNG Y. (1975) Areal extent of the benthic front and variation of the scale height in Pacific deep and bottom waters. *Journal of Geophysical Research*, **80**, 4169–4178.

CRAIG H., Y. CHUNG and M. FIADEIRO (1972) A benthic front in the South Pacific. *Earth Planetary Science Letters*, **16**, 50–65.

DEACON G. E. R. (1933) A general account of the hydrology of the South Atlantic Ocean. '*Discovery*' *Reports*, **7**, 171–238.

DEACON G. E. R. (1937) The hydrology of the southern ocean. '*Discovery*' *Reports*, **15**, 1–124.

ECKART C. (1948) An analysis of the stirring and mixing processes in incompressible fluids. *Journal of Marine Research*, **7**, 265–275.

GORDON A. L. (1967) Structure of Antarctic waters between 20°W and 170°W. *Antarctic Map Folio Series*, no. 6, V. BUSHNELL, editor, American Geographical Society, New York, N.Y.

GORDON A. L. (1975) An Antarctic oceanographic section along 170°E. *Deep-Sea Research*, **22**, 357–377.

GORDON A. L. and E. MOLINELLI (1975) *USNS ELTANIN southern ocean oceanographic atlas, cruises 4–55, June 1962–November 1972.* Lamont-Doherty Geological Observatory and the Department of Geological Science of Columbia University, Palisades, N.Y. (a limited number of copies of the atlas are available on request).

GORDON A. L., H. W. TAYLOR and D. T. GEORGI (1974) Antarctic oceanographic zonation. SCOR/SCAR Polar Oceans Conf., Montreal, Canada, May 1974.

JACOBS S. S. (1974) R/V CONRAD cruises 17-04 and 17-05, southwest Indian Ocean: physical oceanography. *Antarctic Journal*, **9**, 214–219.

KAUFMAN A., R. M. TRIER, W. S. BROECKER and H. W. FEELY (1973) Distribution of 228Ra in the world ocean. *Journal of Geophysical Research*, **78**, 8827–8848.

LYNN R. J. and J. L. REID (1968) Characteristics and circulation of deep and abyssal waters. *Deep-Sea Research*, **15**, 577–598.

McLELLAN H. J. (1957) On the distinctness and origin of the slope water off the Scotian shelf and easterly flow south of the Grand Banks. *Journal of the Fisheries Research Board, Canada*, **14**(2), 213–239.

MANN C. R., A. R. COOTE and D. M. GARNER (1973) The meridional distribution of silicate in the western Atlantic Ocean. *Deep-Sea Research*, **20**, 791–801.

MONTGOMERY R. B. (1938) Circulation in upper layers of southern North Atlantic deduced with use of isentropic analysis. *Papers in Physical Oceanography and Meteorology*, **6**(2), 55 pp.

MONTGOMERY R. B. (1940) The present evidence on the importance of lateral mixing processes in the ocean. *Bulletin of the American Meteorological Society*, **21**(3), 87–94.

MUNK W. H. (1966) Abyssal recipes. *Deep-Sea Research*, **13**, 707–730.

PARR A. E. (1938) Isopycnic analysis of current flow by means of identifying properties. *Journal of Marine Research*, **1**(2), 133–154 (1937–38).

REID J. L. and W. D. NOWLIN (1971) Transport of water through the Drake Passage. *Deep-Sea Research*, **18**, 51–64.

REID J. L., W. D. NOWLIN and W. C. PATZERT (1977) On the characteristic and circulation of the southwestern Atlantic Ocean. *Journal of Physical Oceanography*, **7**, 62–91.

ROSSBY C. G. (1936) Dynamics of steady ocean currents in the light of experimental fluid mechanics. *Papers in Physical Oceanography and Meteorology*, **5**(1).

ROSSBY C. G. (1937) Isentropic analysis. *Bulletin of the American Meteorological Society*, **18**, 201–209.

SEITZ R. C. (1967) Thermostad, the antonym of thermocline. *Journal of Marine Research*, **25**, 203.

WÜST G. (1936) Schichtung und Ziekulation des Atlantischen Ozeans. *Wissenschaftliche Ergebnisse der Deutschen Atlantischen Expedition 'Meteor', 1925–1927*, Bd. VI, TI. 1, 2: Die Stratosphäre.

WYRTKI K. (1962) The oxygen minima in relation to ocean circulation. *Deep-Sea Research*, **9**, 11–23.

Subantarctic Mode Water*

M. S. McCARTNEY

Woods Hole Oceanographic Institution, Woods Hole, Massachusetts 02543, U.S.A.

Abstract—Immediately north of the circumpolar subantarctic front, deep (400–600 m) well-mixed layers are found in late winter. Spring and summer heating isolates (but does not completely erase) these layers beneath the seasonal thermocline as thermostads. The zone in which this active renewal is found is several hundred kilometers wide, but the associated thermostad can be traced much further north—on the order of 2000 km. The thermostads and the often associated dissolved oxygen maxima can be found as far north as the south equatorial current regions of each southern hemisphere subtropical gyre. This water mass formation and spreading process is equivalent to that occurring east and south of the Gulf Stream and Kuroshio currents, where the thermostads are called Subtropical Mode Water (STMW). In light of the association of these southern ocean thermostads with the circumpolar subantarctic front, rather than the subtropical fronts (western boundary currents such as the Agulhas current) the name Subantarctic Mode Water (SAMW) is suggested.

In common with STMW, SAMW contributes substantial volumetric modes to the central water masses, indicating SAMW to be the renewal agent of the high oxygen parts of the main thermocline water of the southern hemisphere subtropical gyres.

Finally, it is noted that the specific types of SAMW formed in the southeast Pacific and Scotia Sea areas are identical in temperature and salinity to the South Pacific and South Atlantic varieties of Antarctic Intermediate Water (AAIW). The renewal process for AAIW is hence indicated as taking place *north* of the polar front zone, in the southeast Pacific and Scotia Sea parts of the subantarctic zone. The actual process is late winter convective overturning of the somewhat warmer and more saline waters advecting into the region from the west along the subantarctic front. The low salinity of AAIW is due to the pronounced excess of precipitation over evaporation in the subantarctic zone. This process is quite different from the traditional concept of circumpolar cross-polar-frontal mixing of Antarctic Surface Water with Subantarctic Surface Water.

1. INTRODUCTION

DEACON's discussions of the subantarctic zone of the southern ocean (1933, 1937, 1963) have focused primarily on the origin and movement of Antarctic Intermediate Water (AAIW): the salinity minimum layer found at depths of a few hundred meters to around 1000 m in the zone, and extending off to the north at around 1000 m. AAIW is the dominant feature of the subantarctic zone in the South Atlantic Ocean, the layer there also being characterized by a dissolved oxygen maximum (DEACON, 1933; WÜST, 1935). In the remainder of the subantarctic zone, there is another feature which has not drawn as much attention as AAIW. In DEACON's (1937) review of southern ocean hydrography, the temperature distributions on Sections 9, 10 and 11 southwest and south of Australia show very thick layers at 8°–10°C lying between the Antarctic convergence and the subtropical convergence. The layers lie beneath a seasonal thermocline and are located immediately north of a pronounced thermal front (subsequently called the subantarctic

* Woods Hole Oceanographic Institution Contribution Number 3773.

103

front). SEITZ (1967) has coined a word for such a layer: a thermostad—a layer with a minimum vertical temperature gradient, i.e. the antonym of thermocline.

Using the much larger data collection now available from the subantarctic zone, this paper will demonstrate that the above configuration—a thermostad at intermediate depths north of the subantarctic front—is an ubiquitous feature of the subantarctic zone. This is of particular significance for our understanding of processes in the world's oceans, because identical configurations are bound in the North Atlantic and Pacific, where thermostads are found south of the Gulf Stream and Kuroshio fronts. It is interesting to find this apparent similarity of processes, since usually the zonally unobstructed southern ocean is taken to have fundamentally different physics from the zonally bounded northern hemisphere oceans.

The formation and spreading processes for the western North Atlantic thermostad has been discussed by WORTHINGTON (1959, 1972a, 1972b, 1976). He notes that a deep well-mixed layer is found in late winter immediately south and east of the Gulf Stream front. Since the thermostad is found all year round, and generally at a temperature within a few tenths of a degree of 18.0°C, he calls the layer 18° water. He also points out that an inflection point at a temperature of near 18°C exists in the temperature–depth curves well to the south of the formation site (the area where the deep well-mixed 18°C layer is found in late winter). WORTHINGTON (1972) views this southward extension of the 18° water at depths of a few hundred meters as the lower half of a meridional cell, with the surface water moving to the north, and cooling and sinking south of the Gulf Stream. MASUZAWA (1969) has identified a similar layer south of the Kuroshio in the western North Pacific, at moderately cooler temperatures, and suggests the more general terminology 'Subtropical Mode Water' (STMW), subtropical referring to the formation region, and mode referring to the substantial volumetric mode this water mass contributes to each ocean's water volume. This shows up as an isolated maximum on a volumetric temperature–salinity diagram (e.g. WRIGHT and WORTHINGTON, 1970, plate 4).

In addition to the similar thermostad/thermal front configuration, it will be demonstrated in the present paper that the formation process is identical, i.e. erosion of the seasonal thermocline during winter leaving the thermostadal layer exposed to the atmosphere in late winter, when convective overturning occurs; and that the spreading equatorward from the formation site via the subtropical gyres also is similar, although with some variation due to the circumpolar character of parts of the southern ocean circulation.

In the light of the similarity in the details of the process, the name Subantarctic Mode Water (SAMW) will be used for this thermostad. Henceforth, the abbreviations SAMW and STMW will be used.

2. THE SUBANTARCTIC MODE WATER THERMOSTAD

Three of the many meridional sections available showing the SAMW thermostad are shown in Fig. 1. Each shows a thermostadal layer extending from immediately north of a deep thermal frontal zone, well to the north through the subtropical gyres, and finally becoming indistinguishable in the south equatorial current region. Except for the associated cooler temperatures, the sections look quite similar to sections across the Gulf Stream and Kuroshio currents. The front itself is circumpolar, and will be called for simplicity the subantarctic front. The associated eastward transport can carry specific types of SAMW

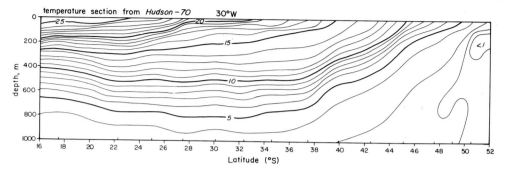

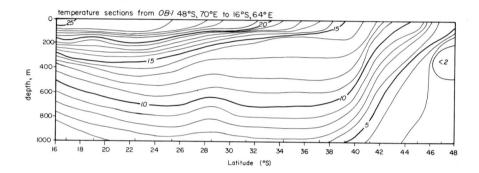

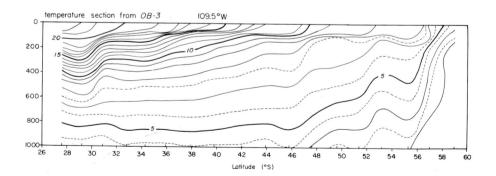

Fig. 1. Meridionally oriented temperature (°C) sections, from the Antarctic zone across the Subantarctic front, through the subtropical gyres and into the south equatorial current zone. Station latitudes are indicated by the tick marks above the latitude axis. The section locations are shown in Fig. 2. The contour interval is 1°C, except in (C), where the 3.5°, 4.5°, 5.5°, and 6.5°C isotherms are indicated by dashed lines to more clearly delineate the thermostad.

from one ocean to another, e.g. from the southeastern Pacific to the southwestern Atlantic. The 30°W section, Fig. 1a, shows the local SAMW thermostad to be centered around 14°C south of 32°S, and somewhat warmer to the north. DEACON (1937, Fig. 12) showed an expanded scale part of his earlier presented (1933) 30°W section. Station 675, at 34°S, shows a somewhat cooler thermostad—perhaps 13.5°C. This suggests a degree of climatic stability similar to 18° water (SCHROEDER, STOMMEL, MENZEL and SUTCLIFFE, 1959).

The circumpolar character of the thermal front is indicated by the stippled thermal frontal zone on the world chart in Fig. 2. The stippled area represents the band of high horizontal temperature gradient at 200 m on the GORDON and GOLDBERG (1970) atlas plate

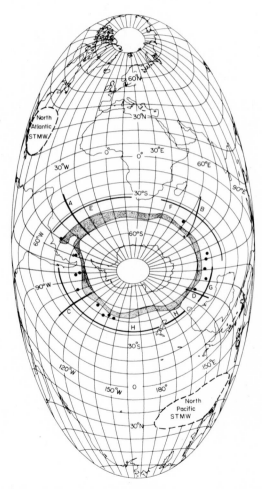

Fig. 2. World Chart: an equal-area (transversed Molleweide) projection. The subantarctic front, as determined from the high temperature gradient areas on the GORDON and GOLDBERG (1970) chart of temperature at 200 m, is indicated by the stippled belt between 39°S and 62°S. The regions of prevalence of STMW in the North Atlantic and North Pacific Oceans are enclosed by dashed lines. The locations of the three meridionally oriented sections presented in Fig. 1 are indicated by curves A, B and C; while that used in Fig. 6 is indicated by curve D. The four zonal sections used in constructing Fig. 4 are indicated by curves E, F, G and H. The fifteen stations used in Fig. 7 and 8 are indicated by the heavy dots.

of 200-m temperatures. This subantarctic front is spiral in nature, starting at close to 40°S in the southwestern Atlantic, and spiraling down to the east to 60°S in the southeastern Pacific. A fairly abrupt turn to the north occurs east of the Drake Passage. The temperatures associated with the front show a corresponding nearly monotonic decrease to the east from the southwestern Atlantic. There is no single indicator that can be used as front indicator, such as 15°C at 200 m is used for the Gulf Stream. Different segments of the front have been given different names in the past, with subtropical convergence being generally used in the Atlantic and southwestern Indian Oceans, and Australasian subantarctic front in the southeastern Indian Ocean (BURLING, 1961). In the Pacific, GORDON (1971) discusses the apparent double polar front zone, the northern boundary of which he calls the primary polar front, whose main characteristic is a pronounced north–south temperature gradient. Also shown on the chart in Fig. 2 are the locations of the three sections in Fig. 1, other stations and sections which will be used to further define SAMW characteristics, and for comparison to SAMW, the regions of prevalence of STMW as indicated by WORTHINGTON (personal communication) and MASUZAWA (1969). Note that the chart in Fig. 2 is an equal area projection (transversed Molleweide).

Since the GORDON and GOLDBERG (1970) atlas only covers the area south of 40°S, it misses most of the western boundary currents. The interaction between the (southward

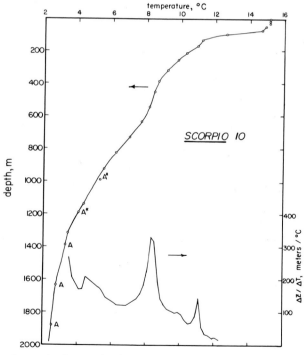

Fig. 3. Example of the three-point Lagrangian interpolated fit to Nansen bottle station data, and the resultant thermostad calculation. The station is number 10 of the SCORPIO expedition (*Eltanin* 28) and was located at 43.2°S, 161.1°E, in the Tasman Sea. The interpolated curve shown does not include the two bottles (labeled A*) from the deep cast (the remaining A bottles) which overlapped in depth the shallow cast (the unlabeled circles). The resulting $\Delta Z/\Delta T$ curve shows a possible weak near surface mode 11.1°, and the strong SAMW thermostad at 8.3°, with strength of well over 300 m/°C.

flowing) Brazil Current and the (northward flowing) Falkland Current in the southwest Atlantic is only partially resolved. This confluence extends off to the east (South Atlantic Current) accounting for the extreme breadth of the thermal frontal zone along 30°W in Fig. 1a. The Agulhas and Agulhas Return Currents lie north of 40°S, 20°E (Duncan, 1970) and account for the southward dip of the thermal frontal zone in Fig. 2 between 10° and 40°E. Two other frontal zones are not shown: the Tasman Current, extending eastward across the Tasman Sea, north of 43°S; and the confluence east of New Zealand along the Chatham Rise (43°S) between the (northeast flowing) Southland Current and the (southward flowing) East Cape Current.

To obtain a more quantitative picture of the geographical variation of thermostad strength and temperature, the following scheme was adopted: for a given hydrographic station a representation of the temperature–depth variation was obtained by using a modified Lagrangian three-point interpolation between observed data points. The resulting curve was finite-differenced to obtain estimates of inverse temperature gradient ($\Delta Z/\Delta T$) as a function of temperature, using $\Delta T = 0.2°C$. Thermostads then stand out as a maximum $\Delta Z/\Delta T$, while thermoclines are a minimum. Figure 3 shows an example of the temperature–depth fit, and the resultant $\Delta Z/\Delta T$ curve. The complete station was made up of a shallow and a deep (labeled A) cast. The deep cast had two bottles (labeled A*) which overlapped in depth with the shallow cast. A fit including these overlapping bottles would be a poor one, having an artificial inflection point in the $T–Z$ curve at 5.4°C. The station falls near the Tasman Current, and 3 hours elapsed between the messengers for the two casts; hence, ship drift in the moderately variable field of the region probably accounts for the differences between the two casts. The fit without the A* bottles looks better, although there is still a weak inflection at the join between the two casts ($T = 3.4°C$). Throughout the thermostad calculations, overlapping bottles in the deep cast were not used, and also one bottle at the join point, if the spacing at the join was less than 50 m.

Figure 4 shows the results of this thermostad calculation for a set of stations taken from four zonal sections, shown on the chart in Fig. 2: 32°S in the Atlantic Ocean (*Atlantis* cruise 247, data obtained from A. R. Miller), 32°S in the Indian Ocean (*Atlantis II* cruise 15, data obtained from NODC), 40°S in the southeast Indian Ocean (*Diamantina* cruise 1/60, data obtained from NODC), and 43°S in the Pacific Ocean (SCORPIO expedition, *Eltanin* cruise 28, data obtained from NODC). A subset of these stations spaced at nominally 5° longitude was used. The results are presented as contours of constant $\Delta Z/\Delta T = 50, 75, 100, 200$ and $300 \, m/°C$ on a temperature versus longitude diagram. It should be noted that these zonal sections lie well to the north of the formation location for SAMW (immediately north of the frontal band), so that the thermostads shown are isolated from the sea surface; that is, the local winter overturning does not extend to the depths where the thermostads are found.

The SAMW thermostad shows up as a pronounced ridge running diagonally across the diagram above the main thermocline valley. The thermostad strength is weakest in the South Atlantic, with maximum strength of about 130 m/°C near 37°W longitude (compare to the 30°W section, Fig. 2a). The thermostad temperature drops progressively to the east from the South Atlantic, while the thermostad strength increases, albeit irregularly. At 65°E the thermostad temperature is 12.7°C (compare to Fig. 2b). South of Australia and the Tasman Sea there is a leveling off of the thermostad temperature at 8–9°C. Again, climatic stability is suggested since Deacon's (1937, plates XIX and XII) sections 10 and 11 from 1932 south of Australia also show 8–9°C thermostads. On the eastern side of New

Zealand, the Chatham Rise is quite shallow, so no data at temperatures colder than 7° or 8°C is found there. East of the Rise, the SAMW thermostad reappears at 7°C, more than a full degree colder than it was west of New Zealand. Finally, just to the west of South America, the thermostad temperature has fallen to 5–6°C, with a maximum calculated thermostad strength of 447 m/°C at temperature 5.5°C, at SCORPIO 65 (*Eltanin* cruise 28, Sta. 65, 93.4°W, 43.2°S). The apparent isolated thermostad 'bubbles' at 4° to 5° are caused by the problems at the join between upper and lower casts previously mentioned. They occur at a temperature lying between that of the deepest bottle of the shallow cast, and the shallowest bottle of the deep cast, and thus are poorly defined and probably not real.

The SAMW thermostad temperature variation of from 14.5°C at 40°W to 5.5°C at 80°W averages out to an east–west temperature gradient on the order 1°C/3000 km. This is the same order of magnitude as STMW, so they have comparable longitudinal variability. STMW appears superficially more homogeneous only because of the relatively small zonal extent of its formation sites, and because it is recirculated in rather small gyres.

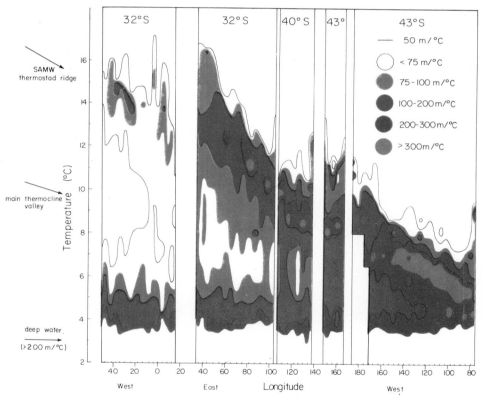

Fig. 4. Plot of the calculated thermostad strength as a function of temperature and longitude, using the four zonal sections whose positions are shown on the world chart in Fig. 2. The SAMW thermostad is the ridge running diagonally across the diagram from 14° to 15°C in the western Atlantic through 8–9°C south of Australia (120°E) and ending at 5° in the eastern Pacific. The main thermocline is the valley lying at temperatures below the ridge. No contours are shown below about 3.5°, in the more homogeneous deep water ($\Delta Z/\Delta T > 300$ m/°C). Also, no contours are shown for temperatures falling shallower than 100 m—in or above the seasonal thermocline.

3. THE FORMATION PROCESS FOR SUBANTARCTIC MODE WATER

The formation process for STMW has been discussed by Worthington (1959, 1972a, 1972b, 1976) and Warren (1972). They agree that the basic mechanism is vertical convective overturning at the end of winter: cooling at the surface in fall and winter, erasing the seasonal thermocline at the formation site (immediately south of the Gulf Stream and Kuroshio) allowing deep convection at the end of winter to depths on the order of 500 m. Spring warming then isolates the thermostad at depth by creating an overlying seasonal thermocline. Both authors thus emphasize the roles that the fundamentally asymmetric cooling and heating cycles play, i.e. cooling causing homogenization by convective overturning, while heating creates gradients. Further to the south, late winter overturning occurs, but not at cold enough temperatures to penetrate to the STMW thermostad. The authors disagree as to whether this process involves a net positive annual heat flux to the atmosphere (Worthington, 1972a) or essentially zero net heat flux (Warren, 1972). They also disagree somewhat as to the details of the recirculation of STMW within the subtropical gyres.

A similar process occurs for SAMW. Figure 5 shows temperature depth profiles from a line along 128°S, south of Australia, taken between 28 August and 6 September, 1968 (late winter). The subantarctic front axis lies near Sta. 873, as can be deduced from the

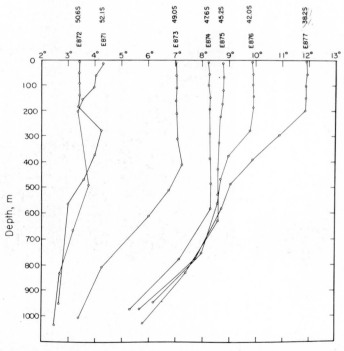

Fig. 5. Temperature–depth profiles for a line of late-winter stations south of Australia along 128°E, taken between 28 VIII 68 and 6 IX 68, on *Eltanin* cruise 35. Stations E871 and E872 lie within the polar frontal zone, while Sta. E873 is actually within the Subantarctic front. Stations E874 and E875 show the SAMW deep mixed layer. Stations E876 and E877 have shallower, warmer mixed layers, and show the SAMW thermostad inflection point at 8–9°C.

plunging of the main thermocline between Stas. E872 and E874. The two stations north of the front (E874 and E875) show deep, well-mixed layers about 600 m in depth. Station E874 is neutrally stable for 580 m, while E875 retains a slight remnant of surface heating (only about 0.01 g/$1\sigma_t$ stability). It should be noted that cooling probably lasted another month with the surface temperature reaching its seasonal minimum around the end of September. However, as WARREN (1972) has pointed out for STMW, the extreme thickness of these mixed layers precludes much change in their temperatures, even if rather extreme cooling were to persist for the entire month. The station spacing suggests a lower bound for the formation site width of $2\frac{1}{2}°$ latitude. A few 750-m XBTs were taken on this section of this cruise (*Eltanin* 35); number 49 was taken at 43.7°S, 128.0°E and showed temperature of 8.7–8.8°C over the upper 640 m. This increases the estimated formation site width to a lower bound of 3.9° latitude (430 km).

Station E873 has a 400-m mixed layer (<0.01 g/$1\sigma_t$). This is not likely to be a volumetric mode, since it is in the middle of the front, where characteristics have a strong north–south gradient. XBTs 39 (49.3°S) and 40 (49.0°S) show temperature profiles similar to station E873, while 38 (50.2°S) has temperature less than 5°C. Thus, the width of the zone showing the station E873 mixed-layer type was greater than 0.3° latitude but probably much less than 2.6° latitude in light of the strong gradient between XBT 38 and Sta. E874.

Further north at Stas. E876 and E877 there are shallower, warmer mixed layers which do not penetrate to the depth of the 8°C thermostadal knee. These shallower mixed layers disappear almost completely in summer, as can be seen in Fig. 6. This shows a temperature section along 132°E, 4° east of the stations in Fig. 5, run in late December 1969, and January 1970 on *Eltanin* cruise 41. The section shows the SAMW in the formation zone (approximately 43° to 48°S) isolated at depth underneath a seasonal thermocline lid. Further north, the SAMW thermostad extends at a depth of 600 m all the way north to the Australian continental shelf, which lies near 34°S. It should be noted that the path of the SAMW thermostad to the Australian shelf is not along the meridional section,

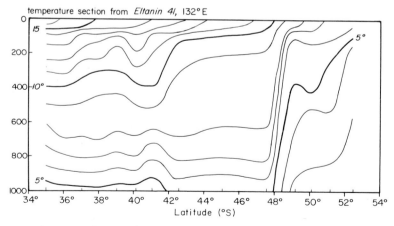

Fig. 6. A summertime temperature section south of Australia along 132°E from *Eltanin* 41. The subantarctic front is at 48° to 49°S. The SAMW north of the front is overlayed by a seasonal thermocline above 200 m, but persists as a pronounced thermostad between 200 and 600 m. North of 42°S the SAMW thermostad is still easily detectable, although not as pronounced as within the formation site between 43° and 48°S. The Australian continental shelf lies north of 35°S.

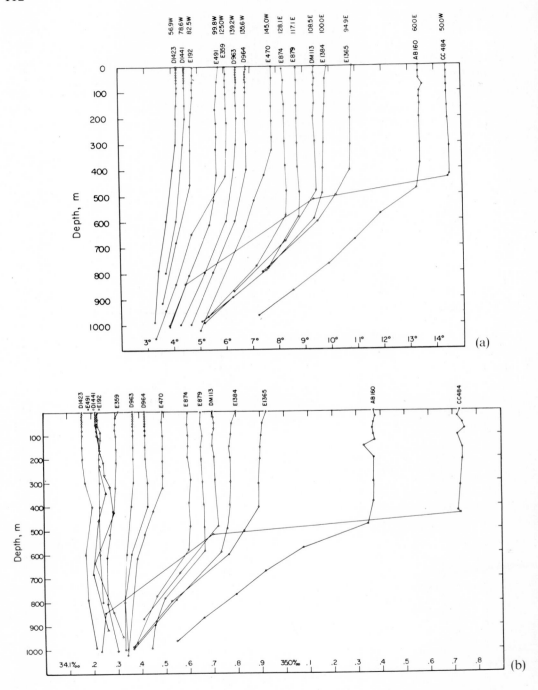

Fig. 7.   Temperature–depth (a) and salinity–depth (b) profiles for a group of fifteen late-winter stations from the zone immediately north of the subantarctic front. The station locations are shown on the world chart of Fig. 2; the station particulars are summarized in Table 1 in the Appendix.

but rather around the weak anticyclonic gyre within the South Australian Basin; that is, first east, then north near Tasmania, then west back to the 132°E section.

The formation zone width of a few degrees, with a formation temperature meridional variation of around $\frac{1}{2}$°C seems to be typical for the SAMW formation process circumpolarly. The longitudinal variation in characteristics within the formation zone in late winter is illustrated in Fig. 7. The figure shows temperature–depth and salinity–depth data for fifteen stations, whose particulars are tabulated in the Appendix; their locations are shown on Fig. 2 by the heavy dots. The stations are selected from all available cool-season extreme stations, which as far as sea surface temperature is concerned means September and October (Sta. E874 was taken on 31 August). When several stations on a given section showed the deep mixed layer, only one is shown. The fifteen stations were chosen to maximize longitudinal coverage. There are about sixteen more *Eltanin* and *Discovery* stations showing this process in the southeast Pacific, about fourteen more *Eltanin* and *Diamantina* stations in the southeast Indian Ocean, one *Anton Bruun* station in the southwest Indian and two *Capitán Cánepa* stations in the southwest Atlantic. These stations do not, however, fill any of the large longitudinal gaps in the late winter data coverage. There is a clear need for more late winter data, particularly from the southwest Pacific (southeast and southwest of New Zealand), southwest Indian (south and southeast of Madagascar, northeast and northwest of Kerguelen Island), and almost the entire South Atlantic. There is not enough data from any given region for looking in detail at year-to-year variability in formation characteristics.

The stations in Fig. 7 show mixed layers of various depths and degrees of homogeneity. The deepest and most homogeneous layers seem to be those in the southeast Indian Ocean. In general the higher the temperature of the mode, the higher the salinity; this relation making density variation rather mild (σ_t from 26.6 to 27.2, δ_t from 145 to 88). The relative homogeneity of a mixed layer is probably dependent on how recently the latest deep convective event took place, and the overall intensity of the winter during that particular year. Station E470 shows a trace of thermostadal knee at 500–600 m and 6.5–7°C, suggesting that 1965 was a mild winter with overturning not penetrating all the way through the previous year's thermostad. This is further suggested by Sta. E491, from the same year, which is fairly well mixed over 300 m, but shows a high salinity layer underneath. These two stations also have rather anomalous temperature–salinity relations, which will be discussed below.

Further south along the section from which Sta. D1441 was taken, the subantarctic front (GORDON's (1971) 'primary polar front') is crossed, south of which the temperature field exhibits a temperature minimum layer at around 400 m at two stations (*Discovery* Stas. 1444 and 1445). Above this layer are somewhat shallower (200–300 m) mixed layers at temperatures of 3.4–3.5°C, salinities of 34.14‰. A similar structure was observed 2 years earlier (Deacon, 1937, plate XLI). These stations were not included in Fig. 7 because they fall in the polar frontal zone rather than in the subantarctic zone. In addition, the structure may not be found every year: the stations from *Eltanin* cruise 10 along 83°W south of Sta. E192 in Fig. 7 show a single front separating the 4° subantarctic mode from the Antarctic zone, with no intervening 3° modes.

In Fig. 8, on a temperature–salinity diagram, the data from the mixed layers of the fifteen stations of Fig. 7 are shown. The symbols enclose the scatter of points within the mixed layer. The 1965 anomalous Sta. E470 and E491 stand out, the former (at 7.8°C, 34.49‰) lying on the warm fresh side of the curve defined by the rest of the data, and the

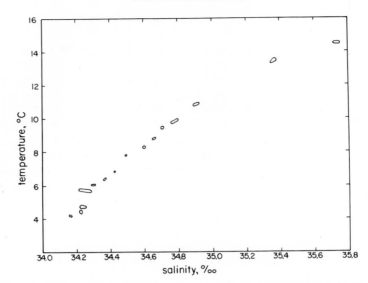

Fig. 8. Temperature–salinity diagram for the mixed layers of the stations shown in Fig. 7. The various symbols enclose the scatter of data within each of the mixed layers. The particular mixed layer depth used for each station is included in the station summary in Table 1 in the Appendix.

latter (at 5.7°C) being elongated in salinity. Stations E1365 and E1384 show compensating T–S variability, the data lines up with constant σ_t lines. The CC484 station at 14.5°C is elongated in salinity, and as a whole appears rather anomalously salty: the 14° to 15°C mode which appeared in Fig. 4 in the South Atlantic had associated salts of 35.4‰ to 35.55‰.

For comparison to Fig. 8, Fig. 9 shows the interpolated temperature–salinity data for the SAMW thermostad ridge seen in Fig. 4. The T–S arc for the 0.2° or 0.4°C range defining the maximum SAMW thermostad strength for each station has been drawn on a temperature–salinity diagram. The general agreement between the two T–S diagrams is good, except for CC484, as mentioned above.

The near linearity of groups of the data in Fig. 9 is rather remarkable. The data from the south Indian Ocean and part of the South Pacific south of the Tasman Sea (South Pacific warmer than 8°C) lies within less than 0.04‰ of the line from 8.0°C, 34.54‰ to 13.5°, 35.33‰. The data from east of the Chatham Rise (east of New Zealand) lies within 0.04‰ of the line from 5.0°C and 34.22‰ to 8.0°C and 34.51‰. The South Atlantic data seem to be divided into two groups, eastern and western, by the mid-ocean ridge at 15°W. The western group data lie within 0.02‰ of a line connecting 13.0°C, 35.25‰ and 15.0°C, 35.55‰ with the indicated thermostads having temperature between 13.2°C and 14.8°C. The eastern group data lie within 0.04‰ of a line connecting 12.0°C, 35.04‰ and 14.0°C, 35.35‰ with the indicated thermostads having temperatures between 11.2°C and 14.0°C. Thus the eastern group data are about 0.05‰ fresher than the western, in their overlapping temperature range (13°C to 14°C), and also have rather weaker thermostad strength.

The late winter data from the Pacific and Indian Oceans show a little more scatter about these lines, but this is perhaps to be expected since these data are from several

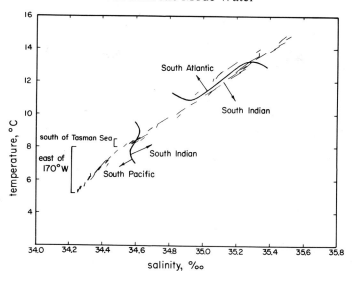

Fig. 9.   Temperature–salinity diagram for the ridge defining the SAMW thermostad in Fig. 4. Each linear segment corresponds to the 0.2°C (occasionally 0.4°C) increment which has the maximum associated value of $\Delta Z/\Delta T$. The salinities were also determined by the modified three-point Lagrangian interpolation scheme.

different cruises of several different ships, with various salinity determination methods. The *Capitán Cánepa* South Atlantic station (CC484) lies 0.25‰ saltier than the western South Atlantic line. This station is rather close to the thermal front, and the high salinity may represent an effect of the southward-flowing Brazil current bringing saltier water into this area.

The isolated thermostad bubbles at 4–5°C which appeared in the South Pacific on Fig. 4 have not been included in Fig. 9, since they are probably artifacts of the calculation procedure as discussed in Section 2. Corroborating evidence for rejecting these particular indicated thermostads is that the associated data do not show the dissolved oxygen maximum which generally characterizes the SAMW thermostad. In the formation zone, the vertical overturning leads to a very high dissolved oxygen content throughout the deep mixed layer, typically 95% of saturation value or greater. North of the formation zone, the SAMW thermostads retain these high oxygen values, typically being recognizable as dissolved oxygen maximum as far north as the south equatorial current system (the northern part of the southern subtropical gyres).

4. CONCLUSIONS

The primary purpose of this paper has been to call attention to the circumpolar character of the subantarctic thermostad, the influence of the thermostad on the central water masses of the southern hemisphere oceans, and the direct analogy to processes in the northern hemisphere oceans. There have been several earlier references to local examples of specific SAMW types. DEACON (1937, p. 53) briefly mentions a 'belt of mixed water' in connection with his sections 9, 10 and 11 south and southeast of Australia. SVERDRUP,

Johnson and Fleming (1942, p. 608) used Deacon's (1937) section 10 in their discussion of Antarctic water masses. They noted the thickness of the 8–9° layer in this section south of Australia, and suggested the name Subantarctic Upper Water; although they made no further comment on it. A related observation was Wyrtki's (1962) 'Subantarctic Intermediate Water', defined as the oxygen maximum stratum found well above the Antarctic Intermediate Water salinity minimum found south of 30°S in the Tasman Sea. This seems to be the oxygen maximum associated with the 8°C SAMW in this area, so his inference that "It is likely that these values indicate water sinking in the subantarctic region" is correct.

In discussing the South Pacific sections, Deacon (1937, p. 56) gives a single sentence description of the essence of the formation process: "The observations made in the southern part of the zone in the Pacific Ocean show that in winter the water is practically uniform down to a depth of 400 m, and probably because of intense vertical mixing the surface water has almost the same properties as the subsurface and intermediate waters." Stations D963 and D964 in Fig. 7 are two of the stations on which he based this statement. In the same area of the Pacific, Mittun and Natvig (1957), using data from summer 1947 along 90°W, noted a 'quasi-homogeneous water mass' they called 'sub-Antarctic sub-surface water' lying between 100 and 400 m at a temperature of about 5.2°C.

Burling (1961) discussed the hydrography south of New Zealand, and defined two types of subantarctic water. Although not making any direct statement about the thickness, he defined an 'Australasian Subantarctic Water' at a temperature of about 8°, which is seen to be the 8–9° variety of SAMW found south of Australia (see Fig. 4). He indicates its domain to be restricted to west of the Campbell Plateau south of New Zealand, i.e. west of 170°E, agreeing with the SAMW thermostad distribution in Fig. 4. The other type, 'Circumpolar Subantarctic Water', he indicates as being colder than 8°C below the surface layer, and salinities less than 34.5‰. This would seem to be identical with the saltiest, warmest SAMW thermostads to the east of the 170°W part of Fig. 9, but this is not completely clear: the boundary between the two types is described as being the Australasian Subantarctic Front, which is simply the section of the subantarctic front south of Australia, the Tasman Sea and New Zealand. SAMW has been shown to be formed north of the subantarctic front, so there seems to be a conflict. There is very little data southeast of New Zealand that can be used to clarify the water mass and circulation picture.

There are two related items which will be pursued in later works. The first is the use of SAMW as a core layer to describe the circulation in the upper kilometer of the southern hemisphere subtropical gyres. Thermostads made perfect tracers: once isolated equatorward of the formation/renewal zone, the thermostad is a thick layer of relatively homogeneous water which, according to the meridional sections, apparently persists as it advects around the various subtropical gyres. Knowing the longitude at which a particular type (T, S) of SAMW is formed, it can be tracked around the gyre to get an indication of the upper layer circulation. A confirmation of the potential of such a calculation is the closeness of the groups of nearly linear SAMW thermostad T, S data in Fig. 9 to the linear 'central water mass' T, S curves of the subtropical southern Pacific, Atlantic and Indian Oceans, described by Sverdrup, Johnson and Fleming (1942). It is suggested that the agent of renewal of these central water masses is in fact the SAMW thermostad. Evidence for this statement is that the SAMW thermostad T, S curves in Fig. 9 overlie the high volumetric ridges found in each of the volumetric T, S diagrams for the southern hemisphere oceans of COCHRANE (Pacific, 1958), POLLAK (Indian, 1958; more recently WYRTKI, 1971) and MONTGOMERY (Atlantic, 1958). Indeed, for the World Ocean, MONTGOMERY (1958) noted 'a strange mode' at 8.25°C, 34.65‰, which probably partially is accounted for by the rather large region of 8° SAMW south of Australia and the Tasman Sea.

The second item which would seem to warrant further study is the relation of the coldest, least saline types of SAMW to AAIW. In the southeast Pacific the deep mixed layers at the end of winter have salinities near 34.2‰ and temperatures ranging from less than 4° to about 5.5°C. These values span the range of AAIW types actually observed within the subtropical gyres. For example, JOHNSON (1972) studied the AAIW core in the South Pacific, defined as $\sigma_t = 27.10 \, \mathrm{g \cdot l^{-1}}$. On this surface he shows a tongue at salinities between 34.2‰ and 34.3‰ (temperatures of 4.5–5°C) extending north and west from the southeast Pacific subantarctic zone. The SAMW colder than 4.5° formed in the extreme southeast Pacific and Scotia Sea appears to pass through the Drake Passage and turn north past the Falkland Islands into the South Atlantic, where it forms a main contribution to the lower salinity, lower temperature variety of AAIW found in the southwest Atlantic. A major entry point of this coldest SAMW into the South Atlantic subtropical gyres seems to be the confluence region of the Brazil and Falkland currents. This can be seen in the data of Sta. CC484 in Fig. 7. At the surface is the 14°C variety of SAMW formed to the north of the confluence. At 800–1000 m is the cold (4°C), 34.2‰ variety of SAMW which has passed through the Drake Passage, turned to the north with the Falkland Current, and, being heavier, has ended up below the warmer SAMW in the confluence zone. The superposed SAMW types are separated by a very sharp pycnocline. The process of AAIW renewal just briefly described is quite different than that traditionally envisioned (e.g. SVERDRUP, JOHNSON and FLEMING, 1942, p. 619), in which AAIW formation occurs circumpolarily by cross-polar-frontal mixing of low salinity Antarctic Surface Water and Subantarctic Surface Water. The mixing product, pure AAIW, is supposed to be 2.2°C and 33.8‰ in the South Atlantic (WÜST, 1935): as it moves northward further strong, rapid mixing with the saltier waters above and below is postulated to bring the salinity up to the 34.2–34.3‰ values generally found in the subantarctic zone. In the present scheme, the renewal and recirculation of AAIW is indicated as taking place predominantly with the subantarctic and subtropical regions. The waters actually being cooled and overturned to form AAIW are just the warmer, saltier types of SAMW advecting in from the west along the subantarctic front. The low salinity of AAIW and, indeed, the entire SAMW trend of progressively lower salinities associated with the west to east trend of progressively cooler temperatures is a result of the large excess of precipitation over evaporation within the subantarctic zone, rather than injection of less saline Antarctic Surface Water across the front.

Acknowledgements—The author would like to express his thanks to VAL WORTHINGTON, TERRY JOYCE and BRUCE WARREN for several helpful comments and discussions. This work was supported by the Office of Naval Research, Contract No. N00014-74-CO262 NR083-004, and by the National Science Foundation, International Southern Ocean Studies Grant No. OCE76-00390.

REFERENCES

Burling R. W. (1961) Hydrology of circumpolar waters south of New Zealand. *New Zealand Department of Scientific and Industrial Research Bulletin* 143, 66 pp.

Cochrane J. D. (1958) The frequency distribution of water characteristics in the Pacific Ocean. *Deep-Sea Research*, **5**, 111–127.

Deacon G. E. R. (1933) A general account of the hydrology of the South Atlantic Ocean. *Discovery Reports*, **7**, 171–238.

Deacon G. E. R. (1937) The hydrology of the Southern Ocean. *Discovery Reports*, **15**, 1–124.

Deacon G. E. R. (1963) The Southern Ocean. In: *The Sea*, M. N. Hill, editor, pp. 281–296. Interscience, 554 pp.

Duncan C. P. (1970) The Agulhas Current, unpublished Ph.D. dissertation, University of Hawaii, 76 pp.

Gordon A. L. (1971) Antarctic polar front zone. In: *Antarctic Oceanology*, I, J. L. Reid, editor, Antarctic Research Series, **15**, American Geophysical Union, pp. 205–221.

Gordon A. L. and R. D. Goldberg (1970) Circumpolar characteristics of Antarctic waters. In: *Antarctic Map Folio Series*, V. Bushnell, editor, American Geographical Society, **13**.

Masuzawa J. (1969) Subtropical mode water. *Deep-Sea Research*, **16**, 463–472.

Mittun L. and J. Natvig (1957) Pacific Antarctic waters. *Scientific Results of the Brategg Expedition, 1947–48*, no. 3, 130 pp.

Montgomery R. B. (1958) Water characteristics of Atlantic Ocean and of world ocean. *Deep-Sea Research*, **5**, 134–148.

Pollak M. J. (1958) Frequency distribution of potential temperatures and salinities in the Indian Ocean. *Deep-Sea Research*, **5**, 128–133.

Schroeder E., H. Stommel, D. Menzel and W. Sutcliffe (1959) Climatic stability of eighteen degree water at Bermuda. *Journal of Geophysical Research*, **64**(3), 363–366.

Seitz R. C. (1967) Thermostad, the antonym of thermocline. *Journal of Marine Research*, **25**, 203.

Sverdrup H. U., M. W. Johnson and R. H. Fleming (1942) *The Oceans: their physics, chemistry and general biology*, Prentice-Hall, 1087 pp.

Warren B. A. (1972) Insensitivity of subtropical mode water characteristics to meteorological fluctuations. *Deep-Sea Research*, **19**, 1–19.

Worthington L. V. (1959) The 18° water in the Sargasso Sea. *Deep-Sea Research*, **5**, 297–305.

Worthington L. V. (1972a) Negative oceanic heat flux as a cause of water-mass formation. *Journal of Physical Oceanography*, **2**, 205–211.

Worthington L. V. (1972b) Anticyclogenesis in the oceans as the result of outbreaks of continental polar air, 169–178. In: *Studies in Physical Oceanography*, A. L. Gordon, editor, Gordon Breach, **1**, 194 pp.

Worthington L. V. (1976) On the North Atlantic circulation. In: *The Johns Hopkins Oceanographic Studies*, R. S. Arthur, D. F. Carritt, R. B. Montgomery, D. W. Pritchard and R. O. Reid, editors, The Johns Hopkins University Press **6**, 110 pp.

Wright W. R. and L. V. Worthington (1970) The water masses of the North Atlantic Ocean: a volumetric census of temperature and salinity. *Serial Atlas of the Marine Environment*, W. Webster, editor, folio 19.

Wüst G. (1935) Die Stratosphäre des Atlantischen Ozeans. In: *Wissenschaftliche ergebnisse der Deutschen Atlantischen Expedition auf dem Forschungs und Vermessungsschiff, Meteor 1925–1927*, Gruyter & Co., **6**(1), 109–288.

Wyrtki K. (1962) The subsurface water masses in the western South Pacific Ocean. *Australian Journal of Marine and Freshwater Research*, **13**, 18–47.

Wyrtki K. (1971) *Oceanographic Atlas of the International Indian Ocean Expedition*, National Science Foundation, 531 pp.

APPENDIX

In Table 1, the details of the fifteen late winter stations used in Fig. 7 and Fig. 8 are summarized.

Table 1. *Summary of details of the examples of late-winter hydrographic stations illustrating SAMW formation*

Ship	Station	Longitude (W or E)	Latitude (°S)	Date	Depth (m)	Max. temp. (°C)	Min. temp. (°C)	Min. salin. (‰)	Max. salin. (‰)	Min. σ_t (g/l)	Max. σ_t (g/l)
								Mixed layer			
Discovery II	D1423	56.9W	53.9	25-IX-34	300	4.23	4.16	34.16	34.17	27.12	27.13
Discovery II	D1441	78.4W	55.7	26-X-34	400	4.52	4.32	34.22	34.23	27.13	27.16
Eltanin	E192	82.5W	55.1	13-X-63	458	4.83	4.61	34.224	34.256	27.10	27.13
Eltanin	E491	99.8W	52.0	23-X-65	420	5.82	5.69	34.216	34.286	26.98	27.05
Eltanin	E359	125.0W	54.9	5-IX-64	425	6.09	6.03	34.311	34.288	27.01	27.02
Discovery II	D963	139.2W	52.0	14-IX-32	400	6.48	6.34	34.37	34.36	27.02	27.02
Discovery II	D964	135.6W	49.7	15-IX-32	400	6.86	6.81	34.42	34.43	26.99	27.01
Eltanin	E470	145.0W	51.0	26-IX-65	323	7.83	7.78	34.494	34.491	26.92	26.93
Eltanin	E874	128.1E	47.6	31-VIII-68	580	8.33	8.23	34.596	34.604	26.93	26.94
Eltanin	E879	117.1E	43.0	14-IX-68	585	8.83	8.76	34.650	34.665	26.90	26.91
Diamantina	D 113	108.5E	45.3	1-X-62	484	9.48	9.38	34.699	34.724	26.83	26.85
Eltanin	E1384	100.0E	44.1	18-X-71	493	9.93	9.70	34.760	34.800	26.82	26.84
Eltanin	E1365	94.9E	42.2	29-IX-71	404	10.89	10.75	34.892	34.926	26.75	26.76
Anton Bruun	AB160	60.0E	40.9	12-IX-63	478	13.44	13.28	35.345	35.378	26.61	26.62
Capitán Cánepa	CC484	50.0W	39.4	12-IX-62	435	14.44	14.53	35.72	35.74	26.65	26.69

Some thoughts on the age and descent of the waters of the northeast Atlantic Ocean

R. JOHNSTON

Department of Agriculture and Fisheries for Scotland, Marine Laboratory, P.O. Box 101, Aberdeen AB9 8DB, Scotland.

Abstract—The age of ocean water is a measure of turnover rate. Age is also related to the descent or origin and subsequent development of the component water masses. In the past, age determinations using a few data points have presupposed a simple model of ocean circulation. Using rates of biological oxygen utilization relative to depth it is possible to make use of a large amount of existing oxygen data and to deduce a detailed vertical and horizontal age distribution. Oxygen along with T,S is also useful in highlighting the structure and development of component waters.

Age determination without reference to the survival of component water identity is of dubious merit. When age and descent are considered together accepted ideas concerning continuity in water masses and ocean models look inadequate. Indeed all age concepts for ocean waters can be interpreted as anthropocentric and misleading; they discourage rational thinking about ocean exchange and renewal processes.

1. INTRODUCTION

AGE of ocean water has proved in the past a fickle problem and some hydrographers and chemical oceanographers have dealt rather harshly with it to their later regret. This paper looks at the concept of age, evolves an attractive solution to age determination of apparently wide application yet despite these advances it concludes by rejecting age as a feasible concept in relation to ocean renewal. The fruitful parts of this paper relate to the pattern of oxygen uptake and the mechanisms of water renewal in the northeast Atlantic Ocean. This paper does not achieve the resolution of the joint problem of "age and descent" looked at side by side; perhaps it merely repeats a familiar insoluble problem in different form.

2. THE CONCEPT OF AGE

Two approaches have been favoured in determining the age of ocean waters or time for circulation and renewal. Both presuppose unchanging patterns in water structure.

The mathematical model juggles with a few 'boxes' and relatively few observations; the laborious field study musters and evaluates voluminous observations, some of dubious reliability, according to some simple code or organisation. Physicists, chemists and biologists have all at times been attracted to the challenging problem of resolving the circulation and renewal of the vast ocean waters into some intellectually satisfying pattern of movement in time and space. This study by a chemist is one of an odd sequence of such ventures that has fallen to commemorative volumes.

121

The age of the oceans might be regarded as that time dimension since the earth cooled and the ocean formed in its initial state or it may be interpreted as the brief cycling rhythms of ocean regeneration. Palaeogeological time is so vast that viewed in this way the world ocean can be said to have reached a steady state. Yet today there is concern about man's impingement on the fringes of the seas and also an awareness that even now day-to-day events in the ocean can modify sizeable areas (such happenings as the eruption of volcanic islands and the injection of helium-rich water from the earth's interior). Such intrusions may be promulgated in a diffuse fashion or with decreasing effect along some preferred physical, chemical or biological pathway.

The sun and its seasonal effects over wide zones of sea and land is the driving power for ocean currents, circulation and renewal. It also dictates rhythms of growth over the oceans and on land which may in turn affect the seas and ocean. The time dimension of ocean change ranges from the first real blink of sunshine that brings the phytoplankton to bloom to the epochs of sunspot years and climatic cycles that alter the balance of ocean water movements and of life in the oceans.

Any method of age determination must acknowledge the existence of this vast time scale of local and general change. For a time-base to be useful effects of much higher frequency of occurrence must be individually insignificant and those of much lower frequency imperceptible or capable of being adequately offset within the framework of the technique used. On the basis of earlier studies (for example, RILEY, 1951) the expected age and descent of major ocean water bodies might be set against a time scale not longer than 5000 years and during which time measurable events of frequency shorter than one or several years are of rare occurrence or of inconsequential magnitude.

The best review I have found of the unbelievably ragged evidence on the circulation of deep ocean waters is to be found in W. S. BROECKER's book *Chemical oceanography* (1974) which marshalls the cycles of many marine chemical elements to serve the topics of age and descent. All the best literature extant on the topic may be found there. He, and the many others quoted, demonstrate the limitations inherent in the pure temperature with salinity approach in determining the identity and continuity of water bodies. There is much wider variability available for exploitation in any number of associated parameters especially certain minor elements, their normal and radio-isotopes and the common vital elements such as oxygen, phosphorus and carbon pre-eminently involved with biological processes. Frequently the differences discernible in the distribution of these minor components may be many-fold greater than those apparent in salinity which serves as a measure of the total salts (BROECKER, 1974).

Each of these single component elements has characteristic drawbacks as a parameter from which age and descent may be derived. For the least abundant elements and isotopes, the complexities of sampling and analysis preordains a limited number of observations. In consequence the only hope of interpreting the observations in a wider sense must rely on some kind of model, frequently a simple one. Consequently if the deductions are to be meaningful the selected ocean area must be of exemplary integrity and of well-defined structure. Sampling should then be arranged best to represent and test the components of the model adopted.

In stark contrast in seeking to interpret temperature and salinity distributions it is boundary or gradient conditions that are sought where water masses may be clear cut and actual or relative water movements are better quantifiable. It is remarkably difficult to build a convincing model that satisfies accurately (a) the turbulent ocean in the limited areas of water mass formation and (b) the remote vast expanses of oceanic drift and meander.

Nutrient elements, especially carbon, oxygen, phosphorus and silicon, are utilized by marine life in the shallow photic layer of the ocean creating there gradients of perhaps 100-fold variation in concentration over a short depth scale. Simultaneously the sinking of moribund and decaying tissues and other downwards migrations set up a further mechanism of segregation. This pattern of shallow uptake and deeper regeneration can be used as rather a blurred time dimension for unravelling water mass, age and descent.

The biological role of any parameter not just carbon, oxygen or a nutrient element is germane to its value as a time base. For example, during the process of tissue decay phosphate, lipids and certain small organic molecules are freed long before decay is complete. The various froms of nitrogen are remineralized and the carbon fully oxidized only slowly from the cell's macromolecules. Durable tissues such as calcareous or silicaceous shells, teeth, otoliths and large bony structures release their organic components very slowly indeed, probably at the sea bottom. This prolonged regeneration time allows a series of separate and probably significant element separations over the time scale. To some extent therefore all the known methods of studying the time dimension in the ocean really ought to be circumscribed with cautions and qualifications. The degree to which these limitations colour and affect the accuracy of age studies has seldom been given adequate attention.

From a number of physical oceanographic studies indirect estimates of rates of turnover of vital elements such as oxygen, carbon, phosphorus and silicon have emerged. These estimates can boast no biochemical basis, but if such rates 'look reasonable' the mathematical model is taken to be in some degree justified or strengthened. At best, rates deduced in this way can only be related to a highly generalized interpretation of oceanographic features (e.g. McGILL, 1964; WRIGHT, 1969).

In this article it is proposed to examine meaningfully but not in every detail the use of dissolved oxygen results to study the renewal processes and circulation of the main water bodies contributing to the northeast Atlantic Ocean, highlighting the basic concepts and their shortcomings. Major difficulties arise later in relating these findings to the general development of these water masses.

3. AGE DISTRIBUTION DERIVED FROM OXYGEN CONTENT

General features

Oxygen dissolved in the sea is required for respiration by all forms of aerobic life and also for chemical and microbial degradation processes. Despite these demands the degree of oxygen saturation in surface waters is always and everywhere close to complete saturation, and where primary productivity is very high it may temporarily exceed saturation. Hence for a limited upper layer the rate of oxygen transfer into the sea mostly exceeds that of utilization. At progressively deeper levels the rate of downwards transfer of oxygen becomes ever less than the rate of utilization. The depth of the layer attaining oxygen equilibration with the atmosphere is determined by vertical stability of the water column. When the water column becomes unstable, usually in September in the latitudes 50–60°N, dissolved oxygen is more or less equally distributed throughout the mixed layer at typically 90–95% saturation for water depths of 100 m in the North Sea. As 100%

saturation is approached the rate of oxygen uptake becomes very slow and complete saturation is commonly not reached before thermal stratification becomes established again. Approximately 100 litres of oxygen are transferred through each square metre of sea surface in about 165 days in these latitudes.

In boreal regions deep water is formed during prolonged periods of water column instability by a combination of processes including (i) the plummeting of cold dense water to the bottom, (ii) the slow descent of thin denser layers as fingers and (iii) less clearly defined mixing processes. The manner of sinking of these waters creates variably oxygen saturated bottom water because of (a) incomplete air–sea oxygen interchange, (b) downwards 'mixing' of well-saturated water from near the surface into a variety of deeper unsaturated waters or (c) oxygen utilization in the resultant sump of bottom water aggregated over its residence period. Probably all three mechanisms work simultaneously, the dominant being determined by local conditions.

According to Broeckers' two-box model of horizontal segregation of elements in the deep ocean, bottom water formation is confined to North Atlantic Deep Water (NADW) containing a substantial component from the Norwegian Sea which ultimately replenishes the deep waters of all the oceans assisted by recooling of deep water in the Antarctic. It is important to emphasize that the deep waters of the Norwegian Sea, of the Greenland Sea (CARMACK and AAGARD, 1972) and of the Weddell Sea (GORDON, 1972) all emerge with a substantial measure of oxygen undersaturation. The age at birth of these waters cannot be retraced to a conventional oxygen saturation content but to a certain level of Apparent Oxygen Utilization (AOU) which is the difference (ml $O_2\,l^{-1}$) between the theoretical Oxygen Saturation and the observed oxygen content. There is some uncertainty in the theoretical oxygen saturation because of narrow variation in the temperature and salinity of NADW due to seasonal and climatic changes and there must be associated variation in the observed oxygen content so that this AOU is likely to vary somewhat.

Whereas sinking at NADW is confined to a relatively small part of the Atlantic Ocean the return process of deep water via surface water to the Arctic occurs over the entire ocean (BROECKER, 1974). Consequently although NADW can be regarded as becoming progressively older as it moves away from its source, the middle and upper waters below the depth of seasonal overturn may or may not be expected to reflect a consistent pattern arising from NADW influence depending on a number of factors such as the proportion of upwelling NADW, the comparative rates of oxygen utilization at various depths in various waters and the proportion of downwards mixing from above the thermocline. The 'return' NADW component in the upper layer subject to seasonal effects will be greatly altered in its oxygen content and virtually undistinguishable from other surface waters.

If the concept of age and descent in ocean water is to be developed beyond existing highly simplistic approaches one has to grapple with three main problems: (1) deduce or model a vertical scale of oxygen utilization rate with depth, (2) recognize, resolve and quantify the progressive development of the component water masses and (3) find some means of reconciling (1) and (2) to produce realistic residence times within the oceans.

The vertical profile of oxygen utilization

Everywhere in the oceans there is a changing profile of rate of oxygen utilization with depth. The problem that has to be resolved is whether this profile changes significantly over the time lapse of ocean circulation and to what extent there can be adopted an average profile which is applicable over a large part of the oceans. As already indicated seasonal oxygen generation belongs only to the effective photic zone, at most 40 m. A high rate of oxygen requirement associated with zooplankton production and the associated fall out of organic detritus must be expected within the immediately underlying layer to perhaps 200 m. At greater depth these effects become smaller as the available food resource decreases. Any shear in the relative water movements, say as between the 0–200-m and 200–400-m layers, will tend to blur local variations. Diurnal vertical migration of certain zooplankton animals and perhaps associated migrations of their predators are sufficiently rapid movements to give rise only to blurring effects. Longer period migrations during reproductive phases may well create some minor redistribution in oxygen requirement in water levels down to about 400 or 500 m. Hence there is a likelihood of some minor perturbations of the profile of rate of oxygen utilization significant certainly on the scale of 1 year in the upper 200 m and perhaps extending to the upper 500 m with progressively decreasing effect. At greater distance and greater depth water shear would attenuate or obscure any imposed local perturbations. There is no suitable body of primary and secondary data to test the spatial variability in the profile of oxygen uptake. It is known that there is poor growth in some central oceanic areas and rich growth at ocean boundaries but the general impression is of relatively uniform mediocrity over the large areas in between. Probably for any particular year there will be some kind of observable pattern within these wide areas, but if water movement through these wide areas takes several years or more this provides reasonable grounds for the adoption of a mean profile. Water bodies moving into a uniform area from a particularly rich or poor area would need separate treatment. Clearly the situation improves the deeper and more extensive the body of water being considered and a more and more complex solution must be sought for water layers at decreasing depths and of minor extent.

Oxygen utilization rate for any small volume of water has several components of which the respiration of phytoplankton and photosynthesis are dominant but for this study are regarded as equal so that they mutually offset one another. Zooplankton respiration, chemical and microbiological breakdown of dead organic matter are of next importance. The respiration of predators will have a minor effect and the demands of higher animals (large crustaceans, large fish and mammals) are, except momentarily, an insignificant component.

Zooplankton

For this study the complexities of the zooplankton are set aside and their distribution regarded simply as biomass (the dry weight of tissue per unit volume) for given depths or depth ranges. There is a substantial amount of evidence that for many purposes biomass is distributed on average as a decreasing smooth curve with increase of depth (YASHOV, 1960; ZENKIVITCH and BIRSTEIN, 1956). The subject is reviewed by BANSE (1964) and other data is given by JOHNSTON (1962). As a broad generalization the curve for biomass against depth is fairly constant for the wide open ocean and the major effect on this curve relates to varying primary productivity between the ocean areas. The curve can be handled conveniently as the best fitting linear regression between the logarithm of the biomass against a linear scale of depth. This approach offers a practical method of determining the profile of oxygen utilization for large areas having uniform productivity.

It is admitted, however, that the vertical distribution of zooplankton as revealed by detailed sampling may be highly complex, for example as seen from the bathyscaph (BERNARD, 1958). It is sometimes possible also to match this degree of complexity with similarly complex short-term details of temperature and salinity depth profiles (e.g. PINGREE, 1971), and possibly also for oxygen, from continuously recording probes. Many sets of data would be needed to explore the wider context of such detailed structures and to reach a valid long-term mean distribution. At the moment the main pool of physical, chemical and biological observations for the open oceans would not sustain such detailed treatment but profiles of this type remain a goal for more advanced interpretations in future. This study was begun many years ago and it would not be feasible to update all parts of it; indeed it will eventually become clear that there are more serious problems.

From a careful search of the literature to 1960 the amounts of zooplankton at various levels in the oceans have been summarised in Table 1. Apart from records for the Pacific Ocean, zooplankton abundance with depth conforms fairly well to a single pattern—if allowance is made for the difference in the primary productivity of the sample areas. Some of the data (Data used in Table 1) will be unfamiliar, other information has been often tabulated and reference is made only to the source most readily available to the writer.

The curve to be used must be unique for the region but the additional support from other parts of the world adds credibility to its application over a wide region. Preliminary statistical tests using data for plankton in the Bay of Biscay showed that a straight-line relationship between depth and logarithm of the biomass parameter was a satisfactory approximation. Regression lines were therefore fitted by the usual statistical method to the eight most complete sets of data (Table 1). For northwest Japan Sea near surface values have been omitted from the calculation since they distort the useful parts of the depth range.

An analysis of all the slopes together and of the goodness of the fit of each regression showed 4, 6, 8, 10 and 11 are of a uniform type with 5 as a solitary curve of less steep slope. Of these set 1 based on the *Research* results has been selected to represent the zooplankton distribution in the northeast Atlantic Ocean.

For the shallow seas, to which virtually all of the food studies relate, MULLIN (1969) concludes that both laboratory approaches and field measurements fail to provide conclusive practical solutions to measuring the open-sea production of zooplankton. There are, however, various useful approximations. The parameters of greatest relevance here are (a) the overall efficiency of food chain utilization of primary production and (b) the magnitude of the average total zooplankton community in the water column at any one time. (a) However deep the water column there will always be some input of organic matter to the bottom; the input to the bottom being a function of biomass distribution with depth. The ratio of total annual zooplankton production to net annual primary production is generally accepted to be in the region of 0.2 to 0.3 for shelf waters with the proportional input to the bottom adding a further 0.2 to 0.3. In the deep ocean, because of greater dispersion with depth, the pursuit of food becomes more difficult and it would be false to conclude that a greater depth of water column must necessarily give rise to higher overall efficiency. ZENKEVITCH and BIRSTEIN (1956) and later calculations given here set the benthic input at about 0.05 times primary production. Additional food sources are organic matter bound to sedimenting inorganic particles of marine and terrestrial origin, also the slow decay of hard organic tissues, which together might help solve the apparent benthic food shortage (SPÄRCK, 1956) in abyssal waters. (b) It is unlikely therefore that the average total zooplankton biomass can exceed 0.35 times the net primary production.

Table 1. Depth distribution of marine animals

m.	1	2	3	4	5	6	7	8	9	10	11
				0–50		*0–50*			*0–100*		
0	11.7	*0–100*		378		54,000	11,700	497.6	234	1650	
		176.23		*50–100*		*50–100*					
100	11.0			144		29,800	5040	320.3			
						100–200			*100–200*		
200	9.6	*100–200*		*100–600*	221	3401	320	246.6	236	1300	
300	7.3	49.14		45	175	*200–500*					
									200–500		
400	13.0	*200–500*			105	2734	84	228.0	155	1450	
500	10.4	37.99			130	*500–*					
									500–		*500–*
600	8.8	*500–1000*		*600–*	150	*1000*					
									1000		*1000*
800	6.6	54.08		*1000*	207	1154	65	59.3	105	740	5.4
1000	5.3	*1000–*	36	77	77	*1000–*		*1000–*	*1000–*		*1000–*
1200	4.6	*1500*		*1000–*	41	*4000*		*2000*	*2000*		*2000*
1400	4.6	32.99		*2000*	39	303		21.8	58	290	1.6
1600	5.0			45	69						
1800	4.8				31						
2000	3.8		33		45						*2000–*
2200	2.8				26						*3000*
2400	1.9		40		28						0.8
2600	1.4				6			*2000–*	*2000–*		*3000–*
2800	1.2				3			*4000*	*4000*		*4000*
3000	0.9		31		5			9.3	34	135	0.3
3200	0.7				2			*4–6000*			
								2.64			*4000–*
								6–8000			*5000*
4000	0.1		33					0.48			0.1

1. $\hat{y} = 1.1974 - 0.0004217\bar{x}$
4. $\hat{y} = 2.228 - 0.0004426\bar{x}$
5. $\hat{y} = 2.5168 - 0.0006035\bar{x}$
6. $\hat{y} = 3.536 - 0.0004353\bar{x}$

8. $\hat{y} = 2.3738 - 0.0004050\bar{x}$
9. $\hat{y} = 2.324 - 0.0002912\bar{x}$
10. $\hat{y} = 3.230 - 0.0004574\bar{x}$
11. $\hat{y} = 0.980 - 0.0004391\bar{x}$

where $\hat{y}$ is the logarithm (base 10) of the measure of animal life per unit volume and $\bar{x}$ is the depth in metres.

Data used in Table 1:

1. H.M.S. *Research*, 1900, area sampled 47°29′N to 46°43′N × 8°18′W to 7°17′W in the Bay of Biscay, vertical hauls, mostly two to five hauls each range, mesoplankton net 36 or 45 meshes per inch for which no major differences in catching power were observed. Volume of 100m haul = 25.53 m³. Number of copepods (all species together) per haul (recalculated in metres from fathoms) (FARRAN, 1926).

2. R.R.S. *Discovery*, fourteen stations in the area 41°15′N–29°52½′N × 9°29′W–20°17½′W, vertical hauls with a closing net, filtered volume 38.5 m³ per 100m haul. Displacement volume, dry weight, ash were measured. (FOXTON, 1958, personal communication.)

3. Research Ship *Dana*, Carlsberg Expedition, 1928–1930, and earlier records. Eastern North Atlantic Ocean, horizontal tows with 1½- and 2-m stramin nets taking five to twelve hauls at each depth. Haul depth assumed to be 0.5 × warp out; filtered volume for 1½-m net 6550 m³ h⁻¹, 2-m net 11,650 m³ h⁻¹. Data refer to displacement volume (JESPERSEN, 1935).

4. Weather Ship *M*, 1948–1949, Norwegian Sea, Nansen closing net forty vertical hauls each range. Volume of 100m haul 38.5 m³. Complete analysis of numbers of each plankton species by subsample recalculated as equivalent number of *C. finmarchicus* stage V (after ØSTVEDT, 1955).

5. Western Atlantic Ocean Horizontal tow, stramin net, mean of nine hauls each depth. Volume of haul 10⁴ m³ h⁻¹. Displacement volume and detailed analysis. (LEAVITT, 1938.)

6. Northwest Japan Sea. Calanoids, number per m³ (BRODSKII, 1960), omitting surface values.

7. North Pacific Ocean, Calanoids, number per m³ (BRODSKII, 1955).

8. Kurile–Kamchatka Trench. Plankton, dry weight in mg m⁻³ (ZENKEVITCH and BIRSTEIN, 1956).

9. Tropical waters. Plankton, dry weight in mg m⁻³ (BOGOROV, 1957).

10. Boreal waters. Plankton, dry weight in mg m⁻³ (BOGOROV, 1957).

11. North Equatorial and Canary Currents. plankton, dry weight in mg m⁻³ (YASHINOV, 1960).

Fish

At depths greater than, say, 600 m, the dominance of the total biomass by phytoplankton, herbivores and benthos relaxes to include a greater proportion of omnivores and predators including fish. With some daring it is possible to derive the approximate weight of fish catch per unit trawl area from the rather extensive old observations from *Talisman* and *Travailleur* (VAILLANT, 1888) from 106 trawl hauls in depths of 30 to 5005 m in the Azores region (Table 2). The promised description of the trawl gear was never fulfilled but NYBELIN (1951) believes that the gear was either a 2- or 3-m-wide sledge trawl. The probable towing period was 4 hrs so that the swept area was 15 to 20 × 10^3 m^2. The wet weight of fish caught per haul has been calculated assuming that deep-water fishes are either "short and deep" or "long and thin". The main stout fishes are *Capros aper*, *Sebastes dactylopterus*, *Hoplostethus mediterraneus* for which wet weight = 0.0131^3; the thin fishes are mainly eels for which 0.6 × length × breadth × depth was used. The data are summarized in Table 2. The regression for depth and $\log_{10}$ weight of fish is $\hat{y} = 0.0002835\bar{x}$. The probable biomass of fish is even more difficult to assess than that of zooplankton. From the above data the average fish would weigh about 50.6 g and on the basis of a North Sea type food chain would consume 100 g wet wt. food annually and respire about 20 l of oxygen. Virtually nothing is known about the age of deep-sea fish but the constraints of food supply and habitat would suggest slower growth and greater mean age, perhaps 6 years compared to 3 years in the North Sea. The data of DEITRICH and KALLE (1957, p. 235) would indicate about 5.8 g dry wt. organic matter as fish per m^2 for the North Sea which on a basis of comparative primary productivity would suggest about 4 g dry wt. organic matter per m^2 for the northeast Atlantic Ocean. As a bold approximation 20 g m^{-2} wet wt. fish, mean age 6 years will be assumed as the standing stock maintained by the zooplankton, representing about 0.04 of primary production which is slightly higher than indicated by HARVEY (1950) and about the conventional one-ninth of the zooplankton production. On this basis the average efficiency of the sledge trawl is about 17% which looks a reasonable figure.

Table 2. Depth distribution of fishes (based on data from Vaillant, 1888)

Depth range (m)	Trawl hauls (n)	Fish per haul (N)	Calculated wet wt. N (g)	Mean temp. (°C)	Respiration (fish at rest) (ml O_2 kg^{-1} h^{-1})
400–600	7	211	4360	12.0	60.1
600–800	8	34	2490	10.4	51.9
800–1000	13	28	3230	9.3	46.9
1000–1200	11	45	3245	8.6	44.0
1200–1400	12	43	2400	7.8	40.9
1400–1600	9	25	1730	6.3	35.6
1600–1800	2	6	345	5.3	32.5
1800–2000	no observations				
2000–2200	8	7	1390	3.9	28.6
2200–2500	10	6	1015	3.3	27.1
2500–3000	5	3	210	3.0	26.3
3000–4000	4	4	930	2.6	25.4
4000–5000	6	2	370	2.5	25.2

$\hat{y} = 3.6302 - 0.0002835\bar{x}$ (for weight of fish).
$\hat{y} = 2.390 - 0.0003668\bar{x}$ (for calculated respiration).

One is left with some uncertainty about pelagic fish species but these are unlikely to be important at depths below 500 m and also about the other large pelagic animals such as shrimps and squids which may fill a niche in the deep ocean food chain. The uncertainties about the distribution and abundance of these species precludes any assessment of their involvement which to put an upper limit to it is unlikely to exceed 0.04 of primary production vertically distributed like demersal fish because they are feeding as carnivores.

The general behaviour of what have been designated here as demersal fish and pelagic animals should not be taken as delimiting one population found only on the bottom and another only in mid-water. Even the method of capture (sledge trawl) does not exclude the possibility that some fraction of each haul was captured during the long descent of the net or its long retrieval haul. It is also much more convenient to apply only one age-determining formula considering both sets of animals to be effectively pelagic. Benthic animals considered in the next section are not pelagic and to include their oxygen requirements would involve much complexity. It is enough at this stage of the study to determine the magnitude of this role and to acknowledge considerable difficulty in incorporating their contribution into the wider scheme. Fish renewal equivalent oxygen requirement (cf. zooplankton, Table 4) is regarded as already met in the zooplankton budget and is not included in the total oxygen requirement.

Benthos

The organic component of particulate matter collected in depths greater than 250 m is resistant to rapid biodegradation and leaching, hence its food value further decreases only slowly at greater depth. By far the dominant deep-water benthic species are sea cucumbers (Holothuria) and starfishes (Asteroidea). In addition certain isopods, amphipods and decapods are found in considerable variety. There are proportionately few molluscs and many fewer lamellibranchs than in shallow waters.

Extensive collections in the Gulf of Gibraltar were made by dredge from *Talisman* (1880–1883) supplemented by *Blake* under the direction of A. Milne-Edwards. The holothurians are described by REMY PERRIER (1901) and the echinoderms by EDMOND PERRIER (1894). The results are summarized in Table 3. Taking all the catches together it is evident that occasional hauls at infrequent intervals yield exceptionally large numbers of individuals, but except at shallow depths there was no clear pattern of spatial variability in numbers and species. Numbers of starfishes and sea cucumbers tended to remain constant with depth.

SPÄRCK (1951) gives values of about $1.85 \, \text{g m}^{-2}$ wet wt. for grab samples taken off the coast of West Africa; the *Galathea* results show a wide range (up to $12 \, \text{g m}^{-2}$) with an average of 1 g for all deep hauls (SPÄRCK, 1956). Taking together the larger epibenthos sampled by dredge and the smaller animals represented in grab collections the average abundance of benthos is about $3 \, \text{g m}^{-2}$ wet wt. (equal to $0.053 \times$ primary productivity).

Respiration

Having established the distribution of biomass of the zooplankton and fish the next step is to estimate the likely oxygen requirements of these populations. Each population is envisaged as an average standing stock biomass which has a typical rate of wastage which is balanced by growth. Oxygen requirement is regarded as the sum of a basal rate and of a rate linked with food consumption, conversion and growth to make good losses. It is recognized that water pressure may alter the metabolic parameters of deep-water species

Table 3. Deep-sea echinoderms

(Results of *Talisman* and *Blake* together, data from E. PERRIER (1894) and R. PERRIER (1901))

Depth (m)	Asteroidea			Holothuria		
	Hauls	Individuals	Catch per haul	Hauls	Individuals	Catch per haul
−100	11	30	2.7	—	—	—
100–200	33	217	6.6	17*	38	2.2
200–300	29	212	7.3	—	—	—
300–400	18	48	2.7	13	6	0.5
400–500	7	57	8.1	—	—	—
500–600	12	32	2.7	21	36	1.7
600–700	5	20	4.0	—	—	—
700–800	5	6	1.2	23	11	0.5
800–900	8	33	4.1	—	—	—
900–1000	6	23	3.8	21	64	3.0
1000–1100	7	49	7.0	—	—	—
1100–1200	9	60	6.7	15	63	4.2
1200–1300	8	95	11.9	—	—	—
1300–1400	4	20	5.0	18	63	3.5
1400–1500	7	41	5.9	—	—	—
1500–1600	4	10	2.5	14	42	3.0
1600–1800	5	18	3.6	3	0	0
1800–2000	2	5	2.5	6	34	5.7
2000–2500	23	154	6.7	27	53	2.0
2500–3000	1	13	13.0	7	0	0
3000–3500	3	5	1.7	5	12[a]	2.4
3500–4000	2	10	5.0	4	5	1.25
4000–4500	2	34	17.0	5	40[b]	8.0
4500–5000	2	15	7.5	2	7	3.5

* 0–200, 200–400 m, etc., to 1600 m.
[a] One 'catch not recorded' taken as 2.
[b] Exceptionally large catch (56) taken as 3.

relative to the more familiar shelf species but no allowance has been attempted other than to apply a conventional temperature-related response. A typical temperature profile is given in Table 2. The rate of stock renewal has been set at 5% for 0–100 m for zooplankton and 1% for fish at 70 m and these figures have been varied by applying the same factor derived from the temperature profile. Thus at greater depth respiration and renewal (or main-tenance) rates per unit body weight are lower. To test whether this model meets the constraints of food available at each depth and over the water column the oxygen utilization has also been converted into food units. For zooplankton (Table 4) it is found that renewal plus standing crop needs add up to about $50 \, \mathrm{g \, C \, m^{-2} \, year^{-1}}$ compared to the primary production of $55 \, \mathrm{g \, C \, m^{-2}}$ year. It is commonly held that entire phytoplankton crop is consumed by zooplankton. For fish (Table 5) the total food requirement is about 17 g dry wt. $\mathrm{m^{-2} \, year^{-1}}$ or $8.5 \, \mathrm{g \, C \, m^{-2} \, year^{-1}}$ (about one-sixteenth of the zooplankton production) and fish production is $0.64 \, \mathrm{g \, C \, m^{-2} \, year^{-1}}$ or 0.1 times primary production. The main parameters satisfy reasonably well the basic requirements of a balanced system in total and at various depth levels.

The sum of zooplankton respiration (standing crop + replacement) and fish standing crop respiration given in Table 5, last column, will be used in assessing the 'age' of ocean waters.

Table 4. *Zooplankton respiration and food requirement*

Depth range (m)	STANDING STOCK Stock (g dry wt m^{-2})	Resp rate (ml O$_2$ kg^{-1} wk^{-1})	Oxygen (ml yr^{-1})	Food (g m^{-2} yr^{-1})	REPLACEMENT OF LOSSES Renewal rate (% day)	Replacement (dry wt m^{-2} yr^{-1})	Food (g m^{-2} yr^{-1})	Oxygen (ml yr^{-1})	BOTH Total oxygen (ml yr^{-1})
0–200	0.566	1890	0.0469	10.69	5.0	10.33	14.76	0.083	0.130
200–400	0.464	1688	0.0343	7.83	4.5	7.55	10.78	0.060	0.094
400–600	0.415	1474	0.0268	6.12	3.9	5.91	8.44	0.047	0.074
600–800	0.304	1316	0.0175	4.00	3.5	3.83	5.48	0.031	0.049
800–1000	0.257	1226	0.0138	3.15	3.3	3.05	4.36	0.024	0.038
1000–1200	0.216	1170	0.0111	2.53	3.1	2.45	3.50	0.020	0.031
1200–1400	0.175	1103	0.0085	1.93	2.8	1.82	2.60	0.015	0.024
1400–1600	0.144	990	0.0062	1.43	2.6	1.37	1.96	0.011	0.017
1600–1800	0.119	923	0.0048	1.10	2.4	1.07	1.52	0.0085	0.0133
1800–2000	0.097	878	0.0037	0.85	2.3	0.82	1.17	0.0066	0.0103
2000–2200	0.082	844	0.0033	0.69	2.2	0.66	0.95	0.0053	0.0088
2200–2400	0.067	810	0.0024	0.54	2.15	0.53	0.75	0.0042	0.0066
2400–2600	0.056	799	0.0020	0.45	2.1	0.43	0.62	0.0035	0.0055
2600–2800	0.045	788	0.0016	0.35	2.1	0.35	0.50	0.0028	0.0044
2800–3000	0.035	776	0.0012	0.27	2.05	0.26	0.37	0.0021	0.0033
3000–3200	*0.029*	765	0.0010	*0.22*	2.0	*0.22*	0.31	0.0018	0.0028
Totals:	3.071 g dry wt m^{-2}			42.15 g dry wt m^{-2} ≡ 21 g C		40.65 g dry wt m^{-2} ≡ 20 g C	58.07 g dry wt m^{-2} ≡ 29 g C		

Table 5. *Fish respiration and food requirements also total oxygen requirement*

Depth range (m)	STANDING STOCK Stock (wet wt m^{-2})	Resp. rate (ml O$_2$ l^{-1} yr^{-1})	Food (g dry wt m^{-2} yr^{-1})	REPLACEMENT OF LOSSES Renewal rate (% per day)	Replacement (g wet wt fish (yr^{-1}))	Food (g dry wt m^{-2} yr^{-1})	BOTH Total food (g dry wt)	TOTAL O$_2$ REQUIREMENT Zooplankton and fish per year
0–200	2.58	0.0114	1.316	0.96	1.511	2.720	4.036	0.141
200–400	2.26	0.0084	0.970	0.80	1.103	1.985	2.955	0.102
400–600	1.98	0.0063	0.728	0.68	0.821	1.478	2.206	0.080
600–800	1.74	0.0047	0.542	0.59	0.626	1.127	1.669	0.054
800–1000	1.53	0.0036	0.416	0.52	0.485	0.873	1.289	0.042
1000–1200	1.34	0.0030	0.346	0.48	0.392	0.706	1.052	0.034
1200–1400	1.18	0.0025	0.288	0.44	0.317	0.571	0.859	0.027
1400–1600	1.03	0.0019	0.220	0.40	0.251	0.452	0.672	0.019
1600–1800	0.91	0.0015	0.174	0.35	0.194	0.349	0.523	0.015
1800–2000	0.80	0.0012	0.138	0.33	0.161	0.290	0.428	0.012
2000–2200	0.70	0.0010	0.116	0.31	0.132	0.238	0.354	0.010
2200–2400	0.61	0.0008	0.092	0.29	0.108	0.194	0.286	0.007
2400–2600	0.50	0.0007	0.080	0.28	0.085	0.153	0.233	0.006
2600–2800	0.47	0.0006	0.070	0.28	0.080	0.144	0.214	0.005
2800–3000	0.41	0.0005	0.058	0.28	0.070	0.126	0.184	0.004
3000–3200	0.37	0.0004	0.046	0.27	0.061	0.110	0.156	0.003
Totals:	20 g wet wt fish (mean survival 6 years)				6.397 g wet wt fish production		17.116 g dry wt ≡ 8.5 g C (fish production ≡ 0.64 g C)	

Table 6. Factors used in Tables 4 and 5

Zooplankton

Standing crop

  (criteria for 0–200 m)   respiratory rate: 1.31 μl oxygen per mg dry wt. per hour at 10°C (a)

                                  food requirement: 5% per day on a dry wt. basis   (b)

Renewal   renewal rate: 5% per day at 100 m   (c)

                                  food requirement: 1 g new tissue from 1.4 g wet wt. food   (d)

Sources: (a) Marshall and Orr (1955): mean of 5° and 15°C results is 0.295 μl O$_2$ per hour for 0.225 mg dry wt. *Calanus*. This is the lower set of values cited.

        (b) The mean of food requirement cited by Marshall and Orr and by Harvey (1950).

        (c) Harvey shows 7–10% daily renewal required to offset predation in a water column 70 m deep; 5% is estimated as the probable renewal rate over 0–200 m.

        (d) Harvey shows 70% of food assimilated goes into tissue.

Fish community

Standing crop

  (criteria for 0–200 m)   respiratory rate: 50 ml oxygen per kg fish per hour at 10°C (e)

                                  food requirement: 60 g food per kg fish per week   (f)

Renewal   renewal rate: 1% per day at 70 m   (g)

                                  food requirement: 1 g new tissue derived from 9 g food   (h)

    (e) Based on Sundnes (1957) and Nicol (1960).

    (f) Based on Brown (1957) and Holliday (unpublished) herring 64 g food kg^{-1} week^{-1} at 11.4°C.

(g, h) Harvey (1950).

4. THE AGE OF SEA WATERS

Vertical profile of age

It is apparent from the previous sections that there are almost innumerable qualifications and complexities that might be incorporated into the determination of age of sea water based on any biologically altered distributions of chemical elements. In the past, age assessments have been proposed on the simplest of bases relating to generalities of water circulation with little or no detail of the associated biological involvement. The biological timing model about to be used cannot claim to be highly accurate and it is certainly not a true simulation of the real ocean. We are dealing with an intermediate type of model which is not too far out of scale taken overall but which being smoothed may impart significant error in any selected segment of the area. Given this timing model, it is very easy to scale it up or down for other ocean areas on the basis of primary productivity and other relevant factors or indeed deduce a better fitting model using the best observations from the area under consideration. There is now a great data bank of observations on the deep water of the Norwegian Sea and on the northeast Atlantic Ocean but at the time this model was evolved the best and latest available section across the North Atlantic Ocean providing complete data on temperature, salinity and dissolved oxygen had been worked by R.R.S. *Discovery II* between Grand Banks and the shelf off southwest England (Stas. 3509–3548, Worthington, 1958). Using these data, oxygen-saturation values were determined for each point and hence the Apparent Oxygen Utilization (AOU). The AOU divided by the appropriate rate from Table 5 gives the age of the sea water (Fig. 1). This intriguing figure brings to mind many questions the answers to which test the age-determining model and if the outcome is satisfactory certain principles should emerge. These should include familiar general features and perhaps fresh discoveries. To be of advantage the age

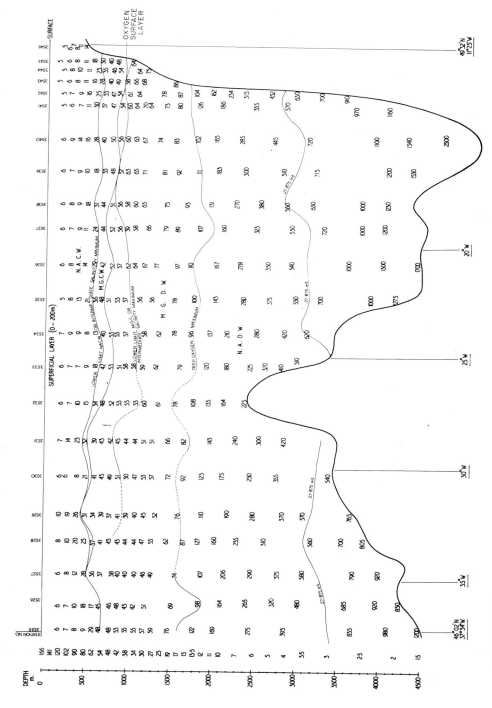

Fig. 1. The depth profile of the age of sea water derived from the rates of oxygen consumption, ml m^{-3} yr^{-1}, given in the left-hand column.

profile should highlight water structure better than ordinary interpretations based on T, S and O individually.

Features of Fig. 1 apparent on inspection are as follows:

(a) The scale of age imposes a more pronounced gradient than is found for $S\%_{00}$ or O_2, the general pattern seems harmonious progressing smoothly from young near surface to old at the bottom. (This would indicate that the profile of oxygen utilization rate is not crazy.)

(b) Figure 1 resembles a density pattern more than a T, S or O distribution but whereas deep-water density structure is very slight the age gradients are most marked in deep water.

(c) Age along horizontal planes is rather uniform with waters of the western basin younger throughout than those of the eastern basin.

(d) The interfaces between the dominant water masses remain clearly discernible with the boundary between North Atlantic Central Water and Gulf of Gibraltar Water showing 30 years, at the core of Gulf of Gibraltar Water 54 to 58 years and at the lower limit of this water 100 to 120 years.

(e) The age at the oxygen minimum level increases from 30 years (600 m) in the west to 58 years (950 m) in the east. The decreasing age of Gibraltar Water as it spreads appears at first sight to be disturbing but progressive admixture with younger waters provides an adequate explanation.

(f) To the west of the mid-Atlantic ridge and above this sill there is little age difference locally between west and east; below this sill the deep waters become increasingly different until at 3500 m there is an age gap of 570 years. The age pattern shows reasonably well the eastwards spread of the younger Labrador Sea Water described by LEE and ELLETT (1965). The very old deep water layer in the eastern basin may be derived from Antarctic Bottom Water as suggested by COOPER (1961) for his station *Cavall*.

(g) To test the indicated age against distance travelled the upper boundary of deep water is an appropriate example. Assuming a smoothly curved track parallel to the mid-Atlantic Ridge deep water descending from the Iceland–Faroe Ridge (*ca.* 550 m) to 1800 m at 48°N covers a distance of 1200 nautical miles* in 56 years (assuming an initial age of 44 years, see later), hence 21.4 miles per year or 0.13 cm s^{-1}.

This velocity is appropriate for the mean rate of advancement of deep water. In volume terms assuming deep water between 1800 and 3050 m at 48°N transport across this latitude would be about 43×10^3 km^3 year^{-1} or about two-thirds of the Arctic outflow (TAIT, 1957). One might hazard that the remaining one-third is entrained northwards with Gulf of Gibralter Water and North Atlantic Central Water.

The residence time for NADW between 40°N and 32°S was estimated as 700 years by BROECKER, GERARD, EWING and HEEZEN (1960) on the basis of ^{14}C dating but WRIGHT (1969) using simultaneous conservation equations for mass, heat, salt, ^{14}C and oxygen found about 100 years—the oxygen depletion rate (based on box models) used in the latter was $15 \pm 5 \times 10^{-4}$ ml O$_2$ l^{-1} yr^{-1}, an order of magnitude smaller than here.

It is of interest to see how Redwood Wright's box model copes with higher oxygen depletion rates (Table 7). Using rates of 50 and 100×10^{-4} ml O$_2$ l^{-1} yr^{-1} has a dramatic effect on the ocean circulation, flow from the Mediterranean is reversed and the waters of the South Atlantic retreat even quicker. Although WRIGHT (1969, p. 437) produces a considerable number of estimates of depletion rates in the range -1 to -43×10^{-4} ml O$_2$ l^{-1} yr^{-1}

---

* 1 nautical mile = 1.853 km.

and finds some grouped about -10 to -20×10^{-4} ml O_2 l^{-1} yr^{-1}, all these estimates are linked with advection and diffusion and have no biological grounding. If these dynamic calculations are of the approved order of magnitude accepting -10 to -20 units is it possible and reasonable to conclude that there must be a mechanism which counteracts an indicated oxygen-depletion rate of -50 to -100 or even ml l^{-1} yr^{-1}? Turbulence which effectively introduces 'younger waters' is one likely factor. Indeed Wright's flow rates imply residence times of 100 years for Deep Water and about 50 years for Antarctic Bottom Water, more in keeping with arguments to be presented in the later section on 'descent'.

Table 7. *Solution of conservation equations formulated by* WRIGHT (1969) *extended to higher values* $(10^{-4}$ ml O_2 l^{-1} yr$^{-1})$

	Oxygen-depletion rates				
	-10	-15	-20	-50	-100
Deep water, 40°N	10.9	10.4	9.9	5.9	1.9
Antarctic Intermediate Water	1.7	1.6	1.6	1.15	0.4
Mediterranean Water	1.6	1.0	0.4	-3.2	-9.2
Antarctic Bottom Water	8.9	7.1	5.2	-6	-25
Deep water 32°S	23.1	20.1	17.1	-0.9	-30.9

Horizons of age

The verticle profile for age based on a single section made by one ship is a much different test of this age method from a set of age horizons based on data drawn from cruises many years apart and by many ships. To reduce the scatter likely to be found in the very old chemical results only observations made since 1947 have been included.

0–200 m

The spatial variation of age in the near surface waters (0–200 m) has not been shown. The range of age indicated was 0–14 years, mainly 0–5 years. The data available were drawn from surveys between 1947 and 1959, irrespective of the month of sampling. Since the flux of the usual annual oxygen cycle in these waters is of the order of $1-1.5$ ml O_2 l^{-1}, most of the variability simply reflects seasonal sampling factors. For this layer therefore seasonal variability greatly exceeds any pattern of age.

At 200 m

At 200 m, only an apparently random display of age variation in the narrow range 3 to 14 years is found. This depth marks the minimum depth of normal winter mixing (excluding exceptional overturn) and annual or seasonal variability would be large enough to distort any age pattern.

At 400 m (Fig. 2a)

It is still difficult to be certain that an age pattern exists. There are hints of patches of older than average water in the west and against the continental shelf.

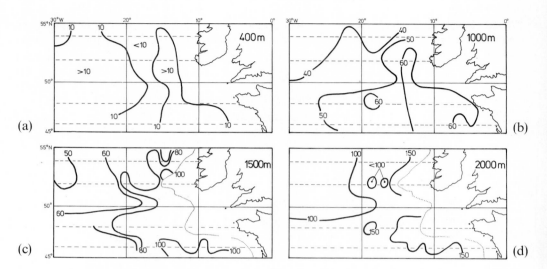

Fig. 2.   Horizon of age at (a) 400 m, (b) 1000 m (800 m and 1200 m are similar), (c) 1500 m (d) 2000 m.

At 600 m

At this depth the pattern becomes clearer with older water (a) in the west and (b) as a layer extending northwards from the Bay of Biscay. The former can be linked (a) with residual Gulf Stream inflow (e.g. SVERDRUP, JOHNSON and FLEMING, 1942) and (b) with Gulf of Gibraltar Water and some kind of upwelling or deeply mixed water (compare COOPER, 1952a).

At 800, 1000, 1200 m (Fig. 2b)

These contours conform to the age pattern at 600m with the Gulf of Gibraltar influence becoming more pronounced and more widespread.

At 1500 m (Fig. 2c)

The general age pattern reflects a declining influence of Gulf of Gibraltar Water.

At 2000 m (Fig. 2d)

Some trends in the age of the deep waters may be discerned. The core of the renewing North Atlantic deep water (predominantly Labrador Sea Water) can be identified but other parts of the deep-water structure do not show a clear-cut pattern of progressive ageing along an axis of travel. Various explanations might be offered: for example, that the deep-water circulation becomes vague and meandering, that the oxygen results lack accuracy and consistency, or that deep water is not as invariable in properties (T, S, O and others) as is commonly assumed.

Horizontal age patterns emerge as broadly consistent with the expected boundaries between the main water masses: as may be confirmed later. The 2000-m horizon also shows that Worthington's speculative age determination (1954) on the basis of a 0.30 ml l^{-1} oxygen shift over 20 years can be explained by a change of about 60 to 100 miles northwards in the '150-year' age front. MONTGOMERY (1955) and COOPER (1961) show small-scale temperature changes occur in periods of 26 days or less and SWALLOW and HAMON (1960) short-term deep current fluctuations. Deep-water micro-layering could also generate changes of appropriate magnitude. Therefore it is unlikely that any age-determining process can ever justify a precise meaning for a few determinations treated in isolation.

Error, bias and other relevant considerations

Age determinations bring with them identifiable and reckonable error and bias; many of the identifiable factors already discussed are difficult to quantify and cannot be disregarded. Their effects are to distort the age pattern especially in the upper and intermediate waters; on average over wide ocean areas their effects will be lessened. There are two types of analytical error relating to oxygen: (a) uncertainties relating to the absolute values for oxygen saturation and (b) uncertainties of the routine chemical analysis. At the time this part of the work was in hand there was conflict between Fox's tables of oxygen-saturation values and the then recent work of TRUESDALE and LOWDEN (1955). The positive bias of about 3% is small and generates a correspondingly small bias in the AOU. At the 5th Meeting of the Joint Panel of Experts on Oceanographic Tables and Standards (December 1969) it was decided to use tables based on the oxygen-saturation formula of WEISS (1970) as standard. On this basis the oxygen-saturation values are 0.22–0.26 ml l^{-1} higher than those used here based on FOX (1907) as reported by JACOBSEN, ROBINSON and THOMPSON (1950) for the range of temperature and salinity. (c) The analytical error is probably of the order ± 0.03 ml O$_2$ l^{-1} = $2\frac{1}{2}$ years at 2000 m, but interlaboratory differences might stretch this to 10 years. COOPER (1960) reports sporadic much larger errors but these would not greatly alter an interpretation based on a large number of samples. It should be noted that the oxygen results for *Discovery* Stas. 3509–3548 were subsequently reported to carry a small systematic error. Quantifiable errors like these are therefore of no consequence.

Much more important are the assumptions in the model relating to the degree of oxygen saturation attained by the various water bodies as they enter the framework of the study area. It is easiest perhaps to consider first the new body of deep water entering the North Atlantic; the concepts relating to which are fundamental to this and all oceanic models. This entry takes place over the Faroe–Iceland Ridge, Faroe Bank channel and Iceland–Greenland ridge. In tracing the fate of new deep water it is essential to study AOU in relation to density (σ_t) rather than depth horizons as just discussed. These horizons are easier to extract from the raw data and served sufficiently well at this stage to demonstrate continuity for a relatively small bit of the northeast Atlantic Ocean. Equal density surfaces together with oxygen core considerations form a more acceptable basis for tracking bodies of water over the much longer pathway of deep water southwards to the equator.

Studies by TAIT (1967) and STEELE (1967) have gone far to clear up the concepts relating to the formation of North Atlantic Deep Water and its overflow components from the Iceland–Faroe Ridge and through the Faroe Bank Channel (CREASE, 1965). It is difficult to specify this 'pure new' Iceland–Faroe deep water accurately because of the sporadic nature of its occurrence but an analysis of T/O and AOU/δ_t curves indicates water of about $-0.5°C$, $34.92‰$ and 6.95 ml $O_2 l^{-1}$ giving an AOU of 1.33 ml $O_2 l^{-1}$. Labrador Sea Water requires $3.4°C$, $34.89‰$ and oxygen close to 6.0 ml l^{-1}. Depending on the strength of overflow and the structure of the recipient water mass the resulting deep water varies somewhat in its character and in its depth range. Quite clearly the initial AOU is not zero. A search was made for deep water of least AOU at all data points south of Iceland beyond the shelf. This revealed a mean AOU of 1.10 ml $O_2 l^{-1}$ corresponding to 85.4% saturation and applied to waters deeper than 1200 m. It is probable that in this area winter overturn extends to at least 600 m (confirmed also from the Denmark and Davis Straits by WATTENBERG, 1929; MEINCKE, 1967) resulting in about 96% oxygen saturation when thermal stratification is restored. Between 600 m and 1200 m the residual modified North Atlantic Waters are about $10–15\%$ undersaturated for oxygen. Hence not only new deep water but also recirculating waters at intermediate depths fail to attain complete oxygen equilibration in the northern Atlantic Ocean.

If the observed initial AOU values are applied to Figs. 1 and 2a–d then intermediate waters (approx. $1000–1500$ m) would be about $15–30$ years younger and deep water (centred at $1800–2200$ m) about 120 years younger assuming Labrador Sea Water, about 110 years younger for Iceland–Faroe overflow and about 90 years younger for the South Iceland reference point.

Such considerations of oxygen saturation and water renewal point to the conclusion that for this large region partial circulation may be a much more important factor than general advection (replacement) for water bodies such as the Gulf Stream and North Atlantic Deep Water. Hence as a corollary, are there really any effective water types and water masses as defined in the textbooks?

5. THE PARENTAGE OF THE WATER MASSES OF THE NORTHEAST ATLANTIC OCEAN

The descent of the bodies of water that make up the northeast Atlantic Ocean has been previously examined with great care, detail and imagination by COOPER (1952 a, b, c, d) for the region between the Celtic Sea and the Rockall plateau. His approach relied firmly on the discrimination of water masses and their parent water types using temperature and salinity as primary criteria (HELLAND–HANSEN, 1916) backed up with density and oxygen. The elaboration of $T–S$ distribution on the basis of σ_t and oxygen logically lead him to speculate about a number of unusual features of the water structure of the area west and south of the Celtic Sea. His general approach was highly innovative but progress was hindered by among other causes data gaps and uncertainties in the observations.

Perhaps a more serious constraint on progress in descriptive oceanography is one of language. Ideally, and as widely used in many models (for example, BROECKER (1974) with regard to age), waters above and below the oceanic thermocline have been separated as two water types, surface and deep, each devoid of precise descriptive parameters, lacking in detailed vertical structure and undergoing very generalized change along some migratory route.

Table 8. Conventional descriptions of northeast Atlantic water bodies
(after COOPER, 1952) (mainly 46° to 54°N, east of the mid-Atlantic ridge)

COOPER (1952a)	*Gulf of Gibraltar Water.* Water type 9.00°C, 35.70‰ salinity. [This water type is only present in the above area admixed with other waters.]
	North Atlantic Deep Water. Water mass 2.7–4.1°C, 34.85–35.00‰ salinity, oxygen 6.7–5.0 ml O_2/l.
	Mixed Gulf of Gibraltar–North Atlantic Deep Water. Water mass, found at intermediate depth, identifiable by strongly decreasing salinity and strongly increasing oxygen with depth.
	Mixed Gulf of Gibraltar–North Atlantic Central Water. Intermediate water mass lies above the preceding water, characterized by sharply increasing salinity with depth, of irregular and uncertain occurrence. Oxygen content always low.
COOPER (1952b)	*North Atlantic Central Water.* Upper water mass, bounded by 19°C, 36.70‰ and 8°C, 35.10‰. This water mass suffers much vertical mixing and becomes blended with coastal waters.

Water types as defined by SVERDRUP, JOHNSON and FLEMING (1942, pp. 141–146) may also be precisely described with appropriate T, S, O, etc., with the same attendant generalizations about integrity and survival as applies to oceanic models. COOPER regards outflowing Mediterranean water as a water type and Gulf of Gibraltar Water as the resulting water type in the Atlantic mixing pot. Similarly the overflowing water from the Norwegian Sea into the northern Atlantic to form North Atlantic Deep Water is treated in the foregoing part of this paper as a water type.

COOPER also recognizes certain water *masses*; bodies of water whose temperature and salinity obey a fixed relationship between certain boundary conditions; usually a water mass results from either a process of evaporation or from the mixing of two types of water. The greater the degree of vertical mixing the more confined lies the T, S domain of the water mass; the geometry of the T, S curve defines the nature of the mixing.

There are no comprehensive studies of the region 46° and 54°N, between COOPER's region and the next area, the Rockall Channel, studied by ELLETT and MARTIN (1973), which lies between 56° and 58°N. To a greater extent than farther south the Rockall Channel bathymetry imposes important constraints on waters deeper than 1250 m. ELLETT and MARTIN were able to describe annual and seasonal variability in the surface water and its changes over a 23-year series. Seasonal variability affects the water column to a diminishing extent down to 600 m. Deep waters, that is those waters situated below the deep salinity minimum and corresponding oxygen maximum usually at depths between 1600 and 1900 m, posed difficulty in identification, but the temperature and salinity characteristics (3.5–3.7°C and 34.94–34.96‰) were deemed not unlike those of Labrador Sea Water (3.4°C and 34.89‰). Between 1900 and 2100 m the potential temperature–salinity curves were ascribed to Norwegian Sea Deep Water entering the northeast Atlantic and giving rise to Northeast Atlantic Deep Water (LEE and ELLETT, 1967). Another deep water was considered to be present, tentatively of Antarctic origin, but the data were inadequate for a firm decision. Thus the deep-water picture for 56–58°N loses the sharpness and detail of COOPER's interpretation and this is clearly supported by study of the data for 46–54°N.

Gibraltar Water is much less distinct in the Rockall Channel than in the areas described by COOPER. For the intervening area Fig. 3 shows how the proportion of Gulf of Gibraltar Water as defined by COOPER decreases northwards but with rather little decrease in lateral spread.

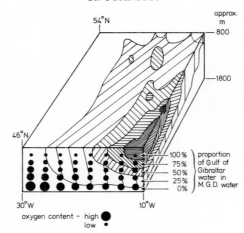

Fig. 3.   Percentage occurrence of Gulf of Gibraltar Water, between 46–54°N and 10–30°W at depths of 800–1800 m.

The change in North Atlantic Central Water, already modified by Cooper as a more dilute form compared to that of Sverdrup, Johnson and Fleming (1942) for the western central Atlantic Ocean, shows marked progressive reduction in its salinity gradient with respect to temperature. For example, Sverdrup's NAC water requires $dS/dz = 0.145 \, dT/dz$

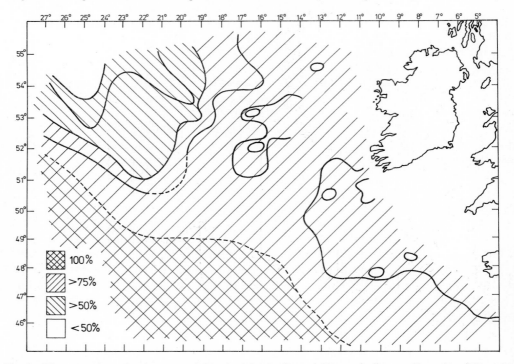

Fig. 4.   The anomalous development of North Atlantic Central Water, isolines -60, -45, -30, -15, $-10 \times 0.001‰$ salinity per 100 m.

and COOPER's NAC water $= 0.214\,dT/dz$ for the ranges encountered. The anomaly of salinity change with respect to temperature change with depth is shown in Fig. 4. Negative anomalies become larger to the north and northwest. This progressive erosion of NACW is confirmed by ELLET and MARTIN for the *Discovery II* section (Stas. 3851–3865) between 56°N, 35°W and Sole Bank. In this study only eight *T–S* curves from over 200 stations were found to be parallel to the NAC envelope for three or more consecutive depths. The occurrence of NACW was nevertheless regarded as a variable feature since the *T–S* curves for Scotia stations in June 1951 fell consistently within the envelope.

As in the ELLETT and MARTIN study, the 0–200-m water layer for 46–54°N was found to be very variable. The onset of vertical stability was about early April and the total breakdown of stratification before mid-November so that the water column was unstable at least to 200 m depth for about 30% of the year.

This brief review of the main component water masses emphasizes the wide departures that occur from classically defined water types and water masses as described by COOPER. These departures are so great that terminology imposes severe restraints and is misleading to the extent that it suggests a constancy in the continuity and structure of the water bodies instead of evolution and dynamic change which are much more typical at least in this region. Possibly it is this labelling process that has led to over-simplification of models of ocean circulation, structure and age, especially where components other than T° and $S\%_0$ are involved.

6. SOME THOUGHTS ON AGE AND DESCENT

Apart from temperature, salinity, density and pressure and other physical effects, the depths of the ocean also experience important gradients in inorganic salts, oxygen and nutrients. In many ways that have been little explored the chemical, biochemical and biological balance of the deep ocean shows very many changes that are unimaginably small in relation to the bulk of water, its transport and its salt balance. Changes in oxygen, conventional nutrients and carbon isotope are grotesquely crude measures of these vital changes. More subtle measures such as the ATP distribution (KARL, ROCK, MORSE and STURGES, 1976) may relate to only transitory conditions, yet the water quality which may directly and to an even greater extent indirectly govern the success of sensitive and delicate systems such as the success and survival of eggs and larvae may well depend on such subtle features. This said, like the many qualifications earlier about detailed hydrography and animal distributions, the major problems raised in this study relate to fundamental concepts of describing the ocean structure, ocean renewal and circulation. Can one accept the great many models based on gross simplifications of what cannot but be recognized as a highly complex medium?

Apart from the local phenomena such as those described by COOPER (1952 a–d) for the northeast Atlantic (and there are others for both adjoining and remoter parts so that the total contribution of 'altered' water is substantial), there is everywhere development and variability in the component waters that defeat the classical concepts inherent in the labels 'water bodies' and 'water masses'. There must be a corresponding mistrust in models based on these concepts and indeed in field measurements that evince advection but not turbulence, dispersal but without regard to longer-term movements, or both without regard to continuity both in a chemical sense as well as in a physical or mathematical sense. The concept of age is similarly suspect and age without a parallel description of development is only half the story.

Surface water is intuitively regarded as young and very deep water as very old yet the H_2O molecules at the surface and at the bottom are equally old. Again, the world ocean (mean depth 3795 m) is essentially a sluggish poorly oxygenated layer of water perturbed by topographical and gravitational forces. Age, fishery resources, coastal pollution and similar topics which most marine researchers are worrying about are the skin on the porridge. This anthropocentric viewpoint is too lax, too familiar and most of all it is entirely misleading. The only scientific way to regard the ocean (at least for the chemist) is to consider the deep water as the norm and to regard processes such as entry of oxygen, tritium or carbon isotopes (or ^{18}O in water) as invasive inputs along with other atmospheric and waterborne inputs. This places all inputs on one rating. No properties other than strictly subsurface temperature and salinity are legitimately subject to a simple mixing diagram approach.

Some of the inputs such as oxygen, $H_2{}^{18}O$ and carbon isotopes are primarily governed by intimate chemical control, namely equilibration between the atmosphere and a surface microlayer but others such as airborne and riverborne particles and indeed many riverborne inputs are largely unidirectional as are many pollutant inputs. The complex cycling of water itself in all its forms at the planet's surface might be regarded as partially subject to intimate control. Most life in the ocean is a skin process. The density, photic and thermal layering in the upper 40 m in European waters are necessary for plant life and the subsequent trophic levels it supports, but the perpetuation of stable surface layering is counterproductive; vertical instability *in situ* or in remote colder regions by lateral transport is vital for life but not for the survival of the ocean.

Discarding age the *in situ* air–sea oxygen flux resolves into, (i) transfer into a turbulent unstable water column (from November to March approximately) and (ii) transfer into or out of a ruffled stable water column for the rest of the year. There will always and everywhere be some input or equilibration at a high rate in the surface wave mixed zone falling off more or less speedily with depth depending on the degree of vertical mixing. This mechanism is predominantly governed by considerations of latitude, increasing to the poles with some important anomalies of low or high column stability related to geographical features.

Aquatic oxygen input resolves into (iii) deep-water overflows, (iv) lateral spread and sinking of more dense well-oxygenated waters beneath the surface and eventually permeating into intermediate waters (e.g. Labrador Sea Water), (v) regionally introduced waters having a separate prior circulation (e.g. Gulf of Gibraltar Water), (vi) the lateral spreading of waters deeply mixed by local geographical anomalies (upwelling and other phenomena).

These then are the parameters needed to assess input. In addition to severe problems in quantifying these parameters (needed to construct a mathematical model) there are problems of simulating the topographical and climatological dispositions, of effectively assessing vertical and lateral spreads and of pursuing the fates of water body introductions.

With regard to improved modelling of the biological system, recent work by THEIL (1975) reviews 12 years of studies on benthic organisms which may put that aspect on a firmer basis. He assesses the biomass and annual production of micro-organisms, meiofauna, macrofauna and megafauna and quotes the biomass of bacteria in the upper 1 cm sediment layer as about 10^6 organisms m^{-2} and an annual production of 8.76 g live wt. m^{-2} (about 0.876 g C m^{-2}). Other estimates gave biomass 4–84×10^6 ml^{-1} and 0.15–17.5 g live wt m^{-2}. Fungi, yeasts and protozoans amounted to about 10^6 organisms m^{-2} but not enough was known about their life cycles to derive the annual production. An indication of the meiobenthos (about 90% nematodes by total number) can be derived from the wet weight of nematodes amounting to $1.1 \, g \, m^{-2}$ (Iceland–Faroe ridge 61° to 63°N) and $0.5 \, g \, m^{-2}$ (Iberian Basin, 42° to 43°N). Net production may be 2.2 and $1.0 \, g$ live wt. m^{-2}. THEIL suggests $1 \, g$ wet wt. m^{-2} macrofauna (Iberian Basin) and an annual production of 0.1–$0.2 \, g \, m^{-2} \, yr^{-1}$. Sampling methods were deemed inadequate for corresponding estimates of megafauna biomass and production. In general for depths greater than 1000 m the benthic biomass decreased only slightly; there was a more marked decrease in relation to the distance from the shore. Decrease with depth was greater for macrofauna than meiobenthos but the major feature was a decrease in overall size of organism. A rate of 10–100 ml $O_2 \, kg^{-1}$ live wt. h^{-1} is quoted which is equivalent to $88 \, ml \, O_2 \, m^{-2} \, yr^{-1}$ for macrobenthos plus meiobenthos.

SMITH and HESSLER (1974) have attempted to measure the *in situ* respiration of bentho-pelagic fishes at 1230 m in the lower oxygen environment ($0.71 \, ml \, l^{-1}$ of the San Diego Trough. If their figures are typical the two species of deep-sea fishes required about $2.3 \, ml \, O_2 \, h^{-1}$ per kg body wt., compared to $55.6 \, ml \, O_2 \, h^{-1}$ per kg body weight for cod (*Gadus morhua*) in well-aerated shelf waters.

Benthic respiration was not included in this oxygen-based age method. THEIL's figures broadly confirm benthos estimates derived here. Since fish respiration is only about one-tenth that of the zooplankton, the change to a very low fish respiratory rate required by the SMITH and HESSLER results (which may not be typical of well-oxygenated waters) would affect the total respiratory rate only marginally. If deep-sea total respiratory rate was out by a similar factor (20 times too great) this would bring the respiratory rate down to about the level required in WRIGHT's (1969) model; it would also help to resolve the mystery of the utterly inadequate food supply for deep-sea life as shown by many writers. But these better agreements cannot be taken as good grounds for confidence in current understanding of the oceans.

If a better synthesis of oxygen input and uptake rates could be achieved, this would be a major advance in understanding the oceans and the broader aspects of the needs of their inhabitants. Prospects for the fulfilment of these hopes are not too remote. It is a significant step forward to see how tolerably adequate solutions of some of the component problems might be achieved and also for the overall picture to be seen the right way up, and somewhat in focus. What applies to oxygen applies in some related way to all the elements and processes capable of framing a time-scale of oceanic events.

What set out as a detailed and intricate outline of age and descent in the oceans has ended as a broad brush caricature. Perhaps no finer picture of the oceans is valid.

REFERENCES

BANSE K. (1964) On the vertical distribution of zooplankton in the sea. In: *Progress in Oceanography*, M. SEARS, editor, Pergamon Press, **2**, pp. 53–125.

BERNARD F. (1958) Plancton et Benthos observés durant trois plongées en bathyscaphe au large de Toulon. *Annales de l'Institut Océanographique*, **25**, 287–326.

BOGOROV V. G. (1957) Regularities of plankton distribution in the northwest Pacific. *Proceedings of UNESCO Symposium of physics Oceanography, Tokyo 1955*, pp. 260–276.

BOGOROV V. G. (1960) Productive regions of the oceans. ICES C. N. 1960, Doc. No. 137 (mimeo).

BOGOROV V. G. and M. E. VINOGRADOV (1955) On the zooplankton in the northwestern part of the Pacific Ocean. *Doklady Akademii Nauk, SSSR*, **102**, 835–38 (in Russian).

BRODSKII K. A. (1955) On the vertical distribution of copepods in the northwestern Pacific Ocean. *Special Scientific Report. Fish O.U.S. Department of the Interior, Fish and Wildlife Series*, No. **192**, 1–6.

BROECKER W. S. (1974) *Chemical oceanography*, Harcourt Brace Jovanovich, New York, 214 pp.

BROECKER W. S., R. GERARD, M. EWING and B. C. HEEZEN (1960) Natural radiocarbon in the Atlantic Ocean. *Journal of Geophysical Research*, **65**, 2903–2933.

BROWN M. E., editor (1957) *The physiology of fishes*, Academic Press Inc., New York, 447 pp.

CARMACK E., and K. AAGARD (1974) The formation of bottom water in the Greenland Sea. *Colloques Internationnaux du C.N.R.S.* No. 215, Paris, 4–7 Octobre 1972.

COOPER L. H. N. (1952a) The physical and chemical oceanography of the waters bathing the continental slope of the Celtic Sea. *Journal of the Marine Biological Association, U.K.* **30**, 465–510.

COOPER L. H. N. (1952b) Factors affecting the distribution of silicate in the North Atlantic Ocean and the formation of North Atlantic deep water. *Journal of the Marine Biological Association, United Kingdom* **30**, 511–526.

COOPER L. H. N. (1952c) Processes of enrichment of surface water with nutrients due to strong winds blowing on a continental slope. *Journal of the Marine Biological Association. United Kingdom*, **30**, 453–464.

COOPER L. H. N. (1952d) Water movements over the continental slope in relation to fisheries hydrography. *Rapport et procès-verbaux des réunions. Conseil permanent international pour l'exploration de la mer*, **131**, 44–50.

COOPER L. H. N. (1961) Vertical and horizontal movements in the ocean. In *Oceanography*, M. SEARS, editor, AAAS Publ. **67**, Washington, D.C., pp. 599–621.

CREASE J. (1965) The flow of Norwegian Sea water through the Faroe Bank Channel. *Deep-Sea Research*, **12**, 143–150.

DIETRICH G. and K. KALLE (1957) *Allgemeine Meereskunde. Ein Einfuhrung in die Oceanographie*, Gebrüder Borntraeger, Berlin. 588 pp.

ELLETT D. J. and J. H. A. MARTIN (1973) The physical and chemical oceanography of the Rockall Channel. *Deep-Sea Research*, **20**, 585–625.

FARRAN G. P. (1926) Biscayan plankton collected during a cruise of H.M.S. *Research*, 1900. Part XIV. The Copepoda. *Journal of the Linnean Society (Zoology)*, **36**, 219–310.

FAO (1956) The influence of hydrographic conditions on the behaviour of fish. *FAO Fisheries Bulletin* **9**, 181–196.

FOX C. J. J. (1907) On the coefficients of absorption of atmospheric gases in distilled water and sea water. I. Nitrogen and oxygen. *Publications de Circonstance* No. 41, 23 pp.

GORDON A. L. (1972) Spreading of Antarctic Bottom Waters, II. In: *Studies in physical oceanography: a tribute to Georg Wüst on his 80th birthday*, A. L. GORDON, editor, Gordon & Breach, New York, 2 vols.

HARVEY H. W. (1950) On the production of living matter in the sea off Plymouth. *Journal of the Marine Biological Association, U.K.*, **29**, 97–137.

HELLAND-HANSEN B. (1918) Nogen hydrografiske metoder. *Forhandlinger ved de Skandinaviska naturforsketes møte, Kristiania*, 10–15 July 1916.

JACOBSEN J. P., R. J. ROBINSON and T. G. THOMPSON (1950) A review of the determination of dissolved oxygen in sea water by the Winkler method. *Publication Scientifique Association Oceanographie Physique*, **11**, 22 pp.

JESPERSEN P. (1935) Quantitative investigations on the distribution of macro-plankton in different ocean regions. *Dana Reports*, **7**, 1–44.

JOHNSTON R. (1962) An equation for the depth distribution of deep-sea zooplankton and fishes. *Rapport et Procès-verbaux des réunions. Conseil permanent international pour l'exploration de la mer*, **153**, 217–219.

KARL D. M., P. A. LA ROCK, J. W. MORSE and W. STURGES (1976) Adenosine triphosphate in the North Atlantic Ocean and its relationship to the oxygen minimum. *Deep-Sea Research*, **23**, 81–88.

LEAVITT B. B. (1938) The quantitative vertical distribution of macro-zooplankton in the Atlantic Ocean basin. *Biological Bulletin, Woods Hole*, **74**, 376–394.

LEE A. J. and D. J. ELLETT (1965) On the contribution of overflow water from the Norwegian Sea to the hydrographic structure of the North Atlantic Ocean. *Deep-Sea Research*, **12**, 129–142.

McGILL D. A. (1964) The distribution of phosphorus and oxygen in the Atlantic Ocean as observed in the International Geophysical Year, 1957–1958. *Progress in Oceanography*, **2**.

MARSHALL S. M. and A. P. ORR (1955) *The biology of a marine copepod*, Oliver & Boyd, Edinburgh, 188 pp.

MEINCKE J. (1967) Die Tiefe der jahreszeitlichen Dichteschwankungen im Nordatlantischen Ozean. *Kieler Meeresforschungen*, **23**, 1–15.

MONTGOMERY R. B. (1955) Characteristics of surface water at weather ship 'J'. *Deep-Sea Research* **3**, Supplement Paper for Marine Biology Oceanography, 331–334.

MULLIN M. M. (1969) Production of zooplankton in the ocean: the present status and problems. *Oceanography and Marine Biology, Annual Review*, **7**, 293–314.

NICOL J. A. (1960) *The biology of marine animals*, Pitman, London. 707 pp.

NYBELIN O. (1951) Introduction and station list. *Reports of the Swedish Deep-Sea Expedition (1947–48)*, **2**, 1–28.

ØSTVEDT O.-J. (1955) Zooplankton investigations from weather ship 'M' in the Norwegian Sea, 1948–49. *Hvalrådets Skrifter*, No. 40, 1–93.

PERRIER E. (1894) *Echinodermes. Expeditions scientifiques du* Travailleur *et du* Talisman, *pendant les années 1880–1883*, Masson, Paris.

PERRIER R. (1901) *Holothuries. Expeditions scientifiques du* Travailleur *et du* Talisman, *pendant les années 1880–1883*, Masson, Paris.

PINGREE R. D. (1971) Analysis of the temperature and salinity small-scale structure in the region of Mediterranean influence in the N.E. Atlantic. *Deep-Sea Research*, **18**, 485–491.

RILEY G. A. (1951) Oxygen, phosphate and nitrate in the Atlantic Ocean. *Bulletin of the Bingham Oceanographic Collection, Yale University*, **13**(1), 1–120.

SMITH K. L. and R. R. HESSLER (1974) Respiration of benthopelagic fishes: *in situ* measurements at 1230 m. *Science*, **184**, 72–73.

SPÄRCK R. (1951) Density of bottom animals on the ocean floor. *Nature, London*, **168**, 112–113.

SPÄRCK R. (1956) *Galathea Report: scientific results of the Danish Deep Sea Expedition (1950–52)*, Danish Science Press, Copenhagen.

STEELE J. H. S. (1967) Current measurement on the Iceland–Faroe Ridge. *Deep-Sea Research*, **14**, 469–473.

STRICKLAND J. D. H. (1972) Research on the marine planktonic food web at the Institute of Marine Resources: A Review of the Last Seven Years Work. In: *Oceanography and marine biology*, H. BARNES, editor, Allen & Unwin, **10**, pp. 349–414.

SUNDNES G. (1957) Notes on the energy metabolism of the cod (*Gadus callarias* L.) and the coalfish (*Gadus virens* L.) in relation to body size. *Fiskeridirektoratets skrifter. Havundersøkelser*, **11**, (9), 10 pp.

SVERDRUP H. U., M. W. JOHNSON and R. H. FLEMING (1942) *The oceans: their physics, chemistry and general biology*, Prentice-Hall, 1087 pp.

SWALLOW J. C. and B. V. HAMON (1960) Some measurements of deep currents in the eastern North Atlantic. *Deep-Sea Research*, **6**, 155–168.

TAIT J. B. (1957) Hydrography of the Faroe–Shetland Channel 1927–1952. *Marine Research Scotland*, **2**, 309 pp.

TAIT J. B., editor (1967) The Iceland–Faroe Ridge Interflow (ICES) 'Overflow' Expedition, May–June, 1960. An investigation of cold, deep water overspill into the northeastern Atlantic Ocean. *Rapport et proces-verbaux des réunions. Conseil permanent international pour l'exploration de la mer*, **157**, 1–274.

THEIL H. (1975) The size structure of deep-sea benthos. *International Revue gesamten Hydrobiologie*, **60**, 575–606.

TRUESDALE G. A., A. L. DOWNING and G. F. LOWDEN (1955) The solubility of oxygen in saline water. *Journal of Applied Chemistry*, 195, **5**, 53–62.

VAILLANT L. (1888) *Poissons. Expeditions scientifiques du* Travailleur *et du* Talisman, *pendant les années 1880–1883*, Masson, Paris.

WATTENBERG H. (1929) Die Durchluftung des Atlantischen Ozeans. *Journal du Conseil*, **4**, 68–80.

WEISS R. F. (1970) The solubility of nitrogen, oxygen and argon in water and seawater. *Deep-Sea Research*, **17**, 721–735.

WORTHINGTON L. V. (1954). A preliminary note on the time scale in North Atlantic circulation. *Deep-Sea Research*, **1**, 244–251.

WORTHINGTON L. V. (1958) Oceanographic data from R.R.S. *Discovery II*, International Geophysical Year. Cruises 1 and 2, 1957. *Woods Hole Oceanography Institute*, pp. 58–130.

WRIGHT R. (1969) Deep water movement in the western Atlantic as determined by use of a box model. *Deep-Sea Research Supplement*, **16**, 433–446.

YASHNOV V. A. (1960) Plankton of the tropical Atlantic Ocean. ICES C.M. 1960, Doc. No. 163 (mimeo).

ZENKEVITCH L. A. and J. A. BIRSTEIN (1956) Studies of the deep-water fauna and related problems. *Deep-Sea Research*, **4**, 54–64.

Trace metals in a section of the North Atlantic Ocean and eastern Caribbean Sea

J. D. Burton,* G. B. Jones*† and H. S. Matharu*‡

Abstract—Surface water samples collected on a section across the North Atlantic Ocean, extending into the eastern Caribbean Sea, were analysed for dissolved copper and molybdenum; mercury was measured on acidified filtered samples. The vertical distributions of molybdenum and mercury were examined at three stations.

Concentrations of copper in surface waters ranged from 0.47 to 1.8 (mean, 0.93) $\mu g\,l^{-1}$, while those of molybdenum in surface and sub-surface samples were in the relatively narrow range of 9.2–13.1 (mean, 11.0) $\mu g\,l^{-1}$, confirming the dominantly conservative behaviour of this element in oceanic waters. The values for mercury ranged from <5–43 (mean, 15) $ng\,l^{-1}$ for surface waters and from <5–23 (mean, 10) $ng\,l^{-1}$ for sub-surface samples. No systematic trends common to the profiles were apparent. The results confirm that the tropical North Atlantic Ocean is a region of generally low concentrations of mercury and show no evidence for enhancement in the bottom waters of this region. It is suggested that the surface enhancement of mercury found for two of the profiles examined here and for a previously reported profile in the North Atlantic Ocean may reflect atmospheric inputs of the element.

INTRODUCTION

Despite the greater attention recently devoted to the subject, our knowledge concerning the distribution of trace metals in oceanic waters remains limited by two main factors. First, the relatively small number of samples which has been examined, particularly from deep waters. Secondly, the indefinite status of many of the values obtained, reflecting uncertainties as to how analytical values may be affected by problems of loss and contamination during sampling and initial treatment, and as to possible differences in the ways in which different methods respond to the speciation of an element, particularly in relation to concentration and separation procedures.

The need for information on the background levels of certain trace metals in environments, including oceanic waters, which have not been subject to significant local pollution has become recognized as an important priority in marine science (International Decade of Ocean Exploration, 1972). There has also been interest, notably in the GEOSECS programme, in trace metals among other potential tracers for water mixing and circulation. The emphases in these programmes are complementary to the marine geochemist's basic concern to understand the processes involved in the movement of elements through the hydrosphere, for which an initial requirement is an adequate description of their distributions.

This paper gives the results of measurements of several trace metals in samples from

---

* Department of Oceanography, The University, Southampton, England.
† Present address: James Cook University of North Queensland, Australia.
‡ Present address: Jawaharlal Nehru University, New Delhi, India.

147

a section across the North Atlantic Ocean, extending into the eastern Caribbean Sea. The primary objective was to obtain information on mercury but samples were also analysed for two other elements of biological interest, namely molybdenum and copper.

METHODS

Sampling

Surface water samples were collected using a plastic bucket from a position forward of any discharges from the moving ship. Subsurface samples were collected using polypropylene water bottles; messengers were coated with polyurethane to reduce the possibilities of contamination of samples. After collection the samples were transferred to polypropylene bottles. The shipboard programme was carried out in late March and early April 1972.

Analysis for copper

Samples of *ca.* 1 litre were filtered through membrane filters of 0.45 μm average pore diameter, acidified to pH 1 and returned to the shore laboratory. After readjustment of the pH to 7.0, copper was concentrated on Chelex-100 chelating ion exchange resin, following Riley and Taylor (1968). After elution of the copper with dilute nitric acid, evaporation of the eluate and dissolution of the residue in 0.1 M nitric acid, copper was determined by atomic absorption spectrophotometry, using an air/acetylene flame. The coefficient of variation of the procedure in this laboratory has averaged 1%. Blank values were below 0.1 μg.

In coastal waters fractions of copper, ranging in some samples up to 40%, have been found to be unavailable to concentration procedures which quantitatively recover inorganic species of the element (see, for example, Foster and Morris, 1971). There is as yet no evidence concerning such organically associated fractions in open oceanic waters. The values reported here are for the analytically available copper fraction which may be lower than the total dissolved copper.

Analysis for molybdenum

Water was filtered as for the determination of copper and analysed by the method of Chan and Riley (1966) which involves concentration by coprecipitation with hydrous manganese dioxide and determination by spectrophotometry using the dithiol complex. Coprecipitation from *ca.* 1 litre was carried out on board ship; the precipitate was filtered off and dissolved in *ca.* 20 ml of 1 M hydrochloric acid, containing 1.5 g of sodium sulphite. The solution was returned to the shore laboratory for completion of the analysis. Experiments with known amounts of molybdenum showed that losses were negligible during storage under these conditions. Replicate determinations in this laboratory have given coefficients of variation in the range of 1–2%. Blank values for the procedure, commenced on board ship, were close to 0.1 μg Mo.

There is no evidence for the presence in sea water of any molybdenum species which are unavailable to this method.

Analysis for mercury

Immediately after collection, 900 ml of the sample were acidified to pH 1 with sulphuric acid and filtered through a Whatman GF/C glass fibre filter, which had been previously heated at 500°C. This procedure was adopted to avoid losses of mercury which can occur in filtering unacidified water. The fraction analysed thus includes any contribution from leaching of particulate material but this is probably negligible for open oceanic waters.

The filtrate was extracted with successive portions (10, 5, 5 ml) of a 12 ppm solution of dithizone in chloroform. The extracts were evaporated, under an infra-red lamp, on *ca.* 50 mg of magnesium oxide, which had previously been heated to 500°C. The dry residues were returned to the shore laboratory, packed in polyethylene vials, and transferred to silica ampoules for neutron irradiation. Experiments using radioactive mercury-203 showed that there was no detectable loss of mercury from residues treated in this way.

Irradiation was usually carried out for 40 hours at a neutron flux of *ca.* 6×10^{12} neutron $cm^{-2} s^{-1}$ and the induced radioactive isotopes of mercury were then radiochemically purified after allowing several days for the decay of short-lived nuclides. The residues were digested under reflux, in the presence of mercury carrier, with sulphuric acid and hydrogen peroxide. After boiling to decompose remaining hydrogen peroxide, mercury was distilled from the solution in hydrochloric acid generated by the addition of glycine and perchloric acid. Chloride ions were removed from the distillate by precipitation with silver nitrate. The mercury was allowed to deposit on copper foil suspended in the solution and the foil was then dissolved in nitric acid and the mercury precipitated as copper ethylenediamine mercuric iodide. The precipitate was filtered, dried and weighed, and the beta radioactivity due to mercury-203 was measured using an anti-coincidence counter, after the elapse of sufficient time for mercury-197m to decay to a negligible level.

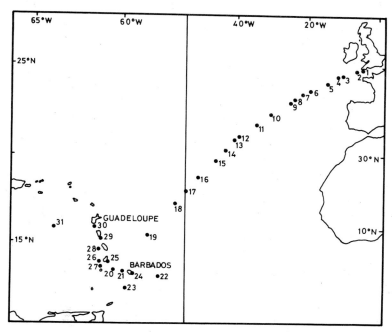

Fig. 1.   Positions of stations.

The radiochemical purity of each source was checked by measurement of the half-life. Standard counting and chemical yield corrections were made. The concentration of mercury was calculated by comparing the corrected counting rates with those from standards, in the form of dilute aqueous solutions, which were irradiated with the samples and subjected to similar purification. Determinations on a reference tissue sample indicated that, over a series of separate irradiations, the coefficient of variation of the neutron activation procedure is *ca.* 15% (Leatherland and Burton, 1974). Blank determinations, commenced on board ship and treated similarly to the samples, gave values close to 2.5 ng of mercury.

A fuller account of the entire method and its evaluation has been given by Leatherland (1974) who showed that the solvent extraction procedure used gives essentially quantitative recovery of mercury in inorganic forms. Methylmercuric ions are also recovered quantitatively; it seems unlikely, however, that this species is of significance in ocean water. Mercury complexed with humic material in a model solution is extractable under the conditions employed (Millward and Burton, 1975).

RESULTS AND DISCUSSION

The positions at which surface samples were collected are shown in Fig. 1. Three vertical profiles were examined, one in the mid-ocean ridge region and two in the Tobago Basin. The results are given in Tables 1 and 2, for the surface samples and vertical profiles respectively, and are discussed below for the individual elements. The salinity values shown are from the ship's oceanographic observations.

Copper

The concentrations of copper in the surface waters ranged from 0.47 to 1.8 μg l^{-1}, with a mean value of 0.93 μg l^{-1}. The mean values for the Atlantic section and the eastern Caribbean were similar (0.91 and 0.96 μg l^{-1}, respectively). The values for the Sargasso region agree well with findings by Spencer and Brewer (1969) for that area and the overall mean is similar to that of 1.0 μg l^{-1} obtained by Chester and Stoner (1974) for surface samples from open parts of the South Atlantic Ocean, although the dispersion of the present values is lower. A very similar range of 0.7–1.6 μg l^{-1} was reported by Riley and Taylor (1972) for surface waters off North-West Africa. The work of Chester and Stoner (1974) suggests relatively uniform mean surface concentrations of copper for the major oceans. Boyle and Edmond (1975), however, have reported substantially lower concentrations, averaging 0.2 μg l^{-1}, for surface waters south of New Zealand. Recent unpublished analyses made in this laboratory on samples from the eastern North Atlantic Ocean between 23°N and 47°N have given similarly low values. More information is needed to establish whether the observed variability is a real environmental feature or a reflection of analytical problems.

Molybdenum

Several workers have reported concentrations of molybdenum for waters in the open East Atlantic Ocean north of the latitudes concerned in this work. Head and Burton

(1970) found a range of 9.4–12.4 (mean 10.9) $\mu g\, l^{-1}$ for fifteen surface and sub-surface samples, and MORRIS (1975) has reported values of 8.8–13.0 (mean 10.7) $\mu g\, l^{-1}$ for seventy-five samples from various depths. The findings indicate rather uniform oceanic concentrations and these workers also found similar concentrations in adjacent coastal waters. The above concentrations are similar to those reported for most other regions which have been examined (see HEAD and BURTON, 1970) and the low dispersion of values accords with the largely conservative behaviour predicted for the molybdate species in a situation in which the supply of the element considerably exceeds the amounts involved in biological processes.

The values for all samples in the region here examined have a range (9.2–13.1 $\mu g\, l^{-1}$) and mean (11.0 $\mu g\, l^{-1}$) which are virtually identical to those for higher latitudes as discussed above. The small variations found in the vertical profiles and the lack of systematic trends with depth emphasize that the flux of the element, between surface and deep waters, maintained by processes of particulate transfer, must be low relative to the total concentration, confirming the small influence of biological activity. Concentrations which show much greater unsystematic variability (2–22 $\mu g\, l^{-1}$) have been reported for upwelling regions (RILEY and TAYLOR, 1972; SOURNIA and CITEAU, 1972). The present work gives no indication of processes which could maintain these variations.

Mercury

FITZGERALD, GORDON and CRANSTON (1974) have summarized the concentrations of mercury reported in eight studies from 1970 onwards. The overall range covered two orders of magnitude. Most values were below 100 $ng\, l^{-1}$, but some higher concentrations were reported. Notably, CARR, HOOVER and WILKNISS (1972) found that in the Greenland Sea concentrations in surface waters were mostly below 50 $ng\, l^{-1}$ but rose to 364 $ng\, l^{-1}$ for near-surface water under ice cover. In contrast with most of the earlier work, the analyses made by FITZGERALD, GORDON and CRANSTON (1974) gave high values, averaging 150 $ng\, l^{-1}$, for an extensive section of surface and sub-surface waters of the northwestern Atlantic Ocean. These measurements were made by a shipboard method and losses of mercury during storage of acidified samples were reported. Such storage procedures were widely employed in earlier work and it was suggested that their use largely invalidated the results obtained previously.

Further considerations regarding the interpretation of analyses of mercury in seawater were raised by LEATHERLAND, BURTON, CULKIN, MCCARTNEY and MORRIS (1973). Using the same technique as in the present work, but on stored acidified samples, they found values from 30–88 (mean 44) $ng\, l^{-1}$ for surface samples from the region 39–47°N, 7–9°W. These were greater than those found by LEATHERLAND, BURTON, MCCARTNEY and CULKIN (1971) on samples, in some cases identical ones, from the same region, using an anion-exchange concentration procedure. It was suggested that the difference was related to the presence of some 50% of the mercury in forms not removed by anion exchange. FITZGERALD and LYONS (1973) demonstrated the presence, in unfiltered coastal seawater, of a fraction of mercury which was available to their analytical procedure only after photo-oxidation of the sample. The existence of associations of mercury with dissolved organic matter in coastal and estuarine water, involving at least 20% of the total dissolved mercury, which ranged from 30–110 $ng\, l^{-1}$, has been demonstrated by ANDREN and HARRISS (1975). MILLWARD and BURTON (1975) inferred the probable environmental

J. D. BURTON, G. B. JONES and H. S. MATHARU

Table 1.   Concentrations of trace metals in surface waters

Station		Salinity (‰)	Cu (μg l^{-1})	Mo (μg l^{-1})	Hg (ng l^{-1})
Number	Position				
1	50°05'N, 04°25'W	35.172	1.39	10.4	—
2	49°43'N, 05°58'W	35.351	1.77	10.8	—
3	48°43'N, 10°02'W	35.588	0.69	10.9	—
4	48°26'N, 10°56'W	35.580	1.48	11.5	—
5	47°28'N, 14°11'W	35.566	0.47	10.7	—
6	45°38'N, 19°19'W	35.594	0.97	9.9	—
7	44°47'N, 21°38'W	35.658	1.12	10.1	—
8	43°25'N, 24°34'W	35.750	1.21	12.6	—
9	42°49'N, 25°56'W	35.703	0.82	10.5	—
10	40°19'N, 30°38'W	36.011	0.68	12.2	—
11	37°47'N, 34°20'W	35.131	0.52	12.6	—
12	34°44'N, 39°21'W	36.423	0.51	12.4	—
13	34°25'N, 40°24'W	36.193	—	12.4	43
14	31°30'N, 43°10'W	36.533	1.14	11.6	<5
15	28°54'N, 46°04'W	36.948	0.78	12.1	21
16	24°13'N, 50°52'W	37.166	0.91	12.3	18
17	20°33'N, 54°03'W	36.783	0.50	12.6	7
18	17°17'N, 56°50'W	36.133	0.61	12.5	14
19	15°25'N, 58°32'W	34.842	0.84	11.4	—
20	13°16'N, 60°25'W	35.941	0.76	12.3	23
21	13°11'N, 59°53'W	35.761	—	—	<5
22	12°58'N, 57°49'W	33.388	0.74	10.4	7
23	12°12'N, 59°45'W	35.708	0.72	11.6	12
24	13°07'N, 59°23'W	35.894	1.46	12.7	7
25	14°03'N, 60°60'W	35.780	0.72	9.8	—
26	13°50'N, 61°08'W	34.532	0.85	9.5	—
27	13°35'N, 61°08'W	34.557	1.50	10.0	—
28	14°34'N, 61°16'W	35.813	0.85	11.1	—
29	15°11'N, 61°19'W	35.789	1.09	10.6	—
30	15°55'N, 61°34'W	35.907	1.08	10.0	—
31	15°50'N, 63°52'W	35.710	0.80	10.0	—

importance of organic associations, from experiments with terrestrial humic materials in model solutions. It seems probable that some of the techniques employed to isolate mercury from seawater could respond differentially to the different mercury species present in natural waters.

There have been other investigations of mercury in surface or near-surface seawater, subsequent to or not included in the compilation of data made by FITZGERALD, GORDON and CRANSTON (1974). TOPPING and PIRIE (1972) found concentrations of 8–55 ng l^{-1} in samples from the North Sea; in their work the initial isolation of mercury was carried out on board ship. A similar range of concentrations has been reported for dissolved mercury in stored samples from relatively unpolluted areas of the Irish Sea (GARDNER and RILEY, 1973). Values around 13 ng l^{-1} were found by ÓLAFFSON (1974) as background levels for waters off the southern coast of Iceland, whereas GARDNER and RILEY (1974) report a range of 12–225 ng l^{-1} for a wider range of samples collected in the vicinity of Iceland. These latter results tended to confirm the findings by CARR, HOOVER and WILKNISS (1972) of high concentrations associated with North Atlantic waters of polar origin. For a station in the northeast Pacific Ocean, WILLIAMS and WEISS (1973) report a near-surface concentration of 270 ng l^{-1}. WILLIAMS, ROBERTSON, CHEW and WEISS (1974) found con-

Table 2.   Vertical profiles of trace metals

Station Number	Station Position	Depth (m)	Salinity (‰)	Mo ($\mu g\,l^{-1}$)	Hg ($ng\,l^{-1}$)
13	34°25′N, 40°24′W	0	36.193	12.4	43
	(bottom depth 3517 m)	245	35.989	12.2	19
		493	35.443	10.7	11
		742	35.118	11.9	10
		1220	34.961	9.2	<5
		1472	—	10.2	<5
		2254	—	10.8	8
		2976	34.731	10.9	15
		3380	34.697	10.9	12
		3482	—	—	<5
20	13°16′N, 60°25′W	0	35.941	12.3	23
	(bottom depth 2207 m)	50	36.022	12.6	14
		149	36.234	12.4	19
		532	34.819	12.2	8
		716	34.606	11.3	12
		1379	34.974	11.4	14
		1810	34.576	11.3	6
21	13°11′N, 59°53′W	0	35.761	—	<5
	(bottom depth 1737 m)	99	35.815	11.3	18
		149	35.959	11.3	5
		297	35.476	13.0	10
		486	34.761	13.1	—
		1203	34.782	12.5	—
		1673	34.745	12.7	6

trasting values for near surface waters at other stations in the northeast Pacific Ocean (16–$50\,ng\,l^{-1}$) and those of southern polar regions (53–$112\,ng\,l^{-1}$). Data cited by these workers indicate a range of 5–90 (mean 20) $ng\,kg^{-1}$ for shelf waters in the area of the Beaufort Sea and Alaska North Slope. The most extensive data are those obtained by ROBERTSON for GEOSECS stations in the western Atlantic Ocean and cited by CARR, JONES and RUSS (1974). These indicate generally low concentrations (2–$40\,ng\,l^{-1}$) between 15°N–35°S, with a region of high concentrations (up to $400\,ng\,l^{-1}$) between 25°–50°N. Concentrations of $10\,ng\,l^{-1}$ have been reported (JOHNSON and BRAMAN, 1975) for surface waters of the Sargasso Sea. In eastern Atlantic continental shelf waters between 30°–33°N, WINDOM, TAYLOR and WAITERS (1975) found an overall range of 5–$370\,ng\,l^{-1}$, a pronounced seasonal variation being apparent.

Similarly, contrasting findings have been reported for the distribution of mercury in vertical profiles. At several stations in the area of the Ramapo Deep in the northwest Pacific Ocean (HOSOHARA, 1961), and in the Atlantic Ocean at 47°N, 8°W (LEATHERLAND, BURTON, CULKIN, McCARTNEY and MORRIS, 1973) significantly increased values have been observed in deep water. For a considerable number of stations in the North Atlantic Ocean (LEATHERLAND, BURTON, McCARTNEY and CULKIN, 1971; FITZGERALD, GORDON and CRANSTON, 1974), the Greenland Sea (CARR, HOOVER and WILKNISS, 1972), the northeast Pacific (WEISS, YAMAMOTO, CROZIER and MATHEWSON, 1972; WILLIAMS, ROBERTSON, CHEW and WEISS, 1974) and southern polar waters (WILLIAMS, ROBERTSON, CHEW and

Weiss, 1974) no systematic variations in concentration with depth have been apparent, even though the average values for these profiles differ substantially. One of the profiles given by Weiss, Yamamoto, Crozier and Mathewson (1972) shows several pronounced maxima. Several profiles have also been reported by Williams and Weiss (1973) in the northeast Pacific Ocean and by Carr, Hoover and Wilkniss (1972) in the Greenland Sea, which show pronounced maximum concentrations in surface layers; in the latter case an influence of melting sea ice was suggested. Williams and Weiss (1973) cite data they have obtained for other stations at which average concentrations of $40 \, \text{ng} \, \text{l}^{-1}$ for surface waters compare with those of $25 \, \text{ng} \, \text{l}^{-1}$ at greater depths. High mercury concentrations occur intermittently in near-bottom waters of the FAMOUS area of the mid-Atlantic ridge (Carr, Jones and Russ, 1974; Carr, Jones, Warner, Cheek and Russ, 1975). Median concentrations on different sampling occasions have varied from 67 to $1080 \, \text{ng} \, \text{l}^{-1}$, with a range for individual values from 12 to $1420 \, \text{ng} \, \text{l}^{-1}$. The enhancement of concentrations of mercury by processes associated with submarine vulcanism was also suggested by Williams, Robertson, Chew and Weiss (1974) as an explanation for the generally high levels in southern polar waters and is substantiated by findings of concentrations up to $480 \, \text{ng} \, \text{l}^{-1}$ in waters influenced by the Heimaey eruption (Ólafsson, 1975).

The values reported here for surface waters range from < 5 to $43 \, \text{ng} \, \text{l}^{-1}$ with a mean of $15 \, \text{ng} \, \text{l}^{-1}$. The shipboard concentration technique eliminated the need to store water samples. A comparison between these values and others obtained by analysis of acidified stored samples collected in the same general area (Chester, Gardner, Riley and Stoner, 1973) or at somewhat higher latitudes (Leatherland, Burton, Culkin, McCartney and Morris, 1973) does not support the contention of Fitzgerald, Gordon and Cranston (1974) that analyses are invalidated by such storage procedures. This is not, however, to argue against the view that shipboard procedures are preferable. The results agree well with the findings by Chester, Gardner, Riley and Stoner (1973), and those of the GEOSECS investigation discussed above, that the low latitude region of the North Atlantic Ocean is one of generally low concentrations of mercury.

For sub-surface waters, concentrations varied from < 5–23 (mean 10) $\text{ng} \, \text{l}^{-1}$. There were no systematic trends common to the three profiles or even to the two profiles in the Tobago Basin. The surface concentration exceeded the average for the water column at two stations but was markedly lower at one of the Caribbean stations. Near-bottom water, including that in the mid-ocean ridge region, showed low concentrations of < 5–$6 \, \text{ng} \, \text{l}^{-1}$ and similarly low concentrations characterized water in the depth range of 1200–1500 m in the mid-Atlantic region at 34°N.

Although the analytical uncertainties which can affect the comparability of data have not yet been entirely clarified, the existence of wide spatial variations in environmental concentrations is clear from the present measurements and other recent work. There is sufficient evidence to indicate that the vertical distribution of mercury is not determined primarily by transference through the biosphere but is primarily controlled by the major pathways of its input to the ocean and subsequent effects of water circulation and mixing, and scavenging by particles. The periodic local significance of effects associated with submarine vulcanism has recently become apparent and glacial processes may affect levels of mercury in some water masses. The enhanced surface concentrations shown in two of the profiles reported here and also observed by Leatherland, Burton, Culkin, McCartney and Morris (1973) somewhat further north suggest the possible importance of atmospheric pathways and sea-surface processes. Williams and Weiss (1973) considered that an atmospheric input might account for the enhancement in surface waters which they observed but they later regarded this as invalid in the light of further measurements (WILLIAMS, ROBERTSON, CHEW and WEISS, 1974). Marked seasonal variations which occur in coastal waters of the eastern Atlantic Ocean may be explained by varying atmospheric inputs from terrestrial sources (WINDOM, TAYLOR and WAITERS, 1975). While the influence of such a pathway will be more readily apparent in coastal waters it is not necessarily limited to such regions. A marked latitudinal and temporal variation in the pattern of injection of mercury to the open sea from natural and anthropogenic sources could arise in this way and might partly explain the latitudinal variations in surface water concentrations in the North Atlantic Ocean.

Acknowledgements—This work, which formed part of the United Kingdom Contribution to CICAR 1971–1972, was supported by the Natural Environment Research Council. One of us (H.S.M.) wishes to thank The University, Southampton, for the award of a research studentship. The authors gratefully acknowledge the facilities and assistance provided by the Hydrographer of the Navy and the first two named thank especially Commander R. A. G. NESBIT, R.N. and his officers and crew for their help and hospitality throughout their work aboard H.M.S. *Hecla*.

REFERENCES

ANDREN A. W. and R. C. HARRISS (1975) Observations on the association between mercury and organic matter dissolved in natural waters. *Geochimica et Cosmochimica Acta*, **39**, 1253–1258.

BOYLE E. and J. M. EDMOND (1975) Copper in surface waters south of New Zealand. *Nature*, **253**, 107–109.

CARR R. A., J. B. HOOVER and P. E. WILKNISS (1972) Cold-vapor atomic absorption analysis for mercury in the Greenland Sea. *Deep-Sea Research*, **19**, 747–752.

CARR R. A., M. M. JONES and E. R. RUSS (1974) Anomalous mercury in near-bottom water of a Mid-Atlantic Rift valley. *Nature*, **251**, 489–490.

CARR R. A., M. M. JONES, T. B. WARNER, C. H. CHEEK and E. R. RUSS (1975) Variation in time of mercury anomalies at the Mid-Atlantic Ridge. *Nature*, **258**, 588–589.

CHAN K. M. and J. P. RILEY (1966) The determination of molybdenum in natural waters, silicates and biological materials. *Analytica Chimica Acta*, **36**, 220–229.

CHESTER R. and J. H. STONER (1974) The distribution of zinc, nickel, manganese, cadmium, copper, and iron in some surface waters from the World Ocean. *Marine Chemistry*, **2**, 17–32.

CHESTER R., D. GARDNER, J. P. RILEY and J. STONER (1973) Mercury in some surface waters of the World Ocean. *Marine Pollution Bulletin*, **4**, 28–29.

FITZGERALD R. A., D. C. GORDON, JR. and R. E. CRANSTON (1974) Total mercury in sea water in the northwest Atlantic Ocean. *Deep-Sea Research*, **21**, 139–144.

FITZGERALD W. F. and W. B. LYONS (1973) Organic mercury compounds in coastal waters. *Nature*, **242**, 452–453.

FOSTER P. and A. W. MORRIS (1971) The seasonal variation of dissolved ionic and organically associated copper in the Menai Straits. *Deep-Sea Research*, **18**, 231–236.

GARDNER D. and J. P. RILEY (1973) The distribution of dissolved mercury in the Irish Sea. *Nature*, **241**, 526–527.

GARDNER D. and J. P. RILEY (1974) Mercury in the Atlantic around Iceland. *Journal du Conseil. Conseil Permanent International pour l'Exploration de la Mer*, **35**, 202–204.

HEAD P. C. and J. D. BURTON (1970) Molybdenum in some ocean and estuarine waters. *Journal of the Marine Biological Association of the United Kingdom*, **50**, 439–448.

HOSOHARA K. (1961) Mercury content of deep-sea water. *Journal of the Chemical Society of Japan*, **82**, 1107–1108.

INTERNATIONAL DECADE OF OCEAN EXPLORATION (1972) Baseline studies of pollutants in the marine environment and research recommendations: the IDOE Baseline Conference, May 24–26, 1972, New York.

JOHNSON D. L. and R. S. BRAMAN (1975) The speciation of arsenic and the content of germanium and mercury in members of the pelagic *Sargassum* community. *Deep-Sea Research*, **22**, 503–507.

LEATHERLAND T. M. (1974) Analytical and environmental studies on the marine chemistry of mercury and some other trace metals. Thesis, Southampton University, 215 pp.

LEATHERLAND T. M. and J. D. BURTON (1974) The occurrence of some trace metals in coastal organisms with particular reference to the Solent region. *Journal of the Marine Biological Association of the United Kingdom*, **54**, 457–468.

LEATHERLAND T. M., J. D. BURTON, F. CULKIN, M. J. MCCARTNEY and R. J. MORRIS (1973) Concentrations of some trace metals in pelagic organisms and of mercury in Northeast Atlantic Ocean water. *Deep-Sea Research*, **20**, 679–685.

Leatherland T. M., J. D. Burton, M. J. McCartney and F. Culkin (1971) Mercury in North-eastern Atlantic Ocean waters. *Nature*, **232**, 112.

Millward G. E. and J. D. Burton (1975) Association of mercuric ions and humic acid in sodium chloride solution. *Marine Science Communications*, **1**, 15–26.

Morris A. W. (1975) Dissolved molybdenum and vanadium in the northeast Atlantic Ocean. *Deep-Sea Research*, **22**, 49–54.

Ólafsson J. (1974) Determination of nanogram quantities of mercury in sea water. *Analytica Chimica Acta*, **68**, 207–211.

Ólafsson J. (1975) Volcanic influence on seawater at Heimaey. *Nature*, **255**, 138–141.

Riley J. P. and D. Taylor (1968) Chelating resins for the concentration of trace elements from sea water and their use in conjunction with atomic absorption spectrophotometry. *Analytica Chimica Acta*, **40**, 479–485.

Riley J. P. and D. Taylor (1972) The concentrations of cadmium, copper, iron, manganese, molybdenum, nickel, vanadium and zinc in part of the tropical north-east Atlantic Ocean. *Deep-Sea Research*, **19**, 307–318.

Sournia A. and J. Citeau (1972) Sur la distribution du molybdène en mer et ses relations avec la production primaire. *Compte Rendu Hebdomadaire des Séances de l'Académie des Sciences*, **275**, 1299–1302.

Spencer D. W. and P. G. Brewer (1969) The distribution of copper, zinc and nickel in sea water of the Gulf of Maine and the Sargasso Sea. *Geochimica et Cosmochimica Acta*, **33**, 325–339.

Topping G. and J. M. Pirie (1972) Determination of inorganic mercury in natural waters. *Analytica Chimica Acta*, **62**, 200–203.

Weiss H. V., S. Yamamoto, T. E. Crozier and J. H. Mathewson (1972) Mercury: vertical distribution at two locations in the eastern tropical Pacific Ocean. *Environmental Science and Technology*, **6**, 644–645.

Williams P. M. and H. V. Weiss (1973) Mercury in the marine environment: concentration in sea water and in a pelagic food chain. *Journal of the Fisheries Research Board of Canada*, **30**, 293–295.

Williams P. M., K. J. Robertson, K. Chew and H. V. Weiss (1974) Mercury in the South Polar Seas and in the northeast Pacific Ocean. *Marine Chemistry*, **2**, 287–299.

Windom H. L., F. E. Taylor and E. M. Waiters (1975) Possible influence of atmospheric transport on the total mercury content of southeastern Atlantic Continental Shelf surface waters. *Deep-Sea Research*, **22**, 629–633.

Note on interpreting e-folding depths[*]

BRUCE A. WARREN

Woods Hole Oceanographic Institution, Woods Hole, Massachusetts 02543

Abstract—Exponentials are often fitted to observed distributions of conservative properties vs. depth in the deep ocean in order to estimate ratios of vertical diffusion coefficient to vertical velocity from the derived e-folding depths. This technique rests on an assumed balance between vertical advection and vertical diffusion of the property in question, and can give misleading results if lateral mixing is important in the property balance. Two illustrative examples are given: one, of horizontal variation in the e-folding depth despite a constant value for the above ratio, and the other, of a uniform value for the e-folding depth despite large horizontal variation in the ratio.

INTRODUCTION

IN TRYING to understand the distributions of conservative properties in the deep ocean, oceanographers commonly fit exponentials to observed property-depth curves, and interpret the derived e-folding depths as ratios of vertical diffusion coefficients (k) to vertical velocities (w), on the assumption of local balance between vertical advection and diffusion of the property in question (e.g., WYRTKI, 1962; MUNK, 1966; CRAIG, 1969, 1971). Since this balance requires that properties vary in the vertical as

$$\int dz \left[\exp \left\{ \int \frac{w}{k} dz \right\} \right],$$

it necessarily limits a strict exponential depth-dependence to layers in which w/k is constant. For instance, even where boundary effects are unimportant, as long as there is meridional flow, the planetary vorticity balance ($\beta v = f w_z$, usual notation) demands vertical shear in w, and therefore a somewhat different form to the property–depth curve. To be sure, unless w varies wildly with depth, the resulting curve will still 'look' exponential-like and, given the usual contamination of station data by transient events, one can probably achieve as good a fit to many data sets with exponentials as with any other form of curve. The point, however, is not the preciseness of the fit, but the lack of significance to fitting data by an inappropriate model.

When exponential fits are restricted to interior regions where the meridional flow (and hence the vertical shear in w) is small in some sense (perhaps certain oxygen-minimum layers, for instance), then the calculated e-folding depths may give useful information about the magnitudes of w and k—if, of course, the assumed vertical balance really holds. It is so very difficult to assess the distribution of vertical velocity in the ocean, and so important to do so to sort out large-scale circulation dynamics, that one is tempted to

---

[*] Contribution No. 3654 from the Woods Hole Oceanographic Institution. This paper was written with support from the U.S. Office of Naval Research under Contract N00014-74-C0262 NR 083-004.

go further: to suppose k constant, and use horizontal variations in e-folding depths to gauge regional variations in mean vertical velocity.

It has become fairly clear, however, that, even on an ocean-wide scale, some deep property distributions cannot be well rationalized without assuming the existence of lateral mixing strong enough to be parameterized with a mixing coefficient $K \sim 10^7 \, \mathrm{cm^2 \, s^{-1}}$ (e.g. KUO and VERONIS, 1973; NEEDLER and HEATH, 1975). One might suspect, therefore, that horizontal property variations brought about by variable w in the vertical balance model could be modified or even erased if lateral mixing were permitted. This is indeed the case, and the purpose of this note is to demonstrate the need for caution in interpreting e-folding depths by means of two simple illustrative examples: (i) a case of uniform w (and k) where lateral mixing nevertheless introduces a variation in apparent e-folding depth; and (ii) a case of variable w where lateral mixing maintains a uniform e-folding depth.

EXAMPLE I

Consider the distribution in some layer of a conservative property (ϕ) that varies in the vertical (z, positive upwards) and in one horizontal direction, which, for definiteness, take as the eastward (x) direction. For negligible horizontal velocity, and constant diffusion coefficients, the steady-state conservation equation is:

$$w\phi_z = k\phi_{zz} + K\phi_{xx}. \tag{1}$$

Let w be uniform in both the x and z directions. If the coördinates are scaled by $\zeta \equiv zw/k$ and $\xi \equiv x\pi/L$, where L is an imposed scale of horizontal variation, then (1) becomes:

$$\phi_\zeta = \phi_{\zeta\zeta} + \lambda\phi_{\xi\xi} \tag{2}$$

where $\lambda \equiv (\pi^2 kK)/(L^2 w^2)$. For k, $K = 1$ and $10^7 \, \mathrm{cm^2 \, s^{-1}}$, and $w = 10^{-5} \, \mathrm{cm \, s^{-1}}$ (a likely global average), then $\lambda = 4$ if $L = 5000 \, \mathrm{km}$, a typical oceanic dimension; significant effects of lateral mixing can therefore be anticipated even on such large scales, notwithstanding that horizontal property differences are usually substantially smaller than vertical differences. Let a periodic variation in ϕ be imposed at the top of the layer ($\zeta = 0$), of the form $\cos \xi$, where the scale L is now the half-wavelength of the variation. Then a solution of (2) that decays with depth and satisfies this condition is:

$$\phi = A + Be^\zeta (1 + \delta \cos \xi e^{\gamma\zeta}) \tag{3}$$

where A, B, δ are arbitrary constants, and $\gamma \equiv \frac{1}{2}[-1 + (1 + 4\lambda)^{\frac{1}{2}}]$.

The shape of the ϕ–ζ curve varies with ξ. If the depth-dependence had been strictly exponential, then the e-folding depth (d) could have been evaluated as:

$$d^{-1} = \frac{1}{\zeta}\ln\left[\frac{\phi(\xi, \zeta) - \phi(\xi, -\infty)}{\phi(\xi, 0) - \phi(\xi, -\infty)}\right].$$

The same quantity for ϕ as given by (3) is:

$$d^{-1} = \frac{1}{\zeta}\ln\left[\frac{e^\zeta(1 + \delta \cos \xi \, e^{\gamma\zeta})}{1 + \delta \cos \xi}\right].$$

If $\delta \ll 1$ and $\gamma\zeta \ll 1$ (equivalent to $\lambda \ll 1$, whereby $\gamma \approx \lambda$), then $d^{-1} \approx 1 + \gamma\delta \cos \xi$. Thus in this approximation the solution behaves exponentially in the vertical with an e-folding depth that varies slightly in the horizontal; but it varies this way in consequence of the lateral mixing, and not because of any variation in vertical velocity. The local value of the dimensional e-folding depth is not therefore a good measure of w, although its average over a distance L would give the correct value. For the ocean, the approximation $\delta \ll 1$ is appropriate, since horizontal differences are generally much smaller than vertical ones, but taking $\gamma\zeta \ll 1$ is not realistic (e.g. for $\lambda = 4$, $\gamma = 1.56$). The inexactness of most curve-fitting, however, is such that one could probably not distinguish a curve of the form (3) with $\gamma\zeta \sim 1$ from a simple exponential with ξ-varying amplitude and e-folding depth. These qualitative conclusions are not limited, of course, to the specific sinusoidal form of variation imposed on ϕ, or to horizontal diffusion in only one direction.

At the other extreme, if $\gamma \gg 1$ ($\lambda \gg 1$) the horizontal variation is entirely confined to a thin boundary layer of thickness $1/\sqrt{\lambda}$ [dimensional thickness $(L/\pi)(k/K)^{\frac{1}{2}}$] near $\zeta = 0$, the essential balance there being between the two diffusive terms in (2). Below the boundary layer the dimensional e-folding depth has the constant value k/w. In this limit, features where horizontal diffusion is important are of such small horizontal and vertical dimensions as not to be of interest, at least in the present context.

EXAMPLE II

Another way to estimate time-mean vertical velocities away from boundaries is to integrate the planetary vorticity balance vertically, assume zero vertical velocity at the bottom, and use the total meridional volume transport per unit zonal distance below a certain level as a measure of the vertical velocity at that level. An attempt to do so is illustrated in Fig. 1: the upper panel gives geostrophic estimates of meridional transport per unit distance below 2000 m for each station interval along the *Scorpio* section at lat. 28°S in the interior of the southwest Pacific Basin. Station 141 is near (but east of) the deep western boundary current of the South Pacific, and Sta. 117 is near the summit of the East Pacific Rise. For reasons given by WARREN (1973), transports are estimated assuming negligible horizontal velocity at 2000 m depth, and the corresponding vertical velocities at 2000 m are to be obtained by multiplying the transport values by $\beta/f = 2.9 \times 10^{-9} \, \text{cm}^{-1}$ at lat. 28°15'S.

As is customary with such estimates, there is a good deal of unevenness from one station pair to the next, which cannot be taken as characteristic of the mean field: the problem is not one of observational 'error', but of interpretation, in that transient disturbances registered in the instantaneous density data at a particular station lead to an erroneously high estimate of the mean transport between the one station pair of which it is a member, and an erroneously low value between the other pair. Smoothing needs to be done, but there is no very satisfactory method for doing so when, as here, the estimates for single station pairs are comparable to, or much greater than, the large-scale mean. In the middle panel of Fig. 1, the transport values are recast as averages over successive groups of four station pairs, and the unevenness is much reduced, although it is not clear to what extent the resulting pattern may be an artifact of the crude smoothing method. In this presentation there seem to be broad zones of predominantly northward and southward flow—southward between Stas. 133–141 and 117–125, north-

ward between Stas. 125–133—that suggest regions of upward and downward motion
at 2000 m, but one cannot have much confidence in the zone width of 1500 km that
emerges through this averaging. The interpretation of the transports in terms of vertical
velocities is further uncertain, moreover, because the ocean bottom is not all flat here.
Stations 117–135, in fact, run down the flank of the east Pacific Rise, and it would not
require a very large cross-isobath flow to produce vertical velocities at the bottom which
are comparable to those indicated by the averaged meridional transports. Nevertheless,
it seems worthwhile to explore the implications for property-depth curves of a variation
in w such as suggested here.

The 2000 m level is close to the core of the oxygen-minimum layer in the central South
Pacific, and there is some reason to suppose that there is a depth interval here where w
can be approximated as constant in the vertical. The bottom panel of Fig. 1 gives rough
estimates of e-folding depths near 2000 m for *Scorpio* Stas. 117–141, as calculated simply
from exponential fits to interpolated values of potential temperature at 1500 m, 2000 m
and 2500 m. There is considerable scatter in these estimates, and there may be a small
eastward trend in values discernible above the noise level, but overall the variation is not
more than 50%. Certainly there is no correlation with the transport distribution in the
middle panel and, in particular, there is no change in sign to match the reversal of

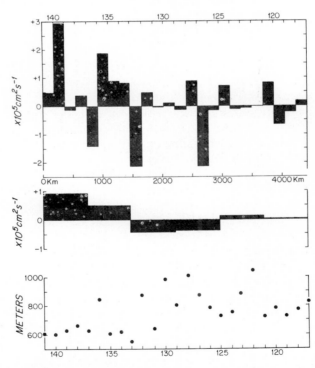

Fig. 1.   Upper panel: estimated meridional geostrophic transport per unit zonal distance below 2000 m along
lat. 28°S in the central South Pacific, *Scorpio* Stas. 117–141, 22 June–8 July 1967; positive values indicate
poleward (southward) flow, negative values, equatorward (northward) flow. Middle panel: same estimates as
above averaged over successive groups of four station pairs. Lower panel: e-folding depths calculated from
exponential fits to interpolated potential temperatures at 1500, 2000 and 2500 m for *Scorpio* Stas. 117–141.
Station numbers along top and bottom, distance scale between upper and middle panels.

net flow at Stas. 125–133. If the transport pattern does indeed represent variations in vertical velocity, then the two sets of calculations are totally inconsistent in terms of the one-dimensional balance between vertical advection and vertical diffusion. The following simple model, however, illustrates how addition of lateral mixing to the balance might reconcile them.

In (1) let the constant w be changed to

$$w = w_0 \left(1 + r \cos \frac{\pi x}{L}\right)$$

where w_0 is a constant, and the scale L is now the half-wavelength of the variation in w. With λ and the scaling of z defined in terms of w_0 rather than w, (2) becomes:

$$(1 + r \cos \xi)\Phi_\zeta = \phi_{\zeta\zeta} + \lambda\phi_{\xi\xi}. \tag{4}$$

Consistent with the rough uniformity of e-folding depths in Fig. 1, (4) admits separable solutions of the form $\phi = A + B\psi(\xi)\exp(\alpha\zeta)$, with α independent of ξ, and A, B undetermined constants. As in the previous example, a possible term linear in ξ and independent of ζ has been disregarded. The equation satisfied by $\psi(\xi)$ is:

$$\psi'' + (a - b \cos \xi)\psi = 0, \tag{5}$$

where $a \equiv (\alpha^2 - \alpha)/\lambda$ and $b \equiv r\alpha/\lambda$. This is Mathieu's equation, and only its periodic solutions need be considered in this example, because imposed conditions will be assumed to be symmetric in ξ. A solution leading to horizontal differences which are small compared with vertical differences (as in the ocean) can be obtained by taking b small and expanding a and $\psi(\xi)$ in powers of b: $a = a_0 + a_1 b + a_2 b^2 + \ldots$, and $\psi(\xi) = \psi_0(\xi) + b\psi_1(\xi) + b^2\psi_2(\xi) + \ldots$. Substituting into (5) and setting to zero the sums of coefficients of like powers of b yields, to order b^2:

$$\psi(\xi) = 1 + (1 - \cos \xi)b + (\tfrac{1}{8}\cos 2\xi - \cos \xi + \tfrac{7}{8})b^2, \qquad a = -\tfrac{1}{2}b^2, \tag{6}$$

where constants in the integrations for ψ_0, ψ_1, ψ_2 have either been set $= 0$ to keep the solution periodic, or have been chosen to normalize $\psi(\xi)$ such that $\psi(0) = 1$ for all b; a_1 was set $= 0$ to keep the solution periodic, and a_0 to zero to make variations in ψ of order b rather than order one.

The definitions of a and b, combined with the derived relation between them, specify the e-folding depth $\alpha^{-1} = 1 + r^2/(2\lambda)$. For $r \lesssim 1$ and $\lambda \ll 1$ (large horizontal mixing, tending to smooth out variations), or $\lambda \gtrsim 1$ and $r \ll 1$ (only slight variation in w), its dimensional value (k/w_0) is given by the mean vertical velocity, as would be expected. On the other hand, if $r \gg 1$, and $\lambda \sim 1$, the vertical scale must be large in order that the advective term in (4) be small enough to be comparable with the horizontal diffusive term. Since this necessarily makes the vertical diffusive term much smaller than the advective term, it reduces (4) to a balance between advection and horizontal diffusion, which leads (by $a = -\tfrac{1}{2}b^2$ above) to $\alpha^{-1} = r^2/(2\lambda) \gg 1$, consistent at this approximation with the general expression for α^{-1}. By a similar analysis one can derive the same balance and the same value of α^{-1} for the case $\lambda \ll 1$, $r \sim 1$, although to have the approximate solution (6) valid in this instance, it is necessary that r be sufficiently large to keep b small. These latter results are discouraging for attempts to interpret observed e-folding depths as values of k/w (however averaged), because they represent a balance of terms different from vertical diffusion and advection, a balance which requires e-folding depths that are un-related to k/w, and are, in fact, much larger.

It might be noted that the case $\lambda \ll 1$ discussed here, although appropriate to small horizontal diffusion, can *not* be derived as a small-K perturbation to a field satisfying the vertical advection–diffusion balance. That perturbed field would consist of an interior region where the vertical balance holds [with vertical scale given by $k/w(x)$], completed by thin lateral boundary layers where horizontal diffusion is important. The present example is of a different kind of solution to (1), in which the vertical scale is so much greater than k/w that horizontal diffusion, however small, is important throughout the fluid, and is balanced only by the vertical advection except, perhaps, in top and bottom boundary layers.

The distribution given in the middle panel of Fig. 1 probably corresponds to none of the extremes above, but to $\lambda \sim 1$, $r \sim 1$. The magnitudes and scales are not at all well determined there, but, for what they are worth, the average estimated volume transport per unit distance between *Scorpio* Stas. 125 and 141 is $0.14 \times 10^5 \text{ cm}^2 \text{ s}^{-1}$ poleward, equivalent to a mean vertical velocity $w_0 = 4 \times 10^{-5} \text{ cm s}^{-1}$ (upward); and they suggest $r = 4$, $L = 1500 \text{ km}$. Estimates of values for the diffusion coefficients are even less satisfactory, of course, but with $K = 10^7$ and $k = 1 \text{ cm}^2 \text{ s}^{-1}$, $\lambda = 2.78$, $\alpha^{-1} = 3.88$, $b = 0.37$ (not really $\ll 1$), and the dimensional e-folding depth $k/(w_0 \alpha) = 970 \text{ m}$, which is somewhat too large (lower panel, Fig. 1). On the other hand, $k/w_0 = 250 \text{ m}$, which is certainly too small. The agreement is not good enough to be persuasive [which is not very surprising, given, among other things, the omission from (4) of meridional diffusion], but it is suggestive of how a fairly uniform e-folding depth (lower panel, Fig. 1) might be reconciled with highly varying vertical velocities (middle and upper panels) through consideration of lateral mixing effects.* The example shows, moreover, that a corrugated field of vertical velocity can lead to a vertical scale that is related as much to the characteristics of the variation as to the regional mean value; and since the deep geostrophic flow is channeled by basins, ridges and passages, one can readily imagine an associated distribution of vertical velocity having scales of variation that could significantly affect property balances and the required e-folding depths.

CONCLUSION

It is difficult to construct positive guidelines for the interpretation of e-folding depths on the basis of these two examples. Rather, the conclusion to be drawn from them is largely negative: that, for realistic conditions in the deep ocean, use of the vertical advection–diffusion balance in making such interpretations can give misleading results, both by fabricating variations in vertical velocity that do not exist, and by leaving unseen some large variations that do exist. Obviously there are many circumstances in which the vertical balance is a useful tool, but its application needs to be examined in the context of the regional property distributions before its product can really be accepted with confidence.

---

*The agreement with observation on the particular *Scorpio* section actually looks much worse, in that (6) requires potential temperature (say) to increase eastward, whereas in fact it decreases (e.g. WARREN, 1973); this solution to (4), however, could be supplemented by a term linear in ξ and independent of ζ that would do much to correct the large-scale trend.

REFERENCES

CRAIG H. (1969) Abyssal carbon and radiocarbon in the Pacific. *Journal of Geophysical Research*, **74**(23), 5491–5506.

CRAIG H. (1971) Son of abyssal carbon. *Journal of Geophysical Research*, **76**(21), 5133–5139.

KUO H. H. and G. VERONIS (1973) The use of oxygen as a test for an abyssal circulation model. *Deep-Sea Research*, **20**(10), 871–888.

MUNK W. H. (1966) Abyssal recipes. *Deep-Sea Research*, **13**(4), 707–730.

NEEDLER G. T. and R. A. HEATH (1975) Diffusion coefficients calculated from the Mediterranean salinity anomaly in the North Atlantic Ocean. *Journal of Physical Oceanography*, **5**(1), 173–182.

WARREN B. A. (1973) Transpacific hydrographic sections at Lats. 43°S and 28°S: the SCORPIO Expedition-II. Deep water. *Deep-Sea Research*, **20**(1), 9–38.

WYRTKI K. (1962) The oxygen minima in relation to ocean circulation. *Deep-Sea Research*, **9**(1), 11–23.

An attempt to test the geostrophic balance using Minimode current measurements*

J. C. SWALLOW

Institute of Oceanographic Sciences, Wormley, Godalming, Surrey, U.K.

Abstract—Some closely spaced float trajectories and other observations made during the Mid-Ocean Dynamics Experiment in 1973 have been combined in an attempt to test the geostrophic balance in relatively weak mid-ocean currents. No significant departures from geostrophy could be resolved. In the most favourable conditions, r.m.s. errors of about $\pm 0.7\,\mathrm{cm\,s^{-1}}$ remained in the estimates of observed and geostrophic current difference through the main thermocline. At the same time, the observed accelerations could have accounted for a difference, observed minus geostrophic, of $0.2\,\mathrm{cm\,s^{-1}}$, well below the noise level attained in the first-order balance. Some intercomparisons of 4-day mean velocities measured with floats and moored current meters showed close agreement, discrepancies being $0.5\,\mathrm{cm\,s^{-1}}$ at 500 m and $0.4\,\mathrm{cm\,s^{-1}}$ at 2900 m. In the deep water (1600–2900 m) again no significant departure from geostrophy could be detected, with a noise level of about $\pm 0.3\,\mathrm{cm\,s^{-1}}$.

INTRODUCTION

LARGE-SCALE slowly varying ocean currents are believed to be nearly geostrophic; that is, to a good approximation there is a balance between the horizontal pressure gradient and the Coriolis force acting on a unit volume of water. For a readable account of the geostrophic relationship, and the conditions in which it can be expected to apply to the ocean, see Chapter 3 of STOMMEL's (1965) book on the Gulf Stream. The classical test of this relationship was made by WÜST (1924) using measurements in the Florida Current. In most of the ocean, where currents are much weaker, geostrophy is often assumed, although less well substantiated by direct observation. The mid-ocean currents observed in the *Aries* cruises (CREASE, 1962), though strikingly variable, were sufficiently large in scale and long in period that they could be expected to be nearly geostrophic. Typical length and time scales were tens to hundreds of kilometers, and several weeks. Where tests could be made, directly observed differences of current did indeed agree with geostrophic differences, calculated from the observed density distribution, within the estimated instrumental accuracy (SWALLOW, 1971). However, the accuracy was low, no better than $\pm 1\,\mathrm{cm\,s^{-1}}$ even in the deep water, and the density data were insufficient to allow all the likely sources of error to be evaluated. With those *Aries* data, no satisfactory geostrophic test could be made across the main thermocline, where usually the largest differences of current were found.

One aim of the Mid-Ocean Dynamics Experiment (MODE-1) in March–June 1973 was to improve the accuracy of comparisons of observed and geostrophic current differences, ideally to allow real departures from geostrophy to be measured in mid-ocean eddies.

*MODE Contribution Number 75 (MODE-1).

Several different approaches were tried, and the results have been summarized by Hogg (1975). None of them, including this one, has achieved much better than 10% accuracy in the first-order balance between the pressure gradient and the Coriolis force per unit volume, whereas estimates of the magnitude of departures from geostrophy, from the observed accelerations, are of order 1%. Even so, because the geostrophic relationship is widely used in studying ocean circulation, it seems worthwhile to show how close a test can be made with these data, and to see which sources of error are important. In addition, these data allow some close intercomparisons to be made between velocities measured by moored current meters and neutrally buoyant floats.

In one of the above-mentioned approaches, recently described in more detail, Bryden (in press) compared differences of current observed with moored current meters, through the main thermocline, to differences of geostrophic current calculated from temperatures recorded on the same moorings. The relatively long duration of the records allowed thirty-two comparisons to be made between 4-day averages of observed and geostrophic currents. By time-averaging the temperature records, the effects of internal waves on the calculated geostrophic currents could be removed. However, errors were incurred through the need to assume linear variation of velocity between moorings 50 to 100 km apart, the assumption of an exact temperature–salinity relationship so that observed temperatures could be converted into densities, and the need to use trapezoidal integration between these density values spaced vertically some hundreds of metres apart, in calculating dynamic heights.

Horton and Sturges (unpublished manuscript) have compared the differences of currents at a single mooring with the differences in geostrophic current in the same depth intervals calculated from smoothed dynamic heights at eleven of the station positions in the MODE density survey. The chosen stations were spaced around the mooring position at radial distances of 35 to 70 km. At each station, vertical profiles of temperature and salinity were taken once every 12 days, on average, for $2\frac{1}{2}$ months. This comparison avoided the disadvantages of the coarse trapezoidal integration for dynamic heights and the assumption of an exact T–S relationship, but at the cost of less effective removal of internal waves and heavy dependence on the currents at a single mooring.

The observations used here are relatively short but more closely spaced. During MODE-1, groups of neutrally buoyant floats were tracked for two periods of 3 weeks each, using the 'Minimode' float tracking system (Swallow, McCartney and Millard, 1974). The floats were mainly at four levels, nominally 500, 1500, 3000 and 4000 m, four floats at each level, the initial spacings being 5, 10 and 20 km between neighbours. Hydrographic stations using a CTD probe (Brown, 1974) and Rosette multisampler were occupied simultaneously with the float tracking. The set of observations made in April 1973 over the abyssal plain near 28°N, 70°10′W are more concentrated spatially than those made in May, in a region of rougher topography to the east, and are more suitable for the present purpose.

The first tests described below use portions of float trajectories at nominal depths of 500 and 1500 m, together with density data collected in the same area. The most serious weakness of this test is the sparseness and uneven spatial distribution of the density data. The stations were not repeated often enough to permit effective smoothing of the displacements of density surfaces due to internal waves. This situation is improved by using dynamic heights derived from the temperature records at two moorings, as Bryden (in press) did, but correcting them by reference to neighbouring CTD stations. Currents recorded by meters in the same moorings are used in addition to those from the float trajectories, to give as complete a direct measure as possible of the components of velocity between the two chosen moorings. This test is slightly more accurate than the first, and approaches the limits set by uncertainties in the mooring positions and even in the equation of state of seawater. Finally, a geostrophic test is made in the deep water using velocities from floats at nominal depths of 1500 and 3000 m and densities from nearby CTD stations.

GEOSTROPHIC TESTS IN THE MAIN THERMOCLINE

For a first attempt at a geostrophic test, one may take the marked 4-day portions of trajectories in Fig. 1, one at each of the nominal depths of 500 and 1500 m, which nearly coincide in time and horizontal mid-position, and compare the observed velocity difference with that deduced from fitting a plane to the appropriate dynamic height differences at

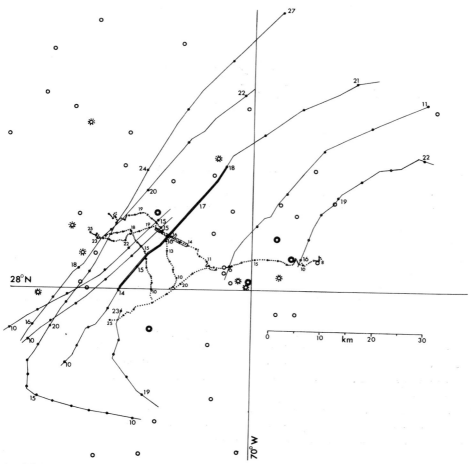

Fig. 1.   Minimode float trajectories at nominal depths of 500 m (solid lines) and 1500 m (dotted) observed in April 1973. The numbered points are interpolated noon GMT positions. The two thickened portions of trajectories are referred to in the text. The open circles are CTD station positions.

neighbouring CTD stations. The actual mean depth attained by the '500 m' floats was 580 m. From the CTD stations, lists of dynamic heights are available at 100 decibar intervals of pressure. The error incurred by using dynamic height differences between 600 and 1600 dbar (approximately 594 and 1582 m equivalent depth), instead of values corresponding to the mean float depths, is small compared to those from other sources.

In the chosen 4-day period, 14–18 April, of the fifteen CTD stations occupied only five (marked with double circles in Fig. 1) were within 30 km of the mid-points of the float trajectories. Fitting a plane to their dynamic height differences (600–1600 dbar) leads to estimates $\delta u_g = 12.4\,(\pm 2.4)\,\mathrm{cm\,s^{-1}}$, $\delta v_g = 9.8\,(\pm 1.6)\,\mathrm{cm\,s^{-1}}$, for the east and north components of geostrophic current difference. These are plotted at (a) in Fig. 2. From the chosen sections of float trajectories, the observed values are $\delta u = 7.8\,\mathrm{cm\,s^{-1}}$, $\delta v = 6.1\,\mathrm{cm\,s^{-1}}$.

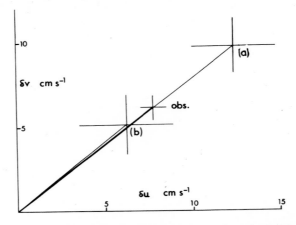

Fig. 2.   Observed difference of current for the marked portions of trajectories in Fig. 1, geostrophic current difference using five stations (a), geostrophic current difference using thirteen stations (b).

Their r.m.s. error is estimated to be approximately $\pm 0.6\,\mathrm{cm\,s^{-1}}$, made up as follows. The velocities are calculated from interpolated noon GMT positions for the floats. Actual fixes were taken at irregular intervals, whenever possible, the average interval between fixes being 17.4 hours for the trajectories in Fig. 1. About one-third of the fixes were done on close passes and have a position accuracy of $\pm 0.3\,\mathrm{km}$, the remainder are long range fixes with estimated accuracies of $\pm 1\,\mathrm{km}$ at 500 m nominal depth and $\pm 0.5\,\mathrm{km}$ at 1500 m. For details of actual fix positions and times, refer to two data reports by Caston, Strudwick and Swallow (1974 a,b). Since the float fixes were too infrequent for suitable low-pass filtering to be applied, a further error arises due to parts of inertial, tidal and other oscillations remaining in the observed displacement—about 1 km at 500 m and 0.5 km at 1500 m.

There is a large disagreement between the magnitudes of these observed and geostrophic current differences (Fig. 2). One possible source of the discrepancy is that the five CTD stations used are not well distributed around the region occupied by the float tracks—their centroid is 12 km southeast of the mid-position of the sections of trajectory used. A better distribution can be got by accepting CTD stations taken in a longer time interval. There are eight more stations that are usable (marked with dotted circles in Fig. 1) in the 12-day interval centred on 16 April. Using all thirteen stations gives the result marked (b) in Fig. 2, for the geostrophic current difference, $\delta u_g = 6.2\ (\pm 2.7)\ \mathrm{cm\,s^{-1}}$, $\delta v_g = 5.2\ (\pm 1.8)$ $\mathrm{cm\,s^{-1}}$, which agrees more closely with the observed difference though the estimated errors are even bigger than before.

From the spatial differences in velocity that can be seen in the float trajectories, it is clearly not satisfactory simply to fit a plane to the dynamic heights, implying a spatially constant velocity difference, even within a radius of 30 km, if we want to make close comparisons of velocities, within $1\ \mathrm{cm\,s^{-1}}$ or better. It seems more appropriate to fit a second-order surface to the dynamic height distribution and derive a geostrophic velocity difference that varies linearly in the horizontal. For comparison, linear gradients of velocity have been fitted to the portions of float trajectories shown in Fig. 3. To cover

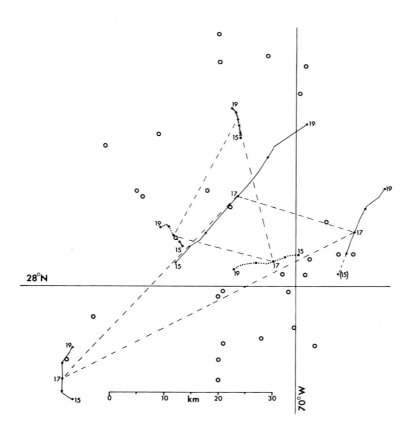

Fig. 3.   Portions of float trajectories at nominal depths of 500 m (solid lines) and 1500 m (dotted) used for fitting velocity gradients. The northernmost 1500 m trajectory is from a Sofar float. Open circles indicate adjusted CTD station positions.

the most suitable 4-day interval, 15–19 April, one trajectory at '500 m' had to be extrapolated back for 18 hours. At the '1500-m' level, the Minimode floats were not well distributed for fitting velocity gradients. Fortunately, one of the MODE Sofar floats (ROSSBY, VOORHIS and WEBB, 1975) was being tracked at nearly the same depth just to the north of them, and a 4-day portion of its trajectory has been used. The resulting expressions for the east and north components of observed velocity difference are:

$$\delta u = (4.93 - 0.0165x + 0.2323y)\,\mathrm{cm\,s^{-1}},$$
$$\delta v = (4.60 + 0.0068x + 0.1448y)\,\mathrm{cm\,s^{-1}},$$

where x and y are measured in kilometres east and north of 28°N, 70°10′W.

A comparable expression for the geostrophic current differences has been derived from the dynamic heights at all the usable CTD stations occupied in the whole observing period, 6 to 27 April. Since the current pattern was changing with time, as can be inferred from the relatively high speed of the earliest '500 m' float trajectory in Fig. 1, some caution is needed in using stations much earlier or later than the chosen current measurements. From other observations on a larger scale in MODE, it appeared that the pattern of currents and density distribution in the thermocline maintained a fairly constant shape whilst moving generally southwestward, during April, through the area with which we are concerned here. The CTD station positions were therefore adjusted to synchronize them with the current measurements, using a mean pattern velocity of 2 km per day towards 220°, inferred from a series of maps of isotherms derived from moored temperature recorders (personal communication from Dr. N. HOGG). After adjustment, the twenty-nine stations within 40 km of 28°06′N, 70°07′W, shown in Fig. 3, were considered usable. The resulting expressions for geostrophic velocity differences are:

$$\delta u_g = (6.77 - 0.0955x + 0.0907y)\,\mathrm{cm\,s^{-1}},$$
$$\delta v_g = (3.30 + 0.2358x + 0.0955y)\,\mathrm{cm\,s^{-1}}.$$

These are compared with the directly observed current differences in Fig. 4. In the central part of the area of observations there is fairly good agreement, within 2 cm s^{-1} in each component, but differences increase rapidly beyond 10 km radius. The errors in the

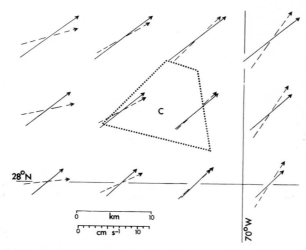

Fig. 4. Interpolated values of observed current difference (solid arrows) and geostrophic current difference, 600–1600 dbar (dashed arrows), inferred from the trajectories and CTD stations in Fig. 3. The region bounded by dotted lines is that common to the two triangles marked in Fig. 3, where the interpolated values of observed currents should be most accurate. C is the centroid of the CTD station positions.

observed velocity differences, within the central area, are approximately the same as were estimated for the previous example, i.e. $\pm 0.6\,\text{cm}\,\text{s}^{-1}$. For the geostrophic current differences, fitting a second-order surface gives very little improvement. The r.m.s. residual in dynamic height difference at a single station is ± 6 dyn. mm for either a quadratic or plane fit, equivalent to $\pm 0.9\,\text{cm}\,\text{s}^{-1}$ in δu_g and $\pm 1.3\,\text{cm}\,\text{s}^{-1}$ in δv_g.

This is consistent with the noise in the dynamic heights being mainly in fluctuations of short period or small horizontal scale. Looking at the distribution of dynamic height differences, 600–1600 dbar, the r.m.s. difference between values observed at close pairs of stations (separated by less than 10 km, and less than 5 days) is 6.7 dyn. mm. As in all the preceding cases, the dynamic heights used are mean values from the 'up' and 'down' CTD records on each station. Not much noise has been removed by this averaging, though; the r.m.s. 'up–down' difference (600–1600 dbar) is 4.5 dyn. mm, and the mean difference 0.9 dyn. mm, with a 'down–up' time interval of about 2 hours for these dynamic heights on most of these stations.

To reduce the effect of this noise in the dynamic heights, and to avoid the uncertainties involved in using data from CTD stations spread over a long period of time, in the next test the dynamic heights from filtered time-series of temperature and pressure at moorings are used [as did BRYDEN (in press)] but systematic errors are corrected by reference to neighbouring CTD stations. Data from at least two moorings are needed; the ones that have most neighbouring CTD stations are moorings 1 and 3. For directly observed components of current through the section defined by these two moorings there are, in addition to the Minimode floats near mooring 3, current meter records from depths of 391 m in mooring 1 and from 427 m and 728 m in mooring 3, and two more Sofar float trajectories not far from mooring 1 (Fig. 5). It is convenient in this case to correct the observed currents

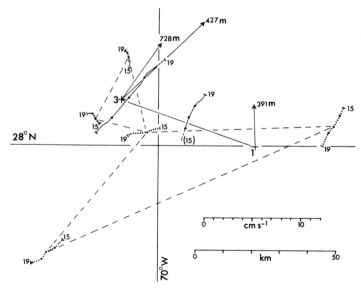

Fig. 5. Currents observed from 15 to 19 April 1973, used for comparison with geostrophic current, 500–1500 dbar, between moorings 1 and 3. There are two float trajectories at 500 m nominal depth (solid lines), two Minimode and three Sofar float trajectories at 1500 m (dotted) and three 4-day mean current vectors from moored current meters (solid arrows). The speed scale applies equally to the length of the arrows and to 4-day displacements along the trajectories.

to mean depths of 500 m and 1500 m, and to evaluate geostrophic current differences for comparison in the pressure interval 500–1500 dbar.

Daily mean dynamic height differences were provided by Mr. J. RICHMAN (personal communication), derived from the records of temperature and pressure at each mooring. Successive daily values rarely differ by more than 1 or 2 dyn. mm. Mooring 3 was near the centre of the float tracking work in April 1973, and eleven CTD stations were occupied within 15 km of its position, during 20 days. The regional gradient of dynamic height difference near the mooring was determined by fitting a plane to the differences between CTD station dynamic heights and the appropriate daily mean values at the mooring. This regional gradient was then used to transfer the CTD station dynamic heights to the mooring position. The mean correction at mooring 3 was −40.6 dyn. mm, the standard deviation of the mean being ±1.5 dyn. mm. There was no evidence for significant change in the correction through the 20-day period. At mooring 1, the mean correction was −14.0 (±1.2) dyn. mm based on forty-one CTD stations within 15 km. These rather large and differing corrections come about because of the necessarily large vertical spacing of the temperature records, different at the two moorings, between which the daily mean dynamic heights for the pressure interval 500–1500 dbar had to be calculated by linear interpolation. At mooring 3, temperature records were available from instruments at 427, 531, 728, 915, 1428 and 3948 m, with those at 531, 915 and 3948 m also recording pressure. Mooring 1 was more plentifully instrumented, providing temperature and pressure records at 491, 697, 898, 1095 and 1896 m, and additional temperature records at 691 and 1392 m.

In this geostrophic test, 4-day mean values of corrected dynamic height difference at each mooring are used, centred on 17 April. For the directly observed currents near 500 m, the small corrections needed to transfer them to that depth are based on the local mean geostrophic profile near each mooring; at 1500 m, where the vertical gradient of velocity is small, no depth corrections were made. At 500 m, the components of velocity from the float trajectories were then transferred to the section by projecting perpendicularly from their mid-positions onto the line joining the positions of moorings 1 and 3. At 1500 m, the current at mooring 3 was estimated by linear interpolation between the three float trajectories enclosing it; at the other end of the section, the linear gradient fitted to the velocities at the three easternmost float trajectories is adopted in that part of the section falling within the triangle formed by their mid-positions (Fig. 5).

For each of the observed components of velocity at 500 m (Fig. 6) the r.m.s. error is estimated to be ±0.6 cm s^{-1}, made up from ±0.5 cm s^{-1} for the observed value and

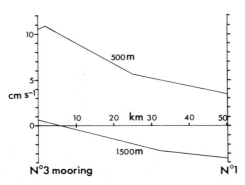

Fig. 6.   Observed components of current between moorings 1 and 3, 15–19 April 1973.

± 0.3 cm s^{-1} for the depth correction. For each observed component at 1500 m, the estimated r.m.s. error is ± 0.4 cm s^{-1}. For the mean difference, 500 m–1500 m, in observed components of current it is assumed that the r.m.s. error is no less than that for the difference between a single pair of observations, i.e. ± 0.7 cm s^{-1}.

Since the daily mean dynamic heights at the moorings vary smoothly through the 4-day period, with a noise level less then 1 dyn. mm, it is assumed that the main contribution to error in the geostrophic current differences comes from the uncertainty in the corrections derived from neighbouring CTD stations, i.e. ± 1.9 dyn. mm r.m.s., equivalent to ± 0.6 cm s^{-1}.

The 4-day mean differences of current through the section are then

$$\delta v \text{ (observed)} = 8.3 \, (\pm 0.7) \text{ cm s}^{-1},$$

$$\delta v \text{ (geostrophic)} = 7.9 \, (\pm 0.6) \text{ cm s}^{-1}.$$

Thus, there is no evidence for departure from geostrophy, within the accuracy of these observations.

Relative to these limits, though, the observed accelerations are insignificant. They have been calculated along the float trajectories, from the change in 2-day mean velocity during 4-day intervals. At 500 m, the component along the section is approximately 1.1 cm s^{-1} per day, towards mooring 1. At 1500 m the component is negligible, 0.1 cm s^{-1} per day towards mooring 3. Divided by the Coriolis parameter to convert these accelerations into velocity units for comparison, their difference is equivalent to 0.2 cm s^{-1}, in the direction to reduce the gap between the observed and geostrophic components listed above.

COMPARISON OF MEAN VELOCITIES FROM FLOATS AND CURRENT METERS

It may have been noticed that the '500-m' float that passed within 2 km of mooring 3 had a component of velocity agreeing fairly well with the one obtained from the current meters in the mooring (Fig. 6). These observations are compared more clearly in Fig. 7, for two different averaging times, taken between actual fixes on the float and chosen to be nearly symmetrical about the mooring position. The geostrophic profile derived from CTD stations within 15 km of mooring 3 is used for interpolation between the current meter

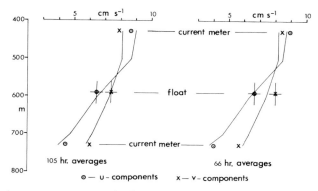

Fig. 7.   Comparison between components of velocity of a float at 590 m and those at two current meters in No. 3 mooring, using the local geostrophic profile for interpolation.

velocities. Averaged over 66 hours, velocity differences (float minus current meter) were $\delta u = +0.1$, $\delta v = +0.5\,\mathrm{cm\,s}^{-1}$. Averaged over 105 hours, the differences were $\delta u = -0.3$, $\delta v = 0.0\,\mathrm{cm\,s}^{-1}$. Uncertainties in fixing the float and determining its depth $(590\pm30\,\mathrm{m})$ would account for $\pm0.5\,\mathrm{cm\,s}^{-1}$, so the observed differences are not significant.

A close comparison between float and current meter velocities can be made at a deeper level, where depth differences may be less serious since the vertical gradient of mean velocity appeared to be much smaller than in the main thermocline. Four floats were tracked at a mean depth of 2865 (±50) m, within 30 km radius of mooring 3, which had a current meter at 2945 m depth. Mean currents and linear gradients have been fitted to successive sets of 4-day portions of the three trajectories drawn with solid lines in Fig. 8, and a current vector interpolated at the mooring position for comparison with that from the current meter, for each of the 4-day periods. The r.m.s. magnitude of the vector difference was $0.4\,\mathrm{cm\,s}^{-1}$, which is not significantly greater than the $0.3\,\mathrm{cm\,s}^{-1}$ obtained in testing the linear fit on the remaining float trajectory, shown dotted in Fig. 8. The current meters were Vector Averaging Current Meters (VACMs) in subsurface moorings. It is particularly reassuring to find such close agreement in these cases, since more than half of the VACM records collected during MODE were at least partly degraded by mineral deposition in the sensor bearings (DEXTER, MILLIMAN and SCHMITZ, 1975). The current meter records used here were ones identified as 'good' from internal evidence, which seems to have been an effective test of quality, at least for low-frequency currents.

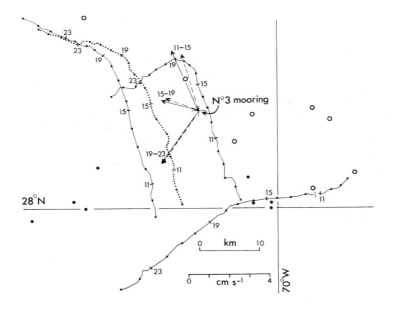

Fig. 8.   Float trajectories at 2865-m depth observed in April 1973, marked with interpolated noon GMT positions. The solid arrows at No. 3 mooring position represent 4-day mean vectors from the current meter at 2945 m depth. Dashed arrows represent vectors interpolated at the mooring position using the three float trajectories drawn with solid lines, for corresponding 4-day periods. The speed scale applies equally to the length of the arrows and to 4-day displacements along the trajectories. CTD stations are marked by circles. Those marked with solid circles form the two groups referred to in the text.

A GEOSTROPHIC TEST IN THE DEEP WATER

Using the linear velocity distributions fitted to 4-day sections of float trajectories at mean depths of 1585 m and 2865 m, differences of observed current can be obtained for each of three 4-day periods, centred on 13, 17 and 21 April. There is another set of float trajectories, at a mean depth of 3770 m, but the CTD stations occupied during the float tracking were not taken deeper than 3000 m because of winch limitations. At some stations, deeper casts of water bottles were taken, but not often enough for a critical geostrophic test to be made. In the pressure interval 1600–2900 dbar, dynamic height differences within close pairs of CTD stations (twelve pairs at less than 10 km, 5 days separation) show an r.m.s. difference of 2.3 dyn. mm, which is equivalent to 1.1 cm s^{-1} between two stations 30 km apart. To reduce this noise level, dynamic heights at several stations need to be averaged. However, only four or five CTD stations were occupied within each 4-day period, within 30 km of 28°05'N, 70°10'W; too few for useful averaging and unevenly distributed in space. If the observed velocity differences had remained steady in time, these groups of unevenly spread CTD stations could have been lumped together for better averaging, but there were appreciable changes. The northward component of observed velocity difference at 28°05'N, 70°10'W increased by 1.25 cm s^{-1} between the first and third 4-day periods, for example. Observed velocities were, however, relatively steady until 15 April, though before 11 April there were too few observations for gradients to be fitted. The sixteen CTD stations occupied up to that date are marked in Fig. 8. Their north–south distribution is poor, but averaging within each of two groups of four stations, near 28°01'N, 70°21'W and 28°01'N, 70°02'W, marked with solid circles in Fig. 8, allows the east–west gradient of dynamic height difference to be estimated fairly accurately. Using both the 'up' and 'down' records at each CTD station, the difference between the mean dynamic heights of the two age groups is 1.8 (± 0.7) dyn. mm, corresponding to a southward velocity component at 1600 dbar relative to 2900 dbar, $\delta v_g = -0.85$ (± 0.33) cm s^{-1}. The observed component at a point midway between the two groups of CTD stations, from the fitted values for the period 11–15 April, is $\delta v = -0.77$ (± 0.3) cm s^{-1}. Again, there is no evidence in these measurements for significant departure from geostrophic balance.

DISCUSSION

Although the accuracy of geostrophic tests could be improved, perhaps by a factor of 2, by more intense sampling on similar lines to those used here, it seems most unlikely that real departures from geostrophy in weak mid-ocean currents could be resolved by such means.

By tracking floats relative to near-bottom transponders, the accuracy and frequency of fixing could be increased to yield daily mean velocities with errors well below 0.1 cm s^{-1}, but then systematic errors, due to the failure of floats to follow high-frequency vertical motions accurately, might become significant. Acoustic current meters may be expected to respond more linearly to high-frequency motions than rotor-vane instruments, and hence be capable of recording mean currents more accurately, but if they are used in compliant moorings there can still be errors due to depth variations being partly correlated with the velocity being measured.

Much greater accuracy in geostrophic currents seems equally difficult to attain. In these tests, corrections to dynamic height differences at moorings were estimated with standard deviations (of the mean) of less than 2 dyn. mm, equivalent to 2 ppm in mean density of the water column through the main thermocline. FOFONOFF and BRYDEN (1975) find that the most precise laboratory determinations of density of seawater as a function of temperature and salinity at atmospheric pressure, have a standard deviation about a fitted polynomial of 1 to 3 ppm, over most of the range. With this as an example of the uncertainty in some of the basic data defining the equation of state of seawater, it seems unreasonable to expect to push the field observations much further on their own.

Although their representativeness is not clear, the observed r.m.s. differences in dynamic heights between close pairs of stations (6.7 dyn. mm in the interval 600–1600 dbar, 2.3 dyn. mm between 1600 and 2900 dbar) may serve as an indication of the noise levels to be expected in dynamic heights along hydrographic sections in mid-ocean.

Acknowledgements—Many people contributed to collecting and processing the data used here, and made un-published material freely available. The author is grateful to all of them, and particularly to J. CREASE and others responsible for processing the MODE density observations.

REFERENCES

BROWN N. (1974) A precision CTD microprofiler. *Ocean 74, IEEE International Conference on Engineering in the Ocean Environment*, **2**, 270–278.

BRYDEN H. (in press) Geostrophic comparisons from moored measurements of current and temperature during MODE. *Deep-Sea Research*.

CASTON G. F., W. K. STRUDWICK and J. C. SWALLOW (1974a) Neutrally buoyant floats serial nos. 242–265, April 1973. *Institute of Oceanographic Sciences, Data Report No. 1* (unpublished report).

CASTON G. F., W. K. STRUDWICK and J. C. SWALLOW (1974b) Neutrally buoyant floats serial nos. 266–293, May 1973. *Institute of Oceanographic Sciences, Data Report No. 2* (unpublished report).

CREASE J. (1962) Velocity measurements in the deep water of the Western North Atlantic. *Journal of Geophysical Research*, **67**(8), 3173–3176.

DEXTER STEPHEN C., JOHN D. MILLIMAN and WILLIAM J. SCHMITZ, JR. (1975) Mineral deposition in current meter bearings. *Deep-Sea Research*, **22**, 703–706.

FOFONOFF N. P. and H. BRYDEN (1975) Specific gravity and density of seawater at atmospheric pressure. *Journal of Marine Research*, **33**, Supplement, 69–82.

HOGG N. (1975) Balance of terms and integral balances. Chapter 11A in *Dynamics and the Analysis of MODE-1* (unpublished report, Massachusetts Institute of Technology).

HORTON C. and W. STURGES. A two-month geostrophic experiment during MODE (unpublished manuscript, Florida State University).

ROSSBY T., A. D. VOORHIS and D. WEBB (1975) A quasi-Lagrangian study of mid-ocean variability using long range SOFAR floats. *Journal of Marine Research*, **33**, 3, 355–382.

STOMMEL H. (1965) *The Gulf Stream*, 2nd edition. University of California Press, 248 pp.

SWALLOW J. C. (1971) The *Aries* current measurements in the western North Atlantic. *Philosophical Transactions of the Royal Society of London*, A, **270**, 451–463.

SWALLOW J. C., B. S. MCCARTNEY and N. W. MILLARD (1974) The Minimode float tracking system. *Deep-Sea Research*, **21**, 573–595.

WÜST G. (1924) Florida und Antillenstrom. Eine hydrodynamische Untersuchung. *Veröffentlichungen des Instituts für Meereskunde an der Universität Berlin, Neue Folge*, A, **12**, 1–48.

A cyclonic ring formed by the Gulf Stream, 1967*

F. C. FUGLISTER

Woods Hole Oceanographic Institution, Woods Hole, Massachusetts 02543.

Abstract—Tracking of a Gulf Stream cyclonic ring for a period of 8 months in 1967 is described. Drogued surface buoys serve to identify the ring and also to show the trajectories of the surface layer currents at various radii within the ring. The marked changes in the surface temperature and salinity during the 8 months are contrasted with the very slight changes beneath the mid-depth of the thermocline.

INTRODUCTION

IN 1967 the study of cyclonic rings formed by the Gulf Stream (FUGLISTER, 1972) was continued on a series of nine cruises, see Table 1. The first cruise, in March, was a shakedown run to test out various pieces of new equipment; a program of searching for, tagging with a drogued surface buoy, and then tracking a newly formed cyclonic ring was to begin with the May cruise.

The first cruise is mentioned here because, as it turned out, while the new equipment was being tested near Bermuda, word was received from the U.S. Navy Oceanographic Office that they were observing, from their aircraft, a cyclonic ring being formed by the Gulf Stream, north of Bermuda (WILKERSON, BRATNICK and ATHEY, 1969). Consequently, on 21 March an expendable bathythermograph section (450 m XBT) was made from the *Crawford* on its return trip to Woods Hole. This section supports the conclusion that a ring was forming but, unfortunately, because of the ship schedules, the program of tagging the ring could not start at that time. Although the navy aircraft continued to observe the cold water of the ring until the 6 April flight, they, in turn, had to discontinue this work because of other schedules. The various positions of this ring and the *Crawford's* track are shown in Fig. 1; the XBT section in Fig. 2.

In May, on *Crawford* cruise 155, an attempt was made to find another newly formed ring. Using the towed V-Fin (FUGLISTER and VOORHIS, 1965) the Gulf Stream was tracked from 38°N, 69°W east to 40°N, 60°W but no ring-forming meander was found. The area south of the Stream was then surveyed and on 13 May at 35°50′N, 65°W a 2-knot current setting to the east was encountered.† This was 125 miles south of the Gulf Stream; to check the assumption that this was the south part of a cyclonic ring, an XBT section was made toward the northwest and then, using only the continuous recordings of the surface (~ 1 m) temperature and salinity, the cool, fresh core of the ring was circum-

---

* Contribution No. 3735 from the Woods Hole Oceanographic Institution.
† 1 knot = 0.51 m s^{-1}.

Table 1.   *Ring cruises in 1967*

Ship	Cruise no.	Dates	Station no.	BTs
1 *Crawford*	152	4–23 March	2032–2036	129
2 *Crawford*	155	4–23 May	2044–2048	246
3 *Crawford*	156	28 May–15 June	2050–2062	264
4 *Crawford*	157	22 June–10 July	—	245
5 *Crawford*	158	14 July–3 August	2064–2080	217
6 *Atlantis II*	35	17 Aug.–10 Sept.	1116–1138	255
7 *Atlantis II*	37	19 Sept.–6 Oct.	1139–1149	228
8 *Atlantis II*	38	11–31 October	1150–1165	247
9 *Atlantis II*	39	4–21 November	—	84

navigated. The position of this elongated pear-shaped core is shown on Fig. 1, as is the track of the XBT section shown in Fig. 3.

In the core of a newly formed cyclonic ring the temperature and salinity structure is much the same as that in the slope water at that time. Here the surface water is definitely warmer and slightly more saline than what was observed in the slope water a week earlier. This, and the position and shape of the cold core, suggests that this is the ring that had been observed in March and April.

On 15 May a radio transmitting buoy with a parachute drogue at a depth of 100 m was placed in, what was estimated to be, the center of this ring. From this time until 29 October at least one such buoy was in some part of this ring. During this 170-day period the *Crawford* and *Atlantis II* were in the ring a total of 102 days tracking drogued buoys, a number of neutrally buoyant (Swallow) floats and collecting hydrographic station data, XBT and continuous surface temperature and salinity records.

On 29 October four drogued buoys were left in the ring near 36°30′N, 67°W. On the next, and last cruise of this series (*Atlantis II* cruise 39, 4–21 November) no buoys were located. Whether or not a remnant of the ring, in process of being absorbed by the Gulf Stream, was observed is open to question.

Several months after this, it was learned that the U.S.N.S. *Gilliss* had made a thorough XBT survey of the area bounded by 36° and 40°N latitude, 65° and 69°W longitude. The area was covered twice, 17–18 October and 31 October–9 November; the 450-m temperatures for these two time periods are shown in Fig. 4. The first phase clearly shows the ring where the *Atlantis II* left it on the 29th; the second phase shows no cyclonic ring in the area but the Gulf Stream had shifted its position considerably. The southeastward extension of Stream might indicate that the ring was reabsorbed there.

The track of the *Atlantis II*, 5–6 November, is shown on the phase II part of Fig. 4. Not having this overall picture at that time, when the cold water was encountered at the position marked A, it was considered to be in the ring, well south of the Gulf Stream. But a survey with XBTs showed that, if it was the ring, it was very much distorted; the cold water stretched far to the northwest and, when none of the identifying buoys were found, the decision was made to search elsewhere for the ring.

Continuously recording the 200-m temperature, a search was made as far south as 35°30′N lat., but no signs of the ring, that had been so clearly defined 10 days earlier, were found. It was, therefore, concluded that the ring must have become reabsorbed by

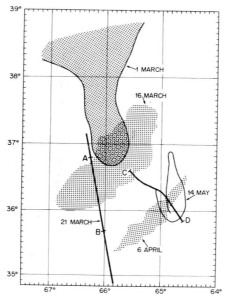

Fig. 1. Formation of the cold core of a cyclonic ring, as observed from aircraft (shaded areas). From *The Gulf Stream* monthly summary, U.S.N.O.O. vol. 2, 4, 1967. Also, track of the R. V. *Crawford* on 21 March and 14 May 1967.

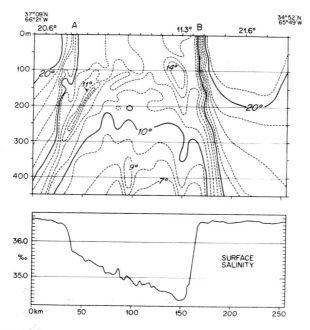

Fig. 2. First XBT section through the ring, 1500/21 to 0300/22 March. Track of the *Crawford* and points A and B shown on Fig. 1.

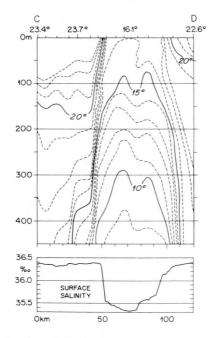

Fig. 3.   Second *Crawford* XBT section through the ring, 1540—2300/13 May 1967. Ship's track, D to C, shown in Fig. 1.

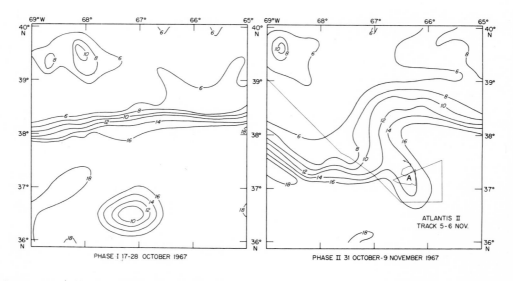

Fig. 4.   Composite 450-m temperature analyses, U.S.N.S. *Gilliss* data. Track of the *Atlantis II*, 5–6 November superimposed on phase II plot.

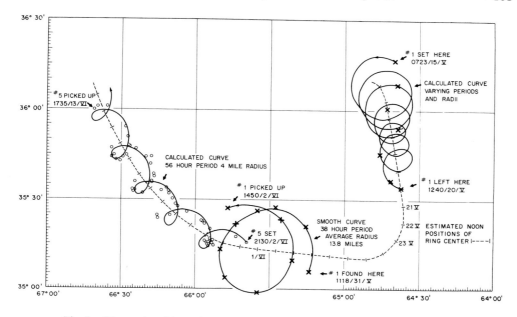

Fig. 5.   Observed positions of buoys 1 and 5 in the period 15 May to 16 June 1967.

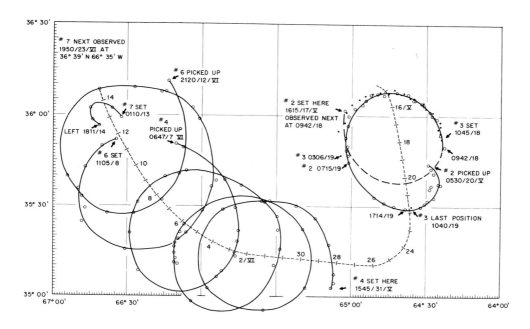

Fig. 6.   Observed positions of buoys 2, 3, 4, 6 and 7 in the period 17 May to 14 June 1967.

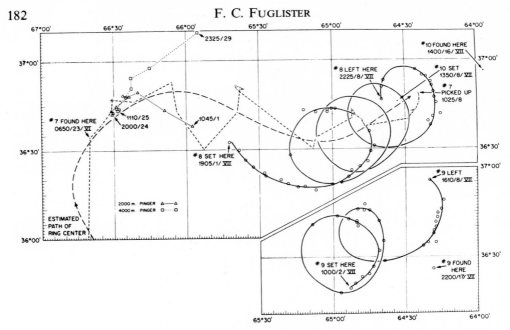

Fig. 7.   Observed positions of buoys 7, 8, 9, 10 and two deep neutrally buoyant floats in the period 23 June to 8 July 1967.

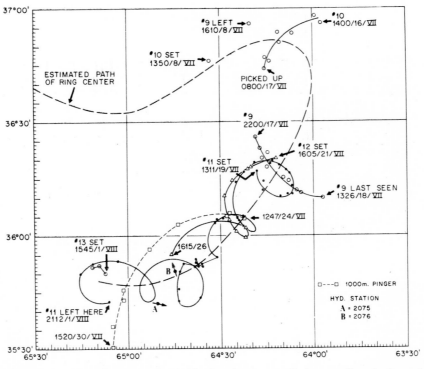

Fig. 8.   Observed positions of buoys 9, 10, 11, 12, 13 and one neutrally buoyant float in the period 16 July to 1 August, 1967.

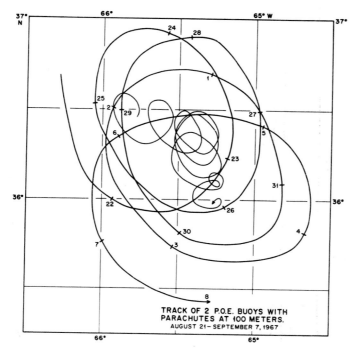

TRACK OF 2 P.O.E. BUOYS WITH
PARACHUTES AT 100 METERS.
AUGUST 21 – SEPTEMBER 7, 1967

Fig. 9.   Trajectories of buoys 11 and 15. See also Fig. 10.

the Gulf Stream and the marker buoys carried away by the more intense surface currents.

If this was indeed the case, a question remains: why did this still vigorous ring, that twice before had been observed moving north toward the Stream only to veer away again, this time, so rapidly disappear into the Stream? In any event, this loss of the ring put an end to the 1967 cyclonic ring observational program.

Although the author, to date, has not published the results of this ring study, he has described the work in numerous lectures; the hydrographic station data and XBTs have been available from the National Oceanography Data Center since 1968 and a number of scientific papers have been based, in part, on some of these data; for instance, BARRETT (1971), MOLINARI (1970) and PARKER (1971).

BUOY TRAJECTORIES AND THE RING TRAJECTORY

On 15 May 1967, for the first time, a buoy was placed in a Gulf Stream ring. It was drogued at a depth of 100 m with a cargo parachute so that its movement should approximately represent the horizontal velocity of the surface layer of the ring. PARKER (1972) gives a detailed description of the drogues.

In the 1965–1966 ring study there was occasionally some question as to the identity of the rings observed; in this experiment the drogued buoy, left in the ring at the end of each cruise, was intended to serve as a marker to identify the ring when it was found on the following cruise. Also, the radio signal from the buoy was intended to help in relocating the ring at the beginning of each cruise. Some of the buoys used had only a flag, a light and radar reflectors but there was always at least one radio-transmitting buoy in the ring.

The experiment proved successful; the buoys did not spin out of the ring, at least there was no evidence of it, on the other hand some were badly damaged and demasted and no attempt was made to continue tracking them. The observed positions (Loran A navigation) from 15 May to 29 October, of the buoys and neutrally buoyant floats are shown in Figs. 5 to 11. Because of the complexity of the data obtained on *Atlantis II* cruise 35—see Figs. 9 and 10—each observed position is not shown, only the smoothed curves.

Two curves (or buoy trajectories) drawn in Fig. 5 are different from all the others in that they are calculated on the bases of a number of assumptions and not smooth curves through the observed positions. One of these is the curve through the first seven positions of buoy no. 1. This may appear rather fanciful and can be disregarded; however, after studying all of the data and estimating the speed and trajectory of the ring, it is evident that this buoy was not at the 'center' as it was intended to be but was rotating around it. Using only the surface temperature and salinity apparently gave a distorted picture of the size and shape of the ring and consequently buoy no. 1 was

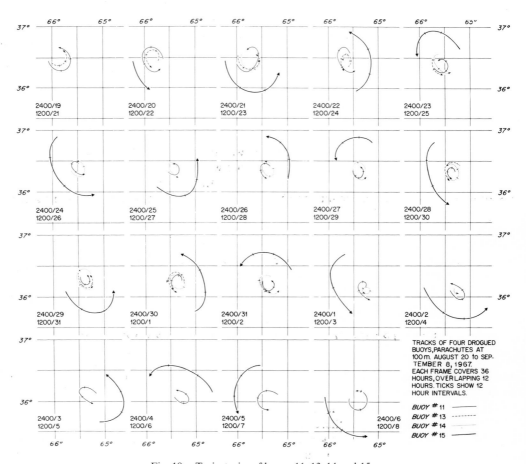

Fig. 10.   Trajectories of buoys 11, 13, 14 and 15.

Table 2. *Averaged buoy speeds and distances from the 'center' of ring*
(N.B.F.: neutrally buoyant 'pinger' floats)

Buoy number	Drogue depth (m	Dates 1967	Hours	Speed curve (knots)	Speed curve (cm/s)	Speed straight (knots)	Speed straight (cm/s)	Radius nautical miles	Radius km
1	100	15, V–20, V	125.3	(2.11)	(109)	0.38	20		
1	100	20, V–31, V	262.6			0.16	8		
1	100	31, V–2, VI	51.5	2.48	128			14	26
2	100	17, V–18, V	17.5			1.94	100		
2	100	18, V–20, V	43.8	2.42	125			15	28
3	10	18, V–19, V	16.4	3.35	172			19	35
3	10	19, V	7.6			3.81	196		
4	100	31, V–7, VI	159.0	2.79	144			21	39
5	100	2, VI–13, VI	260.1	0.55	28			4	7
6	100	8, VI–12, VI	106.2	2.48	128			24	44
7	100	13, VI–14, VI	41.0	0.51	26			5	9
7	100	14, VI–23, VI	217.6			0.20	10		
7	100	23, VI–8, VII	350.6			0.50	26		
8	100	1, VII–8, VII	171.3	2.02	104			13	24
9	500	2, VII–8, VII	150.2	1.20	62			12	22
9	500	8, VII–17, VII	221.8			0.14	7		
9	500	17, VII–18, VII	15.4	1.62	83			17	32
9	500	18, VII–24, IX	1631.3			0.05	3		
10	100	8, VII–16, VII	192.2			0.16	8		
10	100	16, VII–17, VII	18.0	1.22	63			14	26
11	80	19, VII–26, VII	157.8	0.48	25			6	11
11	80	26, VII–27, VII	23.0			0.65	33		
11	80	27, VII–1, VIII	139.2	0.65	33			6	11
11	80	1, VIII–19, VIII	420.1			0.16	8		
11	80	19, VIII–8, IX	473.4	1.00	51			7	13
12	250	21, VII–26, VII	120.1	0.46	24			6	11
13	100	1, VIII–19, VIII	425.5			0.14	7		
13	100	19, VIII–1, IX	316.9	0.62	32			8	15
14	100	20, VIII–3, IX	329.6	0.84	43			7	13
15	100	21, VIII–8, IX	426.1	1.89	97			29	54
16	500	25, IX–4, X	206.6	0.76	39			10	19
16	500	4, X–29, X	608.4			0.14	7		
17	35	27, IX–3, X	144.8	0.61	31			8	15
17	35	3, X–24, X	494.4			0.16	8		
18	150	2, X–4, X	35.7	0.81	34			10	19
18	150	4, X–20, X	388.0			0.14	7		
18	150	20, X–28, X	199.9	0.86	44			10	19
19	150	2, X–4, X	34.7			0.23	12		
19	150	4, X–21, X	426.3			0.17	9		
19	150	21, X–29, X	183.3	1.18	61			12	22
20	100	19, X–21, X	55.0			0.29	15		
20	100	21, X–23, X	27.7	0.65	33			6	11
20	100	23, X–29, X	157.3			0.11	6		
21	150	21, X–29, X	197.4	1.26	65			11	20
N.B.F.	1000	24, VII–30, VII	146.5	0.36	19			11	20
N.B.F.	2000	24, VI–1, VII	158.7			0.24	12		
N.B.F.	4000	25, VI–29, VI	108.2			0.37	19		

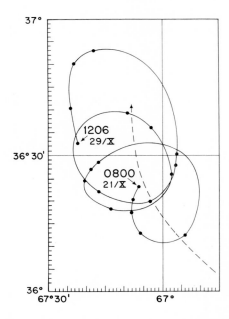

Fig. 11.   Observed positions of buoy 21 in the period 21–29 October 1967.

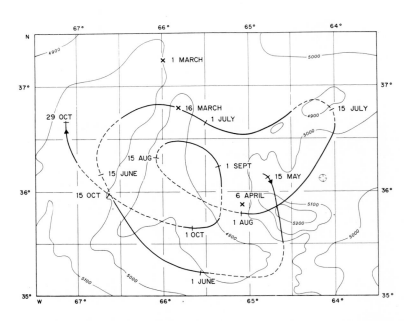

Fig. 12.   Trajectory of the ring (estimated path of the ring 'center') over the period 15 May to 29 October 1967. Interpolation between cruises—broken lines. Bathymetry in meters.

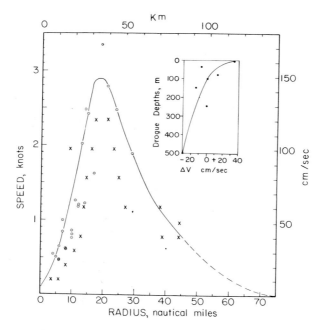

Fig. 13. Averaged buoy speeds (●) and calculated surface current speeds (×) plotted against estimated distances from the center of the ring.

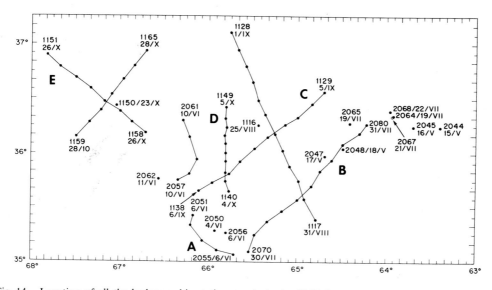

Fig. 14. Location of all the hydrographic stations made in the 1967 ring. Letters refer to sections shown in Figs. 15–19.

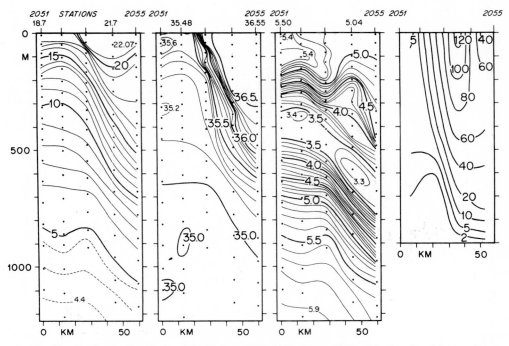

Fig. 15. Section A. *Crawford* cruise 156; Stas. 2051–2055; 6 June 1967. $T°C$; $S‰$; O_2 ml/l; V cm/s.

placed approximately 10 miles (18 km) north of its 'center'. During the first 5 days it may have moved somewhat closer to the 'center' but at the end of 16 days, when it was again observed and followed, its radius was 14 miles (26 km).

The other calculated curve—for buoy no. 5—is a curve traced by a point on the circumference of a circle of 4 miles radius, rotating with a 56-hour period around the estimated path of the ring 'center'. Only the initial observed position was used in drawing this curve. As with the no. 1 buoy, this buoy was intended to be in the ring 'center' and when these observations were made it seemed as though the ring was translating in a jerky fashion, periodically slowing down to a stop and then speeding up again, but using all of the data from the first seven buoys (Table 2 and Figs. 5 and 6) to determine the translatory speed and path of the ring 'center', indicates that buoy no. 5 was 4 or 5 miles away from the 'center'. The translation rate of the ring at this time was 6.1 miles per day (9.7 km day^{-1}), see Table 3. The trajectory of the ring, over the period from 15 May to 29 October, is plotted in Fig. 12; the crosses on this figure show the approximate centers of the cold core as observed from the aircraft on 1 and 16 March and 6 April. It is not at all obvious what path the ring may have taken through this period up to 15 May when it was first tagged with a buoy.

Except for a short period in early July, the path taken by the ring is anticyclonic; after traveling 703 miles (1303 km) in 168 days it is only 110 miles (204 km) west of its position on 15 May. Its speed of advance varies from over 6 miles/day to less than 2 miles/day with no apparent regard for the age of the ring, its direction of movement or the bottom topography. The movements of this tagged ring and the estimated, complex

Table 3.   *Translation speeds of the ring along*
the curve shown in Fig. 12

Dates, 1967	miles/day	km/day
15 May–15 June	6.1	9.7
15 June– 1 July	5.9	9.3
1 July –15 July	5.6-	8.9
15 July – 1 Aug.	5.1	8.1
1 Aug.–15 Aug.	4.3	6.8
15 Aug.– 1 Sept.	2.6	4.1
1 Sept.– 1 Oct.	1.6	2.5
1 Oct. –29 Oct.	3.6	5.7
*29 Oct. – 6 Nov.	6.2	9.8

* Assuming that the ring 'center' moved to the point A shown in Fig. 4.

movements of the two rings observed in 1965–1966 (FUGLISTER, 1972) are in sharp contrast to the idea that these cyclonic rings move steadily in a westerly and southerly direction as implied by RICHARDSON (RICHARDSON, STRONG and KNAUSS, 1973) and others. In order to know the path taken by these cyclonic rings, it seems obvious that frequent—at least weekly—observations are needed, preferably of some identifiable object floating in the surface layer of the ring. Some satellite observations of a drogued buoy placed in a cyclonic ring are being made now (February 1976) that show the ring moving toward the northeast (P. L. RICHARDSON, personal communication).

BUOY SPEEDS

The speed with which the buoys moved in the ring are measured from fix to fix along the curved line through these positions. Where the positions are widely separated, particularly between cruises, the measurements are along a straight line. The periods over which these speeds were averaged and the estimated, average radius, or distance of the buoys from the moving 'center' of the ring are given in Table 2.

The buoy speeds vary with their distance from the 'center' and with the depth of the drogues. The speeds increase in a nonlinear fashion from the center outward, reaching a maximum at a radius of 18 to 20 miles (33–37 km). Only three buoys are more than 20 miles from the 'center' and their average speeds, taken together with the geostrophic calculations from several sections made through this ring, indicate a ring radius of approximately 75 miles (140 km), see Fig. 13. The curve in this figure is based primarily on the data from buoys with drogues at 100 m; at zero radius it is the average translatory speed of the ring.

The speeds of buoys nos. 2 and 3, with drogues at 100 m and 10 m respectively, set next to each other on 18 May, clearly indicate a shear in this layer. Unfortunately the other drogues at different depths were not put out in pairs. However, taking into consideration the different radii, the mean velocities of the buoys with drogues at depths different from 100 m can be compared with the curve shown in Fig. 13. This relationship is shown in the insert to the figure; these few estimates of the shear show the buoys drogued at 500 m moving 30 cm s^{-1} slower than the 100-m buoys and the 10-m buoy going 35 cm s^{-1} faster. These speeds, of course, do not represent the current speeds at these

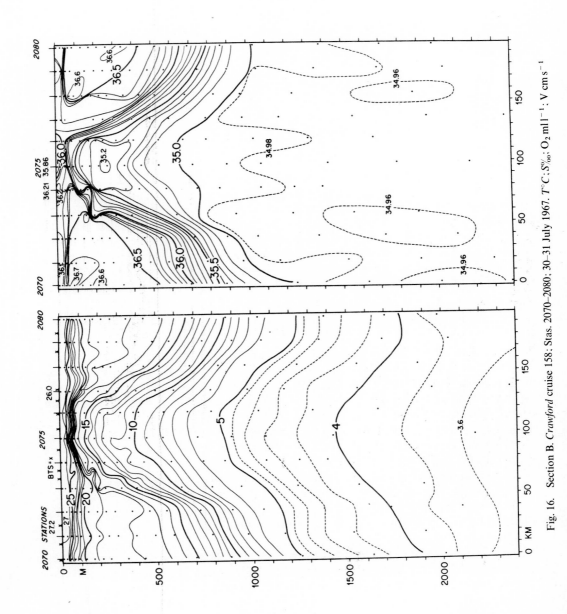

Fig. 16. Section B. *Crawford* cruise 158: Stas. 2070–2080; 30–31 July 1967. $T°C$; $S‰$; $O_2.ml\,l^{-1}$; $V\,cm\,s^{-1}$

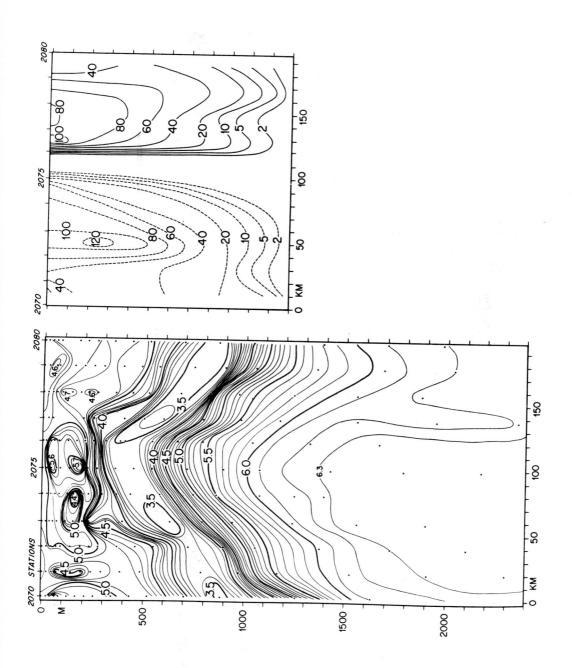

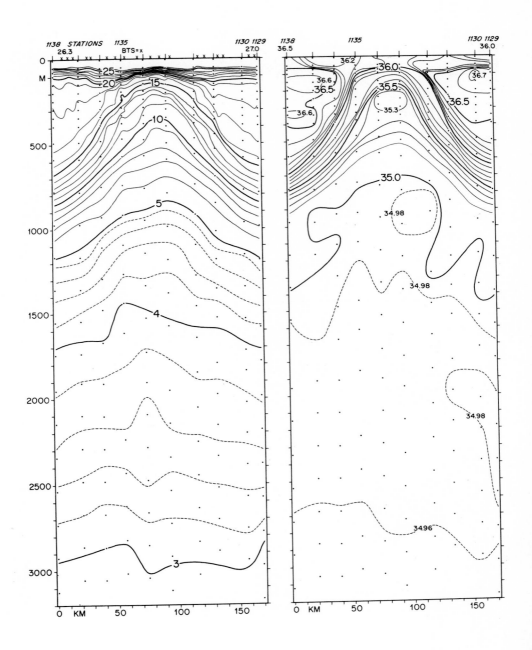

Fig. 17.   Section C. *Atlantis II* cruise 35; Stas. 1129–1138; 5–6 September 1967. $T°C$; $S‰$; O_2 ml l^{-1}; V cm s^{-1}.

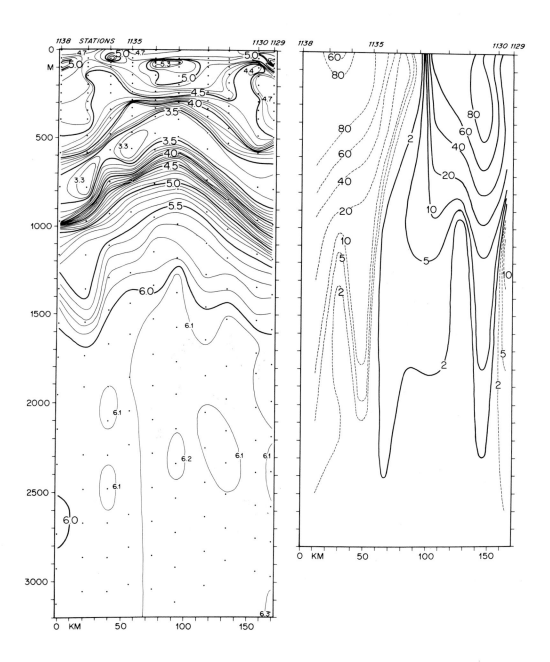

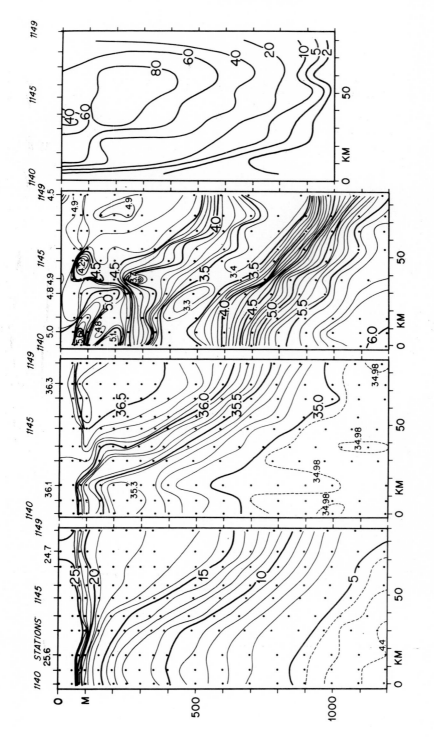

Fig. 18. Section D. *Atlantis II* cruise 37: Stas. 1140–1149: 4–5 October 1967. $T°C$; $S‰$; O_2 ml l^{-1}; V cm s^{-1}

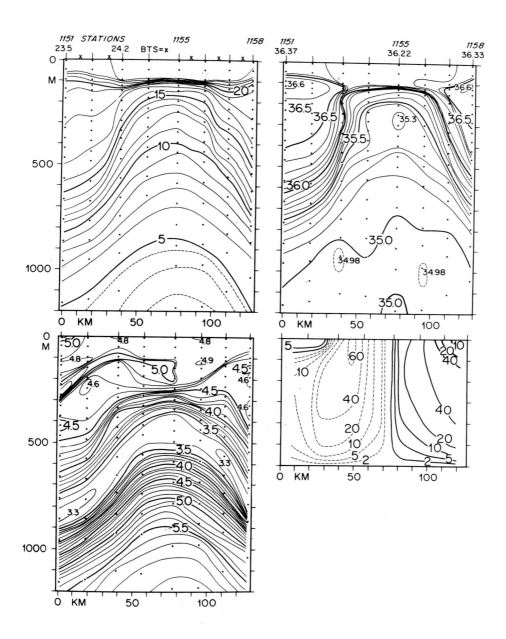

Fig. 19.   Section E. *Atlantis II* cruise 38; Stas. 1151–1158; 26 October 1967. $T°C$; $S‰$; O_2 ml l^{-1}; V cm s^{-1}.

depths but rather some integrated value from the surface to the drogue depth. The neutrally buoyant float at 1000 m, that does represent the current speed at that depth, moved $50 \, \mathrm{cm \, s^{-1}}$ slower than 100-m drogues.

HYDROGRAPHIC SECTIONS

Eight sections of hydrographic stations were made in the ring, three of these are half sections, that is, from the center outward. The location of all the stations is shown in Fig. 14; the station numbers >2000 are R. V. *Crawford* stations, those <2000 are from the *Atlantis II*. The temperature (°C), salinity (‰), dissolved oxygen (ml l⁻¹ and computed,

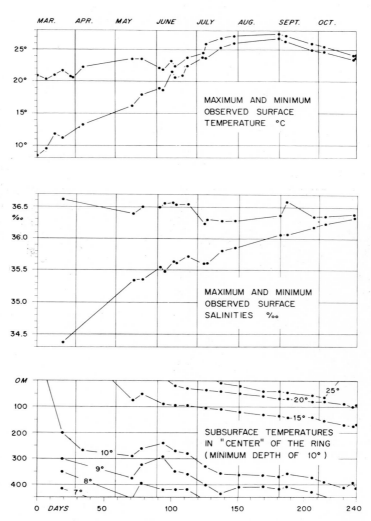

Fig. 20.   Changes in the surface temperature and salinity ranges and in the depth of the thermocline from 1 March to 29 October 1967.

geostrophic velocities—normal to the sections ($cm\,s^{-1}$ for sections A, B, C, D and E are shown in Figs. 15–19. The dots show the depths at which the observations were obtained.

Although sections A and D are called half sections, none of the sections actually go completely across the ring; certainly none show a radius of as much as 75 miles (140 km), as is suggested in the extrapolated velocity curve in Fig. 13. Since the geostrophic velocities are the components normal to the sections and relative to the depth of the shallowest station in each section, the current speeds plotted for each section should be considered minimal. They show a gradual decrease in speed, approximately halved, over the 5-month period. The calculated surface speeds for various radii are plotted as ×s in Fig. 13; all except one, at a radius of 20 km on section B, fall well below the averaged buoy speeds for the same radius.

Over the 5-month period, the only marked change that appears in the sections, A to E, occurs in the temperature and salinity of the surface 100-m layer. There is a decrease in the dissolved oxygen in the ring center but it is not as obvious as the increased temperature and salinity. In the last sections, in October, there is no indication of a 'cold core' ring at the surface, actually it is $\sim 1°C$ warmer in the center. The salinity differences across the ring surface still show less saline water in the core but only $\sim 0.1\%_{\circ}$ less.

Beneath the surface layer there is relatively little change over this 5-month period. There is the gradual subsidence of the thermocline in the core; PARKER (1971) gave a rate of 0.6 m per day for the sinking of the 17°C isotherm. Although it is obvious that the upper part of the thermocline deepens at about that rate (see Fig. 20) the sections A to E show very little, if any, change in the bottom part of the thermocline; the 5°C isotherm remains at approximately $830\,m \pm 10\,m$, this is 400 m shallower than its 'normal' depth for this region. The same applies to the salinity minimum between 200 m and 400 m and to the oxygen minimum layer centering around 400 m; they are virtually unchanged over this period.

The pattern of the dissolved oxygen in these sections, in the region of the maximum currents, is similar to the pattern observed in sections across the Gulf Stream (FUGLISTER, 1963). Below the oxygen minimum layer, these sections are the same as those across the Stream but above it the shallower secondary minimum is gradually disappearing.

Since BARRETT (1971) calculated the available potential energy in these cyclonic rings and found that the decay rate of this energy gave a ring life time of 3 to 5 years, it is not surprising that there are so few changes over a 5-month period in this ring.

CONCLUSIONS

Surface buoys, with drogues in the surface 500-m layer, placed in the central portion of a cyclonic ring, stay in the ring for at least several months and serve to identify and show the path taken by the ring.

The trajectories of these rings vary considerably both in speed and direction so observations made months apart do not show the actual trajectories.

Observations of surface temperatures do not always indicate the presence or the size and shape of a ring.

Finally, the author admits that this work should have been published years ago.

198

Acknowledgements—This investigation was supported by the Office of Naval Research under contract Nonr-2196(00) NR 083-004 and N00014-66-C-0241; NR 083-004. All of the work at sea was carried out under the excellent supervision of Mr. CHARLES PARKER and Mr. MARVEL STALCUP. This manuscript was prepared with support from the National Science Foundation, Grant No. OCE74-01282 (formerly DES74-01282).

REFERENCES

BARRETT J. R. (1971) Available potential energy of Gulf Stream rings. *Deep-Sea Research*, **18**, 1221–1231.
FUGLISTER F. C. (1963) Gulf Stream '60. In: *Progress in oceanography*, M. SEARS, editor, The Macmillan Company, **1**, pp. 265–373, 383 pp.
FUGLISTER F. C. (1972) Cyclonic rings formed by the Gulf Stream, 1965–66. In: *Studies in physical oceanography —a tribute to Georg Wüst on his 80th birthday*, A. L. GORDON, editor, Gordon & Breach, **1**, pp. 137–168, 194 pp.
FUGLISTER F. C. and A. D. VOORHIS (1965) A new method of tracking the Gulf Stream. Alfred C. Redfield 75th anniversary volume, *Limnology and Oceanography*, Supplement to **10**, R115–R124.
MOLINARI R. (1970) Cyclonic ring spin-down in the North Atlantic. Thesis, Texas A & M University.
PARKER C. E. (1971) Gulf Stream rings in the Sargasso Sea. *Deep-Sea Research*, **18**, 981–993.
PARKER C. E. (1972) Some direct observations of currents in the Gulf Stream. *Deep-Sea Research*, **19**, 879–893.
RICHARDSON P. L., A. E. STRONG and J. A. KNAUSS (1973) Gulf Stream eddies: recent observations in the western Sargasso Sea. *Journal of Physical Oceanography*, **3**, 297–301.
WILKERSON J. C., M. BRATNICK and G. L. ATHEY (1969) Aircraft observations of a cyclonic eddy south of the Gulf Stream. Informal report No. 69–41. U.S. Naval Oceanographic Office, Washington, D.C.

On the stability of internal wavetrains

S. A. Thorpe

Institute of Oceanographic Sciences, Wormley, Godalming, Surrey

Abstract—A uniform train of finite amplitude internal gravity waves is propagating horizontally in a fluid of arbitrary Brunt–Väisälä frequency confined between horizontal planes. The effect of finite amplitude on the shape and frequency of the waves is examined. It may increase or decrease the phase speed depending on the density stratification and wavenumber of the waves considered. A small modulation in the direction of the wavetrain may, in some circumstances, increase in size, the wavetrain becoming unstable, and the available theory for surface gravity waves is used to show that stable groups of waves may then occur, although in rather special conditions.

1. INTRODUCTION

IN THE autumn of 1975 Sir George Deacon arranged a course on the Physics of Oceans and Atmosphere at the International Centre for Theoretical Physics in Trieste. He personally arranged and attended the whole of the course and ran it with great enthusiasm, much to the appreciation of the students, many of whom were from developing countries. I was kindly invited to give some lectures on internal waves, and it was as a consequence of reconsidering what little is known of finite amplitude waves, aided by the pleasant and stimulating atmosphere of the course, that this work came to be done.

LAFOND's (1962) studies of internal waves off Mission Beach in California showed that their shape may be unsymmetrical. If the thermocline is near the sea floor, the wave crests are (like surface waves) narrower than the troughs. The converse is true when the thermocline is near the surface. These observations were explained in a theoretical study of the wave shape (THORPE, 1968, hereafter referred to as I) although it was later noticed that the shape depended both on the density profile and also on the mean (Eulerian) current which could in some circumstances reverse the trend in shape (THORPE, 1974, Appendix C).

Knowledge of finite amplitude effects has been extended by the examination, both theoretical and by laboratory experiments, of resonant interactions between triads of internal waves and between internal waves and surface waves. (References are given by THORPE, 1975.) The theory has been successful in providing an explanation of the modulation of surface waves by a train of internal waves observed, for example, in Vancouver Bay (GARGETT and HUGHES, 1972) and in the Straits of Gibraltar (CAVANIE, 1972). The effect of this modulation is to change the appearance of the sea surface, and this has been used in satellite studies by APEL. BYRNE, PONI and CHARNELL (1975) to detect the presence and propagation characteristics of packets of internal waves which seem to be generated at the continental slopes and isolated topographic features by the barotropic tides. These packets of waves appear at first sight to be similar to those observed in Massachusetts Bay (HALPERN, 1971) or in the Straits of Gibraltar (ZIEGENBEIN, 1969; CAVANIE, 1972) or to the waves which follow internal surges in lakes (Hunkins and Fliegel, 1973; Thorpe, 1974) but further observational study of the properties of the waves and a firm theoretical base on which to establish a similarity is needed.

Groups of internal waves have also been reported in the thermocline in various parts of the world by Sabinin (1973) and Brekhovskikh, Konjaev, Sabinin and Serikov (1975). Groups will occur whenever a pair of wave trains of slightly different frequencies are travelling in the same direction, and may be expected when the frequency spectrum is broad band, as it is observed to be for internal waves. However, the observations prompt the questions, when will a uniform train of internal waves be unstable and may a group structure be stable? The question of instability when the wave train is one of a resonant triad of internal waves was discussed by Davis and Acrivos (1968), who were successful in explaining some laboratory observations of the disintegration of a first-mode wavetrain. In general, however, their explanation requires the coexistence of waves not only of the given mode, but of two others, and since the wave modes are discrete (the frequency–wavenumber curves of the dispersion relation for each mode being separated by a finite amount at any point other than the origin), the wave frequencies and numbers required for the two others cannot be obtained by a small perturbation of the given wavetrain (a condition usually required in stability considerations). We shall consider a small modulation of the given wavetrain, here limiting the analysis to modulations in the direction of the wavetrain when the fluid is of finite depth and since we limit the analysis to wavenumbers and frequencies close to the original, waves belonging to modes which differ from the original will not be found. When the modulation is found to increase, the wavetrain is said to be unstable, although in practice what may happen is that the amplitude of groups of waves may vacillate, with the train of waves periodically returning to its initial regular form.

In the analysis which follows, we shall use results which have been obtained in studies of surface waves. The stability of a uniform train of surface waves was studied by Benjamin and Feir (1967) and Whitham (1967). It was discovered that a uniform Stokes wavetrain was unstable if $kh > 1.363$, where h is the water depth and k is the wavenumber of the waves, the train being observed in experiments to break up into groups of waves (Benjamin, 1967a). More recent studies indicate how the groups of waves develop. Hasimoto and Ono (1972) showed that the evolution of a modulation in the direction of the waves is governed by a non-linear Schrödinger equation, from which the stability criterion of Benjamin and Feir, and Whithan, can be deduced. Zakharov and Shabat (1972), studying the solutions of this equation, have demonstrated the evolution of 'solitons', apparently *stable* groups of waves with envelopes having a shape like that of solitary waves, in conditions in which the uniform train of waves is unstable. This theory appears to successfully predict the development of groups of surface waves in the laboratory (Yuen and Lake, 1975). Hayes (1973) and Davey and Stewartson (1974) have extended the theories of Whitham and Hasimoto and Ono to include the modulation of wavetrains by *oblique* disturbances and have found that the region of instability is then extended. The existence of three-dimensional stable groups of waves has not, however, been established, nor will it be here.

In Section 2 we describe the theory of the non-linear modulation of an internal wavetrain. It is convenient, as part of the development of the theory, to consider the effect of finite amplitude on the phase speed of a uniform train of waves. Yih (1974) and Samodurov (1974) found that, for waves in a fluid of constant Brunt–Väisälä frequency, the contribution is negative, the phase speed decreasing as the amplitude increases, but this conclusion is not true for more general density distributions. The finite amplitude contribution has already been calculated for a two-layer fluid (HUNT, 1961; I) and so that a comparison may be made, the full equations are used to determine the contribution for more general density profiles, although later the Boussinesq approximation is used to simplify the equations. Results for some particular density profiles are given in Section 3, and effects of making the Boussinesq approximation are demonstrated. Although the theory is quite general, in the examples we concentrate attention on waves of the first (gravest) internal mode, because observations show that this mode tends to dominate.

2. THEORY

We consider an incompressible, inviscid, stably-stratified, fluid lying between rigid horizontal boundaries at $z = 0, H$, in a Cartesian frame of reference with z vertically upwards. The effect of allowing a free upper boundary will usually be very small in practical applications and is neglected in this treatment. We shall not immediately apply the Boussinesq approximation, so that some results may be obtained which allow the limit of a very thin interface between two homogeneous layers—a two-layer fluid—to be approached. We suppose that the motion is two-dimensional so that we may define a stream function $\psi(x, z, t)$ such that the velocity $u = (\partial\psi/\partial z, 0, -\partial\psi/\partial x)$. The vorticity equation is

$$\rho\left[\frac{\partial}{\partial t}\nabla^2\psi + J(\nabla^2\psi, \psi)\right] + \frac{\partial\rho}{\partial z}\left[\frac{\partial^2\psi}{\partial t\partial z} + J\left(\frac{\partial\psi}{\partial z}, \psi\right)\right] + \frac{\partial\rho}{\partial x}\left[\frac{\partial^2\psi}{\partial t\partial x} + J\left(\frac{\partial\psi}{\partial x}, \psi\right)\right] = g\frac{\partial\rho}{\partial x}, \quad (1)$$

where J is the Jacobian with respect to x and z, and ρ is the density of the fluid. The continuity equation is

$$\frac{\partial\rho}{\partial t} + J(\rho, \psi) = 0. \quad (2)$$

We suppose that the stratification of the fluid at rest is given by $\rho = \rho_0(z)$ and that at time $t = 0$ a progressive internal wave of wavenumber, k, is established on the horizontally infinite layer. (One might imagine, for example, that the wave is established by an infinite periodic moving pressure or stress pattern at a horizontal surface or by body forces, which are then removed.)

We look for a solution at subsequent times in the form

$$\psi = \sum_{n=-\infty}^{\infty} \phi_n E^n, \qquad \rho = \sum_{n=-\infty}^{\infty} \rho_n E^n \quad (3)$$

where $E = \exp\left[i(kx - \sigma t)\right]$, $\phi_{-n} = \tilde{\phi}_n$, $\rho_{-n} = \tilde{\rho}_n$ (a tilde denotes a complex conjugate) and σ is the wave frequency. We also write

$$\phi_n = \sum_{j=n}^{\infty} \varepsilon^j \phi_{nj}, \qquad \rho_n = \sum_{j=n}^{\infty} \varepsilon^j \rho_{nj}, \quad (4)$$

where ϕ_{nj}, ρ_{nj} are functions of ξ, z, τ only and ε is an expansion parameter. Here

$$\xi = \varepsilon(x - c_g t), \qquad \tau = \varepsilon^2 t \quad (5)$$

and c_g is the (horizontal) group velocity of the waves, $\partial\sigma/\partial k$. We suppose that $\rho_{00} = \rho_0(z)$ and that $\phi_{00} = 0$, there being no mean Eulerian flow in the absence of the waves. (The expansion technique follows DAVEY and STEWARTSON, 1974.)

We substitute the expressions for ψ and ρ into (1) and (2) and use the method of multiple scales to obtain a series of differential equations for the functions ϕ_{nj} and ρ_{nj}, which are solved by equating to zero coefficients of $\varepsilon^j E^n$ in sequence and using the results of (2) to remove the ρ_{nj} from (1) to obtain ordinary differential equations for the ϕ_{nj} in terms of known ρ,ϕ with smaller suffix values. These equations are solved for ϕ_{nj} and the equation derived from (2) used to find ρ_{nj}. The procedure follows closely that used in I. The equation of a constant density surface $z + \eta(x, t, \xi, \tau)$ may be found by expressing η in a double series in ε and E similar to (3) and (4), and expanding $\rho(x, z+\eta, t) - \rho_0(z) = 0$ as a Taylor series, and comparing coefficients to find the η_{nj}.

At order $\varepsilon E'$ we find

$$\phi_{11} = A\Psi(z), \qquad \rho_{11} = -\frac{A}{c}\rho_0'\Psi \text{ and } \eta_{11} = \frac{A}{c}\Psi$$

where

$$\rho_0\Psi'' + \rho_0'\Psi' - \rho_0\Psi\left(k^2 + \frac{g\rho_0'}{\rho_0 c^2}\right) = 0, \tag{6}$$

with $\Psi(0) = \Psi(H) = 0$, and $c = \sigma/k$ is the phase speed and $A = A(\xi, \tau)$. A dash denotes derivatives with respect to z. Equation (6) is the standard eigenfunction equation describing the vertical model structure $\Psi(z)$ of the waves.

For a given k there exists an infinite sequence of eigenvalues $\{c_n\}$ for the phase speed corresponding to the various eigenfunctions $\{\Psi_n\}$ of the internal wave modes. From this sequence we select one, specified as c and Ψ in the analysis which follows. If Ψ is normalized in some appropriate manner, A/c may be regarded as the (first order) amplitude of the internal waves.

At order $\varepsilon^2 E^2$ terms are found which contribute to the shape of the waves. The equation for the stream function is $\phi_{22} = A^2\Phi(z)$ where $\Phi(z)$ is the particular solution of

$$\rho_0\Phi'' + \rho_0'\Phi' - \rho_0\Phi\left(4k^2 + \frac{g\rho_0'}{\rho_0 c^2}\right)$$

$$= -\frac{1}{2\sigma^2}\left\{\frac{2gk^2}{c}\rho_0'\Psi^2 + [\rho_0'\sigma k(3k^2\Psi^2 - 2\Psi\Psi'' - \Psi'^2) - 2\rho_0''\sigma k\Psi\Psi']\right\} \tag{7}$$

subject to $\Phi(0) = \Phi(H) = 0$. (If the Boussinesq approximation is made, the second term on the left and the terms in square brackets on the right are omitted and the equation reduces to that found in I.) Substituting in (2) we find

$$\rho_{22} = \frac{A^2}{2c^2}(\rho_0''\Psi^2 - 2c\rho_0'\Phi)$$

and   (8)

$$\eta_{22} = \frac{A^2}{c^2}(\Psi\Psi' + c\Phi).$$

In Section 3 we shall present some solutions for η_{22} and show how the neglect of the Boussinesq approximation can affect the solution. The solution so far is valid whether or not there is ξ variation. We shall now proceed as if there were no ξ variation, that is we shall consider for the moment a uniform wavetrain, and later consider what changes are made when the train has a ξ variation and may be modulated. We continue, therefore, with no Boussinesq approximation, and find

$$\varepsilon E^0: \quad \phi_{01} = 0, \quad \rho_{01} = 0, \qquad \eta_{01} = 0, \tag{9}$$

$$\varepsilon^2 E^1: \quad \phi_{12} = 0, \quad \rho_{12} = 0, \qquad \eta_{12} = 0, \tag{10}$$

$$\varepsilon^2 E^0: \quad \phi_{02} = 0, \quad \rho_{02} = \frac{|A|^2(\rho_0' \Psi^2)'}{2c^2}, \quad \eta_{02} = 0, \tag{11}$$

These solutions were given in I. At order $\varepsilon^2 E^0$ the ρ_{02} term may contain an additional function of z which is not determined by (1) or (2). The condition $\eta_{02} = 0$, however, implies that the volume of fluid lying below a constant density surface is constant and the arbitrary function is then determined. (This condition is imposed here so that comparison may be made with existing results for a two-layer fluid, but will be relaxed later—see the remarks following equation (22).)

After some algebra, the equation (1) for the $\varepsilon^3 E^1$ terms may be written down:

$$\rho_0 \phi_{13}'' + \rho_0' \phi_{13}' - \rho_0 \phi_{13}\left(k^2 + \frac{g\rho_0'}{\rho_0 c^2}\right) = -2\chi(z)|A|^2 A - \frac{2i}{\sigma} A_\tau [(\rho_0 \Psi')' - k^2 \Psi \rho_0] \tag{12}$$

with the boundary conditions $\phi_{13}(0) = \phi_{13}(H) = 0$, where $\chi(z)$ is a function of z and $A_\tau \equiv \partial A/\partial \tau$. The operator on the left-hand side of (12) is identical to that in (6) and is self-adjoint. A solution of (12) satisfying the boundary conditions exists if the following orthogonality condition holds:

$$\int_0^H \Psi \left\{ \chi(z)|A|^2 \cdot A + \frac{iA_\tau}{\sigma} [(\rho_0 \Psi')' - k^2 \Psi \rho_0] \right\} dz = 0 \tag{13}$$

(see Stuart, 1960). Using (6) to simplify (13) we find

$$iA_\tau = \sigma_2 |A|^2 A \tag{14}$$

where

$$\sigma_2 = \frac{\sigma}{4c^2 \int \rho_0' \Psi^2 \, dz} \left\{ \int_0^H \left[\rho_0' \Psi^2 (\Psi'^2 + \Psi\Psi'') + \frac{c^2}{g} \rho_0' \Psi\Psi'(\Psi'^2 - 2\Psi\Psi'') \right] dz \right.$$
$$\left. - 2c \int_0^H \Phi \left[\rho_0'' \Psi^2 + \frac{c^2}{g}(\rho_0'(3k^2\Psi^2 - 2\Psi\Psi'' - \Psi'^2) - 2\rho_0'' \Psi\Psi') \right] dz \right\}.$$

The solution of (14) is $A = A_0 \exp(-i\sigma_2 A_0^2 \tau)$ and so, recalling the definition of ϕ_n, (3), and A, (6), the result of finite amplitude is to increase the frequency of the uniform wavetrain to $\sigma + \sigma_2$. The term $\sigma_2|A|^2$ represents the effect of finite amplitude, and we shall present some calculations of σ_2 for particular density distributions in Section 3. The terms in c^2/g in σ_2 disappear if the Boussinesq approximation is made. Inspection of the expression for σ_2 shows that the approximation may be valid if the wavelength of the waves is much less than the scale height of the fluid.

We now return to consider the effect of ξ variation. Including variation with ξ and comparing terms in (1) of order $\varepsilon^2 E^1$ we find

$$\rho_0\phi_{12}'' + \rho_0'\phi_{12}' - \rho_0\phi_{12}\left(k^2 + \frac{g\rho_0'}{\rho_0 c^2}\right) = 2iA_\xi\left[c_g(\rho_0\Psi')' - \frac{kg}{\sigma}\rho_0'\Psi - \rho_0(c_g + c)k^2\Psi\right], \quad (15)$$

with $\phi_{12}(0) = \phi_{12}(H) = 0$ and, as before in (12), a solution exists if and only if the orthogonality condition

$$\int_0^H \Psi\left[c_g(\rho_0\Psi')^1 - \frac{kg}{\sigma}\rho_0'\Psi - \rho_0(c_g + c)k^2\Psi\right]dz = 0 \quad (16)$$

is satisfied. Now integrating (6) from $z = 0$ to $z = H$ we have

$$\sigma^2 k^2 \int_0^H \rho_0\Psi^2\,dz - \sigma^2\int_0^H (\rho_0\Psi')'\Psi\,dz + k^2\int_0^H g\rho_0'\Psi^2\,dz = 0, \quad (17)$$

and differentiating (17) with respect to k and using (17) to simplify the resulting equation we find

$$\frac{\partial\sigma}{\partial k} = c_g = c\left(1 + \frac{\sigma^2\int_0^H \rho_0\Psi^2\,dz}{\int_0^H g\rho_0'\Psi^2\,dz}\right). \quad (18)$$

(Since ρ_0' is negative, the integral in the denominator is negative whilst that in the numerator is negative, and so $c_g \leqslant c$.) Using (18) it is easily verified that (16) is satisfied, and a solution

$$\phi_{12} = iA_\xi\mathscr{F}(z) \quad (19)$$

for ϕ_{12} exists, where $\mathscr{F}$ satisfies (15) (with iA_ξ removed from the right) and $\mathscr{F}(0) = \mathscr{F}(H) = 0$. (If ρ_0' is constant and the Boussinesq approximation is made, we find $\mathscr{F} = 0$.) Substituting into (2) we find

$$\rho_{12} = iA_\xi\rho_0'\left[\frac{\Psi}{\sigma}\left(1 - \frac{c_g}{c}\right) - \frac{\mathscr{F}}{c}\right]. \quad (20)$$

As before ϕ_{01} and ρ_{01} are zero.

Further progress with the full equation (1) is made extremely tedious and liable to error by the length of the algebra, and since we have no independent estimates with which to check our conclusions we therefore make the Boussinesq approximation in deriving the following solutions. This approximation consists of neglecting the second and third terms in square brackets on the left of (1) and taking $\rho_0 = \bar{\rho}_0$ constant in the first term. To obtain the equation for ϕ_{02} from (1) it is necessary to retain terms of order $\varepsilon^4 E^0$ and after some algebra and two integrations with respect to ξ we find

$$\phi_{02}'' - \frac{g\rho_0'}{c_g^2\bar{\rho}_0}\phi_{02} = \frac{g}{c^3 c_g^2}\left[\frac{\rho_0'}{\bar{\rho}_0}\Psi\Psi'(cc_g + c^2 - 2c_g^2) - c_g^2\frac{\rho_0''}{\bar{\rho}_0}\Psi^2\right]|A|^2. \quad (21)$$

In general a solution exists and $\phi_{02} = \mathscr{I}(z)|A|^2$ where $\mathscr{I}$ satisfies (21) with $|A|^2$ removed from the right-hand side and $\mathscr{I}(0) = \mathscr{I}(H) = 0$. Equation (2) at order $\varepsilon^3 E^0$ now gives after integration with respect to ξ,

$$\rho_{02} = -\left[\rho_0'\frac{\mathscr{I}(z)}{c_g} + \frac{\rho_0''\Psi^2}{4cc_g} - \frac{1}{4cc_g}\left(1 - \frac{c_g}{c}\right)(\rho_0'\Psi^2)'\right]|A|^2. \quad (22)$$

[For a fluid of constant Brunt–Väisälä frequency these results, equations (21), (22), reduce to those found by McINTYRE, 1973.] It is not longer possible in general to apply the condition that $\eta_{0'2} = 0$. The mean density calculated over any wavelength is now changed from $\rho_0(z)$ although it is possible to keep the mean density constant on $\eta =$ constant (following a point moving with the group velocity) by subtracting an appropriate function of τ, z from (22). ϕ'_{02} is also non-zero and horizontal currents are generated. A comparable problem is encountered in the corresponding analysis for modulated surface waves although it is somewhat less significant since the density is uniform. The perturbation is, of course, of second order in ε but is none the less important in the analysis.

We once again have sufficient terms to substitute into (1) and to equate coefficients of $\varepsilon^3 E^1$. This gives

$$\bar{\rho}_0 \phi''_{13} - \bar{\rho}_0 \phi_{13}\left(k^2 + \frac{g\rho'_0}{\bar{\rho}_0 c^2}\right)$$

$$= -2(\Lambda(z) + \chi(z))|A|^2 A - 2i\frac{A_\tau}{\sigma}\left[(\rho_0 \Psi')' - k^2 \Psi \bar{\rho}_0\right] + \frac{2\mathscr{H}(z)}{\sigma^2} A_{\xi\xi}, \qquad (23)$$

with $\phi_{13}(0) = \phi_{13}(H) = 0$, where

$$\sigma\mathscr{H} = \bar{\rho}_0(\sigma k^2 c_g + \sigma^2 k) + kg\rho'_0 \mathscr{F} - \sigma c_g(\rho_0 \mathscr{F}')^1 + c_g^2(\rho_0 \Psi')'$$
$$- \bar{\rho}_0 \Psi(4\sigma k c_g + \sigma^2 + k^2 c_g^2) - g\rho'_0 \Psi,$$

and $\Lambda(z)$ is a function of z. As before a solution exists only if the orthogonality condition holds. After some simplification this condition reduces to

$$iA_\tau + \tfrac{1}{2}\frac{\partial c_g}{\partial k} A_{\xi\xi} = \nu|A|^2 A, \qquad (24)$$

where

$$\nu = \sigma_2 + \frac{\sigma}{4\int \rho'_0 \Psi^2 \, dz}\left[\frac{(c-c_g)(c+2c_g)}{cc_g^2}\int_0^H \mathscr{J}\rho'_0 \Psi\Psi' \, dz\right.$$

$$\left. + \frac{1}{c}\int_0^H \mathscr{J}\rho''_0 \Psi^2 \, dz + \frac{2}{c_g^2}\int_0^H \rho'_0 \Psi^2 \Psi'^2 \, dz\right].$$

(24) is the required non-linear Schrödinger equation, similar in form to the principal equation of HASIMOTO and ONO (1972). ($\mathscr{F}$ does not appear, having cancelled identically in the algebraic derivation.) Using their results it may immediately be concluded that the wavetrain is unstable if

$$\nu\frac{\partial c_g}{\partial k} < 0. \qquad (25)$$

Now differentiating (18) with respect to k it may easily be shown that

$$\frac{\partial c_g}{\partial k} = 3\sigma c_g c \frac{\int_0^H \rho_0 \Psi^2 \, dz}{\int_0^H g\rho'_0 \Psi^2 \, dz},$$

and this is negative, since the denominator is negative, unless $c_g < 0$. However, multiplying (6) by Ψ, integrating by parts from $z = 0$ to H, and substituting in (18) we find

$$c_g = -c^3 \frac{\int_0^H \rho_0 \Psi'^2 \, dz}{\int_0^H g\rho_0' \Psi^2 \, dz}, \quad > 0,$$

since the denominator is negative, so $\partial c_g/\partial k < 0$ and the wavetrain is unstable if $v > 0$. There is neutral stability if $v \leqslant 0$.

3. EXAMPLES

3.1. *The constant density gradient*

Results may be obtained most easily for a fluid in which the Brunt–Väisälä frequency, N, is constant. For this fluid it is already established (YIH, 1974; SAMODUROV, 1974) that the phase speed decreases with increasing wave amplitude. Substituting the solution for $\Psi(z) = \sin(n\pi z/H); n = 1, 2, \ldots; c^2 = N^2 H^2/(n^2\pi^2 + k^2 H^2)$ we find that

$$v = \frac{n^2}{8c_g^2}(c - c_g)\left[c + c_g - \frac{c(c - c_g)(c + 2c_g)^2}{c^3 - 4c_g^3}\right],$$

which is positive if $c^3 < 4c_g^3$. [The singularity at $c^3 = 4c_g^3$ arises in the solution of (21) and at this point the present solution is invalid.] Hence waves with frequency $\sigma < 0.608N$ (or $k < 0.766n\pi/H$) are unstable. This result has also been found by GRIMSHAW (1976) who considered wavetrains in a fluid with constant Brunt–Väisälä frequency modulated by a disturbance in a general direction. It is interesting that the long internal waves are unstable, a conclusion contrasting with that for surface waves (see Section 1), for it is the short surface waves which are unstable when subjected to a modulation of this nature. The weak conoidal internal waves in this case, which were considered by BENJAMIN (1966), thus appear to be unstable as suggested in the discussion of YIH (1974). However, since c approaches c_g as k tends to zero, v tends to zero for long waves, and the terms neglected in making the Boussinesq approximation may ultimately become important in determining the sign of v at very small k. The stability of very long waves is thus uncertain.

3.2. *Numerical studies*

Analytical results for other density profiles are more difficult to obtain and we have resorted to numerical calculations to obtain some further results. The 'tanh' density profile, $\rho = \bar\rho_0(1 - \Delta \tanh az)$, was chosen for study since this not only represents some of the typical oceanic profiles, but tends to the constant N profile as $a \to 0$ with $a\Delta$ remaining finite, and there are analytic solutions for the wave shape, and for σ_2 and v when the fluid depth is infinite. Numerical solutions were compared with the available analytical solutions to verify the accuracy of numerical method. Equation (6) was solved by the numerical method of FJELDSTAD (1935) using 200 points between the boundaries, and the interpolations were made using Simpson's rule.

3.3. *The wave shape*

It was shown by HUNT (1961) that the equation of the interface of a two-layer fluid, $\rho = \rho_2, 0 < z < h_2; \rho = \rho_1, h_2 < z < H = h_1 + h_2$, disturbed by internal waves is

$$\eta = AE^1 + \frac{A^2 k[\rho_2 t_1^2(3-t_2^2)-\rho_1 t_2^2(3-t_1^2)]}{t_1^2 t_2^2(\rho_1 t_1+\rho_2 t_2)}E^2 + 0(E^3) \tag{26}$$

where $t_i = \tanh kh_i$ (I, equation 2.1.3), and A is taken to be real.

The shape of the first mode internal wave in infinitely deep 'tanh' fluid when the Boussinesq approximation is made is determined by

$$\eta_{11} = A \operatorname{sech}^{k/a} az, \qquad \eta_{22} = \frac{-3A^2 k^2}{2(3k+2a)} \operatorname{sech}^{2k/a} az \tanh az, \tag{27}$$

(I, equation 3.3.21). The approximation is valid provided k is not much less than $a\Delta$. At $z = 0$, $\eta_{22} = 0$ whereas the coefficient of E^2 in (26) is in general non-zero, suggesting that the form of η_{22} should tend from an antisymmetric form in deep fluids, $k \sim a\Delta$, to a symmetric form in shallow fluids with small interface thickness, $aH \gg 1$, $k \ll a\Delta$. Figure 1 shows the numerical calculations of the form of η_{11} and η_{22} as k decreases with $aH = 20$ and for four different values of h_2/h_1. The ratio $(\max \eta_{22})/(\max \eta_{11})^2$ corresponds closely with the appropriate values obtained from (26), (27), and the trend in form of η_{22} is as

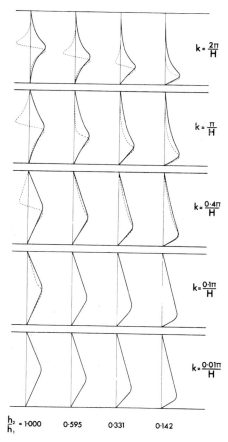

Fig. 1. The curves η_{11} (full lines) and η_{22} (dashed) which define the wave shape for the first wave mode in a density profile $\rho = \bar{\rho}_0(1-0.001 \tanh 20z/H)$ where H is the fluid depth when the boundaries are at $z = h_1$, $-h_2$ for various values of h_2/h_1 and wavenumber k.

expected. It should be noticed that for the two-layer fluid with $h_2/h_1 = 1$, η_{22} is proportional to the difference in densities between the layers (26), and is consequently small, and terms in E^3 determine the wave shape when Ak is much larger than $(\rho_2 - \rho_1)/(\rho_1 + \rho_2)$. In the cases shown in Fig. 1 for $h_2/h_1 \neq 1$ [when $(\rho_2 - \rho_1)/(\rho_1 + \rho_2)$ is not important in determining the coefficient of E^2 in (26)], the calculated solutions for η_{11} and η_{22} with and without the Boussinesq approximation cannot be distinguished, so that the approximation is, in these cases, appropriate, at least to second order.

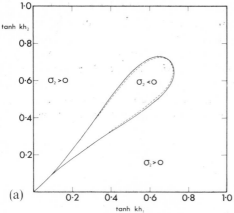

(a)

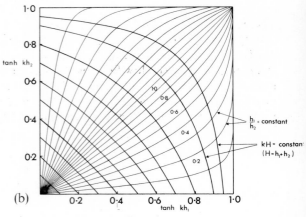

(b)

Fig. 2(a). The locus of points at which the finite amplitude does not affect the frequency of the waves, for internal waves in a two-layer fluid. Within the closed loop, $\sigma_2 < 0$, and outside $\sigma_2 > 0$. The full curve is for $(\rho_2 - \rho_1)/(\rho_1 + \rho_2) = 0.001$ and the dashed line is the curve for $(\rho_2 - \rho_1)/(\rho_1 + \rho_2) \to 0$.

Fig. 2(b). The curves $kH = $ constant and $h_2/h_1 = $ constant.

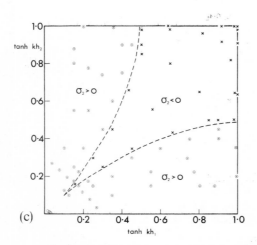

(c)

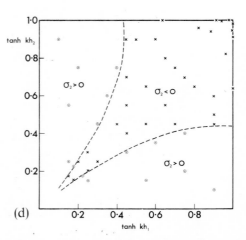

(d)

Fig. 2(c,d). The sign of σ_2 computed at various points in the $\tanh kh_1$, $\tanh kh_2$ plane for the first mode waves in a fluid density $\rho = \bar{\rho}_0(1 - \Delta \tanh az)$ confined between horizontal boundaries at $z = h_1$, $-h_2$. The crosses are points where $\sigma_2 < 0$ and the dots points where $\sigma_2 > 0$. The dashed line shows the inferred position of the line $\sigma_2 = 0$. In (c) $a = 20/(h_1 + h_2)$, $\Delta = 0.001$, and in (d) $a = 6/(h_1 + h_2)$, $\Delta = 0.0033$.

3.4. *Finite amplitude correction to the phase speed*

For waves in the two-layer fluid, HUNT (1961) showed that, when the mean value of the interface displacement is zero,

$$\sigma_2 = \frac{A^2 k^2 \sigma t_1 t_2}{4(\rho_1 t_2 + \rho_2 t_1)} \left\{ \frac{\rho_1}{t_1^3}(2t_1^2 - 1) + \frac{\rho_2}{t_2^3}(2t_2^2 - 1) \right.$$
$$\left. + \frac{1}{4(\rho_1 t_1 + \rho_2 t_2)} \left[\frac{\rho_1}{t_1^2}(3 - t_1^2) - \frac{\rho_2}{t_2^2}(3 - t_2^2) \right]^2 \right\}, \qquad (28)$$

with $\sigma^2 = gk(\rho_2 - \rho_1)t_1 t_2/(\rho_1 t_2 + \rho_2 t_1)$.

If the interface is allowed to change its mean position in the presence of the waves, the depth of the lower layer is increased to

$$h_2 + \frac{A^2 \sigma^2}{4g(\rho_2 - \rho_1)} \left[\frac{\rho_1}{t_1^2} - \frac{\rho_2}{t_2^2} - (\rho_2 - \rho_1) \right]$$

and σ_2 is reduced by an amount

$$\frac{A^2 k^2 \sigma t_1^2 t_2^2}{8(\rho_1 t_2 + \rho_2 t_1)^2} \left[\rho_2 - \rho_1 + \frac{\rho_1}{t_1^2} - \frac{\rho_2}{t_2^2} \right]^2,$$

and, when $\rho_1 \to 0$, the solution approaches that for surface gravity waves given by BENJAMIN, 1967a.)

Figure 2a shows the curve $\sigma_2 = 0$ in the tanh kh_1, tanh kh_2 plane for $(\rho_2 - \rho_1)/(\rho_1 + \rho_2) \to 0$ and for $(\rho_2 - \rho_1)/(\rho_1 + \rho_2) = 0.001$. Near $h_1 = h_2 = H/2$ and tanh$(kH/2) < 0.707$, σ_2 is negative, and the frequency and phase speed of the waves decrease as the amplitude increases. (The limitations to the validity of the approximation very near $h_2/h_1 = 1$ mentioned in Subsection 3.3 should, however, be recalled.) Figure 2b shows the contours $kH = $ constant and $h_2/h_1 = $ constant. By comparing Figs. 2b and 2a, it can be seen that the curve $\sigma_2 = 0$ is intersected twice by the curves $h_2/h_1 = q$, $0.75 < q < 1.33$, and so there exists a limited range of wavenumbers for which the phase speed decreases or, if h_2/h_1 lies outside the range, none at all.

Figures 2 c,d show the sign of σ_2 at various points in the plane tanh kh_1, tanh kh_2 for 'tanh' density profiles with $aH = 20$ and $aH = 6$ respectively. The general trend for long waves (small values of tanh kh_1, tanh kh_2) corresponds to that followed by the curve $\sigma_2 = 0$ in Fig. 2a, but there are short waves for which $\sigma_2 < 0$ in the tanh profiles while $\sigma_2 > 0$ in the corresponding two-layer fluid. (In the limit as $aH \to 0$ the constant density gradient, $\sigma_2 < 0$, case is recovered.) The distribution of points in Figs. 2 c,d suggests that for any fixed depth ratio h_2/h_1 there is a single 'critical' wavenumber at which $\sigma_2 = 0$. For wavenumbers exceeding the critical, the phase speed decreases with increasing wave amplitude, while for wavenumbers less than critical, the phase speed increases.

In the Boussinesq approximation it may be shown analytically that $\sigma_2 < 0$ (for all k/a) for the infinitely deep 'tanh' profile fluid, with the first mode solution, but that in the second mode $\sigma_2 > 0$ as $k/a \to 0$ whilst $\sigma_2 < 0$ for both $a = k$ and $a/k \to 0$.

It has thus been demonstrated that in certain fluids of specified density distributions, there are waves which increase in speed as their amplitude increases ($\sigma_2 > 0$) and others which decrease in speed as their amplitude increases ($\sigma_2 < 0$). This is in contrast to surface gravity waves, which always increase in speed with increased amplitude at this order.

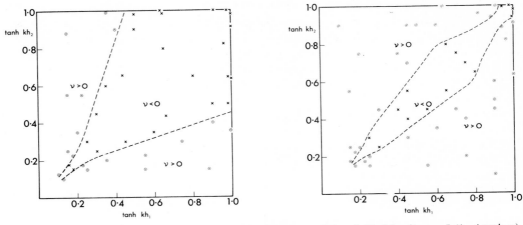

Fig. 3. The sign of the stability parameter v for waves of the first mode in a fluid of density $\rho = \bar{\rho}_0(1 - \Delta \tanh az)$ confined between horizontal boundaries at $z = h_1, -h_2$. The dots are unstable wave trains ($v > 0$) whilst the crosses are waves stable to a modulation in the direction of the waves. The dashed line shows the inferred position of the line $v = 0$. In (a), $a = 20/(h_1 + h_2)$, $\Delta = 0.001$ and in (b) $a = 6/(h_1 + h_2)$, $\Delta = 0.0033$.

3.5. *The stability of the wavetrain*

Two analytic solutions for v have been found for the deep fluid 'tanh' profiles. In general the solution of (21) presents difficulties, unless the right-hand side is identically zero (when $\mathscr{J} = 0$) or when $\mathscr{J}$ is of the form $\mathscr{J} = J_0 \operatorname{sech}^{2 + 2k/a} \tanh az$. The former case has $k/a = 0.28$ and $v > 0$ and is thus unstable, whilst the latter has $k/a = 0.964$ and $v > 0$ and is also unstable.

The results of numerical calculations for the tanh profiles with $aH = 20$ and $aH = 6$ are shown in Fig. 3 a,b, the points indicating stability or instability. It may be shown that v scales with Δ and so, provided Δ is so small that the Boussinesq approximation is valid, the curve $v = 0$ should not be affected by variation in Δ. The results indicate a considerable increase in the extent at the unstable region with decreasing aH. In general the trends suggest that in a given fluid instability will occur, if at all, for long waves and that the shorter waves will be stable.

4. DISCUSSION

Whilst general conclusions for an ocean with arbitrary density distribution may be drawn only with caution, the examples described in Section 3 indicate that trains of internal waves, unlike surface waves, may become unstable if the wavelength of the waves is sufficiently long. Shorter waves appear to be more prone to instability in a shallow thermocline than in a thermocline of similar thickness nearer mid-depth, and as the thickness of the thermocline increases whilst its depth is held constant, shorter and shorter waves become unstable.

The similarity of the controlling equation (24) to that found by HASIMOTO and ONO (1972) implies that stable groups of waves ('solitons') may evolve during the break-up of unstable trains of waves, as described by ZAKHAROV and SHABAT (1972).

Regular trains of internal waves, even 'quasi regular' trains, seem to be rare in the ocean. The packets of waves observed by APEL, BYRNE, PRONI and CHARNELL (1975), or the trains of waves following internal surges mentioned in the Introduction, are some of the most regular which are known. Making appropriate estimates of what the density variation might be (or *is* in those cases where it is known), we conclude that even these waves appear to be unstable to the particular modulation considered in this paper.

The analysis has been restricted to modulations in the direction of the wavetrain. The study of DAVEY and STEWARTSON (1974) and others, referred to in the Introduction, indicates that a yet broader region of instability may be found when modulations inclined to the direction of the waves are included, and the present results may be regarded as demonstrating those areas in which instability is certain, but not those in which stability is ensured. In this connection the results of GRIMSHAW (1972), CHO, LIU and YEH (1975) and LEONOV and MIROPOLSKI (1975) for internal waves in an unbounded fluid of constant Brunt–Väisälä frequency should be mentioned. Waves in certain directions were found to be unstable, whilst stable wavetrains could exist in other directions.

The Konteweg–de Vries equation can be obtained from the non-linear Schrödinger equation (24) in the shallow water limit (HASIMOTO and ONO, 1972, para. 7) as might be expected (BENJAMIN, 1966). However, it does not appear possible to obtain the equivalent equation found by BENJAMIN (1967b) for long waves when the density variation is confined to a narrow region and the depth is infinite, and this configuration requires further attention.

Added in proof—Since preparing this I have discovered that BORISENKO VORONOVICH, LEONOV and MIROPOLSKIY (1976) have examined the stability of an internal wave train in conditions similar to those studied here.

Acknowledgement—I am grateful to Professor K. STEWARTSON for drawing my attention to Dr. GRIMSHAW's work.

REFERENCES

APEL J. R., H. M. BYRNE, J. R. PRONI and R. L. CHARNELL (1975) Observations of oceanic internal and surface waves from the Earth Resources Technology Satellite. *Journal of Geophysical Research*, **80**, 865–881.

BENJAMIN T. B. (1966) Internal waves of finite amplitude and permanent form. *Journal of Fluid Mechanics*, **25**, 241–270.

BENJAMIN T. B. (1967a) Instability of periodic wavetrains in nonlinear dispersive systems. *Proceedings of the Royal Society*, A **299**, 59–75.

BENJAMIN T. B. (1967b) Internal waves of permanent form in fluids of great depth. *Journal of Fluid Mechanics*, **29**, 559–592.

BENJAMIN T. B. and J. E. FEIR (1967) The disintegration of wavetrains on deep water. Part 1. Theory. *Journal of Fluid Mechanics*, **27**, 417–430.

BORISENKO Yu. D., A. G. VORONOVICH, A. I. LEONOV and Yu. Z. MIROPOLSKIY (1976) Towards a theory of non-stationary weekly nonlinear internal waves in a stratified fluid. *Izvestiya Academii nauk USSR, Atmosphere and Oceanic Physics*, **12**, 174–179 (English Edition).

BREKHOVSKIKH L. M., K. V. KONJAEV, K. D. SABININ and A. N. SERIKOV (1975) Short-period internal waves in the sea. *Journal of Geophysical Research*, **80**, 856–864.

CAVANIE A. G. (1972) Observations de fronts interne dans le Détroit de Gibraltar pendant la campagne océanographique otan 1970 et interpretation des results par un modèle mathématique. *Mémoires de la Société royale des sciences de Liège*, **2**(6), 27–41.

CHO H. R., C. H. LIU and K. C. YEH (1975) Nonlinear internal waves in the atmosphere. *Journal of the Acoustical Society of America*, **57**, 14–19.

DAVEY A. and K. STEWARTSON (1974) On three-dimensional packets of surface waves. *Proceedings of the Royal Society of London*, A **338**, 101–110.

DAVIS R. E. and A. ACRIVOS (1967) The stability of oscillatory internal waves. *Journal of Fluid Mechanics*, **30**, 723–736.

FJELDSTAD J. E. (1935) Internal waves. *Geofysiske Publikasjoner*, **10**, 1–35.

GARGETT A. G. and B. A. HUGHES (1972) On the interaction of surface and internal waves. *Journal of Fluid Mechanics*, **52**, 179–191.

GRIMSHAW R. (1972) Nonlinear internal gravity waves in a slowly varying medium. *Journal of Fluid Mechanics*, **54**, 193–207.

GRIMSHAW R. (1976) The modulation and stability of an internal gravity wave. *Mémoires de la Société royale des sciences de Liège*, 6e Series, X, 299–314.

HALPERN D. (1971) Semidiurnal internal tides in Massachusetts Bay. *Journal of Geophysical Research*, **76**, 6573–6581.

HASIMOTO H. and H. ONO (1972) Nonlinear modulation of gravity waves. *Journal of the Physical Society of Japan*, **33**, 805–811.

HAYES W. D. (1973) Group velocity and nonlinear dispersive wave propagation. *Proceedings of the Royal Society*, A, **332**, 199–221.

HUNKINS K. and M. FLIEGEL (1973) Internal undular surges in Seneca Lake: a natural occurrence of solitons. *Journal of Geophysical Research*, **78**, 539–548.

HUNT J. N. (1961) Interfacial waves of finite amplitude. *La Houille Blanche*, **16**, 515–531.

LAFOND E. C. (1962) Internal waves, Part 1 in *The Sea*, M. N. HILL, editor, Interscience Publishers, **1**, 731.

LEONOV A. I. and YU. Z. MIROPOLSKY (1975) Short-wave approximation in the theory of non-linear steady internal gravity waves. *USSR Physics of Oceans and Atmospheres*, **11**, 1169–1178.

MCINTYRE M. E. (1973) Mean motions and impulse of a guided internal gravity wave packet. *Journal of Fluid Mechanics*, **60**, 801–813.

SABININ K. D. (1973) Certain features of short period internal waves in the ocean. *Izvestiya Akademii nauk USSR, Atmospheric and Oceanic Physics*, **9**, 32–36 (English Edition).

SAMODUROV A. S. (1974) Plane non-linear gravity waves in a stratified fluid. *Izvestiya Akademii nauk USSR, Atmospheric and Oceanic Physics*, **10**, 555–556 (English Edition).

STUART J. T. (1960) On the non-linear mechanics of wave disturbances in stable and unstable parallel flows, Part I. *Journal of Fluid Mechanics*, **9**, 353–370.

THORPE S. A. (1968) On the shape of progressive internal waves. *Philosophical Transactions of the Royal Society of London*, Series A, **263**, 563–614.

THORPE S. A. (1974) Near-resonant forcing of a shallow two-layer fluid: A model for the internal surge in Loch Ness? *Journal of Fluid Mechanics*, **63**, 509–527.

THORPE S. A. (1975) The excitation, dissipation and interaction of internal waves in the deep ocean. *Journal of Geophysical Research*, **80**, 328–338.

WHITHAM G. B. (1967) Non-linear dispersion of water waves. *Journal of Fluid Mechanics*, **27**, 399–412.

YIH C. S. (1974) Progressive waves of permanent form in continuously stratified fluids. *Physics of Fluids*, **17**, 1489–1495.

YUEN H. C. and B. M. LAKE (1975) Nonlinear deep water waves: theory and experiment. *Physics of Fluids*, **18**, 956–960.

ZAKHAROV V. E. and A. B. SHABAT (1972) Exact theory of two-dimensional self-focusing and one-dimensional self-modulation of waves in non-linear media. *Soviet Physics: Journal of Experimental and Theoretical Physics*, **34**, 62–69.

ZIEGENBEIN J. (1969) Short internal waves in the Strait of Gibraltar. *Deep-Sea Research*, **16**, 479–487.

On the steady-state nature of the Mediterranean outflow step structure

A. J. ELLIOTT* and R. I. TAIT†

Abstract—An approximate relationship for the diffusive lifetime of a thermohaline step layer is derived which shows the lifetime to be about 10% of the value obtained for an inversion of comparable dimensions. Calculated lifetimes are typically less than 20 hours.

In view of the experimental evidence for the very long lifetimes of established step-layer systems, it is concluded that the stratification is maintained against diffusive erosion by a stirring mechanism within the mixed layers, and that a steady-state condition prevails. A simple numerical model of salt finger transports is applied to the step layers observed beneath the Mediterranean outflow. The results obtained are consistent with the layers being in a steady state with respect to the double-diffusive fluxes of heat and salt.

INTRODUCTION

DURING the past decade, instruments capable of obtaining continuous profiles of salinity and temperature against depth within the ocean have revealed the complexities of the thermohaline properties of water masses. One particular example is to be found beneath the core of the outflowing Mediterranean water in the northeast Atlantic, where continuous profiles revealed the presence of well-defined layers within the depth range 1200 m to 1800 m (TAIT and HOWE, 1968, 1971).

TURNER (1967) proposed that the layers arose, and were maintained, as a result of double-diffusive convection, since the vertical gradients of temperature and salinity were in the right sense for the salt finger mechanism. However, further work (STERN and TURNER, 1969) showed that the observed interface thicknesses were an order of magnitude greater than could be explained theoretically. Some recent observational results (WILLIAMS, 1974, 1975) have supported the suggestion that salt fingers may exist within the layer interfaces and this paper examines the possibility that the layers, once created, are maintained in a steady state by salt finger fluxes. By deriving an expression for the lifetime of a step layer subject to the influence of diffusion, it is shown that the layers could only have persisted in the absence of a stirring mechanism if the effective vertical diffusivities were much smaller than those thought to exist in the ocean (MUNK, 1966). For the gravitationally opposite case of cool fresh water overlying warm salty water, HUPPERT and TURNER (1972) have shown that a thermohaline process could be responsible for the generation and maintenance of layers in Lake Vanda.

* Chesapeake Bay Institute, The Johns Hopkins University, Baltimore, MD 21218.
† SACLANT ASW Research Centre, 19026 San Bartolomeo, La Spezia.

THE FLUX EQUATIONS

Turner (1967) showed that the salt finger flux of salt, F_s, could be related to the solid plane flux, F_*, by a relationship of the form

$$F_s = f(\alpha \Delta T / \beta \Delta S) F_* \tag{1}$$

where

$$F_* = 1.3 \times 10^{-7} (\Delta S)^{4/3}, \tag{2}$$

ΔS being the salinity step across the interface. For this paper, α and β are *in situ* values, evaluated as

$$\alpha = -\frac{1}{\rho} \frac{\partial \rho}{\partial T} \quad \text{and} \quad \beta = \frac{1}{\rho} \frac{\partial \rho}{\partial S}.$$

Linden (1973), using dimensional analysis, suggested that f should be of the form

$$f(\alpha \Delta T / \beta \Delta S) \propto (\alpha \Delta T / \beta \Delta S)^{-1/3}.$$

However, in the case considered here, of layers beneath the Mediterranean outflow, $1 < \alpha \Delta T / \beta \Delta S < 2$ and it is considered sufficient to make a linear approximation to f using Linden's experimental results. Thus, the function f was taken to be of the form

$$F_s = 5(10 - \alpha \Delta T / \beta \Delta S) F_*. \tag{3}$$

Using (2), equation (3) for the vertical salt flux becomes

$$F_s = 6.5 \times 10^{-7} (10 - \alpha \Delta T / \beta \Delta S)(\Delta S)^{4/3}. \tag{4}$$

From Turner (1967), the heat flux is given as

$$\frac{\alpha H_s}{\beta F_s} = 0.56,$$

independent of $\alpha \Delta T / \beta \Delta S$, and hence

$$H_s = 3.04 \times 10^3 F_s. \tag{5}$$

Consider the situation shown schematically in Fig. 1, where a horizontal layer of thickness h is between two layers of infinite thickness. Let S_u, T_u be the values of salinity and

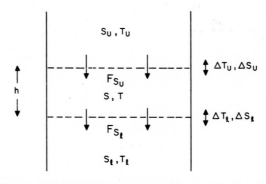

Fig. 1.   Salt finger transports across a layer of thickness h.

temperature in the upper layer, S, T be the corresponding values in the middle layer, and S_l, T_l be the values in the lower layer. Since the fluxes are to be driven by salt finger transports, it will be required that

$$T_u > T > T_l \quad \text{and} \quad S_u > S > S_l.$$

Differential equations for the salinity and temperature of the middle layer are:

$$\frac{dS}{dt} = \frac{10^3}{\rho h}(F_{S_u} - F_{S_l})$$

(6)

and

$$\frac{dT}{dt} = \frac{3.04 \times 10^3}{\rho h Cp}(F_{S_u} - F_{S_l}),$$

(7)

where

$$F_{S_u} = 6.5 \times 10^{-7}(10 - \alpha \Delta T_u/\beta \Delta S_u)(\Delta S_u)^{4/3}$$

(8)

and

$$F_{S_l} = 6.5 \times 10^{-7}(10 - \alpha \Delta T_l/\beta \Delta S_l)(\Delta S_l)^{4/3}.$$

(9)

Cp was the *in situ* value of specific heat, taken as $0.93 \, \text{cal g m}^{-1}\,°\text{C}^{-1}$.

Equations (6)–(9) were applied to the system of layers described by ELLIOTT, HOWE and TAIT (1974). From those data, eleven well-defined and vertically adjacent layers in the depth range 1360–1700 m were selected and mean values of temperature, salinity and thickness were derived for each layer. (Since the coherence of the layers was established, from closely spaced stations, over a horizontal scale of order 20 nautical miles,* layer parameters could be averaged horizontally within each layer. For a given layer, the standard deviation about the horizontally derived mean was of the order of 0.01°C for temperature and 0.008‰ for salinity.)

Using the observed values as initial conditions and keeping the values within the top and bottom layers fixed as boundary conditions, (6)–(9) were solved numerically to determine the steady-state values within the interior layers. The results are presented in Table 1. In almost all cases, the difference between the observed and predicted value was less than the natural variability within each layer (estimated above as being 0.01°C and 0.008‰). Even though the layers near the top and bottom of the sequence were forced by the influence of the imposed boundary conditions, there is an extremely good agreement between observed and predicted values near the middle of the sequence away from the influence of the boundaries. The average time taken for a layer to reach steady state was of the order of 12 days.

THE DIFFUSIVE LIFETIME OF A STEP LAYER

In an early microstructure paper, STOMMEL and FEDEROV (1967) presented a formula for calculating the lifetime of a layer. The formula was subsequently applied by TAIT and HOWE (1968) and ZENK (1970) to layers beneath the Mediterranean outflow. This section

---

* 1 nautical mile = 1.853 km.

shows that the formula referred to by Stommel and Federov (1967) is applicable to the problem of an idealized inversion; an analogous definition is then given for the lifetime of a step layer.

The inversion layer

Consider an idealized temperature (salinity) inversion layer in a uniform ocean at rest. The ocean is supposed to be of infinite thermal (salt) capacity so that it acts as an absorbing media, a situation which is analogous to the heat transfer problem of an infinite slab with absorbing boundaries shown schematically in Fig. 2. If the origin of the depth axis is arbitrarily taken to lie in the upper interface of the inversion, and if the background temperature of the ocean above the upper interface is taken as the origin of the temperature scale, then the problem requires solution of the diffusion equation.

$$\frac{\partial \phi}{\partial t} = k \frac{\partial^2 \phi}{\partial z^2},$$

subject to the initial condition $\phi = T$, for $0 < z < L$ and the boundary conditions that $\phi = 0$ at $z = 0$ and $z = L$ for all t. The problem can be solved by separation of variables to give

$$\phi = \frac{4T}{\pi} \sum_{m=0}^{\infty} \left[\exp\left(-k \left\{ \frac{(2m+1)\pi}{L} \right\}^2 t \right) \cdot \frac{1}{(2m+1)} \cdot \sin\left\{ \frac{(2m+1)\pi z}{L} \right\} \right].$$

(Chapman, 1960, p. 110.)

If we consider the temperature at the centre of the layer, then $z = L/2$ and

$$\phi = \frac{4T}{\pi} \sum_{m=0}^{\infty} \left[\exp\left(-k \left\{ \frac{(2m+1)\pi}{L} \right\}^2 t \right) \frac{1}{(2m+1)} (-1)^m \right].$$

By putting $t = h^2/2k$, where h is the *half thickness* of the layer, we obtain

$$\phi = \frac{4T}{\pi} \sum_{m=0}^{\infty} \left[\exp\left(-\frac{(2m+1)^2}{8} \pi^2 \right) \frac{1}{(2m+1)} (-1)^m \right]$$

$$= \frac{4T}{\pi} \left[e^{-(\pi^2/8)} - \tfrac{1}{3} e^{-(9\pi^2/8)} + \tfrac{1}{5} e^{-(25\pi^2/8)} \ldots \right]. \quad (10)$$

Expression (10) gives the temperature at the centre of the layer after a time of $L^2/8k$; summing just the first two terms of the series gives $\phi = 0.371T$. This differs by less than 1% from $\phi = e^{-1}T$, i.e. $L^2/8k$, or $h^2/2k$ where $h = L/2$, is the time taken for the temperature at the centre of an idealized inversion to reach e^{-1} of its initial value. Taking values of $L = 20$ m and $k = 5$ cm^2 s^{-1} gives a lifetime of 10^5 s (27.8 hours).

The step layer

The temperature (salinity) profile for an idealized step layer is shown in Fig. 2. The diffusion problem cannot be solved immediately by separation of variables because of the non-zero boundary condition at $z = L$. However, the profile shown in Fig. 2 is equivalent to the superposition of the two profiles shown in Fig. 3.

Table 1.   Observed and predicted salinity (‰) and temperature (°C) values within the layers. (Elapsed time 300 hours.)

S_{obs}	S_{pred}	T_{obs}	T_{pred}	$S_{obs}-S_{pred}$	$T_{obs}-T_{pred}$
35.690	35.690	8.26	8.26	—	—
35.666	35.662	8.11	8.10	0.004	0.01
35.637	35.631	7.87	7.85	0.006	0.02
35.600	35.599	7.69	7.69	0.001	0.00
35.559	35.567	7.45	7.48	−0.008	−0.03
35.536	35.536	7.28	7.28	0.000	0.00
35.506	35.506	7.08	7.08	0.000	0.00
35.480	35.476	6.91	6.90	0.004	0.01
35.443	35.446	6.68	6.69	−0.003	−0.01
35.414	35.418	6.48	6.49	−0.004	0.01
35.392	35.393	6.36	6.36	−0.001	0.00
35.372	35.367	6.23	6.22	0.005	0.01
35.341	35.341	6.03	6.03	—	—

Distribution 3(a) has the time independent solution

$$\phi = 2Tz/L, \tag{11}$$

while 3(b) has solution

$$\phi = \frac{2T}{\pi} \sum_{m=1}^{\infty} \left[\frac{1}{m} \exp\left(-k\frac{4m^2\pi^2}{L^2}t\right) \cdot \sin\left(\frac{2m\pi z}{L}\right) \right]. \tag{12}$$

By linearity, the solution to the step problem is obtained by adding (11) and (12).

Lifetime of a step layer

There are a number of ways in which the lifetime of a step layer can be defined. Consider Fig. 4 which shows part of a diffusing step plotted against normalized axes. OCD represents the original profile and OAD is the steady-state distribution. The lifetime will be defined as being the time taken for AB to equal e^{-1} AC. (Other definitions are possible, for example the projection onto OAD of the point on the curve OBD at which the curve is parallel to OAD could be considered the fundamental measure.)

The diffusion of a step layer with $L = 20$ m and $k = 5$ cm^2 s^{-1} was solved numerically using (11) and (12), and the ratio AB/AC computed. It was found that the ratio equalled e^{-1} after 9.9×10^3 s (2.75 hours), thus using this definition the lifetime of the step was less than 10% of the lifetime of an inversion with similar dimensions.

Assuming that the lifetime of a step layer can be expressed in the form

$$t = h^2/(nk) = L^2/(4nk),$$

then n can be found; using the values from the above example we find that

$$n = 10^3(49.5)^{-1} \simeq 20.$$

In terms of the geometric definition, we have from equations (11) and (12)

$$\frac{\phi}{2T} = \frac{z}{L} + \frac{1}{\pi} \sum_{m=1}^{\infty} \left[\frac{1}{m} \cdot \exp\left(-\frac{m^2\pi^2}{n}\right) \cdot \sin\left(\frac{2m\pi z}{L}\right) \right]$$

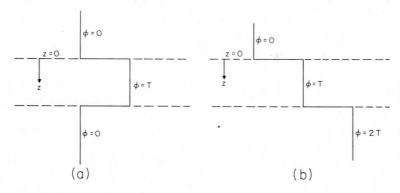

Fig. 2.   Initial distributions for: (a) an idealized inversion layer, (b) an idealized step layer.

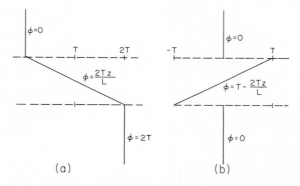

Fig. 3.   Decomposition of the initial step profile.

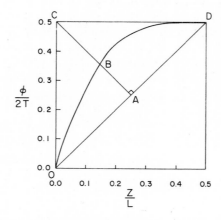

Fig. 4.   Geometric definition of a step-layer lifetime.

while from Fig. 4 the values of $\phi/2T$ and z/L corresponding to $AB/AC = e^{-1}$ are:

$$\frac{\phi}{2T} = 0.342 \quad \text{and} \quad \frac{z}{L} = 0.158.$$

Thus,

$$0.184 = \frac{1}{\pi} \sum_{m=1}^{\infty} \left[\frac{1}{m} \cdot \exp\left(-\frac{m^2\pi^2}{n}\right) \cdot \sin(0.316m\pi) \right],$$

putting $n = 20$ and summing the first two terms gives a value of 0.183 for the r.h.s. of the above equation.

Thus for lifetimes in the two cases

$$t = h^2/2k \quad \text{for an inversion,}$$

and

$$t = h^2/20k \quad \text{for a step layer.} \tag{13}$$

A step layer 20 m thick, with $k = 5 \, \text{cm}^2 \, \text{s}^{-1}$, and assuming no replenishment from above, has a lifetime of less than 3 hours. This is about 10% of the value obtained using an expression for the lifetime of an inversion, and less than 3% of the time estimated by TAIT and HOWE (1968). The step layers reported by ELLIOTT, TAIT and HOWE (1974) had a lifetime of at least 20 days, possibly significantly longer, implying using (13) that $k \leqslant 3 \times 10^{-2} \, \text{cm}^2 \, \text{s}^{-1}$. For the Tyrrhenian Sea step structure, MOLCARD and TAIT (1977) have obtained evidence for a lifetime in excess of 3 years. It is most likely therefore that the persistence of the stratification is due not to a low diffusivity but to a stirring mechanism, within the homogeneous layers, which maintains the interfaces against the effects of diffusion. In this case, steady-state conditions could prevail, a situation not incompatible with the double diffusive hypothesis as we have shown. Since the observations by WILLIAMS (1975) have supported the suggestion that salt fingers may exist within the layer interfaces, the stability analysis made by HUPPERT (1971) should also be considered as a means of defining layer lifetime.

Acknowledgements—This study was made at the University of Liverpool, Department of Oceanography, and was supported by a grant from the Natural Environment Research Council.

REFERENCES

CHAPMAN A. J. (1960) *Heat transfer*, Macmillan, 452 pp.

ELLIOTT A. J., M. R. HOWE and R. I. TAIT (1974) The lateral coherence of a system of thermohaline layers in the deep ocean. *Deep-Sea Research*, **21**, 95–107.

HUPPERT H. E. (1971) On the stability of a series of double-diffusive layers. *Deep-Sea Research*, **18**, 1005–1021.

HUPPERT H. E. and J. S. TURNER (1972) Double-diffusive convection and its implications for the temperature and salinity structure of the ocean and Lake Vanda. *Journal of Physical Oceanography*, **2**, 456–461.

LINDEN P. F. (1973) On the structure of salt fingers. *Deep-Sea Research*, **20**, 325–340.

MOLCARD R. and R. I. TAIT (1977) The steady state of the step structure in the Tyrrhenian Sea. In *Voyage of Discovery*, M. V. ANGEL, editor, *Deep-Sea Research*, supplement to Vol. **24**, 211–233.

MUNK W. H. (1966) Abyssal recipes. *Deep-Sea Research*, **13**, 707–730.

STERN M. E. and J. S. TURNER (1969) Salt fingers and convecting layers. *Deep-Sea Research*, **16**, 497–511.

STOMMEL H. and K. N. FEDEROV (1967) Small scale structure in temperature and salinity near Timor and Mindanao. *Tellus*, **19**, 306–325.

Tait R. I. and M. R. Howe (1968) Some observations of thermohaline stratification in the deep ocean. *Deep-Sea Research*, **15**, 275–280.

Tait R. I. and M. R. Howe (1971) Thermohaline staircase. *Nature, London*, **231**(5299), 178–179.

Turner J. S. (1967) Salt fingers across a density interface. *Deep-Sea Research*, **14**, 599–611.

Williams A. J. (1974) Salt fingers observed in the Mediterranean outflow. *Science*, **185**, 941–943.

Williams A. J. (1975) Images of ocean microstructure. *Deep-Sea Research*, **22**, 811–829.

Zenk W. (1970) On the temperature and salinity structure of the Mediterranean water in the Northeast Atlantic. *Deep-Sea Research*, **17**, 627–631.

The steady state of the step structure in the Tyrrhenian Sea

R. Molcard and R. I. Tait

SACLANT ASW Research Centre, V. le S. Bartolomeo, 400, 19026 La Spezia, Italy

Abstract—A deep step structure, in which mixed layers alternate with high gradient interfaces, is a characteristic feature of the Tyrrhenian Sea. Three oceanographic cruises (May 1972, May 1973 and October 1974), supported by the SACLANT ASW Research Centre, allow a precise description of this phenomenon.

Between 600 and 1500 m, under the maximum of temperature and salinity produced by the Levantine Intermediate Water, ten homogeneous layers were always found in the deepest area of the Tyrrhenian Sea.

The θ–S characteristic of the ten homogeneous layers are constant to within 0.01°C and 0.01‰ from one year to another. This remarkable temporal stability of the stratification observed over 3 years modifies the common hypothesis of an isolated double diffusive system which is expected to evolve in time.

INTRODUCTION

A CHARACTERISTIC feature of the Tyrrhenian Sea is the presence of a persistent thermohaline step structure covering a depth range 600 to 1500 m (MOLCARD and WILLIAMS, 1975; JOHANNESSEN and LEE, 1974). At the Strait of Sicily the Levantine Intermediate Water flows westward, beneath the inflowing surface current, as a relatively homogeneous water mass with potential temperature, θ, ≈ 14°C and salinity, S, ≈ 38.7‰. It sinks to its level of stability at about 400 m and spreads out above the colder and fresher ($\theta \approx 12$°C; $S \approx 38.4$‰) deep water of the western Mediterranean thus producing a maximum in the temperature and salinity profile. Over a wide area of the Tyrrhenian Sea the transition between the warm saline water and the deep water takes the form of a succession of homogeneous layers separated by relatively sharp interfaces. Similar hydrological conditions are found west of Gibraltar beneath the Mediterranean outflow (TAIT and HOWE, 1970; ZENK, 1970) and although there are significant differences in the stratification found in the two areas, the phenomena are clearly related.

Three oceanographic cruises, in May–June 1972, May 1973 and October 1974, have been made by the SACLANT ASW Research Centre R/V *Maria Paolina G.*, to study the step-layer zone of the Tyrrhenian Sea. A comparison of the observations from year to year has shown that the stratification as a whole remains essentially static, indicative of a lifetime greater than 3 years and therefore of steady-state conditions with respect to the transports of heat and salt.

Figure 1 shows the station positions for each year of observation. Temperature and salinity profiles were taken with Plessey 9040 STD which was calibrated in the laboratory

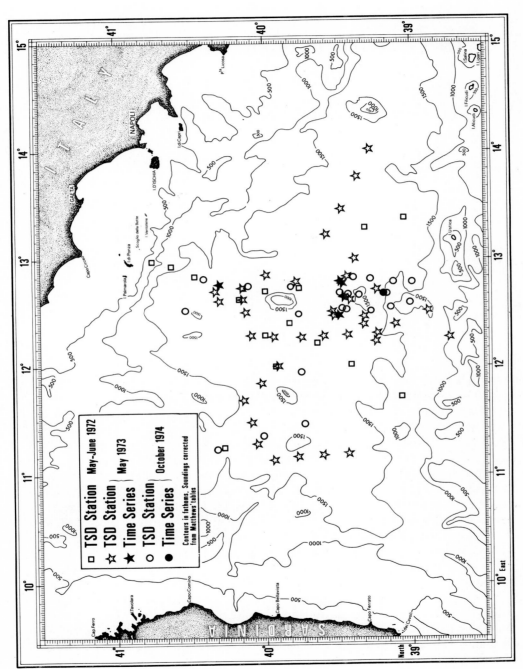

Fig. 1.   Station positions for each year of observation.

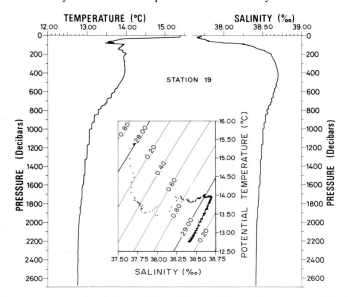

Fig. 2.   Potential temperature and salinity profiles with θ–S diagram for Sta. 19, 1973 (39°27′N: 12°49′E).

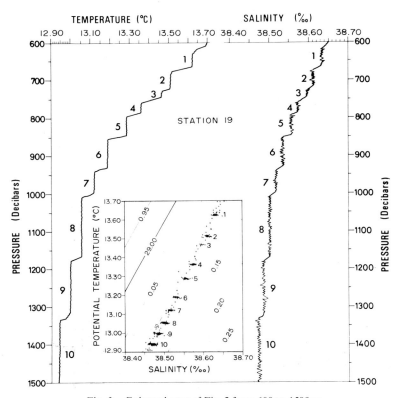

Fig. 3.   Enlarged part of Fig. 2 from 600 to 1500 m.

Table 1. θ–S histogram based on a $0.01°C \times 0.01\%_0$ window, with the choice of layer windows in which mean temperature and mean salinity were calculated in Table 2.

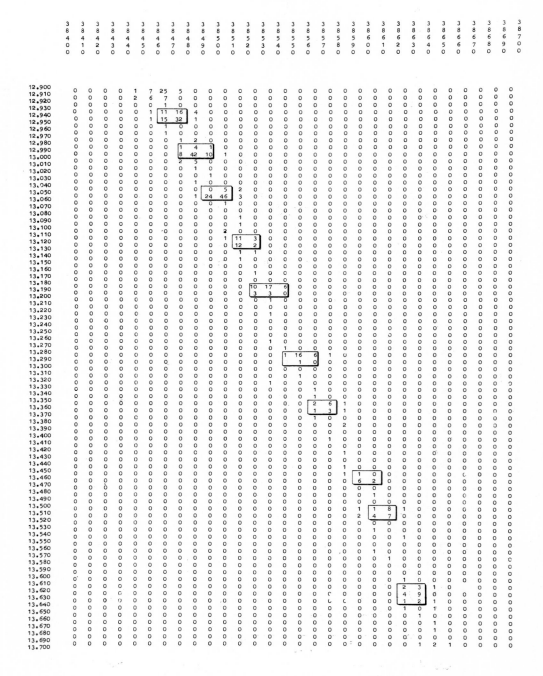

Table 2. *Mean potential temperatures* $\bar{\theta}_i$ *and salinities* $\bar{S}_i$ *with their standard deviations* $\hat{\sigma}$ *for each layer i*

N_i is the number of data points within each layer and $\triangle\theta$ and $\triangle S$ are the differences in temperature and salinity between consecutive layers

Position 39° 27′N 12° 49′E

MAY 1973 (4 MAY 73) Intensity 7.1

θ-S interval		$\bar{\theta}_i$ °C	$\hat{\sigma}\times10^3$ °C	$\triangle\theta\times10^3$ °C	$\bar{S}_i$ ‰	$\hat{\sigma}\times10^3$ ‰	$\triangle S\times10^3$ ‰	σ_θ	N_i	i
12.930	38.460	12.941	3,2		38.472	4,4		29.112	74	10
12.950	38.480									
				54			14			
12.980	38.475	12.995	2,8		38.486	4,3		29.112	61	9
13.000	38.495									
				58			16			
13.040	38.490	13.053	2,5		38.502	4,0		29.112	75	8
13.060	38.510									
				67			15			
13.110	38.510	13.120	2,4		38.517	3,7		29.110	28	7
13.130	38.530									
				67			17			
13.180	38.525	13.187	2,7		38.534	4,6		29.109	36	6
13.200	38.545									
				98			23			
13.270	38.545	13.285	2,3		38.557	4,1		29.106	22	5
13.290	38.565									
				74			15			
13.350	38.560	13.359	2,0		38.572	4,1		29.102	12	4
13.370	38.580									
				105			25			
13.450	38.590	13.464	2,5		38.597	3,9		29.100	9	3
13.470	38.610									
				46			14			
13.500	38.600	13.510	3,4		38.611	4,4		29.101	20	2
13.520	38.620									
				112			21			
13.615	38.620	13.622	3,5		38.632	4,0		29.092	19	1
13.635	38.640									

both before and after each cruise, and checked *in situ* with Nansen casts. Details of the calibrations are given in the table below:

STD calibration

Cruise	Temperature error (°C)	Salinity error (‰)
May 1972	0.0±0.01	+0.05±0.02
May 1973	0.0±0.01	0.0 ±0.02
October 1974	0.0±0.01	0.0 ±0.02

All the data, about 200 profiles, were edited for errors due to salinity spikes filtered and subsampled every 2 dbar.

Figure 2 shows a typical potential temperature and salinity profile together with the corresponding θ–S diagram. The ten most pronounced layers, numbered 1 to 10, are described below.

CHARACTERISTICS OF THE LAYERS

Expanded sections of the Fig. 2 profiles are given in Fig. 3. The temperature gradient in the homogeneous layers was essentially adiabatic within the resolution of the STD. On the θ–S diagram the cross accumulations (1 cross every 2 dbar) represent the homogeneous

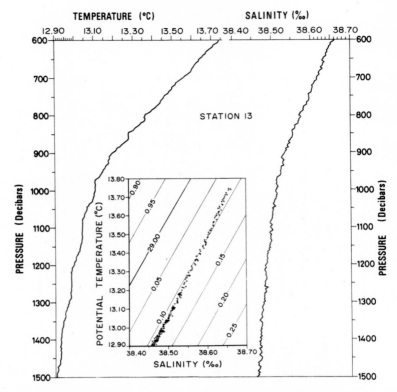

Fig. 4.   Potential temperature and salinity with θ–S diagram from 600 to 1500 m for Sta. 13, 1973 (39°46′N; 11°10′E).

layers. Table 1 shows an histogram, in θ–S space, which records the number of measurements, again at 2-dbar intervals, within a 0.01‰ × 0.01°C 'window'. The ten homogeneous layers are easily identified as an accumulation of higher numbers enclosed by rectangles. A larger window, 0.02‰ × 0.02°C, was then determined in which mean values of temperature $\bar{\theta}_i$ and salinity $\bar{S}_i$ were calculated for each layer i, together with the standard deviations $\hat{\sigma}$ and the number of points N_i within each window. The layer thickness is approximately $2N_i$ metres. The difference in temperature $\triangle\theta$ and salinity $\triangle S$ between adjacent layers, and the potential density σ_θ for each layer were also determined. These data are given in Table 2.

Intensity of layering

Let us define a 'perfect' step-layer system as one in which the interfaces have a thickness less than 2 dbar, and within the homogeneous layers, $d\theta/dz = 0$. In order to quantify, in terms of this criterion, the quality of the stratification observed on any profile, the following procedure was adopted.

The sum of temperatures at 2 dbar increments was calculated from 500 to 1500 dbar giving a quantity Q (proportional to the total amount of heat in the water column). Then

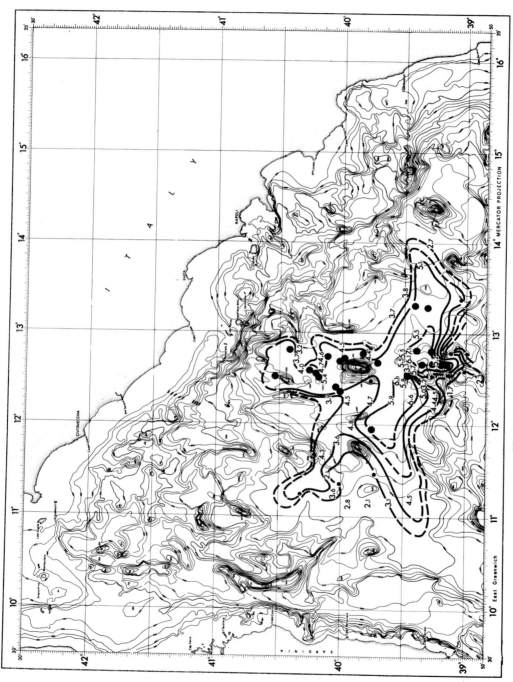

Fig. 5. Iso-intensity contours. Intervals of 1 intensity unit.

Table 3. Layer parameters corresponding to the profile shown in Fig. 4

Position 39°46′N 11°10′E

MAY 1973   (9 MAY 73)   Intensity 2.2

θ-S interval	θ̄ᵢ°C	σ̂×10³ °C	Δθ×10³ °C	S̄ᵢ‰	σ̂×10³ ‰	ΔS×10³ ‰	σθ	Nᵢ	i
12.930 38.460 / 12.950 38.480	12.936	5,1		38.470	4,2		29.112	28	10
			49			13			
12.980 38.475 / 13.000 38.495	12.985	5,2		38.483	4,6		29.112	25	9
			64			16			
13.040 38.490 / 13.060 38.510	13.049	5,4		38.499	5,2		29.111	11	8
			71			16			
13.110 38.510 / 13.130 38.530	13.120	6,6		38.515	2,1		29.108	4	7
			66			19			
13.180 38.525 / 13.2C0 38.545	13.186	5,0		38.534	3,5		29.110	4	6
			89			19			
13.270 38.545 / 13.290 38.565	13.275	4,6		38.553	4,1		29.105	11	5
			86			19			
13.350 38.560 / 13.370 38.580	13.361	3,9		38.572	4,6		29.102	10	4
			98			23			
13.450 38.590 / 13.470 38.610	13.459	3,8		38.595	2,6		29.099	6	3
			47			11			
13.500 38.600 / 13.520 38.620	13.506	5,3		38.606	4,9		29.097	5	2
			116			26			
13.615 38.620 / 13.635 38.640	13.622	5,5		38.632	2,1		29.093	5	1

Table 4. Mean parameters for all the stations made in 1973 with intensity greater than 3

MAY 1973 (88 Stations)

θ-S interval	θ̄ᵢ°C	σ̂×10³ °C	Δθ×10³ °C	S̄ᵢ‰	σ̂×10³ ‰	ΔS×10³ ‰	σθ	Nᵢ	i
12.930 38.460 / 12.950 38.480	12.938	3.8		38.471	4.6		29.112	4127	10
			55			14			
12.980 38.475 / 13.000 38.495	12.993	4.1		38.485	4.5		29.112	4875	9
			59			15			
13.040 38.490 / 13.060 38.510	13.052	3.6		38.500	4.5		29.111	6306	8
			65			17			
13.110 38.510 / 13.130 38.530	13.117	3.4		38.517	4.0		29.111	2058	7
			68			16			
13.180 38.525 / 13.200 38.545	13.185	3.7		38.533	4.4		29.109	2305	6
			99			22			
13.270 38.545 / 13.290 38.565	13.284	4.1		38.555	4.8		29.105	1437	5
			78			17			
13.350 38.560 / 13.370 38.580	13.362	4.3		38.572	4.5		29.102	993	4
			99			25			
13.450 38.590 / 13.470 38.610	13.461	4.6		38.597	4.6		29.100	543	3
			50			11			
13.500 38.600 / 13.520 38.620	13.511	4.8		38.608	4.8		29.098	565	2
			112			22			
13.615 38.620 / 13.635 38.640	13.623	5.3		38.630	4.6		29.091	604	1

Table 5. Stability calculation for adjacent layers which characteristics are determined from Table 4

Layer No.	Pressure (db) P	$\sigma_{\theta,S,P}$	$\Delta\sigma$ $\times 10^{+3}$
10	1335	34.989	2
9	1335	34.987	
9	1170	34.270	3
8	1170	34.267	
8	1000	33.528	3
7	1000	33.525	
7	940	33.263	2
6	940	33.261	
6	850	32.867	6
5	850	32.861	
5	800	32.642	4
4	800	32.638	
4	750	32.419	4
3	750	32.415	
3	730	32.327	2
2	730	32.325	
2	675	32.084	8
1	675	32.076	

for the ten layers, the sum of the products $\bar{\theta}_i N_i$ was determined and divided by Q to give a number representative of the quality of the layer system observed: a perfect step structure would have a value of 1 while a profile devoid of steps would have a much lower number. We define the layer intensity I as this quantity multiplied by 10. Thus:

$$I = \frac{10}{Q} \sum_{i=1}^{i=10} \theta_i N_i.$$

The profile shown in Fig. 3 and analyzed in Tables 1 and 2 has an intensity of 7.1 indicating a good series of step layers. In contrast Fig. 4 and Table 3 show the situation for an ill-defined profile with an intensity of 2.2. Salinity may also be used to find I, and values so obtained agree with those calculated using temperature to within ± 0.2. However, as the salinity profiles were subject to higher noise levels, temperature was used for all estimates of intensity.

The most extensive survey of the Tyrrhenian Sea was made during the 1973 cruise, when ≈ 60 stations positions were worked over a wide area (Fig. 1). For each station the intensity was calculated as outlined above. Iso-intensity contours were then drawn on a bathymetric chart of the area (Fig. 5). As corroborated by other measurements, the step-layer zone was confined to the deepest part of the Tyrrhenian Sea extending over an area of about 4000 square miles, a similar result to that reported for the Mediterranean outflow step layers (TAIT and HOWE, 1970). Even though the deepest layers were some 1500 m from the bottom the correlation of intensity with the bottom contours is striking, particularly near the boundaries.

Within the high-intensity zone the layers were found to have a high horizontal coherence, and no discontinuities were observed, indicating a continuous stratification extending over distances of the order of 60 miles. This result is again in general agreement with the Mediterranean outflow observations (ELLIOTT, HOWE and TAIT, 1974).

Table 4 gives the mean values obtained from all of the 1973 stations whose intensity was greater than 3. The agreement with Table 2 illustrates the uniformity of θ and S in the horizontal plain.

Position 39°47'N 12°46'E

MAY 1972   (11 MAY 72)   Intensity 6.8

θ-S interval		$\bar{\theta}_i$ °C	$\hat{\sigma}\times10^3$ °C	$\Delta\theta\times10^3$ °C	$\bar{S}_i$‰	$\hat{\sigma}\times10^3$ ‰	$\Delta S\times10^3$ ‰	σ_θ	N_i	i
12.930	38.500	12.934	2,3		38.507	2,3		29.141	24	10
12.950	38.520									
				53			14			
12.980	38.510	12.987	1,6		38.521	2,9		29.141	89	9
13.000	38.530									
				62			17			
13.040	38.530	13.049	1,2		38.538	2,5		29.141	82	8
13.060	38.550									
				71			17			
13.110	38.550	13.120	0,8		38.555	2,1		29.140	36	7
13.130	38.570									
				69			19			
13.170	38.560	13.189	0,4		38.574	2,6		19.140	22	6
13.190	38.580									
				102			23			
13.280	38.580	13.291	2,5		38.597	1,3		29.137	24	5
13.300	38.600									
				81			18			
13.370	38.600	13.372	1,3		38.615	2,7		19.133	22	4
13.390	38.620									
				90			22			
13.450	38.630	13.462	1,5		38.637	2,4		19.131	17	3
13.470	38.650									
				68			16			
13.520	38.640	13.530	1,8		38.653	2,8		29.129	15	2
13.540	38.660									
				113			27			
13.630	38.670	13.643	3,3		38.680	1,8		29.125	14	1
13.650	38.690									

Position 40°25'N 12°49'E

OCTOBER 1974   (7 OCT 74)   Intensity 7.7

θ-S interval		$\bar{\theta}_i$ °C	$\hat{\sigma}\times10^3$ °C	$\Delta\theta\times10^3$ °C	$\bar{S}_i$‰	$\hat{\sigma}\times10^3$ ‰	$\Delta S\times10^3$ ‰	σ_θ	N_i	i
12.940	38.470	12.957	1,3		38.475	2,5		29.112	66	10
12.960	38.490									
				55			14			
13.000	38.480	13.012	1,3		38.489	2,5		29.111	63	9
13.020	38.500									
				54			13			
13.050	38.490	13.066	1,9		38.502	2,9		29.110	76	8
13.070	38.510									
				68			17			
13.130	38.510	13.134	1,9		38.519	2,9		29.109	32	7
13.150	38.530									
				63			16			
13.180	38.530	13.197	1,5		38.535	2,5		29.108	41	6
13.200	38.550									
				99			21			
13.290	38.545	13.296	1,6		38.556	2,4		29.103	27	5
13.310	38.565									
				81			19			
13.370	38.565	13.377	2,1		38.575	3,8		29.101	28	4
13.390	38.585									
				112			25			
13.480	38.590	13.489	1,1		38.600	3,1		29.096	18	3
13.500	38.610									
				42			10			
13.520	38.600	13.531	1,7		38.610	3,1		29.095	16	2
13.540	38.620									
				116			27			
13.630	38.625	13.647	1,8		38.637	3,3		29.091	23	1
13.658	38.645									

Table 6. Layer parameters typical of the situation of each year of observation, corresponding to the composite θS diagram of Fig. 6. (See also Table 2)

Data points from the 1972 and 1974 cruises are also included in Fig. 5 as solid circles. The larger circles represent stations whose intensity is greater than or equal to 5 while the smaller ones correspond to $5 > I > 2$. The general agreement with the 1973 contours indicates the overall permanency of the step layer zone.

INTERFACIAL STABILITY

At the depths involved, potential density is an inaccurate indicator of the static stability (ELLIOTT, HOWE and TAIT, 1974). In order to evaluate stability more accurately the *in situ* densities, $\sigma_{S\theta P}$, were calculated for each pair of adjacent layers at the same pressure as their common interface. The values S and θ were taken from Table 4. Confidence intervals for the estimation of the means are less than 10^{-3}°C and 10^{-3}‰ and the corresponding confidence intervals for $\sigma_{S\theta P}$ are of the same order. The third decimals on the sigma values listed in Table 5 are therefore significant. (The use of the potential temperature does not affect the calculation of $\triangle\sigma$ which depends mainly on $\triangle\theta$.) We can conclude that all the layers are statically stable. Differences of density for adjacent layers, adjusted to the same pressure, are of the order of 4×10^{-6} g/cm^3, which for an average interface thickness of 10 m gives a Brünt–Väisälä frequency of (1 ± 0.2) cycle/hour.

Comparison over a 3-year period

The observations made in 1972, 1973 and 1974, in which step layers of appreciable intensity could always be found within the general area delineated in Fig. 5, have enabled the situation to be compared from year to year. For each year's results no significant

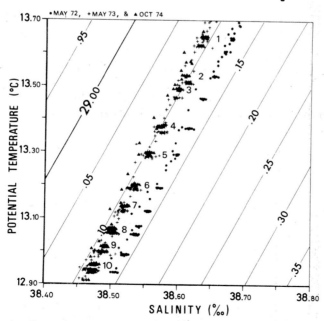

Fig. 6.   Composite θ–S diagram of three stations, covering a period of 3 years.

spatial change in θ–S was observed (compare Tables 2 and 4), hence essentially any set of values representing one station from each year could be used for comparison.

Table 6 gives two such sets where the results for 1972 and 1974 may be compared with the 1973 station already discussed. It will be noted that the θ–S windows differ slightly between the three data sets. This is because in each case the window was chosen to encompass the mixed layer, but these differences are small and of the same order as the experimental error.

A comparison of the θ values shows the mean difference over the 3 years to be 0.013°C, i.e. within the experimental error and less than the interfacial temperature difference by a factor of about 6. This implies that we are indeed looking at the same mixed layers from year to year. Bearing in mind that the salinity figures for 1972 require a $-0.05‰$ calibration correction, a similar result is obtained using the S values. In fact between 1973 and 1974 the salinity differences are negligible.

Figure 6 gives the 3-year composite θ–S diagram where the salinity offset for 1972 is clearly shown. From the point of view of density a more appropriate and legitimate correction for the 1972 salinity values would be $-0.04‰$. The marked accumulation of points representing the ten homogeneous layers illustrates the persistence of the stratification.

Two conclusions may be drawn from the above results: either the layer system remains essentially unchanged over a period of at least 3 years, or, if the stratification does not persist from year to year, then the layers must re-form at preferred values of temperature and salinity.

DISCUSSION

If the Tyrrhenian Sea step structure has an observed lifetime of 3 years it is likely to persist over a much longer period. In any case, the layers must be considered to be in equilibrium with respect to the transports of salt and heat. Much evidence has been put forward in support of salt fingering as the basic mechanism responsible for the step structures (Turner, 1967; Stern and Turner, 1969). Using an optical technique, evidence for the existence of salt fingers, associated with the step structure, was in fact obtained during the 1973 cruise. It has been shown by Molcard and Williams (1975) and Williams (1975) that they are correlated with the microstructure within the interfaces. Hence whereas the double diffusion theory may adequately describe the situation within the interfaces its correlation with the deep convective layers wherein the temperature remains constant has yet to be established.

During the 1974 cruise, vertical and horizontal current measurements were also made. The results of this work, shortly to be reported, are expected to show that the advection processes, associated with internal wave activity, play a significant role in the maintenance of these homogeneous layers.

Acknowledgements—The authors wish to thank Messrs. F. De Strobel and A. Chiarabini for the technical assistance, Messrs. L. Toma and P. Giannecchini for the programming and Messrs. G. Tognarini and A. Carrara for their help in the preparation of the diagrams.

REFERENCES

ELLIOTT A. J., M. R. HOWE and R. I. TAIT (1970) The lateral coherence of a system of thermohaline layers in the deep ocean. *Deep-Sea Research*, **21**, 95–107:

JOHANNESSEN O. M. and O. S. LEE (1974) Thermohaline staircase structure in the Tyrrhenian Sea. *Deep-Sea Research*, **21**, 629–639.

MOLCARD R. and A. J. WILLIAMS 3rd (1975) Deep step structure, in the Tyrrhenian Sea. *Memoires de la Société Royale des Sciences de Liège*, 6 eme Serie, **7**, 191–210.

STERN M. E. and J. S. TURNER (1969) Salt fingers and convective layers. *Deep-Sea Research*, **16**, 497–511.

TAIT R. I. and M. R. HOWE (1971) Thermoline staircase. *Nature, London*, **231**, (5299), 178–179.

TURNER J. S. (1973) *Buoyancy effects in fluids*, Cambridge, University Press, 367 pp.

WILLIAMS A. J. 3rd (1975) Image of ocean mocrostructure. *Deep-Sea Research*, **22**, 811–829.

ZENK W. (1970) On temperature and salinity structure of the Mediterranean water in the North-east Atlantic. *Deep-Sea Research*, **17**, 627–631.

Observations and theories of Langmuir circulations and their role in near surface mixing

R. T. POLLARD

Institute of Oceanographic Sciences, Wormley, Godalming, Surrey, England

Abstract—Langmuir circulations are helical roll vortices usually aligned within a few degrees alongwind, that are frequently observed in the surface layers of lakes and oceans. Their rapid appearance (within a few minutes to tens of minutes) after the onset of winds greater than $3\,\mathrm{m\,s}^{-1}$ makes them a potentially important mechanism for the downward transfer of wind-generated momentum and consequent mixing of heat and momentum through the surface layers. Observational data on the structure of Langmuir circulations are surveyed in this review paper and summarized in Fig. 1 and Table 1. Wind and consequently surface waves are necessary conditions for the existence of the circulations. Surface cooling and surface contaminants are not.

Theories of Langmuir circulations are reviewed and summarized in Table 2. The essential feature that has to be explained is the generation of alongwind vorticity. Inviscid wave theories cannot explain this. Theories which draw vorticity from the shear of the surface drift or Ekman current (GAMMELSRØD, 1975; CRAIK and LEIBOVICH, 1976) predict circulations whose structure appears to differ from that which is observed, but more detailed observations are required. A recent theory by GARRETT (1976) involving wave refraction by currents and preferred wave breaking in enhanced current zones is able, qualitatively at least, to predict all known features of Langmuir circulations.

1. INTRODUCTION

SINCE LANGMUIR (1938) first produced evidence of organized circulations in the surface layers of lakes and oceans, numerous investigators have confirmed their existence, and produced theories of their generation. As yet, no theory has received unanimous acceptance, so in this survey I shall first summarize much of what is known observationally about Langmuir circulations without reference to theories. I shall then review the theories that have been put forward to date, mentioning further observations that are particularly relevant to one theory or another, and finally discuss their importance to mixing processes in the upper ocean.

2. OBSERVATIONS

The cellular structure (Fig. 1) that Langmuir described is well supported by all subsequent observations, in both oceans and lakes (Table 1). Langmuir circulations consist of alternate left- and right-handed helical roll vortices, aligned more or less alongwind, with alongwind surface velocities strongest in the convergence zones. The lines of convergence are often made visible by particulate matter or oil films that collect there, but they can also exist in the absence of surface pollutants, as has been shown by scattering cards, pieces of paper, or dye on the surface (ASSAF, GERARD and GORDON, 1971; FALLER, 1964; HARRIS

Table 1. *Observations of Langmuir circulations*

Author	Ocean or lake	Row spacing (R) (m)	Comments	Vertical velocity (cm/s)	Down or up	Forward velocity of convergence zone (cm/s)	Reorientation time (min)	Other comments
LANGMUIR (1938)	Ocean	100–200					20	
	Lake George	5–10 / 15–25	May–June Oct.–Nov.	1–2	Down (average 0–6 m)	>0	20	
				2–3 / 1–1.5	Down (at 2 m) and up			
WOODCOCK (1950)	Ocean			>3	Down			
STOMMEL (1951)	Ponds on Cape Cod						1–2	Less than 30 cm deep
SUTCLIFFE, BAYLOR and MENZEL (1963)	Ocean			3–6	Down			
WELANDER (1963)	Baltic Sea	8					10	
FALLER and WOODCOCK (1964)	Ocean	20–50						Wind W and row spacing R correlated, R = W × 4.8 sec.
KATZ, GERARD and COSTIN (1965)	Ocean	2–10 / 10–45	Surface 0–7 m deep			>0	30	Wind and R correlated R = W × 4 sec

Owen (1966)	Ocean	1.5					Calm, thermally driven, and only 60 cm deep
Ichiye (1967)	Ocean	20–200	Actual quote 'tens to hundreds of metres'			17	
Scott, Myer, Stewart and Walther (1969)	Lake George			4–7	Down		'almost instantaneous' R and depth to first stable layer correlated
Gordon (1970)	Ocean	5		1.5	Up	1–3	
Assaf, Gerard and Gordon (1971)	Ocean	3–6	Calm			10	
		5–12 30–50 90–300	Hierarchy in 5–15 m s^{-1} wind				
Maratos (1971)	Ocean	6–16	$R = 0.5$ to $1.5 \times$ depth to thermocline	2–4			W and R well correlated $R = 0.1 + W \times 2.8$ sec
Myer (1971)	Lake George	1–5	wind 4 m s^{-1}	2.2	Down		Estimated from sinking rate of fine sand
		2–11	wind 8 m s^{-1}	0.8	Up		
Harris and Lott (1973)	Lake Ontario	3–4		2–3	Down (stable)	≦6	Stable and unstable refer to surface heating and cooling respectively
				5	Down (unstable)		
				2–9	Down		

and LOTT, 1973; ICHIYE, 1967; KATZ, GERARD and COSTIN, 1965; MARATOS, 1971). The difference between down- and upwelling velocities observed (Langmuir, Maratos) suggests that the circulations are asymmetric, with downwelling velocities concentrated in a zone under the streaks that occupies a third or less of the cell width. MYER (1971) found that the onset of a circulation was characterized by downwelling under a newly formed streak in a 'jet' up to 1.0 m wide which could penetrate through stable stratification (Fig. 2).

Unfortunately, only a small part of MYER's thesis has been published (SCOTT, MYER, STEWART and WALTHER, 1969), although it describes the most comprehensive set of observations on Langmuir circulations attempted to date. As it is necessary to reference the thesis extensively, I shall here briefly summarize Myer's experiments, part of the program of measurements on Lake George initiated by Langmuir. The data Myer describes were collected from August to December 1967 (on 15 days), July to October 1968 (10 days) and July to December 1969 (20 days). A wide range of conditions was encountered and thousands of streaks sampled.

Measurements were made from the floating probe described by SCOTT, MYER, STEWART and WALTHER (1969), which could be anchored at any position in the lake, and could also be pushed through the water with a rigid yoke 5 m in front of a barge. In addition, on some occasions, three-component wind measurements were taken on either a fixed tripod or floating support frame, and wave height could be measured from transportable tripod towers in water less than 5 m deep. Event markers on all recorders were used to record visual observations of streaks. Although recording current meters were not used, dye and drogues were frequently used to observe the general pattern of the flow and correlate it with streaks and surface waves.

2.1. Scales of motion

The most variable reported feature of Langmuir circulations is the row spacing (Table 1) ranging from 2–25 m in lakes and 2–300 m in the ocean. Cells of several different scales can

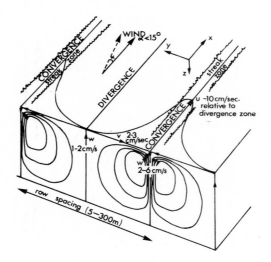

Fig. 1.   Diagrammatic representation of the main features of Langmuir circulations.

exist together. LANGMUIR (1938), at sea, reported 100–200 m streak spacings with smaller streaks between them, and in Lake George he recorded that between the well-defined streaks there were numerous smaller and less well-defined streaks. ASSAF, GERARD and GORDON (1971) reported a hierarchy of two or three cell sizes coexisting in the ocean off Bermuda, and HARRIS and LOTT (1973), state that, in Lake Ontario, the distance between streaks (3–4 m) increased with time and new streaks formed between streaks already marked. SCOTT, MYER, STEWART and WALTHER (1969) said that one to three poorly defined streaks frequently appeared between long well-defined streaks, and that the poorly defined streaks disappeared quickly when the wind died. FALLER and WOODCOCK (1964) may have observed the same phenomenon, though they assumed that the larger spacings they observed were actually two or three cell widths apart.

The factors which control the circulation scale have not been unambiguously determined. LANGMUIR (1938) found larger (15-25 m) spacings in October and November than in May and June (5-10 m), suggesting a correlation with the depth to the seasonal thermocline. SCOTT, MYER, STEWART and WALTHER (1969) support this, finding significant correlation between the streak separation and the depth to the first stable, but FALLER and WOODCOCK (1964) found the same correlation to be not significant.

FALLER and WOODCOCK, on the other hand, and MARATOS (1971) found that the correlation between wind speed and row spacing was significant while SCOTT, MYER,

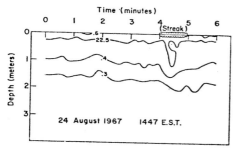

Thermal structure of the near-surface layer for a case in which streaking was just beginning. First streaks were observed at 3 min after time zero.

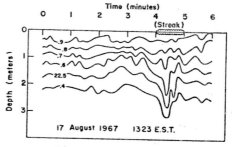

Thermal structure of the near-surface layer for a case in which streaking had been observed for 3 min. Streaks were first observed at 1 min past time zero.

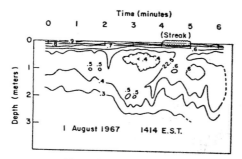

Thermal structure for a case in which streaking had been observed for more than 10 min.

Fig. 2. Downwelling velocities under streaks breaking through stable stratification, from SCOTT, MYER, STEWART and WALTHER (1969).

STEWART and WALTHER (1969) found that it was not. MYER (1971) found that the mean spacing increased slightly with increasing wind speed, and the spread of row spacings on histograms tended to become greater.

The streak spacings observed at the surface are not necessarily a good measure of the dominant scales slightly deeper. KATZ, GERARD and COSTIN's (1965) Fig. 6 supports this viewpoint showing, for the same wind, smaller row spacing when measured at the surface than when measured below the surface. Myer compared autocorrelations from temperature probes pushed acrosswind at depths of 0.35 and 1.0 m, and found that the dominant scale at 1.0 m was twice that at 0.35 m. This was supported by slight depressions of the isotherms frequently observed at 0.35 m over regions of general upwelling approximately midway between two streaks. Elsewhere in his thesis, Myer remarked that the secondary thermal structures (which were accompanied by secondary surface streaks) were often noted for cases of surface cooling. Drogue experiments confirmed the increase of circulation scales with depth.

To the author's knowledge, only MYER has estimated the depth of Langmuir circulations and hence their depth to width ratio. Taking the depth of the circulations to be the depth at which isotherm displacement was no longer observable, he found penetration depths of 2–7 m under stable (heating) conditions. From Myer's Fig. 16, the ratio (streak spacing/penetration depth) appears to be about 1-3 under stable conditions, but is typically 0.2-0.3 under unstable conditions.

2.2 *Velocities*

There is unanimous agreement that the surface velocity in the direction of the wind is larger in the convergence or streak zones than out of them. LANGMUIR (1938) laid a cord on the surface perpendicular to the streak lines and noted that it developed well-defined waves, forwards (in the direction of the wind) in the streaks and backwards out of them. ASSAF, GERARD and GORDON (1971), GORDON (1970), HARRIS and LOTT (1973), ICHIYE (1967) and KATZ, GERARD and COSTIN (1965) have all noted the same effect using computer cards and dye, and have variously estimated the shear between the flow in the convergence and divergence zones as 1–3, 5–10, 6 and 17 cm s^{-1}.

There have been several attempts to measure the downwelling velocity in the convergence zones, and the corresponding upwelling in between them. LANGMUIR (1938), SCOTT, MYER, STEWART and WALTHER (1969), SUTCLIFFE, BAYLOR and MENZEL (1963) and HARRIS and LOTT (1973) used drag plate current meters. LANGMUIR also watched the downward motion of dye in a convergence zone, while GORDON estimated the upwelling velocity from the rate of divergence of dye. MYER (1971), in addition to dye studies, estimated vertical velocities from the displacement of isotherms (Fig. 2). He found the downward motion to be concentrated in 'jets', narrow (0.2–1 m) regions under streaks, with w maximum at about half the depth of the circulation (the depth at which isotherm displacement was no longer observable). From repeated crossings of the same streak, maximum downwelling speeds in the jet were estimated at 2–3 cm s^{-1} in stable conditions, and 5 cm s^{-1} or more in unstable conditions. WOODCOCK (1950) measured the vertical velocity necessary to submerge the pelagic *Sargassum* that accumulate in streaks. Maratos (1971) compared the sinking rates of fine sand in and between streak zones with the sinking rates in still water.

SUTCLIFFE, BAYLOR and MENZEL (1963) have reviewed the data on downwelling (prior to 1963), correlating the vertical velocity w with windspeed W by the line $w = 0.85 \times 10^{-2}W$, or about $1\,\mathrm{cm\,s^{-1}}$ per $1\,\mathrm{m\,s^{-1}}$ wind. HARIS and LOTT's data can be fitted by the same line, though all the data show a scatter of $1-2\,\mathrm{cm\,s^{-1}}$ about the line.

Very few observers have measured upward velocities. Langmuir estimated that the upwelling rate between dye streaks was $1-1.5\,\mathrm{cm\,s^{-1}}$, about one-half the downdwelling velocities he observed. Gordon's estimate, from dye observations, was similar.

A few references to the horizontal velocities perpendicular to the wind can be found. Langmuir noted that leaves placed on the water half-way between streaks travelled at $2-3\,\mathrm{cm\,s^{-1}}$ to reach the streaks in about 5 minutes and WOODCOCK (1944) noted that drift bottles laid in lines perpendicular to the wind always re-formed into lines parallel to the wind. Langmuir also noted on one occasion that a neutrally buoyant drogue at $5\,\mathrm{m}$ deep gradually drifted under a streak, but one at $10\,\mathrm{m}$ had no tendency to do so. HARRIS and LOTT mention that on many occasions drogues placed in the downwelling zones drifted many metres sideways on reaching a depth of $5-6\,\mathrm{m}$. STOMMEL (1951) on small ponds on Cape Cod, Massachusetts, confirmed surface convergence into the streak lines but found no cellular motions at $30\,\mathrm{cm}$ or deeper.

2.3. *Response to winds*

Langmuir circulations do not normally exist in winds less than $3\,\mathrm{m\,s^{-1}}$ (FALLER and WOODCOCK, 1964; HARRIS and LOTT, 1973; ICHIYE, 1967; LANGMUIR, 1938; MYER, 1971; SCOTT, MYER, STEWART and WALTHER, 1969; WELANDER, 1963) though KATZ, GERARD and COSTIN (1965) plot rib spacings in dye patches down to zero wind speed, and ICHIYE remarked that striations in dye patches can exist even in calm seas when there are pronounced swells. MYER observed circulations on a few occasions with smaller winds than $3\,\mathrm{m\,s^{-1}}$, but almost always under cooling conditions.

After the onset of winds larger than $3\,\mathrm{m\,s^{-1}}$, circulations appear within a few tens of minutes. KATZ, GERARD and COSTIN record that an initial rib pattern consisting of three large ribs was broken up when the $5\,\mathrm{m\,s^{-1}}$ wind shifted through $70°$ in 30 minutes and a number of smaller ribs developed aligned with the new wind direction. LANGMUIR recorded that lines of seaweed realigned themselves within 20 minutes when the wind shifted by $90°$. SCOTT, MYER, STEWART and WALTHER report that streaks form almost instantaneously when the wind rises rapidly to speeds greater than $3\,\mathrm{m\,s^{-1}}$. MARATOS (1971) estimated that windrow orientation responded within minutes to major wind shifts. WELANDER suggested that the reorientation is initially confined to a very thin surface layer. He found that 10 minutes after a shift in a $9\,\mathrm{m\,s^{-1}}$ wind, surface streaks had rearranged themselves in the new wind direction and surface floats converged into the streaks but floats at 1 and $2\,\mathrm{m}$ depth continue to move in the original direction.

2.4. *Biological significance*

The presence of circulations, with corresponding convergence and divergence zones, affects the distribution of particulate matter and living organisms in the surface layers. SUTCLIFFE, SHELDON, PRAKASH and GORDON (1971) found that the concentration of particulate matter was larger in convergence than divergence zones. The concentration also increased with increasing wind speed. STAVN (1971) discussed the extent to which the magnitude of the circulations influenced the positioning of zooplankton relative to the circulations, and compared his findings with an earlier theory of STOMMEL (1949).

3. THEORIES

Although LANGMUIR (1938) gave a thorough description of the circulations now named after him, he gave no account of how they were generated. He clearly believed that they were wind driven, and explained the larger forward velocities in the convergence zones by pointing out that the water there had been on the surface since it rose in the divergence zone and had therefore been accelerated by the wind for longest. The lack of a comprehensive explanation has taxed scientists ever since, and a large number of theories have been proposed (Table 2), most of which have several points against them. Theories up to 1971 have been reviewed by SCOTT, MYER, STEWART and WALTHER (1969), CRAIK (1970) and FALLER (1971), so will be only briefly summarized here together with more recent theories.

3.1. *Thermal convection*

The knowledge that convective plumes in the atmosphere can be aligned into rows by wind shear (see review by KUETTNER, 1971) naturally led investigators to try to attribute oceanic circulations to the same mechanism. However, evidence that thermal convection is not a primary mechanism in the ocean is overwhelming. STOMMEL (1951), FALLER and WOODCOCK (1964), CSANADY (1965), ICHIYE (1967) and MYER (1969) have all observed Langmuir circulations in conditions of stable stratification. MYER (1971) (summarized in SCOTT, MYER, STEWART and WALTHER, 1969) observed tongues of warm surface water downwelling under surface streaks on a number of occasions (Fig. 2), and estimated the rate (of order $2 \, \text{erg} \, \text{cm}^{-2} \, \text{s}^{-1}$ at which energy had to be supplied to the circulations to enable them to break down the stable stratification at the rate observed. Several of the above authors calculated surface heat fluxes, but found no correlation between heat fluxes and slick spacing or mixed layer depth. BAYLOR and SUTCLIFFE (reported by FALLER, 1969) observed that the surface temperature in lines of surface convergence could be relatively warm, indicating a downward flux of heat. HARRIS and LOTT (1973) compared down-welling velocities in stable and unstable surface heat flux conditions but found no significant difference.

Although thermal instability is not a necessary condition for the formation of circulations in water, it can certainly enhance circulations generated by other means. This is clear from many of Myer's observations, some of which have been referred to in Sections 2.1 and 2.2. Paradoxically, the only example of naturally occurring thermally driven rolls in the ocean seems to be that of OWEN (1966), in calm *heating* conditions. Convection can occur in this case because the radiative heat input is distributed through a few metres, while cooling takes place right at the surface. However, the rolls OWEN observed (Table 1) were significantly smaller than most Langmuir circulations and confined to the top 60 cm. The streaks seen by STOMMEL (1951) are possibly due to the same cause.

3.2. *Coupling with atmospheric rolls*

The possibility that oceanic roll vortices are coupled to atmospheric rolls can be quickly disposed of, since the patterns in the air move too fast across the surface of the water to be coupled to it. Observations of sea smoke or haze above lakes by LANGMUIR (1938), STOMMEL (1951), and SCOTT, MYER, STEWART and WALTHER (1969), all support this

Table 2. Theories of Langmuir circulations

Theory	Origin	Present status
Convective instability, rolls aligned by wind	Analogy with atmospheric boundary layer, see review by KUETTNER (1971)	Not a primary mechanism, as cells often observed to grow in stable conditions and break down stable stratification
Coupling with atmospheric rolls	Unknown, mentioned by STOMMEL (1951)	Discounted, atmospheric vortices move too fast over ocean surface
Modification of wind over surface slicks	WELANDER (1963)	Discounted, atmospheric vortices move too fast over ocean surface; energy supply 100 times too small (MYER, 1971)
Instability of the Ekman spiral	FALLER (1964)	Not a primary mechanism, cannot account for observed growth rates
Damping of capillary waves in slicks provides radiation stress to drive rolls	KRAUS (1967)	Discounted, as cells may exist in the absence of surface contaminants. Also energy supply too small to explain observed growth rates.
Interaction of two linear wave trains	STEWART and SCHMITT (1968)	Discounted, cannot provide vorticity
'Eddy pressure' of surface waves	FALLER (1969)	Discounted, cannot provide vorticity
Interaction of pairs of inviscid wave trains in a shear flow	CRAIK (1970)	Discounted by LEIBOVICH and ULRICH (1972), inviscid theory creates vorticity of wrong sign
Instability of shear flow in a rotating system	GAMMELSRØD (1975)	Appears unlikely, basic state doubtful, predicted cell structure in conflict with observations
Interaction of pairs of viscous wave trains in a shear flow	CRAIK and LEIBOVICH (1976)	Appears unlikely, predicts maximum wave amplitudes in divergence zones in conflict with observations
Interaction of waves and surface current, with wave dissipation	GARRETT (1976)	Qualitatively, can explain all observed features of circulations. Requires quantitative testing

conclusion. WELANDER's (1963) hypothesis of a feedback mechanism involving the modification of the surface wind over streaks can be similarly discounted. MYER (1971) did find some modification to the air flow over streaks (cold updrafts), but it was clearly induced by the streaks, and not vice versa, as the observed wind convergence was two orders of magnitude smaller than required to supply the energy of the Langmuir circulations.

3.3 Surface films

KRAUS (1967) suggested the first of several wave interaction mechanisms, that the damping of capillary waves as they approach a slick generates a radiation stress that will enhance the slick. KRAUS' hypothesis can be discounted for several reasons. First, while slicks may be caused by Langmuir circulations, surface contamination is not a prerequisite for generation of the circulations (Section 2). Second, there is general agreement that Langmuir cells do not exist in winds less than 3 m s^{-1} (Section 2.3) though capillary waves do. Third, it is doubtful whether capillary waves can supply energy fast enough to account for the rapid generation of Langmuir circulations. For example, MYER (1971) estimated that even if capillary waves typical of those he observed were entirely dissipated in the slicks, and gave up all their energy to the circulations, they could still only supply energy at a rate of about $0.1\,\mathrm{erg\,cm^{-2}\,s^{-1}}$, only 10% of the rate needed to drive the circulations. Finally, GARRETT (personal communication) has suggested that the KRAUS mechanism would squeeze slicks into lines perpendicular to the wind rather than parallel to it, as the secondary currents set up by the capillary waves as they are damped will be primarily alongwind, and strongest on the upwind side of the slick.

3.4 *Ekman instability*

FALLER (1964) suggested that instability of the Ekman spiral (FALLER, 1963; FALLER and KAYLOR, 1966) could be the cause of Langmuir circulations. FALLER supported his hypothesis with observations that showed windrows lying at about 15° *cum sole* to the wind. This does not seem to be a general result. KATZ, GERARD and COSTIN (1965) found similar angles to Faller's in four out of six experiments, but WELANDER (1963) found angles less than 2°, as did WALTHER (1967) on Lake George, and also MYER (1971), MARATOS (1971) found angles within 3° of the wind in sixty-one out of sixty-six estimates, the angle being 0° for twenty-two of these. WELANDER pointed out that observed deviations are no proof of the correctness of FALLER's mechanism. In steady wind conditions in which Ekman flow can be set up, the Langmuir circulations, however generated, may be deflected by the shear flow away from the wind direction.

The main objection to FALLER's theory (WELANDER, 1963; FALLER, 1971) was that it did not seem possible for a rotation-based theory to account for the rapid onset of circulations in response to a sudden change in wind (Table 1) in a time much less than f^{-1}. However, GAMMELSRØD (1975a) has recently suggested that in the presence of a strong shear s, the relevant time scale is $(sf)^{-\frac{1}{2}}$ which may be much smaller than f^{-1}. GAMMELSRØD starts with a steady alongwind shear flow, the coriolis force on which is balanced by friction (or a pressure gradient, in the atmosphere). A vertical perturbation velocity will accelerate the alongwind perturbation flow by vertical advection of the mean shear. The resultant along-wind velocities are then converted to cross-stream velocities by the coriolis effect. Thus any cross-stream vorticity $\partial u/\partial z$* is converted by the coriolis effect into downwind vorticity.

The cellular structure GAMMELSRØD predicts differs from that of Fig. 1 in several ways (deducible from GAMMELSRØD, 1975a, but more clearly presented in GAMMELSRØD, 1975b). Alongwind velocities are zero at the surface, with maxima at the same depths as those at which the vertical velocity has maxima. There may be many vertical modes, allowing circulation cells stacked one above the other. The cell boundaries (straight lines marked w on Fig. 1) are not vertical but at some angle to the vertical.

It is hard to reconcile such features with observations. One can argue that the first mode is the only one observed because it has the fastest growth rate (as GAMMELSRØD predicts). Alongwind surface velocities certainly exist, though they could possibly be inserted into GAMMELSRØD's theory by non-linear terms involving a slightly non-uniform mean shear acted on by crosswind perturbation velocities. There is no evidence that the cells are slanted

---

* Wherever the symbols (x, y, z) for distance and (u, v, w) for velocity are introduced, they refer to (alongwind, acrosswind, vertical) axes as in Fig. 1.

in the vertical. MYER's (1971) data (Fig. 2) indicate vertical alignment, but much more detailed observations of cell structure are desirable.

The major objection to GAMMELSRØD's theory when applied to lakes and oceans, however, may well be found, not in observations, but in the underlying assumption of a friction term (in the ocean) invariant with time which maintains the mean shear. The theory requires that vertical perturbation velocities should displace the mean shear (but not the friction), setting up perturbation horizontal velocities. The Coriolis force on these horizontal alongwind perturbation velocities is assumed unbalanced by friction, and must remain so for long enough that the coriolis force generates acrosswind velocities. But in practice the friction must arise from Reynolds stresses maintained by small-scale turbulence. This field of turbulence will itself be displaced vertically by the perturbation velocities, so can continue to balance the Coriolis term on the perturbation horizontal velocities, and prevent the growth of acrosswind velocities.

3.5. *Waves and wind drift currents*

All the remaining theories that have been advanced rely on winds or wind-driven waves as the driving force for Langmuir circulations. This is consistent with all observations, including LANGMUIR's own, and it is certain that winds are the primary mechanism. It is not at all clear, however, how the winds can set up the circulation, and most of the theories so far suggested can be shown to be inadequate.

The inviscid wave theories of STEWART and SCHMITT (1968) and FALLER (1969) cannot account for the vorticity of the circulations. The vorticity in a similar theory by CRAIK (1970), involving the interaction of two wave trains, was supplied by the vertical shear of a mean current, but LEIBOVICH and ULRICH (1972) point out that the circulation generated would have maximum forward velocity (due to the Stokes drift) in the upwelling zones, which is contrary to all observations (Section 2.2).

CRAIK and LEIBOVICH (1976) have recently modified CRAIK's (1970) theory to include viscosity (and their theory has in turn been extended by LEIBOVICH (1976) to predict the development and structure of the wind-drift current). A pair of wave trains, propagating at equal angles on either side of the wind direction, is accompanied by a Stokes drift (u_s) alongwind that varies in the crosswind direction y. Such a velocity can rotate the y-vorticity present in the wind-drift current $u(z)$ into the x-direction, setting up a Langmuir-like set of vortices (LEIBOVICH and ULRICH, 1972, fig. 1) whose wavelength in the y-direction is half the wavelength of the waves in that direction.

Up to this point CRAIK's and CRAIK and LEIBOVICH's theories are similar, and both forecast *upwelling* velocities where the Stokes drift (and the wave amplitude) is largest. However, CRAIK and LEIBOVICH argue that the surface current in the wind direction is usually dominated, not by the Stokes drift of the waves but by a second-order current associated with the cellular motion, which is largest in the convergence zones, as required by observations. Such a current u_0 arises when a steady solution of the non-linear vorticity and momentum equations is sought. It does not contribute to the vortex rotation argument given above as long as it is independent of x, because in those circumstances the x-component of curl $u_0 \cdot \nabla u_0$ is identically zero.

As with GAMMELSRØD's theory, it is CRAIK and LEIBOVICH's underlying assumption on the flow structure that is hardest to accept. The theory can only work if the Stokes drift varies in the y-direction but is only slowly varying with time. That is, y-variations in the wave drift must remain fixed in space for long enough (several minutes or longer) for the circulations to become established. Mathematically, this requires that wave pairs (the two waves lying at equal and opposite angles to the wind direction) be phaselocked, and a finite amount of wave energy must be concentrated into such wave pairs, as the Stokes drift would not vary in the y-direction for a continuous spectrum. The author is aware of no observations that support the hypothesis of a discretized spectrum, and it seems unlikely except possibly in fetch-limited situations, as CRAIK and LEIBOVICH point out. However, phase-locking between the waves and the secondary circulations can occur if the waves are themselves modified by the circulations, as in the following theory. (See note at end.)

3.6 Wave modification by currents

The wave theories discussed in Section 3.5 assume that the surface waves are not affected by the perturbation currents they attempt to generate. GARRETT (1976) points out that any current will modify the wave field in order to conserve wave action (BRETHERTON and GARRETT, 1968). As waves propagate into a zone of enhanced surface current their propagation speed perpendicular to the current decreases, and then increases again as they leave the zone. To accomplish this, momentum must be transferred from the waves to a current as the waves slow down, and back again as they speed up again. Thus this mechanism causes convergence towards the zone of large current, supplying alongstream vorticity of the sign required by Langmuir circulations.

Since the waves slow down on entering the current zone, their amplitude must increase, and will be maximum where the current is maximum. GARRETT (1976) argues that this will encourage preferential wavebreaking there, transferring wave momentum to the current, which is thereby enhanced. Thus a feedback loop exists, and the cells are generated by an instability mechanism. In terms of vorticity, GARRETT argues that the vertical vorticity $\partial u/\partial y$ of the alongwind current is rotated into x-vorticity of the right sign by the vertical shear of the Stokes drift.

The mathematics of Garrett's paper can be criticized. Use of the WKB approximation is only appropriate for waves whose wavelength is smaller than the zones of enhanced current (typically not more than a few metres wide). The *ad hoc* assumption that wave dissipation is proportional to wave height may need modification. The linear theory is only valid for very small secondary flow currents. But the underlying physics is sound, and it is the reviewer's opinion that Garrett's is the most plausible hypothesis so far suggested to explain the occurrence of Langmuir circulations.

Both CRAIK and LEIBOVICH's (1976) and GARRETT's theories should be amenable to experimental verification without requiring difficult and precise measurements of dissipation or the directional wave spectrum. Both theories predict acrosswind variation of the waves and the Stokes drift. In CRAIK and LEIBOVICH's theory, the Stokes drift is maximum in the divergence zones of the Langmuir circulations. In GARRETT's theory the opposite is the case.

MYER (1971) made observations (reproduced in Fig. 3) of wave height relative to streak zones which show clear wave maxima in the streaks. LEIBOVICH (personal communication) has cautioned that streaks caused by breaking waves could be expected to show such a correlation, and that direct current/wave correlations were not attempted. But MYER showed on numerous occasions that streaks did mark convergence zones and records also

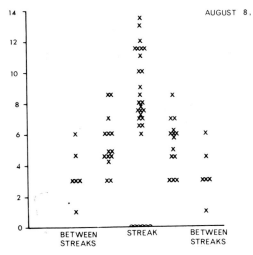

Fig. 3. Wave height as a function of position relative to streaks. Given by MYER (1971) as a plot typical of those obtained during streaking.

that use of dye confirmed that the areas of relative maximum transport agreed 'quite well' with the location of streaks. Thus MYER's observations appear to give strong support to GARRETT's theory at the expense of CRAIK and LEIBOVACH's. However, further observations are desirable in view of their importance to both theories.

4. MIXING DUE TO LANGMUIR CIRCULATIONS

Reynolds stresses associated with Langmuir circulations are potentially very efficient at redistributing momentum through the mixing layer from the thin surface layer in which it is initially deposited by surface wave breaking. This is the primary reason why such circulations are important to mixed-layer dynamics.

Since the alongwind and vertical velocities u and w are in phase (Fig. 1), it is clear that $\overline{uw}$, averaged over a cell width, is non-zero, giving a downward transport of alongwind momentum. For example (from Fig. 1), if the difference between u in the convergence and divergence zones is $6\,\mathrm{cm\,s^{-1}}$ and the difference between down- and upwelling velocities is $4\,\mathrm{cm\,s^{-1}}$ at a depth of $2\,\mathrm{m}$, say, then if $y = 0$ at a line of convergence, one may take

$$u = 3\cos ky\,\mathrm{cm\,s^{-1}}$$
$$w = -2\cos ky\,\mathrm{cm\,s^{-1}}$$

whence

$$\overline{uw} = -3\,\mathrm{cm^2\,s^{-1}}$$

corresponding to a downward stress of 3 dynes $\mathrm{cm^{-2}}$, GORDON (1970), in a similar calculation, estimated the stress as 2–3 dynes $\mathrm{cm^{-2}}$, which was more than enough to transport downwards all the momentum input by the surface stress.

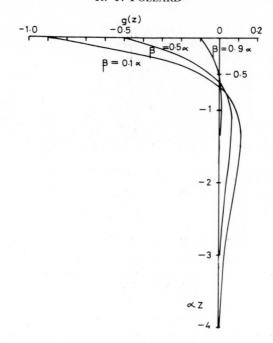

Fig. 4.   Vertical redistribution of momentum by the Reynolds stresses associated with Langmuir circulations. The function $g(z)$ is proportional to the vertical divergence of the Reynolds stresses $(\partial/\partial z)\,\overline{uw}$, so giving a measure of the local acceleration of the flow caused by the Reynolds stresses. The total acceleration $\int_{-\infty}^{0} g(z)\mathrm{d}z$ is of course zero. The definitions of α and β are given in Section 4 and more fully in Garrett (1976).

As an example of the redistribution of momentum with depth consider Garrett's (1976) solution

$$u = u_0 \mathrm{e}^{\alpha z} \cos \beta y\, \mathrm{e}^{qt},$$
$$w = w_0 (\mathrm{e}^{\alpha z} - \mathrm{e}^{\beta z}) \cos \beta y\, \mathrm{e}^{qt} (\beta < \alpha),$$

where α^{-1} is a length scale related to the viscosity and the surface stress. Garrett estimated α^{-1} very crudely as 3 m, and found that larger scales $\beta \ll \alpha$ dominate after sufficient time. For this solution, the rate of change of momentum due to the Reynolds stresses is given by

$$-\frac{\partial}{\partial z}\overline{uw} \propto (\alpha + \beta)\,\mathrm{e}^{(\alpha + \beta)z} - 2\alpha\,\mathrm{e}^{2\alpha z}$$

$$\equiv \alpha g(z)$$

which is sketched in Fig. 4 for $\beta = 0.9\alpha$, 0.5α, and $0.1\,\alpha$.

For all β, the circulations remove momentum from depths less than about $0.5\alpha^{-1}$ and redistribute it rather evenly from that depth down to about $2\alpha^{-1}$ and exponentially below that. The larger scales (small β) are the most efficient at removing momentum from the topmost layer. With $\alpha^{-1} = 3$ m, momentum is redistributed from the top 1–2 m down to 6 to 10 m. While these are only very crude estimates, they suggest the possibility that Langmuir cells may not always be able to mix through the entire mixing layer. This suggestion is supported by data of Scott, Myer, Stewart and Walther (1969), who found that, while Langmuir circulations could break through the diurnal thermocline on Lake George, they usually did not mix the heat right down to the seasonal thermocline (at about 10 m) but formed a secondary thermocline a few metres above it.

STEWART (1970) includes several references to stress calculations, and himself estimated that the stress necessary to overcome the near surface thermal stability on Lake George was 0.6 dyne cm^{-2}. He suggested that the wind speed needed to supply this stress could be the explanation of the threshold speed of 3–4 m s^{-1} below which Langmuir circulations are not observed.

5. CONCLUSIONS

Although the existence of wind streaks does not necessarily mean that Langmuir circulations are present (MCLEISH, 1968), the existence of fairly regular circulations such as were originally described by Langmuir is well documented. Organized circulations quite often exist, and when they do may have to be taken into account both theoretically, in further development of mixed-layer theories (NIILER and KRAUS, 1976), and observationally, in ensuring that surface measurements are not biased by the tendency of any floating platform to seek convergence zones.

If, as existing observations suggest, Langmuir circulations appear rapidly after the onset of winds greater than 3 m s^{-1}, it is likely that they, rather than small-scale turbulence generated by breaking waves, control the downward diffusion and redistribution of wind generated momentum through the surface layers. Detailed measurements of the circulation structure are still lacking, however, and doubts about the importance of organized circulations relative to random wave and wind-driven turbulence will remain until many difficult observational questions have been answered. Do organized circulations *always* form for winds greater than 3 m s^{-1}, and if not, why not? How long do they last, and what causes them to break up? Can they form or continue to exist in strong winds? How rapidly do the circulations penetrate down from the surface, and how deeply can they penetrate? What is the vertical shear of the mean horizontal current (POLLARD, 1976) and how is it modified by the Reynolds stresses associated with the circulations? What are the correlations between the circulations and the surface wave field? How do the organized circulations interact with smaller-scale random turbulence (compare BROWN, 1970)? Detailed and difficult observations of the velocity field, the spatial variation of wave dissipation, and the Reynolds stress field will be needed to answer such questions, and it is likely to be some years before adequate answers are forthcoming.

While surface cooling can enhance Langmuir circulations when it occurs, their driving mechanism is undoubtedly wind. Since much of the energy and momentum input at the sea surface goes initially into surface waves (HASSELMANN, 1974) it is likely that Langmuir circulations are driven by waves rather than by the wind directly. Observations (ICHIYE, 1967; MYER, 1971) support this view. Viscous dissipation of the surface waves must be invoked if the vorticity of the Langmuir circulations is to be explained (LEIBOVICH and ULRICH, 1972). FALLER (1971) suggests that several mechanisms are involved in generating the hierarchy of Langmuir circulation sizes that are observed, with the smallest scales being generated by wave breaking, larger scales by non-linear interaction with smaller scales, and the largest scales by Ekman instability. GARRETT's (1976) instability model, based on the first mechanism, can account for a hierarchy of cell sizes with different growth rates, their relative amplitudes depending on the form of the initial perturbation.

Acknowledgements—I am grateful to ALEX CRAIK, CHRIS GARRETT, SID LEIBOVICH and RORY THOMPSON for their stimulating correspondence on an earlier draft of this paper. Many of their comments have been incorporated into the final version, but the conclusions are my own, and are not unanimously agreed.

REFERENCES

ASSAF G., R. GERARD and A. L. GORDON (1971) Some mechanisms of oceanic mixing revealed in aerial photographs. *Journal of Geophysical Research*, **76**, 6550–6572.

BRETHERTON F. P. and C. J. R. GARRETT (1968) Wavetrains in inhomogeneous moving media. *Proceedings of the Royal Society*, A, **302**, 529–554.

BROWN R. A. (1970) A secondary flow model for the planetary boundary layer. *Journal of Atmospheric Sciences*, **27**, 742–757.

CRAIK A. D. D. (1970) A wave-injection model for the generation of windrows. *Journal of Fluid Mechanics*, **41**, 801–821.

CRAIK A. D. D. and S. LEIBOVICH (1976) A rational model for Langmuir circulations. *Journal of Fluid Mechanics*, **73**, 401–426.

CSANADY G. T. (1965) Windrow studies, Report No. PR26, Great Lakes Institute, University of Toronto, Ontario, 82 pp.

FALLER A. J. (1963) An experimental study of the instability of the laminar Ekman boundary layer. *Journal of Fluid Mechanics*, **15**, 560–576.

FALLER A. J. (1964) The angle of windrows in the ocean. *Tellus*, **XVI**, 363–370.

FALLER A. J. (1969) The generation of Langmuir circulations by the eddy pressure of surface waves. *Limnology and Oceanography*, **14**, 504–513.

FALLER A. J. (1971) Oceanic turbulence and the Langmuir circulations. *Annual Review of Ecology and Systematics*, R. F. JOHNSTON *et al.*, editors, **2**, 201–236.

FALLER A. J. and R. E. KAYLOR (1966) A numerical study of the instability of the laminar Ekman boundary layer. *Journal of Atmospheric Sciences*, **23**, 466–480.

FALLER A. J. and A. H. WOODCOCK (1964) The spacing of windrows of Sargassum in the ocean. *Journal of Marine Research*, **22**, 22–29.

GAMMELSRØD T. (1975a) Instability of Couette flow in a rotating fluid and origin of Langmuir circulations. *Journal of Geophysical Research*, **80**, 5069–5075.

GAMMELSRØD T. (1975b) Instability of linear flow in a rotating fluid and origin of Langmuir circulations. Unpublished manuscript presented at the UGGI Assembly in Grenoble, September 1975.

GARRETT C. J. R. (1976) Generation of Langmuir circulations by surface waves—a feedback mechanism. *Journal of Marine Research*, **34**, 117–130.

GORDON A. L. (1970) Vertical momentum flux accomplished by Langmuir circulations. *Journal of Geophysical Research*, **75**, 4177–4179.

HARRIS G. P. and J. N. A. LOTT (1973) Observations of Langmuir circulations in Lake Ontario. *Limnology and Oceanography*, **18**, 584–589.

HASSELMANN K. (1974) On this spectral dissipation of ocean waves due to white capping. *Boundary-layer Meteorology*, **6**, 107–127.

ICHIYE T. (1967) Upper ocean boundary-layer flow determined by dye diffusion. *Physics of Fluids*, Supplement 5270–5277.

KATZ B., R. GERARD and M. COSTIN (1965) Responses of dye tracers to sea surface conditions. *Journals of Geophysical Research*, **70**, 5505–5513.

KRAUS E. B. (1967) Organised convection in the ocean surface layer resulting from slicks and wave radiation stress. *Physics of Fluids*, Supplement 5294–5297.

KUETTNER J. P. (1971) Cloud bands in the earth's atmosphere. *Tellus*, **XXIII**, 404–425.

LANGMUIR I. (1938) Surface motion of water induced by wind. *Science*, **87**, 119–123.

LANGMUIR I., J. T. SCOTT, E. G. WALTHER and W. X. ROZON (1966) Langmuir circulations and internal waves in Lake George, Lake George Studies Report 1, Publication 42, Atmospheric Science Research Center, State University of New York, Albany.

LEIBOVICH S. (1976) On the evolution of the wind-drift Langmuir current system in the ocean. Part I: theory and the averaged current, Cornell University College of Engineering Energy Program report EPR-76-1, unpublished manuscript, 47 pp.

LEIBOVICH S. and D. ULRICH (1972) A note on the growth of small-scale Langmuir circulations. *Journal of Geophysical Research*, **77**, 1683–1688.

MARATOS A. (1971) Study of the near shore surface characteristics of windrows and Langmuir circulation in Monterey Bay. M.Sc. thesis, U.S. Naval Postgraduate School, Monterey.

McLEISH W. (1968) On the mechanisms of wind-slick generation. *Deep-Sea Research*, **15**, 461–469.

MYER G. E. (1969) A field investigation of Langmuir circulations. *Proceedings of 12th Conference on Great Lakes Research*, Ann Arbor, Michigan, pp. 625–663.

MYER G. E. (1971) Structure and mechanics of Langmuir circulations on a small inland lake. Ph.D. dissertation, State University of New York, Albany.

NIILER P. P. and E. B. KRAUS (1976) One-dimensional models of the upper ocean. In: *Modelling and prediction of the upper layers of the ocean*, E. B. KRAUS, editor, to be published by Pergamon Press.

OWEN R. W. Jr. (1966) Small-scale, horizontal vortices in the surface layer of the sea. *Journal of Marine Research*, **24**, 56–65.

POLLARD R. T. (1976) Observations and models of structure in the upper ocean. In: *Modelling and prediction of the upper layers of the ocean*, E. B. KRAUS, editor, to be published by Pergamon Press.

SCOTT J. T., G. E. MYER, R. STEWART and E. G. WALTHER (1969) On the mechanism of Langmuir circulations and their role in epilimnion mixing. *Limnology and Oceanography*, **14**, 493–503.

STAVN R. (1971) The horizontal–vertical distribution hypothesis: Langmuir circulations and *Daphnia* distributions. *Limnology and Oceanography*, **16**, 453–466.

STEWART R. (1970) On surface stress and Langmuir circulations. *Journal of Geophysical Research*, **75**, 7635.

STEWART R. and K. SCHMITT (1968) Wave interaction and Langmuir circulations. *Proceedings of 11th Conference on Great Lakes Research*.

STOMMEL H. (1949) Trajectories of small bodies sinking slowly through convection cells. *Journal of Marine Research*, **8**, 24–29.

STOMMEL H. (1951) Streaks on natural water surfaces. *Weather*, **6**, 72–74.

SUTCLIFFE W. H., E. R. BAYLOR and D. W. MENZEL (1963) Sea surface chemistry and Langmuir circulations. *Deep-Sea Research*, **10**, 233–243.

SUTCLIFFE W. H. Jr., R. W. SHELDON, A. PRAKASH and D. C. GORDON Jr. (1971) Relations between wind speed, Langmuir circulation and particle concentration in the ocean. *Deep-Sea Research*, **8**, 639–643.

WALTHER E. G. (1967) Wind streaks. M.Sc. thesis, State University of New York at Albany.

WELANDER P. (1963) On the generation of wind streaks on the sea surface by action of surface film. *Tellus*, **XV**, 67–71.

WOODCOCK A. H. (1944) A theory of surface water motion deduced from the wind-induced motion of the *Physalia*. *Journal of Marine Research*, **5**, 196–205.

WOODCOCK A. H. (1950) Subsurface pelagic *Sargassum*. *Journal of Marine Research*, **9**, 77–92.

Note added in proof—The restriction to phase locked wave pairs in CRAIK and LEIBOVICH's (1976) theory has been removed in recent papers by LEIBOVICH (1976a) and CRAIK (1976).

LEIBOVICH S. (1976a) Convective instability of stably stratified water in the ocean, Cornell University College of Engineering Energy Program report EPR-76-5, unpublished manuscript, 43 pp.

CRAIK A. D. D. (1976) The generation of Langmuir circulations by an instability mechanism, University of St. Andrews Mathematical Institute Preprints, unpublished manuscript, 26 pp.

Vertical circulation at fronts in the Upper Ocean

J. D. Woods,* R. L. Wiley† and Melbourne G. Briscoe‡

Abstract—A series of XBT sections cutting through fronts in the central Mediterranean reveals distortions of isotherms in the centre of the thermocline involving vertical displacements of up to ± 50 m from the mean depth.

A kinematic model of the mesoscale structure of these fronts is presented. The model is characterized by large amplitude waves. The horizontal acceleration suffered by water particles in the frontal jet when negotiating the curves created by these waves is comparable with the Coriolis acceleration. This alternating acceleration produces a corresponding isopycnal displacement leading to tongue-like folds alternately up and down the sloping density surfaces, consistent with the observed isotherm distortion.

Some implications of the model for vertical transport through the thermocline are discussed.

1. INTRODUCTION

THE three-dimensional description of a front presented in this paper is based on the analysis of temperature sections derived from expendable bathythermographs (XBT) dropped from high-speed Royal Air Force launches and sea surface temperature maps constructed from airborne radiation thermometer (ART) surveys by the Meteorological Research Flight during summer expeditions to Malta from 1969 to 1971. Preliminary reports of some aspects of the study have been published in papers by WOODS and WATSON (1970), WOODS (1972, 1974b).

BRISCOE, JOHANNESSEN and VINCENZI (1974) have published some results of a parallel series of investigations aimed primarily at exploring the large-scale distribution of temperature and salinity in the broad baroclinic zone centred on the continental slope which forms the eastern edge of the Malta sill, some 80 km east of the archipelago. We have not attempted to explore the large-scale distribution, preferring to concentrate on the meso-scale temperature structure of portions of fronts encountered within the broad baro-clinic zone. The horizontal spectral windows of our sections and maps are typically bounded by 1 and 30 km, whereas those of Johannessen and his co-workers are typically bounded by 20 and 300 km, so there is little overlap in the structural information contained in these two investigations.

2. EXPERIMENTAL METHOD

2.1. *Airborne radiation thermometer maps*

The experimental method differed only in minor detail from that described by WOODS and WATSON (1970). The Barnes PRT-4 radiometer, mounted in the Meteorological

---

* Institut für Meereskunde an der Universität Kiel.
† Meteorological Office, Bracknell.
‡ Woods Hole Oceanographic Institution.

Research Flight Varsity aircraft, was calibrated in flight during each turn in the surveying using a 'black-body' target whose temperature was adjusted successively to three values straddling the sea surface temperature. Tests published by Woods and Watson (1970) showed that the random error in the resulting measurements made at an altitude of 30 to 50 m was less than 0.1°C after averaging for 0.5 km along the flight track.

Most of the flights were carried out during the forenoon, when the sea surface temperature was rising due to the absorption of solar radiation. In calm, cloudless weather the spatially averaged rise approximated to a linear increase of 0.3°C/hour; in windy or cloudy weather the average rate was reduced. The mean rate of rise (assuming spatial homogeneity) was calculated for each flight and the data were corrected to a common time (usually 10 or 14 GMT) by adding or subtracting the appropriate amount, interpolated to the nearest 0.1°C at the mid-time of each straight leg of the survey and held constant throughout each leg.

Prior to 1971, navigation was based on visual siting of markers consisting of 2×1 m fluorescent, numbered boards lying flush with the sea surface and attached to drogues set at 5 m depth. The markers were designed to follow the surface current, which was typically in the range 35–45 km day^{-1} south to southeast. Errors due to current shear between the surgace and 5 m depth, and the wind and waves, are estimated[*] to be less than 2 km day^{-1}. Measurement of the horizontal displacement of a line of markers (Woods and Watson, 1970) leads to estimates of ± 5 km day^{-1} for the maximum horizontal variation of surface current within a survey area (20–50 km across). Maps of sea surface temperature distribution based on measurement using this moving frame of reference do not include significant distortion due to the mean surface flow, but the positions of data points have errors of up to about 1 km due to the horizontal shears, for which no corrections were applied.

In 1971 navigation was based on LORAN-C, and the aircraft flew straight legs along LORAN lines. Before contouring the sea surface temperature distributions, the data positions were adjusted to an equivalent synoptic position by adding a vector equal to the difference in measurement time from the synoptic time multiplied by the current velocity determined from internal evidence in the data (i.e. overlapping tracks or patterns seen in successive maps) or, failing that, put equal to the climatological value of 50 cm s^{-1} to the ESE. The distortion in these maps is estimated to be not greater than 1 km.

2.2. *Expendable bathythermograph sections*

Sippican expendable bathythermograph (XBT) probes were dropped at intervals of 0.5 or 1 km from a Royal Air Force launch moving at 10–15 m s^{-1} or, in 1971, H.M.A.F.V. *Sea Otter* moving at 7–10 m s^{-1}, so that a typical section 20 km long was completed in about half an hour. The XBT system was calibrated in the normal way against a dummy probe at the start and end of each section. In drawing isotherms in the XBT sections, attention was paid to the sharp edges of intrusions detected in the XBT traces and in a parallel series of temperature microstructure measurements.

Navigation for the XBT sections was based on the same surface markers as those used by the aircraft in 1969 and 1970, and on LORAN-C in 1971, when a Hewlett-

---

[*] Based on unpublished trials by Dr. P. M. Saunders, 1974.

Packard 9100B calculator was used to control the ship heading and plot the ship track over the ground.

3. PRELIMINARY OBSERVATIONS

Before describing the model of frontal structure based on our investigation, it is appropriate to present a sea surface temperature map and related section which illustrate the two main features of the model: (i) strong horizontal curvature and (ii) vertical displacement of isotherms.

3.1. *Waves on the surface outcrop of the front*

The surface outcrop of the front, lying between the 25°C and 26°C isotherms in Fig. 1, has a wavy form with radii of curvature as small as 2 km. The impression given by the shape of the contours is that the front is distorted by waves of wavelength 8 km, with perhaps some higher harmonics to explain the waveform being not exactly sinusoidal. Of course, the information content of the ART survey and in particular its bandwidth along the front is quite inadequate to permit one to reject the possibility that the wave in Fig. 1 is the result of aliasing unresolved shorter waves. But this seems unlikely since it would imply that even sharper curvature is present: the 8-km waves offer a minimum curvature interpretation of the data. Equally the presence of these 8-km waves does not exclude the possibility of other longer waves on the fronts, as suggested by BRISCOE, JOHANNESSEN and VINCENZI (1974). However, the curvature in these longer waves is less than for the 8-km waves shown in Fig. 1.

We emphasize the curvature since it produces a horizontal centripetal acceleration which is comparable with the Coriolis acceleration for water flowing along the front at $10 \, \mathrm{cm \, s^{-1}}$, as deduced from the relative motion of drogues (WOODS and WATSON, 1970).

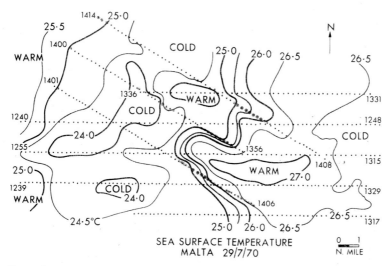

Fig. 1.   Sea surface temperature distribution based on airborne radiation thermometer measurements.

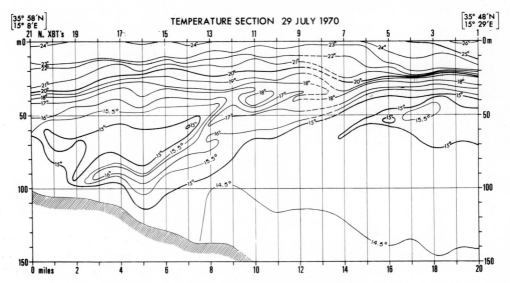

Fig. 2(a).   Vertical temperature section running approximately WNW–ESE along the central of the three parallel lines of ART data positions in Fig. 1.

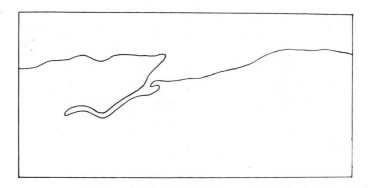

Fig. 2(b).   16°C isotherm redrawn from Fig. 2(a).

For a radius of curvature of $R = 2\,\mathrm{km}$, flow of $v = 10\,\mathrm{cm\,s^{-1}}$, Coriolis parameter $f = 10^{-4}\,\mathrm{s^{-1}}$, Coriolis acceleration, $fv = 10^{-3}\,\mathrm{cm\,s^{-2}}$ and the curvature acceleration $v^2/R = \frac{1}{2} \times 10^{-3}\,\mathrm{cm\,s^{-2}}$. As the water flows along the front the curvature acceleration acts alternately to reinforce, then to oppose the Coriolis acceleration. The flow is strongly ageostrophic.

3.2. *Vertical displacement of isotherms*

The section reproduced in Fig. 2 was drawn from XBT measurements made while the aircraft was mapping the sea surface temperature shown in Fig. 1. The plan was to make a section cutting orthogonally across the mean orientation of the front, but it is clear from Fig. 1 that the actual alignment was approximately 30° clockwise off this target.

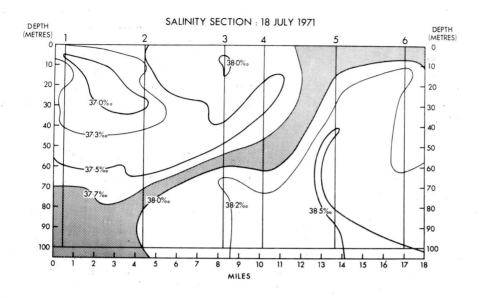

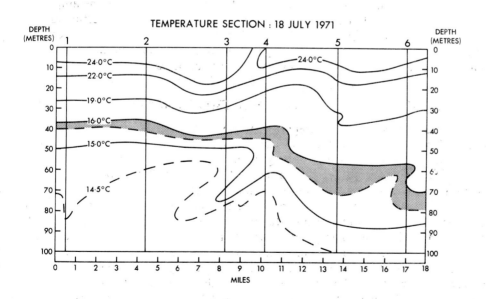

Fig. 3. Zonal temperature and salinity sections through a front drawn from STD stations at the relative positions shown (Sta. 1:15°E, 36°2′N).

The most striking feature in Fig. 2 is the fold in the 16°C isotherm which is redrawn for clarity in Fig. 2b. In this region the water to the east is hotter and more saline than the water to the west at all levels in the thermocline, so the horizontal gradients of salinity and temperature provide opposing contributions to the horizontal density gradient, with the former dominating. Thus at a front, the isotherms tend to be lower and the isohalines and isopycnals higher to the east;* the front slopes down towards the west as shown in Fig. 3. So the fold in the 16°C isotherm does not introduce static instability.

4. THE CAUSE OF ISOTHERM FOLDING AT FRONTS

The first condition necessary for folding is that the isotherms are inclined to the direction of motion, so that the flow advects them. In the thermocline, for scales larger than billows (< 1 m; WOODS, 1974a), the flow is generally aligned rather closely along density surfaces, so that the first condition is met if the isotherms are inclined to the isopycnals. In this situation, which DEFANT (1961, p. 308) calls *thermoclinic*, there is a gradient of temperature along a density surface, with a compensating salinity gradient. This occurs most strongly at locations where water masses with markedly different $T–S$ relationships meet. As was shown in Fig. 3 the upper ocean around Malta is strongly thermoclinic, with isotherms, isohalines and, therefore, isopycnals all strongly inclined to one another.

The second condition for folding is that there should be a maximum in the velocity profile. When such a velocity profile acts on an isotherm inclined to the direction of motion the isotherm is distorted into a fold. If the motion is steady the amplitude of the fold will increase indefinitely as the isotherm is advected, just as a vertical dye streak is distorted into an ever-increasing fold by a maximum in the velocity profile (WOODS, 1968, Fig. 10). But if the flow reverses periodically, then the isotherm fold will also grow to a maximum amplitude, then decay and reverse periodically.

5. ISOTHERM FOLDING IN A TWO-DIMENSIONAL FRONT

WOODS and MACVEAN (in preparation) have shown that weak isotherm folding is produced during the formation of a two-dimensional front, when a deformation field acts on an originally broad density gradient, with thermoclinity characteristic of the Malta area. The model is believed to be valid for the first 3 days of frontogenesis, up to the moment when a density discontinuity forms at the surface. At this stage the isotherm fold has a horizontal amplitude of approximately 1 km and this is increasing as the front continues to sharpen. But, because the model does not include instabilities and mixing which tend to limit the sharpening when the Richardson number falls to small values, it cannot be extended to the final stages of frontal development. So it is impossible to predict the final amplitude of the fold produced by two-dimensional frontogenesis. However, the width of the pressure variation along the density surface at a depth of 50 m in the final stage of the frontogenesis model is approximately 20 km, which is the same, within experimental error, as that for the frontal section shown in Fig. 3. On this evidence, we conclude that the fronts observed off Malta have approximately the same sharpness as those produced after 3 days in the Woods–MacVean model.

---

* Large meanders and eddies may, of course, locally rotate fronts away from this mean orientation.

This conclusion carries the implication that two-dimensional frontogenesis does not achieve more than about 10% of the (over 10 km) isotherm folding observed in our Malta sections. If this is correct, then we must seek an additional process to explain the observed large amplitude folds. This demands an increase in the cross-front motion along density surfaces, beyond that predicted in the frontogenesis model. It seems unlikely that the incorporation of mixing in an improved two-dimensional model would lead to increased cross-front flow: rather the opposite. So we conclude that large amplitude folding cannot be achieved by a two-dimensional model of thermocline fronts. In the next section we shall propose a three-dimensional structure for fronts which is consistent with our observations and provides the additional flow needed to achieve large amplitude folds.

6. A THREE-DIMENSIONAL MODEL .

The basis for assuming two-dimensionality during frontogenesis is the relatively large scale of the controlling deformation field. In searching for cross-front velocity fluctuations capable of inducing the observed isotherm folding, we must concentrate on scales much shorter than this. In this section we argue that baroclinic instability of the front, leading to large amplitude waves at wavelengths of around 10 km, provides the most likely mechanism for the required cross-front flow and present pictures of the resulting isotherm patterns.

ORLANSKI (1968) has considered the stability of a front and concludes that, for parameters appropriate to the fronts off Malta, the fastest growing baroclinic waves will have wavelengths of about 9 km (ORLANSKI, private communications). So far, the finite amplitude growth of these waves has not been analysed, but we suppose they grow to large amplitude ($ka \sim 1$) in a time that is long compared with the 3 days required for frontogenesis but

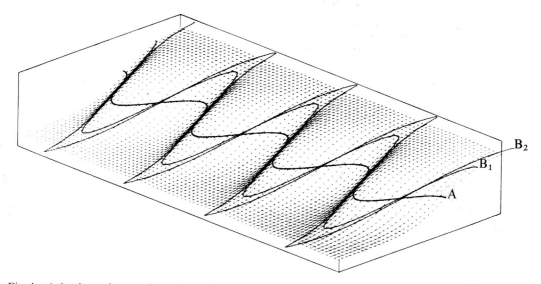

Fig. 4. A density surface at a front distorted by waves, showing the position of the greatest horizontal pressure gradient along the surface (A) and two possible positions (B_1 and B_2) for the isotherm that would coincide with A if it were not displaced in the cross-front direction by the curvature effect.

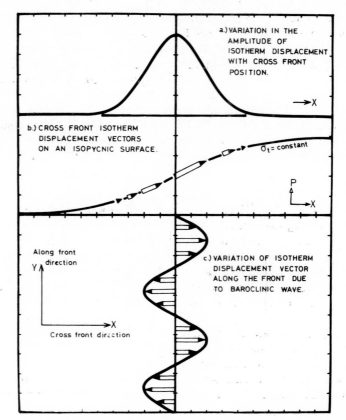

a.) VARIATION IN THE AMPLITUDE OF ISOTHERM DISPLACEMENT WITH CROSS FRONT POSITION.

b.) CROSS FRONT ISOTHERM DISPLACEMENT VECTORS ON AN ISOPYCNIC SURFACE.

σ_t = constant

Along front direction

Cross front direction

c.) VARIATION OF ISOTHERM DISPLACEMENT VECTOR ALONG THE FRONT DUE TO BAROCLINIC WAVE.

Fig. 5. The assumed form of the amplitude of cross-front isotherm displacement along a density surface. The amplitude reaches a maximum where the slope of the density surface is greatest, and decreases smoothly on either side with a scale length similar to that of the slope.

short compared with the (month or so) lifetime of the 100 km deformation field deduced from SAUNDERS' (1972) observation in the Ionian Sea. As was pointed out in Section 3.1, they will then distort the frontal surface with curvatures so large that water flowing along the axis of the front as an otherwise geostrophic jet will experience a periodic cross-front centripetal acceleration whose amplitude is comparable with the Coriolis acceleration. This wave acceleration will displace the jet and therefore isotherms back and forth relative to the locus of maximum cross-front pressure gradient on a density surface, dragging the isotherms with it (Fig. 4). It is assumed that the cross-front displacement along the density surface has a maximum at the axis and decreases smoothly and rapidly with increasing cross-front distance from it (Fig. 5). The form of this distribution will have to be determined from dynamic theory; for the purposes of illustration we have assumed that it is an error curve, with a width chosen (arbitrarily) to be equal to half the horizontal scale of pressure variation along the density surface (see Appendix 1). Isotherm folding in this model is illustrated by displacing isotherms along density surfaces from an undisturbed profile at the wave nodes, where the shape of the isotherm matches that formed by the two-dimensional frontogenesis model. An important feature is that the width of the

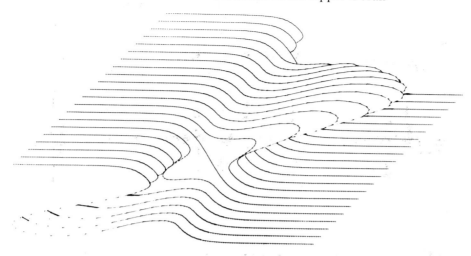

Fig. 6.   Isometric projection of the shape of an isothermal surface in the model front. The surface is depicted in terms of a series of vertical sections orthogonal to the axis of the front, starting at the wave minimum, where the downward displacement is greatest and running through the next maximum, with maximum upwelling, to the following node, where the isotherm is not displaced.

pressure distribution along the isotherm is an order of magnitude less than the corresponding width for a density surface at the same mean depth. The amplitude of isotherm displacement is modulated sinusoidally along the axis of the front in phase with the wave, giving rise to a series of alternating upwelling and downwelling tongues.

The isotherm patterns produced by this model in vertical sections orthogonal to the mean axis of the front are shown in Fig. 6. We believe that this is the explanation of the large amplitude folds seen in our XBT sections. If this is the case, the folds will vary periodically along the frontal surface in phase with the baroclinic wave. The primary motivation of our 1971 expedition was to explore this wave-acceleration idea by searching for the predicted variation of isotherm folding along the fronts.

7. COMPARISON WITH OBSERVATIONS

The kinematic model presented above makes arbitrary assumptions concerning the forms of the density surface, the undisplaced isotherm (at the nodes) and the form of the isotherm displacement along the density surfaces. In due course we hope to develop a dynamic model that will predict these forms in terms of a limited parameter set (wave amplitude and thermoclinity). Meanwhile, it is appropriate to test whether our model is broadly correct (though not, of course, dynamically so), by comparing the shapes of isotherms in our XBT sections with the shapes of isotherms predicted by the kinematic model. The test would be considered satisfactory if the choice of adjustable parameters needed to optimize the simulation were consistent with the totality of our observations of fronts in the Malta area. It would fail (a) if it were impossible to generate isotherm patterns that resembled those in our sections, or (b) if acceptable simulations could only be achieved by adopting parameters that are statistically unlikely because they lie too far outside the ranges encountered during our investigation.

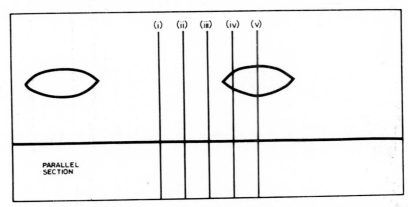

Fig. 7(a).   Vertical section parallel to the model front, showing closed isotherms produced by upwelling tongues.

As our XBT sections do not contain density information, it is impossible to test one of the principal features of the model; namely, the fact that the axis of the isotherm fold has a shallower slope than that of the front (defined by the locus of maximum horizontal pressure gradient on each density surface). What can be compared are the horizontal and vertical dimensions of the isotherm folds and their variation along the front.

Model fitting starts with an inspection of the XBT sections for two symptomatic features: (1) closed isotherms above or below a continuous isotherm at the same temperature, (2) an inclined fold in the isotherm.

(i) For a section aligned precisely parallel with the axis of the front (Fig. 7a) the model produces a series of closed isotherms spaced one wavelength apart and lying above or below the continuous isotherm, according to whether the section is displaced to the upper or lower side of the front.

(ii) For a section aligned precisely orthogonal to the axis of the front (Fig. 7b) the model produces a folded isotherm with a tongue running with a greater or lesser displacement up or down the density surfaces, according to the location of the section relative to the phase of the wave.

(iii) For a section cutting obliquely across the axis of the front (Fig. 7c) the model produces a combination of upper and/or lower closed isotherms, and/or isotherm folds, depending upon the precise orientation, displacement and phase of the section relative to the front and its wave.

Taking these various possibilities into consideration in the initial inspection of the section, it should prove possible to roughly locate the sections relative to the front as in Fig. 7. If the section runs nearly parallel with the axis of the front it should then be possible to estimate the wavelength of the frontal wave responsible for isotherm folding; while, from sections running roughly orthogonal to the axis, it should be possible to estimate the thickness and amplitude of the isotherm tongues displaced along density surfaces. Finally, by examining the variation of isotherm form in oblique sections it should be possible to estimate the amplitude of the wave (i.e. of the periodic horizontal displacement of the density structure), which is expected to be much smaller than the amplitude of isotherm displacement relative to the wave.

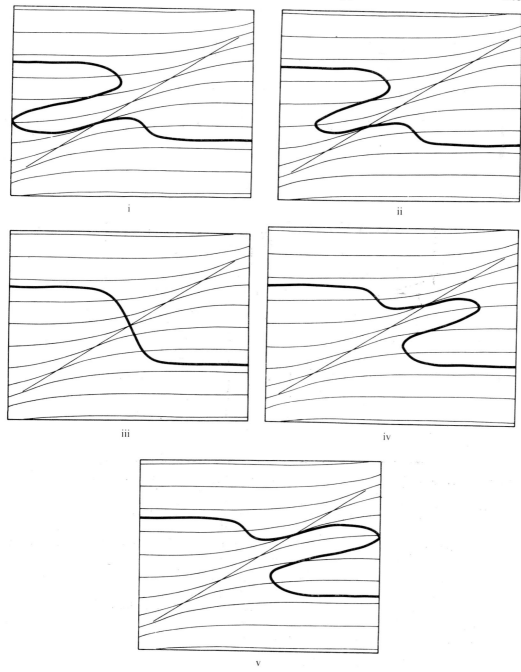

Fig. 7(b). Vertical sections orthogonal to the model front showing the isotherm (bold line) and isopycnals (thin lines). (i) Maximum downwelling ($\phi = -\pi/2$), (ii) intermediate downwelling ($\phi = -\pi/4$), (iii) undisplaced isotherm ($\phi = 0$), (iv) intermediate upwelling ($\phi = +\pi/4$), (v) maximum upwelling ($\phi = +\pi/2$). ϕ is the along-front phase relative to the horizontal isotherm displacement in the model.

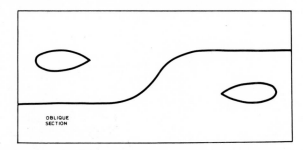

Fig. 7(c).   Vertical section aligned obliquely to the axis of the model front showing closed upper and lower
isotherms caused by upwelling and downwelling tongues respectively.

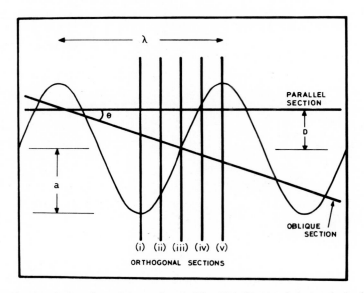

Fig. 7(d).   Plan showing the locations of the sections in Figs. 7(a), (b), (c), relative to the horizontal projection
of the isotherm on a density surface as depicted in Fig. 5. λ = wavelength ($= 2\pi/k$), a = amplitude of horizontal
displacement of isotherm, θ = orientation of section relative to front axis, ϕ = phase relative to horizontal
isotherm displacement, D = distance between section and closest node.

To sum up, a comparison of a single isotherm in the XBT sections and generated by
the model should yield estimates of the following parameters:
   isotherm displacement—amplitude thickness and height,
   density displacement—amplitude and wavelength.
By repeating the process for several isotherms, it should further be possible to estimate:
   the slope of the front,
   variation of the above parameters with depth.

8. RESULTS

The isotherms in the XBT sections presented in this paper were contoured by hand
soon after the end of fieldwork (1970 and 1971 respectively) and do not therefore represent

an attempt to fit the data to the model which was developed some years later. We have resisted the temptation to recontour the data in the light of our model, especially where alternative interpretations as to contour shape between XBT profiles would be equally acceptable (e.g. between 7 and 8 in the section on 11 August 1971). We do not believe that our conclusions are sensitive to any eccentricities of contouring in these sections.

8.1. *11 August 1971* (Fig. 8)

(i) *Location of sections relative to the front*

The 16°C upper closed isotherm between XBTs 9 and 10 shows that this section is aligned roughly parallel to the axis of the front and displaced to the upper side (i.e. southeast) relative to the nodal intersection of the 16°C isotherm and the front. The 17°C isotherm cuts the XBT 7 twice at approximately 42 m, suggesting either the northeast edge of a second upper closed loop, which may encompass the 16°C isotherm beyond the section, or alternatively (as drawn) a closed isotherm only at a temperature just above 17°C. If the latter interpretation were correct, then the section would have been orientated clockwise relative to the frontal axis, whereas the former interpretation would be consistent with a more closely parallel alignment.

The second section (XBTs 1–6) exhibits folds in the isotherms between 17°C and 15.5°C, with the largest amplitude probably between 15°C and 16.5°C. This fold is consistent with the section cutting orthogonally through the front at the position of maximum upward isotherm displacement. It is also consistent with the observation that the intersection between the two sections lies in the centre of the 16°C upper closed isotherm between XBTs 9 and 10. The section parallel to the front is estimated, on the basis of extrapolation beyond XBT 1, to lie 15±3 km from the nodal intersection of the 16°C isotherm with the front.

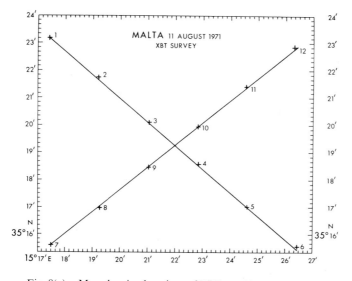

Fig. 8(a).   Map showing locations of XBTs on 11 August 1971.

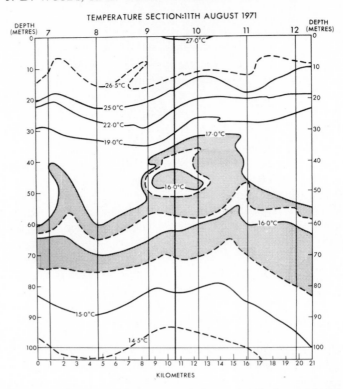

Fig. 8(b).   Vertical section aligned roughly parallel to the front axis as in the model section of Fig. 7(a).

(ii) *Model parameters*

The general shape of the 16°C and 16.5°C isotherms are consistent with the model. Assuming the locations estimated above, model parameters are estimated as follows:

16°C isotherm displacement   amplitude   16 ± 3 km

                                               thickness   7 ± 2 m

Density displacement   wavelength   10 ± 3 km*

      (* depending upon orientation of section 7–12)

Slope of the front   1:400 ± 50

Slope of the 16°C upwelling tongue   1:650 ± 50

8.2. *12 August 1971* (Fig. 9)

An attempt was made to make a second survey of the front sampled on 11 August, using three sections.

(i) *Location of sections relative to the front*

The upper closed isotherms at 17°C and 16.5°C provide evidence that section 7–12 is aligned approximately parallel to the axis of the front, but displaced to the upper (southeast) side relative to the nodal intersection of the 17°C isotherm and the front.

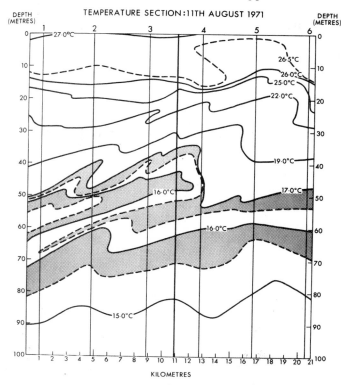

Fig. 8(c). Vertical section aligned roughly orthogonal to the front axis as in the model section of Fig. 7(b)(v).

As the 16.5°C closed isotherm near XBT 8 is smaller than near XBT 10, it is concluded that the section is rotated slightly counterclockwise relative to the front. If the front were precisely periodic, as in the model, this rotation can be estimated from the differences between the distances of the tips of the 16°C tongues from the intersection points ($\simeq 1$ km) and their along-front separation (~ 8 km), thus $\tan^{-1}(1/8) \simeq 7°$. This rotation is too small to produce significant error in estimating our model parameters and will be ignored, i.e. section 7–12 will be assumed to be parallel to the axis of the front.

In the other two sections, the folds in the 16°C isotherms are consistent with the interpretation that they lie orthogonal to the frontal axis at successive maximum upward displacements. The intersection points between these sections and the parallel section (midway between XBTs 8 and 9 and XBTs 10 and 11 respectively) lie approximately 1 km northeast of the centres of the closed loops, which is within the contouring uncertainty of being at the centres of the upwelling tongues. The nodal intersection of the 16°C isotherm with the front lies just beyond XBT 13, and is estimated by extrapolation to be approximately 10 km from section 7–12.

(ii) *Model parameters*

The general shape of the 16°C contour is consistent with the model. Assuming the

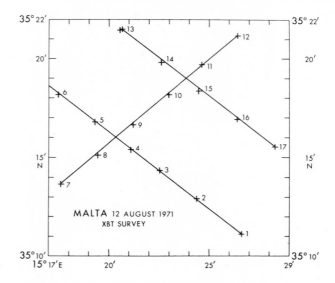

Fig. 9(a).   Map showing location of XBTs on 12 August 1971.

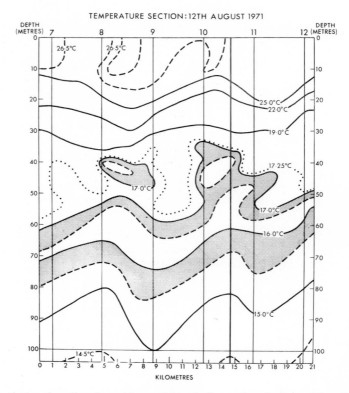

Fig. 9(b).   Vertical section aligned roughly parallel to the front axis as in the model section of Fig. 7(a).

locations estimated above, model parameters are estimated as follows:

16°C isotherm displacement   amplitude 8±2km
                                        thickness   5±2m

Density displacement   wavelength 8±1km
Slope of the front   1:300±50
Slope of the 16°C upwelling tongue   1:650±50

8.3. *13 July 1971* (Fig. 10)

This was an east–west section, which is interpreted as cutting diagonally across the front, such that the 16°C isotherm has an upper closed loop at XBT 30 and a lower closed loop near XBT 35. In this case there is no section parallel to the upwelling tongues from which to estimate their amplitude, but the fact that they extend far enough out on either side of the axis to intersect a diagonal section suggests that, as in the previous two cases, they almost certainly extend more than the wavelength, which is estimated to be less than $2 \times$ (spacing between upper and lower closed isotherms) $\times \sin 45°$, i.e. $\lambda < 14$ km.

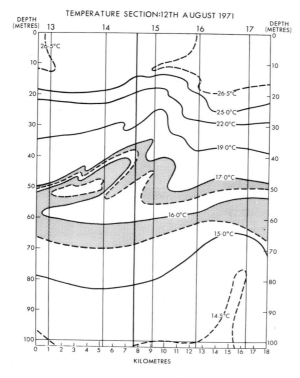

Fig. 9(c). Vertical section aligned roughly orthogonal to the front axis as in the model section of Fig. 7(b)(v).
$\phi = \pi/2$.

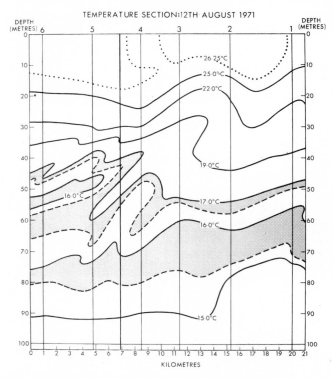

Fig. 9(d).   Vertical section aligned roughly orthogonal to the front axis as in the model section of Fig. 7(b)(v). $\phi = 5\pi/2$.

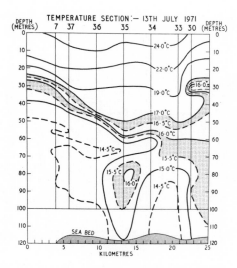

Fig. 10.   Vertical section through a front interpreted as being aligned obliquely to the front axis as in the model section of Fig. 7(c), but with θ negative.

8.4. *Comments*

Contouring uncertainties and isotherm undulations which may be related to internal waves limit the resolution of the data with the result that it is impossible to detect any improvement in model fit by varying the amplitude of the wave on the density surfaces. The results are consistent with the prediction that the isotherm amplitude is much greater than the density amplitude, but, because the latter is too small to be measured, provide no information about their relative phase.

9. CONCLUSIONS

The simple kinematic model presented in this paper, while somewhat speculative, appears to be capable of synthesizing the principal features of isotherms observed in our XBT sections, with parameters that are consistent with other evidence. For example, the wavelengths deduced from vertical sections are close to those deduced from sea surface temperature maps and (within experimental error) of the wavelength of the fastest growing baroclinic disturbance according to ORLANSKI (private communication). The model interprets the large amplitude isotherm folds as being due to centrifugal advection along density surfaces; a logical extension of the frontogenetic isotherm advection along density surfaces observed in the dynamic theory of WOODS and MACVEAN (in preparation). The present model does not predict a relationship between the amplitudes of the centrifugal advection and of the waves that produce it; that will be one of the aims of a dynamic theory currently being developed at Kiel.

It is interesting to consider the possibility that the large amplitude displacements of water by isentropic centrifugal advection at wavy fronts will contribute to vertical 'turbulent' transport through the thermocline. It has been shown that the vertical amplitude of the displacement may be as large as 50 m, giving a peak to peak displacement of up to 100 m, which is comparable with the thickness of the seasonal thermocline. In our model, the vertical displacements are periodic and do not lead to irreversible transport of water. But there is the possibility that the amplitude of the displacements may become so large that closed eddies (analogous to Gulf Stream rings) may become shed from the frontal jet. This process would produce irreversible isentropic transports of momentum as well as heat, salinity, etc. It would also contribute to the generation of microstructure in the seasonal thermocline.

But, even if the frontal waves do not 'break' in this way, the periodic vertical displacements can produce a vertical heat transport by increased absorption of solar radiation near the surface. WOODS (1974b) has shown that this corresponds to the observed mean internal development of the thermocline during the summer.

There may also be a significant biological effect because the mechanism brings water that lies on average below the compensation depth into the euphotic zone for periods of about half a day (the time for water to flow along the front between wave nodes), thereby possibly raising the primary production rate and producing a corresponding upward nutrient flux. This implies that primary production should be more rapid in the vicinity of fronts. In this connexion we note that the Ushant front studied by PINGREE, HOLLIGAN and HEAD (1977) exhibits both curvature and upwelling/downwelling tongues consistent with the model presented in this paper.

Acknowledgements—The fieldwork in Malta was sponsored by the Meteorological Office, Bracknell, where J.D.W. held a research fellowship. M.G.B.'s participation was supported by the NATO SACLANT ASW Research Centre, La Spezia, Italy, where he was affiliated at the time of the fieldwork; the XBT equipment was also provided by that organization.

The authors gratefully acknowledge the assistance of members of the following organizations:
Royal Air Force Marine Craft Unit, Marsaxlokk, Malta,
H.M.A.F.V. *Sea Otter*,
Meteorological Research Flight, Farnborough,
Hydrographer, Royal Navy.

APPENDIX 1. THE ISOTHERM DISPLACEMENT MODEL

The model presented in this paper is intended to demonstrate that isotherm folds of the kind observed can be explained in terms of a simple process, namely the cross-front displacement of water along density surfaces, modulated by (1) frontal waves and (2) the distance from the maximum horizontal pressure gradient on the density surface. In order to synthesize isotherm shapes as they would appear in vertical sections through the model, we have made reasonable assumptions about the form of these modulations, hoping that in the future they may be derived from formal dynamic analysis. The isotherms in Figs. 6 and 7 have been drawn using a HP 9820 calculator programmed to the following model equations:

Density surfaces	$P = P_0 + p \cdot \tanh((X - X_0)/A)$
Slope of the front	$S = P_0/X_0$
Undisplaced isotherm	$P = P_0 + q \cdot \tanh((X - X_0)/B)$
Isotherm displacement	
along density surfaces	$\eta = \eta_0 \sin(ky) \cdot \exp-((X - X_0)/C)^2$

The parameters used in drawing Figs. 6 and 7 had the following ratios

$$A = 5B = 2C = \eta.$$

$$2a = \lambda$$

$$p = -2q$$

APPENDIX 2. FRONTS AS A FEATURE OF THE GENERAL CIRCULATION

J. D. Woods

This paper is concerned with the mesoscale* structure of fronts in the upper ocean, but it is appropriate to speculate a little about the fronts we have studied in the central Mediterranean and the larger-scale hydrography and circulation in the area. The fronts are a feature of the water in and above the seasonal thermocline: they are not associated with bottom mixing as are those studied in the Irish Sea by Simpson and

---

* Following established meteorological terminology, and Soviet practice, I have chosen to apply the name 'synoptic' to those motions at the spectral peak of kinetic energy, and mesoscale to motions between this spectral peak and the Ozmidov scale, which is approximately 1 m in the seasonal thermocline.

HUNTER (1974) and in the English Channel by PINGREE, FOSTER and MORRISON (1974). We prefer to explain their growth in terms of the classical process for atmospheric fronts, in which a synoptic* scale deformation field concentrates a weak horizontal density gradient into a sharp quasi-geostrophic interface in about 2 days (e.g. HOSKINS and BRETHERTON, 1972). SAUNDERS (1972) has analysed synoptic scale (50–500 km) structure in the upper ocean to the east of Malta. His current distributions contain synoptic scale deformation rates of up to $10^{-5}\,s^{-1}$. WOODS and MACVEAN (in preparation) have shown that sharp fronts form in less than 3 days in such conditions. Assuming that the synoptic scale structures analysed by Saunders are part of a spectral cascade with negligible flux divergences of turbulent kinetic energy down to the small scale billows, WOODS (1974a, 1975) has estimated the average lifetime of the synoptic deformation in Saunders's data to be of the order of 1 month. This gives ample time for an initial stage of two-dimensional frontogenesis (lasting 2 to 3 days) then the growth of 9-km baroclinic waves to large amplitude, before the convergence sustaining the front decays.

Our data investigations were not designed to explore the distribution of fronts in the central Mediterranean, but during three seasons' study we encountered them over a wide area from the centre of the Ionian Sea 500 km east of Malta to some 50 km west of Malta; most frequently near the eastern edge of the Malta sill. The wide scatter suggests that fronts are a common feature of synoptic scale eddies in the upper ocean. The universal occurrence in our sea surface temperature data of a spectrum of form close to k^{-2} over horizontal scales ranging from 30 km to 1 m at widely different locations in the central Mediterranean (and even as far away as the Ligurian Sea) is consistent with the flow in the upper ocean being everywhere permeated by two-dimensionally (i.e. horizontally) isotropic turbulence. It seems reasonable to suppose that fronts formed by these eddies are an equally widespread feature of the upper ocean and there is no obvious reason for supposing that their meso-scale dynamical structure differs systematically with location. But around Malta, the presence of exceptionally strong thermoclinity leads to the large isotherm folds seen in our sections.

The picture of fronts as a feature of synoptic scale turbulence in the upper ocean differs significantly from that painted by BRISCOE, JOHANNESSEN and VINCENZI (1974) on the basis of their larger scale surveys of temperature–salinity distribution around Malta. They prefer to represent the fronts, whose meso-scale structure we have sampled, as being parts of a continuous frontal zone extending over hundreds of kilometres making a barrier between waters of the Ionian and Western Mediterranean seas, rather as the Polar front of the old Norwegian school of meteorologists separated Polar and Temperate air masses. Much of the apparent continuity drawn attention to by JOHANNESSEN, DE STROBEL and GEHIN (1971) in their analysis of the *Maria Paolina G.* data results from their failure to correct for the 1 knot southerly drift in the area. MORRICE (1974) has shown that in such a current the ship's ground track in Fig. 6 of BRISCOE, JOHANNESSEN and VINCENZI (1974) would transform as shown in Fig. A1, with the result that the ship passed back and forth through the same patch of water, rather than surveyed a volume of water 300 km in meridional extent. For these reasons, we feel that Johannessen's picture of an extensive continuous front is no more strongly supported than our picture of synoptic scale fronts imbedded in a turbulent upper ocean. We look forward to hearing of new experimental data that can discriminate between these two opposing hypotheses concerning the relationship between fronts and general circulation in the upper ocean.

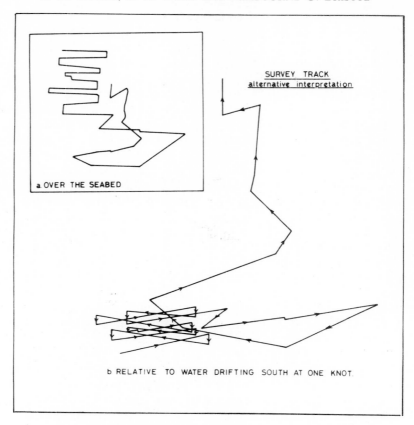

Fig. A1. Morrice's (1974) reinterpretation of the *Maria Paolina G.* ship track, as published in Briscoe, Johannessen and Vincenzi (1974) after allowing for southward drift of the water.

REFERENCES

Briscoe M. G., O. M. Johannessen and S. Vincenzi (1974) The Maltese front: a surface description by ship and aircraft. *Deep-Sea Research*, **21**(4), 247–262.

Hoskins B. J. and F. P. Bretherton (1972) Atmospheric frontogenesis models: mathematical formulation and solution. *Journal of the Atmospheric Sciences*, **29**, 11–37.

Johannessen O. M., F. de Strobel and C. Gehin (1971) Observations of an oceanic frontal system east of Malta in May 1971 (MAY FROST). *Saclantcen Technical Memo*, **169**.

Morrice A. (1974) Analysis and interpretation of temperature distributions in the vicinity of upper ocean fronts. M.Sc. dissertation, Southampton University.

Orlanski I. (1968) Instability of frontal waves. *Journal of the Atmospheric Sciences*, **25**(2), 178–200.

Orlanski I. (1972) Private communication.

Pingree R. D., G. R. Forster and G. K. Morrison (1974) Turbulent convergent tidal fronts. *Journal of the Marine Biological Association of the United Kingdom*, **54**(2), 469–479.

Pingree, R. D., P. M. Holligan and R. N. Head (1977) Survival of dinoflagellate blooms in the western English Channel. *Nature*, **265**, 266–269.

Saunders P. M. 1972. Space and time variability in the upper ocean. *Deep-Sea Research*, **19**, 467–480.

Simpson J. H. and J. R. Hunter (1974) Fronts in the Irish Sea. *Nature*, **250**, 404–406.

Woods J. D. (1968) Wave-induced shear instability in the summer thermocline. *Journal of Fluid Mechanics*, **32**(4), 791–800.

Woods J. D. and N. R. Watson (1970) Measurement of thermocline fronts from the air. *Underwater Science and Technology Journal*, **2**(2), 90–99.

WOODS J. D. (1972) The structure of fronts in the seasonal thermocline. *Proceedings of the Conference "Strait of Sicily"*, La Spezia, pp. 144–152.

WOODS J. D. (1974a) Space-time characteristics of turbulence in the upper ocean. *Mémoires de la Société royale des Sciences de Liège*, **6**, 109–130.

WOODS J. D. (1974b) Diffusion due to fronts in the rotation sub-range of turbulence in the seasonal thermocline. *La Houille Blanche*, **7/8**, 589–597.

WOODS J. D. (1975) The local distribution in Fourier space-time of variability associated with turbulence in the seasonal thermocline. *Mémoires de la Société royale des Sciences de Liège*, 6 ser., **7**, 171–189.

WOODS J. D. and M. K. MACVEAN. Numerical upper ocean frontogenesis models (in preparation).

Heat budget considerations in the study of upwelling

K. F. BOWDEN

Oceanography Department, University of Liverpool, P.O. Box 147, Liverpool L69 3BX, England

Abstract—When upwelled water is brought into the surface layer, its temperature and salinity become modified by the air–sea exchanges of heat energy and by evaporation. Using data obtained during a cruise to the upwelling region off northwest Africa, estimates of these effects have been made. In the narrow upwelling zone near the coast, the rate of evaporation was so small that its effect on the salinity of the upwelled water was negligible on a time scale of 10–20 days. The salinity could thus be treated as a conservative property in tracing the source and movement of the upwelled water. On the other hand, the net gain of heat, of the order of 300 to 500 cal cm^{-2} day^{-1}, was sufficient to cause an appreciable rise of temperature, and corresponding decrease in density, of the upwelled water. This information, used in conjunction with the T–S diagram, can help to identify the source of the upwelled water, to estimate its rate of upwelling and to indicate its movement subsequent to reaching the surface layer.

1. INTRODUCTION

WHEN upwelled water is brought into the surface layer, its temperature becomes modified by exchanges of heat energy across the air–sea interface and its salinity by evaporation or precipitation. Estimates of the rate of heat exchange have been used by several investigators to supplement other data on the processes and effects of upwelling. Heat budget computations were used by SMITH, PATULLO and LANE (1966) in one method of estimating the rate of offshore transport in the upwelling area off the Oregon coast. PAK, BEARDSLEY and SMITH (1970) made use of similar data in their study of a temperature inversion in the same area. Using data from a time series station, also in the Oregon area, REED and HALPERN (1975) deduced that changes in the heat content of the surface layer at a fixed location were influenced more by horizontal advection and diffusion than by the surface heat exchange. From observations off Cap Blanc in the upwelling area off northwest Africa, JONES (1972) suggested that the relatively simple configurations of the isotherms in a vertical section indicated that solar heating of upwelled water took place more rapidly than mixing with adjacent water. BARTON (1973) calculated the heat budget terms and related them to the observed changes in heat content at two time series stations off northwest Africa. At a position near Cap Blanc the advective changes completely predominated but further north, off Cabo Bojador, the influence of the surface heat exchange could be distinguished although advection again had the greater effect, at least on a time scale of several days. No other quantitative estimates of the heat budget effects in this upwelling area appear to have been published.

This paper describes a study of changes in temperature and salinity in relation to heat exchanges, based on data obtained during a cruise by R.R.S. *Discovery* in the upwelling region off Spanish Sahara in July–August 1972. A general account of the investigations during this cruise has been published by HUGHES and BARTON (1974) and a more detailed account of the circulation off Cabo Bojador by Johnson, Barton, Hughes and Mooers (1975). The area of the cruise is shown in Fig. 1. In the first leg, five sections were worked perpendicular to the coast. Between sections the track followed approximately the 500-m isobath, along which continuous recordings of temperature and salinity were made. In the second leg, more detailed observations were made in an area, approximately 20 nautical miles square, off Cabo Bojador.*

2. COMPUTATION OF HEAT BUDGET TERMS

In addition to the oceanographic observations, data were available from which to compute the heat-exchange terms. From continuous recordings, hourly mean values were derived of the solar radiation, dry and wet bulb air temperatures and the sea surface temperature. During the latter part of the cruise, hourly values of wind speed and direction were also available from recordings. For the earlier part, when this system was not working satisfactorily, hourly values of wind velocity were estimated from visual readings of the anemometer while the ship was on station. A comparison of the two methods for the overlapping period showed that the visual method gave a reasonably good approximation to the recorded values.

The net gain of heat by the water through unit area of sea surface is given by

$$Q = Q_s - Q_b - Q_c - Q_e \tag{1}$$

where Q_s = heat absorbed from solar radiation,
Q_b = net loss of heat by back radiation,
Q_c = loss of heat by conduction to the atmosphere,
Q_e = loss of heat by evaporation.

The data available from observations on board ship were the hourly mean values of
Q_R solar radiation recorded by solarimeter, in $\mathrm{cal\,cm^{-2}\,hr^{-1}}$,
t_0 sea surface temperature, in °C,
t_a dry bulb air temperature, in °C
t_w wet bulb air temperature, in °C,
W wind speed, in $\mathrm{m\,s^{-1}}$

and c cloudiness factor, i.e. fraction of sky covered by cloud, as estimated by the ship's officers.

The values of the terms in equation (1) were computed using standard methods, according to the equations:

$$Q_s = (1-r)Q_R \, \mathrm{cal\,cm^{-2}\,hr^{-1}} \tag{1a}$$

where r is the reflectivity of the sea surface, taken as 0.33 for the first hourly mean after sunrise and the last before sunset, 0.10 for the second hourly mean after sunrise and the last but one before sunset and 0.06 for the rest of the hours of daylight.

$$Q_b = 0.985\sigma T_0{}^4(0.39 - 0.05e_a{}^{\frac{1}{2}})(1 - 0.6c^2) \, \mathrm{cal\,cm^{-2}\,min^{-1}} \tag{1b}$$

as used by Kraus and Rooth (1961), where $\sigma T_0{}^4$ is the rate of black body radiation at the absolute temperature T_0 of the sea surface, e_a is the vapour pressure in the air, obtained from

---

* 1 nautical mile = 1.853 km.

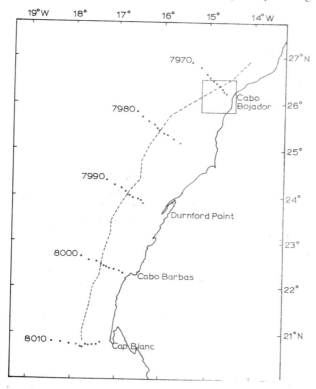

Fig. 1. Station chart for Leg 1 of *Discovery* cruise, 19–27 July 1972. The broken line is the 500-m contour.

equation

$$e_a = e_w - 0.675(t_a - t_w)$$

where e_w is the saturated vapour pressure over water at the wet bulb temperature t_w and e_w are in millibars.

$$Q_c = \rho_a c_p C_{10}(t_0 - t_a) \; W \, cal \, cm^{-2} \, s^{-1} \tag{1c}$$

where ρ_a is the density of the air, c_p is the specific heat of air at constant pressure and C_{10} is the drag coefficient in the wind stress equation for wind speed measured at a height of 10 m above the sea surface.

$$Q_e = \rho_a L C_{10}(q_0 - q_a) \; W \, cal \, cm^{-2} \, s^{-1} \tag{1d}$$

where L is the latent heat of evaporation, q_0 is the saturated specific humidity at sea surface temperature t_0 and q_a is the specific humidity in the air temperature t_a.

The wind drag coefficient C_{10} was taken as 1.3×10^{-3}, following KRAUS (1972). Inserting the appropriate numerical values of ρ_a, c_p and L and expressing $q_0 - q_a$ in terms of $e_0 - e_a$, equations (1c) and (1d) become

$$Q_c = 0.135(t_0 - t_a) \; W \, cal \, cm^{-2} \, hr^{-1}, \tag{1e}$$

$$Q_e = 0.202(e_0 - e_a) \; W \, cal \, cm^{-2} \, hr^{-1}. \tag{1f}$$

Table 1. Average daily values of heat budget terms

	Q_s	Q_b	Q_c	Q_e	Q	E
		(cal cm^{-2} day^{-1})				(g cm^{-2} day^{-1})
Leg 1 (19–27 July 1972)	558	119	−1	124	316	0.21
Leg 2 (2–13 Aug. 1972)	554	122	−10	165	267	0.28
Overall average	550	121	−6	147	288	0.25

Daily values of the terms in the heat budget were computed for all the days worked in the area. Table 1 shows the mean daily values for the 9 days of Leg 1 and 12 days of Leg 2 of the cruise and also the overall mean, in each case irrespective of the actual position of the ship. The wind blew consistently from a direction between north and northeast with an overall average speed of 10.2 m s^{-1}. Individual daily values of Q_s varied with variations in cloudiness, which also affected the values of Q_b, to a lesser extent. Loss by conduction Q_c was small and often negative. Loss by evaporation showed the greatest variability and was more dependent on position, since this term is very sensitive to the difference between air and sea surface temperature. The overall mean values may be taken as representative of the whole area, from 27°N to 21°N and up to 140 km from the coast.

3. EFFECTS ON TEMPERATURE AND SALINITY

In order to estimate the effects of heat exchange and evaporation on temperature and salinity, the simple model shown in Fig. 2 is considered. A column of water depth H, extending through the surface layer, is taken and is assumed to be drifting with the mean velocity of the water so that advective effects may be neglected. Horizontal diffusive effects are also neglected. Q is the net gain of heat and E is loss of mass of water, each per unit area of surface per unit time. It is assumed that water of temperature T_1 and salinity S_1 is upwelling through the base of the column with velocity w.

The equation of continuity of volume of water gives

$$H \operatorname{div} \mathbf{V} = w \tag{2}$$

where $\mathbf{V}$ is the horizontal velocity vector, averaged over the depth H, and the vertical velocity associated with E is assumed negligible compared with w. The equations of conservation of heat content and mass of salt then lead to

$$H\frac{dT}{dt} = \frac{Q}{\rho c_p} - w(T_2 - T_1), \tag{3}$$

$$H\frac{dS}{dt} = \frac{ES}{\rho} - w(S_2 - S_1) \tag{4}$$

where T, S and ρ are the mean temperature, salinity and density of the water within the column, t is time, c_p is the specific heat at constant pressure and T_2 and S_2 are the temperature and salinity of the water being lost to the column by horizontal divergence. For simplicity, in Fig. 2 the effect of the divergence is represented by a velocity u' through one face of the column only, i.e. div $\mathbf{V}$ is represented by $u'/\triangle x$.

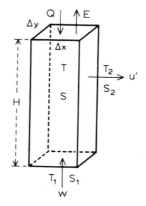

Fig. 2.   Effects of net heat flux and evaporation on the temperature and salinity of a column of water.

Two special solutions of equations (3) and (4) may be considered:
(1) If upwelling is no longer occurring, then $w = 0$ and

$$\frac{dT}{dt} = \frac{Q}{\rho c_p H}. \tag{5}$$

$$\frac{dS}{dt} = \frac{ES}{\rho H}. \tag{6}$$

(2) If, on the other hand, a steady state of upwelling exists, $d/dt = 0$ and

$$T_2 - T_1 = \frac{Q}{\rho c_p w}, \tag{7}$$

$$S_2 - S_1 = \frac{ES}{\rho w}. \tag{8}$$

From overall average figures of Q and E given in Table 1, i.e. $Q = 288 \, \text{cal cm}^{-2} \, \text{day}^{-1}$, $E = 0.25 \, \text{g cm}^{-2} \, \text{day}^{-1}$, equations (5) and (6) give, for a column of water 20 m deep, a temperature increase at the rate of 0.14°C per day and a salinity increase of 0.0045‰ per day.

4. STATIONS OFF CABO BOJADOR

During Leg 2 of the cruise, a grid of stations at approximately 5 nautical miles spacing covering the 20-mile square off Cabo Bojador was worked twice with an interval of 10 days. Hourly values of the heat loss $Q_c + Q_e$ were computed for the stations n the two grids, but only the data for the second grid, during which there was evidence of more active upwelling, are considered here. The station positions for this grid are shown numbered in Fig. 3 and the values of $Q_c + Q_e$ are plotted against the station positions in Fig. 4. For most stations there was only one hourly value available but if more than one was available the average is shown. The hourly values of wind speed varied from 8 to 13 m s^{-1} during this survey.

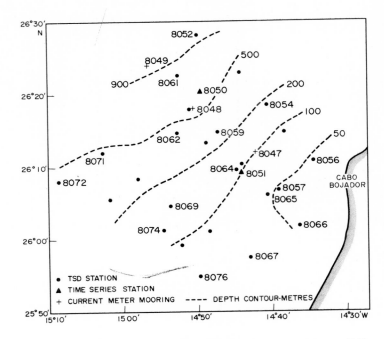

Fig. 3.   Station positions in Cabo Bojador area: Grid 2, 12–13 August 1972.

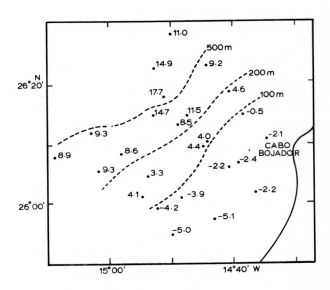

Fig. 4.   Heat loss $Q_c + Q_e$ in cal cm^{-2} hr^{-1} at Cabo Bojador stations: Grid 2.

Table 2. Effects of heating and evaporation

Depth of column (m)	All stations		Cabo Bojador: shallow stations	
	$\dfrac{dT}{dt}$ ($^\circ$C day^{-1})	$\dfrac{dS}{dt}$ (‰ day^{-1})	$\dfrac{dT}{dt}$ ($^\circ$C day^{-1})	$\dfrac{dS}{dt}$ (‰ day^{-1})
10	0.29	0.009	0.51	-0.001
20	0.14	0.0045	0.25	-0.0005

Immediately prior to the survey a time series station was worked at position 8051 and 48 consecutive hourly values of $Q_c + Q_e$ gave an average of 3.8 cal cm^{-2} hr^{-1} with a standard deviation of 2.8 cal cm^{-2} hr^{-1}. Prior to this, 48 hourly values at the time series station 8050 gave an average $Q_c + Q_e$ of 6.8 cal cm^{-2} hr^{-1} with a standard deviation of 3.7 cal cm^{-2} hr^{-1}. The complete survey of the stations in Grid 2 was completed in 30 hours.

Since only one hourly value of $Q_c + Q_e$ was available for most of the stations in Fig. 4, the precision is limited, but a clear systematic variation of $Q_c + Q_e$ with distance from the coast is apparent. The values decrease as the coast is approached and within a line which follows the 100-m contour approximately they become negative. Taking the average values for nine stations, all within the 100-m contour, at which $Q_c + Q_e$ is negative, the mean values of Q_c and Q_e are:

$$Q_c = -2.5\,\text{cal cm}^{-2}\,\text{hr}^{-1} = -60\,\text{cal cm}^{-2}\,\text{day}^{-1},$$
$$Q_e = -0.7\,\text{cal cm}^{-2}\,\text{hr}^{-1} = -17\,\text{cal cm}^{-2}\,\text{day}^{-1}.$$

With the overall average values of Q_s and Q_b given in Table 1, the net gain of heat through the surface, Q would be 506 cal cm^{-2} day^{-1} for these nine stations, compared with the value of 288 cal cm^{-2} day^{-1} derived for the whole area of the cruise. The corresponding rates of heating in a mixed layer of thickness $H = 10$ m or 20 m would be as shown in Table 2. The values for the Cabo Bojador stations, relating to a narrow band of coastal water of low surface temperature, are 70% higher than those for the whole area.

It is also indicated in Table 2 that the small, negative rate of evaporation would produce a decrease in the salinity of a layer 10 m deep of 0.001% per day, which is negligible. In fact the validity of the equation for evaporation with negative values of humidity gradient is doubtful but it seems reasonable to deduce that, under these conditions, the rate of evaporation and its effect on salinity would be negligibly small. It thus seems unlikely that evaporation has any significant effect on the salinity of upwelled water, at least on a time scale of 10–20 days.

The heat exchange, on the other hand, would appear to have a substantial effect on the temperature of the upwelled water after it has reached the surface layer. In a mixed layer 20 m deep it would cause an increase in temperature of the order of 1.4 to 2.5°C in 10 days.

5. *T–S* RELATION OF UPWELLED WATER

From the above results it appears that, under the conditions observed in the Cabo Bojador area:

1. The salinity of upwelled water remains unchanged, to a first approximation, since

the evaporation was very slight, there was no appreciable rainfall during the period and run-off may be assumed to be negligible.

2. The temperature of upwelled water is increased gradually by the net gain of heat through the surface.

These two hypotheses are qualitative and in an attempt to make quantitative deductions a third hypothesis was made:

3. Assuming that a column of upwelled water moves as a whole, the amount of heating since the water reached the surface layer may be obtained from the change in heat content, as shown by the deviation of temperature from the T–S relation for deeper water.

Thus

$$\Delta Q = \rho c_p \int_0^H (T - T_0) \mathrm{d}z \qquad (9)$$

where $\triangle Q$ is the change in heat content, T is the temperature at depth z, T_0 is the temperature given by the deep water T–S relation for water of the same salinity as that at depth z and the integration is carried to a depth H below which T does not differ significantly from T_0.

Figure 5 shows the T–S diagram for three of the near-shore stations of Grid 2 near Cabo Bojador and also, by the broken line, the mean T–S curve for water deeper than 100 m at stations beyond the shelf. The most striking evidence of upwelling is shown at the innermost station, 8056, where only water shallower than 12 m shows any significant heating. The water at 12 m and below has the same temperature and salinity values as at a depth of 200–250 m offshore and has, presumably, upwelled from that depth. The change in heat content of the column is 658 cal cm^{-2} which corresponds to only 1.3 days' heating at

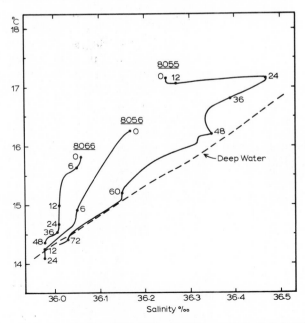

Fig. 5. T–S curves for Stas. 8055, 8056 and 8066 near Cabo Bojador: Grid 2.

the rate deduced above. At another inshore station, 8066, about 16 km to the south, there is some evidence of heating to at least 36 m and the corresponding heating is 4.4 days. On the alternative hypothesis of steady-state upwelling, the data for Sta. 8056 correspond to upwelling at a rate of 4 m per day (5×10^{-3} cm s^{-1}) through the bottom of a column 10 m deep. The data for Sta. 8066 correspond similarly to upwelling at 10 m per day (1.2×10^{-2} cm s^{-1}) at a depth of 30 m.

The T–S curve for Sta. 8055, which is 10 km offshore from 8056, is interesting in showing a layer, about 12 m deep, of low salinity water above water of higher salinity whose properties, at depths below 24 m, differ relatively little from the T–S relation for deeper water. It may be inferred that the low salinity surface water is water which has moved offshore after upwelling nearer the coast, where its properties were similar to those at station 8056. The water below 24 m is probably water which has upwelled from depths of 150–200 m offshore but which has not penetrated the surface layer. Applying equation (9) to the upper 12-m layer only, the increase in heat content corresponds to 4.1 days' heating. Comparing this with the value for Sta. 8056 indicates that the offshore drift would have taken at least 2.8 days to cover the 10 km: a velocity of 3.6 km per day or 4 cm s^{-1}. This time is a lower limit, since not all the heat absorbed through the surface may have been retained in the upper 12 m.

Further evidence for this explanation is seen in the vertical section in Fig. 6, which is typical of others normal to the coast in the Cabo Bojador region. The salinity section shows an indication of a low salinity layer moving offshore, causing a salinity inversion with depth at some distance from the coast. A similar inversion in temperature does not occur, presumably because of the increase in temperature by surface heating as the water moves offshore. The low salinity layer first becomes apparent with a thickness of about 15 m in a water depth of about 100 m and deepens to 20–30 m in a depth of about 300 m, but is not distinguishable further offshore.

Only three stations are shown in Fig. 5 and discussed in detail above but the same treatment has been applied to all those Grid 2 stations which were in depths less than 100 m. At Stas. 8057 conditions were similar to those at Sta. 8056, with only the layer down to 15 m showing any evidence of heating. Four other stations, all in a depth of 50–100 m, indicated the presence in the surface layer of water which had upwelled from depths of 150–250 m off the shelf. The heat content of these stations corresponded to from 8 to 20 days' heating, which had extended to a depth of at least 50 m. None of these stations, shallower than 100 m, showed a salinity inversion like that at Sta. 8055, although inversions occurred at eight other stations in depths between 100 and about 300 m.

Similar considerations of heat budget effects on temperature and salinity were applied to the stations of Grid 1. These confirmed that, as earlier studies of the data by HUGHES and BARTON (1974) and JOHNSON, BARTON, HUGHES and MOOERS (1975) had shown, upwelling was less intense at that time than during the Grid 2 survey, 10 days later. Station 8026, nearest to Cabo Bojador in a position close to Sta. 8056, showed evidence of surface water which had upwelled from a depth of 175–200 m and had been subjected to about 5 days' heating, extending to a depth of 24 m.

6. WATER MOVEMENTS

The circulation in the upwelling region off Cabo Bojador has been described by JOHNSON, BARTON, Hughes and MOOERS (1975), primarily on the basis of time series of

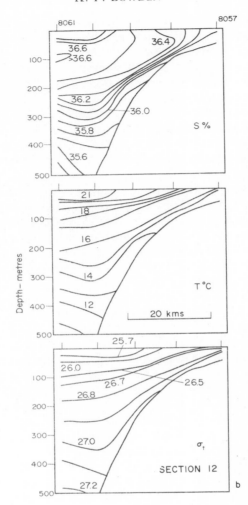

Fig. 6.   Vertical section of S, T and σ_t perpendicular to coast off Cabo Bojador: Grid 2.

measurements by a profiling current meter and several moored current meters, made during the period between the two hydrographic surveys of Grids 1 and 2. From the onshore–offshore current profiles they deduced the existence of a double-celled flow pattern over the continental slope and shelf. The presence of a seaward flow in the surface layer was indicated only tentatively by the profiling current meter but the authors point out that the proximity of the steel-hulled ship probably affected the measured current directions in the upper 10-m layer. They deduced that if an Ekman transport existed, it was confined to the upper 10 m and was probably weak. The measurements indicated a pronounced shoreward flow, centred about the level of maximum static stability, which occurred at about 35 m at a slope station (near Sta. 8060, depth 500 m) and at about 20 m at a shelf station (near Sta. 8058, depth 100 m). Below the shoreward flow there was an offshore flow at mid-depths, from 35 to 80 m at the shelf station and from 150 to at least 230 m at the slope station. A persistent onshore flow along the bottom was found at the shelf station.

From the above considerations of the *T–S* relations and heat budget effects, as at Stas. 8056 and 8066, it appears that the double-celled pattern did not extend into water shallower than about 50 m, where the surface water was continuous in its properties with water which had upwelled from depths of 200–250 m offshore. The inference by JOHNSON, BARTON, HUGHES and MOOERS of seaward flow in a surface layer about 10 m deep is consistent with the *T–S* curve at Sta. 8055 and with the interpretation of the subsurface inversion of salinity found at a number of stations. The formation of upwelled water of sufficiently low density for it to move offshore as a surface layer is dependent on surface heating. It seems probable that the offshore flow at mid-depths, found by the above authors, is fed by water which has upwelled inshore to depths of, say, 15–30 m, but which has not reached the surface layer. Its density would have been reduced to some extent by heat penetrating downwards or by diffusive mixing but not sufficiently for it to penetrate the surface layer.

From the current measurement JOHNSON, BARTON, HUGHES and MOOERS found that the alongshore component of flow was equatorward, reaching a maximum surface velocity of the order of 80 cm s^{-1} over the shelf. Since this is of the order of 10 times the offshore component of surface velocity, one has to visualize a column of surface water as drifting about 100 km southwards parallel to the coast while moving 10 km offshore.

7. STATIONS FROM CABO BOJADOR TO CABO BARBAS

The above considerations have also been applied to the first four hydrographic sections of Leg 1 of the cruise, the positions of which are shown in Fig. 1. The fifth section, off Cap Blanc, cannot be brought into the scheme as the water masses present there are different from those in the other sections. Charts of surface temperature and salinity based on these data, given by HUGHES and BARTON (1974), show that water of salinity less than 36.4‰ occurred at the inshore end of each section. On section 1, off Cabo Bojador, the temperature of this water was about 17.0°C but it increased southwards to about 18.5°C off Cabo Barbas. Figure 7 shows the salinity at depths down to 300 m at the inner stations of each of the four sections and Fig. 8 gives the *T–S* curves for selected stations. On the *T–S* diagram the mean curve for stations beyond the continental shelf is also shown for comparison.

The *T–S* curve for the innermost station of section 1 (Sta. 7979) shows the occurrence of water of salinity less than 36.35‰ in the surface layer and, from the continuity of the isohalines shown in the section, it could have upwelled from a depth of about 200 m beyond the shelf. At 25 m the properties of the water are almost unchanged but at lesser depths there is evidence of heating. Applying equation (9), the heat content of the water column corresponds to 4.5 days' heating at the rate found for Cabo Bojador stations. On the alternative hypothesis of steady-state upwelling, equation (7) gives a rate of upwelling of 5 m per day through the bottom of a 25-m column. These values are similar to those found for Sta. 8026 of Grid 1 but indicate less intense upwelling than at the time of the Grid 2 survey. Consideration of the other stations in section 1 indicate that it was only at the innermost station, 7979, that deep water had upwelled into the surface layer within a period of about 10 days prior to the observations.

At the two innermost stations of section 2, represented by Sta. 7989 in Fig. 8, a thin low salinity surface layer (15-20 m thick) was present overlying a layer in which the salinity

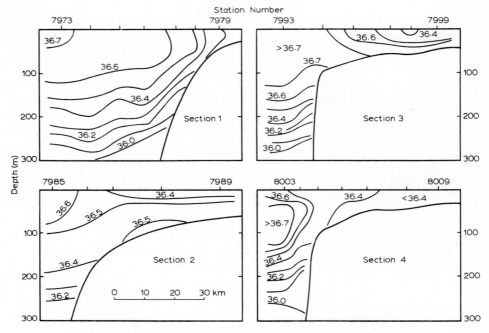

Fig. 7. Vertical distribution of salinity: sections 1–4 of Leg 1 of cruise.

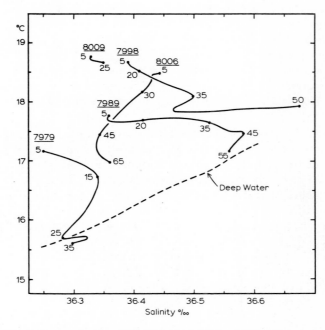

Fig. 8. *T–S* curves for Stas. 7979, 7989, 7998, 8006 and 8009 of Leg 1 of cruise.

increased to a maximum of over 36.55‰. From its salinity, 36.36–36.39‰, the surface under the conditions observed at the time, because of the presence of the high salinity layer with a maximum at about 50 m. The inference is that the surface water upwelled further north, possibly in the vicinity of Sta. 7979 of section 1, and then moved southwards along the coast with some spreading offshore, while its temperature was gradually raised by surface heating. Similar reasoning may be applied to the three inner stations of section 3, only one of which, Sta. 7998, is shown in Fig. 8. The salinity in a thin surface layer is relatively low, 36.39–36.45‰, but below this the salinity increases to more than 36.60‰.

The conditions are somewhat different at the stations on the shelf in section 4. At the first four stations from the coast, all in a water depth shallower than 80 m, the surface water is of relatively low salinity, 36.31–36.45‰. Two of these stations, 8006 and 8009, are shown in Fig. 8. Although there is a slight increase in salinity below the surface layer at the two innermost stations, the salinity does not exceed 36.45‰ at any depth at the four stations. From the T–S relation it seems possible that all the water on the shelf in section 4 could have upwelled from further offshore within the section. The relatively high temperature of the surface water, however, suggests that it is more likely that the surface layer, at least, is composed of water which had upwelled further north, possibly in the vicinity of the innermost station of section 1. The heat content of the two inner stations in section 4, relative to the T–S relation in deep water, corresponds to 15 and 18 days' heating respectively, even at the higher rate of 506 cal cm^{-2} day^{-1}, found for stations in the vicinity of Cabo Bojador during the Grid 2 survey. At the lower rate of 288 cal cm^{-2} day^{-1}, derived for the whole area of the cruise, the times would be 27 and 32 days respectively.

If, in fact, the water at the inshore stations on section 4 had originated by upwelling into the surface layer off Cabo Bojador and had taken about 20 days to drift the distance of about 450 km between the two sections, this would correspond to a drift of 22.5 km per day or 26 cm s^{-1}. The longer time of 30 days would correspond to 15 km per day or 17 cm s^{-1}. Such estimates cannot be more than rough approximations since a column 30–40 m deep would be affected by velocity shear and its heat content changed by horizontal advective and diffusive processes, as well as by surface heating over a distance of several hundred kilometres. However, it is not inconsistent with the maximum alongshore velocity of 80 cm s^{-1} found by JOHNSON, BARTON, HUGHES and MOOERS (1975) off Cabo Bojador, as the velocity of the current would probably be reduced considerably when it flowed over the broad shelf area to the south of Cabo Bojador.

8. CONCLUSION

The heat budget considerations have indicated that in water shallower than about 50 m off Cabo Bojador, the entire water column was composed of water which had upwelled from a depth of about 200 m beyond the shelf. Taking into account the cross-shelf circulation described by JOHNSON, BARTON, HUGHES and MOOERS (1975), this water had probably flowed up the slope on to the shelf as a bottom current, 10–20 m thick. A surface layer, 10–20 m deep, reduced in density by surface heating, appeared to feed the offshore Ekman transport. A study of the sections of Leg 1 of the cruise suggested that the surface layer parallel to the coast, as far south as Cabo Barbas, had probably originated in this way, by upwelling in the vicinity of Cabo Bojador. However, it is not possible from the data of this cruise to distinguish clearly between temporal and spatial variations in upwelling. Intense upwelling may normally be concentrated in the vicinity of Cabo Bojador, associated with topographical features of the area, or it may be that events such as the one observed there during this cruise occur at other positions along the coast at other times.

A more general conclusion from this study is that heat budget calculations, used in conjunction with the T–S diagram of water in the upwelling region, can form a useful addition to other methods of investigating upwelling processes. In conditions similar to those encountered in this cruise, when the effects of evaporation are negligible, salinity may be treated as a conservative property during upwelling. The change in temperature in relation to the rate of heating through the surface then enables an estimate of the upwelling velocity to be made.

The effects of heat exchange, or buoyancy flux, through the sea surface have not been taken into account in theoretical models of upwelling, although the constraint of a given surface temperature distribution has sometimes been applied (HSUEH and KENNEY, 1972; HSUEH and OU, 1975). The importance of including diffusive as well as advective processes in a realistic model is now recognized (e.g. TOMCZAK, 1973; THOMPSON, 1975; HAMILTON and RATTRAY, 1977). It appears that, in its effect on the dynamics of the surface layer, buoyancy flux across the surface may be as important as diffusive mixing.

Acknowledgements—The author wishes to express his appreciation to the Captain, officers and crew of R.R.S. *Discovery* and to his scientific colleagues from the Institute of Oceanographic Sciences and the Oceanography Department, University of Liverpool, for their co-operation and assistance during the cruise. I also wish to thank members of the staff of the Institute of Oceanographic Sciences for the primary processing of the meteorological data and Dr. P. HUGHES, University of Liverpool, for valuable discussions while preparing this paper. The work was supported by N.E.R.C. research grant GR3/1818.

REFERENCES

BARTON E. D. (1973) Upwelling off the coast of North West Africa. Ph.D. Thesis, University of Liverpool, 204 pp.
HAMILTON P. and M. RATTRAY (1977) A numerical model of the depth dependent wind driven upwelling circulation on a continental shelf. *Journal of Physical Oceanography* (in the press).
HSUEH Y. and R. N. KENNEY (1972). Steady coastal upwelling in a continuously stratified ocean. *Journal of Physical Oceanography*, **2**, 27–33.
HSUEH Y. and H-W. OU (1975) On the possibilities of coastal, mid-shelf and shelf-break upwelling. *Journal of Physical Oceanography*, **5**, 670–682.
HUGHES P. and E. D. BARTON (1974) Physical investigations in the upwelling region off North West Africa on R.R.S. *Discovery* cruise 48. *Tethys*, **6**, 43–52.
JOHNSON D. R., E. D. BARTON, P. HUGHES and C. N. MOOERS (1975) Circulation in the Canary Current upwelling region off Cabo Bojador in August 1972. *Deep-Sea Research*, **22**, 547–558.
JONES P. G. W. (1972) The variability of oceanographic observations off the coast of N.W. Africa. *Deep-Sea Research*, **19**, 405–431.
KRAUS E. B. (1972) *Atmosphere–ocean interaction*, Clarendon Press, Oxford, 275 pp.
KRAUS E. B. and C. ROOTH (1961) Temperature and steady state vertical heat flux in the ocean surface layers. *Tellus*, **13**, 231–238.
PAK H., G. F. BEARDSLEY and R. L. SMITH (1970) An optical and hydrographic study of a temperature inversion off Oregon during upwelling. *Journal of Geophysical Research*, **75**, 629–636.
REED R. K. and D. HALPERN (1975) The heat content of the upper ocean during coastal upwelling: Oregon, August 1973. *Journal of Physical Oceanography*, **5**, 379–383.
SMITH R. L., JUNE G. PATULLO and R. K. LANE (1966) An investigation of the early stage of upwelling along the Oregon coast. *Journal of Geophysical Research*, **71**, 1135–1140.
THOMPSON J. D. (1975) The role of mixing in the dynamics of upwelling systems. Third International Symposium on Upwelling Ecosystems, Kiel, Federal Republic of Germany.
TOMCZAK M. (1973) Note on diffusion in coastal upwelling. *Journal of Physical Oceanography*, **3**, 162–165.

On upwelling in the Arabian Sea

Robert L. Smith and Joseph S. Bottero

School of Oceanography, Oregon State University, Corvallis, Oregon 97331

Abstract—Extensive upwelling occurs in the northwestern Arabian Sea during the southwest monsoon. Hydrographic and wind measurements obtained by the R.R.S. *Discovery* in June and July 1963 (*Discovery*, Cruise 1) off the Arabian coast are used to estimate the vertical velocities in the upper several hundred meters by a method previously applied to the Peru Current by Wyrtki (1963). Upwelling into the surface layer occurs along 1000 km of the Arabian coast and extends at least 400 km seaward. The upwelling velocities are greatest over the continental shelf ($>3 \times 10^{-3}$ cm s^{-1}) but remain nearly as strong (1 to 2×10^{-3} cm s^{-1}) for at least 400 km offshore. The considerable width is a result of the strong southwest monsoon winds, blowing nearly parallel to the coast and increasing in intensity with distance offshore, causing, in effect, a superposition of coastal upwelling and open ocean upwelling.

INTRODUCTION

THE major coastal upwelling regions are located along the eastern boundaries of the oceans, with the notable exception of the northwestern Indian Ocean. During the southwest monsoon season (summer) strong upwelling occurs over a large area along the Somali and Arabian coasts and is manifest by cool sea surface temperatures (Bruce, 1974). From May through September the winds blow nearly parallel to the coastline causing an appreciable offshore Ekman transport. The upwelling along the Somali coast is also closely interrelated to the strong western boundary current (the Somali Current) and its separation from the coast near 10°N (Warren, Stommel and Swallow, 1966; Bruce, 1973). The currents along the Arabian coast are relatively weaker; a broad band of upwelling occurs along 1000 km of the coast (15°N to 22°N) as a result of wind induced divergence in the surface layer flow.

In this paper we estimate the vertical field of motion in the upper layers of the Arabian Sea, from the coast to 400 km seaward between 15°N and 21°N, during the southwest monsoon. This study is based on wind and hydrographic data obtained by the R.R.S. *Discovery* during 1963 as part of the International Indian Ocean Expedition. Detailed sea surface salinity and temperature maps from this cruise (*Discovery*, Cruise 1) have been presented by Bruce (1974) and a discussion of the nutrient and biological data can be found in Currie, Fisher and Hargreaves (1973).

The hydrographic stations used (Fig. 1) were occupied by the R.R.S. *Discovery* between 25 June and 18 July 1963. The wind observations obtained during the cruise and the hydrographic data at these stations enabled us to determine the absolute topography of the sea surface by a method previously applied to the Peru Current by Wyrtki (1963). In this method it is assumed that the horizontal flow is in a steady state and is geostrophic, determined by the topography of the sea surface and the density distribution,

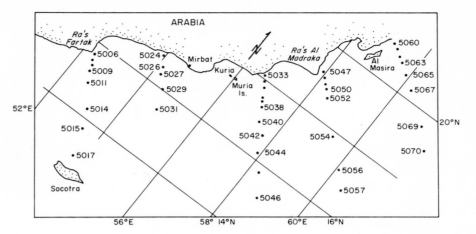

Fig. 1.   The western Arabian Sea, showing stations occupied by R.R.S. *Discovery* during June and July of 1963.

but with an Ekman transport superimposed in the surface layer. The oceanic region under consideration is divided into a number of closed contiguous cells extending from the ocean surface to the bottom, and the equation of continuity is applied. An historical antecedent is STOMMEL's (1956) treatment of the North Atlantic circulation in which he required the net transport across a line of latitude, spanning the North Atlantic, to vanish.

DESCRIPTION OF THE PROCEDURE

The method used in this paper assumes a steady-state ocean and synoptically collected data. The near-surface dynamic topography in the western Arabian Sea changes appreciably from winter (the northeast monsoon season) to summer (southwest monsoon) but not nearly so dramatically as off the Somali coast where the Somali Current develops with the southwest monsoon (BRUCE, 1968). The southwest monsoon begins in April and blows strongly from June through September. July through September is a period of "quasi-consistent conditions" (DUING, 1970) when the monsoon and the associated ocean circulation are well developed. The observations used in this paper were made in the first weeks of this period. Hence, the steady-state assumption is not unreasonable.

The field of Ekman transport for the region was determined from some 200 wind velocity measurements made with shipborne anemometers. The winds during the cruise were "remarkably constant in direction" (ROYAL SOCIETY, 1963) and increased with distance offshore on each line. The wind stress, $\vec{\tau}$, was computed from wind velocity $\vec{u}$ according to the equation

$$\vec{\tau} = \rho_{\text{air}} C_D |\vec{u}| \vec{u} \qquad (1)$$

using 1.3×10^{-3} for the dimensionless drag coefficient, C_D. Wind stress values computed from individual wind observations were smoothed by fitting least-squares planes to τ_x and τ_y, the eastward and northward components, making these functions of longitude and latitude. The smoothed wind stresses at representative stations are given in Table 1.

Table 1.   *Smoothed wind stresses. τ is the magnitude of the stress, and θ is*
the direction toward which the wind blows

Station		τ (dynes cm^{-2})	θ (degrees)	Station		τ (dynes cm^{-2})	θ (degrees)
5008	line 1	1.1	17	5049	line 4	1.9	30
5014		1.9	30	5052		2.2	32
5017		2.7	36	5054		2.9	36
				5057		3.9	39
5026	line 2	1.3	21				
5029		1.7	27	5063	line 5	1.7	27
5031		2.0	31	5067		2.3	32
				5069		2.9	36
5037	line 3	2.1	31	5070		3.4	38
5040		2.5	34				
5044		3.0	37				
5046		3.8	39				

Table 2.   *Wind stress during the southwest monsoon*

Coordinate	τ_x, τ_y (our data)	τ_x, τ_y (HELLERMAN, 1967)
12.5°N, 52.5°E	1.3, 1.9 dynes cm^{-2}	1.7, 1.7 dynes cm^{-2}
12.5°N, 57.5°E	2.9, 3.4 dynes cm^{-2}	3.4, 4.2 dynes cm^{-2}
17.5°N, 62.5°E	2.8, 3.4 dynes cm^{-2}	2.2, 2.1 dynes cm^{-2}

The general wind picture revealed in Table 1 agrees well with that given by RAMAGE (1965) for this region. As a further test of the representativeness of the smoothed wind values, τ_x and τ_y were computed (by extrapolating the least-squares equations) for the center points of the nearest 5° squares for which HELLERMAN (1967) gives seasonal wind stress values. A comparison between our values and HELLERMAN's is made in Table 2 and shows remarkably good agreement, especially considering that our values are from a 3-week period in 1963 and HELLERMAN's represent a seasonal mean based on many years.

Hydrographic stations occupied the corners of the cells (Fig. 2) in the network used in the application of the continuity equation. Consider a cell formed by three hydrographic stations (e.g. Sta. 5040, 5052 and 5054). Let v_n be the horizontal velocity normal to the cell walls which extend from the surface to the bottom, and let s be the horizontal coordinate along the wall. Then:

$$\rho f v_n = -\frac{\partial p}{\partial s} + \frac{\partial \tau_s}{\partial z} \tag{2}$$

where ρ is density, f is the Coriolis parameter, p is pressure, τ is the stress on a horizontal surface, τ_s is the component of τ along the cell wall, and z is measured positive upward. (The flow along the Arabian coast is similar in magnitude to that in eastern boundary currents, and considerably weaker than the Somali Current, which displays the dynamics typical of western boundary currents. The present assumptions, such as neglect of lateral turbulent stresses, would not be appropriate in the latter case.) The horizontal mass transport entering a cell between the sea surface, h, and some depth, z_0, is given by:

$$T(z_0) = \int_{z_0}^{h} \oint_s \rho v_n \, ds \, dz = \int_{z_0}^{h} \oint \left(-\frac{g}{f} \int_z^h \frac{\partial \rho(\zeta)}{\partial s} \, d\zeta - \frac{g}{f} \rho(h) \frac{\partial h}{\partial s} + \frac{1}{f} \frac{\partial \tau_s}{\partial z} \right) ds \, dz \qquad (3)$$

where the dependence on the internal density field, sea surface topography and the wind (the Ekman transport; $\partial \tau / \partial z$ is assumed to vanish at depth) are shown explicitly. Vertical motion results from the convergence, or divergence, of the horizontal flow and is assumed zero at the sea surface ($z = h$). The vertical motion at any intermediate depth, z_0, can be obtained by dividing $T(z_0)$ by the area enclosed by the curve s.

If we ignore exchange with the atmosphere, the equation of continuity requires that the net horizontal mass transport into the cell, which extends from the sea surface to the bottom, be zero, i.e. the right-hand side of equation (3) must be zero if the lower limit of integration is the sea bottom ($z_0 = b$) since no flow can enter through the bottom. The density and wind fields are available from observations, and the only unknown in equation (3) is $\partial h / \partial s$, the slope of the sea surface. In principle one can set $T(b) = 0$ for each cell of the network in turn and solve for $\partial h / \partial s$.

The hydrographic stations available did not extend to the ocean bottom and, moreover, it was not easy to allow for the complicated bottom topography. We implemented the above method in two different ways. In the first, each analysis cell extended to the deepest level for which hydrographic data were available, and the vertical velocity at that level was assumed to vanish. Since we wished to extend the computations as close to shore as possible over the sloping topography, the cell depths could not be the same over the entire region. The cell depths varied from 3000 m in the northeast offshore region to 900 m in the central coastal region. Stations over the continental shelf and inner slope were not used in these computations. The order in which the cells were considered minimized the effect of variations in cell depth. The second procedure used only those stations which could be extended to 2000 m and that depth was used as the bottom for all cells involved. This is equivalent to assuming that there was no vertical flow at 2000 m, or that the flow was everywhere horizontally divergenceless below 2000 m. The different procedures altered the computed flow, both horizontal and vertical, only slightly in the upper 300 m.

In either case, to initiate the procedure for computing the absolute topography and the vertical velocities, we began with Cell 1 (Fig. 2), Stas. 5040, 5052 and 5054, and

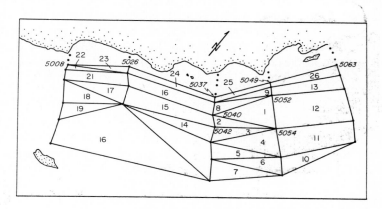

Fig. 2.   Network of cells and order of computations.

assumed that there existed a level of no horizontal motion (LNHM). The net geostrophic transport into Cell 1 was calculated as a function of the reference depth (letting the latter vary from the sea surface to the bottom of the cell). The reference depth that caused the net transport (geostrophic plus Ekman) to vanish was selected as the actual LNHM. (For Cell 1 depth = 3000 m: LNHM $\sim$ 1600 m; for Cell 1 depth = 2000 m: LNHM $\sim$ 1400 m.) Taking this depth as a level surface, we then calculated the absolute heights of the 0-db surface at the three stations, arbitrarily assigning the value $h = 0$ for Sta. 5040. The next cell considered had two stations in common with the first cell and it was necessary only to find the height of the third station (Sta. 5042). This was done by solving the equation $T(b) = 0$ for the cell. Successive cells were handled in essentially the same way and thus the height of each station in the network was calculated. Only for the first cell was the LNHM assumed to be the same on all sides of the cell. For all other cells a LNHM could be found *a posteriori* for each side of a cell, but it was not necessary to do so to compute velocities because the absolute topography of the sea surface was already determined for the stations considered.

In treating the most inshore cells two options were available: the cell walls nearest to the coast and approximately parallel to it could be treated as all other walls with free flow allowed through each cell wall, or these walls could be assumed closed to flow. The latter option was chosen by WYRTKI (1963) in his Peru Current study where he assumed that the proximity of the deep stations to the coastline allowed the assumption that the shoreward boundary of the network represented the coastline across which there could be no transport. The shoreward cell walls in our case (formed by Stas. 5008, 5026, 5037, 5049 and 5063) are the order of 50 km from the coastline, and an assumption of no flow through these cell boundaries seemed unjustified. Direct measurements of the flow in a region (Northwest Africa near 22°N) with similar continental shelf and slope topography and wind intensity showed significant shoreward flow at depths appreciably below the depth of the shelf break yet only a few kilometers away from it (MITTELSTAEDT, PILLSBURY and SMITH, 1975). Therefore the inshore boundaries of the cell network off Arabia were considered open to flow. In any case, the computations for the coastal cells were done last, and the choice of open shoreward boundaries for the coastal cells affected only the dynamic topography of the nearshore stations.

ABSOLUTE TOPOGRAPHY OF THE SEA SURFACE AND HORIZONTAL FLOW

The absolute topography of the sea surface, computed by the method described above and using variable cell depths, is shown in Fig. 3a. The topography has been extrapolated shoreward from the network of cells (see Fig. 2) to give the topography of the stations over the continental slope and shelf. The flow into and out of the region close to the coast (between 5006 and 5008 and between 5061 and 5063) may play a significant role in the mass balance of the inshore coastal upwelling region. This nearshore flow is not determined by the above procedure since the deep network of stations ended relatively far from the coast. The alongshore flow over the inner continental slope and shelf is known to be strong and quasi-geostrophic in most coastal upwelling regions (HUYER, SMITH and PILLSBURY, 1974; MITTELSTAEDT, PILLSBURY and SMITH, 1975); where sea surface slopes can be estimated and compared with current speeds, good agreement has been found for

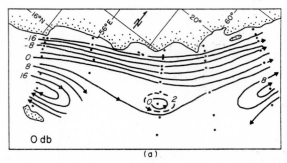

Fig. 3a. Absolute dynamic topography of the sea surface. $h = 0$ at Sta. 5040. The units are dynamic cm with contour interval of 4 dynamic cm. The geostrophic volume transport between adjacent contours (in a layer 1 m thick) is about $10^4 \, \text{m}^3 \, \text{s}^{-1}$.

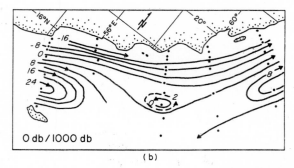

Fig. 3b. Dynamic topography of the sea surface relative to 1000 db. $h = 0$ at Sta. 5040.

the continental slope and shelf regions (SMITH, 1974; REID and MANTYLA, 1976). The geopotential topography of the shallow stations is referenced to the network in the following manner. The level of no motion was determined for the sections of the network perpendicular to the coast by computing the geostrophic velocity from the absolute sea surface topography and the density distributions. The level of no motion varied smoothly near the coast, ranging from 700 db near Sta. 5009 to 1200 m near Sta. 5063, and a level of no motion was assigned for the shoreward part of each of the five sections. Starting with an offshore pair of stations in the network, e.g. 5008 and 5011, a gradient of the geopotential anomaly, relative to the level of no motion, is determined for the deepest pressure surface in common with the first station (5007) shoreward of the network. The gradient at the next deepest common pressure surface is found from 5008 and 5007 and extrapolated to 5006, and so forth if more stations exist shoreward. This procedure is equivalent to extrapolating isopycnals inshore using the last observed offshore slopes of the isopycnals. The method has been used by REID and MANTYLA (1976) off the Oregon coast and gives steric sea levels that compare very well with the tide gauge sea-level records.

   The computed absolute topography of the sea surface (Fig. 3a) can be compared with the dynamic topography relative to the 1000-db surface (Fig. 3b). The absolute topography computed using a constant 2000-m bottom for the network cells is even more similar to the relative topography of Fig. 3b. The differences are not significant and both agree

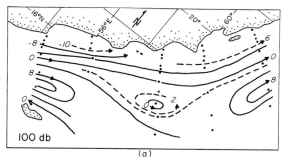

Fig. 4a. Absolute dynamic topography of the 100db surface. $h = 0$ at Sta. 5040. The units are dynamic cm with contour intervals of 4 dynamic cm.

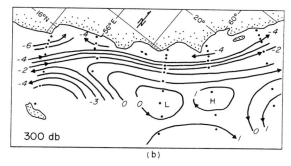

Fig. 4b. Absolute topography of the 300db surface. $h = 0$ at Sta. 5040. Contour interval is 1 dynamic cm.

with the sea surface topographies (relative to 1000db) shown by DUING (1970) and WYRTKI (1971) for the May–June periods. The general rise of sea level offshore is a recurring summer feature, as was noted by BRUCE (1968). The anticyclonic gyre just north of Socotra is clearly seen in the May–June topographies in DUING (1970) and WYRTKI (1971), which are not based solely on data from this cruise; both also show an anticyclonic gyre in the northeast, but the position differs somewhat in the three studies.

Figures 4a and 4b show absolute dynamic topographies of the 100- and 300-db pressure surface, respectively. The 500-db topography (not shown) is similar to that of the 300-db level. These figures indicate that much of the water that passes through the region, beneath the surface layer, is supplied from the southeast quadrant, perhaps as an extension or branch of the Somali Current, as is suggested in the map of the 300-db surface relative to 1000db, for July and August, presented by WYRTKI (1971). This northward-moving stream branches, as it passes Socotra, with part of it turning northeastward along the Arabian coast, and part passing between Socotra and the mainland into the Gulf of Aden. An interesting feature of the 300-m and 500-m topographies (the latter is not shown) is the hint of a deep equatorward (toward the southwest) countercurrent adjacent to the Arabian coast. Several subsurface current measurements, using neutrally buoyant floats with acoustic pingers, were made over the continental slope in the weeks following the hydro-graphic survey (J. C. SWALLOW, personal communication). Although of only short duration, several of the float trajectories (between 17°N and 21°N) indicate flow towards the southwest at 75 m and 210 m close to the shelf break. Poleward undercurrents in upwelling regions associated with eastern boundary currents have commonly been observed (WOOSTER and REID, 1963). The existence of the undercurrents in upwelling regions has not yet been fully explained. However, the existence of a deep equatorward countercurrent off the Arabian coast is consistent with the explanation proposed by PEDLOSKY (1974): the presence of the bottom slope (1) reduces the role of the lower Ekman layer in the upwelling mass balance, and (2) introduces a barotropic boundary layer which can give rise to a deep countercurrent.

VERTICAL FLOW

Vertical velocities at any specified depth can be obtained by applying equation (3) and the equation of continuity to the network of cells as discussed in the section on procedure. We have assumed that the Ekman transport is confined to within 50 m of the surface. Direct current measurements in the upwelling regions off Oregon and Northwest Africa have shown that the Ekman layer is shallower than 50 m (SMITH, 1974; HALPERN, 1976; MITTELSTAEDT, PILLSBURY and SMITH, 1975). In Fig. 5 the cells shown in Fig. 2 have been combined into larger cells to provide some averaging. Vertical velocities are shown for depths of 50, 100, 300 and 700 m. The values given are from computations in which the depths of cells varied; computations with a constant 2000-m cell bottom depth gave values at 50 m and 100 m that differed by less than 10% from those shown in Fig. 5.

The picture revealed here (Fig. 5) is of an upwelling region of broad extent and relatively high velocities. At 50 m and 100 m the motion is upward over the entire area, with speeds on the order of 10^{-3} cm s^{-1}. At 300 m and 700 m this picture is modified near the coast by a band of downwelling 50 km wide and extending parallel to the coast for 1000 km. In applying the same method to the Peru Current region, WYRTKI (1963) found that the vertical flow near the coast was actually downward through the 100-m level, indicating that ascending motion due to the divergence of the Ekman layer was concentrated at depths less than 100 m. Further offshore the vertical flow at 100 m was upward and these ascending motions were found progressively deeper at greater distances from the Peru coast; the vertical velocities at 700 m several hundred kilometers off the Peru coast

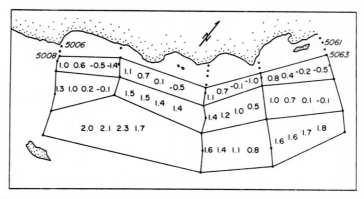

Fig. 5.   Vertical speeds at 50 m, 100 m, 300 m and 700 m. The four numbers in each block are the speeds at 50 m, 100 m, 300 m and 700 m respectively, reading left to right. The units are cm s^{-1} × 10^{-3}.

were about 0.4×10^{-3} cm s^{-1}. A similar pattern is obtained for the Arabian coast, but with upward flow through 50 m and 100 m everywhere, and speeds at these levels almost an order of magnitude larger than off Peru. (The wind stress off the Arabian coast was nearly an order of magnitude greater than off Peru.) In the offshore region of the Arabian Sea vertical velocities decreased with depth and increased with distance offshore: at 700 m the vertical speeds were generally upward and were comparable in magnitude to those found in Peru.

The reason for the rather high vertical speeds and large extent of the upwelling off the Arabian coast can be found in the strength of the monsoon and the spatial distribution of wind stresses shown in Table 1. Wind stress increased with distance offshore and this produced a divergence in the Ekman layer over the entire region. The procedure employed in computing the absolute topography of the sea surface assumes a corresponding convergence in the geostrophic flow within and below the Ekman layer. In a wind-driven ocean away from boundaries the vertical speed at the bottom of the Ekman layer is given by

$$w = \frac{1}{\rho f} (\operatorname{curl} \tau - \beta M) \tag{4}$$

(cf. YOSHIDA and MAO, 1957) where M is the integrated meridional mass transport (geostrophic plus Ekman) between the sea surface and the bottom of the Ekman layer, and β is the change of the Coriolis parameter with latitude: $\beta = \partial f/\partial y$. This equation follows directly from the vertically integrated vorticity equation and the equation of continuity, and is exactly equivalent to our calculation procedure. It is important to realize that the coastal boundary is not invoked to explain the upwelling displayed in Fig. 5— although the boundary may influence the parameter M indirectly through the hydrography. As discussed below, the direct effect of the coastline on vertical motion does not extend beyond about 100 km from shore.

The scale of open ocean upwelling revealed in Fig. 5 is determined by the large-scale wind system and is often referred to as large-scale upwelling (YOSHIDA and MAO, 1957; RODEN, 1972; HIDAKA, 1972). Vertical velocities at the base of the 'mixed' or surface layer are typically the order of 10^{-4} to 10^{-5} cm s^{-1} (HIDAKA, 1972). However, higher values are found off many coastal upwelling regions: 10^{-3} to 10^{-4} cm s^{-1} off California (YOSHIDA and MAO, 1957); 10^{-3} cm s^{-1} off northwest Mexico (RODEN, 1972); and 10^{-4} cm s^{-1} off Peru (WYRTKI, 1963). The high values in the Arabian Sea result from a relatively large wind-stress curl in equation (4). The strength and persistence of the wind stress during the southwest monsoon in the northwestern Indian Ocean are higher than are found in any other mid- or low-latitude region (cf. HELLERMAN, 1967).

ESTIMATES OF COASTAL UPWELLING VELOCITIES

The vertical velocities shown in Fig. 5 are not for coastal upwelling *per se*, by which we mean the upwelling induced by the coastal boundary causing a divergence in the Ekman flow. Coastal upwelling in this sense is limited to a narrow band adjacent to the coast with a scale given by the baroclinic radius of deformation (YOSHIDA, 1955; HURLBURT and THOMPSON, 1973; and MOOERS, COLLINS and SMITH, 1976). The analysis procedure described above could not be used with any confidence in the shallower water over the continental slope and shelf. The network of cells, therefore, ended about 50 km from the

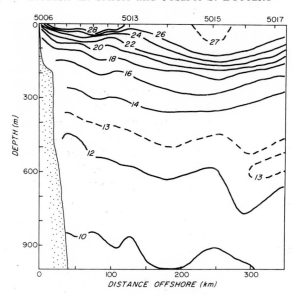

Fig. 6.   Isotherms in the vertical section running S–SE from the Arabian coast to Socotra (Stas. 5006 to 5017).

coast, which is greater than the baroclinic radius of deformation. It is clear from sea surface temperature maps (BRUCE, 1974) and from section data, such as shown in Fig. 6, that appreciable upwelling occurred close to the coast, shoreward of the network. The vertical velocity in the region between the shoreward boundary of the cell network and the coastline can be estimated by computing the divergence of the flow in the surface layer (0–50 m) inshore of the network, using the Ekman and geostrophic transports through the coastal perimeter defined by Stas. 5006 and 5008, the inshore edge of the network (Stas. 5008 to 5063), and Stas. 5063 and 5061. (In making the geostrophic transport estimates through the sections 5006–5008 and 5061–5063, we used the extrapolated sea surface topography discussed above. The transports through the inshore edge of the network had already been computed.) The assumption is made that the net horizontal divergence above 50 m, through this coastal perimeter, is balanced by "coastal upwelling" through the 50-m level inside the perimeter. The net divergence was about $2.5 \times 10^6 \, \mathrm{m^3 \, s^{-1}}$. This yields an average coastal upwelling velocity through 50 m of the value $3.4 \times 10^{-3} \, \mathrm{cm \, s^{-1}}$.

There is no *a priori* constraint in the above computation that the divergence in the upper layer of the coastal region (inshore of the network) be balanced by subsurface convergence in the geostrophic flow. A near balance is obtained with a net onshore transport of $1.6 \times 10^6 \, \mathrm{m^3 \, s^{-1}}$ between 50 m and 300 m through the inshore boundary of the network and an additional $0.4 \times 10^6 \, \mathrm{m^3 \, s^{-1}}$ supplied through the extrapolated coastal sections. Such a balance is consonant with our ideas about coastal upwelling and gives credence to the computations and assumptions. Using the absolute topography based on a uniform 2000-m cell bottom, the coastal upwelling velocity is $3.7 \times 10^{-3} \, \mathrm{cm \, s^{-1}}$. In this case, the resulting subsurface geostrophic flow is insufficient to balance the surface layer divergence, with a deficit greater than $1 \times 10^6 \, \mathrm{m^3 \, s^{-1}}$.

Table 3. *Vertical speeds (w) at 50m predicted by* YOSHIDA's *(1955) theory.* x *denotes distance offshore*

x	w, line 1	w, line 5
10 km	6.1×10^{-3} cm s^{-1}	12.0×10^{-3} cm s^{-1}
20	4.5×10^{-3}	5.8×10^{-3}
40	2.7×10^{-3}	1.9×10^{-3}
60	1.9×10^{-3}	1.2×10^{-3}
80	1.6×10^{-3}	1.1×10^{-3}
100	1.5×10^{-3}	1.1×10^{-3}

To obtain other estimates of nearshore vertical speeds, we turned to a theory due to Yoshida. YOSHIDA (1955; see also HIDAKA, 1972) developed a two-layered coastal up-welling model in which, for $\partial \tau_p / \partial x$ constant, the vertical velocity at the base of the upper layer is given by

$$w = \frac{k}{\rho f} \tau_p e^{-kx} + \frac{1}{\rho f} \frac{\partial \tau_p}{\partial x} \qquad (5)$$

where k^{-1} is the baroclinic radius of deformation, i.e. $k = f(gh\Delta\rho/\rho)^{-1/2}$. Here, ρ is the density of the upper layer, $\Delta\rho$ is the density difference between the two layers, h is the thickness of the upper layer, τ_p is the component of τ parallel to the coastline, and x is the distance offshore. We applied this equation to the area immediately adjacent to the Arabian coast at two locations near the inshore ends of the network. Characteristic values of ρ and $\Delta\rho$ from Stas. 5007 and 5062 were used. The results of the calculations are summarized in Table 3. The vertical speeds calculated should be representative of 50 m. τ_p and $\partial \tau_p / \partial x$ were obtained from the smoothed data in Table 1. The vertical speeds are the order of 10^{-2} cm s^{-1} at the coast and decrease with distance offshore to about 10^{-3} cm s^{-1} at 100 km. These speeds match the speeds given in Fig. 5 fairly well at the inshore boundary of the network.

DISCUSSION AND SUMMARY

During the summer the oceanic regime southeast of the Arabian peninsula is marked by extensive upwelling of cool water into the surface layer and a general rise of the sea surface away from the coast. Both factors can be explained as responses to the southwest monsoon wind system. The southeast monsoon begins in April and by late June the circulation patterns are well developed with horizontal surface velocities of 25 to 50 cm s^{-1} computed from geostrophic calculations. Over the continental shelf and inner continental slope there was a net divergence of the Ekman transport in excess of 3×10^6 m^3 s^{-1}. In the region extending from the inner continental slope to 400 km offshore there was an additional divergence in the Ekman transport of 7×10^6 m^3 s^{-1}. This is partly compensated by a convergence of the geostrophic flow in the surface layer but additional subsurface convergence of the geostrophic flow and upwelling into the surface layer is required to balance the flow.

Using wind and hydrographic data collected during late June and the first half of July we estimated the vertical velocities. The vertical velocities (at depths below the Ekman layer) offshore of the continental shelf were computed from a procedure equivalent to using:

$$w(z) = \frac{1}{\rho f}\left(\text{curl}\,\vec{\tau} + \beta\frac{\tau_x}{f} - \beta M_y(z)\right) \tag{6}$$

where the Ekman transport and the meridional geostrophic transport between the surface and depth z are shown explicitly. The assumptions are that the flow was geostrophic with an Ekman transport, specified by the observed wind distribution, superimposed in the upper 50 m. The use of a reference level of no motion for the geostrophic mass transport is avoided by requiring that the vertical velocity vanish both at the sea surface and at depth (either 2000 m, or as deep as the hydrographic data would allow). By altering the sea surface topography, or equivalently, the slope of any arbitrary isobaric reference surface, M_y can be adjusted to make $w(z = \text{bottom}) = 0$ in equation (6). Thus, the absolute topography of the sea surface is estimated.

The sources of error in the calculations are primarily three: (1) the assumption of no vertical velocity at the cell bottoms; (2) uncertainties in the estimation of the wind stress; and (3) the errors in computations of geopotential anomaly. The vertical velocity at depth, especially over variable topography, is uncertain. However, the computations were made with a variety of cell depths, and the vertical velocities estimated in the upper 300 m changed very little (the order of 10^{-4} cm s^{-1}, which is perhaps a realistic estimate of uncertainty for the vertical velocity near the bottom). Estimates of wind stress from shipboard observations can be unreliable. The data for the Arabian Sea during the monsoon are reasonably consistent among the various cruises and studies, but we have estimated an uncertainty in the wind stress of 50%, not unreasonable considering the uncertainty in the drag coefficient alone (cf. Smith and Banke, 1975). The error in our estimate of the wind stress curl is probably not much more than this, since it is based on differences in a consistent set of measurements. The error in estimates of the geostrophic transport results from uncertainties in the determination of the geopotential anomaly. We use ± 1 cm of sea surface topography as an estimate of the uncertainty. This is consistent with the fluctuations in the absolute topography from varying the computational networks and cell depths used in this paper, and is also consistent with estimates based on uncertainties in the measurements of temperature and salinity (Wooster and Taft, 1958). Equation (6) can be used to estimate the uncertainty in the value of $w(z)$. Uncertainties due to wind stress remain constant throughout the depth range; uncertainties due to errors in geostrophic computations increase with depth, z, since M_y in equation (6) is the integral with depth of the geostrophic flow. Taking typical values for this region, i.e. wind stress of 2 dynes cm^{-2} varying by a factor of 2 over 400 km, and hydrographic stations separated by approximately 50 km with an uncertainty in the surface topography of ± 1 cm, the uncertainty in the vertical velocity at 50 m is 5×10^{-4} cm s^{-1} due to uncertainties in the wind stress and $\pm 10^{-4}$ cm s^{-1} due to uncertainties in the geostrophic computations. By 700 m the uncertainties in the vertical velocity estimates have increased to greater than 10^{-3} cm s^{-1}. We conclude that the measurements in the upper 50 m and 100 m are larger than the uncertainty, but that the vertical velocities computed at 700 m are not meaningful (i.e. they are not significantly different than 0).

In summary, the results indicate that the upwelling in the Arabian Sea during the southwest monsoon is strong and extensive. Strong upwelling into the surface layer occurs not only over the continental shelf and slope but also over the deep ocean. This is a result of the strong steady winds, nearly parallel to the coast, that increase in intensity with distance offshore, resulting in a divergence in the Ekman layer over an area some 1000 km long and at least 400 km wide. This leads to a vertical transport of $8 \times 10^6 \, \mathrm{m^3 \, s^{-1}}$ upward through 50 m. Approximately one-third of the total upwelling in the region is over the continental shelf and the remaining two-thirds over a region extending an additional 350 km offshore. The magnitude of the vertical velocity at 50 m depth in the offshore region (1 to $2 \times 10^{-3} \, \mathrm{cm \, s^{-1}}$) is nearly as strong as the average value over the continental shelf ($3 \times 10^{-3} \, \mathrm{cm \, s^{-1}}$).

Acknowledgements—This paper was started while one of the authors (R. L. S.) spent a year, made especially pleasant by Sir GEORGE DEACON and Dr. J. C. SWALLOW, at the National Institute of Oceanography as a NATO post-doctoral fellow. We thank Dr. J. C. SWALLOW for suggesting the study, providing the data, and encouraging its completion. This is a contribution to the Coastal Upwelling Ecosystems Analysis (CUEA) program of the International Decade of Ocean Exploration supported by National Science Foundation grant OCE 76-00132.

REFERENCES

BRUCE J. G. (1968) Comparison of the near surface dynamic topography during the two monsoons in the western Indian Ocean. *Deep-Sea Research*, **15**, 665–677.

BRUCE J. G. (1973) Large-scale variations of the Somali Current during the southwest monsoon, 1970. *Deep-Sea Research*, **20**, 837–846.

BRUCE J. G. (1974) Some details of upwelling off the Somali and Arabian coasts. *Journal of Marine Research*, **32**, 419–423.

CURRIE R. I., A. E. FISHER and P. M. HARGREAVES (1973) Arabian Sea upwelling. In: *The biology of the Indian Ocean*, B. ZEITZCHEL and S. A. GERLACH, editors, Springer-Verlag, pp. 37–53, 549 pp.

DUING W. (1970) The regime of the monsoon currents in the Indian Ocean. *International Indian Ocean Expedition Oceanographic Monographs*, 1, East–West Center Press, Honolulu, 68 pp.

HALPERN D. (1976) Structure of a coastal upwelling event observed off Oregon during July 1973. *Deep-Sea Research*, **23**, 495–508.

HELLERMAN S. (1967) An updated estimate of the wind stress on the world ocean. *Monthly Weather Review*, **95**, 607–626.

HIDAKA K. (1972) Physical oceanography in upwelling. *Geoforum*, **11**, 9–21.

HURLBURT H. E. and J. D. THOMPSON (1973) Coastal upwelling on a beta-plane. *Journal of Physical Oceanography*, **3**, 16–32.

HUYER A., R. L. SMITH and R. D. PILLSBURY (1974) Observations in a coastal upwelling region during a period of variable winds (Oregon coast, July 1972). *Tethys*, **6**, 391–404.

MITTELSTAEDT E., R. D. PILLSBURY and R. L. SMITH (1975) Flow patterns in the Northwest African upwelling area. *Deutschen Hydrographischen Zeitschrift*, **28**, 146–167.

MOOERS C. N. K., C. A. COLLINS and R. L. SMITH (1976) The dynamic structure of the frontal zone in the coastal upwelling region off Oregon. *Journal of Physical Oceanography*, **6**, 3–21.

PEDLOSKY J. (1974) Longshore currents, upwelling and bottom topography. *Journal of Physical Oceanography*, **4**, 214–226.

RAMAGE C. S. (1965) The summer atmospheric circulation over the Arabian Sea. *Journal of Atmospheric Sciences*, **23**, 144–150.

REID J. L. and A. W. MANTYLA (1976) The effect of the geostrophic flow upon coastal sea elevations in the northern North Pacific Ocean. *Journal of Geophysical Research*, **81**, 3100–3110.

RODEN G. I. (1972) Large-scale upwelling off northwestern Mexico. *Journal of Physical Oceanography*, **2**, 184–189.

THE ROYAL SOCIETY (1963) *International Indian Ocean expedition*, R.R.S. Discovery, Cruise Report 1, 24 pp.

SMITH S. D. and E. G. BANKE (1975) Variation of the sea surface drag coefficient with wind speed. *Quarterly Journal of the Royal Meteorological Society*, **101**, 665–673.

SMITH R. L. (1974) A description of current, wind and sea level variations during coastal upwelling off the Oregon coast, July–August, 1972. *Journal of Geophysical Research*, **79**, 435–443.

STOMMEL H. (1956) On the determination of the depth of no meridional motion. *Deep-Sea Research*, **3**, 273–278.

WARREN B., H. STOMMEL and J. C. SWALLOW (1966) Water masses and patterns of flow in the Somali Basin during the southwest monsoon of 1964. *Deep-Sea Research*, **13**, 825–860.

WOOSTER W. S. and J. L. REID, JR. (1963) Eastern boundary currents. In: *The Sea*, M. N. HILL, editor, Interscience, **2**, 253–280, 554 pp.

Wooster W. S. and B. A. Taft (1958) On the reliability of field measurements of temperature and salinity in the ocean. *Journal of Marine Research*, **17**, 552–566.

Wyrtki K. (1963) The horizontal and vertical field of motion in the Peru Current. *Bulletin of the Scripps Institution of Oceanography*, **8**, 313–346.

Wyrtki K. (1971) *Oceanographic Atlas of the International Indian Ocean Expedition.* National Science Foundation, Washington, D.C. 531 numbered pages.

Yoshida K. (1955) Coastal upwelling off the California coast. *Records of Oceanographic Works in Japan*, **2**, 8–20.

Yoshida K. and H. L. Mao (1957) A theory of upwelling of large horizontal extent. *Journal of Marine Research*, **16**, 40–53.

Nutrients as tracers of water mass structure in the coastal upwelling off northwest Africa

Department of Oceanography, University of Liverpool, P.O. Box 147, Liverpool L69 3BX, England

Abstract—A study was made of silicate [Si], phosphate [P] and organic phosphorus nutrients in the upwelling area between the Canary Islands and Cap Vert on *Discovery* Cruise 26 (April/May 1969). The upwelling was most intense at Cap Blanc and Cabo Bojador, so these areas were examined in detail. The [Si], [P], organic phosphorus and [Si/P] data were used to examine:

1. A poleward undercurrent at about 250 m depth along the shelf. Salinity was used originally to identify the position of the undercurrent. However, south of Cabo Bojador the levels of salinity in the undercurrent and in the surrounding water became indistinguishable. The phosphorus nutrient and the [Si/P] levels were found to be sensitive tracers of the undercurrent throughout the survey area. A second weak poleward undercurrent may have existed at 400–500 m down the shelf south of Cap Blanc.

2. The variability in water properties as a function of time at fixed stations off Cabo Bojador and off Cap Blanc. Biological activity, the movement of advective patches past the stations, and oscillations (which appeared to be internal tides) caused the variability.

3. Water masses in the Cap Blanc area, which were found to be very complex due to the upwelling and the merging of two major water masses, with different nutrient properties. A comparison was made between the Cap Blanc water mass structure and the more classical situation observed at Cabo Bojador.

4. A high [P], low [Si/P] surface water mass which was traced between Cap Vert and south of Cap Blanc.

5. Nutrient regeneration at the pycnocline and in shelf sediments, which caused enrichment of the associated waters with dissolved nutrients. Upwelling processes and internal tides tended to bring these nutrients back into the euphotic zone and played a major role in the recycling of nutrients.

6. A longshore section, in which a rise and fall of 500 m in iso-nutrient lines in the deeper layers was observed over a distance of about 340 miles, west of Cap Timiris.

Abbreviations used

NACW	North Atlantic Central Water
SACW	South Atlantic Central Water
AAIW	Antarctic Intermediate Water
MW	Mediterranean Water
NADW	North Atlantic Deep Water
[Si]	Reactive silicate (μg l^{-1})
[P]	Reactive inorganic phosphate (μg l^{-1})
[Si/P]	The ratio of the two nutrients above
$T(\bar{T})$	Temperature — °C (mean temperature)
$\bar{\sigma}_t$	Mean sigma-t (density = $(1 + 10^{-3}\sigma_t)$g cm^{-3})
S.T.D.	Salinity, temperature, depth
$S(\bar{S})$	Salinity—parts per thousand (mean salinity)

* Present address: CSIRO Division of Fisheries and Oceanography, P.O. Box 21, Cronulla, N.S.W. Australia 2230.

INTRODUCTION

It is well known that the wind stress along the northwest African coast is relatively steady and strong throughout the year, and that it is this factor, linked with the earth's rotation, that causes the characteristic vertical water movements and boundary currents that are known as upwelling. The understanding of upwelling areas has recently become a project of prime importance for oceanographic scientists due to the potential of these areas as food sources for the ever-increasing world population. Upwelling causes enrichment of the euphotic zone with nutritious waters often enabling a massive and consistent supply of fish to be maintained.

Recently many national and international programs have been sponsored to investigate in detail upwelling areas, including those off northwest Africa. Many of the surveys carried out have had a biological bias; however, few have been orientated solely towards a combination of marine chemistry and physics. *Discovery* Cruise 26, in April/May 1969, was organized to investigate the variability of the physical and chemical parameters of the water masses in the upwelling region off the northwest coast of Africa. We studied the area from the coast to approximately 100 nautical miles offshore, between the Canary Islands (lat. 28°N) and Cape Verde Islands (lat. 15°N). The whole area was surveyed in the first phase of the cruise and the results used to identify areas of active upwelling which were studied intensively during phase two of the program. The combined results of the nutrient concentration from both phases were analysed after the cruise was completed.

This paper describes the use of the silicate and phosphorus nutrient concentrations, and silicate to phosphate ratios as water mass tracers. These water properties are also used to study small discrete patches of water, internal oscillations in water properties and areas of nutrient regeneration. An attempt has been made to relate the evidence derived from the analysis of the data to the existing information and hypotheses available in the literature. This approach has led to some consolidation of the oceanographic findings in the upwelling area between Cap Vert and Cabo Bojador.

OBSERVATIONS AND METHODS

Cruise details

Figure 1 shows the positions on phase one of the cruise track of the fifty-nine hydrographic stations which were distributed between the nine diagonal sections and arranged to achieve maximum coverage in the time available. The spacing between the stations close to the mainland was between 5–10 miles, while that of the deeper stations (100 miles from the coast) was between 30 and 35 miles. At the five outer stations (nos. 6896, 6905, 6917. 6930 and 6947) S.T.D. records* were obtained to 1500 m or 2000 m and nutrients were recorded to 2500 m; however, at the remainder both S.T.D. dips and bottle casts were restricted to either 500 m or the bottom for more shallow waters. A continuous-flow fluorometer monitored chlorophyll levels in the surface layers. These levels were used as indicators of what might broadly be described as biological activity. Also shown are the

---

* The physical oceanographic data (wind speeds and direction, salinity, temperature and direct current measurements) have been analysed and presented previously (HUGHES and BARTON, 1974 a, b).

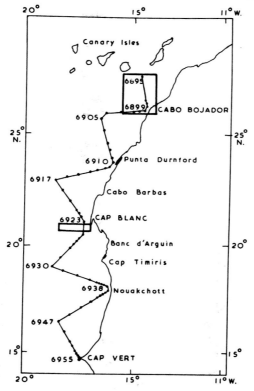

Fig. 1.   Station charts for the *Discovery* Cruise 26 showing cruise track for the first phase of the survey, and the areas examined in detail in phase two of the program.

areas off Cap Blanc and Cabo Bojador selected for further study in the second part of the cruise. A number of additional sections were worked in these upwelling zones. In addition one time-series station was completed in each area. The repeated sampling off Cap Blanc was at a position (20°50′N, 17°45′W) at a depth of 746 m. Further north off Cabo Bojador the position was (26°15′N, 14°43′W) at a depth of 140 m.

Procedure

Water samples for analysis were collected in National Institute of Oceanography water bottles, and analysed immediately for silicate [Si] using a Technicon Autoanalyzer AA1 by the method of BREWER and RILEY (1966) and for phosphate [P] using the semi-automated method described below. Total phosphorus analyses were completed on returning to the laboratory at Liverpool by ultra-violet irradiation and the organic phosphorus calculated from the difference in total phosphorus and [P] concentrations.

Determination of inorganic phosphorus

A multichannel peristaltic pump was used to pump the sample (at laboratory temperature) and the mixed reagent* in the ratios 10:1 into 130-ml glass reaction vessels which had been previously cleaned with HNO_3/H_2SO_4 (1:1). The solutions were well mixed and after not less than 5 minutes and not more than 5 hours, the absorbance values were measured against distilled water at 885 nm in a 10 cm cuvette using a Unicam SP500 spectrophotometer.

Determination of total phosphorus concentrations

Total phosphorus was determined by a method similar to that of STRICKLAND and PARSONS (1972) using treatment with hydrogen peroxide followed by irradiation for 1 hour at 6 cm below a high-intensity 1-kW ultra-violet lamp (GARDNER, 1971). After cooling the samples were made up to 25 ml and the [P] concentrations determined by a modification of the automated method by CHAN and RILEY (1966). The modified technique prevented the coloured complex plating out during transmission and involved (a) the use of glass transmission tubing, (b) a trace of ammonia incorporated in the wash receptacle and (c) omission of the potassium antimonyl tartrate reagent.

Results

The results of the nine sections worked in phase one of the cruise are given in Fig. 2(a–i). Horizontal iso-nutrient lines indicate no upwelling and stable stratification, and are generally seen offshore. Steeply sloping iso-nutrient lines indicate intense upwelling of the deeper water masses to the surface and are seen close to the coast. Figures 2a and 2b show classical upwelling (see, for example, SMITH, 1968), with both the horizontal and steeply sloping iso-lines off Cabo Bojador (sections 1 and 2). On examination of the sections to the south, Fig. 2c, shows iso-lines with less steep gradients while in Figs. 2d, 2e and 2f (sections 4, 5 and 6) they gradually become steeper and more complex. In this area, i.e. Cap Blanc, an increase in nutrient concentrations can be seen between sections 5 and 6. Figures 2g to 2i show a gradual lessening of the slope of the iso-lines. From these observations on board *Discovery* we decided to return to (a) Cabo Bojador to study the water masses of the classical situation in more detail and (b) to Cap Blanc where, although upwelling was indicated, the water mass structure appeared to be highly complex and different to that at Cabo Bojador. GARDNER (1971) has described in detail the sections worked during phase two of the cruise. Table 1 summarizes the ranges, means and standard deviations of the [Si], [P], organic phosphorus concentrations and the [Si/P] levels found for water samples from both the time series stations.

Methods of data analysis

The combined results from both phases of the cruise were subject to a thorough examination. In order to build up a three-dimensional picture, by assuming the finite time required to sample each section or transect was insignificant, plots of isolines were constructed for [Si], [P], [Si/P] and organic phosphorus levels:

---

*100 ml 3% w/v ammonium molybdate solution; 250 ml 5 N H_2SO_4; 200 ml 10.8% w/v ascorbic acid; 50 ml 0.14% w/v potassium antimonyl tartrate.

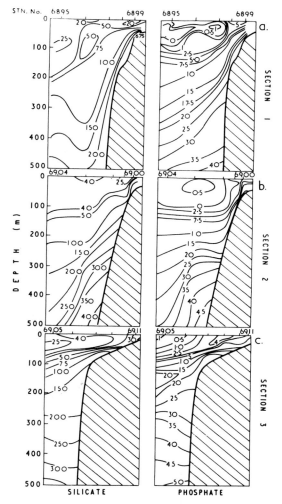

Fig. 2. (a–c) Profiles of iso-silicate ($\mu g[Si]l^{-1}$) and iso-phosphate ($\mu g[PO_4]l^{-1}$) for sections 1–3 in phase one of the survey.

(a) along lines that were diagonal to the coastline versus depth,
(b) along longshore transects at various distances from the coastline versus depth,
(c) as maps at various depths,
(d) using time versus depth for time-series stations.

In addition nutrient concentration versus depth plots were used to show the difference in the properties of the major water masses.

DESCRIPTION OF THE SURFACE WATERS

Wind is the most important force causing deep nutrient-rich waters to upwell along this coast. During phase one of the cruise northeast trade winds prevailed. At first

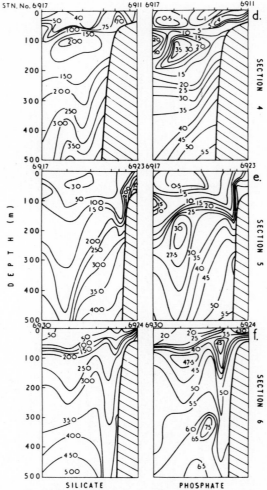

Fig. 2.   (d–f) Profiles of iso-silicate ($\mu g[Si]l^{-1}$) and iso-phosphate ($\mu g[PO_4]l^{-1}$) concentrations for sections 4–6 in phase one of the survey.

(Canaries to Cap Blanc) the wind strength was 20–30 kt.* Later (Cap Blanc to Cap Vert) the winds moderated to ∼10 kt, veering north.

From the steep near-shore parts of the iso-nutrient profiles, intense upwelling was indicated at Cabo Bojador (Figs. 2a and 2b) and Cap Blanc (Figs. 2e and 2f). Relative to offshore conditions ($< 1\ \mu g[P]l^{-1}$ and $< 30\ \mu g[Si]l^{-1}$) the inshore surface waters were enriched with nutrients, with the levels of $> 25\ \mu g[P]l^{-1}$ and $> 200\ \mu g[Si]l^{-1}$.

High [P] concentrations were found in the surface waters as far south as Cap Vert (similar to that found by WEICHART, 1970a, b, 1974). However, the highest [Si] concentrations were confined to Cap Blanc. This suggested that more than one surface water mass was involved in the study area.

Figure 3 shows [Si/P] iso-lines for the surface waters. A division between waters with

---

* $1\,kt = 0.51\,m\,s^{-1}$.

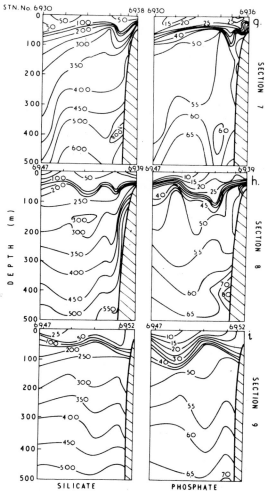

Fig. 2.   (g–i) Profiles of iso-silicate ($\mu g[Si]l^{-1}$) and iso-phosphate ($\mu g[PO_4]l^{-1}$) concentrations for sections 7–9 in phase one of the survey.

high and low [Si/P] levels can be seen at Cap Blanc. FRAGA (1974) also noted differences in nutrients between waters north and south of this point. HUGHES and BARTON (1974a) have shown the upwelled water to the north is North Atlantic Central Water (NACW), while that to the south is South Atlantic Central Water (SACW). At Cabo Bojador the closely spaced iso-lines indicate the position at which the upwelling was strong. To the north of Cap Blanc the iso-lines increase in value seawards to a maximum value of about 100, however south of Cap Blanc the pattern is more complex. For the water mass present in this southern section hugging the coast off the Banc d'Arguin the [Si/P] levels of > 7 are unusually high. Conversely, the tongue of water that intrudes from south of Cap Vert has an unusually low [Si/P] value (1.5). HUGHES and BARTON (1974a) have shown that this surface water mass has a low salinity and also suggest that it flows from the south. Apart from these two discrepancies, however, the [Si/P] levels increase seaward to a maximum of about 5.

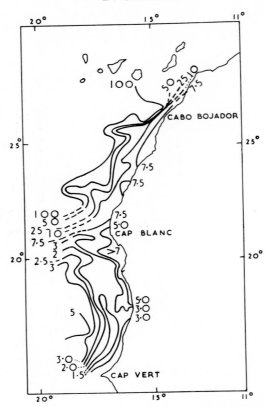

Fig. 3. A contour map of [Si/P] iso-lines for the surface waters showing the division of the properties of the upwelled water masses at Cap Blanc, and the flow of low [Si/P] water from south of Cap Vert.

THE CABO BOJADOR UPWELLING

In phase one intense upwelling was observed off Cabo Bojador, where the surface waters had the same characteristics as waters at 250 m. This feature had disappeared by the time *Discovery* had reached Cabo Bojador in phase two, because of a few days lull in the strong winds, and the waters were calm and clear with negligible fluorescence.

During phase two, three transects were made (GARDNER, 1971), one of which followed the same route as section 1. The [Si/P] levels of the two similar transects (Figs. 4 a, b) show the temporal changes of the nutrient characteristics that occurred between April, when winds were strong, and May when calm weather conditions prevailed. As the upwelling died out, a pycnocline was established. The surface temperature increased while [P] concentrations decreased, and the iso-lines became level in deeper waters. A band of low [Si/P] (< 5) water was well established at the pycnocline, where phosphate concentrations were greater. This feature is usually caused by decomposition of blooms of plankton as they die and their sinking is slowed by the denser water below the pycnocline. As [P] is regenerated from the detritus faster than [Si], the ratio of the two concentrations ([Si/P]) becomes smaller (nutrient regeneration is discussed in more detail later).

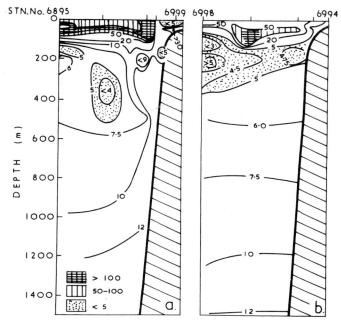

Fig. 4.   [Si/P] iso-lines in the water column of (a) section 1 (phase one) with strong winds—intense upwelling, and its repeat (b) in phase two of the survey with calm weather—negligible upwelling.

The time series at Cabo Bojador

Figure 5a shows the iso-silicate lines for the time series station worked at Cabo Bojador. It shows a simple almost noise-free oscillation with a trough to peak distance of about 50 m and approximately a semi-diurnal period. This behaviour is consistent with the type of oscillations to be expected from internal tides (JOHNSON and MOOERS, 1973; HUGHES and BARTON, 1974b).

At 1200 hours during the time series on 31.4.69 a patch of water depleted in [P] appeared in near-surface layers (Fig. 5c). This patch of water was associated with a rise in temperature to 19°C, and a salinity increase to 36.7‰ (HUGHES and BARTON, 1974b) and a very high organic phosphorus content (Fig. 5b) at the top of the water column.

The overall means at various depths for all the nutrient constituents measured at this time-series station are given in Table 1. It is of interest to note that the mean [P] concentrations increased with depth from 0.92 to 14.5 μg l^{-1} while both the means of the organic phosphorus levels measured at the top and at the bottom of the water column were 21.6 μg l^{-1}.

The distribution of the dissolved organic phosphorus found below the pycnocline (Fig. 5b) might have been caused by the excretory products of the benthos or the first stage in the breakdown of detrital material. It can also be seen in Fig. 5b that there was a finger of increased organic phosphorus concentration from the sediments to about 50 m in the water column, at the times of the maxima of the undulations. It is possible that the postulated internal tide was expending energy into the sediments which caused some concentrated dissolved material to rise into the water column. All the high organic phosphorus levels observed above and below the pycnocline were associated with the

[P] depleted patch of water, and did not exist before or after its appearance. [P] depletion usually indicates a bloom of plankton has fully utilized all the available [P] nutrient. The excretion products from grazing zooplankton populations of fish (WHITLEDGE and PACKARD, 1971) usually contain a high percentage of dissolved organic substances. Consequently it could be argued that a bloom of plankton had recently been consumed by a population of grazing zooplankton or fish, that in turn had released dissolved organics back into the water column.

THE CAP BLANC UPWELLING

During the phase two study off Cap Blanc, the winds were blowing at a speed of 20 kt from NNE.

We have seen that near Cap Blanc there was a division between two surface water masses with different nutrient properties. Consequently it was expected that the nutrient data would show that the pattern of the water mass stratification and structure in this upwelling system would be more complex than that off Cabo Bojador. An examination of the time series would show if this expectation was valid.

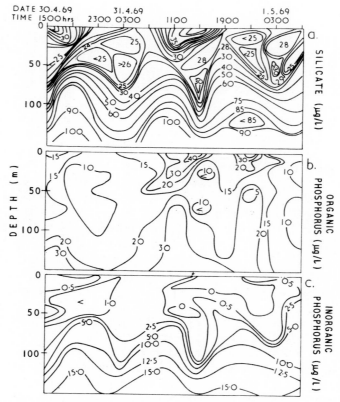

Fig. 5. The time series at *Discovery* Sta. 6985 off Cabo Bojador showing the semi-diurnal internal wave oscillations in deeper waters of (a) silicate iso-lines, (b) organic phosphorus iso-lines and (c) inorganic phosphorus iso-lines, and the differences in the organic and inorganic phosphate concentrations in the surface waters.

The time series at Cap Blanc

A comparison between the almost first mode oscillations (see HUGHES and BARTON, 1974b) of the iso-silicate lines at Cabo Bojador in comparison with the uneven patchy pattern at Cap Blanc (Fig. 6a) immediately indicates the complexity of the area. JONES (1972) and MITTELSTAEDT (1972 and 1974) have commented on similar high-frequency fluctuations found during their cruises in this area. Suggested mechanisms include internal waves or tides, upwellings, downwellings, horizontal turbulence from adjacent waters with different properties and the movement of patches or boluses of water past the point of observation. TOMCZAK (1973) has observed the presence of cold, advective patches of upwelled water moving in this area.

For the present time series the variability outlined in Table 1 will be discussed in relation to semi-diurnal oscillations, advective patches, biological activity and the presence of a poleward undercurrent. Biological activity in near-surface layers was great (high fluorescence) which was obviously a significant contributing factor to the variability. However, the large variability in the surface layers was not due only to biological activity removing nutrients from the upwelling waters, but was largely caused by the appearance and departure of two advective patches during the time we sampled. A small one on 24.4.69 at 2000 hours was followed by a more extensive patch of water at about 1730 hours on 25.4.69. The latter had completely different water properties to the waters present before and after that time (see Fig. 6c). A comparison of the [Si/P], [P] and [Si] data for the advective patch, with the nutrient properties of the surrounding water masses

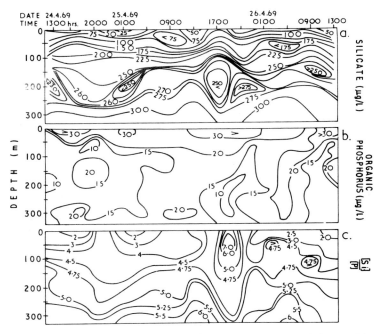

Fig. 6(a). The silicate iso-lines during the time series at *Discovery* Sta. 6970 off the Cap Blanc showing the complex variations in the silicate nutrient with depth. (b) The organic phosphorus iso-lines in the same water column, showing high levels in the surface waters. (c) [Si/P] iso-lines in the same water column, showing the movement past the station of an advective patch with different properties.

(see Figs. 2e, 2f, 7a, b, c) shows that this patch probably originated from the coastline to the east in the vicinity of sections 5 and 6 (Stas. 6920–6925). Its mode of formation and movement to Sta. 6970 is completely speculative but I would like to suggest that it might have been by one of the two following mechanisms.

A convergence occurred between NACW and the surface waters from the south close to the hypothesized point of origin (see Fig. 3). If the conceptual model of coastal up-welling suggested by MOOERS, COLLINS and SMITH (1976) is applied, the [Si/P] values in Fig. 7b and the [Si] and [P] iso-lines in Fig. 2e appear to show that the higher [Si/P] (low [P]) waters from the north may have been sinking at Sta. 6921. The sinking water is seen as a tongue pointing down the shelf in Figs. 7b and 2e. This downwelling was associated with upwelling on the shoreward side of the poleward undercurrent at

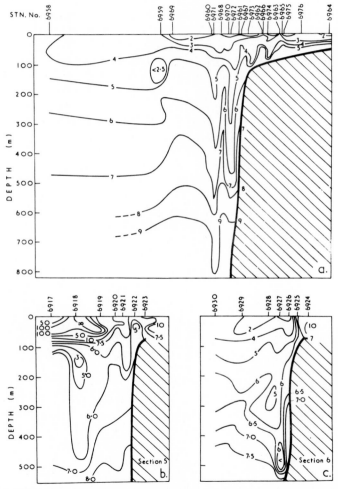

Fig. 7(a).   [Si/P] iso-lines for the combined transects completed off Cap Blanc, showing the effect of the poleward undercurrent along the shelf. (b) [Si/P] iso-lines for section 5, showing the high [Si/P] values in shallow waters for NACW. (c) [Si/P] iso-lines for section 6, showing the low [Si/P] values in the shallow waters for the SACW.

Sta. 6920 (see below) and caused the almost vertical iso-lines seen near the surface in Fig. 2e. Due to the close proximity to the shelf, if the upwelling weakened or if the down-flow veered closer to the shelf, the sinking waters would have cut off the colder newly upwelled waters. This isolated patch could then have drifted offshore with the wind-driven surface flow.

The secondary theory of its place of origin is also feasible. Direct current measurements from this station, using a drogue at 50 m, showed a northerly flow in the opposite direction to the surface winds. As the prevailing winds were from the NNE, there should have been a component of flow due to Ekman transport toward our time series stations from the coast-line at section 6. The advective patch under discussion was high in [Si] compared to the mean value down to 150 m at Sta. 6970, while [P] levels were only slightly lower. SHAFFER (1973) suggested that this type of cold upwelled patch could be due to the pulse-like outflow of high salinity waters from the shallow Banc d'Arguin. These patches might then drift with the northwesterly component of the flow.

The mean concentration (Table 1) of organic phosphorus (0.4–3.8 μg l^{-1}) at the surface for the time series at Cap Blanc was 1.92 times that of the mean [P] concentration. At 200 m, the mean [P] was 0.27. The increased values at the surface could be due to the intensive biological activity. The overall distribution of organic phosphorus with time at Sta. 6970 is given in Fig. 6b. It poses some interesting possibilities on plankton and zooplankton distributions, and nutrient regeneration as did the organic phosphorus distributions at Cabo Bojador. The advective patch discussed above was associated with a layer of high organic phosphorus in the surface water and with low organic phosphorus in deeper water. The finger of > 20 μg l^{-1} organic phosphorus protruding from the surface and observed 12 hours after this patch might have been due to a vertical migration of grazing zooplankton or fish during the night. Alternatively there could have been some connection with the peak maximum of the internal wave found for the deeper waters at that time similar to that observed at Cabo Bojador.

HUGHES and BARTON (1974a) have computed the $\bar{S}$, $\bar{T}$ and $\bar{\sigma}_t$ at the Cap Blanc time-series station. I found an inverse relationship between their $\bar{S}$ and mean [P] and [Si] concentrations. The salinity minimum they found at 300 m was also associated with an increased [Si/P] level, due to the effects of the poleward undercurrent passing some miles to the west.

FEATURES OBSERVED ALONG THE SHELF EDGE

The poleward undercurrent(s)

I have mentioned the existence of a poleward undercurrent in the above discussions. The existence of an undercurrent in areas of upwelling at the eastern boundary of the Atlantic was first reported by HART and CURRIE (1960). A poleward flow at about 300 m is an important feature in upwelling regions and is associated with a surface flow which is generally in the opposite direction, i.e. equatorwards (MOOERS and ALLEN, 1973). TOMCZAK (1973) and HUGHES and BARTON (1974a) have shown a poleward undercurrent is present along most of the northwest African shelf. The latter authors traced slight salinity anomalies found in data from *Discovery* Cruise 26 for 800 nautical miles north of Cap Vert. Unfortunately they lost track of the undercurrent when its salinity and

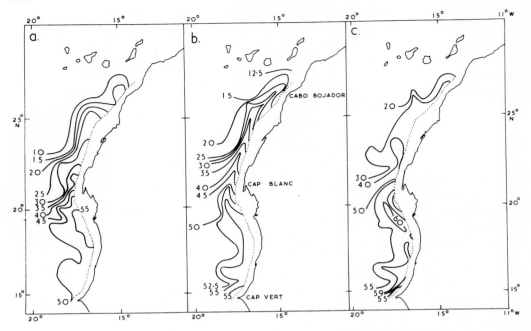

Fig. 8.   Maps of inorganic phosphate ($\mu g\ PO_4^{3-} - P\,l^{-1}$) isolines at (a) 200 m, (b) 250 m and (c) 300 m depth
to trace the poleward undercurrent along the shelf.

temperature characteristics became similar to those of the surrounding water. (Note that
the analogous undercurrent in the Pacific is much easier to trace because its salinity
and temperature properties are significantly different to those of the surrounding waters.)

Often other water properties can be used to trace water masses. The analysis of the
nutrient concentration data and the [Si/P] data showed that in some cases [P] could
be used to trace the undercurrent and in other cases the [Si/P] values could be used.

The salinity anomaly associated with the undercurrent was found at about 250 m using
a S.T.D. probe. No water sample was taken at this depth so an estimate of the nutrient
levels at 250 m was linearly interpolated from the nutrient levels at 200 m and 300 m.
Figure 8b shows a map of these estimated [P] concentrations at 250 m. The core is well
defined north of Cap Blanc due to the great difference in the [P] levels of the intrusive
current of SACW and the less nutritious NACW. Figures 8a and 8c show the iso-line maps
of the 200 m and 300 m [P] data. These figures show only a hint of the existence of the
undercurrent along the shelf.

The core of the undercurrent ([Si/P] $\simeq$ 6–6.5) is indicated in the iso-line map of the
along-shore transect (Fig. 9). The [Si/P] anomaly can be traced along the whole of the
coast surveyed. The intense upwelling at Cabo Bojador (sections 1 and 2) appears to cause
some mixing, and the [Si/P] levels increase slightly.

The profile in Fig. 9 also shows the position at Cap Blanc (between sections 5 and 6)
where the near-surface water properties changed completely. Figure 7a shows the [Si/P]
iso-lines for the combined transects worked there in phase two of the cruise. The effect
of the poleward undercurrent centred at 250 m below. Sta. 6968 is to cause the iso-lines

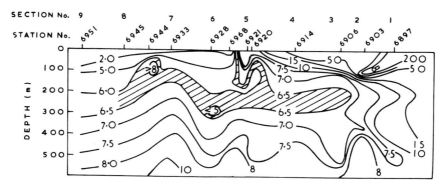

Fig. 9. Longshore section showing [Si/P] ratios along the line of the poleward undercurrent. The hatched area indicates the appropriate position of the tongue of high [P], low [Si/P] water. Also shown is the position near Sta. 6968 at which the water properties change near Cap Blanc.

to rise and fall steeply. The effect at greater depths is seen to be quite considerable, and may be associated with waters downwelling round the undercurrent.

A further anomaly illustrated by the [Si/P] profiles in Fig. 7c is the patch of low [Si/P] water at 400–500 m down the shelf. This anomaly is also seen in [Si/P] plots of all the sections influenced by SACW (not shown). No outstanding salinity–temperature anomalies could be found in the SACW associated with these depths (HUGHES and BARTON, 1974a). The iso-nutrient lines suggest that the anomaly might be caused by regeneration of nutrients from some turbulent disturbance in the shelf sediments. This might be the result of upwelling waters dividing and veering downward due to the intrusive poleward undercurrent at 250 m, or to the breaking of internal waves at the shelf where surplus energy may be absorbed in the sediments (APEL, CHARNEL and BLACKWELL, 1974). Evidence of this anomaly occurring in the Cabo Bojador region is inconclusive probably due to the reduced strength of the 250 m undercurrent, and the increased intensity of the upwelling observed during phase one. Perhaps these data support the two-cell concept of upwelling recently suggested by MOOERS, COLLINS and SMITH (1976) with a second weaker poleward undercurrent at about the depth and position of the anomaly observed south of Cap Blanc.

Regeneration of silicate

Selective regeneration of [P] compared to [Si] at the pycnocline leaves the suspended matter and subsequently the sediments slightly depleted in phosphorus, compared to their silicate component. SHAFFER (1973) has shown that regeneration of the [Si] from the shelf by turbulence and by dissipated energy from internal waves continues even if strong steady winds are not blowing. Consequently replenishment of nutrients can occur more often than would be expected from the wind pattern alone. HART and CURRIE (1960); REDFIELD, KETCHUM and RICHARDS (1963); CALVERT and PRICE (1971), and JONES (1971) have also commented on such a phenomenon.

On the shelf edge off the northwest African coast, we observed a distinct regeneration of the [Si] as the upward-flowing waters were forced along the sediment layers. This can best be seen at the shoreline of sections 5 and 6 (Figs. 2e and f) and also at the shelf edge of section 1 (Fig. 2a). At Cap Blanc there was a stretch of coastal water with a mean [Si/P] value of 7.42 ± 0.76. The results suggest that, over the area covered by Stas. 6922 to 6925, the waters upwelled with a higher than normal concentration of [Si] (mean $184 \mu g l^{-1}$). This is in comparison with the waters slightly to the north, where the [Si] mean is $44 \mu g l^{-1}$ and to the south where the [Si] mean is $40 \mu g l^{-1}$ (see Table 1). At Cabo Bojador, especially at Sta. 6899, the [Si] concentrations were extremely high with [Si] levels of $> 650 \mu g l^{-1}$ in comparison to the $200 \mu g l^{-1}$ iso-silicate line closest to the point of regeneration. As no salinity anomaly was associated with the elevated [Si] levels the most likely explanation is regeneration from sediments. This type of regeneration is vital in areas where [Si] is a limiting nutrient (GOERING, 1974) and often enables the waters to support large diatom blooms.

The iso-line undulation on the shelf and at deeper stations west of Cap Timiris

A further feature of interest became clear during the study of the pathway of the poleward undercurrent along the shelf in Fig. 9. This was the longshore undulation of the iso-lines which had a peak close to Sta. 6933. The undulation caused the core ([Si/P] 6–6.5) to rise by about 100 m. Figure 10 also shows this longshore undulation which affected all section 7 and the peak of which can be seen at Sta. 6930. If this is

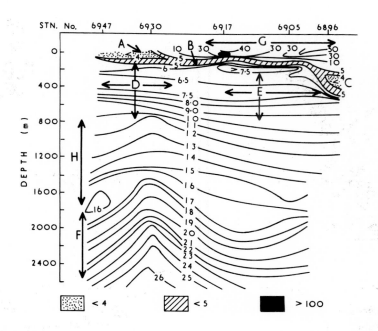

Fig. 10.   Longshore section of [Si/P] iso-lines showing the major water masses and minor features of the seaward deep stations; also shown is the long-shore wave. A = high phosphate surface flow from the south of Cap Vert. B = pycnocline regeneration zone. C = low phosphate, low silicate intermediate water (possibly MW influenced) off Cabo Bojador. D = SACW (300–900 m). E = NACW (300–900 m). F = NADW (> 1500 m). G = biologically worked surface waters (low phosphate) north of Cap Blanc. H = MW (and possibly AIW) influenced intermediate water (low in silicate) (900–1500 m).

assumed to be a wave in the deep waters 'F', it had a half-wave length of about 340 miles and a peak to trough distance of about 600 m. However, these dimensions are too great for natural waves and C. N. K. MOOERS (personal communication) suggests that the feature was due to the interaction of the two major water masses off Cap Blanc which might have caused an offshore current. The finite time required to sample all the stations used in the longshore transect profiles allow only speculations to be made on the nature of this anomaly. However, other chemical and physical parameters also displayed this rise in iso-lines along section 7.

FEATURES OBSERVED AT THE DEEP STATIONS

Figure 10 is a longshore profile of [Si/P] values for the outermost stations which were sampled down to 2500 m. The hatched area contains all levels of [Si/P] < 5. These are divided into three categories: 'A', the high [P] (low salinity) surface water moving in from south of Cap Vert; 'B', the relatively high [P] water lying at the pycnocline; and 'C', a low [Si] water found at the northern-most stations.

HUGHES and BARTON (1974a) have shown that the pycnocline at the deep stations

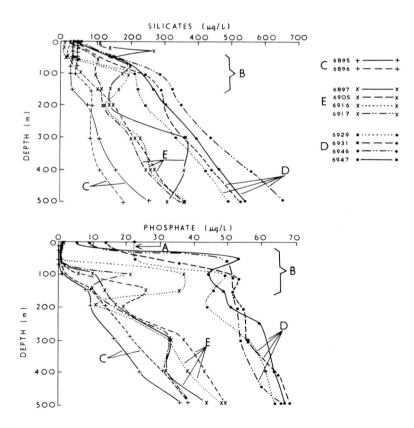

Fig. 11.   Silicate and phosphate concentration versus depth plots to show the nutrient profiles of the differing categories of water for selected stations (A–E as in Fig. 10).

Table 1. *Means, ranges and standard deviations of the means (S.D.) for the silicate and phosphorus nutrient data found at the time series stations on Discovery Cruise 26*

Area	Depth (m)	Silicate [Si] ($\mu g\,l^{-1}$)	Mean (SD) ($\mu g\,l^{-1}$)	Phosphate [P] ($\mu g\,l^{-1}$)	Mean (SD) ($\mu g\,l^{-1}$)	[Si/P]	Mean (SD)	Organic phosp. ($\mu g\,l^{-1}$)	Mean (SD) ($\mu g\,l^{-1}$)
Cap Blanc	0	16– 97	47 (±22)	10 –31	18.5 (±7.3)	1.5–6.9	2.9(± 1.5)	25–40	35.5 (±21.1)
	50	72–192	140 (±42)	27 –48	36.8 (±7.0)	2.7–7.0	3.9(± 1.1)	4–31	19.6 (± 7.6)
	100	176–273	222 (±25)	41 –54	50.1 (±3.4)	4.0–5.4	4.4(± 0.4)	10–25	16.5 (± 3.9)
	150	239–270	256 (±10)	53 –55	54.1 (±0.5)	4.4–5.1	4.7(± 0.2)	11–32	18.4 (± 5.9)
	200	236–283	264 (±13)	52 –55	53.9 (±0.8)	4.6–5.4	4.9(± 0.2)	10–20	14.7 (± 3.4)
	250	254–311	276 (±17)	53 –56	54.4 (±1.0)	4.7–5.7	5.1(± 0.3)	12–25	16.2 (± 3.6)
	300	280–339	307 (±17)	55 –59	56.8 (±1.3)	4.7–6.1	5.4(± 0.4)	9–23	17.0 (± 4.1)
Cabo Bojador	0	22– 38	28.5(± 5.4)	0.0– 2.1	0.9(±0.8)	13.0–∝*	75 (±77)	12–41	21.6 (±12.1)
	20	25– 35	27.6(± 2.7)	0.0– 1.4	0.4(±0.5)	20.0–∝*	108 (±76)	9–33	15.6 (± 6.7)
	50	22– 41	28.5(± 5.5)	0.2– 4.1	0.9(±1.2)	8.5–∝*	102 (±86)	8–30	15.3 (± 5.9)
	100	38– 88	64.3(±20.9)	0.3–13.0	8.7(±4.3)	6.2–93.3	17 (±28)	6–35	17.0 (± 8.5)
	135	85–100	93.0(± 9.4)	12.5–15.4	14.5(±1.5)	5.2–7.5	6.5(± 1)	12–32	21.6 (±10.1)

*200 used for the mean.

Table 2. *Means and ranges of silicate and phosphorus nutrients found on Discovery Cruise 26, for the major water masses and some minor features*

Water mass or feature	Identified in text by:	Depth (m)	Silicate [Si] (μg l^{-1})	Mean (μg l^{-1})	Phosphate [P] (μg l^{-1})	Mean (μg l^{-1})	[Si/P]	Mean
1. NACW[a] (Stas. 6905+6917)	E	300–800	211–625	368	32.0–63.8	48.0	5.8–10.0	7.9
2. SACW[b] (Stas. 6930+6947)	D	300–800	347–839	555	52.0–71.0	63.0	6.3–10.4	8.3
3. MW[c]+AIW[d] (deep stas.)	H	950–1500	637–955	743	48.0–65.0	56.0	11.3–15.2	13.0
4. NADW[e] (deep stas.)	F	>1750	749–1232	930	44.0–48.3	47.3	16.0–20.5	27.2
5. Poleward undercurrent	—	≃250	162–405	308	24.0–56.0	45.0	6.0– 9.2	6.7
6. Surface—north of Cap Blanc	G	0	28–70	44	0.0– 7.0	2.0	4.7–∝*	58.0
7. Stations south of Cap Blanc	A	0	19–60	40	3.0–29.0	17.0	1.2– 4.3	2.5
	—	50	37–215	100	18.0–48.0	28.0	1.2– 4.5	2.7
	—	100	245–264	256	29.0–53.0	43.0	4.7– 5.5	5.0
8. Surface—Cap Blanc	—	0	113–226	184	17.0–30.0	24.4	6.6– 8.3	7.4
9. Regeneration zone (Pycnocline)	B	≃150	38–245	133	2.0–53.0	29.2	1.4– 5.0	3.9
10. Regeneration zone (shelf Sta. 6899)	—	40–50	656–682	669	18.3–18.7	18.5	35.3–35.6	35.5
11. Northern Low Silicate Water (Stas. 6895–6)	C	150–500	26–186	104	7.6–38.8	24.7	3.7– 4.8	4.2

[a]North Atlantic Central Water. [b]South Atlantic Central Water. [c]Mediterranean Water. [d]Antarctic Intermediate Water. [e]North Atlantic Deep Water.
*200 used for the mean.

fell from about 50 m to 100 m from south to north, with a gradual decrease in its stability. On the other hand, the stability of the pycnocline increased with distance from the shore. Consequently 'A' (as above) may have been formed due to the shallowness and instability of the tropical pycnocline close to the shore in waters south of Cap Vert. Strong upwelling would quite readily bring to the surface waters with properties similar to 'A'.

'B' is formed in a similar manner to the low [Si/P] value water at the pycnocline off Cabo Bojador. HERBLAND (1974) has described in some detail how the organic excretion products of phytoplankton seem to induce a bacterial growth which is then increased as the algal cells break down, releasing a great quantity of dissolved products into the sea. The increased [P] values at the pycnocline 'B' can also be seen for selected stations in Fig. 11. Although there are very slight increases in the concentrations of [Si] at 'B', they are not as significant as the maxima seen for [P] concentrations. This difference indicates faster regeneration of [P] than [Si] (hence lower [Si/P]) at the pycnocline.

Figure 11 also shows nutrient level profiles of two stations at which the low [Si/P] water 'C' off Cabo Bojador (Fig. 10) was found. The low [Si] concentrations found in these waters are the main cause for the low [Si/P]. Selective nutrient regeneration similar to that causing 'B' also had a small effect.

The iso-lines slope up from north to south (Fig. 10) due to the higher nutrient concentrations characteristic of the SACW 'D'. Table 2 gives means of both nutrients and their respective [Si/P] values and shows that compared to NACW 'E' in the same depth range (300–800 m), [Si] concentrations are about 50% more, and [P] concentrations about 30% more, giving an overall rise in the [Si/P] levels of about 16%.

HUGHES and BARTON (1974a), using water mass classification by ALLAIN (1970), have shown that the intermediate waters 'H' (Fig. 10) are definitely influenced by high-salinity MW (and possibly slightly by AAIW). The iso-[Si/P] lines spread out in this intermediate water mass ([Si/P] 12–17). Simple calculations indicate that the iso-line spreading is due to a deficiency of [Si] in these waters. This supports the deductions that MW, which is less rich in nutrients ($182.5\ \mu g\,[Si]\,l^{-1}$ and $9.3\ \mu g\,PO_4^{3-} - P\,l^{-1}$) (MCGILL, 1969, and SCHINK, 1967) than the NADW, NACW and SACW, was present. The low [Si] waters 'C' were probably formed in the same way.

WATER MASS CLASSIFICATION OFF NORTHWEST AFRICA

Table 2 summarizes the means and ranges of the [Si] and [P] nutrient concentrations and [Si/P] levels found for selected groups of stations. Each group represents an important water mass or feature that emerged from the analysis of *Discovery* Cruise 26 nutrient data. The outstanding high levels of [Si] and [P] in SACW 'D' compared to those of NACW 'E' can be seen. Also the large variation in the [Si/P] levels in the surface waters 'G', north of Cap Blanc compared to those to the south, 'A'. This feature in the northern surface waters 'G' might be due to any one of the following causes: (a) the great variability in the kinetics of nutrient uptake (COSTE and SLAWYK, 1974; GOERING, 1974; DUGDALE and MacISAAC, 1971); (b) the effects of local winds (CRUZADO, 1974); (c) the size of seed phytoplankton populations and upwelling trajectories (BARBER, 1973); (d) the time lag required to 'condition' newly upwelled waters (BARBERS, DUGDALE, MacISAAC and SMITH, 1971).

CONCLUSIONS

Dissolved silicate [Si], reactive phosphate [P] and organic phosphorus concentrations, and [Si/P] levels were used as water mass tracers in the upwelling area off the northwest African coast.

[Si/P] levels were used to study the poleward undercurrent centred at 250 m which flowed northward along the shelf edge along the whole area surveyed. In the main tongue the [Si/P] levels had a range of 6.0 to 9.2 with a mean of 6.7. The [Si/P] levels also indicated the existence of a surface water mass which appeared to move northward from south of Cap Vert. The [Si/P] level of this tongue of water was low compared to surrounding waters due to its high [P] content. The major upwelled water masses were found to have distinctly different nutrient properties. Consequently [Si/P] levels for surface waters north of Cap Blanc had a mean of 58.0 over a range 4.7 to infinity (zero [P]), while those waters at the surface south of Cap Blanc had a mean of 2.5 over a range 1.2–43. North of Cap Blanc as [P] was often missing it was probably the limiting nutrient. To the south phosphate was constantly replenished by the nutrient-rich, up-welling SACW and the surface water from south of Cap Vert.

Large-scale undulations in the water masses were evident. In particular at Cabo Bojador a time-series showed that there was an organized semi-diurnal oscillation present with a peak to trough distance of about 65 m. At Cap Blanc advective patches and other water movements were superimposed on a semi-diurnal wave form. The overall effect at the latter time-series station was a complex and unorganized series of fluctuations. However, the advective patches could be clearly identified using the [Si/P] ratio technique, due to their differing water properties. A longshore section at the deep stations and the shelf stations showed a long-shore undulation of the iso-nutrient lines throughout the water column. It had a peak to trough distance of 500 m over a distance of 340 miles, and was presumably caused by the interaction of the two water masses merging off Cap Blanc.

Acknowledgements—I would like to thank Captain R. H. A. DAVIES, his officers and crew on the R.R.S. *Discovery* and colleagues from the Department of Oceanography, Liverpool, especially J. MURPHY who assisted with the analyses. Thanks are also due to Professor J. P. RILEY for valuable advice. The assistance during preparation of this manuscript from friends and colleagues at Cronulla was also greatly appreciated. During this programme the author was supported financially by a N.E.R.C. studentship.

REFERENCES

ALLAIN C. (1970) Observations hydrologique sur le talus du Banc d'Arguin en Décembre 1962 (Campagne de la *Thalassa* du 2 Novembre au 21 Décembre 1962). *Rapport et procès-verbaux des réunions. Conseil permanent international pour l'exploration de la mer*, **159**, 86–89.

APEL J. R., R. L. CHARNELL and R. J. BLACKWELL (1974) Ocean internal waves off the North American and African coasts from ERTS-1. *Proceedings of the ninth International Symposium on Remote Sensing of Environment 15–19 April 1974*, pp. 1345–1351.

BARBER R. T. (1973) The relationship between circulation and productivity or the physical components of ecosystem structure in upwelling systems. In: *Coastal Upwelling Ecosystems Analysis*, C. N. K. MOOERS and J. S. ALLEN, editors. Final Report (School of Oceanography, Oregon State University), p. C2.

BARBER R. T., R. C. DUGDALE, J. J. MACISAAC and R. L. SMITH (1971) Variations in phytoplankton growth associated with the source and conditioning of upwelling water. *Investigacion Pesquera*, **35**, 171–193.

BREWER P. and J. P. RILEY (1966) The automatic determination of silicate–silicon in natural waters with special reference to seawater. *Analytica Chimica Acta*, **35**, 514–519.

CALVERT S. E. and N. B. PRICE (1971) Upwelling and nutrient regeneration in the Benguela Current, October 1968. *Deep-Sea Research*, **18**, 505–523.

CHAN K. M. and J. P. RILEY (1966) The automatic determination of phosphate in seawater. *Deep-Sea Research*, **13**, 467–471.

COSTE B. and G. SLAWYK (1974) Structures de Répartitian superficielles des sels nutritifs dans une zone d'upwelling (Cap Corveiro, Sahara Espagnol). *Téthys*, **6**(1–2), 123–132.

CRUZADO A. (1974) Coastal upwelling between Cape Bojador and Point Durnford (Spanish Sahara). *Téthys*, **6**(1–2), 133–142.

DUGDALE R. C. and J. J. MacISAAC (1971) A computation model for the uptake of nitrate in the Peru upwelling region. *Investigacion Pesquera*, **35**, 299–308.

FRAGA F. (1974) Distribution des masses d'eau dans l'upwelling de Mautitanie. *Téthys*, **6**, 5–10.

GARDNER D. (1971) Nutrients in an upwelling area. Ph.D. Thesis (Department of Oceanography, University of Liverpool, England).

GOERING J. J. (1974) Uptake of silicic acid by diatoms. *Téthys*, **6**(1–2), 143–148.

HART T. J. and R. I. CURRIE (1960) The Benguela Current. '*Discovery*' *Reports*, **31**, 123–298.

HERBLAND A. (1974) Activité bactérienne dans l'upwelling Mauritanien relation avec l'oxygène et la matière organique. *Téthys*, **6**(1–2), 202–212.

HUGHES P. and E. D. BARTON (1974a) Stratification and water mass structure in the upwelling area off north-west Africa in April/May 1969. *Deep-Sea Research*, **21**, 611–628.

HUGHES P. and E. D. BARTON (1974b) Physical investigations in the upwelling region of north-west Africa on R.R.S. *Discovery* Cruise 48. *Téthys*, **6**(1–2), 43–52.

JOHNSON D. and C. N. K. MOOERS (1973) Current profiles in the upwelling off Northwest Africa. *CUEA Newsletter*, **2**(5), 5–22.

JONES P. G. W. (1971) The Southern Benguela Current region in February 1966. Part I. Chemical observations with particular reference to upwelling. *Deep-Sea Research*, **18**, 193–208.

JONES P. G. W. (1972) The variability of oceanographic observations off the coast of north-west Africa. *Deep-Sea Research*, **19**, 405–431.

MITTELSTAEDT E. (1972) Der hydrographische Aufbau und die zeitliche Variabilität der Schichtung und Stömung im nordwestafrikanischen Auftriebsgebiet im Frühjahr 1968. '*Meteor*' *Forschungsergebnisse*, A, **11**, 1–57.

MITTELSTAEDT, E. (1974) Some aspects of the circulation in the north-west African upwelling area off Cap Blanc. *Téthys*, **6**,(1–2), 89–92.

McGILL D. A. (1969) A budget for dissolved nutrient salts in the Mediterranean Sea. *Cahiers oceanographiques*, **6**, 543–554.

MOOERS C. N. K. and J. S. ALLEN, editors (1973) Summary of observational knowledge of coastal upwelling dynamics. In: *Coastal Upwelling Ecosystems Analysis*. Final Report (School of Oceanography, Oregon State University), pp. III 1–4.

MOOERS C. N. K., C. A. COLLINS and R. L. SMITH (1976) The dynamic structure of the frontal zone in the coastal upwelling region off Oregon. *Journal of Physical Oceanography*, **6**(1), 3–21.

REDFIELD A. C., B. KETCHUM and F. A. RICHARDS (1963) The influence of organisms on the composition of sea water. In: *The sea*, M. N. HILL, editor, **2** (Wiley Interscience, N.Y.), pp. 26–27.

SCHINK D. R. (1967) Budget for dissolved silica in the Mediterranean Sea. *Geochimica et Cosmochimica Acta*, **31**, 987–999.

SHAFFER G. (1973) Small-scale upwelling variations off north-west Africa. In: *Coastal upwelling ecosystems analysis*, C. N. K. MOOERS and J. ALLEN, editors, Final Report (School of Oceanography, Oregon State University), pp. C32B–B1.

SMITH R. L. (1968) Upwelling. In: *Oceanography and Marine Biological Annual Review*, **6**, 11–46.

STRICKLAND J. D. H. and T. R. PARSONS (1972) Determination of soluble organic phosphorus by ultraviolet oxidation. In: *A practical handbook of seawater analysis. Bulletin 167 of the Fisheries Research Board of Canada* (2nd edition), pp. 141–142.

TOMCZAK M. (1973) An investigation into the occurrence and development of cold water patches in the upwelling region off N.W. Africa (Robbreiten-Expedition 1970). '*Meteor*' *Forschungsergebnisse A*, **13**, 1–42.

WEICHART V. G. (1970a) Kontinuierliche-Registrierung der Temperatur und der Phosphat-Konzentration im oberflächenwasser des nordwestafrikanischen Auftriebswasser-Gebietes. *Deutsche Hydrographische Zeitschrift*, **23**(2), 49–60.

WEICHART V. G. (1970b) Temperatur-und Phosphat Verteilung im nordwestafrikanischen Auftriebsgebiet. *Umschau in Wissenchaft und Technik*, **26**, 856.

WEICHART V. G. (1974) Chemical investigations in the upwelling area off North-West Africa. '*Meteor*' *Forschungsergebnisse Reihe A*, **14**, Seitre 33–70.

WHITLEDGE T. E. and T. T. PACKARD (1971) Nutrient excretion by anchovies and zooplankton in the Pacific upwelling regions. *Investigacion Pesquera*, **35**, 243–250.

The relation of seasonal stratification to tidal mixing on the continental shelf

J. H. SIMPSON, D. G. HUGHES and N. C. G. MORRIS

Abstract—A stratification parameter $\bar{v}$, defined as the amount of mechanical energy required to bring about vertical mixing, has been calculated for an extensive region of the shelf using available data. The distribution of $\bar{v}$ during the summer months is compared with a tidal mixing theory which suggests that the parameter h/u_s^3 (h = depth, u_s = tidal stream amplitude at springs) controls the occurrence of stratification.

The results lend qualitative support to the model, although plots of $\bar{v}$ versus h/u_s^3 show a large degree of scatter which is interpreted as being largely due to variations in wind and wave mixing and surface heat input.

Much of the structure of the $\bar{v}$ and h/u_s^3 distributions is apparent in recently available infra-red images of the sea surface.

INTRODUCTION

In a previous paper (SIMPSON and HUNTER, 1974) a criterion for the occurrence of seasonal stratification in continental seas was proposed. On the assumption that only local tidal mixing opposes the establishment of stratification by buoyancy input at the surface, it was predicted that stratification will occur when the dimensionless parameter ε exceeds a critical value. ε is defined as

$$\varepsilon = A\left(\frac{h}{u_s^3}\right) = \frac{3\pi\alpha Qg}{8k\rho c}\left(\frac{h}{u_s^3}\right) = \frac{h}{L}$$

where h is the depth and u_s^3 is the amplitude of the surface tidal stream at springs, α is the expansion coefficient, c the specific heat and ρ the density of seawater. The quantity A varies slowly with season and latitude through the surface heat input Q and may also be affected by changes in k, the frictional drag coefficient of the bottom. The parameter ε may be interpreted as the ratio of the depth to the Monin–Obhukov length L (TURNER, 1973; p. 131).

The transition from a stratified to a vertically mixed regime is frequently marked by a well-defined boundary front. Examples in the Irish and Celtic Seas have been documented in, for example, SIMPSON (1971, 1976) and PINGREE (1975). Observations of the front in the northwest Irish Sea suggest that the critical value of h/u_s^3 is typically ~ 70. Taking $Q = 40\,\text{cal}\,\text{m}^{-2}\,\text{s}^{-1}$ with $k = 2 \times 10^{-3}$ gives $\varepsilon = 0.0028$ which means that less than 0.5% of the available turbulent energy is used to bring about mixing at the transition point.

A similar approach by FEARNHEAD (1975) also leads to a criterion of the form h^1/u_s^3 where h^1 is the difference between the water depth and the surface mixed layer depth.

THE STRATIFICATION PARAMETER

In this study we have attempted to test the validity of the ε criterion using the large volume of data available from the British Oceanographic Data Service and other sources. As a measure of the stratification we have used the potential energy

$$\bar{v} = \frac{1}{h} \int_{-h}^{0} (\rho - \bar{\rho}) gz \mathrm{d}z; \bar{\rho} = \frac{1}{h} \int_{-h}^{0} \rho \mathrm{d}z$$

where z is the vertical coordinate (positive upwards), h is the depth and ρ is the density with mean $\bar{\rho} \cdot \bar{v}$ is the work which would be done in redistributing the mass in bringing about complete vertical mixing. It is a more meaningful index of stratification than the total surface to bottom temperature contrast $\triangle T$, in that it takes account of the full temperature (and salinity) profile. $\triangle T$, used in SIMPSON (1971) and FEARNHEAD (1975), exhibits more noise due to the formation, at times of low windstress, of shallow transient thermoclines.

Ideally our estimates of $\bar{v}$ should be based on continuous data from surface to bottom, sampled repeatedly in time to eliminate the noise due to internal waves. In practice, of course, we usually have only one vertical profile to assess $\bar{v}$ and in many cases the data does not extend to the full water depth. The fractional error arising from an unsampled bottom region of thickness $\triangle h_2$, for the case of a vertical structure of two layers of uniform density, is just

$$\frac{\triangle \bar{v}}{\bar{v}} = \frac{h_1}{h_1 + h_2} \frac{\triangle h_2}{h_2}$$

where h_1 is the thickness of the upper layer and h_2 that of the lower. In most cases h_1 is much less than $h_1 + h_2$ so that the fractional error in $\bar{v}$ will be small, even when an appreciable fraction of bottom layer is unsampled. Numerical computations of $\bar{v}$ over different depths confirm that this source of error is not serious. It would be preferable, of course, to select only stations for which at least 95% of the column was sampled, but examination of the available stations showed that this would result in the rejection of a large percentage of the data.

THE DATA SAMPLE

The stratification parameter $\bar{v}$ was estimated for the area around the U.K. shown in Fig. 1. The majority of the data was obtained from the data bank of the British Oceanographic Data Service. Supplementary stations were provided by recent observations in the Irish and Celtic Seas made by I.O.S., Bidston and the Marine Science Laboratories. In all a total of more than 10,000 stations were available to us for the investigation of the seasonal variations in $\bar{v}$ in our chosen area. Their distribution, however, both seasonally and geographically, is less than ideal so that we have had to contend with a dearth of data in some areas, particularly in the winter months.

At each station the density profile has been computed from the raw temperature and salinity data. In the case of CTD or TSD data the mean density has been computed over 5-m intervals before forming a summation corresponding to the integral definition of $\bar{v}$. For hydrocast data a depth truncation scheme appropriate to each case has been chosen. The results for each month were plotted in coded form by computer on a mercator

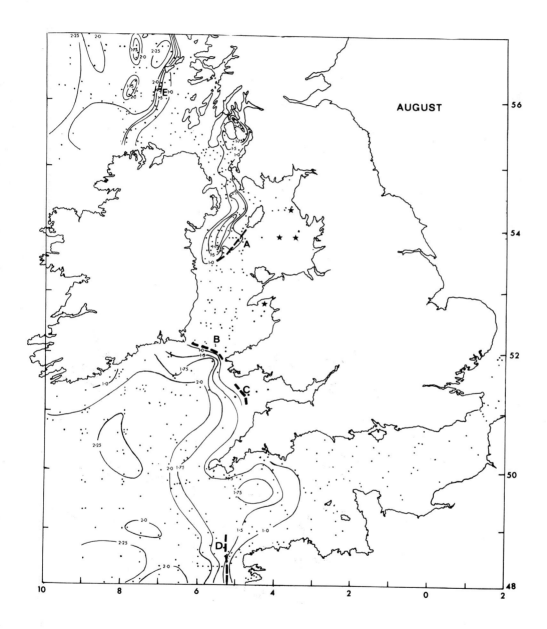

Fig. 1. Contours of $\log_{10} \bar{v}$. The dashed lines indicate known frontal regions reported in Simpson and Hunter (1974), Simpson (1975) and Pingree (1975). The letter E indicates the location of a similar feature which has not yet been surveyed. Stars are shown for the stations in Liverpool Bay and Cardigan Bay referred to in the text. These stations exhibit marginal stratification with $\bar{v} \sim 10\, \text{joules}\,\text{m}^{-3}$. All other stations are shown as dots.

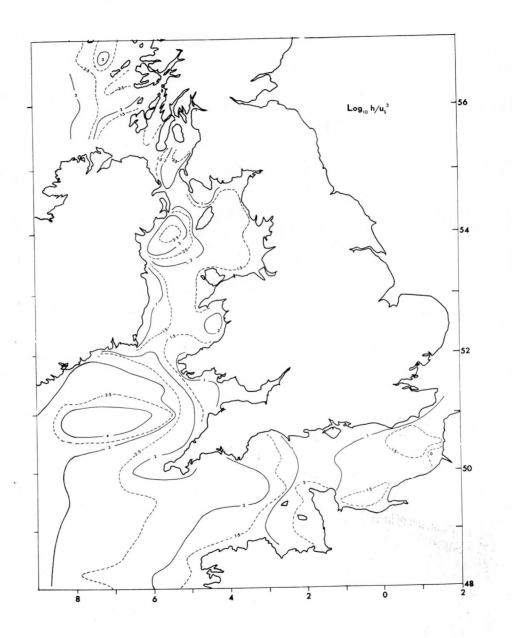

Fig. 2.   Contours of log h/u_s^3 based on all available tidal Atlas data.

projection. Contours of $\bar{v}$ were inserted by visual inspection; their validity may be assessed from the density of data points which are indicated by dots in the diagrams shown here. A not unexpected bias towards more intense sampling in the summer months is evident in the results.

DISTRIBUTION OF $\bar{v}$

Figure 1 shows the distribution of $\bar{v}$ for August, which is the period of maximum stratification and also the best sampled month in the year. $\bar{v}$ varies between ~ 200 joules m^{-3} in regions of high stratification in the southwest and northwest approaches down to virtually zero in large areas of the Channel and Irish Sea. A pronounced isolated maximum in $\bar{v}$ is evident southwest of the Isle of Man.

Overall, the distribution bears a marked resemblance to the h/u_s^3 contours shown in Fig. 2. In particular it is evident that:

(i) The well mixed areas generally occur where h/u_s^3 is rather less than 100. The contour $\bar{v} = 10$, for example, closely tracks the line $h/u_s^3 = 100$.

(ii) The $\bar{v}$ maximum southwest of the Isle of Man coincides with a maximum ih h/u_s^3. There is, however, no strong maximum in $\bar{v}$ in the Celtic Sea where h/u_s^3 is high.

The known frontal regions (A–D) are also marked on Fig. 1. They are seen to occur close to the point where $\bar{v}$ drops from ~ 10 joules m^{-3} to virtually zero. Sections through fronts A and B have shown that this transition may occur in a horizontal distance as small as 1–5 km. Another extensive front of this type (E) is indicated in the $\bar{v}$ contours to the north of Ireland.

In the eastern part of Liverpool Bay, and also in Cardigan Bay some observations show marginal stratification with $\bar{v} \sim 10$ joules m^{-3}. The available data does not permit any generalization about the persistence of stability in either of these areas, though the shallow depth probably means that significant stratification can exist only during periods of low wind stress.

SEASONAL DEVELOPMENT

During the winter months cooling at the surface induces convective mixing which makes $\bar{v}$ small. The distribution of $\bar{v}$ in February (Fig. 3a) shows that it is effectively zero everywhere except in nearshore regions where some stability may be maintained by freshwater run-off.

Significant stratification first starts to develop in April (Fig. 3b) when the heat input at the sea surface becomes strongly positive. Areas with $\bar{v} > 10$ are evident in the Celtic Sea and southwest of the Isle of Man. The degree of stratification at this initial stage is highly dependent on the weather. A large storm may induce sufficient mixing to destroy completely incipient stratification. Consequently the data here shows a large degree of scatter which together with the relatively poor sampling makes it difficult to contour.

More definite stability is apparent in May (Fig. 3c) though much scatter remains. Values of $\bar{v} > 30$ are seen to develop in the areas of high h/u_s^3.

By June the summer pattern of seasonal stratification is generally well established

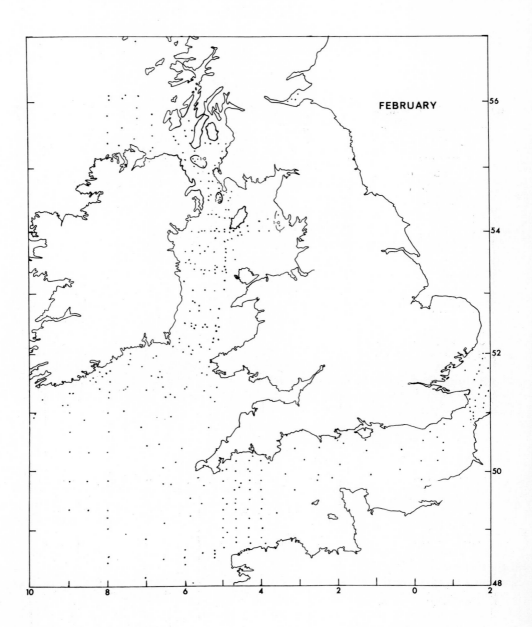

Fig. 3.   Seasonal changes in the distribution of $\bar{v}$; (a) February, (b) April, (c) May and (d) June. The contours are of $\log_{10} \bar{v}$.

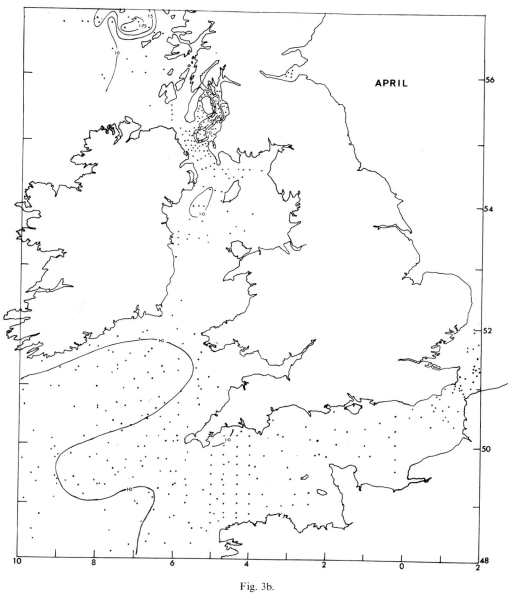

Fig. 3b.

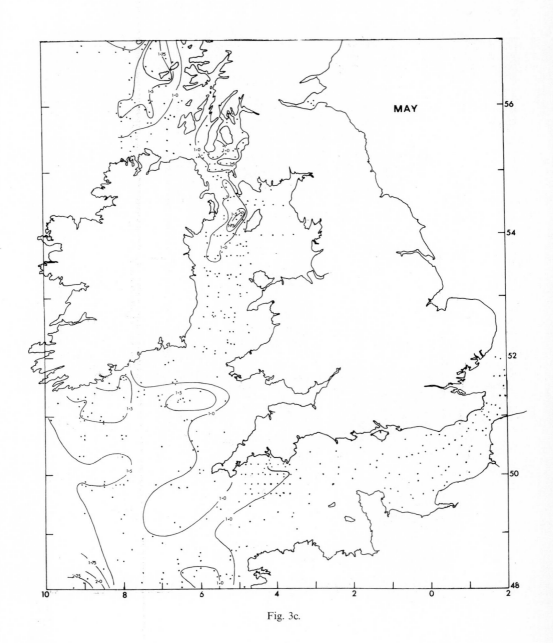

Fig. 3c.

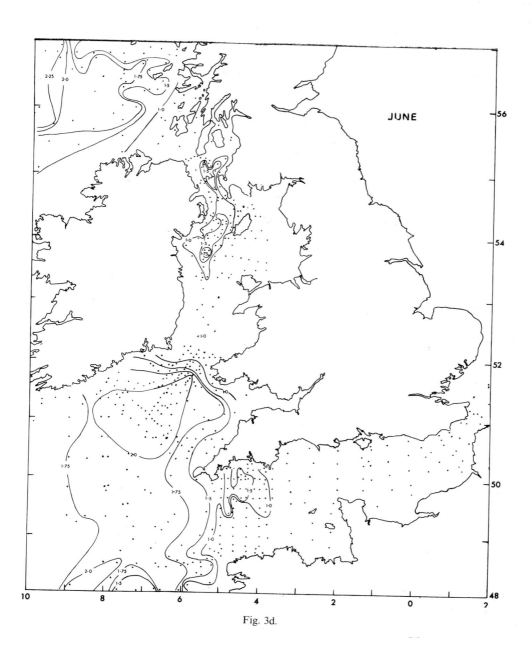

Fig. 3d.

(Fig. 3d) with a second maximum in $\bar{v}$ centred on the region of high h/u_s^3 in the Celtic Sea. This maximum is less apparent in July (not shown) and by August (Fig. 1) has merged into the generally high values of $\bar{v}$ in the western Celtic Sea.

There are generally only relatively small changes in the shape of the $\bar{v}$ contours during the June to September period. The magnitude of $\bar{v}$ in the areas of maximum stratification increases from June to August, but only by a small fraction of the May-June change. After reaching a maximum in August stratification decreases slightly in September and then more rapidly in October and November. Unfortunately data for these months of declining $\bar{v}$ are sparse in comparison with the spring and summer. In addition the onset of convective mixing, like the initial formation process, shows a rather variable pattern from year to year, so no general conclusions can be reliably drawn on the timing of this process until we have a more comprehensive data set.

RELATION BETWEEN $\bar{v}$ AND h/u_s^3

The clear relation between the geographical distributions of $\bar{v}$ and h/u_s^3 has prompted us to compare these two parameters directly for each month's data. Figure 4 shows a

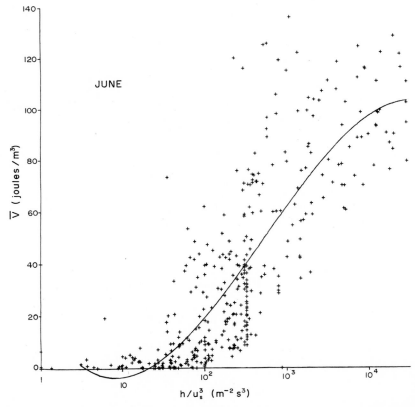

Fig. 4. $\bar{v}$ versus h/u_s^3 for June data for the area 48–54°N, 2–9°W. The curve represents the least-squares fit to a third-order polynomial.

plot for the June data of $\bar{v}$ versus h/u_s^3 for a more restricted range of latitude. At low values of h/u_s^3, $\bar{v}$ is close to zero with little scatter. With increasing h/u_s^3 there is marked rise in $\bar{v}$ and considerably more variability. There are also indications of an upper limit for $\bar{v}$ at high h/u_s^3. This is clear from the polynomial curve which has been fitted to the data by the method of least squares.

Similar curves, summarizing the data from months between February and August for the same area, are shown in Fig. 5. It is seen that the development of stratification is characterized by a marked increase between the May and June results. All the curves intersect the axis at a value of $\log_{10} h/u_s^3 \simeq 1.5$.

During the midsummer months the variation in Q with latitude in the area studied is small ($<10\%$ between 48N and 54N averaged over June, July and August). Separation of the August data into narrower bands of latitude does not indicate any marked increase with latitude, in the critical value of h/u_s^3.

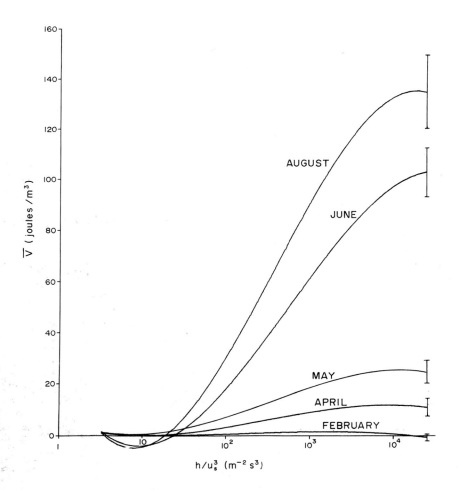

Fig. 5. Curves of $\bar{v}$ versus h/u_s^3 fitted to the data for each month. The bar at the end of each curve indicates the standard deviation of the data points from the best fit.

INTERPRETATION AND DISCUSSION

The form of the $\bar{v}$ dependence on h/u_s^3 suggests a simple theoretical interpretation. In the absence of wind mixing, $\bar{v}$ would increase from zero at low h/u_s^3 to some asymptotic value at high h/u_s^3 where tidal mixing is negligible. The value of this limit will depend on the net heat input at the surface.

Wind mixing will modify this picture by lowering $\bar{v}$ to an extent depending on the wind history. Additional mixing may result from bottom turbulence forced by long waves in regions which are exposed to oceanic swell. The scatter in $\bar{v}$ is then due to the combined year-to-year variability of wind and wave mixing and heat input.

In practice other factors will also contribute to the scatter. Our knowledge of u_s is limited in many areas, so that there may be errors in the computation of h/u_s^3 as large as $\pm 50\%$, particularly in areas of weak tidal streams. In future one would hope that this deficiency may be eliminated by the use of a numerical tidal model to adequately extrapolate u_s for the whole shelf from available data.

In addition to errors in h/u_s^3 some uncertainty in $\bar{v}$ will occur due to the incomplete vertical sampling mentioned already.

A further difficulty is the necessity, imposed by the limited data set, of grouping the data by month which means we ignore the (sometimes substantial) changes which occur within each month. Any significant reduction of this time interval would degrade the spatial coverage unacceptably.

A clear example of year-to-year variability in the distribution of stratification was observed in July 1975 during surveys of the frontal region B following a period of prolonged warm weather in early summer. Stratified water was found at this time penetrating ~ 25 km further northwards into the Irish Sea than the mean position shown in Fig. 1.

A possible technique for the monitoring of such variability, is the use of high resolution infra-red images obtained from orbiting satellites. The pictures shown in Fig. 6 were taken by the NOAA 4 satellite using an infra-red scanner which receives in the 10.5–12.5-μm band and can resolve radiation temperatures to $0.5°$C. At such wavelengths, the temperature measured is essentially that of the top 0.1 mm of the sea surface which may not always be representative of the whole surface layer. Nonetheless it seems likely that the *pattern* of sea-surface temperature will be represented in the radiation temperatures, providing there is not serious interference from cloud cover. The relative immunity of IR observations to low-level cloud and fog is demonstrated by a comparison of the simultaneous IR and visible band images in Fig. 6.

In the infra-red picture, most of the sea surface is visible and the pattern of sea-surface temperature may readily be discerned. Boundaries between the warm (dark) stratified water and the cooler (light) vertically mixed water may be identified with the frontal regions shown in Fig. 1. For example, coincident with the position of frontal region A is a sharp change in image density which has been estimated (from approximate calibration data) as $\sim 2°$C. Similarly the front at B is apparent with a slight intrusion of the stratified water into the Irish Sea already apparent.

Dark areas are also evident in regions of shallow water, for example in the eastern part of Liverpool Bay and Cardigan Bay. With the local exceptions mentioned above, these areas are not generally stratified, so the higher surface temperatures here must be attributed to increasing surface heat input per unit volume with decreasing depth.

In addition to variations in stratification due to weather, a more systematic short-term

Fig. 6. Infra-red and visible band images of the U.K. shelf area. The positions of the frontal regions discussed in the text are clearly discernible in the contrast of the IR image.

fluctuation might result from the variation of tidal range. The ratio of the tidal stream amplitudes at springs and neaps is typically ~ 1.8 for the area under discussion. This implies that the parameter h/u_s^3 changes by a factor of ~ 5.8 with a semi-monthly period. However, this is probably not reflected in movements of the boundary fronts because of the positive feedback inherent in the onset of stratification, i.e. once stratification is established its existence diminishes the capacity of tidally forced turbulence to bring about vertical mixing, so that the front would not adjust to relatively short time scale variations in h/u_s^3.

Acknowledgements—The authors would like to thank the British Oceanographic Data Service for providing much of the temperature and salinity data and assisting with the data processing. Additional CTD profiles were also kindly supplied to us by the Institute of Oceanographic Science, Bidston.

The satellite photographs (Fig. 6) were received from the NOAA 4 satellite by the University of Dundee receiving station, and we wish to thank Mr. P. E. BAYLISS and Mr. R. J. H. BRUSH of the Department of Electrical Engineering and Electronics for making them available to us.

REFERENCES

FEARNHEAD P. G. (1975) On the formation of fronts by tidal mixing around the British Isles. *Deep-Sea Research*, **22**(5), 311–321.

PINGREE R. D. (1975) The advance and retreat of the thermocline on the continental shelf. *Journal of the Marine Biological Association*, **55**, 965–974.

SIMPSON J. H. (1971) Density stratification and microstructure in the western Irish Sea. *Deep-Sea Research*, **18**, 309–319.

SIMPSON J. H. and J. R. HUNTER (1974) Fronts in the Irish Sea. *Nature*, **250**, 404–406.

SIMPSON J. H. (1976) A boundary front in the summer region of the Celtic Sea. *Estuarine and Coastal Marine Science*, **4**(1), 71–81.

TURNER J. S. (1973) *Buoyancy effects in fluids*, Cambridge University Press, 367 pp.

An analysis of 10 years' voltage records from the Dover–Sangatte cable

G. A. Alcock and D. E. Cartwright

Institute of Oceanographic Sciences, Bidston Observatory, Birkenhead

Abstract—Continuous hourly records of electrical potential across the Dover Strait, covering 1956–1964 and part of 1968–1969, are thoroughly analysed for tidal and non-tidal effects. Tidal components are clearly defined at high spectral resolution. There is a remarkable absence of the solar diurnal earth current, attributed to its orientation along the Strait, but higher non-tidal solar harmonics are discernible. Relationship to the difference in tidal elevation between Dunkerque and Dover is rather weak, but at high resolution the annual modulations of the M_2 amplitudes are surprisingly similar. The effect of sea conductivity is variable, but its variations suggest higher values of sea-bed conductivity than previous estimates. About a quarter of the non-tidal variance in e.m.f. is attributable in terms of linear atmospheric pressure gradients across the Southern Bight. Weather conditions associated with the largest positive and negative anomalies in e.m.f. are depicted.

1. INTRODUCTION

TIDAL fluctuations in the voltage recorded on a telegraphic cable across the Dover Strait in 1851 by WOLLASTON (1881) provided the first verification of Faraday's prediction (FARADAY, 1832) that tidal flow in the English Channel (and elsewhere) should produce measurable differences in electrical potential. A 14-hour record of potential on the submarine cable from St. Margaret's Bay near Dover, to Sangatte near Calais, recorded, as on other cross-channel cables, by CHERRY and STOVOLD (1946), was analysed by LONGUET-HIGGINS (1949) to provide an estimate of sea-bed conductivity. Longuet-Higgins' work was part of a series of general investigations of marine electromagnetic phenomena instigated by G. E. R. Deacon at the Admiralty Research Laboratory, Teddington, and later at the National Institute of Oceanography, Wormley (now part of I.O.S.). Longuet-Higgins' theoretical analysis and experiments in various parts of the English Channel inspired Bowden to record the potential on the St. Margaret's Bay–Sangatte cable (and other cables) for several months and to analyse it thoroughly as an indicator of the flow of water through the Strait, again under Deacon's auspices. BOWDEN (1956) identified the main tidal constituents of the cable voltage which are resolvable at one-month level, and showed that they vary seasonally with sea conductivity. Using a "calibration factor" derived from the tidal signal, he also estimated residual flows and correlated them with local winds and sea-level gradients.

Since the time of Bowden's experiments, the potential on that cable has been recorded almost continuously by pen-recording milliammeters installed at Dover by Mr. Norman D. Smith from Deacon's group at Wormley, with the collaboration of the staff at the St. Margaret's Bay Repeater Station and by kind permission of the Post Office Telecommunications Headquarters, London. The records have been used intermittently by various scientists, interested in signals from magnetic storms or in the characteristics of flow into the North Sea affecting fish larvae (RAMSTER, WYATT and HOUGHTON, 1973; TALBOT, 1975). Extensive use of them was made by one of the authors, when at N.I.O. Wormley, to determine the mean difference in sea level between England and France, 1957–1958, in terms of the mean flow through the Strait as estimated by the mean cable potential (CARTWRIGHT, 1961; CARTWRIGHT and CREASE, 1963). The 3-year record of voltage, 1959–1961, was edited and used by Cartwright in an attempt to relate its residual variations to surges in the North Sea, but the results, being largely negative, were omitted from his principal account of that work (CARTWRIGHT, 1968).

By 1968 the recorder had been going for about 14 years. Some of the earliest chart-rolls had been lost, and a fault in one of the circuit components had become apparent since 1966, but there remained a fairly continuous and fault-free stretch of about 9 years, probably the longest record of its kind ever made. It was considered worthwhile to convert to digital form and edit all parts of this 9-year record which had not already been so treated, to collect simultaneous relevant meteorological data, and to apply techniques for analysis of long geophysical time series with strong tidal content which had recently been evolved (e.g. MUNK and CARTWRIGHT, 1966), in an effort to resolve some of the more subtle properties of such records. After several years' delay, caused by pressure of other work, the preparation and analysis of these data have at last been completed and their results are presented and discussed in this paper.

Before proceeding with the main account, we should mention some subsequent developments in voltage recording at Dover. In March 1968 a new recorder was installed by Mr. N. D. Smith, to remedy the fault mentioned above. The first 15 months of its data are also used in the present work. In 1973 the Bidston laboratory (then known as "Institute of Coastal Oceanography and Tides") installed a digital data logger to the same cable circuit, to provide an adjunct to the JONSDAP* data-acquisition project in the southern North Sea. Some preliminary analyses of the JONSDAP cable series have been described by PRANDLE and HARRISON (1975). Further analysis by Prandle, involving the use of a fine-mesh hydrodynamical model of the Dover Strait area, is to follow shortly.

The work just mentioned is related to an important theoretical study recently made by ROBINSON (1976). Robinson's work is a major advance in our understanding of the relationship between potential difference across shallow channels and the pattern of water flow in the vicinity, dispelling the common notion that the p.d. is simply proportional to the integrated flux (see also SANFORD and FLICK, 1975). Another paper by ROBINSON (1977), applying his techniques specifically to the Dover Strait, accompanies our paper in this volume, and illustrates several features which cannot be derived from mere time-series analysis.

Most recently, the Post Office has made known its intention to abandon the continuous cross-channel cables in this area, and to replace them by cables interrupted by under-water repeater units. It is extremely difficult to record the natural potential across the Strait through such cables, so the present records, which are still continuing at the time of writing (March 1976), will soon be terminated for ever.

2. OBJECTIVES OF PRESENT WORK

After the published analyses mentioned in the previous section, to which we may add those of BOWDEN and HUGHES (1961) and of HUGHES (1969) applied to similar cables

---

* JONSDAP = Joint North Sea Data Acquisition Programme.

in the Irish Sea, and analogous work in the U.S.A. (SANDFORD and FLICK, 1975), we should make it clear what we hoped to discover from a 9-year voltage record which cannot be obtained from the much shorter records analysed previously. In brief, the questions we posed were as follows:

(a) How much of the variation in potential can be attributed to variations in the ionosphere at solar daily frequencies, and therefore not related to sea currents?

(b) Can one detect above noise level the long-period lunar tides Mm and Mf, not previously identifiable in sea currents, and in particular the interaction term Msf, which is a useful indicator of the constant e.m.f. set up by the tides?

(c) Could the seasonal modulations of the tidal e.m.f. reported by Bowden and others be partly due to seasonal modulations in the tidal currents themselves?

(d) What is the spectral noise-level of voltage records in comparison with noise-levels of the local tides, and how much of it is attributable to motions induced by local weather?

The rest of the paper describes our attempts to answer these questions. It will be found that no answer is particularly simple, but the processes of their investigation bring out a number of interesting properties which were not previously known.

RECORDING DETAILS AND PREPARATION OF DATA

The simple recording system was similar in principle to that described by BOWDEN (1956), and is sketched in Fig. 1a. In brief, the d.c. potential between the inner conductor of the cable and its screen is measured through a $2.2 k\Omega$ resistor in series with a milliammeter, whose internal resistance was of order $1 k\Omega$. The screen is earthed at both the English and French ends, and at the French end the conductor is also effectively earthed (through inductive elements) as far as low-frequency voltage variations are concerned. Thus the voltage measured at St. Margaret's Bay is effectively the difference in earth potentials between the particular coastal points in England and France. A $0.03 F$ capacitance in parallel with the milliammeter shunts out voltage variations with time scales less than about half a minute. For calibration, a standard voltage cell was switched in place of the cable on occasions.

The cable used for the bulk of the records considered here, before 1968, was that known as 'Number 5'. By the time of the new recorder installation in March 1968, this had been replaced by 'Number 6'. The approximate tracks of both cables across the Dover Strait are shown in Fig. 1, but their recorded potentials should, of course, be independent of their geometry.

The recorder traces go back to July 1954, but when it was decided to digitize the early data about 10 years later, some rolls were missing, notably a 7-week period in November to December 1955. Continuity of time series has advantages in analysis, so the digital series was started at noon, 19 December 1955, and continued through minor lacunae until noon, 9 March 1965, when a 15-day breakdown in the cable occurred. The traces were digitized at hourly intervals, with visual smoothing through passages of high-frequency oscillations which would otherwise be 'aliassed'. About twenty lacunae of 1–3 days were 'filled' by 25-hour interpolations. Six lacunae of 7–15 days were 'filled' by a computer procedure devised to interpolate between monthly tidal analyses of good data immediately before and after each gap.

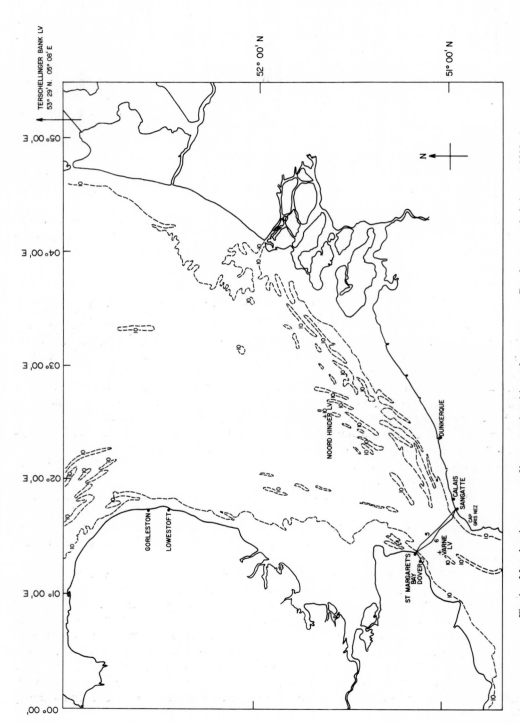

Fig. 1. Map showing sea area of interest and positions of measurements. Depth contours in fathoms (1.8288 m).

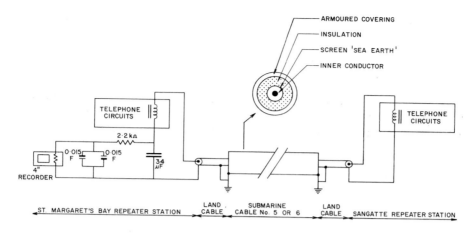

ARMOURED COVERING
INSULATION
SCREEN 'SEA EARTH'
INNER CONDUCTOR

TELEPHONE
CIRCUITS

2·2 kΩ

TELEPHONE
CIRCUITS

0·015
F

10·015
F

3·4
μF

4"
RECORDER

ST. MARGARET'S BAY REPEATER STATION

LAND
CABLE

SUBMARINE
CABLE No. 5 OR 6

LAND
CABLE

SANGATTE REPEATER STATION

Fig. 1a.   Circuit diagram showing principle of recording e.m.f.

Records continue after March 1965, but were not digitized because of an intermittent fault noticed in 1966 (eventually traced to a leaking capacitor) and because 9.2 years makes a useful data set for the type of tidal analysis to be described. Of the records from the later recorder (using No. 6 cable) only the first $15\frac{1}{2}$ months were digitized and edited as above, namely 26 March 1968–11 July 1969, for evidence of continuity with the first series. We thus analysed altogether some 10.5 years of hourly potentials.

For parallel measures of sea temperature (related to conductivity) and local wind conditions, we abstracted the 6-hourly temperatures and barometric pressures from the Noord-Hinder light vessel and the pressures from the Terschellinger-Bank light vessel, all from the annual volumes of the Dutch Royal Meteorological Institute, and the 6-hourly pressures from Gorleston on the East Anglican coast from the U.K. Meteorological Office Daily Weather Report series. Data for sea salinity at the Varne light vessel at 4-day intervals were also extracted from the I.C.E.S.* publications, and expanded to a daily series by cubic spline interpolations. All these data covered the entire span of the cable records.

Some questions, for example (a) and (c) in the previous section, required a measure of the local sea-currents, and this is most easily provided by the differences in sea level across the Strait, again simultaneously with the potentials. The only suitable sea-level series from France at that time was from Dunkerque, of which the period 1957–1958 had been used for a related exercise by CARTWRIGHT and CREASE (1963). The "Service Hydrographique et Océanographique de la Marine" kindly digitized the hourly Dunkerque record for us for a further 3 years, giving us the total period: noon 28 December 1956 to noon 7 March 1962. We confined our analyses of sea-level differences to this 5.2-year period. An extension to the 1957–1958 sea-level series from Ramsgate used by Cartwright and Crease was not readily available for the English station, so we used the hourly series from the Dover tide-gauge, held at IOS-Bidston.

*International Council for the Exploration of the Sea.

3. SPECTRAL AND TIDAL ANALYSIS

(a) *Monthly resolution*

The voltage record $e(t)$ was subjected to the 59-day spectral filters as used for surge and tide analysis by CARTWRIGHT (1968), by the operation:

$$E_r(t) = 2N^{-1} \sum_{n=-N/2}^{N/2} (1 + \cos 2\pi n/N) \exp(2\pi i r n/N) e(t + n\Delta t) \qquad (1)$$

where $\Delta t = 1$ hour, the interval of the digital series, $N = 1416$, so that $N\Delta t$ is 59 days, and the harmonic number $r = 0(1)360$, spanning a little over 6 cycles/solar day (c/d). The rth spectral estimate is centred on $59r$ c/d, and the tidal 'group' (i, j) centred on the frequency

$$(i \text{ cycles/lunar day} + j \text{ cycles/month})$$

is covered by the filter characteristics for $r = 57i + 2j$. 'Side-band leakage' from tidal to non-tidal parts of the spectrum is negligible. For a given time t, the sequence $E_r(t)$ thus represents a set of nearly independent values of the complex spectrum of the 59 days of the potential signal e centred on, and with phases referred to the time t. The set of frequencies determined by r are either centred on tidal groups, which are completely covered by the width of the filter, or are virtually free of tidal energy, where they form estimates of the 'continuum' or noise spectrum of the signal.

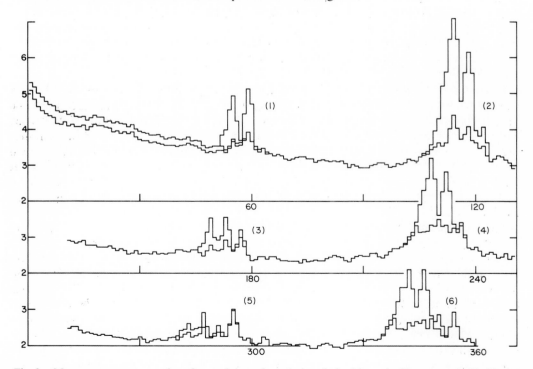

Fig. 2.   Mean energy spectrum of e.m.f. at cycle/month resolution, derived from the filter operator (1). Abscissa shows harmonic number r, equivalent to 59 × frequency in c/day. Bracketed numbers show tidal species. Ordinate is $\log_{10} S$, where S is spectral energy density in (millivolts)^2 (c/day)$^{-1}$. Upper spectral levels represent raw data, lower levels represent tidal and weather residuals.

The central times t were taken in steps of 30 days for the present purpose, giving 110 sample spectra from the $9y$ record. The variance-density spectrum from the ensemble averages $\langle E_r E_r^* \rangle$ for each $r = 1(1)359$, are shown in Fig. 2, scaled to units of millivolts^2 $(c/d)^{-1}$. Where two, or in a few cases, three spectral levels are shown, the highest level is the spectrum of residual series when optimum tidal and weather-dependent syntheses are subtracted from the data by methods to be described later. Peaks corresponding to the tidal groups stand out well above the background noise level, with no more 'cusping' than is common in spectra of typical sea-level records from the area (see, for example, Figs. 5.1, 5.2 of CARTWRIGHT, 1968). Like them, the voltage spectrum shows a sequence of 'overtides' in species 4, 5 and 6, caused by local shallow water. This is to be expected from the strong dependence of the e.m.f. on water flow.

Between the tidal zones, the noise continuum decreases steadily with increasing frequency, as with most geophysical variables. A good approximation to it is given by

$$\log S = 3.4 - 1.5 \log f$$

where $\log S$ is the log spectral density plotted in Fig. 2, and f is the frequency in cycles per day. A power law $f^{-3/2}$ is thus indicated, similar to typical sea-level spectra. The standard sampling error at each value of r is about 9.5%, since the 110 samples at 30-day intervals are effectively independent. This corresponds to a variation of 0.04 in the logarithmic units shown. These figures for variation are probably a little too small, but departures from smoothness much greater than 0.1 may suggest some spectral structures caused by resonances in the sea area. Evidently, there is no very remarkable feature of this nature. There was some interest as to whether the rotational period of sunspots, affecting the ionospheric currents, would give an anomaly in the spectrum near 27-days period ($r = 2$), but there is nothing to support this in Fig. 2.

The tidal residuals shown in Fig. 2 were computed by a 'response' analysis of Munk–Cartwright type. Elaborate details of the analysis would be of little interest, but it should be noted that it included the non-linear interaction terms described by CARTWRIGHT (1968), and some others which have since been added to the computer programs, such as a seasonal modulation of the semi-diurnal tides, which is clearly important in this context. The rising of the residual in the tidal bands, above the smooth trend of the non-tidal continuum, is a normal effect of surge-tide interaction, possibly enhanced here by occasionally imperfect time-keeping of the voltage record, and by the slow variations in amplitude, discussed in Section 5. However, the residual spectral levels are still remarkably low, as shown by their numerical values, listed in Table 1. Corresponding figures from a similar tidal analysis of the Dover 5-year sea-level record are included for comparison. Table 1 shows the variances of data and tidal residuals in frequency bands of width 9 cycles/month centred on and covering each tidal species, and also the ratios (residual variance: data variance). The row labelled 'sum' gives the sum over all the tidal bands above, and 'total' gives the total computed variances of the series without spectral limits. The considerable excess of the overall variances above the summed tidal band variances is mostly accounted for by low-frequency noise below 0.5 c/d. In this respect, the e.m.f. is strikingly noisier than sea level in the 'species 0' band.

Perhaps the most interesting feature shown in Fig. 2 is the series of small but outstanding residual peaks at the solar frequencies n c/d (harmonic number $r = 59m$) for $m = 3, 4, 5$ and 6, which appear to the right of the bands of higher tidal species. The fact that these are not removed, or only slightly reduced, by the tidal convolutions of the 'response'

Table 1. Spectral variances of tidal residuals: e.m.f. and sea level

Tidal species	Cable e.m.f. (mV)^2			Dover sea level (cm)^2		
	Raw data	Tidal residual	Residual data	Raw data	Tidal residual	Residual data
0	10,117	9,848	0.973	97	86	0.886
1	6,842	1,162	0.170	44	18	0.409
2	379,161	1,938	0.005	31,006	46	0.001
3	337	157	0.466	5	4	0.800
4	6,740	498	0.074	617	27	0.044
5	91	64	0.703	3	2	0.667
6	882	111	0.126	74	7	0.149
7	34	30	0.882	2	2	0.999
8	63	50	0.794	10	3	0.300
Sum	404,267	13,858	0.034	31,858	195	0.006
Total	420,640	29,934	0.071	32,063	398	0.012

synthesis shows that they are anomalous to the local tidal regime. We attribute them to the diurnal ionospherically induced variations in the magnetic field, which are known to possess high-order harmonics (MALIN, 1973). The cases of the diurnal S_1 and semi-diurnal S_2 will be considered in subsection (b). The amplitudes of the high harmonics of the residual spectrum are all close to 7 mV. This is in keeping with typical diurnal variations in earth potential of a few millivolts per kilometre (CHAPMAN and BARTELS, 1940). The dependence on the time derivative of the magnetic field no doubt emphasizes the high frequencies.

The part of the residual spectrum shown in Fig. 2 extending from 0 to the diurnal tidal band was derived by subtraction of a 'best' convolution of the barometric pressure data from the cable data. It is indistinguishable from the spectrum of raw data above 1 c/d. It was hoped that this residual would enable some of the small low-frequency tidal components to show above noise level, as was the case with a more elaborate weather convolution described in CARTWRIGHT (1968, Fig. 12), but the reduction in noise level here, about 3 db, was not sufficient for this.

(b) *High resolution spectra*

The series of tidal-group filters $E_r(t)$, $(r = 57i + 2j; i = 0, 1, 2; j = -4(1)4)$, selected from the full spectral series computed by equation (1) at 5-day intervals of t, were themselves analysed spectrally at the highest possible resolution of 1 cycle per 9 years. The computational procedure was similar to that described in CARTWRIGHT and TAYLER (1971). Firstly, each group series (i, j) was 'heterodyned' to a central 'constituent' frequency f_{ijk}, defined by

$$f_{ijk} = i \text{ cycles/lunar day} + j \text{ cycles/sidereal month} + k \text{ cycles/year}$$

with $k = -j$, so that the spectrum of the heterodyned series $E_r'(t)$ at f_{ijk} appears as a (complex) constant of 'zero' frequency. Each series was then analysed by a discrete Fourier transform of the 9×365 days available (the 9.2y of the original series $e(t)$ having been

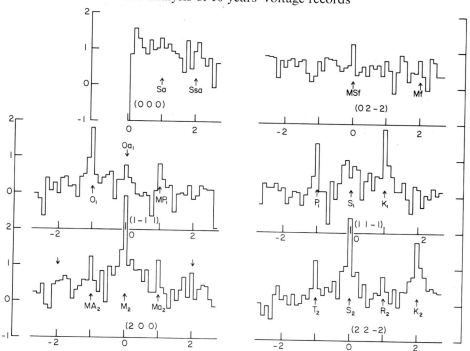

Fig. 3. High resolution (cycle/9y) amplitude spectra of e.m.f. in bands covering selected tidal groups (i, j). Bracketed numbers (i, j, k) represent the tidal 'constituent' frequency at the zero abscissa of each band; abscissa marks frequency scale in cycles/year relative to this frequency. Ordinates are $\log_{10} A$, where A is spectral amplitude in millivolts. Arrowed symbols denote the frequencies of common tidal constituents.

reduced to 9y by the 59d filtering), to cover frequencies from -4 to $+4$ c/y. In this way, all the major tidal 'constituents' (ijk), including those involving the 8.85y perigee cycle (such as N_2) are well resolved by the frequency grid.* The two most interesting results from each of the tidal 'species' 0, 1 and 2 are shown as 'amplitude spectra' in Fig. 3. Here, zero frequency is shown at the heterodyne centre f_{ijk}, with (i, j, k) given in brackets, while frequencies above and below the centre are indicated in cycles per year, at intervals of 9 Fourier components. Only the range of about ± 2.5 c/y is shown, since there were no features of interest outside this range.

The low-frequency groups (species 0) shown in the top pair of diagrams of Fig. 3 are analyses of the weather-residual series mentioned in section (a), since we have shown that this has a 3-db reduction in noise level. In group $(0, 0)$ only the seasonal variation Sa can be seen, but this is only just above noise level, unlike sea-level spectra, where Sa usually stands out clearly in a 9y analysis. The amplitude of Sa is 20 mV, with a maximum about 8 December. The cause of this seasonal change in potential is presumably ionospheric, but the seasonal change in mean sea level could have some influence (ROBINSON, 1976). In any case, the high level of random variations in this frequency range makes its effect rather insignificant.

In group $(0, 2)$ the linear fortnightly tide Mf possibly just shows above the continuum,

* CARTWRIGHT and TAYLER (1971) used 18y analyses, to residual modal components also.

but there are higher peaks nearby which make its amplitude of 6 mV unreliable. (The position of the lunar node is unfavourable for fortnightly tides during 1956–1964.) The monthly tide Mm does not appear at all in group (0, 1), which we have therefore not shown. The one significant peak, however, is clearly at the central frequency, which is that of MSf, an effect of non-linear interaction between the semidiurnal M_2 and S_2 tides. With an amplitude of 16 mV, its phase is within 22° of the difference between the phases of S_2 and M_2, so it may be regarded as giving a positive e.m.f. at spring tides, negative at neaps. It could be caused by a sea current at MSf frequencies, or by a non-linear voltage effect of change of total conductivity of the channel cross-section with tidal sea level, as explained by ROBINSON (1976). However, a similar analysis of the 5y record of Dover sea level showed no energy above noise level at MSf, so a sea current is unlikely to be a cause, and we must attribute the MSf potential variation to the non-linear electrical response. If this is correct, there must also be a *constant* positive e.m.f. generated by the same process, proportional to the squares of the major tidal components, and bearing the ratio $(H_M^2 + H_S^2 + H_N^2 + H_K^2)/2H_MH_S$ to the observed MSf amplitude, where H represents the amplitude of the semidiurnal tidal constituents indicated in, say, the Dover sea level. The above ratio is in fact 1.83, so the constant e.m.f. induced by tides should be about $+20$ mV. This value is in very good agreement with the value calculated by ROBINSON (1977). However, there must also be constant potentials set up by the permanent earth-current system. CARTWRIGHT (1961) deduced a constant e.m.f. of -83 mV, from a comparison of cable e.m.f. with simultaneously measured water flow. His method of deduction may be criticized, but taking that result at its face value, it appears that the tidally induced constant e.m.f. is swamped by a larger negative e.m.f. from earth-currents.

The middle pair of spectra in Fig. 3 include the principal tidal constituents O_1, P_1 and K_1, which all stand out well against the noise continuum with little evidence of line-spreading except for that expected from the 18.6-year nodal variations. The group which includes O_1 also exhibits a peak labelled MP_1, which can only be attributed to non-linear interaction between M_2 and P_1, and a smaller peak labelled Oa_1 which represents the annual modulation of the O_1 line due to seasonal changes in conductivity. The amplitude of Oa_1 is about 5% of the amplitude of O_1, in keeping with the seasonal modulation of M_2, to be discussed later.

The spectral band covering group (1, 0) is not shown here because of its generally low energy levels, but it has some interest in that it confirms the dominance of the line at exactly 1 cycle/lunar day in the cluster of weak lines usually denoted by 'M_1', in accordance with a peculiar property of the tides off western Europe pointed out by CARTWRIGHT (1975). The amplitude of the true M_1 component is 4.6 mV, corresponding to a mean current through the Strait of amplitude about 6×10^{-3} ms^{-1}, with phase lag $G = 335°$.

The most interesting feature of the (1, 1) group is the virtual absence of any line at the precise solar diurnal frequency S_1. This is most unexpected, because it is generally accepted that the non-tidal earth currents are predominantly diurnal, and we have already shown that prominent peaks occur at higher multiples of the S_1 frequency in this signal. LONGUET-HIGGINS (1949, Fig. 2) shows solar diurnal variations comparable in amplitude with lunar diurnal amplitude in potential differences off Plymouth, but since he worked with only one month of data, his 'solar diurnal' variation is indistinguishable from the tidal K_1 variation, so cannot really be attributed to S_1. HUGHES (1969) gives the only published results from cable potentials which are recorded for long enough (one year) for

S_1 to be unambiguously resolved. He quotes amplitudes in units of equivalent water transport; they may be interpreted as 57 mV for S_1, significantly larger than K_1 (46 mV) or O_1 (31 mV). The corresponding figures in the present case are 3, 82, 74 mV for S_1, K_1 O_1 respectively. Hughes's cable (No. 001) is about twice as long as the present one, so one would expect a reduction factor of 2 in earth-current potential difference, but not a reduction of 20 as observed. One may increase the amplitude of 3 mV by allowing for (a) a tidally induced potential from the small S_1 component in the local tides, or (b) a seasonal modulation to K_1. Reasonable estimates for these effects raise the residual S_1 amplitude to about 8 mV, but this is still surprisingly small.

A possible explanation is that the diurnal earth-current is aligned parallel to the Dover Strait and so produces little cross-channel potential gradient, whereas Hughes's cable, between Anglesey and Isle of Man, is roughly parallel to the north–south axis of the Irish Sea. Concentration of earth-currents along the path of high conductivity offered by sea channels has been suggested in other contexts (BARBER, 1948). This explanation is not supported by the presence of $S_3, \ldots, S_8$, but these amplitudes, about 7 mV, are of the same order of magnitude as S_1. They show up only because of the lower noise levels at higher frequencies.

CARTWRIGHT (1961) deduced diurnal earth-current effects of order 50 mV in amplitude between Dover and Sangatte, by assuming a relationship between semi-diurnal and diurnal potentials and sea currents. Those results are now seen to be quite untenable, and may be attributed to over-reliance on dubious assumptions and on the accuracy of figures derived from short spans of data.

Apart from the non-appearance of S_1 it is worth remarking on an enhancement of spectral level in the close vicinity of S_1 in Fig. 3. This spectral feature is equivalent to a quasi-diurnal variation of r.m.s. amplitude 24 mV, whose phase relationship to the mean solar day varies randomly but slowly, with a time scale of a few years. Again one may only speculate as to its origin. It could be caused by a truly diurnal earth current whose alignment along the channel varies slightly due to slow changes in the distribution of sea conductivity, resulting in random changes in the apparent diurnal potential across the channel. In general, it is remarkable how many facts suggest the presence of a diurnal earth-current without a true S_1 line actually appearing in the data.

The lowest panel of Fig. 3 shows the principal lunar and solar semi-diurnal tidal groups. The M_2 line stands out quite clearly, and it is strong enough for its annual modulations Ma_2 and MA_2 (following CORKAN's (1934) notation) to show well above noise level. The nature of this modulation will be discussed in Section 4(c). There is also a suspicion of the second harmonic annual modulation, that is a 2 c/y modulation, marked by downward arrows, particularly at the higher frequency. The annual modulations to S_2 coincide with genuine tidal components T_2 and R_2, so are hardly distinguishable from them. All these components together with K_2 stand out clearly in the bottom right-hand panel. A possible ionospheric contribution to S_2 is discussed in Section 4(b).

4. COMPARISON WITH SEA SURFACE SLOPE

(a) *Dynamics*

We shall use the slope of the sea surface, as measured by differences between tide gauge records at Dunkerque and Dover, as a rough indicator of the characteristics of flow through the Strait, in order to help determine which aspects of the cable e.m.f. signal are due to other causes. The relationship between mean current and slope is complicated by the variable depth and topography of the channel and by the fact that the line joining Dunkerque to Dover is not normal to the general stream direction. In order to identify the principal factors involved, consider the simplified representation of Fig. 4(a), a straight uniform channel containing a uniform tidal current $v(t)$, parallel to the channel's axis and at an angle θ to the normal to the line joining two edge-points D_1 (Dover) and D_2 (Dunkerque), where surface elevations ζ_1 and ζ_2 are respectively measured.

Elementary hydrodynamical theory gives

$$\frac{g}{L}(\zeta_2 - \zeta_1) = fv\cos\theta - \frac{d}{dt}(v\sin\theta) - \frac{F(v)}{D}\sin\theta, \tag{2}$$

where L is the distance between D_1 and D_2, g is gravitational acceleration, f is the Coriolis parameter, $F(v)$ is the bottom frictional stress, and D the depth. If we simplify further by assuming a linear frictional law, $F(v)/D = kv$, and consider a harmonic component of period $2\pi/\omega$, such that $v = v_0\cos\omega t$, then

$$\frac{g}{L}(\zeta_2 - \zeta_1) = (f\cos\theta - k\sin\theta)v_0\cos\omega t + \omega v_0\sin\theta\sin\omega t. \tag{3}$$

If, finally, we assume that the electrical potential induced in the cable, represented by $C_1 C_2$ in Fig. 4(a), can be expressed as

$$e = \alpha(\omega)v_0\cos(\omega t - \varepsilon)$$

where α is a physical factor of proportionality as discussed by BOWDEN (1956) and other authors, and ε is a very small phase correction depending on the positioning of $C_1 C_2$

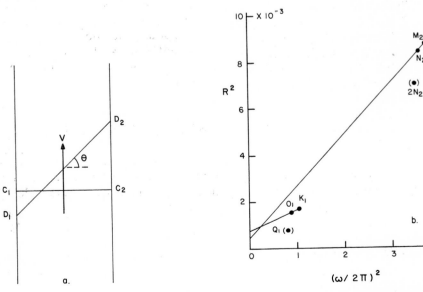

Fig. 4. (a) Simplified representation of channel with cable $C_1 C_2$ and line joining Dover (D_1) to Dunkerque (D_2). (b) Suggested relationships between amplitude ratio R (mV/cm) and frequency ω (radians/day) for dominant diurnal and semi-diurnal tidal constituents, according to equation (5).

Table 2.   *Comparisons of surface slope* ($\zeta_2 - \zeta_1$) *and cable e.m.f. at some major tidal constituents*

Constituent symbol	Group Nos.	Surface slope		Cable e.m.f.		R (mm/mV)	ϕ deg.
		Admittance	Amplitude (mm)	Admittance (mV/mm)	Amplitude (mV)		
Q_1	1–2	0.132	7	0.482	24	0.274	82
O_1	1–1	0.112	29	0.282	74	0.397	85
K_1	1 1	0.092	34	0.226	83	0.407	86
$2N_2$	2–2	0.848	14	1.015	16	0.835	99
μ_2	2–2	2.997	58	1.155	22	2.595	63
N_2	2–1	1.087	132	1.187	144	0.916	85
M_2	2 0	1.180	746	1.266	800	0.932	79
L_2	2 1	3.379	60	2.939	53	1.150	70
S_2	2 2	0.825	243	0.900	265	0.917	89
K_2	2 2	0.906	72	0.930	74	0.975	87
	Radiational components:						
S_2	2 2	17.15	99	10.10	58	1.698	121

with respect to the tidal wave, then the difference in elevation can be expressed as

$$\zeta_2 - \zeta_1 = e_0 R \cos(\omega t - \phi) \tag{4}$$

where

$$R^2 = [L^2/g^2\alpha^2(\omega)][\omega^2 \sin^2\theta + (f\cos\theta - k\sin\theta)^2], \tag{5}$$

$$\phi = \arctan[\omega \sin\theta/(f\cos\theta - k\sin\theta)] - \varepsilon \tag{6}$$

and e_0 is the amplitude of the variation in potential.

Now, at the latitude of Dover Strait, $f = 1.1 \times 10^{-4}\,\text{s}^{-1}$; the tidal streams are on average inclined to the Dover–Dunkerque line by about 45°; and a realistic value for k is

$$0.003 \times (\text{r.m.s. tidal velocity/typical depth}) = 0.003 \times 1.0/30 = 1.0 \times 10^{-4}\,\text{s}^{-1}.$$

Therefore, the expression ($f\cos\theta - k\sin\theta$) may be expected to be of order $10^{-5}\,\text{s}^{-1}$ compared with $\omega \sin\theta$ which is of order $10^{-4}\,\text{s}^{-1}$ at tidal frequencies. We should expect R^2 to be nearly proportional to ω^2 and ϕ to be near 90°. This is different from the situation often considered, when θ is nearly zero, and the term involving f is the most important on the right-hand side of (2).

Table 2 shows values of admittances to the tide-generating potential and the corresponding amplitudes, of both ($\zeta_2 - \zeta_1$) and e, computed from the group-filtered series $E_r(t)$ (equation (1)), and from corresponding series derived from the hourly tide-gauge records at Dunkerque and Dover, by a straightforward correlation procedure. All ensemble-averages were based on the 5-year span for which the tide-gauge records were both available. Most of the tidal constituents selected are those of reasonably large amplitude and not usually disturbed by non-linear effects, but μ_2 and L_2 are also included to show the effect of strong non-linear interactions. The table also shows the resulting values of R and ϕ (equation (4)).

Neglecting anomalous values from the predominantly non-linear terms μ_2 and L_2, we see that the phase lags ϕ are indeed near 90°, and values of R for the diurnal tides are of order half the typical values from the semi-diurnal tides, thus confirming that R is roughly proportional to frequency ω, as suggested by the above speculations.

There is nothing very systematic about the phases other than general magnitude, but the really major tidal lines O_1, K_1, N_2, M_2, K_2 do suggest systematic trends in the ratios R with respect to frequency within the narrow bands of their species. For these lines, values of R^2 are plotted against ω^2 in Fig. 4(b). K_2 is plotted rather than S_2, because S_2 is more likely to contain non-tidal solar variations—see Section 4(b). Values for Q_1 and $2N_2$ are shown bracketed; they do not fit well into the trends by the major terms and we may suppose they are affected either by noise or non-linear effects. The major results for both species support linear relations, with small positive intercepts at the axis $\omega^2 = 0$, in accordance with equation (5). But the lines are now seen not to be identical. This is reasonable, since both $\alpha(\omega)$ and θ can be expected to vary from one species to another according to their different cotidal patterns in the southern North Sea, and according to Robinson's calculations of cable sensitivity.

We shall not burden the rough relation (5) with any deeper interpretation, but the anomalous non-linear terms deserve a few more comments. As is well known, in such shallow waters the linear tidal component μ_2 is swamped by a third-order action between M_2 and S_2 (i.e. $M_2 + M_2 - S_2$), while L_2 is swamped by $(M_2 + M_2 - N_2)$. One would have liked to find that the elevations, currents and e.m.f. are similarly affected, so that the non-linear terms would fit into the same trends as the linear terms, but this is evidently not the case. In retrospect, the differences are not unreasonable. The sea in the vicinity of Dunkerque is much shallower than on the Dover side, so much of the non-linearity will be generated there. The distribution of non-linear current variations across the Strait will therefore be very different from the distribution of the dominant linear tides; understandably from Robinson's work, the corresponding electrical potentials will be different.

(b) *The non-tidal part of* S_2

S_2 is the only solar component for which we do not yet have an estimate of its ionospheric component. Eliminating the part due to gravitational tides is not the only difficulty, because the marine tides also have a non-gravitational part which MUNK and CARTWRIGHT (1966) termed the 'radiational tide'. CARTWRIGHT (1968) found that the radiational tide accounted for about 18% of the total S_2 tide at typical places round the British Isles, and showed evidence to suggest that much of the effect is a result of coupling with the atmospheric tide. The S_2 component of the e.m.f. will therefore contain a gravitational tide, a radiational tide caused by the sea currents, and an ionospheric component. In so far as the ionospheric component is solar in origin, it may be assumed for the present purpose to have a spectral structure similar to the radiational tide itself. We can then estimate its magnitude by isolating the 'radiational part' of the e.m.f. and of $(\zeta_2 - \zeta_1)$, in the semi-diurnal solar band, and subtracting the anomalous part of the former.

The best way to isolate the radiational tide is by the 'response method' of MUNK and CARTWRIGHT (1966). The data series is multiply correlated with the gravitational tide-generating potential, the 'radiational potential', and a function which accounts for the principal non-linear tides. The radiational part is distinguished from the gravitational principally by their different proportions in the components S_2 (purely solar) and K_2 (luni-solar). The radiational parts of S_2 resulting from response analyses of the cable data and the two tide-gauge data are summarized in the lowest line of Table 2. Admittances are referred of course to the radiational instead of the gravitational potential. The anomalous values of R and ϕ are interpreted as being due to the ionospherically induced earth current. The phase lag of the 58 mV radiational component of the e.m.f. is 179° (relative to 0 h GMT). To have the same (R, ϕ) as the gravitational S_2 component, which is similar to that for K_2, the radiational e.m.f. would be 89 mV with phase lag 136°. The vector difference then gives our estimate of the ionospheric S_2 component, namely:

Amplitude 61 mV   Phase lag 276°.

This is considerably larger than S_1 or any of the higher harmonics. However, it is obviously difficult to assess its accuracy.

(c) Annual modulations

We have already noted (Fig. 3) the annual sidebands Ma_2 and MA_2 accompanying M_2. These lines are the spectral manifestations of a cyclic annual change in amplitude, and possibly phase, of the tidal variations in the e.m.f. The modulation is usually attributed (for example, by BOWDEN, 1956) to the seasonal change in the conductivity of the sea, but this is not necessarily the entire cause, because the tides themselves have small seasonal modulations (CORKAN, 1934; CARTWRIGHT, 1968). The practice of making a correction factor based on the predicted tidal range for a local port does not remove the tidal effect, because common tide predictions do not include the seasonal modulation. Besides this, it is not known whether the tidal flow has the same degree of modulation as the elevation. Following the plan of the previous parts of this section, we shall use the surface slope as an indicator of the annual modulation in the flow, and hence assess how far it is responsible for the modulation in e.m.f.

Spectral analysis gives the modulated M_2 component in the form

$$X(t) = a_0 \cos(\omega t - \phi_0) + a_1 \cos[(\omega + v)t - \phi_1] + a_{-1} \cos[(\omega - v)t - \phi_{-1}], \tag{7}$$

where ω is the M_2 frequency, v is $2\pi y^{-1}$, and the notation for the three amplitudes and phases is obvious. Following CARTWRIGHT (1968), it is more informative to express the four annual parameters in terms of amplitude- and phase-modulations, that is,

$$X(t) = a_0 \lambda(t) \cos[\omega t + \mu(t) - \phi_0], \tag{8}$$

where it can be shown that, provided $r_1 = a_1/a_0$ and $r_{-1} = a_{-1}/a_0$ are both small,

$$\begin{aligned}
\lambda(t) &\approx 1 + (r_1 \cos \Delta_1 + r_{-1} \cos \Delta_{-1}) \cos vt \\
&\quad + (r_1 \sin \Delta_1 - r_{-1} \sin \Delta_{-1}) \sin vt \\
&= 1 + \delta \cos(vt - \alpha),
\end{aligned} \tag{9}$$

and

$$\begin{aligned}
\mu(t) &\approx -(r_1 \sin \Delta_1 + r_{-1} \sin \Delta_{-1}) \cos vt \\
&\quad + (r_1 \cos \Delta_1 - r_{-1} \sin \Delta_{-1}) \sin vt \\
&= \varepsilon \cos(vt - \beta)
\end{aligned} \tag{10}$$

where

$$\Delta_i = \phi_i - \phi_0.$$

(CARTWRIGHT (1968) used the opposite signs for α, β.)

Table 3.   *Parameters of annual modulation*

	Units	Dunkerque sea level	Dover sea level	Slope $(\zeta_2 - \zeta_1)$	Cable e.m.f.
r_1	—	0.0054	0.0025	0.0174	0.0173
r_{-1}	—	0.0160	0.0110	0.0175	0.0223
Δ_1	deg	79.0	6.3	12.9	10.0
Δ_{-1}	deg	59.2	59.4	4.9	−47.8
δ	—	0.0125	0.0123	0.0345	0.0375
ε	rad.	0.0203	0.0103	0.0054	0.0137
α	days	71	65	119	149
β	days	277	280	303	127

Table 3 shows the numerical values of the parameters in equations (9) and (10), resulting from analysis of the three data series. The phase lags α and β are given in days from the Vernal Equinox (21 March); their calculation depends on the phase origins used, 13 July 1959 for the sea-level records, 17 July 1960 for the e.m.f. record. The degree of amplitude and phase-modulation in the two sea levels is between 1 and 2%, similar to values reported from other places, but it is very interesting to note that the amplitude modulation in the slope, and presumably in the flow, is much larger, approximately $3\frac{1}{2}$%, while the phase modulation ε is quite small. These characteristics match those of the e.m.f. remarkably well, while the phase lag α of the amplitude modulations differ by only 30 days in the yearly cycle. (Ironically, the phase lags β differ by nearly half a year, but the phase modulations are small and presumably subject to random errors.)

Thus, a remarkable fact emerges, that within the limits of the sampling errors of 5–9-year records, the *strictly annual* modulation of the e.m.f. M_2 component is largely accounted for by the annual variations in water flow, as estimated here. This appears to contradict the hypothesis that these variations are caused by changes in sea conductivity, or at least to reduce its importance. However, the total modulation of all variables covers a much wider band of frequencies. Although the annual lines stand out above noise level at 1/5 or 1/9 cycle/year resolution, their contribution to the total sideband variance over 1 or 2 c/y on either side of M_2 is in all cases fairly small. It will be seen in the next section that the time plots of the total (modulated) M_2 amplitudes of the e.m.f. and surface slope bear no resemblance, and in fact the annual cycle in the M_2 amplitude of the slope is hardly discernible against the random variations. So the results noted above are applicable only to rather small, very narrow-band spectral components.

5. RELATION TO SEA CONDUCTIVITY

We now consider the time history of the M_2 modulations which we discussed in Section 4(c) in spectral terms. The series $E_r(t)$ defined by equation (1) with $r = 114$ well represents the behaviour of the total signal of the (2, 0) tidal group in a band of width 1 cycle/month, with t now every 5 days. The second panel from the bottom of Fig. 5 plots values of

$$H(t) = |E_r(t)|/f(t)$$

for the whole 9 years' record, where $f(t)$ is the conventionally computed function which describes the small modulation of the M_2 tide of period 18.6 years, depending on the

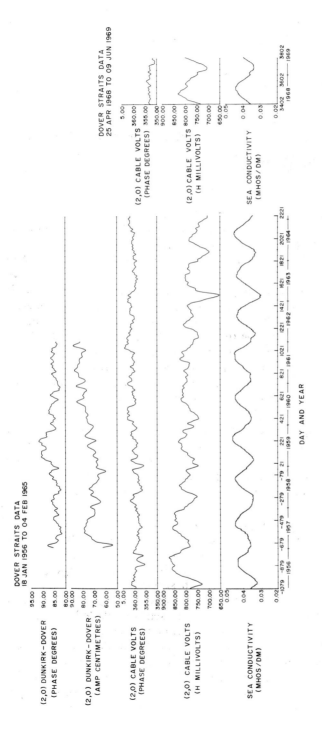

Fig. 5. Top two panels: phase lag and amplitude of 'M₂' spectral band of $(\zeta_2 - \zeta_1)$; 3rd and 4th panels: phase lag and amplitude of E_{114} from cable e.m.f. (equation (1)); bottom panel: sea conductivity in vicinity of Dover Strait. The bottom scales show a continuous day count with 0 arbitrarily at 1 January 1959, and also the years themselves.

position of the Moon's node. (Over the period of the record, $f(t)$ rises from 1.015 to a maximum 1.037 in mid-1959, then falls to 0.988 at the beginning of 1965; its value in 1969 is 0.963.) $H(t)$ is therefore equivalent to $a_0 \lambda(t)$ in equation (8). The panel above plots

$$P(t) = \mathrm{Arg}[E_r(t)] + \omega t + u(t) - \phi_0$$

where ω is the M_2 frequency as before, $u(t)$ is a nodal function associated with $f(t)$, and ϕ_0 is value of $\mathrm{Arg}(E)$ at $t = 0$. $P(t)$ is therefore equivalent to $-\mu(t)$ in equation (8). The top two panels plot analogous functions from the 5-year slope record.

The phase lag $P(t)$ varies randomly about $0(360°)$ with vagaries of order $5°$ which are not consistently related to any other plotted quantity, although it has some resemblance to the corresponding phase of slope (top panel) during the year 1958. $H(t)$ has a much more pronounced annual cycle, as we should expect, although a high degree of randomness is also present, as was suggested in Section 4(c). It also exhibits a steady downward drift in mean value, from about 820 mV in 1956 to about 750 mV in 1964. We at first thought the drift might be due to a slow change in the Earth's magnetic field, but inspection of yearly values of the vertical magnetic component at Hartland for the period showed in fact a very slight increase. The nodal factor $f(t)$ is usually distorted by shallow water, but not by more than 1% or so, and the distortion can hardly produce a steady downward trend. Finally, we were convinced that the drift must be caused by a slowly developing leak in the capacitors placed across the recording instrument (Fig. 1a) which finally broke down about 1966. The calibration used for the whole record had been obtained in 1960. Inspection of earlier and later calibration figures did in fact reveal a decline in sensitivity of about 0.5% per year, similar to that observed in Fig. 5.

The 14-month records in 1968–1969 whose results are added at the right of Fig. 5 were analysed principally to confirm this diagnosis of the cause of the pre-1965 drift. It confirms a return to an average amplitude of about 780, as in the middle of the earlier record. However, the 1968–1969 phases are consistently lower than the earlier set, by about $7°$. We can suggest no explanation for this. The change from cable No. 5 to cable No. 6 (see Fig. 1) should make no difference, since the earth points at St. Margaret's Bay and Sangatte were apparently unaltered. S_2 phase lags also drop by $7°$, but values for O_1 actually increase by about $1°$. There is no question of any consistent timing error. The new cable system behaves as though it were sensitive to tidal currents about 10 miles further into the North Sea, but we have no evidence to suggest why this should be so.

The lowest panel in Fig. 5 shows the time history of local sea conductivity during the entire period. This was calculated from 6-hourly surface temperatures at the *Noord Hinder* light-vessel (51°39'N, 2°34'E) published by the Dutch Meteorological Institute, and from surface salinities recorded every 4 days at the *Varne* light-vessel (50°56'N, 1°17'E) as published by the "International Council for Exploration of the Sea". Salinities were interpolated to the same times as the temperatures, and the resulting conductivities smoothed over a 59-day tapered filter to simulate the same smoothing as in the calculation of equation (1). The annual cycle is fairly regular, with the highest conductivity during the warm summer of 1959 and the lowest during the cold winter of 1963.

There is evidently a fair degree of positive correlation between the e.m.f. M_2 amplitude and conductivity in their annual cycles, but the relationship is by no means uniform from year to year. In the period from mid-1960 to mid-1961 $H(t)$ shows no relationship to conductivity, while the drop in $H(t)$ in February 1963 is much greater and sharper than one would expect from the rest of the record. The M_2 amplitude of surface slope appears irrelevant to $H(t)$, despite the similarity of their strictly annual modulations discussed in Section 4(c). The variations of $H(t)$ must depend on other factors not represented, possibly the finer details of the spatial patterns of conductivity and flow.

BOWDEN (1956) used a formula for the potential difference across a channel of semi-elliptical cross-section to deduce the conductivity of the sea bed κ_0 of the Dover Strait in terms of the M_2 amplitudes of p.d. and water flow and the seawater conductivity, κ_1. The values he derived for κ_0 varied inversely with κ_1, whereas one would expect them to be constant, or to vary slightly but positively with κ_1. Our more extensive measurements make it possible to estimate a *constant* value of κ_0 in terms of the variation of H with κ_1.

Bowden's formula, adapted from one derived by LONGUET-HIGGINS (1949), is

$$H(t) = VZL \left[1 + \frac{\kappa_0}{\kappa_1} \frac{L}{2D} \right]^{-1} \tag{12}$$

where V is the amplitude of the mean velocity in the channel, Z is the vertical magnetic component, and L and $2D$ are the major and minor axes of the hypothetical ellipse. We use the derivative H', where

$$\kappa_1 H' = \kappa_1 \frac{\partial H}{\partial \kappa_1} = VZLQ(1+Q)^{-2}, \tag{13}$$

and

$$Q = \frac{\kappa_0}{\kappa_1} \frac{L}{2D}$$

with the equivalent dimensions for the Dover Strait, $L = 34\,\text{km}$, $D = 45\,\text{m}$, used by Bowden, and $V = 1.06\,\text{m s}^{-1}$ as measured by CARTWRIGHT (1961). Z varied little from a typical value 43.5×10^{-6} Weber m^{-2}.

A mean derivative H' appropriate to the mean sea conductivity $\bar{\kappa}_1$ was estimated for each year separately, in the form of a least-squares regression coefficient of H on κ_1. These varied widely, as will be evident from Fig. 5, and results with low correlation ratio, as for example in 1961, are valueless. The results whose correlation ratios are above 0.80 are listed in Table 4.

Table 4.   *Estimates of sea-bed conductivity*

Year	Corr. ratio	$\bar{\kappa}_1 H'(\text{mV})$	Q	κ_0 (mho m^{-1})
1956	0.94	422	1.18	12×10^{-3}
1957	0.88	270	3.56	37
1960	0.82	245	4.16	44
1963	0.90	322	2.50	35
1964	0.83	184	6.43	66
1968/9	0.93	376	1.52	16
			Average:	33×10^{-3}

The rather wide variation in the six yearly estimates of κ_0 do not, of course, represent real changes in conductivity, but are due to the random elements in the data and the many simplifying assumptions made. However, it is interesting that all the estimates are greater than those listed in table 1 of BOWDEN (1956), which are near 10 in our scale. In fact, direct measurements of rock conductivity from boreholes in the vicinity quoted by ROBINSON (1977) give 20 as a typical value. Thus, our mean estimate of $33 \times 10^{-3}\,\mathrm{mho\,m^{-1}}$ may be nearer the correct value than those derived by Bowden's method.

6. RELATION TO LOCAL BAROMETRIC PRESSURE GRADIENTS

(a) *Empirical correlation*

As a final stage of this investigation, we examine the dependence of our fluctuating e.m.f. signal on those features of the local weather which may be expected to produce a general flow of water through the Strait. The principal objective of BOWDEN (1956) was similar to this. He showed convincingly that large anomalies in the e.m.f. are associated with a combination of northeasterly wind stress up the Channel and southwesterly gradients of sea-surface elevation between say Shoreham and Lowestoft. The former is a direct driver of up-Channel water flow, while the latter is a somewhat independent indicator of storm surges set in motion by remoter weather systems. We did not intend (or succeed) to improve on Bowden's formulation, but merely wished to find how much of the signal could be expressed in terms of a triad of atmospheric pressures, which are in general more reliable than recorded wind vectors, and are more easily available for the 9 years of our cable record.

The principle is similar to that investigated by CARTWRIGHT (1968) in the context of surges in elevation. Storm surges, of which the current through Dover Strait is just one manifestation, are driven by atmospheric pressure gradients and by wind stress. But wind stress is itself related to the pressure gradients in a manner which may be roughly represented by a simple linear formula, so if one can adequately represent the pressure gradient vector in time and space over a relevant sea area, one should expect to find that the given surge variable is related to its parameters in a fairly simple way. Cartwright worked in terms of a spatial array of eight pressure stations; here we attempt a much simpler scheme in terms of only three stations. We ignore, as Bowden did, the possibility of variable response in e.m.f. to different spatial patterns of flow through the Strait. That is the subject of Robinson's work.

Six-hourly pressure data were extracted for the whole period December 1955 to February 1965 from the Dutch Meteorological Institute publications for the light-vessels *Noord Hinder* (51°39′N, 2°34′E) and *Terschellinger Bank* (53°28′N, 5°08′E) and from the U.K. Daily Weather Reports for Gorleston (52°36′N, 1°43′E). All positions are shown in Fig. 1. Although rather to the north of the Dover Strait, they span a large area of shallow water in the Southern Bight which is sensitive to surge-generating winds. All data were carefully checked for translation and other errors by interpolations in both time and space. Axes were chosen to point southeast (x) and southwest (y), as seemed most appropriate to the geographical situation, and pressure triads converted to a time series of pressure gradients, $p_x(t)$, $p_y(t)$ in these directions by means of straightforward linear combinations. A third independent parameter, which could be described as the mean pressure over the area, was ignored as being irrelevant to the present case.

Before correlating to these variables, the tides were removed in an optimum way from the e.m.f. series by subtracting an hourly tidal synthesis computed in the 'response' formalism. This synthesis contained thirty-six complex terms, including a wide range of non-linear interactions, and terms describing annual modulations. The resulting reduction in variance in the tidal frequency bands has been described in Section 2 (see Table 1). The residual series was then convoluted with a low pass filter with unit response from 0 to 2 c/d and cut-off frequency near 3 c/d, in order to eliminate residual tides of species 4 and above which would otherwise be aliassed to low frequencies on thinning the series down to 6-hourly time steps as required for this comparison.

Calling the 6-hourly tide-free residual e.m.f. series $e'(t)$, a covariance matrix was computed from the 9-year span, involving cross-products between time-lagged pairs of the variables $e'(t)$, $p_x(t)$, $p_y(t)$, and enabling $e'(t)$ to be approximated with minimum squared residual by the form

$$e'(t) \approx \tilde{e}(t) = u_0 p_x(t) + u_1 p_x(t-\tau) + u_2 p_x(t-2\tau)$$
$$+ v_0 p_y(t) + v_1 p_y(t-\tau) + v_2 p_y(t-2\tau), \qquad (14)$$

where $\tau = 6$ hours and the coefficients u_n, v_n are evaluated by inversion of the 'normal' equations.

Table 5.   Coefficients in equation (14) and predictable variances

u_0	u_1	u_2	v_0	v_1	v_2	p.v.
1274	1588	−127	751	−1734	−415	7351
1208	1546		763	−2112		7324
2500			−1054			6134

Table 5 shows the values of these coefficients and the 'predictable variance' (p.v.) associated with the complete form $\tilde{e}(t)$ and with its equivalent when the terms with lags 2τ and τ are successively ignored. In this table, numerical values are such that $\tilde{e}(t)$ is in millivolt units and p_x, p_y are in millibars per nautical mile (1 n.mile = 1′ in latitude = 1853 m). Predictable variances are in mV^2. The total variance of the $e'(t)$ series is 28,229 mV^2.

With all six terms, the p.v. supplies only about 0.26 of the total variance, so the physical mechanism of flow generation is rather poorly represented by the variables used. This may be because of the lack of terms to describe the externally generated surge, included by Bowden. The spectral distribution of the p.v. is described by the difference in logarithmic spectral levels at the low-frequency end of Fig. 2. It is confined to the frequency range 0–1 c/d and is ineffective above 1 c/d. This means that the implied mechanism will inadequately represent changes in weather in a time scale shorter than a day.

The rather small increase in p.v. associated with the lagged variables shows that the major part of the mechanism represented is a quasi-static one. The two-term form on the bottom line of Table 5 is almost as good as the six-term form. From the two terms on the bottom line we see that positive potentials (corresponding to flow into the North Sea) are mainly produced by pressure gradients a little to the east of southeast, more precisely, increasing in the direction 112°. Surface winds in this area are typically 14° anti-clockwise of the geostrophic wind, so this pressure gradient produces the most effective wind as coming from some 10° west of south, which is physically appropriate. It might have been thought that a positive southwesterly pressure gradient would also produce flow (more strictly, acceleration) into the North Sea by its direct action on the sea surface, but this effect is evidently swamped by the geostrophic effect on both wind and water.

Since wind stress is proportional to the square of the speed, we attempted to improve the physical realism of equation (14) by replacing p_x, p_y by

$$p_x(p_x^2 + p_y^2)^{1/2}, \qquad p_y(p_x^2 + p_y^2)^{1/2}, \tag{15}$$

and carrying out the same correlation procedure. The resulting p.v. values were 5802 mV² (six terms) and 3798 mV² (two terms), so this form is distinctly inferior to the first formulation. Possibly, the larger stresses which are emphasized by (15) are also more heavily damped by bottom friction, so that at the assumed level of simplicity a linear law works best.

(b) Two selected weather situations

Finally, we depict the two weather situations which produced the greatest positive and negative anomalies in the residual e.m.f. on the cable. Figure 6 shows the 6-hourly

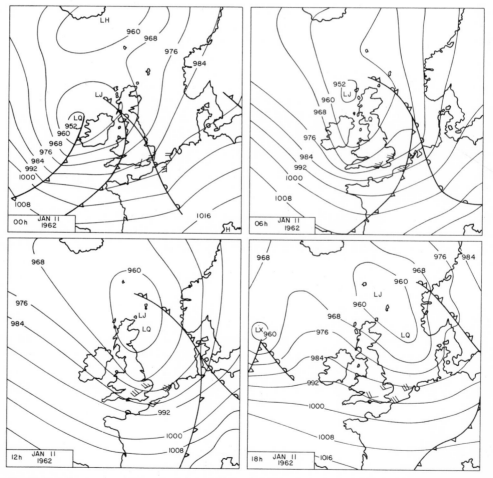

Fig. 6.   Six hourly sequence of weather charts about the time of greatest positive anomaly in cable e.m.f. (1070 mV).

progression of weather on 11 January 1962, copied from the U.K. Daily Weather Reports. A deep and complex depression over Scotland followed by the passage of fronts caused winds up to Gale Force 10 at *Noord Hinder* veering through southerly. Pressure gradients attained largest values in the southeasterly direction at 12 h, and then produced our largest recorded e.m.f. residual of 1070 mV. The relevant data are given in the upper section of Table 6.

Figure 7 and the lower section of Table 6 depict the corresponding situation when an equally large negative residual e.m.f. was recorded, on 18 October 1961. A deep depression over Denmark moved southwestwards towards a high-pressure region in mid-Atlantic. Gale Force 10 was again reached at *Noord Hinder*, but with winds mainly from the north. Pressure gradients were large and negative in p_x, large and positive in p_y, which are the situations we diagnosed as causing negative e.m.f.

The last column of Table 6 shows the values of $\tilde{e}(t)$ corresponding to the given pressure

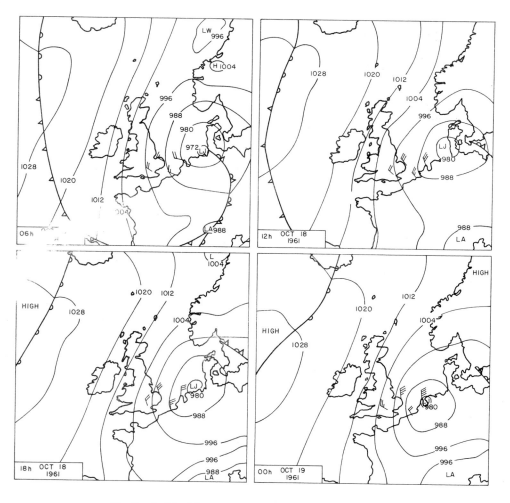

Fig. 7.   As Fig. 6 for greatest negative anomaly (-1040 mV).

Table 6.   Summary of two extreme weather situations

GMT (h)	Wind at N. Hinder		$10^3 p_x$	$10^3 p_y$	$e'(t)$	$\tilde{e}(t)$
	deg.	m s^{-1}				
			11 January 1962			
0	170	19	20	-34	210	-8
6	210	16	59	8	370	165
12	220	26	88	24	1070	221
18	240	19	38	73	460	189
			18 October 1961			
6	330	19	-64	76	-560	-256
12	330	19	-50	82	-540	-264
18	360	23	-69	64	-1040	-283
24	360	26	-81	52	-690	-313

gradients with the six-term expression in equation (14). The largest estimated values were $+220$ mV and -280 mV respectively, barely a quarter of the recorded values. Clearly a simple formalism such as (14) cannot do justice to the complicated mechanics of storm surge generation. A realistic description of the meteorologically induced e.m.f. in the cable requires a large-scale dynamical model of the North Sea of the type developed by HEAPS (1969), for example, coupled to a realistic electromagnetic model of the potential field set up by the flow in the vicinity of the Dover Strait, as developed by ROBINSON (1977). Such a simulation is being investigated by D. Prandle at I.O.S. Bidston.

7. CONCLUSIONS

In this rather discursive examination of an exceptionally long record of electrical potential across the Dover Strait in relation to associated measurements in the vicinity, the following positive facts have emerged.

(a) The tidal signal is about as good in its signal/noise quality as a typical tide gauge record in the area, except at very low frequencies.

(b) Solar frequency earth currents of non-tidal (ionospheric) origin are particularly weak, especially S_1, indicating that these currents are directed nearly parallel to the axis of the Strait. S_2 is possibly stronger, but all higher solar harmonics are detectable at a level of about 7 mV.

(c) The only clearly defined low-frequency tide is the non-linear MSf term. It agrees with calculations of the interaction between the electromagnetic effects of surface elevation and of flow.

(d) Difference in elevation between Dunkerque and Dover is a poor indicator of flow through the Strait because of its obliqueness, but there is some meaningful relationship to the strongest linear tidal components in the e.m.f. However, the non-linear tidal components are not similar.

(e) The finely resolved annual modulations in M_2 are very similar in e.m.f. and Dunkerque–Dover slope, but they are swamped by wider band modulations which show no similarity.

(f) The relationship between M_2 amplitude modulation in e.m.f. and sea conductivity is variable from year to year. Estimates of sea-bed conductivity from the regression coefficients between these two factors are variable, but suggest higher values than those suggested by Bowden.

(g) About a quarter of the non-tidal residual e.m.f. can be accounted for by a convolution in terms of time-lagged pressure gradients across the Southern Bight. Linear pressure gradients produce better correlation than terms which simulate quadratic wind stress. Positive gradients in the direction 112° have greatest tendency to produce positive e.m.f. However, such convolutions are inadequate in reproducing extreme storm conditions.

REFERENCES

BARBER N. F. (1948) The magnetic field produced by Earth currents flowing in an estuary or sea channel. *Monthly Notices of the Royal Astronomical Society, Geophysical Supplement* **5**(7), 258–269.

BOWDEN K. F. (1956) The flow of water through the Straits of Dover related to wind and differences in sea level. *Philosophical Transactions of the Royal Society*, A, **248**, 517–551.

BOWDEN K. F. and P. HUGHES (1961) The flow of water through the Irish Sea and its relation to wind. *Geophysical Journal* **4**(5), 265–291.

CARTWRIGHT D. E. (1961) A study of currents in the Strait of Dover. *Journal of the Institution of Navigation,* **14**(2), 130–151.

CARTWRIGHT D. E. (1968) A unified analysis of tides and surges round north and east Britain. *Philosophical Transactions of the Royal Society*, A, **263**, 1–55.

CARTWRIGHT D. E. (1975) A subharmonic lunar tide in the seas off western Europe. *Nature, London*, **257**, 277–280.

CARTWRIGHT D. E. and J. CREASE (1963) A comparison of the geodetic reference levels of England and France by means of the sea surface. *Proceedings of the Royal Society*, A, **273**, 558–580.

CARTWRIGHT D. E. and R. J. TAYLER (1971) New computations of the tide-generating potential. *Geophysical Journal of the Royal Astronomical Society*, **23**, 45–74.

CHAPMAN S. and J. BARTELS (1940) *Geomagnetism*, Clarendon Press, Oxford, 2 vols., 1007 pp.

CHERRY D. W. and A. T. STOVOLD (1946) Earth currents in short submarine cables. *Nature, London*, **157**, 766.

FARADAY M. (1832) Experimental researches in electricity—second series. *Philosophical Transactions of the Royal Society*, **122**, 163–194.

HEAPS N. S. (1969) A two-dimensional numerical sea model. *Philosophical Transactions of the Royal Society,* A, **265**, 93–137.

HUGHES P. (1969) Submarine cable measurements of tidal currents in the Irish Sea. *Limnology and Oceanography*, **14**(2), 269–278.

LONGUET-HIGGINS M. S. (1949) The electrical and magnetic effects of tidal streams. *Monthly Notices of the Royal Astronomical Society, Geophysical Supplement*, 5, **8**, 285–307.

MALIN S. R. C. (1973) Worldwide distribution of geomagnetic tides. *Philosophical Transactions of the Royal Society*, A, **274**, 551–594.

MUNK W. H. and D. E. CARTWRIGHT (1966) Tidal spectroscopy and prediction. *Philosophical Transactions of the Royal Society*, A, **259**, 533–581.

PRANDLE D. and A. J. HARRISON (1975) Relating the potential difference measured on a submarine cable to the flow of water through the Strait of Dover. *Deutsche Hydrographische Zeitschrift*. **28**(5), 207, 207–226.

RAMSTER J. W., T. WYATT and R. G. HOUGHTON (1973) Towards a measure of the rate of drift of planktonic organisms in the vicinity of the Strait of Dover. (Unpublished MS. presented to the International Council for the Exploration of the Sea, Hydrographical Commission.)

ROBINSON I. S. (1976) A theoretical analysis of the use of submarine cables as electromagnetic flow meters. *Philosophical Transactions of the Royal Society*, A, **280**, 355–396.

ROBINSON I. S. (1977) A theoretical model for predicting the voltage response of the Dover–Sangatte cable to typical tidal and surge flows. In, *Voyage of Discovery*, M. V. ANGEL, editor, *Deep-Sea Research* Supplement to Vol. **24**, 367–391.

SANFORD T. B. and R. E. FLICK (1975) On the relationship between transport and motional electric potentials in broad shallow channels. *Journal of Marine Research*, **33**(1), 123–139.

TALBOT J. W. (1975) Changes in plaice larval dispersal in the last 15 years. *Rapport et procès-verbaux des réunions. Conseil permanent international pour l'exploration de la mer* (in press).

WOLLASTON C. (1881) Discussion of paper by A. J. S. Adams entitled 'Earth currents' (same vol.). *Journal of the Society of Telegraphic Engineers and of Electricians*, **10**, 50–56.

A theoretical model for predicting the voltage response of the Dover–Sangatte cable to typical tidal flows

I. S. Robinson

Institute of Oceanographic Sciences, Bidston Observatory, Birkenhead*

Abstract—A theoretical model is described which is capable of predicting the potential difference experienced across the ends of the submarine telephone cable between Dover and Sangatte, given a particular distribution of velocity in the Dover Strait and adjacent sea areas. The model can cope with seasonally varying sea conductivities and incorporates a three-dimensional description of the electrical conductivity of the rock strata beneath the Strait. It is also designed to include the effect of the tidal heights on the instantaneous response of the cable. It is found that the seasonal variation of the M_2 response due to sea conductivity changes can be modelled very realistically. The model is used to determine the extent to which the southern North Sea and the English Channel contribute to the cable signal.

Using realistic tidal flows derived from a numerical nonlinear dynamical model of the area, the cable response is obtained, both to individual tidal harmonics over a single cycle, and to the major tidal harmonics interacting over a 15-day period. Theoretical calibrations (voltage/water transport) are produced which agree well with the empirical calibration of the cable. It is shown that this calibration is valid over the diurnal and semi-diurnal wave band. However, errors are predicted if this calibration is applied to the shallow-water harmonics, and the principal tidal calibration is shown to be of little value in interpreting the long period and mean voltage signals. The consequences of calibrating the voltage in terms of the mean across the Strait of the depth mean velocity are also explored.

1. INTRODUCTION

A KNOWLEDGE of the flow of water through the Dover Strait is of considerable importance to oceanographers studying the North Sea. The net transport controls the flow of nutrients and sediment, and affects the salinity and temperature of the southern North Sea. At times when there is a large surge produced in the North Sea by suitable wind conditions the surge flow through the Dover Strait can have a direct effect on the surge heights recorded in the southern North Sea (PRANDLE, 1975).

One way of obtaining an estimate of the flow through the Strait is to use submarine telephone cables to measure the voltage produced across the Strait by the movement of the sea in the earth's magnetic field. BOWDEN (1956), CARTWRIGHT (1961) and CARTWRIGHT and CREASE (1963) used this method, assuming that the instantaneous transport through the Strait could be related to the instantaneous voltage by a constant of proportionality which varied only with the seasonal variation of sea conductivity, provided that allowance was made for a constant potential due to electrochemical effects and steady earth currents. Unfortunately there is no unique relationship between the transport of a stream and its generated electric potential field, as LONGUET-HIGGINS, STERN and STOMMEL (1954) and

---

* Present address: Department of Oceanography, University of Southampton.

SANFORD and FLICK (1975) demonstrate. Moreover, ROBINSON (1976) has shown that it is not necessarily correct to assume that the calibration of a cable, made on the basis of observations of M_2 tidal frequency stream currents correlated to M_2 voltage signals, can be applied either to other tidal frequencies or to residual flows. This is borne out by the recent observations of PRANDLE and HARRISON (1975). The voltage observed across the cable is produced by the electric field **E**, which is related to $\mathbf{v} \times \mathbf{B}$, the motionally induced field, and the electric current density **j** (throughout the whole sea and the earth beneath it) by Ohms law,

$$\mathbf{j} = \sigma(\mathbf{E} + \mathbf{v} \times \mathbf{B}).$$

v is the velocity of the conducting medium, σ is the conductivity and **B** the magnetic flux density. BEVIR (1970) has shown that the voltage between two points in the conducting fluid (in our case the ends of the submarine cable) can be expressed as

$$U = \sum \mathbf{v} \cdot \mathbf{W} \, d\tau. \tag{1}$$

W is a weighting vector such that $\mathbf{W} = \mathbf{B} \times \mathbf{J}$ where **J** is a vector dependent only on the topography and conductivity of the sea and the earth beneath. **J** is equivalent in magnitude and direction to the electric current distribution which would result if unit current were injected into the sea at one end of the cable and extracted at the other, in the absence of $(\mathbf{v} \times \mathbf{B})$ induced fields. The volume integral covers the whole of space, and should be evaluated throughout the whole sea volume where **v** is non-zero. In practice, in most topographies the value of the integral is dominated by the contribution from the sea area near to the cable section, and contributions from the sea beyond a distance of about one cable length away from the cable section can be ignored. This weight vector method eliminates the need to evaluate the actual electric currents flowing in the sea and earth and thus differs essentially from the approach of SANFORD and FLICK (1975) to similar problems.

ROBINSON (1976) showed that for cables in the Irish Sea, **W** could be evaluated once for a given sea conductivity distribution. It was then easy to use this **W** distribution with various velocity distributions to show that the theoretical response of the cable varied considerably with different distributions of velocity flow, tidal or non-tidal in origin. In the case of the Dover Strait, the problem is complicated because the tidal range is not negligible in comparison with the mean sea depth. In this situation the weight vector distribution varies over a tidal cycle and produces a modulating effect on the voltage signal. Robinson briefly explored this modulation using crude data to represent tidal elevations and currents and found it to be significant. It is the purpose of this paper to describe a more detailed model used to evaluate the response of the Dover cable to flows of tidal origin, in conjunction with the results of a fine-mesh numerical model of the fluid dynamics of the area, produced by PRANDLE (1977). In a companion paper to this, ALCOCK and CARTWRIGHT (1977) describe the analysis of 10 years of voltage data from the Dover–Sangatte cable, thus providing a means of testing the validity of this model.

The ultimate objective of this work is to evaluate whether the Dover–Sangatte cable is useful for measuring residual drifts and storm surge flows. That part of the residual due to tidal interaction is dealt with in this model. The response to storm surge flows requires coherent surge velocity data from a dynamical model which is not yet available.

It is intended, however, to model the flows of an actual period during the JONSDAP 1973 exercise and to use the present model to predict the theoretical voltages for comparison with the observed voltages over the same period. This will be presented in a later paper.

2. THE MODEL FOR EVALUATING THE WEIGHT VECTOR

For any realistic model of the response of an actual cable situation, an analytical solution of equation (1) is not possible. Once **W** has been defined for a given topography and state of tidal elevation, the numerical solution of (1) is a simple matter. It is not so easy to determine **W** in a model incorporating as much information as possible about the sea depth and the conductivity distribution of the sea and the earth beneath. Basically the problem is to solve numerically for the potential distribution in the sea when unit flux enters the region at one end of the cable and leaves at the other.

Using Bevir's terminology, let the 'virtual current' density be **i** and the 'virtual potential' be ϕ. If the electrical conductivity is σ then

$$\mathbf{i} = -\sigma\nabla\phi \tag{2}$$

Also,

$$\operatorname{div}\mathbf{i} = 0 \tag{3}$$

everywhere except the cable ends, i.e. the source and sink points, at which

$$\operatorname{div}\mathbf{i} = +1 \quad \text{and} \quad -1, \text{ respectively.} \tag{4}$$

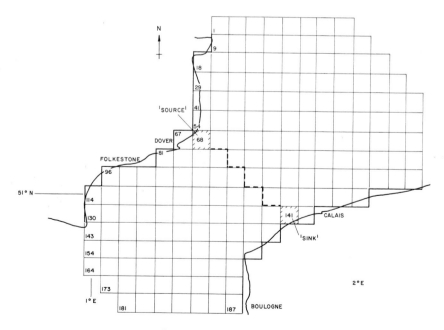

Fig. 1.   The basic sea model.

The sea model

Because the sea depths are very small compared with the horizontal length scales involved, the virtual current density distribution in the sea can be treated as two-dimensional, the vertical component being ignored. The numerical solution for ϕ and $\mathbf{i}$ is therefore split into two parts, a sea model and an earth model as in ROBINSON (1976).

The numerical solution for the sea virtual current $\mathbf{i}_s$ is based on the finite difference grid shown in Fig. 1, parameters being specified in Table 1. This was designed to correspond to Prandle's fine-mesh dynamical model which has grid spacings of half the size, i.e. four grid 'squares' of the dynamical model fill one square of the virtual current model. The fine-mesh dynamical model is itself designed to intermesh with a coarse grid model by Prandle covering the southern North Sea and eastern English Channel as in Fig. 2 (PRANDLE, 1975).

Table 1.   *Parameters used in the virtual current models*

(a) Main sea model

Grid spacing, north–south		4.94 km		
east–west		4.66 km		
No. of grid squares		187		
Vertical component of earth's magnetic field		$4.36 \times 10^{-5}\,\mathrm{Wb\,m^{-2}}$		

(b) Supplementary sea models

Grid spacing, north–south		12.36 km		
east–west		11.66 km		
Sea conductivity		$4.0\,\mathrm{ohm^{-1}\,m^{-1}}$		
No. of grid squares, English Channel model		207		
North Sea model		238		
No. of boundary squares $\{$ English Channel		4		
where ϕ is defined, $\{$ North Sea		5		

(c) Earth virtual current model

Layer	1 (top)	2	3–5	6–8
Horizontal grid spacing, north–south (km)	4.94	4.94	4.94	9.88
east–west (km)	4.66	4.66	4.66	9.32
Depth (km)	0.4	1.0	5.0	10.0

It is convenient to deal with the total virtual current I^E and I^N crossing each grid square side in an easterly and northerly direction respectively. There is also a current C which flows into the earth from each sea square. If the grid notation is as shown in Fig. 3, with ϕ defined at the centre of each square and I defined as shown at the centre of the sides, the finite difference equations are:

$$I_{m,n}^{E} = \left(\frac{\sigma_{m,n} + \sigma_{m-1,n}}{2}\right)(\phi_{m-1,n} - \phi_{m,n})d_{m,n}^{E}\, l_{\mathrm{NS}}/l_{\mathrm{EW}},$$

$$I_{m,n}^{N} = \left(\frac{\sigma_{m,n} + \sigma_{m,n-1}}{2}\right)(\phi_{m,n-1} - \phi_{m,n})d_{m,n}^{N}\, l_{\mathrm{EW}}/_{\mathrm{NS}},$$

and

$$I_{m,n}^{E} + I_{m,n}^{N} - I_{m+1,n}^{E} - I_{m,n+1}^{N} = C_{m,n}.$$

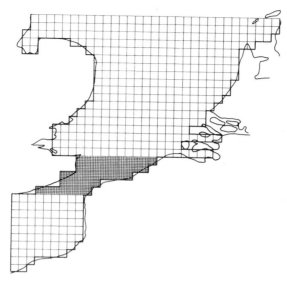

Fig. 2.   The grids for Prandle's dynamical model.

σ is defined at the centre of each square. $d_{m,n}^{E} = h_{m,n}^{E} + \zeta_{m,n}^{E} + D$ where $h_{m,n}^{E}$ is the chart depth at the centre of the appropriate eastern grid side, $\zeta_{m,n}^{E}$ is the tidal elevation obtained from Prandle's model at the same point and D is the height of Prandle's model datum above chart datum. $d_{m,n}^{N}$ is similarly the instantaneous depth on a northern grid side.

Elimination of I produces a set of equations, one for each grid square, which can be written in matrix form

$$A \cdot \phi = c \qquad (5)$$

A is a square matrix composed of terms in σ and d, most of the elements in each row being zero and the sum of the rest being zero. c is a column vector containing C, and to satisfy (4) there is an additional term in c of $+1$ and -1 at the source and sink points respectively. To produce a unique solution for ϕ it is necessary to fix the potential of one grid square, for which $a_{pq} = 0 (p \neq q)$, $a_{pp} = 1$ and $C_p = 0$. It was found most suitable to the numerical solution for this square to be in midstream well removed from the cable location.

Supplementary sea models

In ROBINSON (1976) it was assumed that the boundaries of the model were chosen far enough away from the cable ends for the virtual current passing outside the limits of the model to be negligibly small. In the Dover cable topography, however, such an assumption is not easily justified, since the opposite coastlines diverge away from the cable section, in contrast to the Irish Sea topography where the coastlines tend to enclose the virtual current within a limited area. Moreover, even if only a small fraction of the total virtual current passes outside the limits of the model, it may be distributed across a very large area of the English Channel and southern North Sea with the consequent possibility of contributing significantly to the integral (1). It was therefore

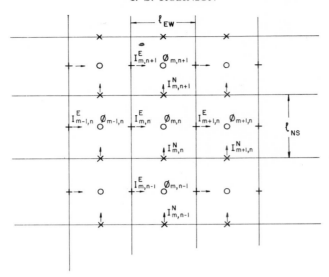

Fig. 3.   Notation of virtual current model.

considered necessary to construct supplementary virtual current models of the North Sea and English Channel, as shown in Fig. 4 (details of parameters in Table 1). These were matched to the large grid model of Prandle to facilitate calculation of the cable voltage due to these parts of the sea.

For simplicity the earth currents are ignored for these models. The potential at the squares bounding the main model is fixed by the potential distribution obtained in the main model. A set of equations of the same form as (5) is obtained, in which c is zero except in the equations referring to the fixed potential squares where it is set at the fixed potential. The elements a of the matrix are dependent on the depth, and for simplicity the conductivity was considered to be uniformly $4\,\mathrm{ohm^{-1}\,m^{-1}}$. The matrix rows corresponding to the fixed potential squares are

$$a_{pq} = 0(p \neq q), \qquad a_{pp} = 1.$$

Now

$$\phi = A^{-1}c.$$

The numerical inversion of A was performed once for each model, using a standard Gauss–Jordan algorithm. It was then possible to express the virtual current flowing through each boundary square out of the main model as

$$k_r = d_r \sum_{s=1}^{N_b} b_{rs} \phi_s, \tag{6}$$

N_b being the total number of boundary squares, ϕ_s the potential in those squares, fixed by the main model, d_r a term depending on the local depth and the conductivity, and where

$$b_{rs} = (a_i)_{r,s}(r \neq s) \quad \text{and} \quad b_{rr} = (a_i)_{rr} - 1,$$

a_i being elements of A^{-1}.

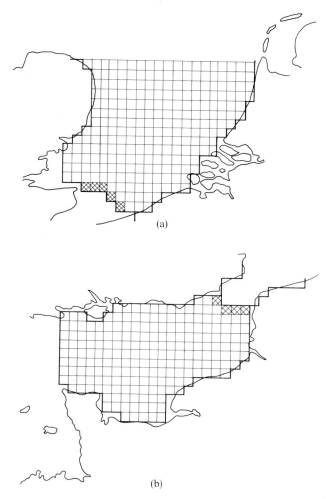

Fig. 4. Supplementary sea models (a) North Sea. (b) English Channel. The shaded squares are those where the potential is defined from the main model.

For each evaluation of the ϕ distribution in the main model, the resulting current leaking out through the boundaries and back again could be evaluated, and incorporated into the main model in the value of C for those grid squares at the boundary of the main model. Consequently the main model could be corrected for the effect of these leakage currents in the course of the iteration procedure devised to compensate for the earth currents.

Earth currents

In the study of the Irish Sea cables, Robinson (1976) included the effect of the induced electric currents circulating through the earth beneath the sea by calculating the virtual current passing into the earth in each grid square of the sea model. This was achieved by a method which required the simplification of asuming the earth conductivity to be uniform. Since the geology of the upper strata in the Dover Strait is relatively well known, it was decided in this case to employ a fully three-dimensional model with provision for earth conductivity varying in space. It is clear from the analytical solutions of LONGUET-HIGGINS (1949) that electric currents due to the motional electric potential are significantly large to a depth of the same order as the width of the sea channel. Therefore the eight-level model shown in Fig. 5 was adopted. Boreholes in the area penetrate to around 1000 m, so that the structure is quite well known to this depth. Between about 400 m and 1000 m there is carboniferous limestone or older rocks of similar conductivity throughout the whole area, and so it is worth while to include the spatial variation of earth conductivity only in the top layer where the conductivity differs considerably between the southwest and northeast. Below 400 m each layer is assumed to have uniform conductivity.

No detailed geological map could be found of the region under the Strait of Dover. However, the sketch of Fig. 6 was produced using the 1 inch to 25 miles geological map of the British Isles as a basis, together with detailed information about the geology of southeast Kent from the I.G.S. 1:50,000 series geological maps (sheet 290—Dover, and sheet 305/6—Folkestone). Knowledge about the French side of the Strait was obtained

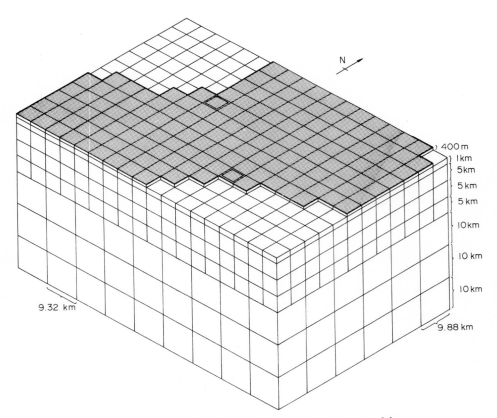

Fig. 5.   The three-dimensional grid for the earth current model.

Table 2.   Electrical conductivities used in the earth model

Rock type	Observed conductivity used to calculate the conductivity of the top layer, ohm^{-1}m^{-1}
Middle chalk	0.01
Lower chalk	0.02
Gault	0.1 (varies from 0.07 to 0.4)
Lower Cretaceous and Jurassic	0.05 (varies from 0.03 to 0.3)
Coal measures and carboniferous limestone	0.03

Model layer	Conductivity assigned to lower layers of model, ohm^{-1}m^{-1}
2	0.02
3–5	0.005
6–8	0.001

from AGER and WALLACE (1966) and WALLACE (1968). This information had to be interpolated under the sea except near the line of the cable itself, where detailed offshore
information is available along the line of the proposed channel tunnel, DESTOMBES and
SHEPHARD-THORN (1971).

Actual resistivity data contained in borehole records obtained by the I.G.S. applied
geophysics unit (A. J. BURLEY, personal communication) indicated the average values of
conductivity for different types of rock in the Dover Strait region, shown in Table 2. Using
these, and the depths of the different strata obtained from Fig. 6, a vertical and a
horizontal conductivity was deduced for each of the grid boxes in the top layer. The
horizontal grid spacing of the top five layers corresponded to that of the sea model.

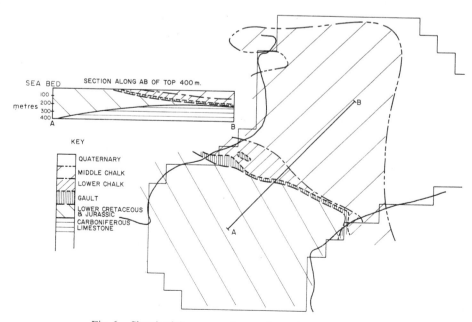

Fig. 6.   Sketch of the near-surface geology of the Dover Strait.

The topmost layer had boundaries matching the sea model but the remaining layers had a rectangular boundary, both for ease of computation and to model the earth currents which flow under the land areas adjacent to the cable ends. The second layer, 1 km thick, was assigned a uniform conductivity of $0.02\,\text{ohm}^{-1}\,\text{m}^{-1}$ to model the carboniferous limestone and Devonian rocks at that depth. Below this, there is little knowledge of actual conductivities and informed guesswork is necessary. Using the information about conductivity of rocks given by BARSUKOV (1967) and ignoring the fact that the lower boundary of the model lies below the Mohorovicic discontinuity (where a rapid increase of conductivity with depth is likely) the values given in Table 2 were assigned to the lower layers, with a view to adjusting them later to tune the model to produce realistic estimates of cable voltage. In fact the initial values chosen produced very satisfactory predictions of M_2 cable response, as will be seen later, and no adjustment was therefore necessary. The lower three layers, being 10 km deep, had twice the horizontal grid spacing of the upper layers.

The model was used to obtain the virtual potential distribution throughout the earth given the potential distribution at the top of the uppermost layer (i.e. at the sea bed— the potential of the sea model) and thence to calculate the virtual current flowing between sea and earth in each top layer grid box. The governing equation is

$$\mathbf{V} \cdot \sigma \nabla \phi = 0. \tag{7}$$

This was solved using a successive over-relaxation method, proceeding a layer at a time starting at the top. The potential ϕ_p was predicted for a grid box using (7) and the most recently obtained values of ϕ in the surrounding boxes. This was compared with the previous value ϕ_0 in that box, and the updated value assigned to that location was $\phi = \phi_0 + f(\phi_p - \phi_0)$ where f, the over-relaxation factor used, was 1.5. With initial values of zero in all boxes (which was approximately the mean of the sea potential distribution) convergence occurred after around fifty iterations for a typical sea potential distribution, and further updating after the sea potential had been adjusted was achieved typically with less than ten iterations.

The overall computational scheme

With the need in this model to recompute the virtual current distribution at every time step, inversion of the matrix A in equation (5) for every step is too costly a method in computer time, and a more economical approach was developed. For the first time step of a particular time series, the matrix A is formed and numerically inverted. A first approximation to ϕ is then obtained by assuming that C is zero for every grid square (i.e. the earth conductivity is zero). From these values of ϕ the appropriate current flowing through the sea boundaries can be evaluated using equation (6), and the earth currents required by the distribution of ϕ in the sea can be obtained using the relaxation model described above. Predicted values, C_p, of the current leaving each sea square can then be obtained. Now in the topography of the Dover Strait, particularly because of the relatively shallow-water depth, nearly half of the total virtual current passing from source to sink travels through the earth. Under these conditions, a simple iterative procedure using C_p to form the column vector c in $\phi = A^{-1} \cdot c$ updating ϕ each time, is found to be unstable. Stability is achieved by forming c from $C = C_0 + f(C_p - C_0)$ where C_0 is the previous value of C and f in this case is 0.3. It is also necessary for stability to add a small correction quantity to each value of C to ensure that the summation of C over all sea squares is zero. Adopting this approach, satisfactory convergence was achieved within eight iterations, and the actual values of I^E and I^N could be calculated.

For subsequent times, as the depth varies, a new matrix A is formed for each time step. The final value of c at the previous time step is used as a first estimate of c in the new time step. Since the variation in A is small from one time step to the next, inversion of the new A can be avoided by using a Gauss–Seidal iterative technique to solve equation (5) for ϕ. The initial values of ϕ used in this numerical algorithm are the final values of ϕ in the previous time step. c can be updated using the earth model, the corresponding ϕ calculated by the Gauss–Seidal method, and the process repeated until convergence is achieved. Typically only four iterations are necessary, and the method is at least an order of magnitude faster than inverting A each time step.

At each time step, the values of I^E and I^N obtained are interpolated linearly to give the weight vector at the centre of each grid square of Prandle's fine mesh model, to enable the integral (1) to be performed numerically, summing over all the grid squares within the boundary of the main cable model. Thus a predicted cable voltage V is obtained, corresponding to the v distribution at that time step. Also, the volume transport, T, through the Strait is calculated at each time step, numerically performing the integral $\int d(\mathbf{v} \times d\mathbf{l})$ across the Strait, where $d\mathbf{l}$ is an element of the line joining the two ends of the cable. It should be noted that d, as well as v, is varying with time. From these the instantaneous calibration of the cable is obtained as V/T.

3. THE SEASONAL VARIATION OF CONDUCTIVITY

It was noted above that nearly half the virtual current passes between the cable ends through the earth rather than the sea. Thus the predicted voltage will be little more than half that which would be expected if the earth were an electrical insulator. Under these conditions the predicted voltage varies considerably with the values of earth conductivity used in the earth current model. The present impossibility of obtaining true values of earth conductivity at all the depths required by the earth model means that this method of modelling cable response is not suitable for actually obtaining a calibration factor by which to interpret observations of voltage in terms of flow. However, if the earth conductivity can be adjusted so that the model predicts voltages which are similar to those observed, then the results of experiments performed with the model can be considered to describe with reasonable accuracy the real response of the actual cable voltages under the same conditions as those being modelled.

It was pointed out in ROBINSON (1976) that a good test of whether the earth conductivity is being suitably estimated is to model the effect of the seasonally varying sea conductivity on the voltage response to the basic M_2 tidal flow. For the areas covered by this model, maps of mean monthly temperature and salinity distributions, averaged over 50 years in the first half of this century, are available in CONSEIL PERMANENT INTERNATIONAL POUR L'EXPLORATION DE LA MER (1962). To simplify the calculations in this and some of the other experiments performed with the model, the tidal variation of elevation was considered to be zero and two velocity data sets were used, consisting of (u_a, v_a) and (u_b, v_b) respectively, for each sea square. The velocities used were obtained

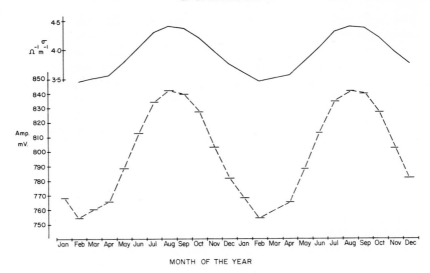

Fig. 7. The seasonal variation of the M_2 signal. Ordinate of upper graph; sea conductivity, $ohm^{-1} m^{-1}$. Ordinate of lower graph; signal amplitude, mV.

from a run of Prandle's model with realistic M_2 and M_4 input, the velocity components which resulted in each grid square being Fourier analysed and the pure M_2 component being represented as $(u_A \cos \omega t + u_B \sin \omega t)$ in the easterly direction and $(v_A \cos \omega t + v_B \sin \omega t)$ in the northerly direction. With no tidal variation of elevation, there is no modulation effect, and the pure M_2 input to the cable model results in a pure M_2 output of predicted voltage.

The model was run for the twelve monthly distributions of sea conductivity, obtaining the response to the cosine and the sine phase of velocity for each month. From these the amplitude and phase of the predicted voltage response to the pure M_2 input were calculated and are presented in Fig. 7 and Table 3 along with the value of sea conductivity at the centre grid square of the Dover Strait.

Table 3.   *Seasonal variation of* M_2 *cable response*

Month	Average sea conductivity ($ohm^{-1} m^{-1}$)	Amplitude of M_2 signal (mV)			Phase (deg.)
		Cos phase	Sine phase	Amplitude	
January	3.60	710	−294	768.5	337.5
February	3.46	697	−289	754.5	337.5
March	3.52	702	−292	760.3	337.4
April	3.57	707	−294	765.7	337.4
May	3.80	728	−303	788.5	337.4
June	4.04	752	−309	813.0	337.7
July	4.30	770	−322	834.6	337.3
August	4.39	777	−325	842.2	337.3
September	4.37	775	−324	840.0	337.3
October	4.20	764	−318	827.5	337.4
November	3.96	741	−309	802.9	337.4
December	3.75	722	−300	781.9	337.4

There is a negligible seasonal variation in the phase relationships between the calculated transport through the Strait and the predicted voltage, which is worth knowing in the analysis of year-long periods of cable data, since it means that any multiplier used to compensate for sea conductivity variation need only be a scalar rather than a vector. The lack of phase variation also indicates that the seasonal variation does not alter the relative distribution of the weight vector in the area. Its effect must therefore be the same whatever velocity distribution is considered. There is, of course, a considerable variation of amplitude response. Since the transport remains constant throughout the year in this experiment, the amplitude is presented in Fig. 7 directly as the predicted voltage. There is a seasonal variation between 754 mV in February and 843 mV in August. Figure 7 should be compared with Fig. 5 of ALCOCK and CARTWRIGHT (1977) which shows the seasonal variations of the amplitude of the M_2 component of actual cable observations. A direct comparison between the model and observation is thus possible, and the agreement is remarkably good, both in the absolute values of voltage predicted, and in the size of the seasonal variation. It therefore seems that appropriate values of earth conductivity have been chosen, and further adjustment of these is unnecessary. It must be borne in mind when comparing the model and observed results that the model is governed by average monthly conductivities and is likely to give a smoother seasonal variation than actually occurs year by year. There is, however, a direct correlation between the sample sea conductivity and the predicted voltage, which could be useful in analysis of data. The model predicts this to be of the form $V = 429 + 94.3\sigma$. Because of the uncertainty of the cause of the long period drift of the data presented by Alcock and Cartwright the model results should be compared with the central years of that data set, which are likely to be the most reliable.

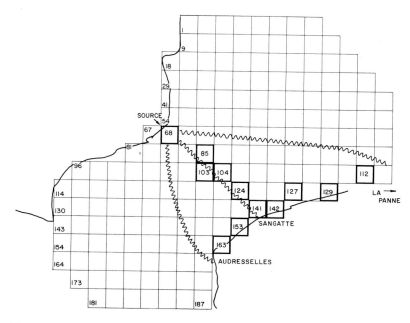

Fig. 8.   Positions of the different cable end locations tested in the model. The locations of three actual cables is also shown.

4. VARIATION OF CABLE LOCATION

From time to time it is necessary, for reasons relating to the primary function of the cables as telecommunication channels, to monitor a different cross-channel cable. The data referred to by Alcock and Cartwright is from the Dover–Sangatte cable, but some recent measurements have been made on the Dover–Audresselles cables. In future, it may be that data is obtained from the Dover–La Panne cable. It is important to know whether these different cable locations produce very different amplitude and phase responses of the cable voltages to the same M_2 flow distribution, since this governs whether data from different cables can be incorporated into the same time series. It also happens that the cable is sometimes broken in midstream, and a different voltage response is observed in the time series until the cable is repaired. It is possible using the present method to model both these situations and gain some knowledge which will assist the analysis of cable records. All that is necessary is to change the position of the source or sink square in the virtual current model, and a new cable location is defined.

Using the June sea conductivity distribution, and the pure M_2 velocity data set with zero tidal elevation described above, the model was run with the source always at St. Margaret's Bay, but the sink point in ten different locations as well as the usual Sangatte location. These are shown in Fig. 8 along with the lines of those Dover Strait cables used as oceanographic flowmeters. The results of these runs are given in Table 4. In Fig. 9 the amplitude and phase of the predicted voltage signal have been plotted against the geographical location of the cable end, in Fig. 9(a) along the French coast and in Fig. 9(b) across the Strait.

In Fig. 9(a), the local variability between having the sink point at squares 141, 142, 127 and 129 is probably due to the limitations of the finite differences grid size. However, the general trend which is evident is that with the French end of the cable at Sangatte (141) the voltage response is a maximum. Moving the cable end southwestward results in a rapid fall off in response, and towards the east a lesser decrease. The phase of the signal increases steadily from southwest to northeast, which would be expected since the tidal phase progresses in a similar way. It may therefore be concluded that the response

Table 4.   Variation of cable response with the position of the eastern end of the cable

Sink square	Amplitude of M_2 signal (mV)			Phase (deg.)
	Cos phase	Sine phase	Amplitude	
85	268	−62	175.1	347.0
103	377	−85	386.5	347.3
104	485	−116	498.7	346.6
124	672	−195	699.7	343.8
141	267	−313	828.4	337.8
112	774	−255	814·9	347.8
129	726	−249	790.6	341.1
127	770	−269	815.6	340.7
142	732	−290	787.4	338.4
141	267	−313	828.4	337.8
153	698	−302	760.5	336.6
163	693	−297	754.0	336.8

of the Audresselles cable is likely to be nearly 10% less than that of the Sangatte cable, and about 1° advanced in phase. If the trend is extrapolated eastwards beyond the limits of the model, then it might be expected that the La Panne cable would have a response about 3% less than the Sangatte cable, and about 5° later in phase. However, to calculate this accurately, it would be necessary to construct a new model extending much further northeastwards to include La Panne and beyond. Even with the cable end at square 112 there is a considerable virtual current flow outside the northeastern boundary, and so a

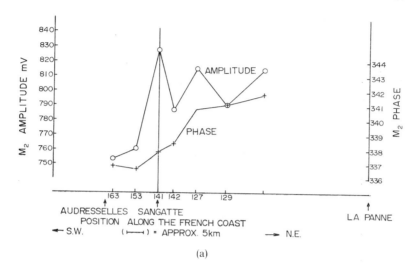

(a)

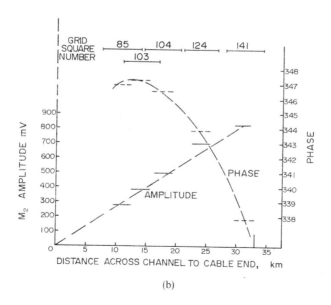

(b)

Fig. 9.   Variation of the M$_2$ signal with cable end position. (a) Along the French coast; (b) Across the Dover Strait.

possibly significant proportion of the integral (1) is being ignored. It may be, in fact, that the response of the La Panne cable is greater than the Sangatte cable.

Figure 9(b) shows an almost linear decrease in cable response with distance across the channel, as would be expected from the simple theories of cables as flowmeters. There is, however, a noticeable increase in phase lag as the east end of the cable moves towards the English coast, presumably due to the phase of tidal currents being later towards the English coast. Thus if the cable were cut in midstream the amplitude would be reduced to about a half, but the phase would be expected to increase by about 9°.

5. THE CONTRIBUTION TO THE CABLE VOLTAGE OF INDUCTION OUTSIDE THE MODEL AREA

It has been noted already that the relative position of the Strait of Dover, opening out into wide seas on either side, casts some doubt on the validity of assuming that only the area of sea covered by the main model contributes significantly to the voltage observed across the cable. In all the experiments described in this paper, virtual current is permitted to leak through the sea boundaries of the model, returning through the boundary closer to the sink point. However, this virtual current is ignored in the evaluation of integral (1) and its contribution is lost. It was therefore considered desirable to take a particular case and evaluate the contribution of the English Channel and North Sea to the predicted voltage at the cable. The supplementary models were supplied with the virtual potential at the boundaries obtained from the main model run which evaluated the response to cosine and sine phases of M_2 velocities, with zero tidal elevation and March sea conductivities. The virtual current distribution in the supplementary models was evaluated, and the integral (1) performed in the area of the supplementary models with the cosine and sine phase velocities obtained from Prandle's large mesh model. For the North Sea, the virtual current model defined the northern boundary for the integration, whilst for the English Channel Prandle's model defined the western limit (see Figs. 2 and 4). The results are presented in Table 5. The greatest contribution comes from the North Sea, with an additional signal having an amplitude 2.5% of that predicted in the main model, and lagging it in phase by 53°. The contribution of the English Channel model is 1% with a 40° phase lead. Their joint effect on the predicted voltage is to raise its amplitude by 0.017 volt (2.2%) and retard its phase by 0.7°. In the context of the investigations of this paper, notably the modulation of the voltage signal and the variation of cable response with frequency, the contribution of the supplementary areas

Table 5.   Contributions to the cable voltage from the supplementary sea models

	Amplitude of M_2 signal (mV)			Phase (deg.)
	Cos phase	Sine phase	Amplitude	
Main model only	716.0	−296.0	775.0	337.6
North Sea model	16.4	9.8	19.3	30.9
English Channel model	3.6	−6.9	7.8	297.6
Sum of three models	736.0	−293.0	792.0	338.3
Comparison of sum with main model only			+2.2%	+0.7°

can safely be ignored. Any tidal modulation of the supplementary signal will be very small because the water is much deeper in those areas. Any variation of the signal due to the different flow distribution of different tidal frequencies in the English Channel and in the southern North Sea is likely to be a similar proportion of the supplementary signal as the variations due to the main model are of the main signal. Therefore any contribution to the overall variation will be of order 2%, which is much smaller than the likely accuracy of the variations calculated in the main model.

However, it is worth noting that as much as 2.2% of the voltage recorded at the cable comes from induction in sea areas so far removed from the cable.

6. THE CABLE RESPONSE TO INDIVIDUAL TIDAL HARMONICS

Having determined that the model was predicting realistic seasonal variations in the cable response, it could now be used to evaluate aspects of the cable response which cannot be explored observationally. Firstly a series of experiments was performed to test the response of the cable to individual tidal harmonics. In each case the velocity and elevation distribution were supplied at each of twenty-five time steps covering one cycle of the particular harmonic. This data was obtained from corresponding runs of Prandle's model in which the boundary conditions were governed by observed elevations of the particular harmonic. Since each harmonic was run separately in this case, the non-linearity in Prandle's model would generate only a mean term and simple overtones of the basic harmonic. For the case of M_2, to ensure sensible results, M_4 was included in the boundary conditions of Prandle's model. Thus the information supplied to the cable model represented the tidally generated mean flow, the tidal flow and higher harmonics which would be present if only one particular harmonic from the gravitational tide generating potential were present in the Dover Strait and surrounding seas. The modulation effect in the cable model produces exactly the same set of frequencies, and so the twenty-five values of predicted voltage, and transport calculated directly from Prandle's model results, could be Fourier analysed to determine the amplitude and phase of the components present.

Four semi-diurnal (M_2, S_2, N_2, K_2) and two diurnal (O_1, K_1) tidal frequencies were run separately. The results are presented in Table 6. A positive value of transport indicates a flow into the North Sea from the English Channel. The most obvious test is to compare the theoretical calibrations for the fundamental frequencies. This shows very clearly that across the diurnal and semi-diurnal tidal harmonics, the calibration of the cable is essentially independent of frequency, being $580 \pm 2 \times 10^{-9} \, \mathrm{V \, s \, m^{-3}}$ with the exception of M_2 for which the value is $565 \times 10^{-9} \, \mathrm{V \, s \, m^{-3}}$. The flow distribution must be so similar for each harmonic that the cable responds in the same way to them all, except for M_2. In view of the uniformity between the other five harmonics, the 2.6% lowering of the calibration for M_2 cannot be ignored as being due to numerical 'noise' within the model, but is probably due to the fact that the non-linear modulation is more significant for M_2 than for the other, smaller amplitude, harmonics, and is sufficiently strong to influence the response to the fundamental frequency. However, the error is unlikely to be important in comparison with the errors incurred in the harmonic analysis of a real set of cable voltage data. Thus it may be concluded that the normal practice of applying the M_2 calibration to the other harmonics in the diurnal and semi-diurnal wavebands is justified.

Table 6. *The response of the cable to individual tidal harmonics*

Velocity input to model	Harmonic	Voltage observed by Alcock and Cartwright (1977) (mV)	Voltage predicted by cable model (mV)	Transport from Prandle's model ($10^4 m^3 s^{-1}$)	Theoretical calibration Amplitude ($10^{-7} V s m^{-3}$)	Phase lag (voltage behind transport) (deg.)
O_1 and non-linear products	mean		4.3	1.0	4.2	
	O_1	74	86.4	14.9	5.80	−1.8
	O_2		16.5	2.9	5.8	−1.0
K_1 and non-linear products	mean		−3.6	−1.0	3.6	
	K_1	83	85.2	14.7	5.79	−2.0
	K_2		12.6	2.3	5.6	−0.9
N_2 and non-linear products	mean		1.5	0.27	5.6	
	N_2	144	110.0	19.0	5.81	0.1
	N_4		9.4	1.5	6.2	0.2
M_2, M_4 and non-linear products	mean		15.0	5.0	3.0	
	M_2	800	820.0	145.0	5.65	−1.0
	M_4		108.0	20.8	5.2	3.2
	M_6		24.8	3.9	6.4	4.6
	M_8		7.8	1.4	5.7	−113.1
Pure M_2	mean		18.1	5.6	3.2	
	M_2	800	800.0	139.0	5.76	−0.4
	M_4		16.9	5.4	3.1	−3.6
	M_6		2.2	0.6	3.5	−4.3
	M_8		0.2	0.2	1.0	−3.1
S_2 and non-linear products	mean		−1.5	0.006	−249.0	
	S_2	265 (+58 radiational)	305.0	52.1	5.82	−0.1
	S_4		10.1	1.9	5.4	172.0
K_2 and non-linear products	mean		1.3	−0.05	−23.5	
	K_2	74	138.0	23.7	5.82	1.2
	K_4		6.8	1.0	7.0	172.5

Comparison between the predicted voltage amplitudes and those extracted from observations by Alcock and Cartwright (1977) shows good agreement (the latter are included in Table 6 where appropriate). That the model predicts absolute values of voltage for the major tidal constituents which differ by only a few percent from observations encourages confidence in those conclusions drawn from the model which cannot be compared with observations.

To understand the cable response to higher harmonics and the mean flow, the results of the model run for M_2 must be examined in detail. It should be remembered that the voltage and transport means and higher harmonics are the result of both (i) the linear operation of the cable model and the transport evaluation on any mean velocity field and higher harmonics of velocity already present in the input data, and (ii) the non-linear modulating effect of the tidal rise and fall operating on the fundamental M_2 velocity field in both the cable model and the transport evaluation. To isolate the latter of these two operations, the input velocity field was harmonically analysed to extract the pure M_2 frequency, and a twenty-five-step time series of velocity data was constructed

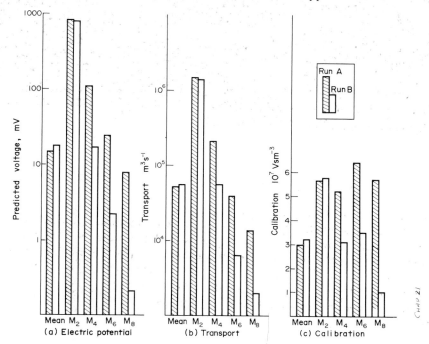

Fig. 10.   Comparison between the cable response to M_2 plus non-linear product velocities (run A) and pure M_2 velocities (run B).

to cover the M_2 tidal cycle, containing only the pure M_2 velocity. The cable model was re-run and the transports re-evaluated with the pure M_2 velocity input, but retaining the elevations appropriate to the original M_2 run of Prandle's model. The comparison between the voltage and transport obtained from the ordinary run with M_2 and M_4, etc., velocity input (A), and the run with pure M_2, velocity input (B), is made in Fig. 10.

The fundamental M_2 frequency is very little changed between the two, both in the absolute magnitudes of voltage and transport and in the calibration factor. In the case of the higher harmonics M_4 and M_6, etc., the voltages and transports in run B are an order of magnitude less than in run A. Also, where the calibration factors in the higher harmonics of A differ from the M_2 calibration by less than 15%, in B they differ a great deal more. This suggests that the cable responds to effect (i) in a way that can be reasonably predicted by the M_2 calibration, but its response to effect (ii) is not predictable in that way. However, because effect (ii) is apparently much less than effect (i), the higher harmonics can probably be reasonably well calibrated in practice by the M_2 factor. The same is not true for the mean. The magnitude of mean transport and voltage are similar for both run A and run B, and in both cases the calibration factor is almost half that for M_2 suggesting that for both runs effect (ii) is very much greater than effect (i). Consequently if effect (ii) (i.e. the non-linear tidal interaction) is the dominant source of mean transport through the Dover Strait, then applying the M_2 calibration factor to an observed mean voltage to deduce the mean transport would result in an estimate which is half the correct value. There are, of course, other causes of mean transport and voltage, arising from a distribution of residual velocities due to wind drift, density currents, etc., and the present model does not attempt to evaluate the cable response to these. None the less a typical estimate of the actual mean transport through the Strait is given by CARTWRIGHT (1961) as $4.9 \times 10^4 \, \mathrm{m^3 \, s^{-1}}$, which is very similar to the $5.1 \times 10^4 \, \mathrm{m^3 \, s^{-1}}$ or $5.6 \times 10^4 \, \mathrm{m^3 \, s^{-1}}$ encountered in the model, suggesting that the transport due to non-linear tidal interaction may well be the largest contribution except for periods of storm surges which are not considered here. The conclusion to be drawn from the model results is therefore that the use of the M_2 calibration factor in calibrating the mean voltage will lead to very large errors in the estimated mean flow through the Strait. It is, of course, not possible to verify these conclusions from the available observations, particularly because the voltage record is subject to zero errors caused by electro-chemical potentials and steady earth currents due to geoelectrical phenomena. However, that part of the mean transport due to non-linear interaction of the semi-diurnal tides will vary in magnitude with the spring neap cycle. The resulting fortnightly period signal is more readily obtained from observational records and contains no ambiguities as does the mean voltage.

Therefore it was considered worth running the model over a full spring neap cycle to obtain a more comprehensive picture of the cable response to all the interactions between the major tidal constituents. This would reveal the modulating effect of the M_2 elevations on the voltage and transport of the other semi-diurnal and diurnal harmonics present.

7. FIFTEEN-DAY MODEL RUN WITH MAJOR TIDAL CONSTITUENTS

The model was run as before, but in this case velocities were specified every hour for 360 hours, and the cable model was run to predict the voltage at every hour. The boundary conditions for Prandle's model specified M_2, S_2, O_1, K_1 and M_4. It was felt that these would represent the major features of the tidal regime. Harmonic analysis of the resulting time series for voltage and transport was by a standard tidal analysis least squares method, and the introduction of any of the other relatively large amplitude semi-diurnal harmonics would have required the total length of the time series to be doubled to enable these harmonics to be resolved. The sea conductivity used in the model was that corresponding to the mean May temperatures and salinities from the ICES atlas. The mean across the cable section of the depth mean velocity was also evaluated at each time step (i.e.

$$1/L \int_0^L \bar{\mathbf{v}} \times d\mathbf{l},$$

where L is the width of the channel and $\bar{\mathbf{v}}$ the depth mean velocity).

The non-linear interactions in both Prandle's dynamical model and the cable model would be expected to produce the input harmonics plus sums, differences and a mean value. The harmonic analysis algorithm was programmed to look for as many of these as could be resolved from 15 days' data. Table 7 indicates which spectral lines were asked for in the analysis, and lists the voltages, transports and mean velocities obtained for each frequency. When the voltages are compared with those obtained by Alcock and Cartwright from analysis of cable records, it is seen that again there is good agreement for M_2 and for S_2. The very much lower voltages predicted for O_1 and K_1 cannot be

Table 7. Results of the 15-day run of the model with major tidal inputs

Tidal constituent	Model results			Observed voltages	
	Voltage (mV)	Transport ($10^4\,m^3\,s^{-1}$)	Mean velocity ($10^{-2}\,m\,s^{-1}$)	PRANDLE and HARRISON (1976) (mV)	CARTWRIGHT (1961) (mV)
Mean	19.6	5.99	0.9		20*
Msf	17.8	4.20	1.8		16*
O_1	25.5	4.43	3.9	81	92
K_1	33.7	5.91	5.1	60	
M_2	855.3	148.30	133.2	720	785
S_2	229.0	39.77	35.4	259	266
$2SM_2$	3.9	0.64	0.6		
MO_3	3.3	0.55	0.4		
MK_3	0.9	0.14	0.2		
SK_3	1.0	0.15	0.1		
M_4	102.5	19.65	14.2	80	86
MS_4	30.3	6.09	4.4	58	54
S_4	4.6	0.97	0.6	3	10
M_6	21.4	3.20	4.0	12	23
$2MS_6$	15.9	2.56	2.9		
$2SM_6$	3.3	0.53	0.6		

* ALCOCK and CARTWRIGHT (1976).

accounted for, except by suggesting that they may be due to underestimation of the diurnals in the dynamical model of the area which had not been fully developed and tested when these velocity predictions were obtained from it. In view of the good values of the calibration factor obtained for the diurnals, it is unlikely that the cable model is producing the discrepancy, and the overall conclusions drawn from this modelling exercise should not be affected.

Looking at the lower frequency interaction terms, the value of 17.8 mV obtained for Msf is in surprisingly good agreement with the 16 mV observed by Alcock and Cartwright, and has a similar phase relationship. Moreover, Alcock and Cartwright predict from their value for Msf that the mean voltage arising from tidal interactions should be about 20 mV, to be compared with the model prediction of 19.6 mV. The picture is not so clear with the higher harmonic interaction terms, since with the exception of M_4 the higher harmonics present contain no contribution at that frequency from input at the boundaries of Prandle's model, and are therefore due entirely to generation within the bounds of the model. Observed values of voltage at these higher harmonics, as reduced by PRANDLE and HARRISON (1975), CARTWRIGHT (1971) and ALCOCK and CARTWRIGHT (1977) are listed in Table 7. It will be seen that the model predicts M_4 too high and MS_4 too low, possibly due to error in resolving the two over a fortnight since the total energy in the two spectral bands is approximately correct. S_4 and M_6 are predicted within the range of observed values. Apart from $2MS_6$, the other interaction terms are all very small.

Overall, apart from the diurnals where the input data is probably at fault, the model appears to be predicting voltages similar to those observed, and consequently we can take seriously the theoretical calibration factors produced by the model for each harmonic. The transport calibration is plotted in Fig. 11. The principal transport calibration, for M_2, is $5.76 \times 10^{-7}\,V\,s\,m^{-3}$. This compares favourably with the figure deduced from observations by CARTWRIGHT (1961) of $6.06 \times 10^{-7}\,V\,s\,m^{-3}$ and $5.60 \times 10^{-7}\,V\,s\,m^{-3}$ due

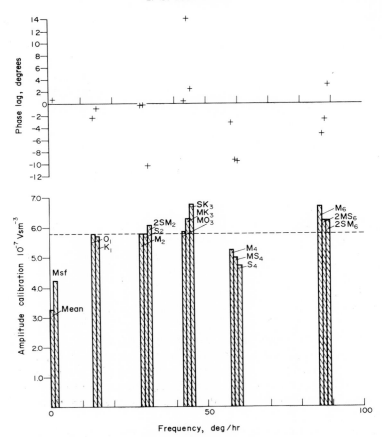

Fig. 11.   Transport calibration from the analysis of the 15-day run.

to PRANDLE and HARRISON (1975). Clearly this cable model is much better than the crude model described by ROBINSON (1976) for which the value was $10.0 \times 10^{-7} \, V \, s \, m^{-3}$.

Consideration of the frequency dependence of the transport calibration shows a clear uniformity across the four major constituents present. The anomalous difference between M_2 and the other harmonics, encountered when each was run separately, has now disappeared. Over the rest of the spectrum some definite trends emerge. The third diurnals are all higher than the diurnal/semi-diurnal principal calibration factor. The quarterly diurnals, much more significant in absolute magnitude, have calibration factors between 10% and 20% lower than the principal calibration, whilst the sixth diurnals are between 8% and 16% greater than the principal calibration. In view of the fair agreement between the predicted and observed voltages at these frequencies, we must assume that the model is accounting for most of the quarter and sixth diurnal motions present in reality. It follows that the model calibration is a realistic one, and so this experiment confirms that the use of the principal calibration factor to predict transport from observed voltages at these frequencies is likely to lead to errors. It is worth noting that PRANDLE and HARRISON (1975)

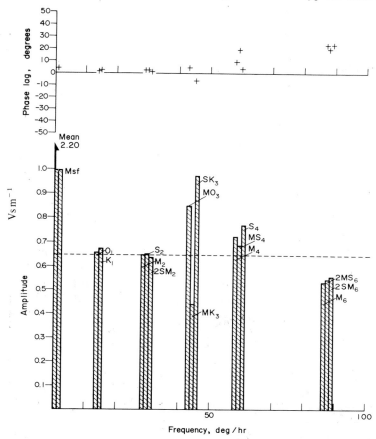

Fig. 12. Velocity calibration from the analysis of the 15-day run.

found a lower calibration than the principal factor for both M_4 and M_6.

The most serious deviation from the principal calibration occurs at the long period end of the spectrum, which is modelled very well in terms of the values of voltage predicted. The fortnightly period harmonic (Msf) has a calibration which is 73% of the principal calibration, whilst the mean calibration is only 57%. Significant underestimation of the fortnightly and mean transport would thus occur if the principal calibration factor were used on real voltage data.

Finally it is worth looking at the velocity calibration (voltage/section mean of depth mean velocity) shown in Fig. 12. This ratio is important because of its relevance to using the voltage observations in the estimation of surface slope across the Dover Strait (see CARTWRIGHT and CREASE, 1963). The variation across the frequency spectrum is more erratic than for the transport calibration, but there is a reasonable uniformity across the diurnal and semidiurnal harmonics. The M_2 value of $0.64\,\mathrm{V\,s\,m^{-1}}$ agrees well with Prandle and Harrison's observed value of $0.67\,\mathrm{V\,s\,m^{-1}}$. At the higher frequencies it is interesting to note that the trend of the variation from the principal calibration is opposite to the trend noted for the transport calibration, the quarter diurnals are higher and the sixth diurnals lower. The same is also true of the fortnightly and mean trends, the latter being 3.4 times the principal calibration. Consequently, where applying the principal calibration to the mean voltage would result in underestimating the transport by nearly half, it would predict a mean velocity which was 3.4 times too large, with implications for the estimation of the mean surface slope across the Strait.

8. CONCLUSIONS

Because it is able to produce a realistic seasonal variation of voltage and is able to predict calibrations for the semi-diurnal harmonics which are close to observed values, it is concluded that the model described in this paper can be considered to be a reliable indicator of the response of the Dover Strait cable voltage to a given velocity field. The implications for the analysis of voltage data are that the M_2 calibrations in terms of transport and of the section mean of the depth mean velocity can be reliably used for the diurnal and semi-diurnal tidal harmonics. Their use for the higher, shallow water, harmonics is likely to result in errors of 10% to 20%. For the fortnightly period tidal signal, the errors would be even greater whilst it is considered that it would be completely wrong to apply the principal M_2 calibration to the mean voltage. However, because it is thought that this method is modelling the major part of the fortnightly, quarter diurnal and sixth diurnal tides present in the real situation, then the calibrations deduced here could be used in data analysis, if they were suitably scaled to make the model-produced principal calibration equal to the empirical calibration derived from semi-diurnal tidal stream observations. On the other hand, the model does not attempt to include the cable response to any residual currents in the Dover Strait due to meteorological effects, variations in sea water density, or the difference between mean sea level in the North Sea and in the English Channel. This is because their spatial distribution is just not known. Therefore the model predictions for the mean response, due only to the tidal interaction mean, should not be considered necessarily to be realistic. The evaluation of mean transport from the records of mean voltage therefore remains an outstanding problem. Some encouragement may be derived from the indication by this model that the principal calibration and the mean calibration should have the same sign, since even this simple assumption has been questioned by the crude model of ROBINSON (1976) and the observations of PRANDLE and HARRISON (1975).

A final conclusion that can be drawn from the model relates to the difference between the calibration in terms of transport and in terms of the section mean of the depth mean velocity. It must be remembered that the voltage is truly related to the spatial distribution of velocity and tidal elevation. Relating it to either of the above two quantities is done in an attempt to extract some useful oceanographic information from easily and cheaply obtainable data. In a situation like the Dover Strait where tidal elevation modulates the transport and the voltage significantly, it is not clear which of the two calibrations is the more appropriate. On balance, it appears from the model that it is most sensible to relate the voltage to the transport. The wide discrepancy between the two types of calibration for the mean flow, however, puts in question the validity of using cable voltage measurements in an evaluation of the mean sea-level slope, this being related to the velocity rather than the transport calibration.

Acknowledgements—The author wishes to acknowledge the co-operation of Dr. D. PRANDLE of I.O.S. in his running of the dynamical model of the southern North Sea to supply the velocity data used in this work. He is also indebted to Dr. A. J. BURLEY of the Applied Geophysics Unit of I.G.S. who supplied information about the electrical conductivities measured in boreholes and gave advice about the geology of the Dover Strait area. Thanks are also due to Mr. J. E. JONES for running some of the computer programmes.

REFERENCES

AGER D. U. and P. WALLACE (1966) History of the Boulonnais, France. *Proceedings of the Geologist's Association*, **77**, 385–417.

ALCOCK G. A. and D. E. CARTWRIGHT (1977) An analysis of 10 years' voltage records from the Dover Sangatte cable. In, *Voyage of Discovery*, M. V. ANGEL, editor, *Deep-Sea Research*, Supplement to Vol. **24**, 341–366.

BARSUKOV O. M. (1967) Electric conductivity of the Earth. In: *International Dictionary of Geophysics*, S. K. RUNCORN, editor, Pergamon Press, pp. 321–324.

BEVIR M. K. (1970) The theory of induced voltage electromagnetic flow meters. *Journal of Fluid Mechanics*, **43**, 577–590.

BOWDEN K. F. (1956) The flow of water through the Straits of Dover related to wind and differences in sea level. *Philosophical Transactions of the Royal Society of London*, A, **248**, 517–551.

CARTWRIGHT D. E. (1961) A study of currents in the Strait of Dover. *Journal of the Institute of Navigation*, **14**, 130–151.

CARTWRIGHT D. E. and J. CREASE (1963) A comparison of the geodetic reference levels of England and France by means of the sea surface. *Proceedings of the Royal Society of London*, A, **273**, 558–580.

CONSEIL PERMANENT INTERNATIONAL POUR L'EXPLORATION DE LA MER (1962) *Mean monthly temperature and salinity of the surface layer of the North Sea and adjacent waters from 1905 to 1954*, Denmark: Charlottenlund Slot.

DESTOMBES J. P. and E. R. SHEPHARD-THORN (1971) *Geological results of the Channel tunnel site investigation 1964–65. Report no. 71/11*, N.E.R.C. Institute of Geological Sciences, 12 pp.

LONGUET-HIGGINS M. S. (1949) The electrical and magnetic effects of tidal streams. *Monthly Notices of the Royal Astronomical Society, Geophysical Supplement*, **5**, 285–307.

LONGUET-HIGGINS M. S., M. E. STERN and H. STOMMEL (1954) The electric field induced by ocean currents and waves, with applications to the method of towed electrodes. *Papers in Physical Oceanography and Meteorology*, M.I.T. Press, **13**, no. 1, 37 pp.

PRANDLE D. (1975) Storm surges in the southern North Sea and River Thames. *Proceedings of the Royal Society of London*, A, **344**, 509–539.

PRANDLE D. (1977) Tidal residuals in the Dover Strait and southern North Sea. *Deep-Sea Research* (In preparation).

PRANDLE D. and A. J. HARRISON (1975) Relating the potential difference measured on a submarine cable to the flow of water through the Strait of Dover. *Deutschen Hydrographischen Zeitschrift*, **28**, 207–226.

ROBINSON I. S. (1976) A theoretical analysis of the use of submarine cables as electromagnetic oceanographic flowmeters. *Philosophical Transactions of the Royal Society of London*, A, **280**, 355–396.

SANFORD T. B. and R. E. FLICK (1975) On the relationship between transport and motional electric potentials in broad, shallow currents. *Journal of Marine Research*, **33**, 123–139.

WALLACE P. (1968) The sub mesozoic palaeogeology and palaeogeography of N.E. France and the Strait of Dover. *Palaeogeography, Palaeoclimate, Palaeoecology*, Elsevier Publ. Co., Amsterdam, **4**, 241–255.

Some effects of finite steepness on the generation of waves by wind*

Michael S. Longuet-Higgins

University of Cambridge. Dept. of Applied Mathematics and Theoretical Physics, and Institute of Oceanographic Sciences, Wormley, Surrey

Abstract—General reasons are given for expecting the localization of the stresses exerted by the wind on the surface of steep gravity-waves. Some recent observations of the phase velocities of wind-generated waves can be very simply interpreted by supposing that the energy of waves at frequencies higher than that of the dominant waves is propagated at an angle θ to the wind, given by $\cos\theta = c/c_0$, where c denotes the phase speed and c_0 the speed of the dominant waves. This in turn is explained by the hypothesis that the stresses are localized on the dominant waves, probably near the wave crests. The hypothesis is similar to the resonance theory of wave generation by a turbulent wind, except that the angle θ is related to the phase speed c_0 of the steep waves, and not to the convection velocity U of the turbulent eddies.

Rough estimates of the energy imparted by the tangential stresses confirm that they could play a significant part in the growth of the waves.

1. INTRODUCTION

THE inadequacy of linear theories of wave generation to explain the observed rates of growth of sea waves under the action of wind has led to the suggestion by STEWART (1967), HASSELMANN (1967) and others of various non-linear mechanisms for wind-wave generation. In this paper we wish to emphasize one likely mechanism that has received little attention, namely the localization of the surface stresses, brought about by the very non-sinusoidal profile of steep gravity waves.

General reasons for expecting the localization of both the normal and the tangential stresses in the neighbourhood of the wave crests are given in Sections 2 and 3. At the same time, in a recent paper (1976) RAMAMONJIARISOA and COANTIC have described some unexpected measurements of the phase-velocities of wind-generated waves which it appears can best be interpreted by assuming that the energy has a bimodal distribution with regard to direction, and hence that the surface stresses are localized at certain phases of the dominant waves (see Section 4).

The normal stress at the air–water interface has usually been regarded as the probable agent for wave generation. But of the two kinds of stress, it is the tangential wind stress which is more likely to become localized. In Section 5 we make estimates which suggest that it may indeed be possible for the tangential wind-stress to account for a significant part of the observed growth of the dominant waves.

Attention is also drawn to the probable existence of strong non-linear interactions in the wave field, associated with instabilities and even breaking at the crests of the dominant

---

* Submitted to *Deep-Sea Research* on 24.6.76.

393

waves. These will tend to give a similar angular distribution of energy. Such a process may be called 'speed-locking'.

2. THE TANGENTIAL WIND STRESS

There are at least two factors affecting the distribution of tangential wind stress over the profile of gravity-waves, namely variations in the relative wind speed and variations in the small-scale roughness of the surface.

Variations in the relative wind speed

By continuity of mass flux we expect the air speed to be generally greater over the wave crests than over the troughs. For low waves, a qualitative estimate is provided by considering the perturbation of a uniform, frictionless air stream, of speed U, flowing over a sinusoidal boundary $z = a\cos(kx - \sigma t)$ propagated with phase speed $c = \sigma/k$. The tangential velocity of the air is given, to first order, by

$$u_1 = U + (U - c)ak\cos(kx - \sigma t). \tag{2.1}$$

When $U > c$, the velocity is greatest at the wave crests. The tangential velocity of the water, however, is given by

$$u_2 = c \cdot ak\cos(kx - \sigma t) \tag{2.2}$$

so that the relative velocity is

$$(u - u_2) = U + (U - 2c)ak\cos(kx - \sigma t). \tag{2.3}$$

This is greater or less at the crests than in the troughs according as $U \gtrless 2c$. In reality, the critical ratio of U/c will be affected by the presence of shear in the mean flow, which also introduces phase differences between u_1 and u_2 (see, for example, Miles, 1957).

At larger wave steepnesses, the sharper curvature of the surface near the crests tends initially to accentuate the difference between u_1 and u_2. This effect will be somewhat modified by shearing of the air stream and by the tendency of the airflow to separate in the lee of sharp corners. Nevertheless, qualitatively we may expect an increase in $(u_1 - u_2)$ at the wave crests, at least for waves of moderate steepness ak and for larger values of U/c. If $U/c \simeq 1$ then it is possible for $(u_1 - u_2)$ to be less at the wave crests than in the troughs, just as for waves of low amplitude.

Variations in surface roughness

Among the factors contributing to a variation in the small-scale roughness of the surface are the following.

(a) The horizontal scale of the roughnesses is reduced by the lateral contraction of the surface near the wave crests. For low waves, this has been discussed quantitatively by LONGUET-HIGGINS and STEWART (1960, 1963). For steep waves, we note that the tangential separation Δs of two neighbouring particles at the surface is simply proportional to the surface velocity q, in the frame of reference travelling with the wave:

$$\triangle s \propto q. \tag{2.4}$$

Hence the relative *contraction* between trough and crest is given by

$$(\triangle s)_{crest}/(\triangle s)_{trough} = q_{crest}/q_{trough} \qquad (2.5)$$

If the roughnesses consist of short waves, whose speed of propagation c' is small compared to q, then the ratio (2.5) gives the relative change in wavelength of the short waves, assuming that they persist throughout the passage of the long wave.

As the crests become sharp, q_{crest} tends to zero, and the scale of the roughnesses becomes so compressed that, if the waves do not break, their speed is increased due to capillarity. The condition $c' \gg q$ is then not met. If the group-velocity c_g of the capillaries exceeds q, then the short waves will tend to accumulate on the forward face of the long waves, in the neighbourhood of the point where $c_g = q$. A general discussion, taking account of the shearing current near the interface, has been given by PHILLIPS and BANNER (1974).

(b) Closely associated with the kinematical effect (a) is the dynamical effect of the surface contraction in doing work against the radiation stress of the short waves, and so increasing the short wave amplitude. For low gravity waves this effect was calculated by LONGUET-HIGGINS and STEWART (1960), but for steep gravity waves the effect is far more pronounced. A detailed calculation would have to take into account the effects of viscous dissipation and possible breaking of the shorter waves, as well as possible replenishment of short-wave energy by the wind.

(c) Because of the variation in wind speed over the longer waves the wind will have a greater capacity to generate short waves at the long wave crests (at sufficiently large values of U/c). If short-wave generation takes place preferentially at the wave crests, and if $q_{crest} > c'_g$, the short-wave energy will tend to be left behind, and increased roughness will be observed on the *rear* slopes of the waves. If on the other hand short-wave generation takes place preferentially in the troughs of the long waves, increased roughness will be found on the forward face of the longer waves, possibly at some critical position not far from the crest.

(d) There is probably a tendency, even in the absence of wind, for steep gravity waves to develop instabilities near the wave crest. One such instability, resulting in the formation of capillary waves ahead of a sharply curved wave crest, was analysed by LONGUET-HIGGINS (1963); see also VANDEN-BROEK (1974). A more extreme instability is apparent in the breaking of steep gravity waves and the formation of white caps on the forward face. When, owing to the advance of a steep wave through a group, a steep wave ceases to break, the whitecap is left behind, and the surface roughness which it represents is distributed over other phases of the wave and possibly reduced by horizontal extension of the elements at the free surface.

Significant variations in the surface roughness have been found experimentally by LAI and SHEMDIN (1971) and by KITAIGORODSKII (1976). The latter reports variations in short-wave energy in plunger-generated waves under the action of wind. The mean-square of the high-frequency roughness $<\zeta^2>$ was found to vary by a factor of order 10 over the phase of the longer waves, with the highest roughness generally occurring near the crests of the longer waves. Further observations of this nature would seem to be very desirable.* Some distinction may be necessary between conditions when the longer waves are essentially passive swell, as in the case just mentioned, and when they are being actively generated by the wind.

The above discussion strongly suggests that the combined effect of the variation in the

---

* Measurements at lower values of U/c_0 have been made by KELLER and WRIGHT (1975).

surface roughness and of the variation in wind speed will be to produce a tangential stress that is strongly localized near the crests of the dominant waves, particularly for the steeper waves and for larger values of U/c_0.

3. THE NORMAL WIND STRESS

Most calculations of the normal stress have assumed that the dominant waves are of sufficiently low amplitude that the perturbations in normal pressure are small and sinusoidal.* In fact, as the waves become steeper and the crests more sharply curved, a separation of the airflow near the crest becomes increasingly likely. This implies a marked difference in pressure between the rear and the forward face of the wave, with the strongest pressure gradient occurring near the crest itself.

The work done by the normal stress, is however, limited by the fact that in a progressive surface wave the surface slope never significantly exceeds $30°$. Thus the normal velocity does not exceed $\frac{1}{2}c_0$. The horizontal component of velocity, on the other hand, may be equal to c_0 if the crest is sharp-angled. This suggests that the steepness of the waves may increase the work done by the tangential stresses in a greater proportion than the work done by the normal stresses.

4. EVIDENCE FROM MEASUREMENT OF PHASE SPEED

Some interesting observations of the phase speeds of waves under the direct action of wind have been published by RAMAMONJIARISOA and COANTIC (1976). In a laboratory wind-wave channel of length 40 m and width 1.6 m and with wind-speeds of 0.5 to 14 m/s, they recorded the surface elevation at two points simultaneously in line with the mean wind, separated by a distance Δx of several centimetres. Using two independent methods, they deduced the apparent speed of each frequency component, in the direction parallel to the wind. Typical results, reproduced in Fig. 1, show that while the speed of the dominant waves agrees fairly closely with the theoretical speed at the frequency corresponding to the energy peak, the speeds of the higher-frequency components are almost constant, and equal to the speed of the dominant waves.

Various explanations may be considered and rejected. The first is that the higher frequencies represent harmonics bound to the dominant waves. If this were so, the effect would appear only in the neighbourhood of certain frequencies, namely integral multiples of the peak frequency. In fact, the speed is remarkably constant at all frequencies to the right of the peak.

Secondly, the effect is not due to a smooth spread of directions among the higher frequencies, the effect of which, for spectral densities varying as $\cos \theta$ or $\cos^2\theta$, is shown in Fig. 1. The constancy of the observed speed is also an argument against this interpretation.

Thirdly, the effect could not be produced entirely by a surface current, unless this were comparable in magnitude to the phase speed. Since surface currents are generally only about

---

*In a recent paper, GENT (1976) has introduced a second harmonic into the wave profile, which does not, however, correspond precisely with water waves.

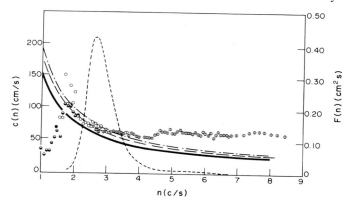

Fig. 1.   Phase velocities parallel to the wind, as a function of frequency, measured in a laboratory wave channel. $U = 5\,\text{m/s}$, $c_0 = 0.6\,\text{m/s}$ (from RAMAMONJIARISOA, 1974).

3% of the wind speed, while the observed discrepancies are much larger, this explanation also is ruled out.

However, we can interpret the observations on the assumption that the waves of frequency greater than the peak frequency are propagated at an angle θ to the wind such that their *apparent* phase speed in the wind direction is just equal to c_0, the speed of the dominant waves. This implies the relation

$$\cos \theta = c/c_0 \tag{4.1}$$

where c is the phase velocity of the waves.

Among the possible physical reasons for this effect are, first, a localized effect at the wave crests, such as wave breaking, which, by non-linearity in the fluid motion, would tend to transfer energy to other wave components. This would not be a *weak* non-linearity, such as that suggested by Phillips and Hasselmann, but rather a *strong* non-linearity, scattering energy from one steep, short-crested wave into an infinity of free wave components. We may call this a phase-locked wave interaction; and the resulting effect 'speed-locking'.

A second possibility is that such a local instability might arise from the boundary layer at the side walls of the tank, where the wave amplitude is likely to be enhanced by reflection. This possibility cannot yet be ruled out, except by experiments in a broader channel or in the open sea. We note, however, that a similar though less pronounced effect has also been reported by YEFIMOV, SOLOV'YEV and KRISTOFOROV (1972) in wave measurements over open water.

A third hypothesis, in the light of the previous discussion, is that the wind stresses themselves are localized on the wave crests. For the crests of the dominant waves are not generally uniform along their length.* Where they are particularly steep we expect a concentration of surface stress, either normal or tangential to the long-wave profile. Moreover, these spots of high stress will travel forward with the local phase velocity of the dominant waves. Hence they will tend to generate waves travelling at an angle with the

---

*The view of the wave surface in fig. 12b of RAMAMONJIARISOA (1974) shows clearly that the waves in this instance are quite short-crested. Hence localization occurs both in the down-wind and cross-wind directions.

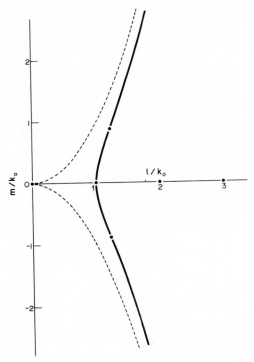

Fig. 2.   Wavenumber locus for gravity waves generated by a localized surface stress, or by strong non-linear wave interactions. k_0 is the wavenumber of the dominant waves.

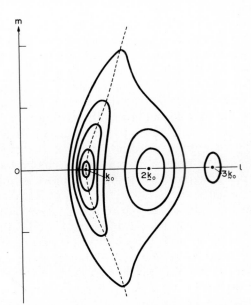

Fig. 3.   Sketch of the contours of spectral density at an early stage of wave generation, when $U/c_0 \gg 1$. On the left is the free-wave energy clustered round the resonance curve. Also shown are subsidiary peaks near $2k_0$ and $3k_0$, representing energy bound to the lowest harmonic.

wind direction, but with the same speed in the κ-direction as the speed of the dominant waves.

The effect is very similar to that proposed in PHILLIPS's theory (1957) of the production of waves by spots of pressure in a turbulent wind stream, except that the patches of stress arise not from the free turbulence but from the existing waves; and hence the angle θ is related not to the wind velocity but to the phase speed of the dominant waves.

Whichever interpretation is chosen, equation (4.1) implies a relation between the wavenumber and the direction of the shorter waves. Let $\mathbf{k}$ be the vector wavenumber with components (l, m) in the (x, y) directions, and let $k = |\mathbf{k}| = (l^2 + m^2)^{\frac{1}{2}}$. Assuming the linear dispersion relation $c = (g/k)^{\frac{1}{2}}$, we have from (4.1)

$$(k_0/k)^{\frac{1}{2}} = \cos\theta = l/k$$

so

$$l^2 = kk_0$$

and hence

$$l^4 = k_0^2(l^2 + m^2). \tag{4.2}$$

This curve is shown in Fig. 2. It passes through the point $(l, m) = (k_0, 0)$ where it has radius of curvature k_0. At large wavenumbers, it is asymptotic to the parabolas

$$m = \pm l^2/k_0.$$

There are two points of inflexion, where

$$(l, m) = \left\{ \left(\frac{3}{2}\right)^{\frac{1}{2}}, \pm \frac{3}{2} \right\} \frac{l^{\frac{1}{2}}}{}\, k_0, \quad k = \frac{3}{2} k_0,$$

and where

$$\frac{dm}{dl} = \pm 2\sqrt{2}.$$

In a continuous spectrum, we may expect the wave energy to lie generally in the neighbourhood of this curve, with some dispersion about it, as in Fig. 3. This portion of the spectrum represents the free-wave energy. Waves bound to frequencies in the spectral peak may be found in the neighbourhood of the wavenumbers $2\mathbf{k}_0, 3\mathbf{k}_0, \ldots$.

As a function of direction, the spectrum is bimodal rather than unimodal. This type of spectrum is to be expected only under conditions of strong wave generation, and possibly only in the earlier stages. At later stages, the spectrum will be slowly modified by dissipation and by weak non-linear interactions, the energy being dispersed from the peak in the characteristic directions $dm/dl = \pm 1/\sqrt{2}$; see LONGUET-HIGGINS (1976) and Fox (1976).

In the observations of RAMAMONJIARISOA and COANTIC (1976) the waves were being strongly generated, with $U/c_0 \simeq 8$ or more. In the field observations of YEFIMOV, SOLOV'YEV and KHRISTOFOROV (1972), where a comparable effect was observed, the ratio U/c was generally less.

It is not difficult to take into account the effect of capillarity on the form of the resonance curve (4.2). Adopting the dispersion relation

$$c = (g/K + T/\rho \cdot k)^{\frac{1}{2}}$$

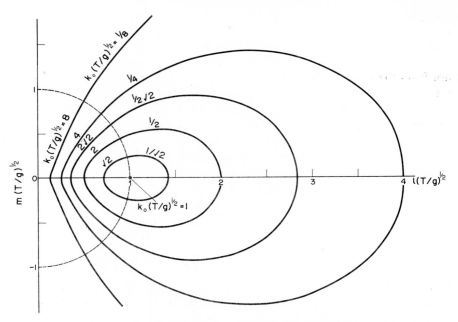

Fig. 4. Wavenumber loci for gravity-capillary waves, showing the effect of capillarity on the resonance curve of Fig. 2.

and choosing units with $\rho = T = g = 1$ we have

$$\cos\theta = \left(\frac{k+k^{-1}}{k_0+k_0^{-1}}\right)^{\frac{1}{2}}.$$

(4.3)

The family of resonance curves is shown in Fig. 4, with k_0 as parameter. The resonance curves corresponding to the longer, gravity-type waves lie inside the circle $k = 1$. The effect of capillarity on these is to bend them round towards the l-axis, so that ultimately they form closed ovals. The parts of the curve lying outside the circle $k = 1$ corresponds to capillary-type waves. Evidently these bend backwards towards the origin, meeting the circle $k = 1$ in points increasingly close to the m-axis. In the special case $k_0 = (1,0)$, corresponding to gravity-capillary waves of minimum phase-velocity, the resonance curve shrinks to zero, as would be expected; there are no other waves with phase-velocities less than c_0.

5. DISCUSSION; RATES OF WAVE GROWTH

In wind-generated waves it is universally observed that the dominant waves occur in groups, the wave amplitude varying both in time and space. This is a well-known property of any stationary stochastic process having a non-zero spectral bandwidth. An important consequence, however, is that the effects of non-linearities in the generating process may be accentuated. For even when most of the individual waves are linear and sinusoidal, there can still be a few which are steep and sharp crested. Hence the bulk of the energy transfer from the wind may be imparted to the steepest waves.

Table 1. *Observed phase angles ϕ between air pressure variations and the dominant surface waves; and the estimated values of τ_n^*/τ^**

Author	ϕ	τ_n^*/τ^*
DOBSON (1971)	$40°$	0.80
ELLIOTT (1972)	$20°$	0.20
SNYDER (1974)	$45°$	0.02

Despite many attempts, the distribution of the normal stress on the surface of a steep water wave in conditions of active wave generation has still not been well determined experimentally. The distribution of the tangential stress is even less well known. We can, however, use the comparatively well-determined value of the total mean wind stress τ^* to estimate the rates of growth due to the wind, according to various hypotheses.

Let us divide the total horizontal wind stress τ^* into two parts: a part τ_n^* corresponding to normal pressures on the dominant waves, and a part τ_s^* corresponding to the 'tangent stress', which may also include normal stresses on the small-scale elements.

The direct observations of surface pressure by DOBSON (1971), ELLIOTT (1972) and SNYDER (1974) have led to widely differing estimates of the ratio of τ_n^*/τ^* (see Table 1). Nevertheless it does appear that the mean tangential stress τ_s^* is generally of the same order as τ^*.

Let us write τ_s for the local tangential stress, so that $\tau_s = \tau_s^*$ and

$$\tau_s = \tau_s^* + \tau'$$

where τ' is a variable stress with mean zero. We may estimate the work done on the waves by the tangential stress under two different hypotheses. First, assume the waves are sinusoidal and τ' varies in-phase with the surface elevation. Then it was shown by LONGUET-HIGGINS (1969a) that the work done on the waves, per unit time and unit surface area, is

$$W = \overline{\tau' u_s}$$

where u_s is the tangential velocity of the particles at the free surface.* in sinusoidal waves, the amplitude $|u_s|$ is equal to $ak \cdot c$, and if we take

$$|\tau'| = K\tau_s^*$$

where K is a constant we have

$$W = \tfrac{1}{2}|\tau'|\,|u_s| = \tfrac{1}{2}K\overline{ak} \cdot c_0\tau_s^*.$$

For waves under the action of wind it is reasonable to take $\overline{ak} \simeq 0.14$ (LONGUET-HIGGINS, 1969a), and since the maximum value of K is not likely to exceed 1 we have

$$W \leqslant 0.07\,c_0\tau_s^* \leqslant 0.07\,c_0\tau^*. \tag{5.1}$$

If, on the other hand, we adopt the extreme hypothesis that all of the tangential stress is concentrated at the sharpest wave crests, where the particle velocity is of order c_0,

---

*In a recent contribution GARRETT and SMITH (1976) have shown that this formula effectively applies also when τ' includes the transfer of momentum from the wind to the *short* waves.

it follows that

$$W \sim c_0 \tau_s^*. \tag{5.2}$$

A large part of this work can be expected to go into the dominant waves. Comparing (5.1) and (5.2) we see that the work done on waves by the tangential stress is increased by an order of magnitude.

Working now with orders of magnitude, if we take

$$\tau_s^* \sim \tau^* \sim C\rho'U^2 \tag{5.3}$$

where ρ' is the density of air and C a drag coefficient of order 2×10^{-3}, and if we take the mean energy density of the waves to be

$$E = \tfrac{1}{2}\rho g a^2 \tag{5.4}$$

where $g = kc_0{}^2 = \sigma c_0$ and $ak = 0.14$, then from the last three equations, the proportional growth of the energy in one wave cycle T_p is given by

$$\frac{T_p}{E}\frac{dE}{dt} = \frac{2\pi W}{\sigma E} \gg C\,\frac{\rho'}{\rho}\frac{4\pi U^2}{ga^2\sigma} \sim 10^{-3}(U/c_0)^2.$$

If $U/c_0 \simeq 4$, say, this leads to a proportional rate of growth exceeding 10^{-2}, which in turn is greater than the observed rates by at least one order of magnitude (see PHILLIPS, 1967).

We have not, of course, allowed for the weak non-linear interactions, or for the probably more important dissipation of energy by wave breaking and viscosity. Nevertheless the above estimate suggests that the action of the tangential stress in generating the dominant waves may well be significant.

As against this it should be pointed out that if the wind speed should fall so that $U/c_0 \simeq 1$, say, then relative velocity of air and water will be greater in the troughs, where the orbital velocity is negative, than at the crests. Though the surface roughness of the troughs may be less, it is nevertheless possible that in such a case the tangential wind stress might act so as to damp the energy of the dominant waves.

REFERENCES

DOBSON F. W. (1971) Measurements of atmospheric pressure on wind-generated sea waves. *Journal of Fluid Mechanics*, **48**, 91–127.

ELLIOTT J. A. (1972) Microscale pressure fluctuations near waves being generated by wind. *Journal of Fluid Mechanics*, **54**, 427–448.

FOX M. J. H. (1976) On the nonlinear transfer of energy in the peak of a gravity-wave spectrum, II. *Proceedings of the Royal Society, London*, A, **348**, 467–483.

GARRETT C. and J. SMITH (1976) On the interaction between long and short surface waves (Preprint, submitted to the *Journal of Physics and Oceanography*)

GENT P. R. and P. A. TAYLOR (1976) A numerical model of the air flow above water waves. *Journal of Fluid Mechanics* (in press).

HASSELMANN K. (1967) Nonlinear interactions treated by the methods of theoretical physics (with application to the generation of waves by wind). *Proceedings of the Royal Society, London*, A, **299**, 77–100.

HASSELMANN K. (1971) On the mass and momentum transfer between short gravity waves and larger-scale motions. *Journal of Fluid Mechanics*, **50**, 189–205.

---

*In a recent contribution GARRETT and SMITH (1976) have shown that this formula effectively applies also when τ' includes the transfer of momentum from the wind to the *short* waves.

KELLER W. C. and J. W. WRIGHT (1975) Microwave scattering and the straining of wind-generated waves. *Radio Science*, **10**, 139–147.

KITAIGORODSKII S. A. (1976) Personal communication.

LAI R. J. and O. H. SHEMDIN (1971) Laboratory investigation of air turbulence above simple water waves. *Journal Geophysical Research*, **76**, 7334–7350.

LONGUET-HIGGINS M. S. (1963) The generation of capillary waves by steep gravity waves. *Journal of Fluid Mechanics*, **16**, 138–159.

LONGUET-HIGGINS M. S. (1969a) A nonlinear mechanism for the generation of sea waves. *Proceedings of the Royal Society, London*, A, **311**, 371–389.

LONGUET-HIGGINS M. S. (1969b) Action of a variable stress at the surface of water waves. *Physics of Fluids*, **12**, 737–740.

LONGUET-HIGGINS M. S. (1976) On the nonlinear transfer of energy in the peak of a gravity-wave spectrum: a simplified model. *Proceedings of the Royal Society, London* A, **347**, 311–328.

LONGUET-HIGGINS M. S. and R. W. STEWART (1960) Changes in the form of short gravity-waves on long waves and tidal currents. *Journal of Fluid Mechanics*, **8**, 565–583.

LONGUET-HIGGINS M. S. and R. W. STEWART (1964) Radiation stresses in water waves; a physical discussion, with applications. *Deep-Sea Research*, **11**, 529–562.

MILES J. W. (1957) On the generation of surface waves by shear flows. *Journal of Fluid Mechanics*, **3**, 185–204; see also **6**, 568–598 (1959), **7**, 469–478 (1960) and **13**, 433–448 (1962).

PHILLIPS O. M. (1957) On the generation of waves by turbulent wind. *Journal of Fluid Mechanics*, **2**, 417–445.

PHILLIPS O. M. (1963) On the attenuation of long gravity waves by short breaking waves. *Journal of Fluid Mechanics*, **16**, 321–332.

PHILLIPS O. M. (1966) *The dynamics of the upper ocean*, Cambridge University Press.

PHILLIPS O. M. and M. L. BANNER (1974) Wave breaking in the presence of wind drift and swell. *Journal of Fluid Mechanics*, **66**, 625–640.

RAMAMONJIARISOA A. (1974) *Contribution a l'ètude de la structure statistique et des mécanismes de generation des vagues de vent*. Thèse, D. es Sc., Université de Provence.

RAMAMONJIARISOA A. and M. COANTIC (1976) Loi experimentale de dispersion des vagues produites par le vent sur une faible longeur d'action. *Comptes Rendns de l'Académie des Sciences, Paris*, **282** B, 111–114.

REECE A. M. and O. H. SHEMDIN Modulation of capillary waves by long waves. *Report*, Coastal and Oceanography Engineering Laboratory, University of Florida, Gainesville, pp. 1–12.

SNYDER R. L. (1974) A field study of wave-induced pressure fluctuations above surface gravity-waves. *Journal of Marine Research*, **32**, 497–531.

STEWART R. W. (1967) Mechanics of the air–sea interface. *Physics of Fluids*, **10**, (Supplement on Boundary layers and Turbulence), pp. S47–S55.

VANDEN-BROECK J.-M. (1974) *Mécanique des vagues de grande amplitude*. Thesis for Ingen. Phys., Université de Liège.

YEFIMOV V. V., YU. P. SOLOV'YEV and G. N. KHRISTOFOROV (1972) Observational determination of the phase velocities of spectral components of wind waves. *Izvestiya Akademii Nauk SSSR*, **8**, 435–446.

An investigation of beach cusps in Hell's Mouth Bay

J. Darbyshire

Department of Physical Oceanography, University College of North Wales

Abstract— Detailed surveys have been carried out of the topography of a small beach at Hell's Mouth Bay off the Llŷn Peninsula, North Wales. Less detailed surveys have also been carried out of a much longer beach adjacent to the first. The surveys often revealed the presence of beach cusps spaced at regular intervals. These cusps changed from time to time due to variation in the sea wave conditions which were recorded and also due to changes in the beach gradients.

The cusps' spacing was always equal to about one-half the length of an edge-wave mode, calculated using the observed beach gradients normal to the shore and assuming the same frequency as the incident waves. Data obtained previously by Longuet-Higgins and Parkin (1962) also gave the same relationship.

Various possible causes for the generation of these cusps are discussed but it appears that the most likely cause is the variation of wave radiation stress along the beach due to the refractive effects of small transient changes in the topography of the beach within about 200 m of the shore-line.

List of symbols

x	Co-ordinate normal to beach
y	Co-ordinate along beach
U	Oscillatory part of volume transport in x direction
$X(x)$	Amplitude of U
V	Oscillatory part of volume transport in y direction
$Y(x)$	Amplitude of V
η	Wave elevation
$Z(x)$	Amplitude of wave elevation
A	Amplitude of wave elevation
k	Wave number
k_x	Wave number component in x direction
k_y	Wave number component in y direction
σ	Wave angular frequency
ρ	Density of water
g	Acceleration of gravity
β	Angle between plane of beach and horizontal plane
$h(x)$	Depth of still water
$\bar{\eta}$	Wave set-up
$d = h + \bar{\eta}$	Depth with wave action
H	Wave height (vertical distance between crest and trough of wave)
γ	Dimensionless constant in relation $H = \gamma d$
S_{xx}	Wave radiation stress in x direction
S_{yy}	Wave radiation stress in y direction

τ_x Stress per unit mass in x direction
τ_y Stress per unit mass in y direction
L_0 Wavelength in deep water ($h > L_0/2$)
θ_0, ζ_0 Angles of incidence of wave rays to line normal to the beach direction
π 3.141592654...

INTRODUCTION

HELL'S MOUTH BAY is situated at the northwestern end of Cardigan Bay (Fig. 1), being just to the southeast of Aberdaron. The name appears to be an ancient one and was used by Lewis Morris in his chart of the Welsh coast, published in 1748. The Welsh name Porth Neigwl is quite different and Neigwl is probably a proper name. The English name was undoubtedly given because this coast is exposed to all the fury of the waves from the Atlantic and in the days of sail, ships were often driven ashore there. There is some mention of this by HUGHES and EAMES (1975) in their book about Portmadoc ships.

This beach has fortunately remained completely unspoilt as it has been rather

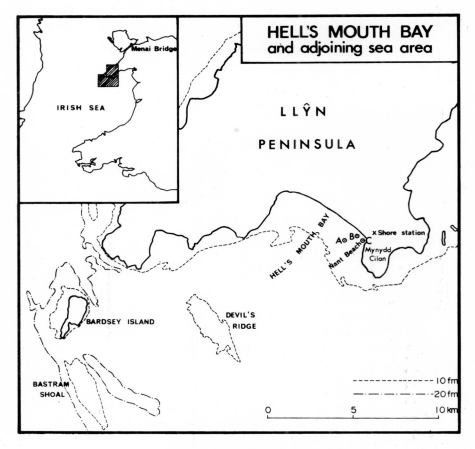

Fig. 1.   Map of Hell's Mouth Bay and adjoining sea area.

inaccessible by motor-car and all 4 miles of beach are completely free of man-made structures. It is therefore an excellent beach to study. An investigation was started by the Department of Physical Oceanography at Menai Bridge in conjunction with IOS Bidston in 1973. Originally the emphasis was on measuring the actual currents caused by tides and waves and on the identification of edge waves. It is hoped to describe this part of the work later.

Early in 1975, however, it was decided to study the beach itself in a very thorough manner as it had become apparent that its topography was changing significantly from one week to the next. These changes were particularly obvious on Nant beach, which lies in the southeastern angle of the bay, bounded by the great headland of Mynydd Cilan and a large spit near the mouth of a small stream about 400 m to the northwest (see Fig. 1). After the great storms of January 1975, there was deposited on this beach a large amount of shingle and loose stones up to about a third of a metre in diameter. Four or five very long ridges, 25 m apart, were cut in the shingle by the waves as can be seen in Fig. 2 which shows the beach at the end of February 1975. The tides in January were so high that even the highest parts of the beach were affected but later tides were not high enough for waves to have any effect on these higher areas. Thus in Fig. 3, taken in July 1975, the upper parts denoted by A are more or less the same as in Fig. 2 but the lower parts are noticeably different. In the area marked B are cusps caused by waves of shorter period and lower height than the January ones and in the area marked C are still smaller cusps caused by waves of 4–5-seconds period which occurred during the last tidal cycle before the photograph was taken. It was clear then that the beach contours

Fig. 2. Photograph of Nant Beach, February 1975, looking towards the northwest, showing cusps spaced about 25 m apart.

Fig. 3.   Photograph of Nant Beach, 4 July 1975, showing A, the long cusps shown in Fig. 2, B, the shorter cusps formed more recently and C, very short cusps during the high tides before the photograph was taken.

reflected the wave climate over several weeks and that detailed surveys taken at frequent intervals combined with a precise knowledge of wave conditions should lead to an increase in our knowledge of beach development processes. A good deal of work has been done on this subject in the past. Longuet-Higgins and Parkin (1962) studied cusps on the Chesil Beach, Dorset, and at Tide Mill between Newhaven and Seaford. The whole subject gained considerable impetus as the result of the theoretical work of Longuet-Higgins and Stewart (1964) on wave radiation stress and the application of this work to beach currents by Bowen (1969, 1972) and Bowen and Inman (1969). Observations on rip currents and beach cusps were taken by Komar (1971) both in a laboratory wave basin and on a real beach. It is therefore of interest to see if the results found for Hell's Mouth Bay are in agreement with those found previously by other workers.

METHODS OF MEASUREMENT

The waves were measured by a wave recorder which initially consisted of a pressure transducer anchored to the bottom at a depth of 15 m and connected by cable to a surface buoy which contained a radio transmitter. At this depth of recording, the response of the instrument is as given in Table 1 (a). It falls to 57% for 8-second waves and so the work was confined to cases where the period was greater than this. The pressure transducer signals were transmitted at 6-hourly intervals by the radio transmitter and picked up by a radio receiver at a shore station.

Table 1. *Response of wave recorder*

Period (secs)	Response (ratio of amplitude of pressure oscillation to that of wave surface elevation)	(Response)^2
(a) Depth 15 m. Position A in Fig. 1.		
15	0.87	0.76
12	0.80	0.64
10	0.72	0.52
8	0.57	0.32
6	0.33	0.11
5	0.17	0.03
(b) Depth 4.5 m. Positions B and C in Fig. 1.		
15	0.96	0.92
12	0.94	0.88
10	0.91	0.83
8	0.86	0.74
6	0.76	0.58
5	0.66	0.44

The position of the buoy is denoted by A and that of the shore station by × in Fig. 1. This system worked from the end of February until the end of September when the transducer disappeared due to unknown causes. In December, a second transducer was set up by laying it on the bottom near the low spring tide mark at position B in Fig. 1. This recorder was connected to the shore by a cable and with one or two breaks worked well until the middle of March 1976 when it was replaced by a similar transducer at Nant Beach, at position C in Fig. 1. In these last two positions, records were only taken at high tide when the depth was 4–5 m. The response of the instrument at a depth of 4.5 m is shown for various periods in Table 1(b); it falls to 66% for 5 seconds so all waves of greater period than this could be considered. The wave profiles, however, are distorted to some extent due to the effect of the shallow depth and harmonics are formed. (See the example for 3 February in Fig. 8.)

In all cases, the pressure variations were sampled every 2 seconds for 1300 seconds and recorded on paper tape which could be fed into the College's ICL 4130 computer. The power spectra were then obtained using a programme based on the work of BLACKMAN and TUKEY (1958), taking 50 lags which with 650 observations give 26 degrees of freedom corresponding to 95% fiducial limits of 0.55 to 1.66 times the observed values. To obtain the power spectral estimates for the wave elevation, the values shown should be divided by the square of the appropriate response factor shown in Table 1.

As regards the beach, it was decided to do a fine survey over a grid of 5 × 2 m over a large part of Nant Beach every fortnight. The part chosen consisted of a strip 160 m long extending from the cliff edge 30 m downbeach. These limits in fact included most of the shingle area. It was thought better to carry out the surveys during neap periods as it was then possible to isolate the effect of the waves which occurred during the intervening springs and which were the only ones that could affect the upper parts of the beach. The levels obtained from the beach were referred to a datum which was the level of the local mean low water springs. Examples of such surveys, where the levels have been contoured at one metre intervals, are shown in Fig. 4. Particular attention was given to

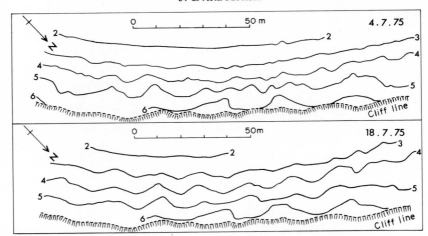

Fig. 4.   Map of Nant Beach, showing contours for 2-, 3-, 4- and 5-m levels above mean low water springs, for 4 July and 18 July 1975.

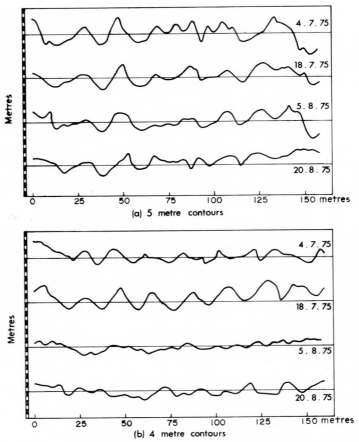

Fig. 5.   (a) Plots of 5-m contours above M.L.W.S. for 4 July, 18 July, 5 August and 20 August 1975. The horizontal lines represent the same position on the beach for each survey. (b) Same for 4 m.

the horizontal changes in the 4-m and 5-m contours. Figure 5a shows 5-m contours for several consecutive surveys and Fig. 5b shows the corresponding 4-m contours. The contour lines are usually wavy in character due to cusp formation and it was thought that their variation could be best described by making a Fourier analysis of the wave pattern. These analyses are not meant to be considered as power spectra but provide a more complete picture of the variation in cusp length than is obtained by taking histograms of cusp length distributions as was done by other workers such as LONGUET-HIGGINS and PARKIN (1962). Thus changes are often much more apparent on the amplitude spectra shown in Fig. 6 than on the original contours. This is the case with those of 18 July 1975, 13 February 1976 and 21 May 1976 (Fig. 17).

RESULTS

An examination of Fig. 6 shows many interesting features. Between 4 and 18 July 1975, there is, for instance, a marked change in the 4-m contour and a smaller effect in the 5-m

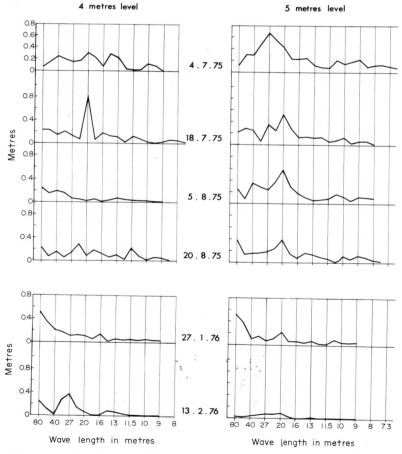

Fig. 6. Amplituce spectra for 4-m and 5-m contour lines for 4 July, 18 July, 5 August, 20 August 1975, 27 January and 13 February 1976.

contour and both show very prominent cusps of 20-m length. These cusps appear very clearly on the 4-m contours shown in Fig. 5b. The sea was very calm during most of this period and there was appreciable swell only on 14 July. This is shown by the plot of daily wave heights and wave periods shown in Fig. 7. The power spectrum of the waves for 14 July is shown in Fig. 8 and indicates a dominant wave period of between 8.5 and 9.0 seconds. Spring tides occurred on the 10th to 11th (Fig. 7) and so the tide on the 14th was high enough to enable the waves to affect the higher levels of the beach. The cusps at 4 m had disappeared by 4 August, probably due to the effect of the waves of 1.5 m height and 7.5 seconds period which occurred, as shown in Figure 7, on 22 July. The cusp spectra for 20 August 1975 show two prominent bands, one centred at 22.85 m and the other at 11.4 m. Between 5 and 22 August there was appreciable swell only on 7 August when the spectrum (Fig. 8) shows two dominant peaks at 12.5 and 8.3 seconds period.

The wave recorder ceased to work after 20 September and wave recording was not

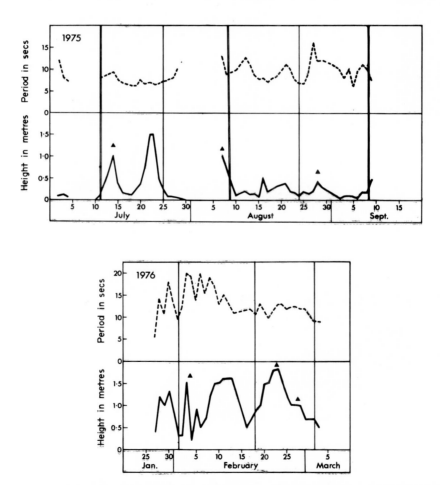

Fig. 7.   Daily wave heights and periods July–September 1975 and January–March 1976. The times of the higher spring tides are indicated by the double vertical lines and those f the lower spring tides by the single lines. The black triangls indicate the times the cusps analysed in Table 3 were formed.

restarted until mid-December, so no examples are available for the intervening period. A very good example of cusp formation occurred on 2 to 3 February 1976. A visit to the beach on 1 February revealed no trace of cusps and the photograph (Fig. 9) taken on 27 January confirms this but by 6 February, as shown by the next photograph (Fig. 10), a series of eight very long ridges, spaced about 30 m apart, had been formed. Figure 12 shows a marked variation in the 4-m contour but negligible differences in the 5 m for this case. This is confirmed by an examination of Fig. 9 and 10 where debris deposited on the upper beach has obviously not moved at all. These cusps could be clearly associated with the swell of 1.5 m height and 20 seconds period which arrived on 2 February. The wave power spectrum is shown in Fig. 8. By 26 February the cusps had largely disappeared as shown in Fig. 11.

The main beach was also surveyed but only in one instance was the grid used as fine as that for the Nant Beach. Cusp formation took place here also though not always at the

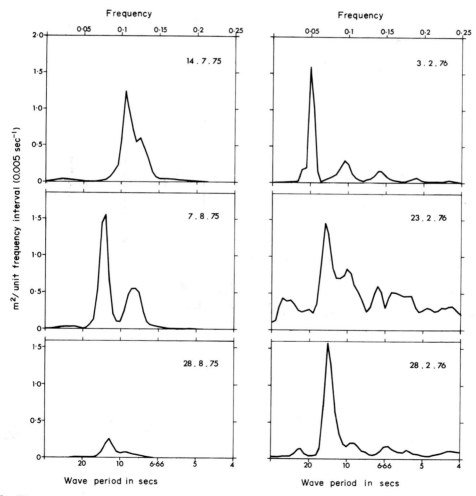

Fig. 8. Wave power spectra for 14 July, 7 August and 28 August 1975, 3 February, 23 February and 28 February 1976.

Fig. 9.   Nant Beach, looking towards northwest, 27 January 1976.

Fig. 10.   Nant Beach, 6 February 1976, showing long cusps spaced about 30 m apart. Note the position of the debris on upper beach is identical with that shown in Fig. 9.

Fig. 11.   Nant Beach, 26 February 1976, showing disappearance of long cusps shown in Fig. 10.

same time as Nant Beach. One such example was observed on 5 September 1975 where light pebbles had been regularly arranged in lines 6 m apart for over a mile of the beach. Similar cusps were observed on the lower parts of Nant Beach on 30 August and 5 September. The sea was again very calm for most of the time immediately preceding this and, as indicated by Fig. 7, the only swell that stands out occurred on 28 August when the height was 0.4 m and the period was 11.75 seconds. The spectrum is shown in Fig. 8. Another example on the main beach was observed between 26 February and 3 March 1976. The photograph, Fig. 13, was taken on 26 February and shows a series of stone ridges spaced about 12 m apart. The photograph, Fig. 14, shows the same ridges on 3 March when they have now increased in height and become more regular. Figure 7 shows that during all this time there was a high swell of period 13.33 seconds. The wave spectrum is shown in Fig. 8. It was possible to make detailed surveys every 2 m on a 160-m-long section of the main beach on 13 May and 21 May 1976. Figure 15 shows the variation of the 3.5-m contour for both times. There has been a marked change between the two and a very clear periodicity is evident on the 21st. The wave spectra for 14, 15, 16, 17, 18, 19 and 20 May are shown in Fig. 16. These show a large increase in wave activity between the 14th and 15th due to the arrival of a heavy swell of 8 seconds period. This persisted until the 17th with the period gradually lengthening to 8.66 seconds. The height decreased on the following days but the period of 8.66 seconds persisted. The cusp spectra are shown in Fig. 17 and show an appreciable change, the 7–9-m cusps originally present on the 13th have been destroyed and by the 21st very pronounced cusps of 13.7 m length are present.

Table 2. Cusp wavelength and sea wave period

Date	Cusp spacing (ms)	Wave period (secs)
Nant Beach		
18 July 1975	20.0	8.5
20 August 1975	22.85 11.42	12.5 8.3
6 February 1976	29.0	19.5
Main Beach		
5 September 1975	6.0	11.75
26 February to 3 March 1976	12.0	13.33
26 May 1976	13.7	8.0

EDGE WAVES

It has been mentioned by previous workers, LONGUET-HIGGINS and PARKIN (1962), BOWEN (1972), KOMAR (1971), and others that beach cusps and rip currents may be associated with edge waves which are orientated along the beach direction. These waves are mentioned by LAMB (1932) and a fuller theoretical treatment is given by URSELL (1952) and an indication of the theory is given by GROEN and GROVES (1962).

Taking axes with x normal to the beach and y along the beach, if U is the volume transport in the x direction and V in the y direction and η the elevation, we have:

$$U = X(x)\exp(iky+\sigma t), \tag{1}$$
$$V = Y(x)\exp(iky+\sigma t), \tag{2}$$
$$\eta = Z(x)\exp(iky+\sigma t) \tag{3}$$

where k is the wave number and σ the angular frequency.

Following GROEN and GROVES, we can obtain from the equations of motion and continuity, ignoring the effect of the Earth's rotation, the following equations:

$$i\sigma X + gh(x)Z' = 0, \tag{4}$$
$$i\sigma Y + igh(x)Z = 0, \tag{5}$$
$$X' + ikY + i\sigma Z = 0 \tag{6}$$

where $h(x)$ is the water depth, taken to be a function of x only.

$$\frac{\partial^2 Z}{\partial x^2} + \frac{\dfrac{\partial Z}{\partial x} \cdot \dfrac{\partial h(x)}{\partial x}}{h(x)} + Z\left(k - \frac{\sigma^2}{gh(x)}\right) = 0. \tag{7}$$

When the gradient is constant, $h(x)$ is a linear function of x and the only convergent solutions of the equation are Laguerre functions with a finite number of terms and these can only exist if:

$$\sigma^2/k^2 = \frac{g(2n+1)\sin\beta}{k} \tag{8}$$

where $h = x\sin\beta$ and $n = 0, 1, 2, 3, \ldots$

When gradient is not constant, the equation can be solved numerically by a step by step

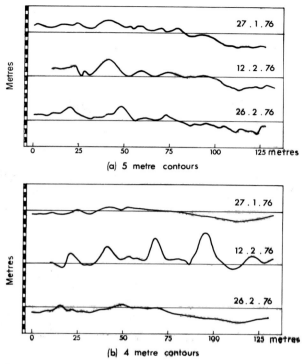

(a) 5 metre contours

(b) 4 metre contours

Fig. 12. Plot of (a) 5-m contours and (b) 4-m contours for 27 January, 12 February and 26 February 1976. Details are the same as for Fig. 5.

process, using the Runge–Kutta method, taking a given value of σ and then using trial values of k until both $\partial Z/\partial x$ and Z approach zero together within a distance of 40 m, as with edge waves of the frequency, mode and wavelength discussed in this paper, the amplitude diminishes to less than 1% of the shore-line value at this distance.

This method was used with all the examples given in Table 2 as the beach profiles were known in each case. There is still, however, some uncertainty about the exact profile to use as the cusps would be formed over a range of tidal levels. Some allowance is made for this. The results are shown in Table 3.

Table 3. *Edge-wave wavelengths and cusp spacings*

Date	Angular frequency	Edge-wave number ($(k)\,\mathrm{m}^{-1}$)	Edge-wave wavelength ($(2\pi/k)\,\mathrm{m}$)	Cusp spacing (m)	Ratio of lengths
18 July 1975	0.698	0.45, 0.133	13.99, 47.24	20.0	2.36
20 August 1975	0.503	0.147	42.74	22.9	1.87
	0.754	0.330	19.00	11.4	1.66
3 February 1976	0.322	0.109	62.83	29.0	2.16
5 September 1975	0.534	0.423	14.69	6.0	2.45
26 February to 3 March 1976	0.471	0.233	27.00	12.0	2.25
21 May 1976	0.785	0.537, 0.212	11.70, 29.64	13.7	2.16

Fig. 13.   Main Beach, looking northwest, 26 February 1976 showing cusps about 12 m apart.

Fig. 14.   Main Beach, looking northwest, 3 March 1976, showing further development of the cusps shown in Fig. 13.

In two cases where there was cusp formation, the cusp spacing was half the wavelength of the first mode edge wave (taking $n = 1$) with the same frequency as the incident waves. In the other cases, the spacing was half the wavelength of the zero mode edge wave (taking $n = 0$). The ratio of the wavelength to the cusp spacing is given in the last column in Table 3.

COMPARISON WITH PREVIOUS WORK

The paper by LONGUET-HIGGINS and PARKIN (1962) has already been mentioned. They measured beach cusps at Tide Mill between Newhaven and Seaford and on the Chesil Beach, Dorset. The mean wave period and mean gradient over the part of the beach where the cusps were formed were determined and also the mean cusp spacing. A histogram of the distribution of values of cusp spacing was also given for each example. There were three examples from Tide Mill and nine from Chesil Beach, the wave period varying from 2.85 seconds to 6.85 seconds, and the beach gradient from 0.09 to 0.20 and the cusp spacing from 9 to 32 ft (2.7 to 9.7 m). The authors only considered edge waves of zero mode and could not find any clear relationship between the cusp spacing and the edge-wave wavelength nor between the cusp spacing and wave period. They did find, however, a linear relationship between cusp spacing and swash length but this did not go through the origin, a cusp spacing of 10 ft (3.03 m) corresponding to zero swash length.

It has been pointed out to me, however, by A. W. LEWIS that if higher-order modes are taken into account, these results do agree with those shown in Table 3 and they are shown in Fig. 18 which shows a plot of half the edge-wave wavelength predicted from the mean gradient and wave period, against the mean cusp spacing. In the graph, numbers indicating the modes are used instead of points. The straight line drawn represents a one to one relationship and all the points fall evenly about it and the scatter is certainly no worse than that obtained by the authors with swash.

KOMAR (1971) describes cusps of spacing 5.8 m at Blue Stone Beach, Crail, Fife, Scotland, but unfortunately no details are given of the wave period and beach gradient.

In Fig. 19, the data of LONGUET-HIGGINS and PARKIN (1962) shown in Fig. 18 are combined with those given in Table 3 and plotted in the same way as in Fig. 18, except that the lengths have been converted to metres. The points again fall evenly about the line and there is clearly a very close relationship between cusp spacing and half the length of an edge-wave mode for all the examples shown.

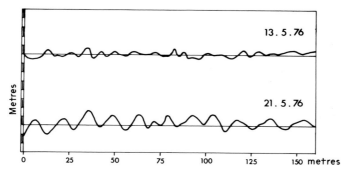

Fig. 15.   Plot of 3.5-m contours with respect to M.L.W.S. for Main Beach, 13 May and 21 May 1976.

Taking both sets together then, there are three separate beaches involved, the wave periods vary from 2.85 seconds to 20 seconds, the cusp spacing from 3 m to 30 m, the beach gradient from 0.07 to 0.20 and the results taken in conjunction prove quite conclusively that the cusp spacing is equal to half an edge-wave mode length. Most of the examples involve zero and first-order modes but there is one example of a second mode.

DISCUSSION

This result, although a simple one, is difficult to explain in terms of present theories.

The formation of beach cusps is closely related to the formation of rip currents and the theory has been worked out by BOWEN (1969) and LONGUET-HIGGINS (1972). The theory

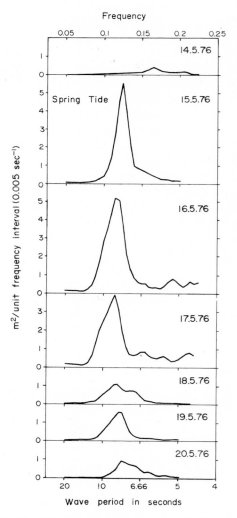

Fig. 16.   Wave power spectra for 14 to 20 May 1976.

ascribes the current formation to the effect of the spatial variation of the wave radiation stress. In the wave direction x in shallow water, the radiation stress is for a progressive wave:

$$S_{xx} = 3/16\,g\rho H^2,\tag{9}$$

where H is the wave height, and in direction y at right angles to the wave direction, the radiation stress is:

$$S_{yy} = 1/16\,g\rho H^2,\tag{10}$$

g and ρ have the usual meanings.

In the breaking or surf zone, H is assumed to vary as the depth d which is the sum of the still water depth h and the wave set-up $\bar{\eta}$.

$$d = h + \bar{\eta}\tag{11}$$

and thus

$$H = \gamma(h + \bar{\eta}).\tag{12}$$

The stresses per unit mass are:

$$\tau_x = -\frac{1}{\rho d}\frac{\partial S_{xx}}{\partial x}\tag{13}$$

and

$$\tau_y = -\frac{1}{\rho d}\frac{\partial S_{yy}}{\partial y}.\tag{14}$$

Hence

$$\mathrm{curl}\,\tau = \frac{\partial}{\partial x}\left(\frac{1}{\rho d}\frac{\partial S_{yy}}{\partial y}\right) - \frac{\partial}{\partial y}\left(\frac{1}{\rho d}\frac{\partial S_{xx}}{\partial x}\right).\tag{15}$$

According to LONGUET-HIGGINS and STEWART (1964), with a stationary wave given by

$$\eta = A\cos ky\cos\sigma t$$

with the wave direction along the line of the beach y, in shallow water:

$$S_{xx} = \tfrac{1}{8}g\rho A^2(1 + 2\cos 2ky)\tag{16}$$

and

$$S_{yy} = \tfrac{3}{8}g\rho A^2.\tag{17}$$

With a stationary wave of this kind, the curl will thus vary in a periodical manner with distance along the beach, giving a series of current and circulation cells of spacing $2\pi/2k$. This analysis does not, however, take account of the shoaling water and of breaking but this has been considered by LIU and MEI (1974) who deal with a stationary wave formed by the total reflection of a progressive wave normally incident on a breakwater extending 350 m out perpendicularly from a shore-line, and where the beach gradient is 0.1. They find that in the surf zone the wave set-up $\bar{\eta}$ varies as $\cos 2ky$ and the current cells set up would have the same spacing.

BOWEN (1972) considers the effect of stationary edge waves on a beach and he distinguishes three cases: (1) stationary edge-waves of wave number k with no incident waves;

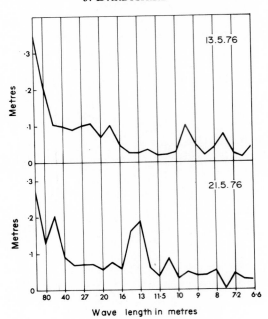

Fig. 17.   Amplitude spectra for 3.5 m contours for 13 May 1976.

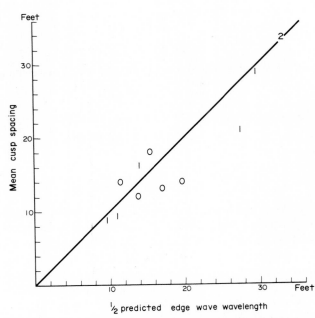

Fig. 18.   Half the edge-wave wavelength predicted from observed wave period and mean beach gradient (for $n = 0$ or 1 or 2), plotted against mean cusp spacing for data obtained by Longuet-Higgins and Parkin (1962). Units in feet. The numbers used as points indicate the value of n used.

(2) stationary edge-waves of wave number k and incident waves of different frequency;
(3) stationary edge-waves of wave number k and incident waves of the same frequency. In the first case, as expected, cusps of spacing $2\pi/2k$ would be set up. In the second case, although there would be some interaction effects, the result would be the same but in the third case a phase relationship would exist between the edge-wave and the incident wave and cusps would be formed with a spacing equal to the wavelength of the edge-wave. In our examples, case (1) is clearly inadmissible and so at first sight is case (2). We are, however, dealing with a narrow spectrum of wave frequencies rather than with a discrete wave frequency and so although the frequencies of the incident and edge-waves would be nearly identical, no permanent phase relationship would exist and so conditions would correspond more to case (2) than case (3).

A more serious objection to a stationary edge-wave theory is that there seems to be no mechanism available to generate such waves. The two headlands shown on the map in Fig. 1 are 6800 m apart and if a stationary wave system were set up between them by the reflection of waves off them, we would be dealing with harmonics of the order 3000 to 4000. The work of LIU and MEI (1972) shows that when 10-second period waves impinge at 45° to a perfectly reflecting breakwater, the cell pattern largely disappears within a distance of 700 m. The headland in our case is not likely to be a perfect reflector but there could be a slight effect at Nant Beach but hardly any effect at all on the Main Beach. No headlands are mentioned in the work described in LONGUET-HIGGINS and PARKIN (1962). It seems, therefore, that the effect of wave reflection off headlands cannot be the principal cause.

The other possible cause is the effect of short-crested waves of form

$$\eta = A \cos k_y y \cos (k_x x - \sigma t).$$

The wave height H would normally be given by $A \cos k_y y$ and vary along the beach. In the surf zone, equation (12) would operate, however, and

$$H = A \cos k_y y \text{ for } \gamma d > |A \cos k_y y| \tag{18}$$

and

$$H = \gamma d \text{ for } \gamma d < |A \cos k_y y|. \tag{19}$$

There is thus a truncated cosine wave variation in the y direction. LIU and MEI (1972) deal with this case when they make waves of 10 seconds period approach at 45° to the breakwater referred to above. Equation (12) is applied by them also. They find that current cells are set up with a spacing of approximately $2\pi/2k_y$. If reflection is ruled out, one must look for a different mechanism to create such waves. They can be formed by two intersecting plane waves and can occur even in mid-ocean but it seems doubtful if the long 20-second period swell arriving at Hell's Mouth would originally have been short-crested. Short crests can be caused by refraction of intersecting wave trains. DALRYMPLE (1975) considers the effect of the refraction of two intersecting wave trains and finds that cusps could be formed with a spacing of:

$$L_0/(\sin\theta_0 - \sin\zeta_0) \tag{20}$$

where L_0 is the original deep sea wavelength and θ_0 and ζ_0 the original angles of incidence of the two trains. The least spacing possible would thus be $L_0/2, (\theta_0 = \pi/2, \zeta_0 = -\pi/2)$, but this is much too high to explain the spacings we have observed. For 20-second waves, for instance, this would give 310 m as against the observed value of 30 m.

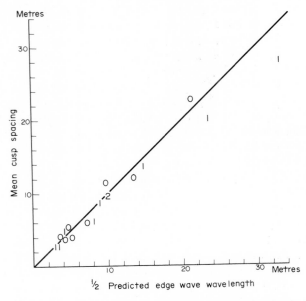

Fig. 19.   Same as Fig. 18 but including data shown in Table 3 and units in metres.

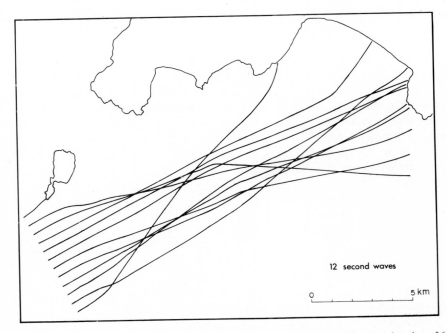

Fig. 20.   Orthogonals for 12-second waves assumed to be originally approaching on a bearing of 61.5°.

Refraction can, however, be effective in another way. Most real beaches, in contrast to artificial models, contain endless irregularities in depth and shape and any wave refraction pattern is very complicated. Thus in Fig. 20 there are orthogonals drawn for 12-second waves progressing originally on a bearing of 061.5°. There is a good deal of convergence in some places followed by divergences in others so there would be an appreciable variation of wave amplitudes along the beach. There are variations over a 'wave-length' of 500 m or so but these are still too large. Figure 20 is, however, based on the Admiralty chart and no account can be taken of minor variations in beach topography near the shore-line which change from day to day. Bearing this in mind, a survey was carried out on 13–14 April 1976 over a part of Nant Beach extending from the cliff edge to 220 m away and over a length of 240 m, taking a 10-m square grid. Refraction diagrams were prepared, based on this survey, for 5, 6, 10 and 12 seconds and these are shown in Fig. 21. Two bearings of approach, 072° and 067°, were taken as these correspond roughly to the bearing the waves shown in Fig. 20 would have in this region. The lower period waves could be generated by the wind in this direction. There is often a marked difference in the convergence pattern for the two bearings although they only differ by 5°. The distance between convergences is now of the order of 20–30 m which is within the range we are interested in. The wave height on the beach would thus be a very involved quasi-periodic function of y which would be synthesized by the superposition of many simple sinusoidal variations with different wave numbers. Those components with wave numbers corresponding to edge-wave modes would grow at the expense of the others. It is not easy to give a clear-cut physical reason for this but a possible cause is that when orthogonals converge, the waves usually lose energy by breaking but if the wave number corresponds to an edge-wave mode, the system can be 'defused' by the setting up of two sets of progressive edge-waves which originally overlap and then leave in opposite directions along the beach, and the stationary wave pattern could be maintained. Waves have in fact been observed progressing along the beach during cusp formation.

The cusps then would not be affected so much by the beach contours in the surf zone but rather by the bottom topography a little way offshore which determines the refraction pattern. Hence it could be explained why some waves create cusps and some do not, since cusp formation depends on the wave refraction pattern which in its turn is determined by the period and direction of the waves as well as bottom topography. A change in any of these conditions would drastically alter the refraction pattern.

CONCLUSION

The spacing of wave cusps at Hell's Mouth Bay is clearly related to the period of the incident swell and the beach gradient. The cusp spacing being half the wavelength of an edge-wave mode. These results are in agreement with those found from the data obtained by LONGUET-HIGGINS and PARKIN (1962). These results are consistent with the existence of stationary waves of the same wave number as that of an edge-wave mode, giving rise to a periodic spatial variation of radiation stress along the beach. It is suggested that these waves are formed by the variation of wave height along the beach due to refraction.

Observations are still being taken and it is intended to record the actual formation of these cusps by time-lapse photography and to study more closely the effects of wave refraction, constructing diagrams like those in Fig. 21 which would apply to the examples studied.

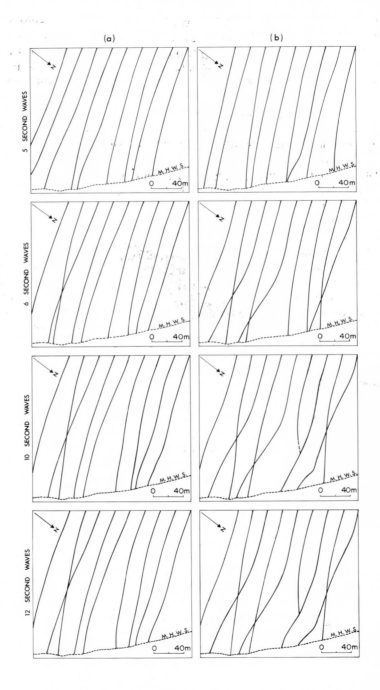

Fig. 21.   Orthogonals for 5-, 6-, 10- and 12-second waves approaching Nant Beach. (a) Original bearing 71.5 and (b) original bearing 66.5°.

Acknowledgements—The author wishes to thank the Natural Environment Research Council for a grant to carry out this work.

He is very greatly indebted to Mrs. E PRITCHARD for plotting and analysing all the surveys and preparing the drawings for this paper, Mr. A. W. LEWIS who assisted with the work until the end of September 1975 for much valuable discussion and for setting up the original wave buoy system, Mr. D. COLLINS who took over in October 1975 and with Mr. ALUN ROBERTS set up the present system, and to Mrs. R. HODDER for typing the manuscript.

REFERENCES

BLACKMAN R. B. and J. W. TUKEY (1958) *The measurement of power spectra*, Dover Publications, New York, 190 pp.

BOWEN A. J. (1969) Rip currents. Part I: Theoretical investigations. *Journal of Geophysical Research*, **74**, 5467–78.

BOWEN A. J. (1972) Edge waves and the littoral environment. *Proceedings, 13th Conference on Coastal Engineering*, Vol. II, chap. 71, p. 1313.

BOWEN A. J. and D. L. INMAN (1969) Rip currents. Part II, Laboratory and field measurements. *Journal of Geophysical Research*, **76**, 8662–8671.

DALRYMPLE R. A. (1975) A mechanism for rip current generation on an open coast. *Journal of Geophysical Research*, **80**, 3485–3487.

GROEN P. and G. W. GROVES (1962) Surges, In: *The Sea*, Vol. I: *Physical Oceanography*, M. N. HILL, editor, pp. 611–646.

HUGHES E. and A. EAMES (1975) *Porthmadog Ships*, Gwynedd Archives Service, 426 pp.

KOMAR D. K. (1971) Nearshore cell circulation and the formation of giant cusps. *Geological Society of America Bulletin*, **82**, 2643–2650.

LAMB H. (1932) *Hydrodynamics*, Cambridge University Press, 738 pp.

LEWIS A. W. Personal communication.

LIU P. L. F. and C. C. MEI (1974) Effects of a breakwater on nearshore currents due to breaking waves. *Report 192*, R. M. Parsons Laboratory, Massachussets Institute of Technology, Cambridge, Mass.

LONGUET-HIGGINS M. S. L. (1972) Recent progress in the study of longshore currents. *Waves on Beaches*, Academic Press, New York and London, pp. 203–248.

LONGUET-HIGGINS M. S. L. and D. W. PARKIN (1962) Sea waves and beach cusps. *Geographical Journal*, **128**, 194–201.

LONGUET-HIGGINS M. S. L. and R. W. STEWART (1964) Radiation stresses in water waves, a physical discussion with applications. *Deep-Sea Research*, **11**, 529–562.

URSELL F. (1952) Edge waves on a sloping beach. *Proceedings of the Royal Society*, A, **214**, 79–97.

Waves at Shambles Light Vessel, Portland Bill, during 1968

Laurence Draper

Institute of Oceanographic Sciences, Wormley, Godalming, Surrey, U.K.

Abstract—Waves have been measured by a Shipborne Wave Recorder[†] fitted to Shambles Light Vessel. The recordings were made systematically over a complete year and the results are presented in standardized ways which have been developed to make them easily usable for many practical purposes.

DESCRIPTION OF THE INVESTIGATION

WAVES have been recorded by a Shipborne Wave Recorder (TUCKER, 1956) placed on the Shambles Light Vessel which was stationed in 15 fathoms of water off Portland Bill, at 50°31′N, 2°20′W. The records for the complete year 1968 have been analysed, mainly following the method of analysis developed by TUCKER (1961) from theoretical studies by CARTWRIGHT and LONGUET-HIGGINS (1956). The method of presentation is that recommended for data for engineering purposes (DRAPER, 1966).

Records were taken for 15 minutes at 3 hourly intervals, and the analysis yields the following parameters:

(a) H_1 = The sum of the distances of the highest crest and the lowest trough from the mean water level.

(b) H_2 = The sum of the distances of the second highest crest and the second lowest trough from the mean water level.

(c) T_z = The mean zero-crossing period (the mean interval between successive crossings in an upwards direction of still water level).

(d) T_c = The mean crest period.

From these measured parameters the following parameters have been calculated, after allowing for instrumental response:

(e) H_s = The significant wave height (mean height of the highest one-third of the waves): this is calculated separately from both H_1 and H_2, and an average taken. The relationship between the parameters is $H_1 = f \cdot H_s$ where f is a factor related to the number of zero-crossings in the record (TUCKER, 1963). A similar relationship is used for the calculation of H_s from H_2.

(f) $H_{max\,(3\,hours)}$ = The most probable value of the height of the highest wave which occurred in the recording interval (DRAPER, 1963).

[†] The calibration of the instrument and the subsequent analysis were made using the Imperial system, and are in feet and seconds.

(g) ε = The spectral width parameter, which is calculated from T_z and T_c (TUCKER, 1961):

$$\varepsilon^2 = 1 - (T_c/T_z)^2$$

RESULTS

The results of these measurements are expressed graphically, divided into seasons thus:

Winter:	January	February	March
Spring:	April	May	June
Summer:	July	August	September
Autumn:	October	November	December

For each season a graph (Figs. 1–4) shows the cumulative distribution of significant wave height H_s, and of the most probable value of the height of the highest wave in the recording interval, $H_{max(3 hours)}$.

The distribution of zero-crossing period is given for each season (Figs. 5–8).

The distribution of the spectral width parameter is given for the whole year (Fig. 9).

Figure 10 is a scatter diagram relating significant wave height to zero-crossing period.

Figure 11 is a persistence diagram for the whole year.

Figure 12 is a plot of $H_{max(3 hours)}$ on probability paper for the whole year.

Figure 13 is a plot of $H_{max(3 hours)}$ on Weibull probability paper, for the whole year.

Plate 1.   Waves at the Shambles Light Vessel.

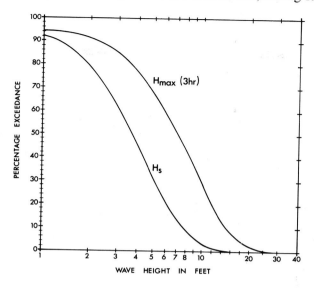

Fig. 1.　Percentage exceedance of H_s and H_{max} Winter 1968, January–March.

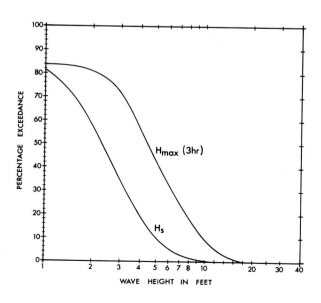

Fig. 2.　Percentage exceedance of H_s and H_{max} Spring 1968, April–June.

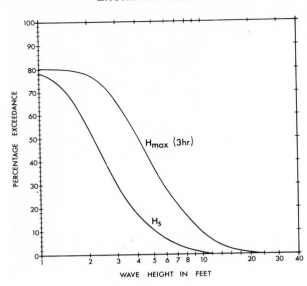

Fig. 3.   Percentage exceedance of H_s and H_{max} Summer 1968, July–September.

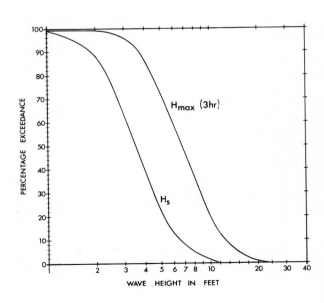

Fig. 4.   Percentage exceedance of H_s and H_{max} Autumn 1968, October–December.

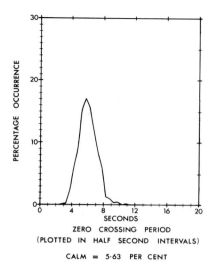

Fig. 5.

Percentage occurrence of the zero-crossing period (T_z) Winter 1968, January–March.

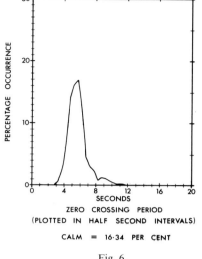

Fig. 6.

Percentage occurrence of the zero-crossing period (T_z) Spring 1968, April–June.

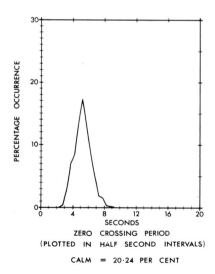

Fig. 7.

Percentage occurrence of the zero-crossing period (T_z) Summer 1968, July–September.

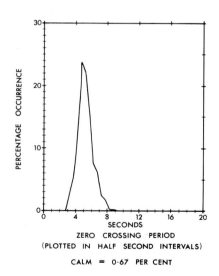

Fig. 8.

Percentage occurrence of the zero-crossing period (T_z) Autumn 1968, October–December.

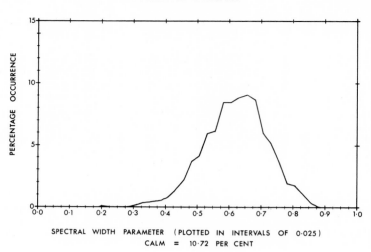

Fig. 9.   The percentage distribution of the spectral width parameter for the whole of 1968.

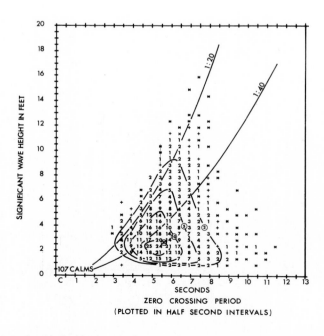

Fig. 10.   Scatter diagram for the significant wave height and zero-crossing period in parts per thousand for 1968.

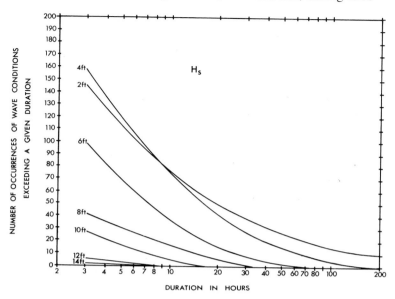

Fig. 11.   Persistence of wave conditions for 1968.

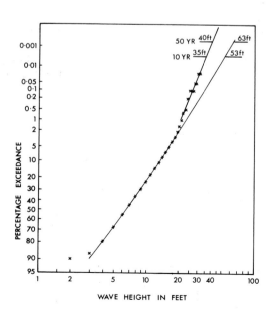

Fig. 12.   Lifetime wave prediction for 1968 assuming normal probability.

DISCUSSION

Wind conditions

Wind conditions during the year of wave measurement are relevant to the conclusions of this report. There is no site near to Portland from which long-duration wind records were available and from where good records were available for 1968. The nearest such site is Hurn airport, Bournemouth, about 24 nautical miles from Shambles Light Vessel.* At Hurn the mean speed in 1968 was 93% of a 9-year mean. On this basis it is reasonable to deduce that the wave heights measured at Shambles were perhaps 10–15% lower than average, and wave periods, at least the locally generated ones, were perhaps 4% lower than average. The periods of the swell waves are, of course, virtually unaffected by the local winds.

Highest waves and calmest months

The highest values of H_1 (after correction for instrumental response) are 26.5 ft on 4 February (zero-crossing period 7.13 seconds), 23.2 ft on 4 February (6.76 seconds) and 22.4 ft on 15 January (7.58 seconds). The steepness of the highest wave (defined as the ratio of wave height: wave length), using zero-crossing period, is 1:9.8.

The calmest months were July and August, with 34.7% and 25% respectively of the records classified as calm. (A record is classified as calm when H_1 is less than 1 ft.)

Percentage occurrence of wave heights

Figures 1–4 indicate for what proportion of time H_s or $H_{max(3\ hours)}$ exceeded a particular height. Comparison of these figures shows, as would be expected, that in the winter and autumn months there were more waves exceeding a particular height value than in the spring and summer months. For example, in the winter the significant height exceeded 6 ft for 21% of the time, whilst in the summer it exceeded 6 ft for only 6% of the time.

Zero-up-crossing periods

Figures 5–8 show relatively little seasonal variation in the zero-crossing period, which lies in the range 3 seconds to 9 seconds (with a few occurrences of periods up to 12 seconds in winter and spring). Taking the year as a whole, the most commonly occurring periods are between 5 seconds and 6 seconds.

Spectral width parameter

Figure 9 shows the spectral width parameter to lie mainly between 0.4 and 0.9, which is the upper half of its possible range; and the most common value is between 0.6 and 0.7. This implies that during 1968 both wide- and narrow-band spectra occurred.

---

* 1 nautical mile = 1.853 km.

Scatter diagram

The scatter diagram (Fig. 10) shows, in parts per thousand, the numbers of occurrences of particular combinations of zero-crossing period and significant wave height. It indicates that, for the whole of 1968, the waves most often encountered at this location had zero-crossing periods of about 5 seconds, and significant heights of about 2 ft. It can be seen from the diagram that there are no waves shown having periods less than 3 seconds. This is because the rapid attenuation with depth of the shorter waves means that the pressure units, which are necessarily situated at about 4.9 ft below mean water level, do not record waves which have a period less than about 3 seconds.

Lines of wave steepness of 1:20 and 1:40 are shown on this diagram. Wave length is calculated from the zero-crossing period using the formula for waves in deep water $L = gT_z^2/2\pi$; the wave height here is the significant wave height. The great majority of the waves represented on this diagram are less steep than 1:14.

Persistence diagram

From the persistence diagram (Fig. 11) may be deduced the number and duration of the occasions in 1 year on which waves persisted at or above a given height. For example, if the limit for a particular operation of a vessel is a significant height of 6 ft, it would have been unable to operate for spells in excess of 10 hours on thirty-seven occasions, or spells in excess of 24 hours on eleven occasions.

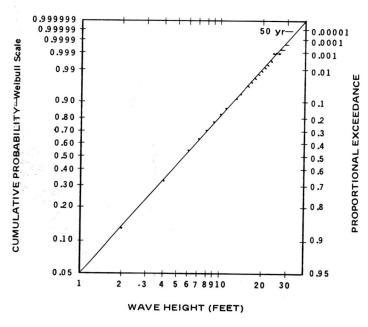

Fig. 13.   Lifetime wave prediction for 1968 using Weibull probability paper.

'Lifetime' wave prediction (DRAPER, 1963)

Values of $H_{max\,(3\,hours)}$ have been plotted on both log-normal probability and Weibull probability paper. The latter yields a 50-year lifetime wave height of 37 ft whilst the former suggests a value of 40 ft. However, on the former, the bulk of the data fits a smooth, almost straight, line very well up to a wave height of 20 ft but falls away severely above that value. It is the values above 20 ft which have been used to extrapolate to the extreme (50-year) value of 40 ft. It is possible that Shambles bank, which the vessel marked, protects the site from the higher waves, and that although the 40-ft value is the appropriate value for the actual site, a value for open water should come from an extrapolation of the gentle curve from the bulk of the data. This yields a 50-year wave of 63 ft. This value is consistent with results from another nearby recorder and with predictions from wind.

Acknowledgements—The author wishes to thank the Corporation of Trinity House for permission to install the equipment on their vessel, the masters and crew for operating it, the Meteorological Office for providing wind data and several colleagues for help with the analysis. The assistance of Mr. BRIAN FORTNUM in the drafting of the paper is much appreciated.

REFERENCES

CARTWRIGHT D. E. and M. S. LONGUET-HIGGINS (1956) The statistical distribution of the maxima of a random function. *Proceedings of the Royal Society*, A, **237**, 212–232.
DRAPER L. (1963) The derivation of a 'design-wave' from instrumental measurements of sea waves. *Proceedings of the Institution of Civil Engineers*, **26**, 291–304.
DRAPER L. (1966) The analysis and presentation of wave data—a plea for uniformity. In: *Proceedings of the Tenth Conference on Coastal Engineering*, Tokyo, American Society of Civil Engineers, 1–11.
TUCKER M. J. (1956) A shipborne wave recorder. *Transactions of the Institution of Naval Architects, London*, **98**, 236–250.
TUCKER M. J. (1961) Simple measurement of wave records. In: *Proceedings of the Conference of Wave Recording for Civil Engineers*, L. DRAPER, editor, National Institute of Oceanography, 22–23.
TUCKER M. J. (1963) Analysis of records of sea waves. *Proceedings of the Institution of Civil Engineers*, **26**, 304–316.

A brief review of sampling techniques and tools of marine biology

M. R. CLARKE

Marine Biological Association, Plymouth

MARINE biology is the study of every aspect of life in salt water, from upper tidal limits to deep sea, from water surface to within the bottom sediments, from bacteria to whales, from chemical composition to species distribution, from the equator to the poles. Such a broad study advances over a wide front by application of mathematics, physics, chemistry, engineering, electronics as well as biology itself. Dependence on non-biological subjects is nowhere more evident than with basic techniques of observation and collection of living organisms. The enormous diversity of problems can only be attacked by a correspondingly enormous diversity of techniques of observation, collection and analysis. Each problem requires a different approach; obviously bacteria and whales cannot be sampled in the same way; less obviously a net fished horizontally, vertically or obliquely will select species in different ways so giving a different picture of what is present in the water. There is no universal method of observation or collection. Each biologist must have a clearly defined aim and must endeavour to find the gear and the method to fulfil this aim. Limitations of the gear must be studied and only questions should be asked which the sampling technique can answer—obvious but sometimes subtly ignored. While a review of techniques and tools of marine biology should perhaps start with a list of aims, such a task is impossible for such a broad subject and we must be content with an appraisal of

OBSERVATION

Techniques of observation, collection and analysis can sometimes be taken over from terrestrial biology. On the sea shore, above low tide, similar methods of observation, counting and collection can be applied. Some of these may also be used without too much difficulty with the aid of aqualungs, to depths of about 30 m. Some observations may be made by merely looking into the sea surface, but even in the clearest waters this is of little practical value deeper than a few metres. Life within the great horizontal and vertical expanse of the ocean and seas cannot be observed without sophisticated apparatus of some kind and collection of organisms involves gear unparalleled in the world above water. the most commonly used methods and their limitations.

Shore and inshore waters

Many terrestrial biological studies are dependent upon pure observation—bird

439

migration, ecological grouping of plants, effect of air pollution on evolution of moths—but how far will observation alone take us in marine biological studies? On rocky shores attached and boring animals can be readily observed at low tide. They can be counted by use of the quadrat or square frame and densities per square metre compared. Even this simplest of observation technique is hampered by subjectivity in site selection on many eroded shores. Sand or mud may be observed more objectively but most associated organisms are subsurface and must be dug out. Statistical methods developed by terrestrial ecologists can often be used to analyse shore observations and reduce, if not entirely remove, subjectively in drawing conclusions and may help in site selection. Some experimental ecology has been carried out on rocky shores by removal of dominant animals or plants to study effects of their removal. General observations on habitat, settlement, growth and feeding methods provide the background knowledge of life style necessary before more detailed studies of many kinds are made (CONNELL, 1974). Observation of deeper water animals which have been stranded in shallow water has proved very useful particularly in the study of cetacean distribution and giant squids (e.g. SHELDRICK, 1976; REES, 1955).

Below low tide levels the diver can observe much of interest on behaviour of mobile, as well as sessile, animals. For slow-moving animals or for plants he can use a quadrat frame or for larger animals count those within a metre or two of a previously pegged-out line of known length, or use photogrammetry (LUNDÄLV, 1971; MACHAN and FREDA, 1975). Indeed diving offers the only method whereby steep rocky slopes and cliffs, caves, canyons and boulders can be studied since adequate observation or sampling from the surface is rarely possible. The principal limitation on diving without resort to elaborate support systems involving habitats and control of gases for breathing are depth and time. Thirty metres is the depth limit for anything but the briefest work and 20–90 minutes at shallower depths on the normal quantities of compressed air. A diving scientist needs to plan his work with great care to achieve useful results in a work day of 1 hour. Visibility can be severely limited over soft sediments.

The time available for work at the bottom has been greatly extended by the use of undersea habitats such as the 'Tektite' series of experiments (MILLER, VAN DER WALKER and WALLER, 1971; HERRNKIND, 1974) in which scientists lived underwater for 2 or 3 weeks, and made numerous excursions totalling as much as 8 hours in one day. While these experiments showed great potential for studying the biology, particularly the behaviour, of animals near the habitat, the costs and hazards involved are serious limitations to the wide use of this technique.

Many techniques have been developed to improve vision underwater, ranging from simply placing clear water-filled bags between the eye and the target to sophisticated image intensifiers (COCKING, 1975).

Deeper on the shelf

Moving offshore, the bottom soon draws out of reach of practical diving depth and observations by divers in midwater have contributed very little to biological studies. The reasons for this are that diving in other than coastal waters involves use of a relatively large and costly ship, several divers in healthy condition and a concisely defined problem which observation will solve. Scientific returns rarely justify the aiming of a cruise toward a diving operation in midwater. Among the exceptions are the fascinating analysis of the fine structure of the thermocline with the aid of dye (WOODS and FOSBERG, 1967), a physical observation which must greatly influence future interpretation of biological data, the special study of gelatinous plankton and their filter feeding structure (HAMNER, MADIN, ALLDREDGE, GILMER and HAMNER, 1975), the study of fish behaviour in relation to trawls during which divers actually hang on to the netting while making observations (HEMMINGS, 1971) and a study of animals associated with floating weed and thermal discontinuities (PINGREE, FORSTER and MORRISON, 1974).

Surface observations

Some useful observations can be made above the surface, or rather while breathing atmospheric air, with much less risk, discomfort and organization than the diver must endure. Some oceanographic ships are equipped with transparent subsurface ports which have proved useful in studying dolphins and fish, but even from the deck of a ship, bird (e.g. BAILEY, 1966), whale and squid distributions (e.g. BAKER, 1960) can be studied. Another variant for the study of oceanic communities which become attracted to drifting objects is the manned raft; a version with underwater ports has produced a fascinating insight into a variety of fishes attracted to it (GOODING and MAGNUSEN, 1967). A large tube with ports has been used for observation under sea ice in the Antarctic and direct observation from a high platform to study shark and cephalopod behaviour has proved useful (HODGSON and MATHEWSON, 1971; YARNALL, 1969). Aircraft cover large areas in a short time and have proved valuable for observing fish stocks (CRAM, 1974), whale behaviour (PINKERTON and GAMBELL, 1968), phytoplankton blooms and pollution. Extension of this remote observation is involving the development of remote sensing systems operating from satellites (DYRING, 1973; JOSEPH and STEVENSON, 1974). The first earth resource satellite ERTS-1 was launched in July 1972 and offers the possibility of operational scanning of biological changes over a long period. Such satellites may provide information on pollution, phytoplankton and marine resources by relaying pictures from a battery of cameras each sensitive to different bands of the electromagnetic spectrum, the solar radiation band being restricted in its use to cloudless days, microwave bands including infra-red and radar without such restriction. For marine biology information from remote sensors on specially launched rockets is possible in the future but at present it seems the expense would hardly be justified by scientific returns.

Deeper in the sea

What of the sea at depths greater than the diver can penetrate? Here, for observation we must rely upon manned (PHLEGER and SOUTAR, 1971; STRASBERG, JONES and IVERSEN, 1968; PARRISH, AKYUZ, ANDERSON, BROWN, HIGH, PERES and PICCARD, 1972) and un-manned submersibles (e.g. ALEXANDER, 1975; TIEMON, 1972), or recording devices such as echo sounders and cameras. Since direct human observation is always superior to even the best undirected recording equipment, submersibles give the most information per observation. During one dive something definite about vertical distribution and attitude of fish or squids may be ascertained, great populations of midwater fish may be identified (BACKUS, CRADDOCK, HAEDRICH, SHORES, TEAL, WING, MEOD and CLARKE, 1968) and distribution of some commercial species of bottom-dwelling fish can be studied (CADDY,

1976). Observations are often only relevant to a specialist in a particular animal group and it is often difficult to plan dives involving the right personnel to make the best of opportunities presented. Observations allowing firm scientific conclusions are few relative to the high cost of submersible operation and the risk to which personnel are exposed. Cost of operation can be reduced by lowering by wire a simple chamber with an air-

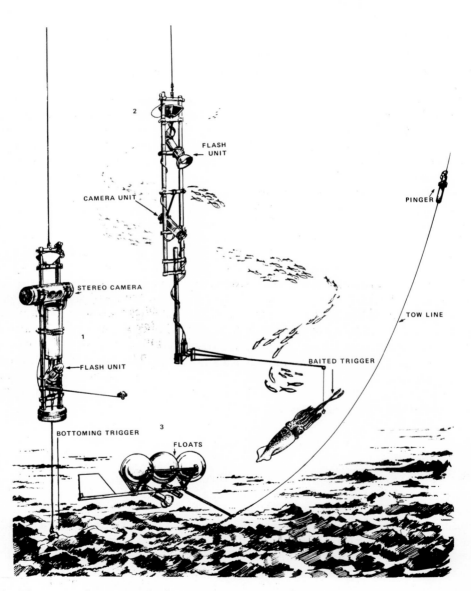

Fig. 1.   Three types of camera application: 1, stereo camera triggered by touching the bottom; 2, midwater camera triggered by animal taking the bait; 3, time-lapse camera takes a continuous series of pictures of the bottom above which it is towed.

bottle supply. This is probably rather more dangerous than the failsafe submersibles which automatically pop to the surface in the event of electrical failure but has been used in observing the behaviour of some commercial fish on the bottom.

Sonar

Since the scattering layers detected by sonar were first shown to be caused by animals, several studies have been devoted to determining the species in oceanic water (CURRIE, BODEN and KAMPA, 1969; GIBBS, ROPER, BROWN and GOODYEAR, 1971).

Little information concerning individual species can be obtained without the use of concomitant sampling or direct observing techniques. However, where the scattering is from a single species, echo-sounding can give useful data on distribution, numbers and diurnal behaviour (BODEN, 1971; BACKUS *et al.*, 1968; REVIE, WEARDON, HOCKING and WESTON, 1974). Arrays of transducers have been used to produce a narrow beam sonar capable of mapping a near-plan view of a large fish shoal (RUSBY, SOMERS, REVIE, MCCARTNEY and STUBBS, 1973; NEWTON and STEFANON, 1975). A similar array set up on a coastline could give useful information on fish migration, although such devices have defence roles which usually preclude biological interpretation. Sound detection arrays have been used in a few instances to track large animals (SCHEVILL, WATKINS and BACKUS, 1964; WATKINS and SCHEVILL, 1971). Sonar which scans a sector continually can give a ciné-film type record of fish shoals and this has proved useful in studying the reaction of fish to trawls (JONES, 1969; CUSHING, 1969).

Fig. 2.   This rare 12-cm medusa *Solmissus incisa* took its own photograph at 1000 m; its own luminescence was picked up by a phototube which triggered the electronic flash. The 2-cm watch provides both scale and time of exposure.

Cameras

Cameras, still, ciné and television are cheaper and less dangerous to use than submersibles. The problem of making cameras, cables and lighting units watertight under immense pressures has long been solved and such units are relatively inexpensive. Cameras may be triggered by midwater animals pulling at a baited hook (Fig. 1), by a weight touching bottom, by a clock at regular intervals or even by a luminescent flash of a passing animal (Fig. 2). In oceanic waters there is little return per cost when hanging a camera in midwater, since much expended ship time produces few pictures; in the northeast Atlantic the majority of pictures turn out to be squids of two species. Bottom photographs can give useful information on density of animals even when taken at random by bouncing the camera along the bottom as the ship drifts (HEEZEN and HOLLISTER, 1971; MARSHALL and BOURNE, 1964; SOUTHWARD, ROBINSON, NICHOLSON and PERRY, 1976) and on rarely caught species such as the cirrate octopods (ROPER and BRUNDAGE, 1972; PEARCY and BEAL, 1973). The approach of the camera is probably detected by many animals and so the active ones may never be included in such surveys. The recent development of pop-up cameras has greatly increased the ratio of photographing time to ship time (ISAACS and SCHWARTZLOSE, 1975). A number of cameras can be dropped from a ship, each with a can of bait to attract fish.

Fig. 3.   A large shark, with a prawn on top of its head, and several smaller fish round a bait at 1890 m off Hawaii.

They will descend to the bottom and photograph or film the area around the can every few minutes or hours, and then will pop to the surface when a command signal is produced by the accompanying ship. This has given truly remarkable pictures of many species and has suggested that many species are not so rare and some have greater depth limits and horizontal distributions than was thought previously (Fig. 3). Unfortunately it is not known how far the animals travel to reach the bait so that absolute estimates of density cannot be made at present (BROWN, 1975). Most difficulties with cameras are those of photographic technique but identification of many oceanic species from photographs is often impossible even by a specialist who may need fin ray or even vertebral counts for certainty! Television

Fig. 4.   Sledge bearing television camera for bottom photography. On deck are video recording equipment, generator for power supply and cable.

can be useful for bottom surveys in shelf waters particularly if used on a towed sledge (Fig. 4). Records can be stored on videotape and analysed later. In deeper water its use has not been fully exploited since television cables to the ship are heavy and, if too long, would break under their own weight. Television used in midwater often provides tantalizing but useless glimpses of out-of-focus animals; so its use is limited to a few biological problems where animals can be held within the field of focus as in studies involving their reaction to nets. However, many new developments are taking place and, in particular, the 'anthropomorphic' TV vehicle in which movements of the television camera are controlled by movements of the operator's head seem very hopeful (MYRBERG, 1974; STRICKLAND, 1973).

The use of holographic techniques for studying fine-scale plankton distribution shows promise (STEWART, BEERS and KNOX, 1974).

COLLECTION

While observation can often supply useful information on distribution and behaviour most biological investigations require collection of specimens or samples. For anatomical and many physiological studies all that is necessary is the collection of a few specimens in good condition. For many ecological studies, however, the exact position of the specimens in time and space needs to be related to the immediate physical or biological environment and such sampling often requires more sophisticated tools and techniques. Difficulties vary with the environment and with the type of animals being collected; the deep sea is more inaccessible than the shore, and very soft-bodied coelenterates and ctenophores are almost impossible to catch in as good condition as muscular fish, but are more readily caught. Different targets need different sights although the intention remains the same.

Shore

Life of the shore may be directly collected and their environments sampled, measured or monitored by instruments placed by hand. Sampling may be made quantitative by using appropriately sized quadrats depending on the organisms being studied. Even in such accessible regions there are many problems to sampling effectively and quantitatively. Rocky shores are characterized by an enormous diversity of habitats so that *typical* conditions cannot readily be defined and categorized; here the microhabitat is all important and to monitor even such things as temperature, salinity and exposure to wave action is difficult without altering the habitat concerned. Other difficulties in shore monitoring are susceptibility of instruments to wave action and corrosion as well as possible interference from inquisitive beachcombers. Sandy and muddy shores can be characterized by the spectrum of particle size which affects the interstitial space, and the organic content of the substrate. Organisms are patchy even in these seemingly homogeneous environments. Larger animals are collected quantitatively by hammering in quadrats with deep metal sides, digging out the sediment to a standard depth and then hand sorting it or passing it through graded sieves. The very small animals—the meiofauna—are usually sampled by using a 1 cm^2 core which may be only 2 or 3 cm long in fine mud but must be perhaps a metre long in sand where the meiofauna extends much deeper. Main problems in handling samples for extraction of meiofauna are the high degree of patchiness and species diversity even within a square metre; there may be more than 100 species of nematode worm, for example, in 1 cm^2, animals may exceed a million per square metre in density. Each type of organism requires different treatment before identification. Crustacea are tough and can be preserved and extracted later while in contrast flat worms are delicate and must be identified while still alive. Aids to the extraction of the meiofauna include the use of ultrasonics (THIEL, THISTLE and WILSON, 1975). Injection of sediments with epoxyresins, thus trapping organisms *in situ* enables close study of their relationships to sediments. Epoxy castes have also been used to study burrows of larger animals.

Fortuitous strandings of large Cetacea, fish and squid, not only supply specimens of some rare species, but their stomachs sometimes contain a wealth of deep sea squids and fish. Such remains also occur in regurgitats of sea birds at their nesting sites and from stomachs of commercially exploited species such as whales. While stomach-content examination requires persistent qualities in the collector, it provides much valuable information on cephalopod (CLARKE and MACLEOD, 1974; ASHMOLE and ASHMOLE, 1967; IMBER, 1975) and fish species (FITCH and BROWNELL, 1968) which are seldom caught by other sampling techniques.

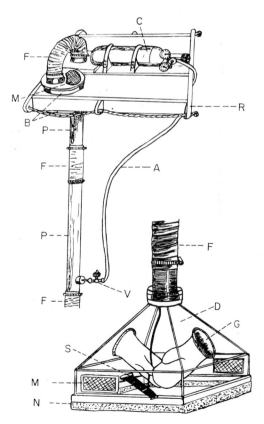

Fig. 5. Finnish suction sampler for hard bottoms: (A) air hose; (B) plastic bucket, with open bottom covered with fine steel mesh; (C) air cylinder; (D) perspex funnel; (F) flexible pipe; (G) latex glove; (M) stainless-steel mesh; (N) foam plastic; (P) rigid plastic pipe; (R) raft; (S) scraper; (V) air valve.

Shallow seas

Below water, chemicals like the toxic rotenone or anaesthetic quinaldine compounds have been used to drive clandestine animals from the protection of holes and crevices (Gibson, 1967) so that they can be captured; at least one new species, an octopus, has been discovered in this way. Explosives have sometimes been used to collect all the fish from coral reefs (Talbot, 1965) but such methods can hardly be recommended from the stand-point of conservation. Estuaries and some beaches are suitable for the siting of stationary nets or traps which depend for efficiency on the design of a 'leader' and the placing of it in the correct position relative to fish movement. Such traps can be very complicated and all parts besides the leader are for preventing the escape of fish or squid entering it. Such large nets (Andreev, 1966) are used more for biological and commercial sampling in the Far East than in western countries. Advantages of diving in depths less than 35 m have been touched upon and various techniques for sampling quantitatively on or near the bottom have been devised (Fager, Flechsig, Ford, Clutter and Ghelardi, 1966; Potts, 1976). Quadrats may be used on hard and soft substrates in conjunction with simple mesh bags for general collection or, slurp guns (Wilcox, Meek and Mook, 1974) but for keeping delicate animals alive impervious plastic bags are better. Hard substrates may be sampled by some device like the Finnish suction sampler incorporating rubber gloves in a perspex cone from which detached plants and animals are drawn up to a raft by an airlift (Fig. 5). Soft substrates are more accurately and comprehensively sampled by divers than by samplers operated from the surface. Cores taken gently by hand have proved essential for sampling meiofauna in fine sediments occupying the top 2 or 3 cm; gravity cores operated from the surface displace top mud by a pressure wave as they fall (Hertweck, 1974; Kirchner, 1974). In one diver-operated sampler a 3 h.p. motor on the surface is used to suck sediment through a 3.2-mm mesh from a metal frame hammered into the sediment (Brett, 1964). In a more advanced device suction is obtained by a 12-volt electric motor and batteries thus giving greater ease of handling (Fig. 6) (Emig and Lienhart,

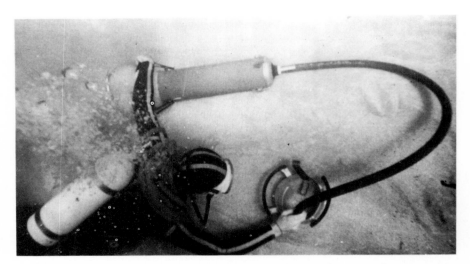

Fig. 6.   Quantitative suction sampler of Emig and Lienhart which contains a propellor powered by a self-contained 12-volt electric motor and batteries and a filter of metal gauze.

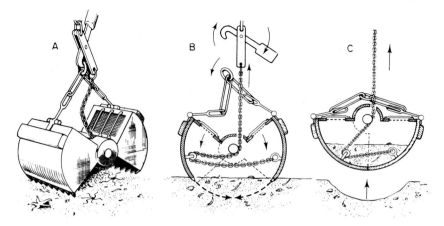

Fig. 7. Petersen grab taking a sample on the sea bed.

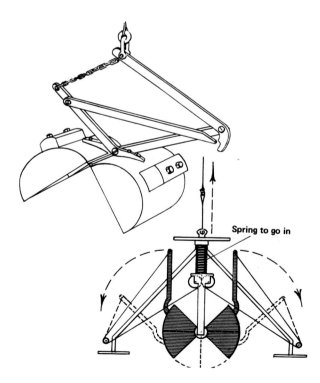

Fig. 8. *Top.* Van Veen grab in which the locking notch disengages when the grab hits the bottom and the long arms are pulled together closing the jaws as the warp is drawn up again. *Bottom.* Smith–McIntyre grab in which the springs drive the bucket into the bottom when the trigger plates touch the sea bed and hauling closes the jaws (drawn from Barnes, 1959, and Smith and McIntyre, 1954).

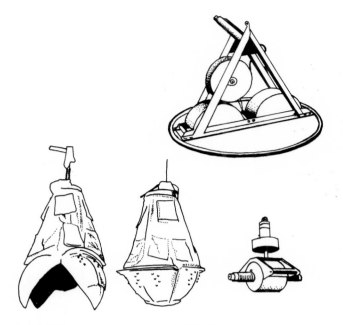

Fig. 9. *Top.* Holme double scoop grab; *lower left and centre*: orange peel grab in open and closed positions; *lower right*: Shipek grab with scoop closed (drawn from photographs in HOLME and MCINTYRE, 1971).

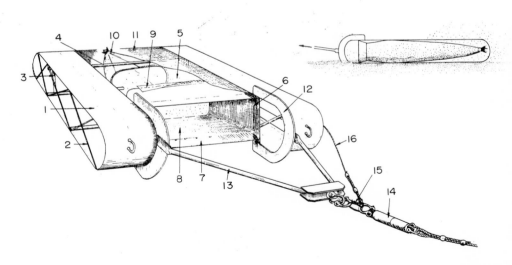

Fig. 10. Epibenthic sledge. 1, runners; 2,3, strengthening members; 4, tubular cross piece; 5, collecting net; 6, side plate at mouth of net; 7, biting edge at top and bottom of net; 8, canvas collar at front of net which is tied by canvas flaps; 9, to the tubular cross pieces and struts; 10, net tied at posterior end; 11, heavy wire screen to protect net (partly cut away in diagram); 12, flange preventing mud entering net from the side; 13, towing yoke; 14, swivel; 15, weak link; 16, safety line. The small drawing shows the mode of operating (from HOLME and MCINTYRE, 1971, after HESSLER and SANDERS, 1967).

1967). A third sampler uses an airlift to suck a cover into the sediment and then to suck the contained sediment into a sieve and collecting bag (BARNETT and HARDY, 1967); this sampler facilitates deep penetration and accurate quantitative work on macrofauna but the positions of animals within the sediment are obscured. Diver-operated samplers have been used to find the efficiency of bottom sampling gear operated from the surface so that the quantitative value of samples taken from depths beyond the reach of divers may be assessed. These remotely controlled samplers have to be relied upon for the vast majority of biological work on the sea bed and they may be grouped into spot samplers like grabs and corers and travelling samplers like dredges and trawls (HOLME and MCINTYRE, 1971; MENZIES, 1972).

The bottom in deeper waters

Spot samplers provide the best method of obtaining quantitative samples and there are many varieties, each having advantages and disadvantages according to the type of substrate, depth of water and size of vessels being used. Corers rely on their weight to enter the bottom sediments. They give reproducible penetration but there is danger of the pressure wave pushing the top layer of sediment aside, the core is nearly always considerably shorter than the penetration depth probably due to it acting as a rod after the first few centimetres of penetration so pushing deeper sediments ahead or to the side. Furthermore, there is danger that some of the core will wash out during the ascent. These difficulties must be allowed for in analysis but can be reduced by use of larger diameter cores and valves to prevent out-wash. Free fall deep-sea corers which pop to the surface with a sample are used extensively in the deep sea by geologists but rarely have they been used by biologists. Most grabs in use sample $0.1 \, \text{m}^2$ or $0.2 \, \text{m}^2$ of the bottom surface. Most commonly used designs are the Petersen (Fig. 7), the Van Veen, the Smith-McIntyre (Fig. 8), the Okean, the orange peel, the Holme and the Shipek (Fig. 9). The principal disadvantage of the Petersen grab is loss of sediment as the grab leaves the bottom; the Van Veen grab is liable to premature triggering during lowering, and the orange peel grab loses sediment between the pointed jaws during closing. The Okean and Smith–McIntyre grabs have overcome most of these problems and are practical for deep-water work. The Shipek is reliable but too small for macrofauna sampling. The Holme grab is useful on the continental shelf; the latest model has two scoops so providing two samples for comparison. None of these grabs are as accurate as cores for quantitative studies since they dig in deeper in the centre of the excavated hole than at the sides but they sample the macrofauna better. Suction samplers overcome this disadvantage. The Knudsen sampler acts as a piston, sucking itself into the sediment, as it is pulled out the strain goes on to an arm which inverts the coring tube so retaining the sample (KNUDSEN, 1927). Its main disadvantage is that it anchors the ship and may prove impossible to pull out. Another suction sampler depends upon a submerged electric motor with power from the surface or a water jet from the surface to suck sediment into a collecting basket; this is limited to shallow water except when used from a submersible (TRUE, REYS and DELAUZE, 1968).

A semi-quantitative method of sampling sediments is the anchor dredge. A shallow water form depends upon dropping a rectangular-mouthed dredge on the bottom and backing off so that the dredge pulls into the sediment and anchors the boat. The boat then moves to above the dredge and lifts it up vertically (FORSTER, 1953). Warp length

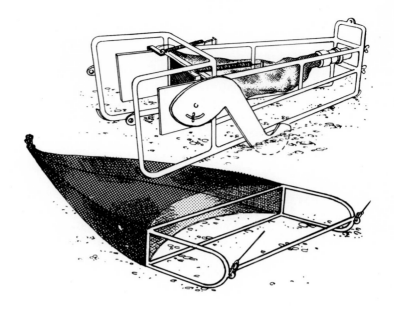

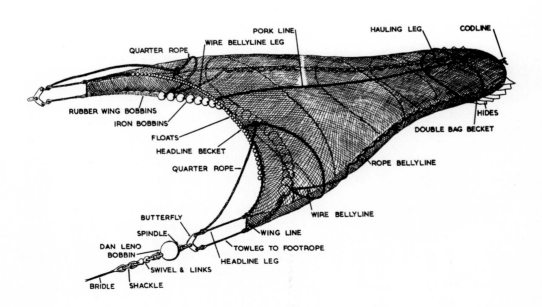

Fig. 11. *Top.* Bossanyi trawl for sampling plankton just above the bottom; the spring-loaded side arms hold the doors shut until they touch the sea bed. *Middle.* Agassiz trawl (after BARNES, 1959). *Bottom.* The Granton otter trawl used for commercial fishing. Each bridle runs to an otter board (not shown) which spreads the mouth horizontally (from *Fishing Manual, World Fishing News*).

makes this manoeuvre impossible to control in deep water. For work in deep water the dredge was first made double edged so that it could work either way up (HOLME, 1964), and then a flat plate was fixed across the mouth so limiting penetration to 11 cm and also directing sediment out of the mouth when the dredge was full. This development probably gives a good quantitative sample if the sediment will allow penetration to 11 cm, but sandy deposits are sometimes too hard and errors in estimates of population density result. Large rectangular or conical dredges are only useful for qualitative collecting but the former may be used to collect rock samples and associated fauna unobtainable in other ways. Sampling of animals on top of sediment involves the use of an epibenthic sledge (Fig. 10) (HESSLER and SANDERS, 1967) for small animals or various trawls for larger and active animals.

Commercial bottom trawls have been used successfully for qualitative biological sampling on the continental shelf (KRISTJONSSON, 1959; F.A.O., 1972, 1974; FOSTER, 1971). There are two basic forms, the beam trawl which has a transverse bar or beam spreading the mouth open, and otter trawls with transversely pulling otter boards on each side giving spread, and floats to elevate the top and weights to depress the ground line (Fig. 11). Beam trawls are limited in size by the maximum length of the bar possible, which is dependent upon strength of materials, drag of the net, bending load on the bar and handling convenience. Its efficiency in capturing animals largely depends upon the

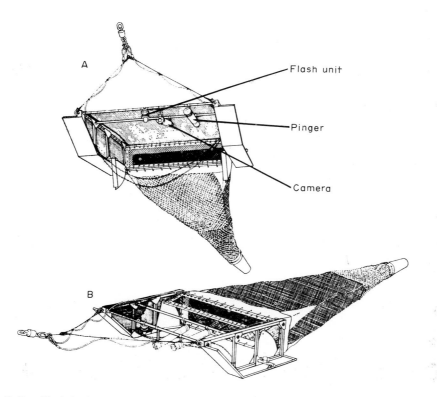

Fig. 12.   Epibenthic sledge in the mid-water attitude (*top*) with the quadrant levers lowered and the blind closed, and in the fishing position (lower) with the levers raised and the blind closed (from *Deep-Sea Research*).

speed of towing, the extent to which the top of the mouth runs ahead of the footrope and the number of tickler chains stirring the sediment ahead of the footrope—for prawns the catch was found to continue to increase with the addition of extra chains to a maximum of seven. One advantage of the beam trawl is that it is easily towed with a single wire and, since a small net has relatively little drag, long wires can be used and deep bottom trawling is possible. Because in deep water, a simple beam trawl could land upside down, the Agassiz trawl which can fish either way up was developed (Fig. 11). This loses the advantage of the headrope being ahead of the footrope. Another possibility is to have heavy skis either side of the mouth so that the net cannot easily turn over. The greatest difficulty with these in deep water is to assess the time the gear is fishing on the bottom so that some quantitative limits can be put upon the sample, and a development by the Institute of Oceanographic Sciences, England, has overcome this by use of pingers and a closing device (ALDRED, THURSTON, RICE and MORLEY, 1976) (Fig. 12). In the Bossanyi trawl (Fig. 11) spring-loaded side arms open doors over the mouth but drag limits such a trawl in size, the mouth has to be clear of the bottom to stand a chance of operating and the design is likely to give too many unsuccessful hauls to justify use in deep water where failure costs so much time. Various ways have been devised to determine

Fig. 13.   A mass of hagfish around a baited wire basket at 1450 m in the Gulf of California.

the length of time or distance that the net is on the bottom (MENZIES, 1964). Wheel-type devices to measure distance are probably much more prone to failure than pingers. Pingers placed above the sampler indicate by the time interval between the directly received sound pulse and its reflection from the bottom when the sampler is on the bottom and a minimum of wire is dragging ahead of it. An improvement is to include a mercury switch in the pinger which changes the pulse rate or length when the trawl touches the bottom.

Otter trawls can be much larger than beam trawls, they have various design features limiting the depth to which such large trawls can be used. With a normal commercial trawl, warps of more than $3\frac{1}{4}$ inch circumference must be accommodated and if a fishing depth of, say, 1500 m is desired, 5000 m of warp are needed together with a very large powerful winch on a large powerful ship. For yet greater depths like those at the bottom of the continental slope exceeding 2000 m, truly enormous winches are required. Headline buoyancy is provided by special metal deep sea floats but these are not reliable deeper than 1500 m and implode with the great pressures; while other types of buoyancy such as microspheres or glass floats are available they are very expensive. Use of such trawls in

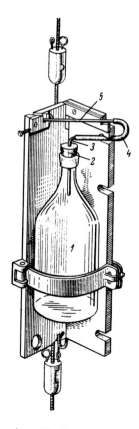

Fig. 14. Zobell's sampler for micro-organisms. The brass messenger in sliding down the wire breaks the glass tube (5) allowing the rubber tube (4) to straighten and ingress of water through the tube (3) to the bottle (1). At the same time a messenger is released to open the bottle on the wire below.

deep water with its very special requirements can only be carried out after major policy decisions involving design of ships.

Other ways of sampling animals on the bottom of the deep sea include lines, traps, net-bearing submersibles and predators. Lines bearing up to 100 hooks can be dropped to the bottom with a weight at each end and attached to a buoy at the surface (FORSTER, 1973). These have proved extremely useful in obtaining large bottom fish in deep water even where the bottom is steep or rough and inaccessible to most other methods of collection. Stomachs of these fish sometimes yield information on other animals which are not caught by any other means (CLARKE and MERRETT, 1972). Traps may be used to catch smaller fish and crustaceans in deep water and these have sometimes proved very successful for particular species (Fig. 13). An advance over the buoyed line is the pop-up device which incorporates a slowly dissolving magnesium link to release a weight and petrol-filled polythene containers to bring the trap to the surface. Such magnesium links are unreliable in very deep water because of temperature and pressure effects and so it is difficult to predict the time the trap appears at the surface and much ship time may be wasted. Devices attached to submersibles are by far the most attractive for sampling since man can then act as a chasing predator. However, the great expense in building and running submersibles prohibits such schemes from being more than opportunistic adjuncts to more predictable sampling programmes. Recent advances of unmanned submersibles have made possible accurate sampling and even *in situ* experiments (ALEXANDER, 1975; SMITH, CLIFFORD, ELIASON, WALDON, ROWE and TEAL, 1976).

Plankton and nekton

So far we have only concerned ourselves with sampling on or near the bottom. But what of the great mass of water of the world's ocean? Biological investigation of life in midwater requires different sampling devices to near-bottom investigations (BARNES, 1959; TRANTER, 1968) and one of the main requirements is to monitor the depth at which animals are caught in this three-dimensional world.

Small Organisms

The smallest organisms, bacteria, yeasts, etc., can be sampled by messenger-operated sterilized bottles on a vertical wire, e.g. the ZoBell and the Niskin samplers (RODINA, 1972). The ZoBell sampler is a glass bottle with sealed tube attached by a rubber tube. A messenger breaks the end of the sealed tube which springs straight and admits water to the bottle (Fig. 14). The tube, however, is not resealed and while the original sampler can only be used at shallow depths, a deep-water version has been developed (SOROKIN, 1962). With the Niskin sampler the messenger releases powerful springs which pull two metal arms apart so inflating a sterilized polythene bag between them; a tube opens to allow ingress of water and this tube is then sealed again by a clamp. It can be used at any depth, and versions collecting up to 5 litres of water have been designed. Microorganisms are concentrated from the samples by membrane filters and then cultured for further studies.

Biomass of the phytoplankton can be estimated by counting or from measurements of chemical components (VOLLENWEIDER, 1974).

An indication of the rate of production of organic matter can be obtained by adding a known amount of radioactive carbon isotope (^{14}C) to a known volume of water and suspending this at a particular depth on a wire below a buoy (STEEMAN NIELSEN, 1964). After several hours the water is filtered and the ^{14}C present in the plants on the membrane filter is found with a geiger counter. Controls with light-proof bottles are paired with transparent bottles so that the carbon assimilated by the plants in ambient light conditions may be found by subtraction. The amount of living plant material, or phytoplankton biomass, is often estimated by measuring chlorophyll concentration after filtration from a known volume of water collected in a large water bottle. The standing stock of phytoplankton is now measured by fluorometry (EGAN, 1974; HORNE and WRIGLEY, 1974; LINCOLN, 1976) and this may be incorporated into pumping systems with particle counters (PLATT, DICKIE and TRITES, 1970). Most tiny organisms obtained must be filtered off and identified and counted while alive since they are not easily preserved but others with hard skeletons such as diatoms can be sedimented by adding Lugols solution to a known volume of water, preserved and then counted with an inverted microscope and a haemocytometer. Special techniques have been developed to collect such small organisms very gently (GRAHAM, COLBURN and BURKE, 1976). The volume and biomass of the phytoplankton can be obtained from counts and sample measurements estimates. Cell numbers, volumes and biomass may also be calculated from electrical conductivity measurements with a Coulter counter. Adaptations of electronic counting devices are used to study the fine distribution of plankton (BOYD, 1973; VON BEHRENS and EDMONTON, 1976).

Small plankton can also be sampled by pumps (BEERS, STEWART and STRICKLAND, 1967).

Slightly larger organisms can be caught in fine-mesh conical ring nets. Up to 1 m^2 in mouth area can be conveniently hauled vertically. By hauling vertically at a steady 2 knots* (1 m s^{-1} closing by messenger and using a flowmeter consisting of a propellor coupled to a depth meter (Bourdon tube), valid comparisons between hauls and quantitative estimates can be made. Unfortunately, many of these nets catch very little because of the small mouth area and the scaring effect of wire bridles in front of the mouth, and as they sink through surface waters animals tend to get trapped in the net so contaminating deep hauls; methods to avoid this have been developed (GRICE and HULSEMAN, 1968).

For certain studies where an indication of relative numbers of phytoplankton and small zooplankton are required to show seasonal, long-term or geographical changes, high-speed samplers operating at ships' speeds of over 5 knots are used. These are metal streamlined instruments with a small sampling hole at the front and incorporating a net or gauze to collect the plankton. The Gulf III sampler used in basic fishery studies can also carry sensors for physical data as well as small phytoplankton samplers (ADAMS, 1976) (Fig. 15). The jet net is designed to reduce the flow of water so that organisms are not extruded by the water pressure through the netting (Fig. 15, 2). The continuous plankton recorder has a propellor which pulls a gauze through the water channel so that organisms caught on the gauze are sandwiched by a second gauze and rolled into a tank of formalin (Fig. 15, 3). This type of sampler has been used from merchant ships for long-term monitoring of changes in the North Atlantic (GLOVER, 1970). To overcome the major disadvantage of this sampler in that it tows at a fixed depth of 10 m, an undulating version has been developed (BRUCE and AIKEN, 1975). In samplers with small mouths messengers may be used to operate simple mechanical opening/closing mechanisms (e.g. Clarke–Bumpus sampler, Fig. 16, 3).

---

* 1 knot = 0.51 m s^{-1}.

Larger animals

All small-mouthed samplers sample a very limited size range of the animal community; for larger animals larger samples are needed. By effectively using a continuous plankton recorder as the cod-end of a net the spatial distribution can be determined for larger animals (LONGHURST, REITH, BOWER and SEIBERT, 1966). Towed horizontally, ring nets catch more than when hauled vertically since much more water can be filtered. Several opening/closing nets may be attached to a warp stretched by a large weight on the end. The simplest opening/closing method is the throttle system operated by a brass messenger sent down the wire (MOTODA, 1971, Fig. 16). In the Leavitt release a narrow messenger opens the net and a broad messenger strikes a second lever which releases the towing point of the net so it is throttled by a rope. As the levers operate they release similar messengers beneath the release gear and thus activate a net below; by this means a whole series of nets on one wire may be operated and fished almost at the same time. Such devices have proved useful for studying the vertical distributions of the smaller zoo-plankton such as copepods and euphausiids (TRANTER, 1968). Principal disadvantages of the simple ring net are that the bridles and the warp preceed the net scaring active animals away, the messenger-operated gear is often unreliable (for example, the efforts of a 3-month cruise with this gear were seriously jeopardized by *Pyrosoma* colonies wrapping round the warp, and acting as a soft cushion which failed to operate the levers!) and handling difficulties limit the practical size of such nets to a diameter of about 2 m. The

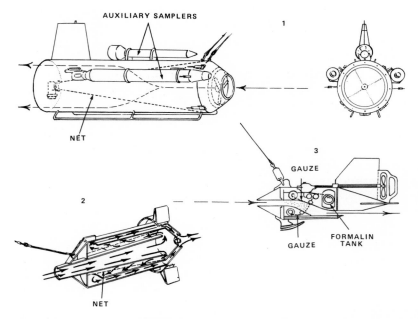

Fig. 15.   (1) Gulf 111 high-speed sampler; water flows from the small mouth through a net encased in a metal housing. (2) The jet net in sectional view; water speed through the net is reduced by its more circuitous path than in the Gulf 111. (3) Continuous plankton recorder; as the recorder passes through the water the propellor winds a continuous length of filtering gauze across the path of the entering water. Plankton is collected on the gauze and is sandwiched by a second strip of gauze before being wound into a tank of preservative (1, after BEVERTON and TUNGATE, 1969; 2, after CLARKE, 1964; 3, after HARDY, 1936).

bridle and warp problem has been tackled in several ways (Fig. 17) and these prevent the use of the usual opening–closing messenger system. In the 'bongo' type two nets balance one another on either side of the warp. Here the mouths are at first covered by canvas hoods, these are released to open the nets which are then closed by throttling (MᴄGᴏᴡᴀɴ and Bʀᴏᴡɴ, 1966). Operation may be achieved by a clock timing device or by a propellor which opens and closes the net after the water has induced a preset number of turns. An alternative system which is effective with 1 m ring nets is the cod-end catch dividing bucket which was first designed to operate by a plunger-type pressure switch (Fᴏxᴛᴏɴ, 1963) but later by an acoustic command signal directed at the net from the ship (Fig. 18). The Bé sampling device also operates by a plunger type of pressure switch but depends upon elastic to pull open the mouths of three small nets within one frame (Fig. 18, Bé, 1962). Another cod-end sampler contains chambers with gates operated by an electric current from the ship (Aʀᴏɴ, Rᴀxᴛᴇʀ and Aɴᴅʀᴇᴡs, 1964; Bᴏᴜʀʙᴇᴀᴜ, Cʟᴀʀᴋᴇ and Aʀᴏɴ, 1966) which, combined with depth sensors, gives good control of sampling. With any cod-end opening–closing system there is always a danger that some of the animals are held

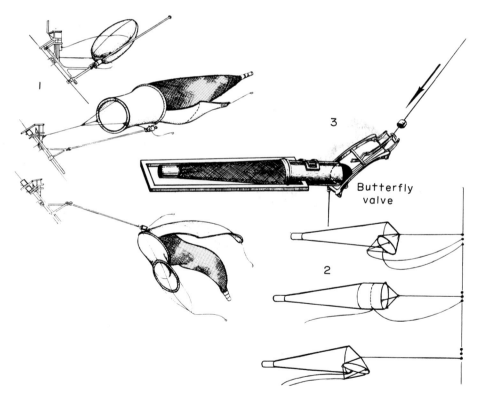

Fig. 16.   Three types of opening and closing systems; 1, one of Mᴏᴛᴏᴅᴀ's methods in which the first messenger releases the net from a canvas flap, and it opens, and the second messenger releases the bridles and the net closes; 2, Leavitt release method; the first messenger releases the throttling rope and opens the net, the second messenger releases the bridles and the net is throttled again; 3, Clarke–Bumpus sampler; the mouth is covered by a butterfly lid which is opened and closed by messengers. All three methods allow the use of several nets on a single wire for at each operation another messenger is released to trigger the net below (1, after Mᴏᴛᴏᴅᴀ, 1967; 3, after Bᴀʀɴᴇs, 1959).

up in front of the opening–closing device because of a reduction in filtration and pass into the wrong sample container after the device operates; a possibility which becomes greater the larger the net.

Special nets of small size have been developed to sample the neuston, the animals forming the community in the top few centimetres of the sea (David, 1975; John, 1976; Zaitzev, 1971). To sample the larger zooplankton and nekton, including decapods, fish and cephalopods, ring nets are ineffective and three kinds of net are widely utilized, the Isaacs–Kidd midwater trawl (IKMT), the rectangular midwater trawl (RMT) and commercial midwater trawls. The IKMT has a large V-shaped depressor and a bar to hold the mouth open (Fig. 19); the weight and handling difficulties of the depressor impose a practical limit to the mouth width of 3 m giving a mouth area of about 7.5 m². Cod-end devices such as the catch dividing bucket and four-chambered sampler have been used on this net but introduce some doubt into the analysis because animals may be held up in front of the device. An attempt to open and close an IKMT at the mouth by means of an additional top panel which drops to change the water flow into a second bucket which is then throttled after the horizontal tow (Isaacs and Brown, 1966) suffers, on throttling, from the same disadvantage as the cod-end device and is mechanically less attractive. The simple rectangular midwater trawl (RMT) consists of a mouth with bars top and bottom and side wires with weights or depressors at the bottom to hold the mouth open. This has been made to open and close either by having three bars and a clockwork

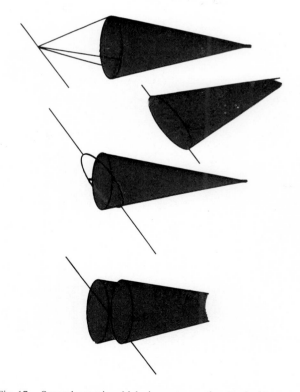

Fig. 17.   Several ways in which ring nets may be attached to a warp.

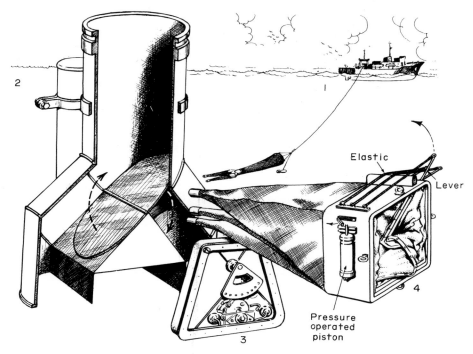

Fig. 18. (1) Catch-dividing bucket fitted at the end of a net allows the catch from different depths to be diverted into different legs. (2) Sectional view of the bucket showing the flap which controls the flow of the catch. (3) Control of the flap which can be operated mechanically, electrically or acoustically. (4) The Bé multiple sampler fishes three nets in one frame. A pressure-sensitive piston allows the successive release of three levers each forming one side of a different net. When released, elastic pulls each lever into a vertical position, thereby closing the mouth of its net and allowing the next net to fish (4, after Bé).

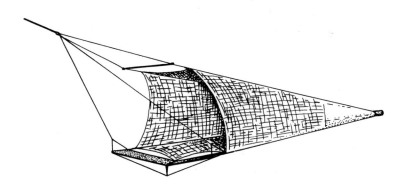

Fig. 19. Isaacs–Kidd midwater trawl.

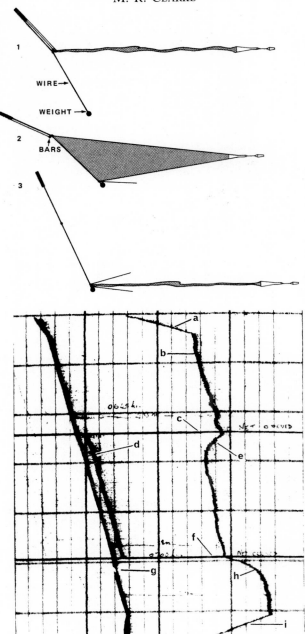

Fig. 20. *Top.* Operation of a rectangular opening and closing trawl; 1, net is lowered closed; 2, net in trawling position; 3, net closed at end of haul. *Bottom.* Record received on the echo-sounder from the 'pinger' monitoring the operation of an acoustically controlled rectangular midwater trawl. At (*a*) the increasing separation of the two traces indicates the net is sinking as the warp is payed out; at (*b*) the winch is stopped and the depth remains constant until an acoustic signal opens the net at (*c*). Opening is indicated by the appearance of a third trace (*d*) and the increased drag of the open mouth causes the net to rise slightly at (*e*). A second signal closes the net at (*f*), the third trace disappears at (*g*) and the decreased drag causes the net to sink at (*h*). At (*i*) the net is hauled up again.

device (DAVIES and BARHAM, 1969) or four bars and an acoustic device (CLARKE, 1969; Fig. 20. The latter system has since been developed and used for a major sampling programme in the North Atlantic (BAKER, CLARKE and HARRIS, 1973). The simple version operates as in Fig. 20. As the net is lowered in the closed position (Fig. 20a, 1) a pinger on the net gives a crystal-controlled 2-second reference pulse and a second pulse which varies in interval from the first pulse according to depth; the interval increases as the net sinks (Fig. 20b). When the winch stops paying out and the depth remains constant the Mufax recorder traces stay the same distance apart (b). An acoustical signal is sent from the ship (shown as a line at c) which activates the release of bridles holding the lower bar and the net opens (Fig. 20, 2). This operation is shown by a third ping (d) and the rise of the net in the water due to increased drag. Closure of the net is achieved by a second signal (f) which causes the release of the bar holding the top of the mouth (Fig. 20, 3), cessation of the third ping (g) and a sinking of the net in the water (h). This acoustically controlled net fulfils all the requirements of a midwater discrete depth sampler, it can be opened and closed effectively at the mouth, estimates of the fishing mouth area can be made, the depth of the net can be accurately monitored during the whole fishing operation, the exact time and depth and the success (electronic and mechanical) of opening and closing can be checked at the time. In addition various other parameters such as temperature and flow can be easily obtained by extra pinging circuits. This type of net has the great advantage over other designs that it is effective in a wide range of sizes; opening–closing versions have been used with $1\,m^2$, $8\,m^2$ and $25\,m^2$ mouths and non-closing versions from $\frac{1}{2}\,m^2$ to $90\,m^2$. In the RMT combination net one release gear is used to operate more than one size of net so sampling a wider size range of animals than is possible with one net (BAKER, CLARKE and HARRIS, 1973).

As larger nets are used opening and closing becomes more difficult but the samples include more rarely caught animals. The $90\,m^2$ RMT collects some of these and some

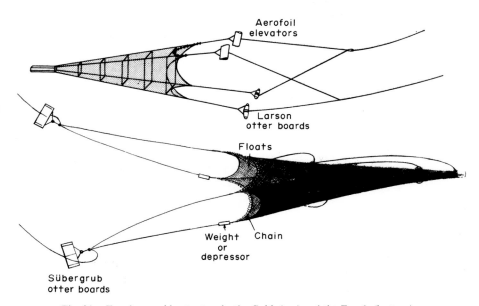

Fig. 21.   Two large midwater trawls, the Cobb (top) and the Engels (bottom).

commercial trawls such as the British Columbia, Engels and Cobb, trawls of immense proportions (Fig. 21), catch greater numbers but without the precision of discrete sampling. However, these giant trawls are of great use in the present early exploratory phase in the study of many large midwater animals.

Sampling with purse seines has usually been restricted to fish and squid of comestible size but providing the animals are in shoals the technique has potential for plankton collection (MURPHY and CLUTTER, 1972).

Various methods have been used to attract animals to traps, pumps or nets but such methods have been exploited more commercially than for biological research. These include attraction by lights (JONES, 1971; OGURA, 1972), sound (WORLD FISHING, 1974) and electricity (DANIULYTE and MALUKINA, 1967).

For some work on the physiology of midwater animals, samplers have been developed which can be closed so that the pressure at which the animal is caught is almost maintained as the animal is brought to the water surface. The sample chamber can then be connected to a high-pressure aquarium and physiological studies be conducted under pressure. This sampling has necessitated the development of highly sophisticated apparatus and its applications are in a relatively narrow but important field (MACDONALD, 1975).

Collection of animals with lines in midwater for research purposes is limited to a few abundant species which are usually of commercial importance such as tuna (MERRETT in HERRING and CLARKE, 1971), sharks and certain squid (MAEDA and ASO, 1974).

To study migrations and populations many types of tag have been developed from simple clips on fish to foot-long cylindrical marks shot into whales. Migrations immediately prior to capture, can sometimes be surmised by the remnants of food in the stomach (CLARKE, 1972) and populations characterized by their parasites. To follow particular animals during migration radio or sonic devices can be attached and tracked by boat or satellite (MARTIN, EVANS and BOWERS, 1971; STASKO, HORRALL and HASLER, 1976) or, with larger animals, recording instruments can be attached and the data collected later (WARTZOK, RAY and MARTIN, 1975).

Finally let us ponder on our deficiencies. We know what gear will catch but, with animals, we do not know what it will not catch; comparisons between many methods helps us to assess this. We have sampling methods to give specific answers, we have no universally useful sampler. We know more types and more larger animals can be caught by increasing the size of nets and by increasing the speed they are pulled through the water but we have little idea of the reasons why a sampling gear will take one species and not another.

A commercial bottom trawl was deemed an engineering failure but a costly redesign with perfect engineering parameters does not catch as many fish. An old Madeiran fisherman baiting his longline hooks was asked by a young oceanographer why he cut the squid in such a mysterious way. "It is *the* way!" he replied. It took the oceanographer several nights of despair to find that *the* way produced luminous lures by bacterial action while the simpler, more obvious way did not! This is the problem we face when we step from the time-tested and moulded sphere of commercially successful techniques into the realm of biological sampling for research. The greatest volume of the oceans lies outside the sphere of fisheries (with a few exceptions such as tuna longlining) and the biologist is in a new world; small wonder he has far to go and much to find.

Acknowledgements—I should like to thank Dr. M. ANGEL, Mr. A. DE C. BAKER and Prof. N. A. MARSHALL, F.R.S., for helpful criticism of the manuscript and the following for permission to use figures; Dr. G. L. CLARKE (Fig. 2), Dr. C. EMIG (6), Dr. N. A. HOLME (4, 7, 10), Professor J. D. ISAACS (3, 13), Dr. A. McINTYRE (5), Dr. T. RICE (12), Deep Sea Research, Pergamon Press Ltd. (12), Blackwell Scientific Publications (7, 10), Morgan-Grampian (Publishers) Ltd. (11), University Park Press (14) and the Institute of Oceanographic Sciences, Wormley (20).

REFERENCES

ADAMS J. A. (1976) The Bridger version of the Gulf III high speed plankton sampler. *Scottish Fisheries Research Report* No. 3, 1–42.

ALRED R. G., M. H. THURSTON, A. L. RICE and D. R. MORLEY (1976) An acoustically monitored opening and closing epibenthic sledge. *Deep-Sea Research*, **23**, 167–174.

ALEXANDER C. M. (1975) Sea floor effectiveness of RUM II. *Marine Journal*, **9**(7), 9–15.

ANDREEV N. N. (1966) *Handbook of fishing gear and its rigging*. Translated from Russian. Israel Program for Scientific Translations, Jerusalem, 454 pp.

ARON W., R. N. BAXTER, R. NOEL and W. ANDREWS (1964) A description of a discrete depth plankton sampler with some notes on the towing behaviour of a 6-foot Isaacs–Kidd midwater trawl and a one-meter ring net. *Limnology and Oceanography*, **9**(3), 324–333.

ASHMOLE N. P. and M. J. ASHMOLE (1967) Comparative feeding ecology of sea birds of a tropical oceanic island. *Peabody Museum of Natural History Bulletin*, **24**, 1–131.

BACKUS R. H., J. E. CRADDOCK, R. L. HAEDRICH, D. L. SHORES, J. M. TEAL, A. S. WING, G. W. MEAD and W. D. CLARKE (1968) *Ceratoscopelus maderensis*—peculiar sound scattering layer identified with this myctophid fish. *Science*, **160**, 991–993.

BAILEY R. (1966) The sea birds of the southeast coast of Arabia. *Ibis*, **108**, 224–264.

BAKER A. DE C. (1960) Observations of squid at the surface in the N.E. Atlantic. *Deep-Sea Research*, **6**, 206–210.

BAKER A. DE C., M. R. CLARKE and M. J. HARRIS (1973) The N.I.O. combination net (RMT 1+8) and further developments of rectangular midwater trawls. *Journal of the Marine Biological Association of the U.K.*, **53**, 167–184.

BARNES H. (1959) *Oceanography and marine biology*, George Allen & Unwin Ltd., London, 218 pp.

BARNETT P. R. O. and B. L. S. HARDY (1967) A diver operated quantitative bottom sampler for sand macro-faunas. *Helgoländer wissennschaftliche Meeresuntersuchungen*, **15**, 390–398.

BÉ A. W. H. (1962) Quantitative multiple opening–closing plankton samplers. *Deep-Sea Research*, **9**, 144–151.

BEERS J. R., G. L. STEWART and J. D. H. STRICKLAND (1967) A pumping system for sampling small plankton. *Journal of the Fisheries Research Board of Canada*, **24**, 1811–1818.

VON BEHRENS W. and S. EDMONDSON (1976) Comparison of techniques improving the resolution of standard Coulter cell sizing systems. *Journal of Histochemistry and Cytochemistry*, **24**, 247–256.

BODEN B. P. (1971) Bioluminescence in sonic-scattering layers. In: *Proceedings of the International Symposium on Biological Sound Scattering in the Ocean*, G. B. FARQUHAR, editor, Warrenton, Virginia, Washington, D.C., 60–66.

BOURBEAU F., W. D. CLARKE and W. ARON (1966) Improvements in the discrete depth plankton sampler system. *Limnology and Oceanography*, **11**(3), 422–426.

BOYD C. M. (1973) Small scale spatial patterns of marine zooplankton examined by an electronic *in situ* zooplankton detecting device. *Netherlands Journal of Sea Research*, **7**, 103–111.

BRETT C. E. (1964) A portable hydraulic diver operated dredge-sieve for sampling subtidal macrofauna. *Journal of Marine Research*, **22**, 205–209.

BROWN D. M. (1975) Four biological samplers: opening–closing midwater trawl, closing vertical tow net, pressure fish trap, free vehicle drop camera. *Deep-Sea Research*, **22**, 565–567.

BRUCE R. H. and J. AIKEN (1975) The undulating oceanographic recorder—a new instrument system for sampling plankton and recording physical variables in the euphotic zone from a ship underway. *Marine Biology*, **32**, 85–97.

CADDY J. F. (1976) Practical considerations for quantitative estimation of benthos from a submersible. In: *Underwater research*, DREW E. A., J. N. LYTHGOE and J. D. WOODS, editors, Academic Press, pp. 295–298.

CLARKE M. R. (1969) A new midwater trawl for sampling discrete depth horizons. *Journal of the Marine Biological Association of the U.K.*, **49**, 945–960.

CLARKE M. R. (1972) New techniques for the study of sperm whale migration. *Nature, London*, **238**, 405–406.

CLARKE M. R. and N. MACLEOD (1974) Cephalopod remains from a sperm whale caught off Vigo, Spain. *Journal of the Marine Biological Association of the U.K.*, **54**, 959–968.

CLARKE M. R. and N. MERRETT (1972) The significance of squid, whale and other remains from the stomachs of deep sea fish. *Journal of the Marine Biological Association of the U.K.*, **52**, 599–603.

COCKING S. J. (1975) Improving underwater viewing. In: *Underwater Research*, DREW E. A., J. N. LYTHGOE and J. D. WOODS, editors, Academic Press, London and New York, 430 pp.

CONNELL J. H. (1974) Ecology: field experiments in marine ecology. In: *Experimental marine biology*, R. N. MARISCAL, editor, Academic Press, New York and London, 373 pp.

CRAM D. L. (1974) Rapid stock assessment of pilchard populations by aircraft-borne remote sensors. *Proceedings of the 9th International Symposium on Remote Sensing of the Environment*, Ann Arbor, Michigan, Willow Run Laboratories, Environment Research Institute of Michigan, 15–19 April 1974, **2**, 1043–1050. Ann Arbor, Michigan.

CURRIE R. I., B. P. BODEN and E. M. KAMPA (1969) An investigation on sonic-scattering layers: the R.R.S. *Discovery* SOND cruise, 1965. *Journal of the Marine Biological Association of the U.K.*, **49**, 489–514.

CUSHING D. H. (1969) The use of echo sounders and scanners in the study of fish behaviour. *Fisheries Papers F.A.O.*, No. **62**(2), 115–130.

DANIULYTE G. and G. MALUKINA (1967) The reaction of some fishes in an electric field. *F.A.O. Fisheries Report*, **62**, (3), 775–780.

DAVID P. M. (1965) The Neuston net. A device for sampling the surface fauna of the ocean. *Journal of the Marine Biological Association of the U.K.*, **45**, 313–320.

DAVIES I. E. and E. G. BARHAM (1969) The Tucker opening–closing microplankton net and its performance in a study of the deep scattering layer. *Marine Biology*, **2**, 127–213.

EGAN W. G. (1974) Measurement of the fluorescence of Gulf Stream water with submerged *in situ* sensors. *Marine Technology Society Journal*, **8**(10), 40–47.

EMIG C. C. and R. LIENHART (1967) Un nouveau moyen de recolté pour les substrats meubles infralittoraux: l'aspirateur sous-marin. *Recueil des Travaux Station Marin d'Endoume*, **58**, 115–120.

DYRING E. (1973) The principles of remote sensing. *Ambio*, **11**(3), 57–69.

FAGER E. W., A. O. FLECHSIG, R. F. FORD, R. I. CLUTTER and R. J. GHELARDI (1966) Equipment for use in ecological studies using SCUBA. *Limnology and Oceanography*, **11** (4), 503–509.

FITCH J. E. and R. L. BROWNELL (1968) Fish otoliths in cetacean stomachs and their importance in interpreting feeding habits. *Journal of the Fisheries Research Board of Canada*, **25**, 2561–2574.

F.A.O. (1972) Department of Fisheries, Fishery Industries Branch, Fishing Gear and Methods Branch, *FAO Catalogue of fishing gear designs*, West Byfleet, Fishing News (Books) Ltd., 155 pp.

F.A.O. FOOD AND AGRICULTURE ORGANIZATION.(1974) *Otter board design and performance*, Rome, FAO (FAO Fishing Manuals) (viii, 91 pp.).

FORSTER G. R. (1953) A new dredge for collecting burrowing animals. *Journal of the Marine Biological Association of the U.K.*, **32**, 193–198.

FORSTER G. R. (1973) Line fishing on the continental slope. The selective effect of different hook patterns. *Journal of the Marine Biological Association of the U.K.*, **53**, 749–751.

FOSTER J. J. (1971) Gear studies at Aberdeen Laboratory. *World Fishing*, **20**(10), 12–14 and 20.

FOXTON P. (1963) An automatic opening–closing device for large midwater plankton nets and midwater trawls. *Journal of the Marine Biological Association of the U.K.*, **45**, 295–308.

GIBBS R. H., C. F. E. ROPER, D. W. BROWN and R. H. GOODYEAR (1971) *Biological studies of the Bermuda Ocean Acre. 1: station data, methods and equipment for cruises 1 through 11, October 1967–January 1971*, Smithsonian Institute, 62 pp.

GIBSON R. N. (1967) The use of the anaesthetic quinaldine in fish ecology. *Journal of Animal Ecology*, **36**, 295–301.

GLOVER R. S. (1970) Synoptic oceanography—the work of the Edinburgh Oceanographic Laboratory. *Underwater Science Technology Journal*, **2**, 34–40.

GOODING R. and J. MAGNUSON (1967) Ecological significance of a drifting object to pelagic fishes. *Pacific Science*, **21**, 486–497.

GRAHAM L. B., A. D. COLBURN and J. C. BURKE (1976) A new simple method for gently collecting planktonic protozoa. *Limnology and Oceanography*, **21**, 336–341.

GRICE G. D. and K. HULSEMAN (1968) Contamination in Nansen like vertical plankton nets and a method to prevent it. *Deep-Sea Research*, **15**, 229–233.

HAMNER W. M., L. P. MADIN, A. L. ALLDREDGE, R. W. GILMER and P. P. HAMNER (1975) Underwater observations of gelatinous zooplankton. *Limnology and Oceanography*, **20**(6), 907–917.

HEMMINGS C. C. (1971) Fish behaviour. In: *Underwater Science*, J. D. WOODS and J. N. LYTHGOE, editors, London: Oxford University Press.

HERRING P. J. and M. R. CLARKE (1971) *Deep Oceans*, Arthur Barker Ltd., London, 320 pp.

HERRNKIND W. F. (1974) Behaviour: *in situ* approach to marine behavioural research. In: *Experimental marine biology*, R. N. MARISCAL, editor, Academic Press, New York and London.

HERTWECK G. (1974) Handstechkasten zur Gewinnung von ungestörten sedimentproben in Taucheinstaz. *Senckenbergiana Maritima*, **6**, 119–127.

HEEZEN B. C. and C. D. HOLLISTER (1971) *The face of the deep*, Oxford University Press, London, 659 pp.

HESSLER R. R. and H. L. SANDERS (1967) Faunal diversity in the deep sea. *Deep-Sea Research*, **14**, 65–78.

HODGSON E. and R. MATHEWSON (1971) Chemosensory orientation in sharks, *Annals of the New York Academy of Sciences*, **188**, 175–182.

HOLME N. A. (1964) Methods of sampling the benthos. *Advances in Marine Biology*, **2**, pp. 171–260.

HOLME N. A. and A. D. McINTYRE (1971) Methods for the study of marine benthos. *International Biological Programme Handbook* No. 16, Blackwell Scientific Publications, Oxford and Edinburgh, 334 pp.

HORNE A. J. and R. C. WRIGLEY (1974) Hunting phytoplankton by remote sensing. *British Phycological Journal*, **9**, 220.

IMBER M. J. (1975) Lycoteuthid squids as prey of petrels in New Zealand seas. *New Zealand Journal of Marine and Freshwater Research*, **9**(4), 483–489.

ISAACS J. D. and D. M. BROWN (1966) Isaacs-Brown opening–closing trawl. *Methods and Technique*, **16**(29).

ISAACS J. D. and R. A. SCHWARTZLOSE (1975) Active animals of the deep-sea floor. *Scientific American*, **233**, 84–91.

JOHN H. C. (1976) Beschreibung eines Zweistufen-Neustonsammlers nach dem Prinzip von Sameoto und Jaroszynski. (Description of a two-net neuston-sampler after the principles of Sameoto and Jaroszynaki.) *Bericht der Deutschen Wissenschaftlichen Kommission für Meeresforschung*, **24**, 342–344.

JONES F. R. H. (1969) Observations on the behaviour of fish, made with the bifocal sector scanner. *Fisheries report of the Fisheries and Agricultural Organisation*, No. **62**, Vol. 3, 667–670.

JONES D. A. (1971) A new light trap for plankton. *Fourth European Marine Biology Symposium*, D. J. CRISP, editor, Bangor, 14–20 September, 1969, Cambridge University Press, pp. 487–493.

JOSEPH J. and M. R. STEVENSON (1974) A review of some possible uses of remote sensing techniques in fishery research and commercial fisheries. In: *Approaches to earth survey problems through the use of space techniques*, P. BOCK, editor, Akademie-Verlag, Berlin.

KIRCHNER W. B. (1974) A SCUBA diver operated cover for determining vertical distribution in the benthos. *Methods and Techniques*, **19**(24).

KNUDSEN M. (1927) A bottom sampler for hard bottoms. *Meddelelser fra Kommissionen for Havundersøgelser, København*, Serie Fiskeri, **8**, 4 pp.

KRISTJONSSON H. (1959) *Modern fishing gear of the world*, Publ. Fishing News (Books) Ltd., London, 607 pp.

LINCOLN A. (1976) The use of a fluorometer to measure the standing stock of marine phytoplankton. *Technical Report of the Fisheries Laboratory Lowestoft*, No. 19, 15 pp.

LONGHURST A. R., A. D. REITH, R. E. BOWER and D. L. R. SEIBERT (1966) A new system for the collection of multiple serial plankton samples. *Deep-Sea Research*, **13**, 213–222.

LUNDALV T. (1971) Quantitative studies on rocky-bottom biocoenoses by underwater photogrammetry. A methodological study. *Thalassia Jugoslavica*, **7**(1), 201–208.

MACDONALD A. G. (1975) *Physiological aspects of sea biology*, Cambridge University Press, London, 450 pp.

MACHAN R. and K. FEDRA (1975) A new towed underwater camera system for wide range benthic surveys. *Marine Biology*, **33**, 75–84.

MAEDA H. and K. ASO (1974) The distribution pattern of squids caught by the automatic powered reel. *Journal of the Shimonoseki University of Fisheries*, **22**, 115–145.

McGOWAN J. A. and D. M. BROWN (1966) *A new opening–closing paired zooplankton net*, University of California Scripps Institute of Oceanography (Ref. 66–23).

MARSHALL N. B. and D. W. BOURNE (1964) A photographic survey of benthic fishes in the Red Sea and Gulf of Aden, with observations on their population density, diversity and habits. *Bulletin of the Museum of Comparative Zoology of Harvard*, **132**(2), 223–244.

MARTIN H., W. E. EVANS and C. A. BOWERS (1971) Methods for radio tracking marine mammals in the open sea. *IEEE '71 Eng. in the ocean environment conf.—44, San Diego, California, September 21–24, 1971*, N.Y. Inst. Electrical & Electronic Engrs., Inc., 1971, pp. 44–49.

MENZIES R. J. (1972) Current deep benthic sampling techniques from surface vessels. In: *Barobiology and experimental biology of the deep sea*, BRAUER, editor, University of Carolina, pp. 164–174.

MILLER J. W., J. VAN DERWALKER and R. WALLER (editors) (1971) Tektite II *Scientists-in-the-sea*, U.S. Dept. of the Interior, Washington, D.C.

MOTODA S. (1971) Devices of simple plankton apparatus, V. *Bulletin of the Faculty of Fisheries, Hokkaido University*, **22**, 101–106.

MYRBERG A. A. JR. (1974) Underwater television—a tool for the marine biologist. *Bulletin of Marine Science*, **23**, 823–836.

MURPHY G. I. and R. J. CLUTTER (1972) Sampling anchovy larvae with a plankton purse seine. *Fishery Bulletin*, **70**, 789–798.

NEWTON R. S. and A. STEFANON (1975) Application of side-scan sonar in marine biology. *Marine Biology*, **31**, 287–291.

OGURA M. (1972) Squid fishing and light. *Bulletin of the Japanese Society of Scientific Fisheries*, **38**, 881–889.

PARRISH B. B., E. F. AKYUZ, J. ANDERSON, D. W. BROWN, W. HIGH, J. M. PERES and J. PICCARD (1972) *Submersibles and underwater habitats*: a review. *Underwater Journal*, **4**, 149–167.

PEARCY W. G. and A. BEAL (1973) Deep-sea cirromorphs (Cephalopoda) photographed in the Arctic Ocean. *Deep-Sea Research*, **20**, 107–108.

PINGREE R. D., G. R. FORSTER and G. K. MORRISON (1974) Turbulent convergent tidal fronts. *Journal of the Marine Biological Association of the U.K.*, **54**, 469–479.

PINKERTON K. J. and R. GAMBELL (1968) Aerial observations of sperm whale behaviour. *Norsk Hvalfangst-Tidende*, **57**, 126–138.

PHLEGER C. F. and A. SOUTAR (1971) Free vehicles and deep-sea biology. *American Zoologist*, **11**, 409–418.

PLATT T., L. M. DICKIE and R. W. TRITES (1970) Special heterogeneity of phytoplankton in a near-shore environment. *Journal of the Fisheries Research Board of Canada*, **27**, 1453–1473.

POTTS G. W. (1976) A diver controlled plankton net. *Journal of the Marine Biological Association of the U.K.* **56**, 959–962.

REES W. J. (1955) On a giant squid *Ommastrephes caroli* Furtado stranded at Looe, Cornwall. *Bulletin of the British Museum of Natural History*, **1**(2), 31–41.

REVIE J., G. WEARDEN, P. D. HOCKING and D. E. WESTON (1974) A twenty-three day twenty-mile echo record of fish behaviour. *Journal du Conseil, Conseil Permanent International pour l'Exploration de la Mer*, **36**(1), 82–86.

RODINA A. G. (1972) *Methods in adequate microbiology*, translated from the Russian, edited and revised by R. R. COLWELL and M. S. ZAMBRUSKI, Baltimore University Park Press and Butterworth & Co. Ltd., 461 pp.

ROPER C. F. E. and W. L. BRUNDAGE (1972) Cirrate octopods with associated deep-sea organisms: new biological data based on deep benthic photographs (Cephalopoda). *Smithsonian Contributions to Zoology*, **121**, 1–46.

RUSBY J. S. M., M. L. SOMERS, J. REVIE, B. S. MCCARTNEY and A. R. STUBBS (1973) An experimental survey of a herring fishery by long-range sonar. *Marine Biology*, **22**, 271–292.

SCHEVILL W. E., W. A. WATKINS and R. R. BACKUS (1964) The 20-cycle signals of Balaenoptera (fin whales). *Marine Bio-acoustics*, pp. 1147–1152.

SHELDRICK M. C. (1976) Trends in the strandings of Cetacea on the British coast 1913–72. *Mammal Review*, **6**(1), 15–23.

SMITH K. L., C. M. CLIFFORD, A. H. ELIASON, B. WALDEN, G. J. ROWE and J. M. TEAL (1976) A free vehicle for measuring benthic community metabolism. *Limnology and Oceanography*, **21**, 164–170.

SOROKIN Y. I. (1962) Problems in the sampling method used for the study of marine microflora. *Okeanologiya*, **5**, 888–897.

SOUTHWARD A. J., S. G. ROBINSON, D. NICHOLSON and T. J. PERRY (1976) An improved stereocamera and control system for close-up photography of the fauna of the continental slope and outer shelf. *Journal of the Marine Biological Association of the U.K.*, **56**, 247–257.

STASKO A. B., R. M. HORRALL and A. D. HASLER (1976) Coastal movements of adult Fraser river sockeye salmon (*Onchorynchus nerka*) observed by ultrasonic tracking. *Transactions of the American Fisheries Society*, **105**, 64–71.

STEEMAN NIELSEN E. (1964) Recent advances in measuring and understanding marine primary production. *Journal of Ecology* (Suppl.), **52**, 119–130.

STEWART G. L., J. R. BEERS and C. KNOX (1974) *Application of holographic techniques to the study of marine plankton in the field and in the laboratory*, University of California (La Jolla) Institute of Marine Resources. Research on the marine food chain. Progress Rep., July 1973–June 1974, California.

STRASBERG D. W., E. C. JONES and R. IVERSEN (1968) Use of a small submarine for biological and oceanographic research. *Journal du Conseil. Conseil Permanent International pour l'Exploration de la Mer*, **31**(3), 410–426.

STRICKLAND C. L. (1973) Underwater television—its development and future. *Underwater Journal*, **5**, 244–249.

TALBOT F. H. (1965) A description of the coral structure of Tutia reef (Tanganyika Territory, East Africa) and its fish fauna. *Proceedings of the Zoological Society of London*, **145**, 431–470.

TIEMON A. (1972) *Exploration by unmanned submersibles*, Oceanology International Conference papers, Brighton, No. 253.

TRANTER D. I., editor, (1968) *Reviews on zooplankton sampling methods*, UNESCO, Monograph on oceanographic methodology.

TRUE M. A., J. P. REYS and H. DELAUZE (1968) Progress in sampling the benthos; the benthic suction sampler. *Deep-Sea Research*, **15**, 239–242.

THIEL H., D. THISTLE and G. D. WILSON (1975) Ultrasonic treatment of sediment samples for more efficient sorting of meiofauna. *Limnology and Oceanography*, **20**, 472–473.

VOLLENWEIDER R. A. (1974) *Primary production in aquatic environments*, International Biological Programme Handbook No. 12, Blackwell, London, 225 pp.

WARTZOK D., G. C. RAY and H. B. MARTIN (1975) A recording instrument package for use with marine mammals. *Rapport et Procès—Verbaux des Réunions du Conseil Permanent International pour l'Exploration de la Mer*, **169**, 445–450.

WATKINS W. A. and W. E. SCHEVILL (1971) Four hydrophone array for acoustics three-dimensional location. *Technical Report of the Woods Hole Oceanographic Institute*, **71–60**, 1–61.

WILCOX J. R., R. P. MEEK and D. MOOK (1974) A pneumatically operated slurp gun. *Limnology and Oceanography*, **19**, 354–355.

WOODS J. D. and G. G. FOSBERG (1967) The structure of the thermocline. *Underwater Association Report*, *1966–67*, 5–18.

WORLD FISHING (1974) Luring fish. *World Fishing*, **23**(1/2), 54 and 64.

YARNALL J. (1969) Aspects of the behaviour of *Octopus cyanea* Gray. *Animal Behaviour*, **17**, 747–754.

ZAITSEV YU. P. (1971) *Marine neustonology*, translated from the Russian. Jerusalem, Israel Program for Scientific Translations, 1971, vi, 207 pp. Originally published as *Morskaya Neustonologiya*, Kiev, Naukova Dumka, 1970.

The effects of a pure carbohydrate diet on the amino acid composition of *Neomysis integer*

M. E. ARMITAGE,* J. E. G. RAYMONT* and R. J. MORRIS†

Abstract—*Neomysis integer* was fed carbohydrate or kaolin (starved) for up to 2 weeks, and the ability of the animal to maintain its free and protein amino acid patterns was examined. Marked effects on the body composition were seen. Total protein was reduced in starch-fed animals and reduced even further in the starved controls. The same was true of certain free amino acids, though the levels of others (glycine, taurine, aspartate and arginine) were not reduced but conserved. The protein was examined in more detail by analysis of hydrolysates. Results showed that the protein fraction was altered not only in quantity but also in quality by the starch diet. Compositional changes occurred in the protein amino acids which indicated that a selective metabolism of particular proteins, rather than general degradation of all body proteins, occurs during periods of deficient dietary intake.

INTRODUCTION

BODY protein is normally regarded as a relatively stable entity in animals, especially mammals. Thus the pattern of amino acids arising from protein hydrolysis would not be expected to show much variation during an animal's life or within one species. For planktonic invertebrates it has also been shown that there is little difference in protein amino acid composition between estuarine and neritic species (RAYMONT, FERGUSON and RAYMONT, 1973) and even between widely differing species of marine animals (RAYMONT, MORRIS, FERGUSON and RAYMONT, 1975). Indications were that the patterns observed were not altered by maturity, environmental depth or geographical locality. However, studies on mammals have suggested different turnover rates for different proteins; for example, liver and plasma proteins have a much faster turnover than skeletal tissue proteins (ROUTH, EYMAN and BURTON, 1969). Possibly proteins associated with DNA and RNA and other fundamental units of cell structure are likely to be conserved at the expense of other cell constituents during times of nutritional stress.

The various invertebrate taxa differ with respect to the type of reserve used for metabolism during periods of starvation (see MARSDEN, NEWELL and AHSANULLAH, 1973, and references therein). Although little is known concerning planktonic invertebrates, IKEDA (1974), reviewing his results and earlier data, suggests that lipid is the constituent most frequently used, but that protein can be substantially reduced, especially in smaller animals with low lipid reserves found more frequently in lower latitudes. Studies on the effect of a pure starch diet on the lipid chemistry of *Neomysis integer* MORRIS, ARMITAGE, RAYMONT, FERGUSON and RAYMONT, 1977) have demonstrated that the mysid is capable of maintaining its phospholipid composition for periods of about 2 weeks, but that pure carbohydrate is not an adequate diet, and during this time changes in the lipid

---

* Department of Oceanography, University of Southampton.
† Institute of Oceanographic Sciences, Wormley, Surrey.

composition occur—the levels of total lipid, triglyceride and long-chain polyunsaturated fatty acids gradually fall. *Neomysis integer* also appears to rely to a considerable extent on protein for oxidative metabolism (RAYMONT, AUSTIN and LINFORD, 1968) so that when mysids are maintained on a starch diet some variation in the pattern of protein amino acids might be expected to occur. These experiments test the ability of *Neomysis* to maintain its normal protein and free amino acid composition using a controlled diet and seek to evaluate the importance of individual amino acids in both fractions.

METHODS

1. *Materials*

Collections of *Neomysis integer* were made at low tide, by hand net, from the head of the estuary of the River Test at Redbridge, Southampton, and from shallow tidal sections of the River Beaulieu at Buckler's Hard, Hants. Samples were returned to the laboratory within the hour where they were maintained at 15°C in 25% (8‰) seawater (mean environmental salinity). When animals were not used immediately they were fed various species of unicellular algae (*Phaeodactylum tricornutum, Monochrysis lutheri, Dunaliella minuta, Tetraselmis tetrathele*) and *Artemia* nauplii. *Neomysis* has been kept successfully in the laboratory under such conditions for several months (RAYMONT, unpublished). Generally speaking, however, samples were kept overnight only and the experiment begun the following day.

The pure starch diet was prepared by washing starch (Analar; BDH) twice with Analar chloroform to remove any lipids. The chloroform was pipetted off after centrifugation and the starch vacuum-distilled to dryness. A starch solution was then made which was heated gently to boiling point and left to set as a gel. A kaolin diet was similarly prepared, but fed as a lipid-free powder.

2. *Experimental*

Mature male specimens of *Neomysis*, mature but not gravid females, and large juveniles (approximately 10–15 mm) were selected for these experiments. The animals, generally about eighty per experiment, were transferred to clean seawater and starved for 24 hours to clear the intestine of residual food. The population was then randomly divided into two groups: (i) an initial control sample of about thirty animals which was sacrificed; (ii) the experimental sample. The experimental animals were isolated in crystallizing dishes containing 600 ml 25% seawater to which small samples of the solid starch had been added. The seawater in the dishes was changed every 2 or 3 days, uneaten food and faeces removed and fresh food added. The experiments were terminated usually after 10 days; one experiment was abandoned before this time because mortality exceeded 60%. The remaining animals were starved for a further 24 hours to clear the gut of starch, then sacrificed. Samples were not normally analysed immediately, but carefully dried on filter paper and stored under nitrogen at −20°C. These are experiments I, II, III, IV and VI (V abandoned).

In some cases further controls were made. The original population was divided into three, instead of two groups, the third group being starved. In order to simulate the energy expenditure due to active feeding which occurred in the starch-fed animals, chloroform-washed kaolin was given to the starving animals, which were seen to ingest it readily and form faecal pellets. Starvation could usually be continued only for 5 or 6 days until the 60% mortality limit had been reached. In these experiments (VII and VIII) the starch-fed and starved animals were sacrificed at the same time. All samples were analysed for total protein, free amino acids and protein amino acids.

In order to ensure that the surviving animals constituted a random sample all dead animals from one experiment were weighed and compared with the weights of the control animals and those in the final sample. The results demonstrated that mortality occurred randomly.

Since the differences in protein hydrolysate values between control and experimental animals were very small (usually less than 5%—see Results) an assessment of the reproducibility of the extraction and analytical technique was necessary. A field collection of fifty animals was randomly divided into five groups of ten animals each; protein was extracted, hydrolysed and analysed for each group. The amino acids, calculated as percentages of the total hydrolysate, showed approximately 5% variability between the five sets.

3. Analytical

Total protein was estimated by the biuret method as reported previously (RAYMONT, AUSTIN and LINFORD, 1964) and the results are expressed as a percentage of wet body weight. Since only a small amount of material was available for analysis at the end of each experiment the specimens used for protein estimation were those previously used for wet–dry weight ratio determinations, reconstituted with distilled water at room temperature. Protein values of oven-dried material tend to be more variable and on average 25% lower than those for fresh mysids (ARMITAGE, unpublished; BAMSTEDT, 1974). It was considered, however, that the variability was not unacceptably high (a maximum of $\pm 2.5\%$ for any analysis). A minimum of three estimations was performed on each sample.

Free amino acids were prepared by homogenizing samples with two washes of 2:1 chloroform-methanol, according to the method of FOLCH, LEES and SLOANE-STANLEY (1957) for the removal of lipids. The homogenates were centrifuged and the supernatants washed three times with distilled water. The lower organic phase containing lipids was discarded. The solid residue was also washed with distilled water and all water–methanol phases combined. TCA was added to a final concentration of 5% and the solution centrifuged again to remove precipitated proteins. The solution of amino acids was then vacuum-distilled to dryness, taken up into HCl to a final concentration of 0.01 N, filtered through a fine glass sinter and stored at $-20°C$. Protein hydrolysates were prepared by refluxing the residue from the free amino acid extraction with 6 N HCl for 24 hours. The amino acid solution was vacuum-distilled to dryness, taken up in 0.01 N HCl and stored at $-20°C$.

The amino acids were separated and measured by ion-exchange chromatography (Technicon Amino Acid Autoanalyser). The chromatographic system employed Type-B Chromobeads and sodium citrate buffers, pH 2.8–5.0. Glutamine and asparagine are not detected by this system and tryptophan is destroyed by the preparation procedures. Cystine occurred normally in trace quantities; it is partly destroyed during hydrolysis and the values have been excluded from the results.

RESULTS

The changes in total protein which occur after periods of starch-feeding and starvation are superimposed on the protein hydrolysate data in Fig. 1 (A) and (B). As expected, the reduction after starvation is more severe than after starch-feeding. The degree of reduction varied between the experiments, reflecting differences in the duration of feeding.

Examination of the protein hydrolysate figures (expressed as μmoles/g wet weight) showed in general a reduction in experimental values from control levels. This was not surprising since the biuret analyses showed a considerable fall in total protein. Figure 1 (A) and (B) show the pattern of amino acids obtained from the protein hydrolysates of control and experimental animals from each experiment; these are expressed as percentages of the total hydrolysate amino acids. Differences between control and experimental patterns are small, usually less than 5%, and may be accounted for by variations in extraction and autoanalysis (see Methods).

Further quantitative examination of the hydrolysate data, expressed in this manner, did not assist interpretation, since any changes were likely to have been masked by the use of percentages. A more revealing approach to the changes in the amino acid patterns was therefore attempted. A Friedman rank sum analysis was applied to both the protein and free amino acid data of the two types of feeding experiments (see Appendix for details). The results were as follows:

Protein hydrolysates:

Control group versus (a) *Starch-fed* (b) *Kaolin-fed*

$\chi^2 = 34.39$ (15 d.f.) $\chi^2 = <15.34$ (15 d.f.)

P $= <0.01$ (99%) P $= 0.5-0.1$ (50%)

Free amino acids:

Control group versus (a) *Starch-fed* (b) *Kaolin-fed*

$\chi^2 = 36.3$ (17 d.f.) $\chi^2 = 25.4$ (17 d.f.)

P $= <0.01$ (99%) P $= 0.1-0.05$ (90%)

The results of the starvation experiments were not statistically significant but the results of only two experiments were available. The results of the starch-feeding experiments showed that there were changes in the amino acid composition of the protein hydrolysates and in the free amino acid pool. Mean ranks plotted on a linear scale can be used to demonstrate qualitative changes in composition. Thus, in the protein hydrolysate (see Fig. 2) the largest reductions are seen in aspartic acid, serine and histidine, and the smallest reductions in alanine and valine.

Results of the free amino acid analyses from four experiments are available and are shown in Fig. 3 (A) and (B); the difference between control and experimental samples has been plotted as a percentage increase or decrease. Asterisks refer to significant changes, and mean values for the percentage change are given below each histogram. The figures show that there was no significant change in the total free amino acids, and in the individual amino acids taurine, glycine, aspartate and arginine, in both the starch-fed and starved animals. The mean rank plot (Fig. 2) also shows that, of the free amino acids, the levels of these four are least changed. Changes in proline, methionine, phenylalanine and ornithine were large but highly variable and therefore were not statistically significant.

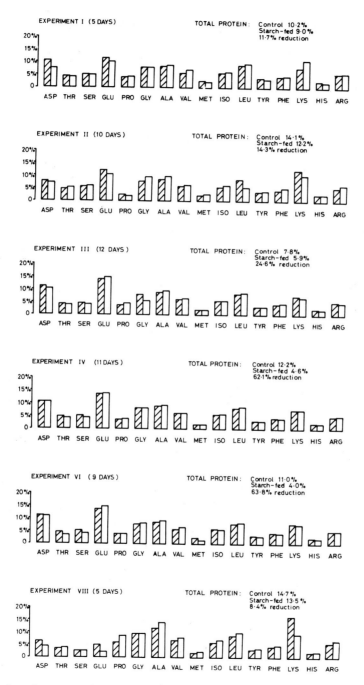

Fig. 1. (A) Protein amino acid composition of *Neomysis integer*, fed pure starch, expressed as percent of total. Hatched areas, control; open areas, experimental. Total protein values expressed as percent wet weight.

DISCUSSION

Results of the total protein estimations show that the body composition of *Neomysis* is partially maintained on a starch diet (cf. similar experiments on lipid metabolism—Morris, Armitage, Raymont, Ferguson and Raymont, 1977), presumably because a certain portion of the animals' energy requirements can be met directly by metabolism of the starch. Some protein degradation during starch-feeding, however, does occur because free amino acids are undoubtedly required for osmotic balance and for other important biochemical processes. Protein appears to be a significant metabolic substrate for oxidative metabolism in *Neomysis* (Raymont, Austin and Linford, 1968). Differences in the duration of feeding gave rise to varying reductions in total protein. Thus, experiments I and VIII (5–6 days) lost only about 10% of body protein. Experiments II and III (10–12 days) lost 15–25%. Experiments IV and VI (9–11 days) lost much more, but these mysids were from a different stock of animals obtained from the Beaulieu River estuary. Results of only two experiments are available for kaolin-fed (starved) animals. These mysids fed actively on the kaolin and produced faecal pellets, so that the results may properly be compared with those for starch-fed animals. As expected, total protein was sharply reduced in these experiments (80% in experiment VII; 60% in VIII) which lasted only 5 or 6 days.

Further reference to Fig. 1 (A) and (B) shows that certain differences exist in the hydrolysate data between the experiments, which may be related to the extent of change in total protein. The two starvation experiments Fig. 1 (B) which suffered the largest reductions in total protein showed the greatest differences in amino acid composition between control and experimental samples. For aspartate and lysine these differences exceed the 5% variability associated with the extraction and analysis. These acids assume greater (aspartate) and lesser (lysine) importance relative to other amino acids in the hydrolysate. Changes in the other acids are less consistent and less than 5%. Although the result of the

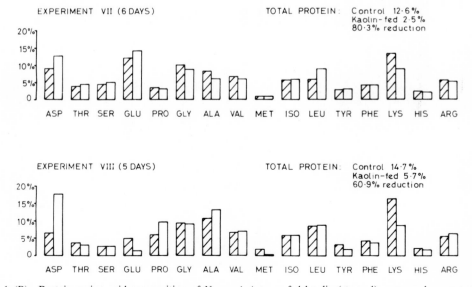

Fig. 1. (B) Protein amino acid composition of *Neomysis integer*, fed kaolin (starved), expressed as percent of total. Hatched areas, control; open areas, experimental. Total protein values expressed as percent wet weight.

rank sum analysis on these data was not significant, it appears to confirm the conclusion that a compositional change occurs in the amino acid spectrum. In the hydrolysates from the starch-fed samples (Fig. 1(A)) the variability in pattern is much smaller, usually less than 5%, and compositional differences may therefore be masked. Further, in these experiments the body protein can be conserved for a greater length of time by metabolism of the dietary starch, and hence one would expect the compositional changes to be smaller. If starvation experiments were performed for increasing lengths of time, increasing reductions in total protein might be reflected in a gradation of compositional changes in the hydrolysate.

The extent to which protein hydrolysate data can be interpreted is limited and for the present, in the absence of further experiments, only tentative explanations can be made for any changes which are seen. Examination of the hydrolysate data by the Friedman method revealed a change in the composition of the body protein during the starch-feeding experiments, as well as a drastic reduction in its amount. The conclusion is that a selective metabolism of proteins of a particular amino acid composition occurred. This, together with the knowledge that *Neomysis* relies heavily on protein in oxidative metabolism, suggests that the body protein might be broadly divided into two types. The first is that which is readily accessible for general metabolic requirements in times of nutritional deficiency, which may be synthesized with this purpose in view, and which, for ease of distinction, is here called the 'general-protein'. The other type of protein is that which cannot be broken down without loss of fundamental biological processes to the animal and eventually death. These proteins may include basic enzyme systems, membranes and essential structural proteins which are likely to be conserved by the animals at all costs. This type may be called the 'basic-protein'. Undoubtedly the distinction cannot be as clear as suggested here—there is likely to be a gradation of proteins between the two types, some being more important, and hence more likely to be conserved.

Changes in the pattern of the individual amino acids may permit some speculation about protein amino acid composition in general qualitative terms. Figure 2 shows that aspartic acid, serine and histidine are reduced most in the hydrolysates. These acids must have been present in the control animals in the general-protein which is degraded during the course of the experiment and is lost from the final starch-fed sample (largely composed of basic-protein only). Thus it can be inferred that, relative to the general-protein, the basic-protein is poorer in aspartate, serine and histidine. At the other end of the scale the smallest changes are seen in alanine and valine (Fig. 2). After the general-protein has been degraded the levels of these two amino acids still remain largely unchanged in the starch-fed animals, implying that the basic-protein is relatively richer than the general-protein in these amino acids. Reference to the free amino acid scale in Fig. 2 indicates that it may be necessary for the animal to metabolize a general-protein, rich in aspartate, since this is one of the acids which is maintained in the free pool when others are clearly reduced. Indeed, the notion that it is conserved in the free amino acid pool, at the apparent expense of the protein aspartic acid, can be seen as supporting evidence for the composition and function of the general-protein. Alternatively, it is possible that particular proteins are degraded readily, not only because of their amino acid composition but because these proteins have high turnover rates and are, therefore, more accessible for metabolism.

Interpretation of the free amino acid data is rather easier. Figure 3 (A) and (B) show that, in contrast to the substantial reduction seen in total protein content, there are only small and non-significant reductions in total free amino acids (19% after starch-feeding; 29% after

5 days starvation). The small changes in the total free amino acids are due to the conservation of certain individual acids, namely taurine and glycine, which are normally present at high concentration in the free amino acid pool, and aspartate and arginine (both in the starch-feeding and starvation experiments). Glycine is important in *Neomysis*, which lives in an estuarine environment, because of its function in osmotic control (RAYMONT, AUSTIN and LINDFORD, 1968). The changing levels of taurine may be due to an increased metabolism of the sulphur-containing amino acids, taurine being commonly thought of as a degradation product of cysteine metabolism, though it has also been shown to be osmotically active (ALLEN and GARRETT, 1972). Aspartate and arginine are likely to be very important biochemically, aspartate because of its central position in the metabolism of several keto and amino acids, and arginine because of its importance as arginine phosphate in energy transfer; Arginine may be largely 'bound' as arginine phosphate. Unlike aspartate, glutamate, which is equally central to metabolism in mammalian systems, is reduced in *Neomysis* by some 40% by both starch-feeding and starvation.

The alternative treatment of the starch-fed data by the rank sum method (Fig. 2) supports these results further, showing not only that a compositional change in the free amino acid pool occurred, but that aspartate, taurine, glycine and arginine had been least reduced. The

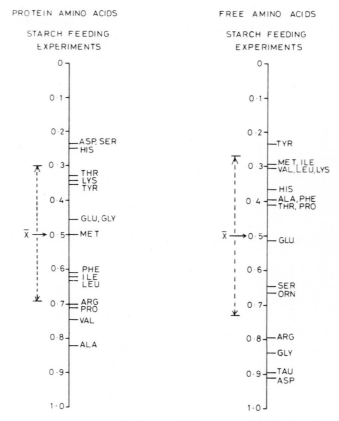

Fig. 2.   Mean rank values of compositional change in protein and free amino acids of *Neomysis integer*, fed pure starch (see text).

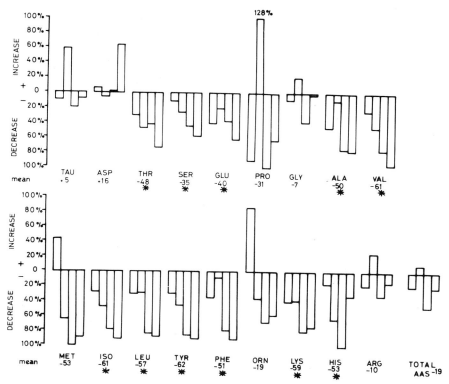

Fig. 3. (A) Percentage change in free amino acid levels in *Neomysis integer*, fed pure starch. Asterisks denote statistically significant changes. Mean percentage change for each amino acid is indicated.

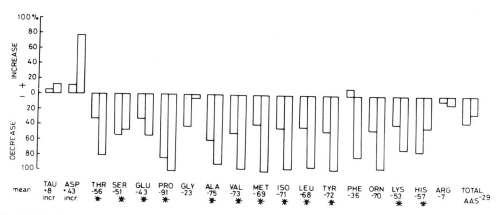

Fig. 3. (B) Percentage change in free amino acid levels in *Neomysis integer*, fed kaolin (starved). Asterisks denote statistically significant changes. Mean percentage change for each amino acid is indicated.

notion that these four amino acids are maintained at the expense of protein is countered by the suggestion that aspartate and glycine in particular undergo a rapid turnover and that these are most easily resynthesized, using amino units derived from the other free amino acids. While this resynthesis is not impossible, it is unlikely that a significant amount of glycine and aspartate will be formed by *de novo* synthesis since the free amino acid pool is small in comparison with the tissue protein pool. It is more probable that, in the absence of an intake of dietary nitrogen, the amino nitrogen in the free amino acid pool is derived largely from amino nitrogen in the protein pool. Other amino acids were reduced slightly more or slightly less than average. Tyrosine appears to have been used more than the others (Fig. 2); this is borne out by reference to Fig. 3 (A) which gives tyrosine the highest percentage reduction. The same is true of isoleucine, leucine, valine and lysine which follow tyrosine in both Figs 2 and 3 (A). The exception is methionine which has more variable individual data, and is therefore not significant, as shown in Fig. 3 (A), but has a mean rank value of 0.3 in Fig. 2, which puts it amongst the more degraded amino acids. The starvation experiments (Fig. 3 (B)) generally support these results. The changes are of much the same order, with tyrosine, leucine, isoleucine, methionine and valine being depleted, together with a high reduction in proline and alanine (which are reduced only slightly more than average in the starch-fed samples (Fig. 2)). The starvation results reinforce the conclusion that the levels of aspartate, glycine, taurine and arginine are maintained, either due to their relative ease of synthesis or because of their special importance to the survival of the animal. Experiments using ^{14}C-labelled amino acids might help to clarify this problem.

APPENDIX: THE FRIEDMAN METHOD

Figure 2 is derived from a rank sum analysis which tests whether there has been any compositional change in the samples during the experiment. A randomization test, which makes fewer assumptions about the normality of the data, is one of the better ways of dealing with these results. By using ranks the test overcomes certain problems in the calculations encountered when using actual values, although it may be argued that it becomes qualitative rather than quantitative.

The difference between controls and experimental values are calculated (% increase or % reduction) and are ranked for each experiment. Ranks are then totalled and an average rank calculated for each amino acid. The null hypothesis is that if no selective metabolism has occurred during the experiments the ranks will be randomly distributed throughout the matrix of ranks for all experiments. There should be no pattern to the ranks produced by the experimental conditions. The test now follows the standard Friedman procedure (COLQUHOUN, 1971; pp. 201–203).

The mean ranks for individual amino acids may be considered further. If no selective metabolism has occurred the mean rank for each amino acid will be close to the mean rank for the whole matrix. Plotted on a linear scale (Fig. 2) it is thus possible to see which amino acids are close to the mean rank ($\pm$ standard deviation of the mean). Because of the difference in the number of amino acids in the free and protein fractions (18 and 16 respectively) the plots are best represented on a zero-to-one scale; the mean ranks for each scale are then in line and comparison of the two is easier.

Acknowledgements—We are grateful to Dr. A. P. M. LOCKWOOD for contributions to the discussion and to J. K. B. RAYMONT for assistance in the experiments.

REFERENCES

ALLEN J. A. and M. R. GARRETT (1972) Studies on taurine in the euryhaline bivalve *Mya arenaria*. *Comparative Biochemistry and Physiology*, **41A**, 307–317.

BÅMSTEDT U (1974) Biochemical studies on the deep-water pelagic community of Korsfjorden, western Norway. Methodology and sample design. *Sarsia*, **56**, 71–86.

COLQUHOUN, D (1971) *Lectures on biostatistics. An introduction to statistics with applications in biology and medicine*, Clarendon Press, Oxford, 155 pp.

FOLCH J., M. LEES and G. H. SLOANE-STANLEY (1957) A simple method for the isolation and purification of total lipids from animal tissues. *Journal of Biological Chemistry*, **226**, 497–509.

IKEDA T. (1974) Nutritional ecology of marine zooplankton. *Memoirs of the Faculty of Fisheries, Hokkaido University*, **22**, (1), 1–97.

MARSDEN I. D., R. C. NEWELL and M. AHSANULLAH (1973) The effect of starvation on the metabolism of the shore crab, *Carcinus maenas. Comparative Biochemistry and Physiology*, **45 A** (1), 195–214.

MORRIS R. J., M. E. ARMITAGE, J. E. RAYMONT, C. F. FERGUSON and J. K. RAYMONT (1977) Effects of a starch diet on the lipid chemistry of *Neomysis integer* (Leach). *Journal of the Marine Biological Association of the United Kingdom*, **57**, 181–189.

RAYMONT J. E. G., J. AUSTIN and E. LINFORD (1964) Biochemical studies on marine zooplankton. I. The biochemical composition of *Neomysis integer. Journal du Conseil international pour l'Exploration de la mer*, **3**, 354–363.

RAYMONT, J. E. G., J. AUSTIN, and E. LINFORD (1968) Biochemical studies on marine zooplankton. V. The composition of the major biochemical fractions in *Neomysis integer. Journal of the Marine Biological Association of the United Kingdom*, **48**, 735–760.

RAYMONT, J. E. G., C. F. FERGUSON and J. K. B. RAYMONT (1973) Biochemical studies on marine zooplankton. XI. The amino acid composition of some local species. *Special publication of the Marine Biological Association, India*, 91–99.

RAYMONT, J. E. G., R. J. MORRIS, C. F. FERGUSON and J. K. B. RAYMONT (1975) Variation in the amino acid composition of lipid-free residues of marine animals from the northeast Atlantic. *Journal of Experimental Marine Biology and Ecology*, **17**, 261–267.

ROUTH J. I., D. P. EYMAN and D. J. BURTON (1969) *Essentials of general organic and biochemistry*, Saunders, London, 718 pp.

The existence of a benthopelagic fauna in the deep-sea

N. B. MARSHALL* and N. R. MERRETT†

THOUGH it is a hundred years since H.M.S. *Challenger* returned from her long and fruitful voyage, there is still much to be learned of the main environments of deep-sea animals. At midwater levels a mesopelagic ('twilight zone') is now distinguished from an underlying bathypelagic ('sunless') zone, but it was not until 1957 that these two strata were 'officially' recognized (HEDGPETH, 1957). In Hedgpeth's synthesis of marine environments (as shown in his fig. 1, p. 18) the transition between mesopelagic and bathypelagic environments is set tentatively at a depth of about 1000 m. Indeed one of the tasks of deep-sea biologists is to investigate the nature of the change from one faunal region to the other and to see how far the depth of this transition varies with regional conditions.

On the deep-sea floor a bathyal fauna is distinguished from a deeper abyssal fauna, and in Hedgpeth's figure the transition between the two is placed, again with a query, at about 4000 m. At midwater levels below this depth there may also be a distinct abyssopelagic fauna (see also BRUUN, 1957). Proper investigation of such a deep midwater fauna will surely be a long and daunting task.

Besides our pleasure in contributing this paper to Sir George Deacon's 70th Anniversary Volume, our purpose is to show that there is growing evidence to substantiate further the existence of a separate and distinct pelagic fauna near the deep-sea floor. The members of this benthopelagic fauna range from millimetre-sized copepods to cephalopods and fishes, some of which exceed a metre in length. After reviewing the evidence for the existence of this fauna, which extends to depths of 6000 m or more, we present findings, including our own, on how the fishes make a living. In our final section we conclude that conditions of life near and on the deep-sea floor, at least in subtropical and tropical regions (which are the headquarters of deep-sea animals), have favoured the evolution of benthopelagic fishes utilizing a mixed diet of benthic and pelagic organisms. The same conclusion is apparent when benthopelagic fishes, which are neutrally buoyant (or nearly so) are similarly compared to the negatively buoyant, and thus more handicapped, benthic fishes.

THE EVIDENCE FOR A BENTHOPELAGIC FAUNA

If benthopelagic and benthic forms are broadly described as bottom-dwelling fishes, close to a thousand species have been taken in nets fished on the deep-sea floor. Bottom-dwelling species include hagfishes (Myxinidae), squaloid sharks, skates (Rajidae), chimaeras, alepocephalids, tripod-fishes (Bathypteroidae), chlorophthalmids, synapho-

---

* Department of Zoology and Comparative Physiology, Queen Mary College, University of London, Mile End Road, London E1 4NS.

† Institute of Oceanographic Sciences, Wormley, Godalming, Surrey, GU8 5UB.

branchid eels, halosaurs, notacanths, deep-sea cod (Moridae), rat-tails (Macrouridae), brotulids, nototheniiforms, sea snails (Liparidae), eel pouts (Zoarcidae), scorpaenids, cottids, flatfishes and anglerfishes. The contrasts between benthopelagic and benthic types are considered later.

During lowering and hauling, bottom nets also catch species that belong to typical mid-water groups of animals, but it is not safe to assume that every such representative has been taken well away from the bottom (see p. 495). Proper sampling of the bottom-dwelling fauna at least requires the combination of a fine-meshed net and a large bottom net. Ideally both nets would be lowered closed, opened for fishing and then closed before hauling. Two IOS nets, the combination midwater trawl (RMT 1 + 8, Baker, Clarke and Harris, 1973) and the 1.5-m^2 epibenthic sledge (Aldred, Thurston, Rice and Morley, 1976), both acoustically monitored and mouth-closing, are capable of separately sampling the smaller elements of the benthopelagic and epibenthic fauna. Yet the simultaneous sampling of a wider spectrum of both faunas is complicated by the disturbing influence of a necessarily large bottom net on the catches of a plankton net in tandem with it.

No doubt with such difficulties in mind, Grice (1972) has used the deep submersible *Alvin* to fish twin closing nets near and well above the bottom. During two dives off New England to depths of 1465–1500 m and 992–1000 m, one net was closed after a tow close to the bottom (about 20 cm away on each occasion) for an hour at 1 knot.* The other net was fished and closed well above the bottom (260 m on the first and 120 m on the second dive). The catches of copepods in the two nets were quite different. Indeed, the near-bottom net had evidently sampled a special fauna of small calanoid copepods (*ca.* 1–3 mm in length) some with parts of the exoskeleton bristling with many small spines. There were, for instance, seven species of *Xanthocalanus*. These planktobenthic copepods, as Grice calls them, are like species known to be benthic in their small size, plump bodies and certain details of limb structure. They are also assignable to genera rarely recorded in plankton samples or to genera with known benthic affinities. He concluded that they live just above the bottom or perhaps in the bottom sediment or overlying flocculent zone. We have found the remains of such copepods in the alimentary tracts of various bottom-dwelling fishes. In the present context, it is interesting that we found so many of these copepods in the tripod-fish *Bathypterois dubius* (see Table 1). Bathypteroids use their tripod under-carriage to rest on the bottom, and we suggest they are well placed to snap up plankto-benthic copepods.

The existence of a benthopelagic fauna of deep-sea fishes is now well documented. One reason is that the adult sizes of these fishes are well within the resolution of deep-sea cameras. Moreover, they are not disturbed by cameras, or indeed by deep submersibles. By 1960 there were enough photographs and observations to support the conclusion that the development of an aptly organized, gas-filled swimbladder in a diversity of bottom-dwelling fishes should give them the buoyancy to hover or move easily over the bottom (Marshall, 1960). Such fishes, which include the macrourids, morids, halosaurs, noto-canths, synaphobranchid eels and brotulids, were later described as benthopelagic ("Ben-thopelagic fishes, and other organisms, are those that swim freely and habitually near the ocean floor" (Marshall, 1965)). If degree of success is judged by species-diversity (*sensu* species richness), ubiquity and biomass, the outstandingly successful fishes are the macrourine rat-tails, which, like the halosaurs, have the right form and fin pattern to adopt

* 1 knot = 0.51 m s^{-1}.

a snout-down posture when hovering or swimming over the sediments. The underslung jaws of these fishes are thus easily able to pick food organisms off the bottom (MARSHALL and BOURNE, 1964, 1967).

A minority of benthopelagic fishes have no swimbladder and nearly all are squaloid sharks, chimaeras and alepocephalids. Representatives of all three groups have been photographed over the bottom. Species of the first two groups have a large liver so charged with the light hydrocarbon squalene (s.g. 0.86) that they are neutrally buoyant (SCHMIDT-NIELSEN, FLOOD and STENE, 1934; CORNER, DENTON and FORSTER, 1969). Like other pelagic deep-sea fishes without a gas-filled swimbladder, alepocephalids have a poorly ossified skeleton and weak, watery muscles, and are thus likely to be close to neutral buoyancy (see DENTON and MARSHALL, 1958). The same is probably true of the ateleopiform fishes. Evidently, there has been strong selection pressure for the retention or the attainment of neutral buoyancy in benthopelagic fishes. This is even true of deep-sea rays (BONE and ROBERTS, 1969). For instance, two specimens of *Raja longirostra* from 930–1600 m had a percentage weight in water of 1.5. Rays from still greater depths (2500 m), such as *Raja richardsoni* and *Breviraja pallida*, which have soft bodies and extremely large livers, were judged to be "... probably near to neutral buoyancy".

Benthic fishes have no swimbladder or other buoyant means. Tripod-fishes, already mentioned, have a firm skeleton and well-formed axial muscles, and are also negatively buoyant. The same is probably true of the related genus *Ipnops* and of *Bathysaurus*,* which have also been photographed resting on the bottom (see ROPER and BRUNDAGE, 1972; HEEZEN and HOLLISTER, 1971). The form, fin pattern and inner organization of chlorophthalmids, deep-sea flatfishes, scorpaenids, cottids, ogcocephalids and notatheniiform fishes also strongly suggest a benthic habit coupled with negative buoyancy. Deep-sea liparids and zoarcids, which are most diverse in cold northern waters (ANDRIASHEV, 1965), are classic types of benthic fishes, though certain species have taken to a pelagic existence.

Lastly, deep benthic photographs have also revealed that cirrate octopods live near the bottom. ROPER and BRUNDAGE (1972) conclude: "Twenty-seven photographs from seven deep-sea localities in the North Atlantic reveal cirrate octopods in their natural habitat. The photographs demonstrate that these octopods are benthopelagic, living just above the bottom at depths of 2500 m to greater than 5000 m. Typical cephalopodan locomotion is exhibited as well as a drifting or hunting phase, and possibly a pulsating phase. Animals range in size from approximately 10 to 128 cm in total length, and up to 170 m across the outstretched arms and webs."

Evidence for a benthopelagic, deep-sea fauna thus comes from both ends of the food chain. Clearly we need to discover the intermediate links between the smallest copepods and the largest fishes and cephalopods. Study of the trophic tendencies of a wide range of benthopelagic fishes should contribute to such a discovery.

SOURCES OF FOOD AND TROPHIC TYPES

Knowledge of the food patterns of benthopelagic animals is almost entirely confined to the fishes. Both pelagic and benthic kinds of food organism are taken. As already seen,

---

* Possible evidence of an approach to neutral buoyancy of *Bathysaurus agassizi* is given from preliminary observations on a preserved specimen of 495 mm standard length. The liver constituted 21% of the total preserved body weight and was rich in lipid (70% of total liver). The liver lipids had a specific gravity of 0.920 and were found to be composed mainly of triglycerids (R. J. MORRIS, personal communication).

Table. 1. *Dominant food organisms of bottom-living fishes caught off northwest Africa, based on preliminary analysis of stomach and intestinal contents*

Species	Number examined	Size range examined (mm)	Buoyancy	Range of capture (mid-depth) (m)	Dominant food organisms
Galeus polli	12	310–410	TL	?L	510	Fish (mainly myctophids), cephalopods
Etmopterus princeps	5	550–655	TL	L	1153–1164	Fish (1 ?myctophid)
Centrophorus granulosus	6	850–1195	TL	L	578–798	Fish (including myctophids)
C. ?foliaceus	18	390–600	TL	?L	578–601	Fish, unidentified? crustaceans
C. squamosus	1	1205	TL	L	985	Cephalopods
Centroscymnus coelolepis	4	655–1015	TL	L	601–1164	Fish
C. ?macracanthus	1	780	TL	?L	1164	?Crustaceans
Deania calcea	8	655–935	TL	L	940–985	Fish (myctophids)
Raja nidarosiensis	1	1150	DW	L	918	Fish
Synaphobranchus kaupi	14	150–465	SL	SB	814–981	Fish, cephalopods
Halosaurus ovenii	6	185–254	GL	SB	981	Mixed crustaceans, fish
Halosaurichthys johnsonianus	5	100–140	GL	SB	933–1017	Copepods
Halosauropsis macrochir	5	145–270	GL	SB	2375–2985	Polychaetes, mixed crustaceans
Aldrovandia phalacra	3	78–121	GL	SB	1510	Mixed crustaceans (mainly amphipods), polychaetes
Notacanthus bonaparti	4	140–160	GL	SB	981	Mixed crustaceans
Alepocephalus rostratus	6	207–300	SL	RO	981	Mysids (Gnathophausia zoea), unid. jelly
A. bairdi	1	362	SL	RO	981	Natant decapod
Chlorophthalmus agassizi	1	162	SL	—	295	Chaetognaths, mixed crustaceans (incl. euphausiids)
Parasudis fraser-brunneri	3	141–185	SL	?L	295	Unid. crustaceans
Bathysaurus agassizi	1	495	SL	—	2375	Fish
Bathypterois dubius	10	148–193	SL	—	981–1261	Copepods
B. longipes	1	150	SL	—	2984	Copepods
Benthosaurus grallator	1	368	SL	—	2375	Mixed crustaceans
Bathytyphlops azoriensis	2	270–280	SL	—	2984	Crustaceans (?natant decapods)

Species	n	Size range	Measure	SB	Food
Chaunax pictus	5	48–206	SL	—	Copepods, natant decapods
Dibranchus atlanticus	6	43–68	SL	—	Copepods, polychaetes
Mora mediterranea	2	490–555	SL	SB	Fish, cephalopods
Laemonema laureysi	2	146–235	SL	SB	Mixed crustaceans, fish
Mixonus laticeps	2	142–165	SL	—	Natant decapods
Trachyrhynchus trachyrhynchus	15	33–155	HL	SB	Mixed crustaceans, cephalopods
Gadomus longifilis	8	22–46	HL	SB	Chaetognaths
Bathygadus favosus	3	30–48	HL	SB	Crustaceans, cephalopods, chaetognaths
B. melanobranchus	23	22–74	HL	SB	Mysids (*Gnathophausia zoea*), chaetognaths
Coelorhynchus occa	1	93	HL	SB	Polychaetes, crustaceans
C. coelorhynchus	2	49–56	HL	SB	Mixed crustaceans, fish (1 myctophid)
Coryphaenoides zaniophorus	14	42–88	HL	SB	Mixed crustaceans (mainly copepods and amphipods), polychaetes
C. macrocephalus	1	215	HL	SB	Natant decapods
Chalinura mediterranea	1	76	HL	SB	Polychaetes, cephalopod
Hymenocephalus italicus	12	11–33	HL	SB	Mixed crustaceans (including copepods, euphausiids and natant decapods)
Sphagemacrurus hirundo	1	32	HL	SB	Polychaetes
Nezumia aequalis	44	17–46	HL	SB	Mixed crustaceans (mainly copepods and ostracods), polychaetes
N. micronychodon	6	30–46	HL	SB	Polychaetes, tanaids
Ventrifossa occidentalis	4	45–58	HL	SB	Crustaceans (?natant decapods)
Hoplostethus mediterraneus	23	51–242	SL	SB	Natant decapods
H. cadenati	1	102	SL	SB	Natant decapods
Helicolenus dactylopterus	87	100–345	SL	—	Ophiuroids
Trachyscorpaea cristulata	10	29–137	SL	—	Natant decapods
Scorpaena maderensis	2	68–107	SL	—	Mixed crustaceans
Epigonus denticulatus	4	75–114	SL	SB	Mixed crustaceans
E. pandionis	3	86–123	SL	SB	Copepods
Bathysolea profundicola	4	98–130	SL	—	Mixed crustaceans

Key: TL: total length, DW: disc width, SL: standard length, GL: gnathoproctal length, HL: head length, L: liver, RO: reduced ossification, SB: swimbladder.

there seems to be a special fauna of benthopelagic copepods, and it is clear that over slope levels, at least, typical members of the midwater zooplankton may approach the bottom and become the food of benthopelagic fishes (pp. 493–495). But, apart from the copepods, we have no knowledge of the special zooplankton at benthopelagic levels. It is likely, though, that peracaridean crustaceans, especially gammarid amphipods, and natant decapod crustaceans may at times leave the bottom and swim over the sediments. Indeed, during a dive to 4160 m off Madeira in the French bathyscaphe *Archimède*, WOLFF (1971) observed '. . . an incredible number of isopods and amphipods swimming close to the bottom (from 0 to 1–2 m above it)". Moreover, decapods and large gammarids certainly find their way to baited cameras (see ISAACS and SCHWARTZLOSE, 1975).

Nearly all the benthic food of bottom-dwelling deep-sea fishes consists of epifaunal species, but small infaunal forms, such as polychaetes and bivalves, may well be more important than we realize. Macrourine rat-tails have the means to process the sediment for such food, and we have seen small polychaetes among their gut contents. But no fish would seem to be capable, unless it is an ooze-eater (see p. 492), of utilizing the members of the meiofauna.

The present endeavour is to integrate previous findings on food patterns in bottom-dwelling deep-sea fishes with our analyses from fishes taken in surveys over parts of the slope off northwestern Africa. A detailed report is in preparation. The work was carried out from R.R.S. *Discovery* during 1972–1975 using a variety of fishing gear; 2.4 m² and 1.5 m² epibenthic sledges (35 tows), 14 m Marinovich semi-balloon shrimp trawl (5 tows), baited bottom traps (18 operations) and bottom longlines (9 operations). The main fishing effort was concentrated between 250–1500 m (54 operations) with a further 13 operations in depths down to 4400 m. Feeding was studied from stomach and intestinal contents analysis. The latter partly overcame the common problem of food loss by stomach eversion among swimbladdered forms. It provided qualitative evidence of the general diet of the species sampled and was advantageous as it lessened the chances of anomalous results due to unnatural net feeding.

The initial analyses of the diets of 405 specimens of 51 species represents about 9% of the total catch and about 35% of the species taken during the surveys (see Table 1 and also Fig. 1 for the depth distributions of the more numerous species concerned). It will be best to consider first the macrourids, whose food patterns are best known (see OKAMURA, 1970; MARSHALL, 1973a; GEISTDOERFER, 1973, 1975; HAEDRICH and HENDERSON, 1974; PEARCY and AMBLER, 1974). Study of the design of their fin patterns, jaw suspension, dentition, gill structure and gustatory system, combined with some preliminary analysis of their diet, indicated that the bathygadine species are adapted for taking pelagic kinds of food, whereas the very diverse and more adaptable macrourines are able to feed on both pelagic and benthic kinds (MARSHALL, 1973a). Our analyses support this conclusion (see Table 1). For instance, *Bathygadus melanobranchus* contained the remains of mysids and chaetognaths. Specimens of *Gadomus longifilis* which were taken at 1510 m contained little but the heads of chaetognaths.

The macrourine species contained both pelagic forms such as copepods and small decapods and benthic species (e.g. polychaetes). Indeed, most of the macrourines also contained the remains of sediment in the gut and there is now cinematographic evidence (ISAACS and SCHWARTZLOSE, 1975) to support earlier conjecture (MARSHALL, 1973a) that these rat-tails take in mouthfuls of sediment, then filter off contained food organisms by blowing the deposit through their gill slits. It is, of course, possible that some of the sediment may come from their prey. GEISTDOERFER (1973, 1975), who has studied the morphology and histology of their alimentary system as well as their diet, also found that representatives of macrourine species from the Atlantic and Mediterranean contained both pelagic and benthic organisms.

Even so, it is not true that all macrourine rat-tails, which, as we have said, are very diverse, have a mixed pelagic and benthic diet. Reference to Table 1 will show that *Hymenocephalus italicus* contained copepods, euphausiids and natant decapods. GEISTDOERFER (1975) has similar evidence of a pelagic diet. Perhaps all species of this genus, which are relatively small with subterminal jaws, small teeth and rather numerous gill rakers, have a marked bias for pelagic food. Species of *Trachyrhynchus*, which MARSHALL (1973a) places in a separate subfamily, have markedly inferior, highly protrusible jaws, minute teeth and numerous gill rakers. Like GEISTDOERFER (1975), we found that *T. trachyrhynchus* has a marked preference for pelagic food. Lastly, in certain species food patterns may change with age. For instance, off the Oregon coast small individuals of *Coryphaenoides leptolepis* and *C. armatus* feed largely on epifaunal crustaceans, whereas large individuals have a marked preference for pelagic forms of fish and squid, which may be taken well away from the bottom (PEARCY and AMBLER, 1974).

There are also contrasts in the régimes of the two main groups of notacanthiform fishes. Halosaurs seem to prefer crustaceans, many of which (e.g. copepods) are pelagic, but also take benthic animals, such as polychaetes (see Table 1 and McDOWELL, 1973). Notacanths also take pelagic and benthic food organisms, but the latter make up most of the food of *Notacanthus* spp. (McDOWELL, 1973 and p. 492).

Like the macrourine rat-tails, benthopelagic species of squaloid fishes are both neutrally buoyant and possess inferior, protrusible jaws. One might suppose then that the sharks have a rat-tail-like ability to seize both pelagic and benthic kinds of food, but this inference is not supported by our data. While the thorough and rapid digestion of sharks makes an inventory of their diet a hazardous task, such evidence as could be obtained indicated clearly the pelagic origin of their food. The most abundant food items found were fish, cephalopods and natant decapods, in order of importance. Little or no evidence was obtained of scavenging on the remains of large organisms as was found by CLARKE and MERRETT (1972). Several species (*Etmopterus princeps*, *Centrophorus granulosus* and particularly, and surprisingly from its morphology, *Deania calcea*) were found to be eating mesopelagic myctophid fishes, such as *Myctophum*, *Symbolophorus*, *Notoscopelus* and *Diaphus* species. Other myctophids (*Notoscopelus* and *Benthosema* species) have previously been reported in *C. granulosus* from the Indian Ocean (FORSTER, BADCOCK, LONGBOTTOM, MERRETT and THOMSON, 1970). There was no evidence to suggest that any of these elasmobranchs were dependent upon the pelagic feeding bottom-dwelling teleosts for food.

STRATEGIES FOR PREDATION

There would seem to be three main trophic types among benthopelagic fishes of the deep sea:

1. Species depending on a mixed diet of pelagic and benthic organisms.
2. Species with a marked preference for pelagic animals.
3. Species with a marked preference for benthic animals.

Considering our earlier criteria for the success of taxonomic groups, what may be said of the above three trophic groups? The macrourids, which have been best investigated, provide useful frames of reference.

The available evidence suggests that most of the species of the subfamily Macrourinae fit into group 1. Species of *Hymenocephalus* and large individuals of certain macrourines (especially of the largest species of *Coryphaenoides*) may be put in group 2 together with species of two other subfamilies, the Bathygadinae and Trachyrynchinae. Though it is to be expected that members of group 1 will sometimes contain a preponderance of benthic food, present evidence suggests that no more than a few species are assignable to group 3.

Group 1, as typified by macrourids with a mixed diet, is by far the most successful. As already stated, the macrourines are very diverse and indeed comprise about 90% of the 250 or more species of the family. Even excluding the species with habitual preferences for pelagic or benthic food and the few species that have taken to a midwater existence, the overall species-diversity criterion is well covered. Concerning their ubiquity and niche diversity, macrourines live from polar to tropical regions and range in depth from upper slope levels to abyssal reaches of at least 6000 m. In terms of biomass, macrourine species easily outweigh all other macrourids in the combined catches of any extensive trawling survey from slope to abyssal levels. The catches we have examined certainly confirm the criterion of biomass, and regarding the other two criteria, reference to MARSHALL (1973 b) will show that off northwest Africa macrourine rat-tails are both more diverse and occupy a wider range of depths than the other macrourids. Discussion of the sequence of species with depth is reserved for another paper, although preliminary evidence of the more numerous species in the present survey is given in Fig. 1.

Apart from these general criteria of success, it is clear that fully grown macrourids are what WILSON (1975) calls 'bonanza strategists'. Over a wide range of depths (750 to 6000 m) in the Pacific, Antarctic and Indian oceans, ISAACS and SCHWARTZLOSE (1975) found that macrourids—and all of their photographs, so far as we know, are of macrourine species— surpassed all other kinds of fishes in finding baited automatic cameras. Where hagfishes occurred, and they are not known below 2000 m, they were the most thorough bonanza strategists in that they sequestered the bait by covering it with copious secretions of mucus from their skin (see ISAACS and SCHWARTZLOSE, p. 89).

It is clear then that macrourids with preferences for benthopelagic food (group 2) are less diverse and abundant than the macrourine species that subsist on a mixed diet. Moreover, they are less widespread and their centres of abundance are restricted to depths above 2000 m (see MARSHALL, 1973 a and b). Perhaps the main factor that keeps them to modest depths is the nature and quantity of their food. Standing stocks of zooplankton are certainly very low at depths below 2000 m (VINOGRADOV, 1968) and the same may be true of benthopelagic copepods. Concerning their geographical distribution, it is striking that bathygadine rat-tails, for instance, are confined to productive waters in the temperate to tropical eastern Atlantic and are most abundant in upwelling regions: they are unknown from the impoverished Mediterranean. Elsewhere, but again only in warm temperate to tropical regions, the same correlation is also apparent. Unlike such macrourine (group 1) species as *Coryphaenoides rupestris* and *Macrourus berglax*, no bathygadine has been able to exploit the seasonal abundance of food in cold northern waters.

The only species reported to feed largely on benthic organisms is *Coelorhynchus coelorhynchus* and GEISTDOERFER (1975) also has evidence that the same applies to *Coryphaenoides zaniophorus*. In his samples GEISTDOERFER found that the main food

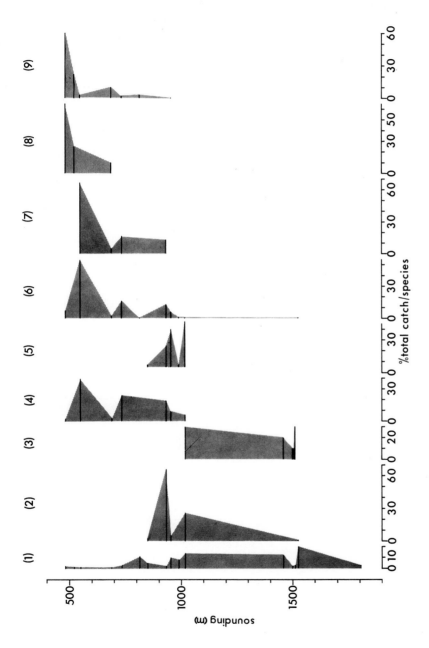

Fig. 1. Sounding ranges of nine species of slope-dwelling fishes based on the catches of seventeen approximately comparable epibenthic sledge tows (distance of bottom contact = *ca.* 1000 m/tow) from an area (21–26° N) off northwest Africa. The percentage catch per species is indicated at the mid-sounding position of relevant tows, with the range of capture shaded:
(1) 98 *Synaphobranchus kaupi*, (2)506 *Halosaurichthys kaupi*, (3) 10 *Bathypterois dubius*, (4) 54 *Trachyrhynchus trachyrhynchus*, (5) 214 *Bathygadus melanobranchus*, (6) 222 *Nezumia aequalis*, (7) 66 *Nezumia micronycholon*, (8) 124 *Hymenocephalus italicus*, (9) 261 *Hoplostethus mediterraneus*.

organisms of *C. coelorhynchus* were polychaetes and amphipods, but the latter may sometimes behave as pelagic forms (WOLFF, 1971). In South African waters *Coelorhynchus fasciolus* certainly feeds on both pelagic and benthic organisms (RATTRAY, 1947) which seems also to be true of *C. occa* (MARSHALL and IWAMOTO, 1973).

Rat-tails feed on freely moving benthic organisms. Indeed, sessile invertebrates are rarely found in bottom-dwelling deep-sea fishes. The exceptions are *Notacanthus* species, which eat considerable quantities of hydrazoans, sea-pens, sea anemones and bryozoans (MCDOWELL, 1973). No doubt the cutting, serrated edge of their close-set premaxillary teeth enable *Notacanthus* spp. to crop sessile invertebrates. Such a dentition and food pattern are evidently unique among bottom-dwelling fishes of the deep sea. Species of the related genus *Polyacanthonotus*, which have terminally hooked and separated teeth in the jaws, feed on a mixture of benthic and benthopelagic organisms, as do halosaurs (MCDOWELL, 1973).

The notacanthiform fishes seem to be unique in one other respect. *Lipogenys gillii*, the sole representative of the family Lipogenyidae, has a toothless, sucker-like mouth, "...a vacuum cleaner for drawing up consolidated sediment", according to MCDOWELL. He found sponge spicules, plant fibres, crustaceans and sand grains in the gut, which has a "long and complexly folded intestine". If, as seems likely, MCDOWELL's conclusions are substantiated, *Lipogenys gillii* will be the only known ooze eater among deep-sea fishes.

When we have more data on the food patterns of benthic deep-sea fishes, it will be interesting to compare and contrast their trophic types and degree of success with those of benthopelagic species. The negatively buoyant species of the former seem at a disadvantage compared to the neutrally buoyant species of the latter, which in subtropical and tropical regions, at least, appear easily to be the more successful. It seems better to 'hover and explore' rather than be confined, when resting, to the interface between sea and land. But this is too sweeping a statement, as shown very well by the tripod-fishes (p. 484). As in neritic waters, the benthic fishes of the deep-sea range in trophic types from classic, 'sit and wait', large-jawed predators to more active searching types. Work on these aspects is in progress.

ECOLOGICAL SIGNIFICANCE OF THE ABUNDANCE OF SLOPE-DWELLING FISHES

Previous studies of the bathymetric distribution of abundance and diversity in oceanic bottom-dwelling fishes have centred on the slope fauna. They indicate a general peak in abundance within the upper 1200 m irrespective of sampling technique or region (e.g. *Trawling*: PECHENIK and TROYANOVSKII, 1971 (North Atlantic); SCHROEDER, 1955 and HAEDRICH, ROWE and POLLONI, 1975 (western North Atlantic); ANON, 1973 and BAKKEN, LAHN-JOHANNESSEN and GJØSAETER, 1975 (eastern North Atlantic). *Lining*: POWELL, 1964 (Pacific islands); FORSTER, BADCOCK, LONGBOTTOM, MERRETT and THOMSON, 1970 (western Indian Ocean); PHLEGER, SOUTER, SCHULTZ and DUFFRIN, 1970 (eastern Pacific). *Submersible observations*: GRASSLE, SANDERS, HESSLER, ROWE and MCLELLAN (western North Atlantic)). In the present survey, off the northwest African slope, the peak of abundance of fishes caught by the I.O.S. epibenthic sledge occurred in the upper 1250 m (Fig. 2).

Evidence of zonation of species assemblages indicates a change in pattern around 1200 m depth. Thus, DAY and PEARCY (1968) found an assemblage of fish species in 594–1143 m, while GOLOVAN (1974) distinguished a zone between 500–1000 (1200) m. HAEDRICH, ROWE and POLLONI (1975) indicated faunal breaks at 300–400 m and 1000–1100 m depth for epibenthic megafauna, including fishes. They related the shallower break to the shelf/slope transition and the deeper to the upper/lower slope transition, while pointing out their considerable horizontal extent and coherence. They also showed that, while peak abundance occurred in 141–285 m, the total weight of fish remained about the same to 1928 m, as individual sizes increased.

An insight into what sustains this abundant and distinct fish fauna can be gained from the preliminary analysis of the dominant food organisms utilized by slope-dwellers off northwest Africa (Table 1), with the tacit realization that such an upwelling area is unusually productive. Most species are benthopelagic in habit with the facility for neutral buoyancy at depth, so lessening the energy requirement in foraging for food. In turn, most of these benthopelagic forms are pelagic feeders from the food items examined; an observation confirmed in a similar study in the same general area by GOLOVAN (1974). He reported upperzone (< 1000 m) fishes (e.g. macrourids, oreosomatids, congrids, scorpaenids and trachichthyids) to contain mostly shrimps and a few myctophid and gonostomatid fishes, while deeper zone (> 1000 m) alepocephalids were eating macroplankton. In particular *Alepocephalus bairdi* was found to be eating mainly *Pyrosoma* and medusae (GOLOVAN and PAKHORUKOV, 1975).

The occurrence of pelagic food in macrourids and its importance has been discussed in the literature (OKAMURA, 1970; PECHENIK and TROYANOVSKII, 1971; HAEDRICH and HENDERSON, 1974; PEARCY and AMBLER, 1974). On the slope off Labrador, PECHENIK and TROYANOVSKII (1971) found that *Coryphaenoides rupestris* were feeding on migrant

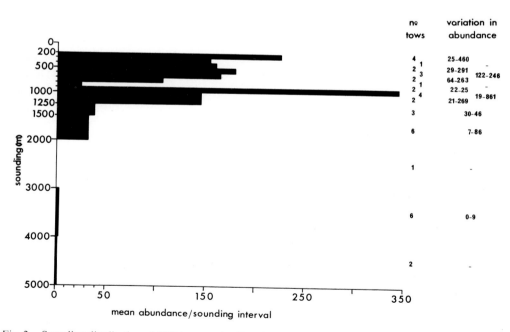

Fig. 2. Sounding distribution of 3748 bottom-dwelling fishes based on the catches of thirty-five approximately comparable epibenthic sledge tows (distance of bottom contact = *ca.* 1000 m/tow) from an area (08–26 N) off northwest Africa. Mean abundance per sounding interval was calculated from the total catch of all tows whose mid-sounding occurred within the indicated range. The number of hauls and catch variation per sounding interval are shown.

zooplankton and they correlated the fluctuations in catch rate with the diurnal vertical migrations of zooplankton and the feeding rhythm of the fish. At night, as the forage organisms migrated towards the surface *C. rupestris* concentrated near the bottom and catches increased. In the morning, the shoals dispersed, the fish beginning to feed actively on the descending zooplankton up to 100–200 m from the bottom, and the catches dropped. HAEDRICH and HENDERSON (1974) proposed extensive migrations to explain food items such as the midwater fish, *Chauliodus*, in the stomachs of the abyssal *Coryphaenoides armatus* and HAEDRICH (1974) reported the pelagic capture of *C. rupestris* up to 1440 m from the sea-bed. Similarly, among deep-sea squaloids, FORSTER (1971) reported the capture of *Centrophorus squamosus* (13) and *Etmopterus princeps* (1) on vertical lines between 222–1900 m above soundings of 1537–3000 m. Thus there is a body of data to indicate that fishes hitherto considered to be bottom dwellers do make excursions well off the bottom and that these, and other benthopelagic species, rely to varying degrees on pelagic prey.

Extensive migrations would be necessary to enable abyssal fishes to prey upon live meso-pelagic food. In oceanic conditions off the Canary Islands, euphausiids were found to be concentrated in the upper 700 m (BAKER, 1970), while the peak of decapod distribution by day was around 800 m (FOXTON, 1970). Nearby, in 30°N, 23°W, the peak of species diversity in mesopelagic fishes was found at 400–900 m by day (BADCOCK and MERRETT, 1976). The peak of abundance by day along a transect, near the 20° meridian, of six stations at approximately 10° intervals from 11°N to 60°N and including the above station, was centred in 300–800 m from 11–53°N (BADCOCK and MERRETT, 1977) and this embraced the majority of myctophid fishes. At the sixth station, 60°N, the overall peak occurred rather deeper. Oceanically, at least, the depths of peak abundance of such potential prey organisms for benthopelagic fishes occurs within the upper 1000 m.

Such excursions may not necessarily be required for upper slope-dwellers. The dominant prey organisms found in many of the species we examined were euphausiids, natant decapods and, in the case of squaloid sharks, myctophid fishes (Table 1). Some indication of the influence of the shoaling water over the continental slope off northwest Africa on these mesopelagic organisms can be gained from the preliminary results of two transects, at right angles to the submarine contours at 23°N and 25°N, from soundings of 2500 m to the shelf (IOS unpublished data). These data were obtained from oblique and horizontal tows of a mouth-closing combination midwater trawl (RMT 1 + 8). The two most abundant euphausiids in the area, *Euphausia krohnii* and *Nematoscelis megalops*, extended shelfwards to beyond the 500 m contour station, about 5 nautical miles* from the slope/shelf break. *Nematoscelis megalops* was most abundant below 150 m and extended close to the bottom in this depth, while *E. krohnii* occurred in peak abundance in 50–150 m and even extended onto the shelf to a station in 50 m depth (A. DE C. BAKER, personal communication). The mysid *Gnathophausia zoeae*, eaten by several of the benthopelagic fishes in the area, occurred shelfwards to close to the bottom in a sounding of 500 m and from one sample over the shelf (P. M. HARGREAVES, personal communication). Certain, as yet unidentified, natant decapods extended close to the bottom into soundings of 500 m also. Among the mesopelagic fishes, species diversity decreased shelfwards, yet at the 500 m contour eleven myctophid and six stomiatoid species, among others, were recorded. Several species extended close to the bottom in daytime. In this connection, it is worth noting the

---

* 1 nautical mile = 1.853 km.

observations of J. E. CRADDOCK (personal communication) on *Alvin* Dive 325 in the western North Atlantic. While the submersible was close to the bottom in about 1025 m and moving at 1 knot, two species of lantern-fish (*Lampadena* sp. and *Lampanyctus macdonaldi*) all *ca.* 6 inches in length, were seen in roughly equal numbers over a period of several hours, distributed singly about 30 m apart and less than 2 m off the bottom. Returning to myctophids occurring in the catches from the transects we are considering, species of *Myctophum*, *Symbolophorus*, *Benthosema* and *Diaphus* were present, which are known to fall prey to several species of benthopelagic squaloid sharks. *Benthosema glaciale*, incidentally, has been reported to be numerous over the slope off Nova Scotia (HALLIDAY, 1970), while this species together with *Ceratoscopelus maderensis*, *Myctophum punctatum* and *Maurolicus muelleri* have been reported to be permanent members of the mesopelagic fauna of the Gulf of Mexico (MUSICK, 1973). Thus, while the depth distribution of mesopelagic organisms may well be modified by the proximity of the bottom over the continental slope, many organisms do extend close to the shelf/slope break where they expose themselves to possible predation by benthopelagic fishes. Widespread though the supplementing of the diet of benthopelagic fish with mesopelagic food may be, the extent to which this is possible may vary widely. Clearly conditions are better in areas of considerable productivity, especially upwelling areas. For instance, GEISTDOERFER (1975) found that *Nezumia sclerorhynchus* took much less pelagic food (copepods, etc.) in the relatively unproductive western Mediterranean than off northwest Africa.

In a study of the interaction of migrant sound scatterers with the sea floor, ISAACS and SCHWARTZLOSE (1965) have suggested that where currents continually sweep from oceanic areas over coastal regions and over shallow seamounts, a large resident benthic, or epibenthic, community may live by feeding on the diurnally migrating plankton. If such an ecologically significant mechanism is widespread in slope regions, with benthopelagic slope-dwelling fishes largely living on mesopelagic organisms extending shelfwards as we find occurs (cf. also PECHENIK and TROYANOVSKII's (1971) findings on the feeding habits of *C. rupestris* off Labrador, pp. 493–494), then the depth of peak abundance in such fishes could be expected to coincide with the similar peak in oceanic mesopelagic organisms. The existing data, we find, indicate that this is indeed the case and that both peaks occur in the upper 1200 m. We may infer from this also that such a food source would more directly favour benthopelagic bottom-dwelling fishes, rather than truly benthic forms.

Acknowledgements—We would like to thank Messrs. J. R. BADCOCK and P. M. DAVID for their helpful criticism on reading the manuscript.

REFERENCES

ANDRIASHEV A. P. (1965) A general review of the Antarctic fish fauna. In: *Biogeography and ecology in Antarctica*, J. VAN MEIGHEM and P. VAN OYE, editors, Dr. W. Junk, The Hague, pp. 491–550.

ANON, (1973) Fish at 625 fathoms. *World Fishing*, **22**, (7), 52–54.

ALDRED R. G., M. H. THURSTON, A. L. RICE and D. R. MORLEY (1967) An acoustically monitored opening and closing epibenthic sledge. *Deep-Sea Research*, **23**, 167–174.

BADCOCK J. and N. R. MERRETT (1976) Midwater fishes in the eastern North Atlantic. I. Vertical distribution and associated biology in 30 N, 23 W, with developmental notes on certain myctophids. *Progress in Oceanography*, **7**, 3–58.

BADCOCK J. and N. R. MERRETT (1977) Aspects of the distribution of midwater fishes in the eastern North Atlantic. In: *Oceanic sound-scattering prediction*, 209–242. N. R. ANDERSEN and B. J. ZAHURANEC, editors, *Marine Science Series*, **3**, Plenum Press.

BAKER A. DE C. (1970) The vertical distribution of euphausiids near Fuerteventura, Canary Islands ('Discovery'
    SOND Cruise, 1965). *Journal of the Marine Biological Association of the United Kingdom*, **50**, (2), 301–342.
BAKER A. DE C., M. R. CLARKE and M. J. HARRIS (1973) The N.I.O. combination net (RMT 1+8) and
    further developments of midwater trawls. *Journal of the Marine Biological Association of the United Kingdom*,
    53 (1), 167–184.
BAKKEN E., J. LAHN-JOHANNESSEN and J. GJØSAETER (1975) Demersal fish on the continental slope off Norway.
    Saertrykk av Fiskets Gang, (34), pp. 557–565.
BONE Q. and B. L. ROBERTS (1969) The density of elasmobranchs. *Journal of the Marine Biological Association
    of the United Kingdom*, **49**, (4), 913–937.
BRUUN A. (1957) Deep sea and abyssal depths. *Geological Society of America, Memoir 67*, **1**, 641–672.
CLARKE M. R. and N. R. MERRETT (1972) The significance of squid, whale and other remains from the
    stomachs of bottom-living deep-sea fish. *Journal of the Marine Biological Association of the United Kingdom*,
    52 (3), 599–603.
CORNER E. D. S., E. J. DENTON and G. R. FORSTER (1969) On the buoyancy of some deep-sea sharks.
    Proceedings of the Royal Society, Series B, **171** (1025), 415–429.
DAY D. S. and W. G. PEARCY (1968) Species associations of benthic fishes on the continental shelf and slope
    off Oregon. *Journal of the Fisheries Research Board of Canada*, **25**, (12) 2665–2675.
DENTON E. J. and N. B. MARSHALL (1958) The buoyancy of bathypelagic fishes without a gas-filled swimbladder.
    Journal of the Marine Biological Association of the United Kingdom, **37**, (3), 753–767.
FORSTER G. R. (1971) Line fishing on the continental slope. III. Mid-water fishing with vertical lines.
    Journal of the Marine Biological Association of the United Kingdom, **51** (1), 73–77.
FORSTER G. R., J. R. BADCOCK, M. R. LONGBOTTOM, N. R. MERRETT and K. S. THOMSON (1970) Results of the
    Royal Society Indian Ocean Deep Slope Fishing Expedition, 1969. *Proceedings of the Royal Society*, Series B,
    175, 367–404.
FOXTON P. (1970) The vertical distribution of pelagic decapods (Crustacea: Natantia) collected on the SOND
    Cruise 1965. II. The Penacidea and general discussion. *Journal of the Marine Biological Association of the
    United Kingdom*, **50** (4), 961–1000.
GEISTDOERFER P. (1973) Régime alimentaire de Macrouridae (Téléostéens, Gadiformes) atlantique et méditer-
    ranéens, en relation avec la morphologie du tube digestif. *Bulletin du Muséum National d'Histoire
    Naturelle, Paris*, 3e série (161), 285–295.
GEISTDOERFER P. (1975) *Ecologie alimentaire des Macrouridae (Téléostéens Gadiformes)*. Thèse de Doctorat D'État
    Ès-Sciences Naturelles, l'Université de Paris VI, 315 pp.
GOLOVAN G. G. (1974) Preliminary data on the composition and distribution of the bathyal ichthyofauna (in
    the Cap Blanc area). *Oceanology*, **14** (2), 288–290.
GOLOVAN G. A. and N. P. PAKHORUKOV (1975) Some data on the morphology and ecology of *Alepocephalus
    bairdi* Goode and Bean (Alepocephalidae) of the eastern central Atlantic. *Voprosy Ikhtiologii*, **15** (1), 51–58.
GRASSLE J. F., H. L. SANDERS, R. R. HESSLER, G. T. ROWE and T. McLELLAN (1975) Pattern and zonation:
    a study of the bathyal megafauna using the research submersible *Alvin Deep-Sea Research*, **22** (7), 457–
    481.
GRICE G. D. (1972) The existence of a bottom-living calanoid copepod fauna in deep water with descriptions of
    five new species. *Crustaceana*, **23** (3), 219–242.
HAEDRICH R. L. (1974) Pelagic capture of the epibenthic rattail *Coryphaenoides rupestris*. *Deep-Sea Research*,
    21, 977–979.
HAEDRICH R. L. and N. R. HENDERSON (1974) Pelagic food of *Coryphaenoides armatus* a deep benthic rattail.
    Deep-Sea Research, **21**, 739–744.
HAEDRICH R. L., G. T. ROWE and P. T. POLLONI (1975) Zonation and faunal composition of epibenthic
    populations on the continental slope south of New England. *Journal of Marine Research*, **33** (2),
    191–212.
HALLIDAY R. G. (1970) Growth and vertical distribution of the glacier lanternfish, *Benthosema glaciale*, in
    the northwestern Atlantic. *Journal of the Fisheries Research Board of Canada*, **27** (1), 105–116.
HEDGPETH J. W. (1957) Classification of marine environments. *The Geological Society of America, Memoir 67*,
    1, 17–28.
HEEZEN B. C. and C. H. HOLLISTER (1971) *The face of the deep*, New York: Oxford University Press, Inc.,
    659 pp.
ISAACS J. D. and R. A. SCHWARTZLOSE (1965) Migrant sound scatters: interaction with the sea floor. *Science*,
    150, 1810–1813.
ISAACS J. D. and R. A. SCHWARTZLOSE (1975) Active animals of the deep-sea floor. *Scientific American*, **233** (4)
    85–91.
McDOWELL S. B. (1973) Order Heteromi (Notacanthiformes). In: *Fishes of the western North Atlantic*.
    Memoir, Sears Foundation for Marine Research, No. 1, part 6, pp. 1–228.
MARSHALL N. B. (1960) Swimbladder structure of deep-sea fishes in relation to their systematics and biology.
    'Discovery' Reports, **31**, 1–122.

MARSHALL N. B. (1965) Systematic and biological studies of the macrourid fishes (Anacanthini, Teleostii). *Deep-Sea Research*, **12**, 299–322.

MARSHALL N. B. (1973a) Family Macrouridae. In: *Fishes of the western North Atlantic. Memoir, Sears Foundation for Marine Research*, No. 1, part 6, pp. 496–665.

MARSHALL N. B. (1973b) Macrouridae. In: *Check-list of the fishes of the north-eastern Atlantic and of the Mediterranean*, J. C. HUREAU and TH. MONOD, editors, UNESCO, Paris, **1**, 287–299.

MARSHALL N. B. and D. W. BOURNE (1964) A photographic survey of benthic fishes in the Red Sea and Gulf of Aden, with observations on their population density, diversity and habits. *Bulletin of the Museum of Comparative Zoology*, Harvard University, **132** (2), 225–244.

MARSHALL N. B. and D. W. BOURNE (1967) Deep-sea photography in the study of fishes. In: *Deep-Sea Photography*, J. B. HERSEY, editor, Baltimore: Johns Hopkins Press, pp. 251–257.

MARSHALL N. B. and T. IWAMOTO (1973) Genus *Coelorhynchus*. In: N. B. MARSHALL, Family Macrouridae. *Fishes of the western North Atlantic. Memoir, Sears Foundation for Marine Research*, No. 1, part 6, 538–563.

MUSICK J. A. (1973) Mesopelagic fishes from the Gulf of Maine and the adjacent Continental Slope. *Journal of the Fisheries Research Board of Canada*, **30** (1), 134–137.

OKAMURA O. (1970) Studies on the macruroid fishes of Japan—morphology, ecology and phylogeny. *Reports of the Usa Marine Biological Station*, **17** (1–2), 1–181.

PEARCY W. G. and J. W. AMBLER (1974) Food habits of deep-sea macrourid fishes off the Oregon coast. *Deep-Sea Research*, **21**, 745–759.

PECHENIK L. N. and F. M. TROYANOVSKII (1971) Trawling resources on the North-Atlantic Continental Slope. Jerusalem: Israel Program for Scientific Translations, 71 pp. Translations of: *Syr'evayabaza tralovogo rybolovstva na materikovom sklone Severnoi Atlantiki*. Murmansk: Murmanskoe Knizhnoe Izdatel'stvo, 1970.

PHLEGER C. F., A. SOUTAR, N. SCHULTZ and E. DUFFRIN (1970) Experimental sablefish fishing off San Diego, California. *Commercial Fisheries Review*, **32**, 31–40.

POWELL R. (1964) Dropline fishing in deep water. *Modern Fishing Gear of the World 2*, pp. 287–291. London: Fishing News (Books) Ltd.

RATTRAY J. M. (1947) Observations on the food-cycle of the South African stockfish, *Merluccius capensis* Cast. off the west coast of South Africa: with a note on the food of the King-klip *Genypterus capensis* (Smith). *Annals of the South African Museum*, **36**, 315–331.

ROPER C. F. E. and W. L. BRUNDAGE (1972) Cirrate octopods with associated deep-sea organisms: new biological data based on deep benthic photographs (Cephalopods). *Smithsonian Contributions to Zoology*, (121), 46 pp.

SCHMIDT-NIELSEN S., A. FLOOD and J. STENE (1934) On the size of the liver of some gristly fishes, and their content of fat and vitamin A. *Kongelige Norske videnskabernes selskabs forhandlinger*, **7**, 47–50.

SCHROEDER W. C. (1955) Report on the results of exploratory otter-trawling along the Continental Shelf and Slope between Nova Scotia and Virginia during the summers of 1952 and 1953. *Papers in Marine Biology and Oceanography, Deep-Sea Research, Supplement to Volume 3*, pp. 358–372.

VINOGRADOV M. E. (1968) *Vertical Distribution of the Oceanic Zooplankton*, 339 pp. Moscow: Nauka. (Translation from the Russian. Jerusalem: Israel Program for Scientific Translation, 1970.)

WILSON E. O. (1975) *Sociobiology*, Harvard University Press, Cambridge, Massachusetts, 697 pp.

WOLFF T. (1971) Archimède Dive 7 to 4160 metres at Madeira: observations and collecting results. *Videnskabelige Meddelelser fra Dansk naturhistorik Forening i Kjøbenhavn*, **134**, 127–147.

Depth distributions of *Hyperia spinigera* Bovallius, 1889 (Crustacea: Amphipoda) and medusae in the North Atlantic Ocean, with notes on the associations between *Hyperia* and coelenterates

MICHAEL H. THURSTON

Institute of Oceanographic Sciences, Wormley, Godalming, Surrey, U.K.

Abstract—Material from two vertical series of nets fished at 60°N, 20°W and 53°N, 20°W was utilized to clarify the status of *Hyperia spinigera*. The species was shown to be a small-scale migrant concentrated at depths of 600–800 m. The vertical distribution and migratory behaviour of medusae from the same hauls were studied in order to investigate associations between these species and *H. spinigera*. With the exception of *Aglantha digitale*, an epipelagic species, most medusae were deep mesopelagic in distribution. *Pantochogon haeckeli, Aeginura grimaldii* and *Periphylla periphylla* were the most abundant species, accounting for 81% of the total specimens. Most species showed small-scale vertical movements at night, but *Atolla vanhoeffeni* and *P. periphylla* migrated about 200 and 150 m respectively. Larval amphipods were found in *P. periphylla* and it was concluded that the relationship was that of an obligative parasite to a host. Examination of amphipod gut contents failed to provide conclusive evidence for a parasitic relationship between non-larval individuals and *P. periphylla* despite the presence of much coelenterate debris. Literature records have been used to delineate the geographical distribution of *H. spinigera*, and to document the association between species of *Hyperia* and medusae.

INTRODUCTION

SINCE 1968 detailed sampling programmes have been operated from R.R.S. *Discovery* in order to investigate the depth distribution and diurnal migration of mid-water organisms. Most of these investigations have been carried out in the eastern part of the North Atlantic Ocean. During April and May 1971 series of horizontal hauls using the RMT 1+8 combination net (BAKER, CLARKE and HARRIS, 1973) were made in the vicinity of 60°N, 20°W and 53°N, 20°W. Amphipods formed a significant part of the catches from these hauls and contained material belonging to the genus *Hyperia* which posed a number of taxonomic and biological problems.

The genus *Hyperia* Latreille, 1823 has been used as a depository for a hotch-potch of morphologically diverse species (see, for example, STEBBING, 1888; BOVALLIUS, 1889; STEPHENSEN, 1925). Successive critical studies of the genus (YANG, 1960; LAVAL, 1968) culminating in the recent report by BOWMAN (1973) have resulted in the transference of many of the previously included species to other genera. Bowman has accepted only eight species in his concept of *Hyperia*. Of these eight species, two, *Hyperia medusarum* (MÜLLER,

1776) and *H. galba* (MONTAGU, 1813), are common elements of the boreal fauna, while *H. leptura* from the seas off Baja California and *H. crassa* from the tropical east Atlantic Ocean were described as new. A fifth species, *H. macrocephala* (DANA, 1853), is large and distinctive and has been recorded only south of the Antarctic Convergence. *H. gaudichaudii* Milne Edwards, 1840, a temperate southern hemisphere species, is morphologically very close to *H. medusarum* but has been maintained as a distinct species by Bowman mainly on zoogeographic grounds. Currently available data are insufficient to determine whether this separation is justified or whether *H. medusarum* should be regarded as a bipolar species with a distribution paralleling that demonstrated by BOWMAN (1960) and KANE (1966) for *Parathemisto gaudichaudi*. The remaining two species, *H. spinigera* BOVALLIUS, 1889 and *H. antarctica* SPANDL, 1927, were, until the recent redescription by BOWMAN (1973), rather poorly known and conflicting opinions as to their validity and relationships have been put forward by various authors (NORMAN, 1900; TATTERSALL, 1906; BARNARD, 1930, 1932). VINOGRADOV (1976) has described an additional species, *H. bowmani*, from the tropical Pacific Ocean.

The association of *Hyperia* with medusae was first reported by STRØM (1762) and has been documented extensively since that date by many authors (see, for example, EDWARD, 1868a; ROMANES, 1877; TATTERSALL, 1906; HOLLOWDAY, 1948; DAHL, 1959 a,b; BOWMAN, MEYER and HICKS, 1963; METZ, 1967; WHITE and BONE, 1972; SHEADER, 1974).

The nature of this association has given rise to considerable speculation. EDWARD (1868a) concluded that, in the adult stage at least, *H. galba* was not parasitic but attached to medusae for resting, protection and feeding, the latter not involving medusan tissue. ROMANES (1877) observed *H. galba* to damage or destroy gonadal tissue of *Aurelia aurita*. SARS (1890), SCHEURING (1915), CHEVREUX and FAGE (1925) and AGRAWAL (1967) have referred to the relationship as parasitic, while STEPHENSEN (1928) and POULSEN (1950) used the term 'semiparasitic'. ORTON (1922), concluded that *Hyperia* was a food parasite, feeding on mucus and entrapped food particles but not on the tissues of the host. HARDY (1956) concurred with this hypothesis. HOLLOWDAY (1948) studied large numbers of *H. galba* found on R*hizostoma octopus*. Both adults and juveniles were present in what was described as a commensal relationship. DAHL (1959 a,b) made a histological examination of *H. galba* taken from *Cyanea capillata* and found evidence to suggest that the amphipods were feeding directly on medusan tissue, and therefore should be regarded as true ecto-parasites. BOWMAN, MEYER and HICKS (1963) considered that further evidence for parasitic behaviour was required, although suggesting that *H. galba* might feed on the tentacles and oral arms of *C. capillata*. METZ (1967) considered *H. galba* to be a true parasite of *A. aurita* as numerous observations were made of direct feeding on the medusan gonads by adult and juvenile amphipods. WHITE and BONE (1972) were able to show by an analysis of nematocysts in the gut that *H. macrocephala* fed preferentially on the manubrium of the medusa *Desmonema gaudichaudi*. It was concluded that the relationship of hyperiid to medusa was that of an obligative parasite to host. SHEADER (1974) found that *H. galba* fed on medusan tissue and on food material trapped by the medusa.

The routine use of electronically monitored, opening/closing nets for sampling midwater organisms is a relatively modern innovation. As a consequence detailed information on the vertical range and diurnal migration is available for a rather limited number of species. Recent summaries of data pertaining to medusae (KRAMP, 1959, 1968a; RUSSELL, 1970) highlight this deficiency. A similar situation exists among the Amphipoda although some data have been published recently (THURSTON, 1976 a,b).

MATERIALS AND METHODS

The material on which this report is based was obtained during R.R.S. *Discovery* cruise No. 39 in April and May 1971. Vertical series of horizontally towed nets were worked at 60°N, 20°W (Sta. 7709) and 53°N, 20°W (Sta. 7711) using the Institute of Oceanographic Sciences RMT 1 + 8 Combination Net (BAKER, CLARKE and HARRIS, 1973) as part of a long-term investigation of vertical distribution and migration of planktonic organisms in the eastern North Atlantic Ocean. Horizontal tows were made by day and by night to investigate the animal distributions in the top 2000 m of the water column. A day and night series, each of sixteen separate hauls were made at the two localities. Hauls were made at 25–10, 50–25 and 100–50 m, through 100 m layers between 100 m and 1000 m, and thereafter at 1250–1000, 1500–1250 and 2000–1500 m. Hauls at depths of 1000 m or greater lasted 4 hours and those at shallower depths were generally of 2 hours duration.

Catches were fixed in buffered 4% formaldehyde on shipboard, sorted on return to the laboratory and subsequently stored in 80% alcohol (*Hyperia*) or a propylene phenoxetol/ propylene glycol/formaldehyde preserving fluid (STEEDMAN, 1974).

Lengths of *Hyperia* have been measured, on straightened animals, between the anterior margin of the head and the distal extremity of the telson. The diagnosis of stages of sexual maturity in male *Hyperia* has been based on the degree of development of antenna 1, and in females on the size of oostegites. Individuals in which the flagellum of antenna 1 is thickened and slightly elongated but unsegmented have been considered as juvenile males. Adult males are characterized by strongly elongate, multiarticulate, filiform antennal flagella. Immature males show an intermediate condition in which the flagella are more or less elongate and multiarticulate, but not filiform. The complete development of oostegites in females requires several moults and the division into juvenile, immature and adult stages is somewhat arbitrary. Individuals in which the oostegites of perapeopods 4 and 5 were less than one-quarter the length of the corresponding branchial lamella have been considered to be juvenile females. In fully adult females the oostegites are broadly expanded, extremely thin, completely transparent and about as long as the branchial lamella. Immature individuals bear oostegites which are rather narrow, at most partly transparent, and not more than half as long as the branchial lamella. Animals lacking genital papillae or oostegites and showing no differentiation of the antennal flagella have been designated juveniles.

In order to minimize the effects of damage, distortion and shrinkage shown by specimens of *Periphylla periphylla*, the diameter of these medusae has been measured across the central disc. In the many cases in which the central disc was slightly oval rather than a true circle, the major diameter has been taken.

The depth range of a species, that is the upper and lower limits of vertical distribution, has been taken as the shallowest fishing depth of the positive haul nearest the surface to the deepest fishing depth of the positive haul furthest from the surface. This method will overestimate the true depth range occupied by a species. The maximum possible error of upper or lower limits is related to the thickness of the stratum, fished by any net, and will increase with increasing depth. In near-surface hauls the error will not exceed 15, 25 or 50 m, while below 1000 m, the error may approach 250 or 500 m.

Cumulative percentage calculations have been performed on the vertical distribution data. Comparisons of the day and night depths above which 10%, 25%, 50%, 75% and 90%

of the population occurred have been used to give a measure of the extent of vertical migration.

The method adopted by RUDJAKOV (1971) based on mean depths of occurrence has also been used to quantify vertical migration and assess the significance of the values so obtained.

RESULTS AND DISCUSSION

Hyperia spinigera BOVALLIUS, 1889

Hyperia spinigera BOVALLIUS, 1889, pp. 191–194, pl. 10, figs. 33–39; TATTERSALL, 1906, p. 22; STEPHENSEN, 1924, p. 82 (part, Bovallius' type specimens only), 1942, pp. 460–462; BARNARD, 1932, pp. 273–274, fig. 160 (part, female = *H. crassa*); SHOEMAKER, 1945, p. 238, fig. 35; HURLEY, 1955, pp. 140–143, figs. 83–95, 1969, p. 33, map 5 (part); BOWMAN, 1973, pp. 20–23, figs. 15–16.
   ? VOSSELER, 1901, p. 58*; DUNBAR, 1942, p. 37; HURLEY, 1956, p. 15; EALEY and CHITTLEBOROUGH, 1956, p. 22; BRUSCA, 1967a, p. 388, 1967b, p. 452.
   not NORMAN, 1900, p. 129 (= *H. galba*); BARNARD, 1930, p. 412; THORSTEINSON, 1941, pp. 87–88, pl. 8, figs. 79–82 (= *H. medusarum*); OLDEVIG, 1959, p. 125 (= *H. medusarum*); DUNBAR, 1963, p. 3.
Hyperia antarctica SPANDL, 1927, pp. 153–156, fig. 2; BOWMAN, 1973, pp. 18–20, figs. 12–14.
Hyperia galba (not MONTAGU): REID, 1955, p. 18.
   ? HURLEY, 1960, pp. 111–112*.

Status

BOVALLIUS (1889) distinguished *H. spinigera* from *H. medusarum* and *H. latreillei* MILNE EDWARDS, 1830 (the latter currently considered synonymous with *H. medusarum*) by the armature of peraeopods 1 and 2, and from *H. galba* by differences in the third uropods. SPANDL (1927) described only the peraeopods of *H. antarctica* and failed to compare the new species with others in the genus except to point out that it belonged to the group of species characterized by unfused peraeon segments 1 and 2. The uropods were stated to be of 'characteristic form'. No additional specimens of *H. antarctica* were recorded in the literature. BOWMAN (1973) re-examined the holotype of Spandl's species and equated with it material from the north-eastern Pacific and Atlantic Oceans. The holotype was also compared with material of *H. spinigera*, including part of Bovallius' type series, and separated from that species by the relative proportions of the mandibular palp, length of maxilliped inner lobe, armature of peraeopods 1–4 and breadth of uropod 3 inner ramus. It should be noted that both the diagnosis and description of *H. spinigera* were based solely on adult male specimens and that the single specimen described as *H. antarctica* was female.

   During the examination of material from the RMT 8 series worked at Sta. 7709, several immature male specimens in a premoult condition were found in which the soft new cuticle of the subsequent stage was clearly visible beneath the existing integument. It was clear from the antennae that the impending moult would give rise to a specimen showing characteristics of the adult stage. The old cuticle of articles 5 and 6 of peraeopods 3 and 4 was smooth lacking setae, and thus agreeing with the condition described for *H. antarctica*

* See note added in proof, p. 536.

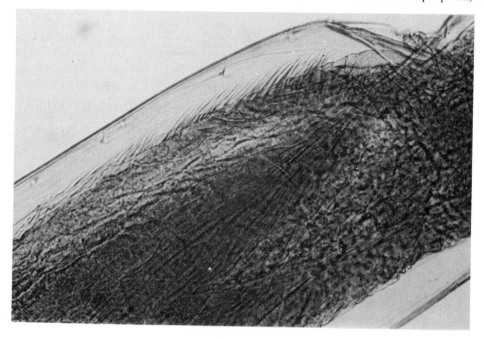

Plate 1.   *Hyperia spinigera*, setation of peraeopod 3 carpus. *a*, Immature male just prior to moult showing underlying adult cuticle. *b*, Adult male.

by BOWMAN (1973). The underlying new cuticle, however, showed clearly a dense armament of short setae along the posterior margin of each article, as has been described in *H. spinigera* (Plate 1). This finding prompted a re-examination of those characters listed by Bowman as separating the two species.

Immature males 7, 10 and 14 mm long and a mature male 13 mm long have been figured to illustrate changes which occur with increasing maturity. The degree of development of each individual can be gauged by the condition of antenna 1 (Fig. 1 a–d). Contrary to the text, Bowman's figures of mandibular palps of male and female show all three articles subequal in each case. Differences lie in the stouter first article and the less strongly falcate article of males. In the present material, the stout first article is particularly apparent in the adult. The third article is uniformly longer than the second in all cases and becomes less falcate with increased maturity. The mandibular palp of females resembles that of the immature male but tends to be more slender (Fig. 1 e–h). The changes in proportion of inner and outer lobes of the maxillipeds is shown in Figs. 1 i–l. The growth of the outer lobes is relatively greater than the inner lobe. With increasing maturity the maxilliped passes from a condition similar to that found in the female in which the length of the

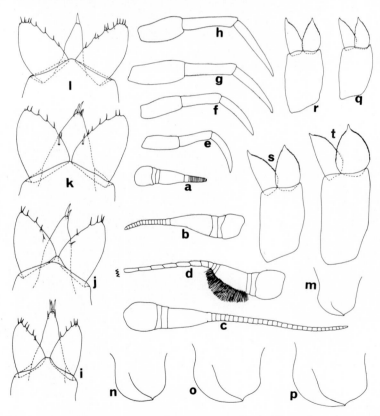

Fig. 1. *Hyperia spinigera*, males. Antenna 1: *a*, 7-mm juvenile; *b*, 10-mm immature; *c*, 14-mm immature; *d*, 13-mm adult. Mandibular palp: *e*, 7-mm juvenile; *f*, 10-mm immature; *g*, 14-mm immature; *h*, 13-mm adult. Maxillipeds: *i*, 7-mm juvenile; *j*, 10-mm immature; *k*, 14-mm immature; *l*, 13-mm adult. Epimeron 3: *m*, 7-mm juvenile; *n*, 10-mm immature; *o*, 14-mm immature; *p*, 13-mm adult. Uropod 3: *q*, 7-mm juvenile; *r*, 10-mm immature; *s*, 14-mm immature; *t*, 13-mm adult.

inner lobe exceeds the outer lobes to the situation where the outer lobes become more prominent, attaining the condition shown by BOWMAN (1973, Fig. 13f). The maxilliped of the 13-mm adult male figured here (Fig. 1l) is aberrant in that the inner lobe bears only a single terminal spine rather than the usual two.

Changes apparent in the distal articles of peraeopods 1 and 2 are of two sorts, relating to the articles themselves and to the setation of these articles. Carpus and propodus of first and second peraeopods remain subequal in length through the developmental series but the carpal process becomes rather more prominent. In the fully mature individual this projection and the posterior margin of the propodus became minutely serrate. The increase in prominence of the carpal process of peraeopod 2 is particularly noticeable, being about one-quarter the length of the propodus in the 7-mm specimen but nearly half the length in the fully mature specimen. The numbers of setae on the fourth, fifth and sixth articles of peraeopods 1 and 2 are relatively constant during development but, contrary to the usual situation, the setae become much reduced in length relative to their respective articles. For example, the longest setae on the carpal process of peraeopod 2 in the 7-mm specimen are about 70% of the length of the carpus whereas in the mature

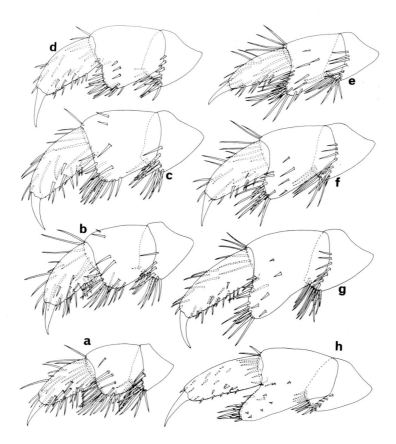

Fig. 2.   *Hyperia spinigera*, males. Gnathopod 1: *a*, 7-mm juvenile; *b*, 10-mm immature; *c*, 14-mm immature; *d*, 13-mm adult. Gnathopod 2: *e*, 7-mm juvenile; *f*, 10-mm immature; *g*, 14-mm immature; *h*, 13-mm adult.

specimen they are less than 30%. Similarly, the setae on the posterior margin of the propodus of peraeopod 2 are nearly 25% as long as the propodus in the juvenile specimen but less than 10% in the 13-mm specimen (Fig. 2). The condition of all females resembles that of the smallest male figured.

The remaining character by which Bowman separated *H. spinigera* and *H. antarctica* was the relative width of the inner ramus of uropod 3. As can be seen from Fig. 1 q–t, the width of the inner ramus of the male increases until at maturity it is more than 60% of the length. In contrast, this ratio remains more or less constant from individuals showing no external sexual differentiation through to fully mature females.

Bowman's concepts of *H. spinigera* and *H. antarctica* are clearly separated from the remaining seven species in the genus by the form of the third epimera. The strongly convex margin is developed at an early age (Fig. 1 m–p) and is equally prominent in female specimens.

Morphological evidence requires that *H. antarctica* be placed as a junior synonym of *H. spinigera*.

In the synonymy given above, the unequivocal assignment of a reference to *H. spinigera* or to another species of *Hyperia* is based on an examination of some or all of the reported material by Dr. T. E. BOWMAN or the author. The only exception to this rule is that reference of HURLEY (1955) which contained illustrations sufficient to confirm the identity of the material available. References questionably assigned to *H. spinigera* are those dealing with specimens which for various reasons have not been available for examination.

The original description of *H. spinigera* was based on a series of adult male specimens from off Spitzbergen, northern Norway and the south coast of England (BOVIALLUS, 1889). STEPHENSEN (1924) reported specimens in the collections of the Copenhagen Zoological Museum determined by Bovallius as '*Hyperia spinosa*', a name never published, and considered them to be part of the type series of *H. spinigera*. BOWMAN (1973) and the present author have subsequently examined these specimens and found that with the exception of the individual from the north of Iceland all were identifiable with *H. spinigera* as described by Bovallius. Bowman has pointed out that none of the localities given for these specimens agrees with those to be found in the original description. It seems likely that material from the original localities belonged to other museums, but as the specimens cannot now be traced, this distribution cannot be confirmed.

NORMAN (1900) placed *H. spinigera* in the synonymy of *H. galba*, dismissing the differences in spination of the gnathopods and the shorter, broader uropods by which Bovallius had characterized his species, as due to age and sex respectively. The specimens from Birturbury Bay and from off Valentia, which Norman referred to the *H. spinigera* facies have been examined both by BOWMAN (1973) and by myself. We agree that they are typical, fully adult *H. galba*. TATTERSALL (1906) disagreed with the fusion of *H. spinigera* with *H. galba* proposed by NORMAN (1900), pointing out that in addition to differences between the gnathopods, the two species were separable by the condition of epimera 3, 'sharply pointed' in *H. galba* and 'rounded' in *H. spinigera*. The two specimens obtained in May 1905 at 55°07′N, 15°06′W, which are in the collections of the British Museum (Natural History), have been examined and found to agree with the characters of *H. spinigera*. STEPHENSEN (1924) was able to examine specimens presumed to be part of Bovallius' type series and, although remaining unconvinced of the validity of *H. spinigera*, was able to write: "I may say at once that in the collections of the 'Thor' there is not a single specimen which can be ascribed to *H. spinigera*." The type specimen of *H. antarctica*

Spandl has been re-examined by BOWMAN (1973). BARNARD (1930) pointed out the marked similarity between SPANDL's (1927) figures of *H. antarctica* and those of *H. spinigera*, and considered that they were one and the same species. He also stated his belief that the characters separating *H. spinigera* and *H. galba* were neither constant nor always real, and followed NORMAN (1900) in placing *H. spinigera* as a junior synonym of *H. galba*. Material obtained by the *Terra Nova* from west of the Falkland Islands and from Melbourne Harbour has been reidentified as *H. gaudichaudii*. Specimens from the tropical Atlantic Ocean and from South Georgia led BARNARD (1932) to reverse his previously held opinion, and reinstate *H. spinigera* as a valid entity. These specimens were characterized by broad third uropods, rounded epimera 3 and the condition of the gnathopods, thus agreeing with Bovallius' figures. Barnard clearly had reservations about the conspecificity of male and female specimens from the tropical Atlantic Ocean. These doubts were justified as BOWMAN (1973) found it necessary to describe the female specimen as a new species, *H. crassa*. Barnard considered that the affinities of *H. antarctica* lay with *H. spinigera* rather than *H. galba*. Through the courtesy of Dr. S. Rohwer of the University of Washington, I have been able to re-examine the material assigned to *H. spinigera* by THORSTEINSON (1941). Two 7.5-mm adult males collected 'in jelly fish' by Kincaid in 1925 were present, although only a simple specimen was mentioned by Thorsteinson. Neither is in good condition and one lacks the third pleon segment and urosome. A comparison of those specimens with the morphological entities described by BOWMAN (1973) showed that they agree most closely with the intermediate form of *H. medusarum* which also came from Friday Harbour. Characters indicating an affinity with *H. medusarum* rather than *H. spinigera* include the short basal article of mandibular palp, distinctly angled carpal process of peraeopod 1, moderately long spines on propodus of peraeopods 1 and 2, subtriangular coxa 3, 3–5 spines on the carpus of peraeopods 3 and 4, and relatively narrow peduncle of uropod 3 which is more than twice as long as the telson. Although the spination of the propodus of peraeopods 1 and 2 is much longer than is found in adult *H. spinigera* (Figs. 2 d,h) it bears a close resemblance to the condition found in the 7-mm juvenile male from the present collections (Figs. 2 a,e). Two further characters are not fully consistent with the morphology of *H. medusarum*. The spination and setation of these specimens has suffered damage since the original fixation, but the carpus of peraeopod 5 on one individual retains not only a few spines characteristic of *H. medusarum hystrix* but also a dense covering of short spinules such as is found in *H. spinigera*. In addition, the posterior margin of epimeron 3 is more strongly convex than has been figured for either *H.m. medusarum* or *H.m. hystrix*, and tends to approach the condition found in the 7-mm male from Sta. 7709. Although most morphological characters of the Friday Harbour material suggest affinity with *H. medusarum*, the possibility that they represent a neritic or relict form of *H. spinigera* cannot be ruled out entirely. Such a situation would parallel that recently demonstrated by SHEADER and EVANS (1974) who showed that *Parathemisto gracilipes* is no more than a neritic warm-water form of *P. gaudichaudi*. SHOEMAKER (1945a) assigned specimens from off Bermuda to *H. spinigera* but Bowman considered them to be a mixture of *H. antarctica* and *H. spinigera*. A single large *Hyperia* was captured by the *Atlantide* in the Gulf of Guinea and Reid (1955) accepting the widely held belief that *H. galba* was a widely distributed, variable species, placed the specimen in this species. It is, in fact, an adult male *H. spinigera*. The single specimen referred to *H. spinigera* by OLDEVIG (1959) has been examined and reidentified as *H. medusarum*. BOWMAN (1973) has critically reassessed *H. spinigera* and *H. antarctica*.

Table 1. *Numbers of* Hyperia spinigera *obtained from day and night hauls with the RMT 1+8 combination net at Sta. 7709 and Sta. 7711*

Depth (m)	Station 7709				Station 7711			
	RMT 8		RMT 1		RMT 8		RMT 1	
	Day	Night	Day	Night	Day	Night	Day	Night
10–0	–	–	–	–	–	–	–	–
25–10	–	–	–	–	–	–	–	–
50–25	–	1	–	–	–	–	–	–
100–50	–	–	–	–	–	–	–	–
200–100	–	–	–	–	–	2	–	–
300–200	–	1	–	1	–	2	–	–
400–300	–	1	–	2	–	–	–	–
500–400	–	5	–	–	–	1	1	–
600–500	2	26	–	4	7	13	2	1
700–600	24	86	2	5	9	43	–	–
800–700	92	228	2	9	2	–	–	–
900–800	75	30	1	3	2	1	–	–
1000–900	13	11	–	–	–	–	–	–
1250–1000	1	1	–	–	–	–	–	–
1500–1250	–	–	–	–	1	–	–	–
2000–1500	–	2	–	–	–	–	–	–
Total	207	392	5	24	21	62	3	1

Following an examination of type material of both species, differences in the mandibular palp, maxillipeds, armature of peraeopods 1–4 and width of inner ramus of uropod 3 were used as diagnostic characters. Data have been presented above to show that the separation of these species is not justified.

VOSSELER (1901)*, DUNBAR (1942), HURLEY (1956), EALEY and CHITTLEBOROUGH (1956) and BRUSCA (1967 a, b) assigned specimens to *H. spinigera*. In each case, the material has been destroyed or mislaid, so that it has not been possible to confirm any of these identifications. HURLEY (1960)* was uncertain of the status of *H. spinigera*, including two BANZARE specimens 'of the *spinigera* type' under *H. galba*. STEPHENSEN (1942), VINOGRADOV (1956) and DUNBAR (1963), treated *H. spinigera* as a dubious species or a junior synonym of *H. galba* on the basis of literature reports. HURLEY (1969) has plotted the geographical distribution of species of *Hyperia* south of lat. 35°S. Eight localities for *H. spinigera* are included. Five of those have been dealt with above (specimens referred to by BARNARD, 1932; HURLEY, 1955, 1960; and EALEY and CHITTLEBOROUGH, 1956). Of the remaining three records, those southeast and north of the Falkland Islands represent specimens assigned to *H. gaudichaudi* by CHILTON (1912) and that in the Drake Passage (which has been incorrectly plotted) the *Challenger* material identified as *H. gaudichaudi* by STEBBING (1888). All this material has been checked and the original identifications found correct.

Since 1968 this Institute has worked extensive series of nets at various localities mainly in the eastern North Atlantic. While an examination of the amphipod material from those sources is not yet complete, the indications are that *H. spinigera* is far less common in lower latitudes than at 60°N. No specimens were found in a complete vertical series at

* See note added in proof, p. 536.

Table 2. *Numbers of* Hyperia spinigera *obtained from the day and night RMT 8 series at Sta. 7709*

Depth (m)	Males ad. D	Males ad. N	Males imm. D	Males imm. N	Males juv. D	Males juv. N	Females ovig. D	Females ovig. N	Females ad. D	Females ad. N	Females imm. D	Females imm. N	Females juv. D	Females juv. N	Juvs. D	Juvs. N	Total D	Total N
25–10	–	–	–	–	–	–	–	–	–	–	–	–	–	–	–	–	–	–
50–25	–	–	–	–	–	–	–	–	–	1	–	–	–	–	–	–	–	1
100–50	–	–	–	–	–	–	–	–	–	–	–	–	–	–	–	–	–	–
200–100	–	–	–	–	–	–	–	–	–	–	–	–	–	–	–	–	–	–
300–200	–	–	–	–	–	–	–	–	–	–	–	–	–	–	–	1	–	1
400–300	–	1	–	–	–	–	–	–	–	–	–	–	–	–	–	–	–	1
500–400	–	3	–	1	–	–	–	–	–	–	–	–	–	–	–	1	–	5
600–500	2	–	–	7	–	7	–	–	–	–	–	6	–	1	–	5	2	26
700–600	1	–	2	5	10	5	–	9	–	2	3	5	2	6	6	54	24	86
800–700	7	5	23	14	17	54	4	7	6	2	9	20	7	44	19	82	92	228
900–800	5	1	4	3	5	2	12	5	6	3	8	4	8	3	27	9	75	30
1000–900	3	–	–	–	–	1	4	1	2	1	1	1	1	1	2	6	13	11
1250–1000	–	–	1	–	–	–	–	1	–	–	–	–	–	–	–	–	1	1
1500–1250	–	–	–	–	–	–	–	–	–	–	–	–	–	–	–	–	–	–
2000–1500	–	–	–	1	–	–	–	–	–	–	–	–	–	–	–	1	–	2
Total	18	10	30	31	32	69	20	23	14	9	21	36	18	55	54	159	207	392

40°N, 20°W, while at 11°N, 20°W a single immature male was found in an RMT 8 haul fished at 800–700 m.

During the course of these investigations a number of samples from various sources have been examined. Material having a direct bearing on the current concept of *H. spinigera* has been covered in this discussion while other determinations have been listed in Appendix 2.

Vertical distribution

Hyperia spinigera was found in catches from large and small nets taken at northern and southern stations (Table 1). The species was most abundant in the RMT 8 series fished at lat. 60°N (Sta. 7709) and the analysis of vertical range and migration has been based primarily on material from those samples.

By day 207 specimens were caught in the RMT 8 between 500 m and 1250 m, with maximum numbers at the 800–700-m level. At night the species was represented by 392 specimens caught between 25 m and 2000 m, with highest numbers again at 800–700 m. The numbers of individuals in the various population categories obtained from each depth stratum have been listed in Table 2. If single individuals are ignored, all categories are confined to vertical depths ranges of 500 m or less, both by day and by night, and the vertical range of the whole population lies between 500 m and 1000 m by day and 400 m and 1000 m by night.

The 25%, 50% and 75% cumulative catch levels and interquartile range for the more

Table 3. *Depths of occurrence and migration of* Hyperia spinigera *based on the RMT 8 series at Sta. 7709*

	Males			Females			Juvs.	Total
	ad.	imm.	juv.	ovig. +ad.	imm.	juv.		
DAY								
Average depth (m)	783	769	734	838	783	794	796	787
Cumulative % 25	–	724	680	785	–	–	739	728
50	786	757	735	839	783	800	807	784
75	–	789	782	886	–	–	857	850
Interquartile range	–	65	102	101	–	–	118	122
NIGHT								
Average depth	630	721	728	743	719	745	724	726
Cumulative % 25	–	596	710	664	660	715	661	674
50	720	718	742	744	735	747	723	733
75	–	773	774	837	780	765	771	776
Interquartile range	–	177	64	173	120	50	110	102
MIGRATION								
Average depth	153	48	6	95	64	49	72	61
Cumulative % 25	–	128	−30	121	–	–	78	54
50	66	39	−7	95	48	53	84	51
75	–	16	8	49	–	–	86	75

abundant categories of the population have been calculated (Table 3). The numbers of adult males and day-caught immature and juvenile females were relatively low, so that no great importance can be attached to small differences in depths of occurrence. By day the interquartile ranges of the various population elements were remarkably consistent. There was a tendency for males to occur at somewhat shallower depths than females of a comparable stage of development. The difference between median depths of occurrence of adult males and females was about 50 m, of immatures about 25 m and of juvenile males and females about 65 m. Although there was a very considerable overlap in vertical range, sexually mature individuals of both sexes were found rather deeper than were younger animals.

By night the interquartile range of the total population was less than that from day hauls, being affected by the large catch from the 800–705-m level. Several of the individual categories, however, were not so concentrated about a particular depth horizon with the result that the interquartile ranges of adult plus ovigerous females and immature males approached 200 m. The partial segregation of developmental stages of each sex found by day was not apparent, the median depths of all categories falling within the range 718–747 m.

H. spinigera was less abundant in the RMT 1 hauls at Sta. 7709, and in the series at Sta. 7711 (Table 1), and an analysis of depth distribution for the population elements was not practical. The overall depth distribution of these series was, however, closely comparable with that from the RMT 8 series at Sta. 7709. With the exception of the RMT 1 series from the southern station, which contained only four specimens, the remaining series demonstrated a similar pattern of vertical distribution. The population maximum lay at 700–800 m at 60°N and 600–700 m at 53°N. This difference in the depth of population maximum conforms with the pattern demonstrated for the decapod *Acanthephyra pelagica* by Foxton (1972) and for some ostracod species by Fasham and Angel (1975).

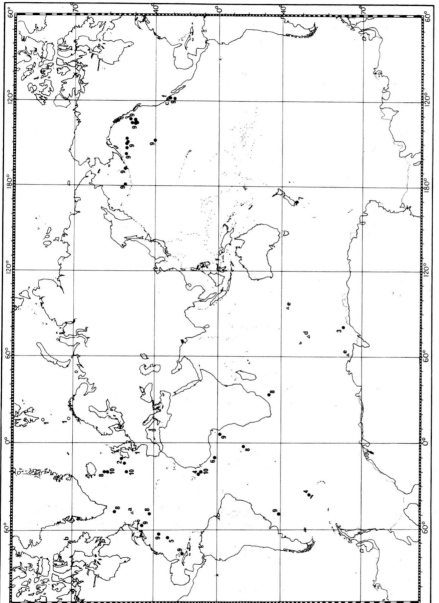

Fig. 3.   *Hyperia spinigera*, distribution records. ○—localities imprecise: 1, BOVALLIUS (1889). ●—identity confirmed: 2, TATTERSALL (1906); 3, SPANDL (1927, as *H. antarctica*); 4, BARNARD (1932); 5, SHOEMAKER (1945a); 6, REID (1955, as *H. galba*); 7, HURLEY (1955); 8, BOWMAN (1973, BOVALLIUS types); 9, BOWMAN (1973, part as *H. antarctica*); 10, this paper. △—identity not confirmed: *a*, VOSSELER (1901)*; *b*, DUNBAR (1942); *c*, HURLEY (1956); *d*, EALEY and CHITTLEBOROUGH (1956); *e*, HURLEY (1960)*; *f*, BRUSCA (1967 a, b).

* See note added in proof, p. 536.

Vertical migration

Differences between day and night quartile depths (Table 3) were considered a better measure of vertical migration than were differences between overall limits. The median depths for all categories except that of juvenile males were shallower at night by 40–100 m, while the upward migration for the whole population was 51 m. The interquartile ranges of the categories by day were roughly comparable (65–118 m), but showed greater differences by night (50–177 m). Changes in the 25% levels of immature males and ovigerous plus adult females were much greater than was the case with the respective 75% levels, implying variations in the extent of migration by individuals in those categories. This situation contrasts with that of juvenile specimens in which the 25%, 50% and 75% levels were 78, 85 and 86 m shallower respectively by night. Corresponding values for the total population were 54, 51 and 74 m. The differences between average depth of occurrence by day and by night was 61 m.

Although the material from the other series was much less abundant than that from the RMT 8 series at Sta. 7709, a similar picture of a small-scale upward migration at night has emerged.

Geographical distribution

The distributional records shown in Fig. 3 represent all the material which has been confirmed as belonging to *H. spinigera* either by Dr. T. E. Bowman or by the present author. Also included are the localities of unchecked material assigned to this species by VOSSELER (1901)*, DUNBAR (1942), EALEY and CHITTLEBOROUGH (1956), HURLEY (1956, 1960)* and BRUSCA (1967 a, b).

H. spinigera is widely distributed in both northern and southern hemispheres. It is not possible to ascertain whether the apparent preponderance of localities in the boreal Atlantic and Pacific Oceans is a genuine indication of higher population levels in those areas, or merely a reflection of fishing effort. On the evidence of unpublished Institute of Oceanographic Sciences collections from the eastern part of the North Atlantic Ocean which have been examined by the author, *H. spinigera* is much more abundant at 60°N and 53°N than in more southerly areas.

In connection with this distribution it is of interest to note that the single specimen obtained at 11°N, 20°W came from a depth of 800–700 m at which the temperature was 6.0–6.5°C. This temperature is only marginally lower than that at which the main concentrations of *H. spinigera* occurred at 60°N and 53°N. FOXTON (1972) has shown that temperature was a major factor in controlling the depth distribution of decapods of the genus *Acanthephyra* at those same stations, and that light, among other physical variables, must also have been involved. A similar limitation in the case of *H. spinigera* might explain the distribution of this species in the North Atlantic Ocean. FASHAM and ANGEL (1975) have figured the temperature profiles obtained during all the vertical series worked between 60°N and 11°N. These data show that temperatures of 6–8°C occurred between 1000 and 1500 m at 30°N and 40°N where *H. spinigera* was not collected rather than 600–800 m at localities where it was found. On a temperature basis, *H. spinigera* should have been recorded from the series worked at 18°N, 25°W. Absence at this locality may be explained by low densities of the species which appear characteristic of low latitudes.

Most of the records of *H. spinigera* in the northeastern Pacific Ocean originated from

* See note added in proof, p. 536.

Table 4.   Numbers and depth distribution of medusae obtained from the day and night RMT 8 series at Sta. 7709

Depth (m)	Chromatonema rubrum D	N	Colobonema sericeum D	N	Pantochogon haeckeli D	N	Crossota alba D	N	Crossota rufobrunnea D	N	Aglantha digitale D	N	Halicreas minimum D	N	Halicera bigelowi D	N	Botrynema brucei D	N	Aegina citrea D	N	Aeginura grimaldii D	N	Nausithoe globifera D	N	Atolla parva D	N	Atolla vanhoeffeni D	N	Atolla wyrillei D	N	Periphylla periphylla D	N	Total D	N
25–10	–	–	–	–	–	–	–	–	–	–	49	25	–	–	–	–	–	–	–	–	–	–	–	–	–	–	–	–	–	–	1	14	50	39
50–25	–	–	–	–	–	–	–	–	–	–	51	28	–	–	–	–	–	–	–	–	–	–	–	–	–	–	–	–	–	–	–	9	52	37
100–50	–	–	–	–	–	–	–	–	–	–	172	70	–	–	–	–	–	–	–	1	–	–	–	–	–	–	–	–	–	–	7	3	179	74
200–100	–	–	–	–	–	–	–	–	–	–	110	65	–	–	37	1	–	–	–	–	–	–	–	–	–	–	–	–	–	–	33	97	143	163
300–200	–	–	–	10	–	–	–	–	–	–	10	40	–	–	41	5	–	–	–	–	–	–	–	–	–	–	–	23	–	–	7	61	54	106
400–300	–	–	2	12	–	–	–	2	–	–	4	203	–	–	110	41	–	–	–	–	–	–	–	–	–	–	–	2	–	–	27	82	75	369
500–400	19	2	27	15	–	2	–	1	–	–	2	20	–	–	1	18	–	–	–	–	–	10	–	10	–	–	–	16	–	3	24	528	163	596
600–500	13	25	8	5	67	5	4	–	–	–	12	2	–	–	–	–	–	–	–	–	–	–	7	3	–	–	15	4	1	11	125	283	235	326
700–600	8	10	8	1	125	440	4	–	–	–	7	1	–	–	–	–	–	–	2	–	35	2	13	1	–	–	15	4	5	12	1379	78	1593	543
800–700	–	4	2	–	416	899	–	–	–	–	10	3	13	23	–	–	–	–	12	2	213	450	1	–	–	–	14	7	10	7	129	101	814	1498
900–800	–	–	–	–	482	453	–	–	2	2	4	5	41	53	–	–	–	–	2	9	1000	630	–	–	2	–	10	14	44	14	60	28	1640	1167
1000–900	–	–	–	1	357	582	–	–	261	63	2	2	99	94	–	–	–	–	3	1	496	560	–	–	–	2	–	–	4	4	14	15	929	1296
*1250–1000	1	–	2	–	360	497	–	–	–	404	2	4	12	63	–	–	100	173	–	2	740	286	–	–	1	–	–	–	5	–	9	9	1482	1301
*1500–1250	–	–	–	–	320	370	–	–	–	6	1	2	10	4	–	–	60	66	–	–	82	62	–	–	–	–	–	–	3	2	2	–	520	670
*2000–1500	–	–	–	–	55	120	–	–	–	–	–	–	–	–	–	–	–	–	–	–	22	11	–	–	–	–	–	–	5	9	7	2	160	214
Total	41	41	49	44	2182	3368	4	3	263	475	439	474	175	227	189	65	160	239	19	15	2588	2011	21	14	2	2	55	49	77	62	1825	1310	8089	8399

*4-hour tow.

R.V. *Brown Bear* stations worked at estimated depths of 225–400 m in late July 1959 (BOWMAN, personal communication). Data from Ocean Station P (50°N, 145°W) for the corresponding period of 1974 (COX and DE JONG, 1975) indicate a temperature of 4–5°C at those depths. Temperatures of the layers occupied by the two populations were thus different, but the horizontal distributions may be governed in the same way.

Medusae

Over 16,000 medusae belonging to sixteen species have been identified from the day and night RMT 8 hauls made at Sta. 7709 (Table 4). KRAMP (1959, p. 241) has shown that most of the oceanic medusae belong to the holoplanktonic Trachymedusae or Narcomedusae. The present material conformed to this pattern as ten of the eleven species of Hydromedusae belonged to these groups and only one, *Chromatonema rubrum*, is an anthomedusan. The remaining five species belong to the Coronatae (Scyphomedusae).

Chromatonema rubrum FEWKES, 1882

This species was taken in rather small numbers between 700 and 1250 m by day and 600 and 1000 m by night. In both cases, maximum population density occurred in the 800–700 m haul. Comparison of the day and night values of median depth of occurrence and interquartile range indicate an upward movement of a few tens of metres by night.

C. rubrum is a deep-dwelling species widely distributed throughout the Atlantic Ocean and extending in the deep warm layer into the Atlantic and Indian sectors of the Southern Ocean (KRAMP, 1968a).

Colobonema sericeum VANHÖFFEN, 1902

C. sericeum was found in rather small numbers between 300 and 800 m, with occasional specimens in catches from below 1000 m. The species was most abundant at 400–600 m. Although no very small individuals were obtained, fully grown specimens 30–35 mm high tended to occur deeper than did 15–20-mm animals with an incomplete complement of tentacles. There was no evidence of diurnal migration.

KRAMP (1968a) listed *C. sericeum* as a bathypelagic species recorded from deep water in the warmer parts of all the oceans of the world. It has not been recorded from the Arctic or Southern Oceans, nor from the Mediterranean Sea. The extensive material from the *Dana* collections was, with very few exceptions obtained from hauls made with at least 1000 mwo (KRAMP, 1959, 1965). From the 'best estimate' of fishing depths (see BRUUN, 1943, p. 15) about 20% of those specimens would have been caught at or about a depth of 300 m and the remainder in deeper layers. This analysis is in broad agreement with the present findings which suggest that *C. sericeum* is a shallow mesopelagic species.

Pantochogon haeckeli MAAS, 1893

P. haeckeli was the most abundant medusa in the RMT 8 hauls made at Sta. 7709. Over 5500 specimens were obtained of which more than 3300 came from night catches.

By day the species was found between 500 m and the lower limit of sampling. If allowance is made for the greater length of hauls made below 1000 m, it is clear that the main concentration of *P. haeckeli* lay between 700 and 1000 m and that maximum numbers occurred in the 900–800 m layer. At night the overall depth range extended from 400 to 2000 m with concentrations between 600 and 1000 m and peak density in the 800–700 m layer. The median depth of occurrence by night was about 30 m shallower than by day, a difference shown to be significant at the 5% level by the method of RUDJAKOV (1971).

P. haeckeli has been found in most parts of the oceans, being absent only from the Arctic Ocean and Mediterranean Sea. It is commoner in the Atlantic Ocean than in the Pacific. Nearly all of the *Dana* material was caught in hauls made with at least 2000 mwo, i.e. from depths of 1000 m or greater. In the northern Atlantic Ocean there is evidence to suggest that, as shown by the present material, this species also occurs at depths of less than 1000 m (KRAMP, 1959, 1965, 1968a).

Crossota alba BIGELOW, 1913

A few specimens of this species were found in hauls made between 400 and 700 m. *C. alba* has been recorded previously from lower latitudes of the eastern Atlantic Ocean and from parts of the Pacific Ocean (KRAMP, 1959, 1968a). Most of the *Dana* material was obtained from hauls made to depths in excess of 1000 m. The present records are thus much shallower and more northerly than previous finds.

Crossota rufobrunnea (KRAMP, 1913)

C. rufobrunnea was the fifth most abundant medusan species in the collection, being represented by 738 specimens of which 263 came from day hauls. By day this species was almost entirely confined to the 1250–1000 m layer. By night most of the specimens were found at the same depth, with some at 1000–900 m. This increase in numbers above 1000 m at night was indicative of a small-scale but significant upward migration.

C. rufobrunnea is confined to the north Pacific Ocean and the Atlantic Ocean north of 30°N (KRAMP, 1959). The specimens under consideration here confirm the bathypelagic status of this species but suggest that the vertical range may be much more narrowly circumscribed than had been realized heretofore.

Aglantha digitale (MÜLLER, 1775)

This species was represented by over 800 specimens and ranked fourth in the RMT 8 series. It was the only species of medusa which was abundant in the surface layers. By day the main concentration was found above 200 m. At night there appeared to be some degree of dispersion, but most of the specimens obtained came from the top 500 m. Both by day and by night, small numbers were found in deep mesopelagic and bathypelagic catches. This distributional pattern conforms with that given by KRAMP (1968a). *A. digitale* is common in the Atlantic and Pacific Oceans north of 35°N and in the Arctic Ocean (KRAMP, 1959).

Halicreas minimum FEWKES, 1882

With 402 specimens, *H. minimum* was the sixth most abundant species in the collection. Both by day and by night the species was confined to depths in excess of 800 m, with maximum numbers occurring at or about 1000 m. The data from these catches suggested that *H. minimum* is non-migratory.

KRAMP (1968a) has listed the distribution of this species as cosmopolitan except in Arctic waters and the Mediterranean. The extensive collections described by KRAMP (1959, 1965) were obtained over a wide depth range, but most came from hauls made with 2000 mwo or more, corresponding to a depth of at least 1000 m.

Halicera bigelowi KRAMP, 1947

Over 250 badly damaged halicreid medusae have been found in RMT 8 hauls from Sta. 7709. Some of the least damaged of those specimens could be assigned, with some degree of confidence, to *H. bigelowi*, but the determination of many others must remain uncertain. All the specimens are assumed to belong to the same species.

By day 189 specimens were found in the four hauls made between 200 and 600 m with maximum numbers at the 500–400 m level. Night catches were much smaller, only sixty-five specimens being obtained in four hauls between 100 and 500 m. Maximum numbers occurred at the 400–300 m level. Those data suggest a small but significant upward movement at night.

H. bigelowi is known from various localities in the North Atlantic and Pacific Oceans but probably has a much wider distribution (KRAMP, 1968a). Previously reported material suggests a rather deeper vertical distribution than has been demonstrated for the present material.

Botrynema brucei BROWNE, 1908

Nearly 400 individuals of this species were obtained in the two deepest catches of both day and night series. In each case, maximum numbers were found in the 1500–1250 m layer.

B. brucei is widely distributed in the Atlantic and Indian Oceans, but appears to be restricted to the north and southwest parts of the Pacific Ocean (KRAMP, 1965). Both the present collection and the *Dana* material demonstrate a bathypelagic distribution for this species. Most Most material from the Atlantic Ocean obtained by the *Dana* was from nets fished with at least 4000 mwo corresponding to depths of 2000 m or more (KRAMP, 1959) although the species can occur much nearer the surface (RUSSELL, 1953).

Aegina citrea ESCHSCHOLTZ, 1829

A. citrea was poorly represented in these collections, nineteen specimens being obtained from four of the five day hauls between 600 and 1250 m, and fourteen from four night hauls made between 700 and 1250 m. A single specimen was also found in the night 100–50 m haul. Both by day and by night, maximum numbers were found in the 900–800 m layer.

KRAMP (1965, 1968a) has shown that this species is widely distributed in the tropical and temperate areas of all the oceans. Near the limits of geographic range *A. citrea* is, as has been found at Sta. 7709, more or less confined to deep water but in warmer areas may occur throughout the water column (KRAMP, 1959).

Aeginura grimaldii MAAS, 1904

A. grimaldii was the second most abundant species in the northern RMT 8 series. A total of 4599 specimens was recovered of which day hauls accounted for 2588 and night hauls 2011.

By day specimens of *A. grimaldii* were present in all samples collected below 600 m. At least 10% of the day total was collected in each of the three samples fished between 800 and 1250 m with a maximum of 38.6% in the 900–800 m haul. At night the shallowest positive haul was that made at 500–400 m, with significant numbers taken in catches made between 700 and 1250 m. The maximum density, 31.3% of the night total, again occurred in the 900–800 m haul. An analysis of the vertical distribution data has shown that the median depth of occurrence was about 930 m and that a small but significant upward movement of 40 m took place at night.

A. grimaldii is widely distributed in the three major oceans but is more abundant in some areas than others. It has not been found in the Arctic Ocean or the Southern Ocean, nor in the Mediterranean, but is particularly abundant in the North Atlantic Ocean (KRAMP, 1959, 1965), a finding confirmed by the present collections. Most of the *Dana* material was obtained in hauls fished to depths of 1500 m or more. The upper limit of vertical distribution as determined by material from Sta. 7709 is much shallower, with the shallowest catches comparable to those obtained at about 650 m by the *Michael Sars* (KRAMP, 1948).

Nausithoe globifera BROCH, 1913

A total of thirty-five specimens of this species was obtained in six of the RMT 8 hauls. By day the three hauls made between 500 and 800 m proved positive, while at night it was found in the 400–300, 500–400 and 800–700 m layers. Numbers were too small for certainty but an upward migration at night may be indicated.

N. globifera has been recorded only from the temperate North Atlantic Ocean (RUSSELL, 1970). Little confirmation is available on vertical distribution but it is, as appears from the present results, most probably a mesopelagic species (BROCH, 1913; RUSSELL, 1956).

Atolla parva RUSSELL, 1958

Four small rather damaged specimens have been assigned to this species, two from the 900–800 m day haul and two from the 1000–900 m night haul.

A. parva has been recorded only from the north and tropical Atlantic Ocean (RUSSELL, 1970). Most of the large collection studied by REPELIN (1966) came from hauls made at depths of 700–750 m.

Atolla vanhoeffeni RUSSELL, 1957

A total of 103 specimens of this species was obtained at lat. 60°N. By day fifty-four specimens were found in catches from the four hauls taken between 500 and 900 m and one additional specimen in the 2000–1500 m haul. At night a total of forty-eight individuals were recovered from five hauls made between 300 and 800 m. The median depth of occurrence by day was 680 m and by night 470 m indicating an upward migration of over 200 m during the hours of darkness.

RUSSELL (1970) has summarized the distribution of *A. vanhoeffeni* showing it to occur in parts of the three major oceans. The species has been found closer to the surface than either *A. wyvillei* or *A. parva*. RUSSELL (1957) and REPELIN (1964) reported specimens caught in hauls made with only 600 mwo, and REPELIN (1966) found this species commoner at 350–400 m than at 700–750 m. Those results are in close agreement with the findings based on the present collections.

Atolla wyvillei HAECKEL, 1880

With 139 specimens this species was the most abundant *Atolla* in the collections. By day specimens were caught in all hauls made below 600 m with maximum numbers in the 900–800 m layer. A single specimen was also found in the 400–300 m catch. All night hauls fished below 500 m combined to give a total of sixty-two specimens. There was no clear peak of abundance. Both by day and by night numbers were lower at depths greater than 1000 m. The available data gave no indication of migratory activity although some degree of dispersal seems indicated.

A. wyvillei is very widely distributed, having been found in all the major oceans from the edge of Antarctica to the north polar basin (KRAMP, 1968b). The vertical distribution of this species has been discussed by a number of authors. BIGELOW (1938) found a marked concentration at depths of about 900 m off Bermuda, while STIASNY (1940) working on the extensive *Dana* collections obtained roughly comparable numbers throughout the water column from 500 to 2500 m. KRAMP (1968b) reported on the *Galathea* material and showed that appreciable concentrations of this species occurred within the depth range 2000–5000 m. Clearly more data are required to reconcile those anomalies.

Periphylla periphylla (PÉRON and LESUEUR, 1809)

P. periphylla was the third most abundant medusa in the series, being represented by 3135 specimens of which 1825 were from day catches and 1310 from night hauls (Table 4). With the exception of the night haul from 1500–1250 m, *P. periphylla* was present in all catches in the series. Individuals were concentrated at mesopelagic depths with the bulk of specimens at 600–900 m by day and at 100–800 m by night. Numerical abundance was highest in the 700–600 m day haul and 500–400 m night haul.

The smallest specimens found were about 2 mm across the central disc and the largest about 140 mm. The depth distribution of individuals in various size groups is given in Table 5. The 10%, 25%, 50%, 75% and 90% cumulative totals have been calculated and plotted in Fig. 4. Although the overall depth range in the two smallest size categories was wide, a large proportion of each category was found at one or two adjacent sampling depths both by day and by night. In the 10 mm and 15 mm size groups this concentration into a narrow depth range was less evident, the median 80% of the population occupying 400–800 m of the water column rather than 100–400 m as in the smaller groups. The distinct downward trend of the 50% cumulative level was continued in all but the largest size category (60–140 mm) by night but not by day. Both by day and by night all medusae in this category were found in the uppermost 300 m of the water column with the exception of single individuals from 800–710 m and 900–810 m day hauls. This pattern of distribution, in which large adult individuals occurred at shallower depths than smaller individuals, is unusual among planktonic and nektonic organisms. The present analysis has confirmed the tendency noted by KRAMP (1968b) most of whose material was obtained from much greater depths in tropical waters. The overall patterns of day and night distribution as shown in Fig. 4 are, with the exception of the 20 mm and 25–50 mm categories, quite similar one to the other. Those two size categories were poorly represented, and day figures were grossly affected by the catch from the 200–110 m haul. Hydrological conditions were monitored during the series using a temperature–salinity–depth probe. Considerable changes occurred during the early part of the sampling programme resulting in isothermal and isohaline conditions in the upper 1000 m of the water column. These changes were attributed to the passage through the sampling area of a polar front. No precise timing for this event can be fixed, but it is possible that the 200–110 m day haul, the first of the series, sampled a different population of *P. periphylla* than did most or all of the remaining hauls in the series.

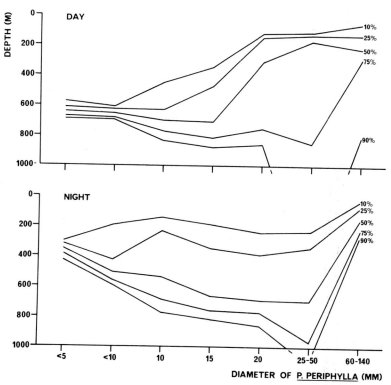

Fig. 4. *Periphylla periphylla*, cumulative percentage depths of occurrence.

Table 5.   Depths of occurrence of size groups of Periphylla periphylla *obtained from the day and night RMT 8 series at Sta. 7709*

Depth (m)	<5 D	<5 N	<10 D	<10 N	10 D	10 N	15 D	15 N	20 D	20 N	25–50 D	25–50 N	60–140 D	60–140 N	Total D	Total N
25–10	1	—	—	—	—	1	—	1	—	—	—	1	—	11	1	14
50–25	1	—	—	3	—	—	—	2	—	1	—	—	—	3	1	9
100–50	1	—	—	—	—	1	—	2	1	—	2	—	3	—	7	3
200–100	—	1	—	37	—	36	1	16	13	3	17	1	2	3	33	97
300–200	—	8	—	11	2	14	2	16	—	5	—	3	3	4	7	61
400–300	—	41	1	16	12	6	12	10	2	6	—	3	—	—	27	82
500–400	—	409	2	103	13	10	8	5	1	1	—	—	—	—	24	528
600–500	115	59	6	170	3	41	1	12	—	1	—	—	—	—	125	283
700–600	720	5	564	13	69	19	16	25	2	12	—	4	—	—	1379	78
800–700	2	—	32	3	70	30	19	50	4	17	1	1	1	—	129	101
900–800	1	—	6	3	27	7	20	12	4	5	1	1	1	—	60	28
1000–900	8	2	4	2	—	2	—	3	1	1	1	5	—	—	9	15
1250–1000	4	3	2	3	2	—	1	—	—	—	—	3	—	—	2	9
1500–1250	—	—	2	—	—	—	—	—	—	—	—	—	—	—	7	—
2000–1500	4	—	—	—	—	—	1	1	—	1	2	—	—	—	7	2
Total	857	528	619	364	198	167	81	155	28	53	24	22	10	21	1825	1310

In a discussion of the extensive collections made by the *Dana*, KRAMP (1968b, p. 178) has pointed out that medusae are 'not good swimmers' and that vertical movements are usually attributable to physical conditions. The use of opening/closing nets fished through narrow vertical layers at accurately known depths enables vertical movements of relatively small magnitude to be detected.

The day median depths of the three smallest size categories are deeper than the corresponding night levels by 150–200 m (Fig. 4). Differences of 50 m and 130 m exist between the day and night levels for the 15-mm and 60–140-mm categories. In the remaining two size groups, the day median depths are shallower than the night levels by about 500 m, but these data are suspect. A number of factors may have affected the observed distributions but it is clear that *P. periphylla* undergoes a diurnal migration during at least part of its life cycle. Additional supporting evidence for an upward movement at night by large individuals was provided by visual observations at the surface, these medusae being common at night but absent during daylight hours.

P. periphylla has a very wide geographical distribution, having been recorded from all oceans except the Arctic (STIASNY, 1940; KRAMP, 1968b; RUSSELL, 1970). KRAMP (1947) discussed *P. periphylla* at some length, and tabulated data on depth distribution from most of the major sources then available including the extensive catches made by the *Dana* in the north Atlantic. The quoted depths at which those catches were made can be at best only educated guesses as non-closing nets and imprecise depth-measuring techniques were used. Despite the degree of uncertainty produced by these factors a strongly marked numerical maximum was apparent for the smallest sized medusae in all five sets of data. Maxima at 660 m (data from KRAMP, 1913) and 500–600 m (BROCH, 1913) coincide with the current findings while those at 750–1000 m (STIASNY, 1934), 800 m (KRAMP, 1924) and 1000 m (KRAMP, 1947) are all deeper. These differences are, in all

Table 6. *Summary of vertical distribution and migration of the common medusae from the RMT 8 series at Sta. 7709*

	DAY				NIGHT				MIGRATION				
	25% level	50% level	75% level	Mean depth	25% level	50% level	75% level	Mean depth	25% level	50% level	75% level	Mean depth	Significance of mean
Chromatonema rubrum	754	813	892	832	733	774	837	789	21	39	55	43	*
Colobonema sericeum	438	484	602	541	409	501	575	503	29	−17	27	38	—
Pantochogon haeckeli	800	936	1211	1017	755	914	1178	999	45	22	33	18	*
Crossota rufobrunnea	1061	1124	1187	1124	1039	1111	1182	1108	22	13	5	16	***
Aglantha digitale	63	126	189	221	181	324	374	331	−118	−198	−185	−100	***
Halicreas minimum	1001	1111	1221	1165	984	1125	1283	1152	17	−14	−62	13	—
Halicera bigelowi	325	415	458	390	325	365	410	367	0	50	48	23	*
Botrynema brucei	1388	1542	1770	1580	1360	1470	1711	1537	28	72	59	43	**
Aeginura grimaldii	846	933	1102	971	811	894	988	931	35	39	114	40	***
Atolla vanhoeffeni	592	683	780	685	353	475	573	477	239	208	207	208	***
A. wyvillei	813	862	1035	998	746	943	1585	1110	67	−81	−550	−112	—
Periphylla periphylla	617	651	684	655	415	476	570	492	202	175	114	163	***

$*p < 0.05.$ $**p < 0.01.$ $***p < 0.001.$

probability, partly artificial rather than real, being a function of the method of estimating fishing depths. It is clear, however, that *P. periphylla* is primarily an inhabitant of the deep mesopelagic zone, and that large individuals, particularly in colder waters, may occur in the surface layers. BIGELOW (1938) has shown that *P. periphylla* is only rarely found in water warmer than 12–13°C, hence in high latitudes the upper limit of vertical distribution may approach or reach the surface.

The vertical range, mean and median depths and migration of the more abundant medusan species are summarized in Table 6. Only *A. digitale* of the twelve species considered is typical of the epipelagic zone, as *P. periphylla*, although reaching the surface, is basically a mesopelagic species. The main concentrations of medusae in terms of numbers of individuals and of species were found in the deep mesopelagic zone.

A striking feature of the day vertical distribution was the marked change in relative abundance and species composition of the total medusan population at the 600–500 m level. The sixteen species recorded at Sta. 7709 have been listed in Table 4, and of those *C. alba* and *A. parva* were represented by very low numbers. Ten of the fourteen more abundant species had vertical distributions which appeared to be limited to a greater or lesser extent at 500–600 m. *H. bigelowi* was found only above this level while *P. haeckeli*, *A. citrea*, *A. grimaldii*, *N. globifera*, *A. vanhoeffeni* and, if the single specimen in the 400–300 m haul is ignored, *A. wyvillei* were all confined to greater depths. Only three species were found both above and below this depth layer. *C. sericeum* and *A. digitale* were much more abundant above the boundary, while *P. periphylla* was concentrated mainly below it. A significant number of amphipod species, including *H. spinigera*, also had vertical distributions which were curtailed or which showed marked changes of abundance at those depths (THURSTON, unpublished data). It was at this depth that the North Atlantic

Central Water began to show signs of mixing with the North Atlantic Deep Water (FASHAM and ANGEL, 1975). Thirteen of the sixteen species of medusae obtained at lat. 60°N had vertical distributions centred in this layer of mixed water.

Of the twelve species which were sufficiently numerous to permit analysis, nine showed a significant vertical displacement at night (Table 6). Eight species moved upward and one, *A. digitale*, moved downward. The extent of vertical migration of most species was small, thus agreeing with KRAMP's (1968b) statement that medusae are not good swimmers. The apparent negative migration of *A. digitale* may not be real but a product of patchy distribution. With the exception of the massive concentration of individuals in the 400–300 m night haul, the overall distributions by day and by night are very similar. Two other species showed an extensive migration. *A. vanhoeffeni* was not very abundant, and although the differences in day and night median depths of occurrence are highly significant, the irregular distribution at night suggests that numbers may have been affected by patchiness. If the strong migratory tendencies are real, this species contrasts sharply with the related *A. wyvillei* for which no significant diurnal movement could be established. The remaining species, *P. periphylla*, migrated on average through a vertical distance of 170 m.

The high proportion of deep mesopelagic and bathypelagic species is in line with the findings of BIGELOW (1938). KRAMP (1959, pp. 242; 253) has listed epipelagic and bathypelagic medusae occurring in the boreal North Atlantic Ocean. Totals of thirteen and sixteen species respectively (of which *A. citrea* and *A. digitale* occur both shallow and deep) are not reflected in the present results. On further analysis, however, it is apparent that several of the epipelagic species have been recorded only once from the area and that most of the remainder rarely occur further north than 50°N. The exceptions are those two species with very wide vertical ranges. In contrast to this situation nine of the sixteen bathypelagic species known from the boreal North Atlantic Ocean were found at lat. 60°N. Of the seven species not obtained five could be excluded on geographical grounds. The species occurrence and the pattern of vertical distribution established by the present material agrees well with the zoogeographic data assembled by KRAMP (1959).

ASSOCIATION OF *HYPERIA SPINIGERA* WITH *PERIPHYLLA PERIPHYLLA*

During the present investigation, over 16,000 medusae belonging to sixteen species have been examined for the presence of larval hyperiids. The only medusan species found to be infected was *P. periphylla* and the only larvae present belonged to the genus *Hyperia*. Associations between adult *Hyperia*, or larval amphipods of other genera, may have occurred, but if so, failed to survive the disturbance of catching and subsequent handling. No larvae other than those of *Hyperia* were found in RMT 1 catches.

A total of 311 *Hyperia* were dissected out of 36 *P. periphylla*, a rate of infection of 1.1 %. Two species of *Hyperia* were involved, 300 specimens from 33 medusae being *H. spinigera* and 11 from 4 medusae being *H. medusarum*. One *P. periphylla* from the 900–800-m night haul contained juveniles of both species. Most of the *Hyperia* (89%) were located in the gastrovascular pouches, but some (11%) were embedded in the mesogloea, usually close to the coronal groove.

The number of *H. spinigera* per medusa varied from 1 to 66 (average 9.09). The smallest

Table 7.   *Numbers of* Periphylla periphylla *from the RMT 8 series at Sta. 7709 containing juvenile* Hyperia spinigera

			Number of *P. periphylla*		
	Haul No.	Depth (m)	Total	Infected	% Infected
DAY	21	800–710	129	7	5.4
	25	900–810	49	7	14.3
NIGHT	36	700–600	78	4	5.1
	28	800–705	101	9	8.9
	22	900–800	24	5	20.8
	17	1000–900	15	1	6.7

specimens found were 1.0 mm long and, by comparison with hatched larvae from female brood pouches, had undergone one moult. The peraeopods were indistinctly articulated and the antennae and abdominal appendages rudimentary or absent. Individuals at this stage of development, corresponding to the protopleon stage of Laval (1965), would possess at most a limited locomotory ability and completely lack swimming powers. In contrast the largest specimens were 4.3–4.6 mm long with well-developed antennae and peraeopods, and multiarticulate, setose pleopods composed of four or five short articles in addition to the long basal article. The degree of development was such as to suggest a fully functional swimming ability. Although a few smaller specimens have been found free in RMT 8 and RMT 1 samples, it is at this size, i.e. 4–5 mm, and this stage of development that *H. spinigera* appears to become capable of an existence at least temporarily independent of the medusan host. Smaller specimens free in samples were probably an artefact of catching and subsequent handling.

Several authors (e.g. KANE, 1963; SHIH, 1969) have used the increasing number of articles in the pleopodal rami to delimit growth stages. The method is of limited value in the present case due to variation in number of articles between rami of one pleopod, between the pleopods of a pair and between pairs of appendages. It is probable, however, that the size range of juveniles found in these medusae represents about five growth stages.

The numbers and depths of occurrence of *P. periphylla* which contained *H. spinigera* have been listed in Table 7. It is clear from a comparison with data in Table 4 that the infected medusae came from depths greater than the levels of maximum density of *P. periphylla* but within the normal range of that species. In contrast, the depths at which amphipod-bearing medusae occurred agreed precisely with the depth range of ovigerous *H. spinigera*. As the range of diurnal migration of the two species is comparable, this agreement tends to be true of both day and night distributions.

The overall size range of the 3135 *P. periphylla* obtained from RMT 8 hauls at Sta. 7709 was 2–140 mm across the central disc. The thirty-three medusae containing young *H. spinigera* fall into two size groups; thirty measured 10–25 mm in diameter, while the remaining three were 51–55 mm in diameter (Table 8). The absence of *H. spinigera* in *P. periphylla* less than 10 mm in diameter, despite this category representing over 75% of the total numbers caught, can be explained on the grounds of size. The situation parallels that found by BUCHHOLZ (1953) who found that free *H. galba* used the anthomedusans *Halitholus cirratus* and *Sarsia tubulosa* as temporary resting places in the spring until *Cyanea capillata* attained a diameter of about 20 mm. No young amphipods

Table 8. *Size of* Periphylla periphylla *from the RMT* 8 *series at Sta.* 7709 *containing juvenile* Hyperia spinigera

Size (mm)	Number of *P. periphylla*		
	Total	Infected	%
<5	1385	0	0
<10	983	0	0
10	365	11	3.0
15	236	14	5.9
20	81	5	6.2
25–50	46	3	6.5
60–140	31	0	0
Total	3135	33	1.1

were found in large *P. periphylla*. The number of medusae with a diameter of 60 mm or more formed only a small proportion of the total population but, in any case, this element was concentrated primarily at depths much shallower than those occupied by ovigerous *H. spinigera*. It is not clear whether the absence of infected medusae of 30–40 mm diameter was real or simply a reflection of the low numbers of this size obtained.

Most of the juvenile *H. spinigera* removed from *P. periphylla* and a substantial proportion of larger individuals found free in samples had full or partially full digestive tracts. Without exception, the gut contents were dark reddish-brown in colour, matching closely the pigment of *P. periphylla*. Samples of material from the gut have been shown to contain protoporphyrin, the presence of which is conclusive evidence of an origin in one or other of the brown-coloured deep water medusae (HERRING, 1972).

A microscopical examination of gut contents of *H. spinigera* taken from *P. periphylla* and found free in samples demonstrated the presence of considerable quantities of cell debris among which were many nematocysts. A large proportion of the nematocysts were damaged or distorted, but commoner types included holotrichous haplonemes of various shapes and sizes and at least one form of microbasic eurytele. These two types of nematocyst, although found in other coelenterate groups, are the only ones known to occur in the Coronatae (WERNER, 1965). The relative proportions of different sizes and types of nematocyst present varied widely from amphipod to amphipod. This was true even for young stages taken from the same medusa. The size and shape of nematocysts in tissue samples from different parts of the anatomy of *P. periphylla* corresponded poorly with those found in amphipod stomachs. The other brown medusae, *A. grimaldii*, *A. vanhoeffeni* and *A. wyvillei*, were also examined as was the abundant *P. haeckeli*. There was no reason to believe that the hyperiids had been feeding on any of these species rather than *P. periphylla*, although evidence for the latter was far from conclusive. In view of the known association of *H. spinigera* with *P. periphylla*, it seems probable that the amphipods were feeding, at least in part, on *P. periphylla* tissue, perhaps from areas such as the lining of the stomach which lack nematocysts. A more extensive survey of the shape, size and distribution of nematocysts in *P. periphylla* would be required to show whether individual variation could explain the apparent lack of agreement between those found *in situ* and those from amphipod gut contents. The possibility that the hyperiids, particularly the smaller juveniles, were feeding on the food of *P. periphylla*, rather than the medusa itself, cannot be substantiated as no information is available concerning the diet of *P. periphylla*. WHITE and BONE (1972) referred to *H. macrocephala* as "... an obligative parasite ... which alternates between an endoparasitic juvenile stage and an ectoparasitic reproductive and dispersive phase ...". Juvenile *H. spinigera* are clearly also obligative parasites of *P. periphylla* as initially they possess only the most rudimentary of locomotory powers and feed either directly on host tissue, or on the host's food.

No data on the number of eggs carried by female *H. spinigera* is available, but the closely related *H. galba* averages 225 per brood (METZ, 1967) and the antarctic species *H. macrocephala* 600 per brood (WHITE and BONE, 1972). Egg size of *H. spinigera* is similar to that of *H. macrocephala* so that although the former species is somewhat the smaller of the two, the brood size of *H. spinigera* is unlikely to be widely different from these two species. The inferred high fecundity of *H. spinigera* coupled with the low numbers of juveniles found in individual medusae (1 to 66, average 9.09) suggests either that a single amphipod places larvae on a number of medusae or that a very substantial loss occurs during or after the transfer process. Once the larvae are in a medusa there seems no reason to suppose that the rate of development of individuals will be widely different, so that marked differences in size and state of development which have been noted among the *H. spinigera* from three of the medusae studied are attributable, in all probability, to deposition by more than one female. Those three cases were in addition to that in which both *H. spinigera* and *H. medusarum* were found in the same medusa.

A comparison has been made between the length of the largest *H. spinigera* from each infected medusa and the diameter of the host (Fig. 5). Although some of the smallest medusae

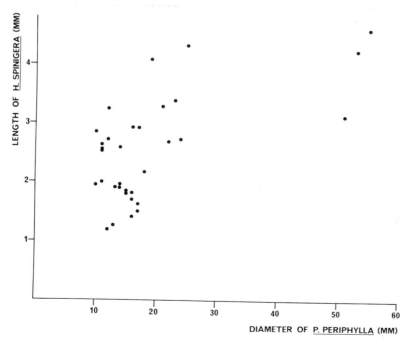

Fig. 5. *Hyperia spinigera*, size of juveniles relative to host *Periphylla periphylla*.

contained fairly large amphipods, there was a tendency for the sizes of parasite and host to be positively correlated, at least among medusae in the 10–25-mm size range. As all the samples were collected within a few days of each other, and no data on growth rates or breeding cycles of either amphipod or medusa are available, the significance of this correlation must remain obscure.

ASSOCIATION OF *HYPERIA* SPECIES WITH MEDUSAE

Many authors have reported species of *Hyperia* in association with other organisms. With the exception of several occurrences on ctenophores (CHUN, 1880; STEPHENSEN, 1923a; SCHELLENBERG, 1942; BUCHHOLZ, 1953; OLDEVIG, 1959) and one dubious record from a salp (VERRILL and SMITH, 1874), all such associations have been with medusae. Although several reports of *Hyperia* on Anthomedusae and Leptomedusae have appeared in the literature (BLANC, 1884; EVANS and ASHWORTH, 1909; TESCH, 1911; STEPHENSEN, 1923a; SCHELLENBERG, 1942; BUCHHOLZ, 1953), the number of original records is small. For example, the observation of *H. galba* on *Melicertum octocostatum* (as *Melicertidium*) given by BLANC (1884) has been repeated at least three times. The work of BUCHHOLZ (1953) has suggested that associations with Anthomedusae and Leptomedusae may be of a temporary nature and abnormal in so far as they result from an absence of larger Scyphozoa, the group with which *Hyperia* is normally found. DAHL (1959 a, b), METZ (1967), WHITE and BONE (1972) and SHEADER (1974) have shown that the relationship between adult *Hyperia* and medusae is parasitic in nature, the amphipod feeding, in part at least, directly on medusan tissue. Evidence is available to show that *H. medusarum*, *H. galba*, *H. gaudichaudi* and *H. macrocephala* are associated with one or more species of Scyphozoa of the orders Semaeostomeae and Rhizostomeae (Appendix 1) and data has been produced in this paper to demonstrate a similar relationship between *H. spinigera* and *P. periphylla*. This latter case is the first to be reported involving a scyphozoan belonging to the order Coronatae.

The young of many hyperiid species hatch and are liberated by the female as larvae with vestigial or non-existent abdominal appendages (MINKIEWICZ, 1909; LAVAL, 1963, 1965, 1968; KANE, 1963; SHIH, 1969; WHITE and BONE, 1972). Hyperiids at this stage of development (the protopleon stage of Laval, 1965) are incapable of swimming and require a substrate on which to live until the pleopods and uropods become functional.

There are few explicit reports of protopleon larvae or very small juveniles in medusae. METZ (1976) and WHITE and BONE (1972), working on *H. galba* and *H. macrocephala* respectively, have shown that the very small stages burrow into the mesogloea or infest the peripheral parts of the gastrovascular system of the host medusa. Either position may be adopted by *H. spinigera* larvae although most occur free in the gastrovascular pouches.

In view of the close morphological similarity among species of *Hyperia*, it seems probable that all members of the genus parasitize Scyphomedusae both as larvae and at later stages in the life history.

Acknowledgements—My thanks are due to Mr. P. M. DAVID for his continued encouragement. I am most grateful to colleagues in various museums and institutions, particularly Dr. R. J. LINCOLN of the British Museum (Natural History) and Dr. T. E. BOWMAN of the Smithsonian Institution, for the loan of specimens and answers to my questions. I extend my thanks to Dr. P. J. HERRING for confirming the presence of protoporphyrin in amphipod gut contents. Mrs. C. E. DARTER kindly inked my pencil drawings of Figs. 1 and 2.

REFERENCES

AGRAWAL V. P. (1967) The feeding habit and the digestive system of *Hyperia galba*. *Symposium of the Marine Biological Association of India*, Series 2, 545–548.

ALVARADO R. (1955) El 'cangrejito' de las medusas. *Boletin de la Real Sociedad Española de Historia Natural*, **53**, 219–220.

BAKER A. DE C., M. R. CLARKE and M. J. HARRIS (1973) The N.I.O. combination net (RMT 1 + 8) and further developments of rectangular midwater trawls. *Journal of the Marine Biological Association of the United Kingdom*, **53**, 167–184.

BARNARD K. H. (1930) Crustacea. Part XI. Amphipoda. *Natural History Report. British Antarctic 'Terra Nova' Expedition 1910*, Zoology, **8**(4), 307–454.

BARNARD K. H. (1932) Amphipoda. *Discovery Reports*, **5**, 1–326.

BASSINDALE R. and J. H. BARRETT (1957) The Dale Fort marine fauna. *Proceedings of the Bristol Naturalists Society*, **29**(3), 227–328.

BATE C. S. (1862) *Catalogue of the specimens of amphipodous Crustacea in the collection of the British Museum*, London, Trustees of the British Museum.

BATE C. S. and J. O. WESTWOOD (1868) *A history of the British sessile-eyed Crustacea*, London, Van Voorst.

BIGELOW H. B. (1926) Plankton of the offshore waters of the Gulf of Maine. *Bulletin of the Bureau of Fisheries, Washington, D.C.*, **40**(2), 1–509 [1924].

BIGELOW H. B. (1938) Plankton of the Bermuda Oceanographic Expeditions, VIII. Medusae taken during the years 1929 and 1930. *Zoologica, New York*, **23**, 99–189.

BLANC H. (1884) Die Amphipoden der Kieler Bucht nebst einer histologischen Darstellung der 'Calceoli'. *Nova Acta Academiae Caesarea Leopoldino-Carolinae Germanicum Naturae Curiosorum*, **47**(2), 37–104.

BOVALLIUS C. (1889) Contributions to a monograph of the Amphipoda Hyperiidea. Part 1:2. The families Cyllopodidae, Paraphronimdae, Thaumatopsidae Mimonectidae, Hyperiidae, Phronimdae and Anchylomeridae. *Kungliga Svenska vetenskapsakademiens handlingar*, **22**(7), 1–434.

BOWMAN T. E. (1960) The pelagic amphipod genus *Parathemisto* (Hyperiidea: Hyperiidae) in the north Pacific and adjacent Arctic Ocean. *Proceedings of the United States National Museum*, **112**, 343–392.

BOWMAN T. E. (1973) Pelagic amphipods of the genus *Hyperia* and closely related genera. (Hyperiidea: Hyperiidae.) *Smithsonian Contributions to Zoology*, No. 136, 1–76.

BOWMAN T. E., C. D. MEYERS and S. D. HICKS (1963) Notes on associations between hyperiid amphipods and medusae in Chesapeake and Narragansett Bays and the Niantic River. *Chesapeake Science*, **4**, 141–146.

BROCH H. (1913) Scyphomedusae from the 'Michael Sars' North Atlantic Deep-Sea Expedition. *Report on the Scientific Results of the 'Michael Sars' North Atlantic Deep-Sea Expedition 1910*, **3**(1), 1–21.

BROWNE E. T. (1895) Report on the Medusae of the L.M.B.C. district. *Proceedings and Transactions of the Liverpool Biological Society*, **9**, 243–286.

BROWNE E. T. (1908) The medusae of the Scottish National Antarctic Expedition. *Transactions of the Royal Society of Edinburgh*, **46**, 223–251.

BRUCE J. R., J. S. COLMAN and N. S. JONES (1963) *Marine fauna of the Isle of Man*, Liverpool, Liverpool University Press.

BRUSCA G. J. (1967a) The ecology of pelagic Amphipoda, I. Species accounts, vertical zonation and migration of Amphipoda from the waters off Southern California. *Pacific Science*, **21**, 382–393.

BRUSCA G. J. (1967b) The ecology of pelagic Amphipoda, II. Observations on the reproductive cycles of several pelagic amphipods from the waters off Southern California. *Pacific Science*, **21**, 449–456.

BRUUN A. (1943) The biology of *Spirula spirula* (L.). *Dana Report*, No. 24, 1–46.

BUCHHOLZ, H. A. (1953) Die Wirtstiere des Amphipoden *Hyperia galba* in der Kieler Bucht. *Faunistische Mitteilungen aus Norddeutschland. Kiel*, **1**(3), 5–6.

CAULLERY M. (1922) *La Parasitisme et la symbiose*, Paris, Libraire Octave Doin.

CHEVREUX E. and L. FAGE (1925) Amphipodes. *Faune de France*, **9**, 1–488.

CHILTON C. (1912) The Amphipoda of the Scottish National Antarctic Expedition. *Transactions of the Royal Society of Edinburgh*, **48**, 455–520.

CHUN C. (1880) Die Ctenophoren des Golfes von Neapel. *Fauna und Flora des Golfes von Neapel*, **1**, 1–313.

COX B. J. and C. DE JONG (1975) Oceanographic observations at Ocean Station P (50°N, 145°W), Volume 61, 2 August–18 September 1974. *Pacific Marine Science Report 75-4*, 1–153 (unpublished).

CROTHERS J. H. (editor) (1966) Dale Fort marine fauna. *Field Studies*, **2**, Supplement, 1–169.

DAHL E. (1959a) The amphipod, *Hyperia galba*, an ectoparasite of the jelly-fish, *Cyanea capillata*. *Nature, London*, **183**, 1749.

DAHL E. (1959b) The amphipod, *Hyperia galba*, a true ectoparasite on jelly-fish. *Universitetet i Bergen Arbok*, 1959, (9) 1–8.

DAHL E. (1961) The association between young whiting, *Gadus merlangus*, and the jelly-fish *Cyanea capillata*. *Sarsia*, **3**, 47–55.

DAHL F. (1893) Untersuchungen über die Tierwelt der Unterelbe. 6. *Bericht der Kommission zur wissenschaftlichen Untersuchungen der deutschen Meere, Berlin*, pp. 149–185.

DALES R. P. (1966) Symbioses in marine organisms. In: *Associations of micro-organisms, plants and marine organisms*, Vol. 1. *Symbiosis*, S. M. HENRY, editor, New York, Academic Press, pp. 299–326.

DANA J. D. (1852) Crustacea. In *United States Exploring Expedition during the years 1838, 1839, 1840, 1841, 1842 under the command of Charles Wilkes, U.S.N.*, **13**(2), 691–1618.

DUNBAR M. J. (1942) Marine macroplankton from the Canadian eastern Arctic. 1. Amphipoda and Schizopoda. *Canadian Journal of Research*, **20**, 33–46.

DUNBAR M. J. (1963) Amphipoda. Sub-order: Hyperiidea. Family: Hyperiidae. *Fiches d'Identification du Zooplankton*, Sheet 103, 1–4.

EALEY E. H. M. and R. G. CHITTLEBOROUGH (1956) Plankton, Hydrology and marine fouling at Heard Island. *Australian National Antarctic Research Expeditions Interim Reports*, No. 15, 1–81.

EDWARD T. (1868a) Stray notes on some of the smaller crustaceans. Note I. On the habits, & c., of the Hyperiidae. *Journal of the Linnean Society of London*, Zoology, **9**, 143–147.

EDWARD T. (1868b) Stray notes on some of the smaller crustaceans. Note II. On the habits, & c., of the Hyperiidae. *Journal of the Linnean Society of London*, Zoology, **9**, 166–170.

EMISON, W. B. (1968) Feeding preferences of the Adélie Penguin at Cape Crozier, Ross Island. In: *Antarctic bird studies*, D. L. AUSTIN, editor, Washington, D.C., American Geophysical Union. *Antarctic Research Series*, **12**, 191–212.

EVANS F. and M. SHEADER (1972) Host species of the hyperiid amphipod *Hyperoche medusarum* (Krøyer) in the North Sea. *Crustaceana*, Suppl. 3, 275–276.

EVANS W. and J. H. ASHWORTH (1909) Some medusae and ctenophores from the Firth of Forth. *Proceedings of the Royal Physical Society of Edinburgh*, **17**, 300–311.

FASHAM M. J. R. and M. V. ANGEL (1975) The relationship of the zoogeographic distributions of the planktonic ostracods in the north-east Atlantic to the water masses. *Journal of the Marine Biological Association of the United Kingdom*, **55**, 739–757.

FOXTON P. (1972) Observations on the vertical distribution of the genus *Acanthephyra* (Crustacea: Decapoda) in the eastern North Atlantic, with particular reference to species of the '*purpurea*' group. *Proceedings of the Royal Society of Edinburgh*, B, **73**, 301–313.

GOORMAGHTIGH E. and M. PARMENTIER (1973) Le crustacé amphipode *Hyperia galba*, "parasite" de la méduse *Rhizostoma octopus*. *Les Naturalistes Belges*, **54**, 131–135.

GOSSE P. H. (1853) *A naturalist's rambles on the Devonshire coast*, London, Van Voorst.

GOULD A. A. (1841) *A report on the Invertebrata of Massachusetts comprising the Mollusca, Crustacea, Annelida and Radiata*, Cambridge, Mass., Folsom, Wells and Thurston.

HAAHTELA I. and J. LASSIG (1967) Records of *Cyanea capillata* (Scyphozoa) and *Hyperia galba* (Amphipoda) from the Gulf of Finland and the northern Baltic. *Annales Zoologici Fennici*, **4**, 469–471.

HAMOND R. (1967) The Amphipoda of Norfolk. *Cahiers de Biologie Marine*, **8**, 113–152.

HARDY A. C. (1956) *The open sea. Part I. The world of plankton*. London, Collins.

HERRING P. J. (1972) Porphyrin pigmentation in deep-sea Medusae. *Nature, London*, **238**, 276–277.

HOLLOWDAY E. D. (1948) On the commensal relationship between the amphipod *Hyperia galba* (Mont.) and the Scyphomedusa *Rhizostoma pulmo* Agassiz, var. *octopus* Oken. *Journal of the Quekett Microscopical Club*, **11**(4), 187–190.

HURLEY D. E. (1955) Pelagic amphipods of the sub-order Hyperiidea in New Zealand waters. I. Systematics. *Transactions of the Royal Society of New Zealand*, **83**(1), 119–194.

HURLEY D. E. (1956) Bathypelagic and other Hyperiidea from Californian waters. *Occasional Papers of the Allan Hancock Foundation*, No. 18, 1–25.

HURLEY D. E. (1960) Amphipoda Hyperiidea. *British Australian and New Zealand Antarctic Research Expedition Reports*, Ser. B, **8**(5), 109–113.

HURLEY D. E. (1969) Amphipoda Hyperiidea. In: *Distribution of selected groups of marine invertebrates in waters south of 35°S latitude*. Antarctic Map Folio Series, V. C. BUSHNELL and J. W. HEDGPETH, editors, Folio 11, 32–34.

JONES N. S. (1948) The ecology of the Amphipoda of the south of the Isle of Man. *Journal of the Marine Biological Association of the United Kingdom*, **27**, 400–439.

KANE J. E. (1963) Stages in the early development of *Parathemisto gaudichaudii* (Guér.) (Crustacea Amphipoda: Hyperiidea), the development of secondary sexual characters and of the ovary. *Transactions of the Royal Society of New Zealand*, Zoology, **3**(5), 35–45.

KANE J. E. (1966) Distribution of *Parathemisto gaudichaudii* (Guér.), with observations on its life history in the 0° to 20°E sector of the southern Ocean. *Discovery Reports*, **34**, 163–198.

KINAHAN G. H. (1859) Notes on dredging in Belfast Bay, with a list of species. *Natural History Review*, **6**(2), 79–85.

KRAMP P. L. (1913) Medusae collected by the 'Tjalfe' Expedition in Greenland waters. *Videnskabelige Meddelelser fra Dansk Naturhistorisk Forening i Kjøbenhavn*, **65**, 257–286.

KRAMP P. L. (1924) Medusae. *Report on the Danish Oceanographical Expeditions, 1908–1910, to the Mediterranean and adjacent seas*, **2**, Biology, H1, 1–67.

KRAMP P. L. (1947) Medusae, Part III. Trachylina and Scyphozoa with zoogeographical remarks on all the medusae of the northern Atlantic. *Danish Ingolf-Expedition*, **3**(8), 1–100.

KRAMP P. L. (1948) Trachymedusae and Narcomedusae from the 'Michael Sars' North Atlantic Deep-Sea Expedition, 1910, with additions on Anthomedusae; Leptomedusae and Scyphomedusae. *Report on the Scientific Results of the 'Michael Sars' North Atlantic Deep-Sea Expedition, 1910*, **15**(9), 1–24.

KRAMP P. L. (1959) The Hydromedusae of the Atlantic Ocean and adjacent waters. *Dana Reports*, No. 46, 1–283.

KRAMP P. L. (1965) The Hydromedusae of the Pacific and Indian Oceans. *Dana Reports*, No. 63, 1–162.

KRAMP P. L. (1968a) The Hydromedusae of the Pacific and Indian Oceans, sections II and III. *Dana Reports*, No. 72, 1–200.

KRAMP P. L. (1968b) The Scyphomedusae collected by the Galathea Expedition 1950–52. *Videnskabelige Meddelelser fra Dansk Naturhistorisk Forening i Kjøbenhavn*, **131**, 67–98.

LAMBERT F. J. (1936a) Jelly-fish. The difficulties of the study of their life history and other problems. *Essex Naturalist*, **25**, 70–86.

LAMBERT F. J. (1936b) Observations on the Scyphomedusae of the Thames Estuary and their metamorphoses. *Travaux de la Station Zoologique de Wimereux*, **12**, 281–307.

LAVAL P. (1963) Sur la biologie et les larves de *Vibilia armata* Bov. et de *V. propinqua* Stebb., Amphipodes Hypérides. *Compte rendu hebdomadaire des séances de l'Académie des sciences*, **257**, 1389–1392.

LAVAL P. (1965) Présence d'une période larvaire au début du développement de certains Hypérides parasites (Crustacés Amphipodes). *Compte rendu hebdomadaire des séances de l'Académie des sciences*, **260**, 6195–6198.

LAVAL P. (1968) Développement en élevage et systématique d'*Hyperia schizogeneios* Stebb. (Amphipoda, Hypéride). *Archives de zoologie expérimentale et générale*, **109**, 25–67.

LAVAL P. (1972) Comportement, parasitisme et écologie d'*Hyperia schizogeneios* Stebb. (Amphipode Hypéride) dans le plancton de Villefranche-sur-Mer. *Annales de l'Institut Océanographique*, New Series, **48**, 49–74.

MANSUETI R. (1963) Symbiotic behaviour between small fishes and jelly fishes, with new data on that between the stromateid, *Peprilus alepidotus*, and the Scyphomedusa, *Chrysaora quinquecirrha*. *Copeia*, **1963**, 40–80.

MARINE BIOLOGICAL ASSOCIATION (1957) *Plymouth Marine Fauna*, Plymouth, Marine Biological Association.

MEINERT F. (1890) Crustacea Malacostraca. In C. G. J. PETERSEN, *Den Videnskabelige Udbytte af Kanonbaaden 'Hauchs' Togter i de Danske Have indenfor Skagen, 1883–1886*, **3**, 147–230.

METZ P. (1967) On the relations between *Hyperia galba* Montagu (Amphipoda, Hyperiidae) and its host *Aurelia aurita* in the Isefjord area (Sjaelland, Denmark). *Videnskabelige Meddelelser fra Dansk Naturhistorisk Forening i Kjøbenhavn*, **130**, 85–108.

MEYER A. H. and K. MÖBIUS (1862) Kurzer Ueberblick der in der Kieler Bucht von uns beobachteten wirbellosen Thiere, als Vorläufer einer Fauna derselben. *Archiv für Naturgeschichte*, **28**(1), 229–237.

MINKIEWICZ R. (1909) Memoire sur la biologie du Tonnelier de mer (*Phromina sedentaria* Forsk.) Chapitre I. La coloration des Phronimes et son développement par migration progressive des chromatophores. *Bulletin de l'Institut Océanographique*, No. 146, 1–21.

NAGABHUSHANAM A. K. (1959) Studies on the biology of the commoner gadoids in the Manx area, with special reference to their food and feeding habits. Ph.D. Thesis, University of Liverpool.

NORDGAARD O. (1912) Faunistiske og biologiske iakttagelser ved den biologiske station i Bergen. *Kongelige Norsk Videnskabernes Selskabs Skrifter*, **1911**(6), 1–58.

NORMAN A. M. (1900) British Amphipoda of the tribe Hyperiidea and the families Orchestiidae and some Lysianassidae. *Annals and Magazine of Natural History*, Ser. 7, **5**, 126–144.

OLDEVIG H. (1959) Arctic, Subarctic and Scandinavian amphipods in the collections of the Swedish Natural History Museum in Stockholm. *Göteborgs Kungl. Vetenskaps- och Vitterhets-Samhälles Handlinger*, Ser. B, **8**(2), 1–132.

ORTON J. H. (1922) The mode of feeding of the jelly-fish, *Aurelia aurita*, on the smaller organisms in the plankton. *Nature, London*, **110**, 178–179.

PIRLOT J. M. (1929) Results zoologiques de la Croisière Atlantique de 'l'Armauer Hansen' (Mai–Juin 1922). 1. Les Amphipodes Hypérides. *Memoires de la Société royale des Sciences de Liége*, Ser. 3, **15**(2), 1–196.

PIRLOT J. M. (1932) Introduction à l'étude des Amphipodes Hypérides. *Annales de l'Institut Océanographique, Monaco*, New Series, **12**(1), 1–36.

PIRLOT J. M. (1939) Sur des amphipodes hypérides provenant des croisières du Prince Albert Ier de Monaco. *Resultats des campagnes scientifiques accomplies sur son yacht par Albert Ier Prince Souvrain de Monaco*, **102**, 1–64.

POULSEN E. (1950) *Krebsdyr i vort lands dyreliv*. Copenhagen, Christiana and Copenhagen.

REID D. M. (1955) Amphipoda (Hyperiidea) of the coast of tropical west Africa. *Atlantide Report*, No. 3, 7–40.

REPELIN R. (1964) Scyphomeduses de la familie Atollidae dans le Golfe du Guinée. *Cahiers de l'Office de la Recherche Scientifique et Technique Outre-Mer*, Ser. océanographie, **2**(3), 13–30.

REPELIN R. (1966) Scyphomeduses Atollidae du Bassin du Guinée. *Cahiers de l'Office de la Recherche Scientifique Outre-Mer*, Ser. océanographie, **4**(4), 21–33.

ROMANES G. J. (1876) An account of some new species, varieties, and monstrous forms of medusae. *Journal of the Linnean Society of London, Zoology*, **12**, 524–531.

ROMANES G. J. (1877) An account of some new species, varieties, and monstrous forms of medusae—II. *Journal of the Linnean Society of London, Zoology*, **13**, 190–194.

RUDJAKOV YU. A. (1971) Details of the horizontal distribution and diurnal migration of *Cypridina* (*Pyrocypris*) *sinuosa* (G. W. Müller) (Crustacea Ostracoda) in the western equatorial Pacific. (In Russian.) In: *Life activity of pelagic communities in the ocean tropics*, M. E. VINOGRADOV, editor, Moscow, Izdatelstvo Nauka. (English translation: Jerusalem, Israel Program for Scientifique Translations, 1973, pp. 240–255.)

RUSSELL F. S. (1953) *The Medusae of the British Isles. Anthomedusae, Leptomedusae, Limnomedusae, Trachymedusae and Narcomedusae*, Cambridge, Cambridge University Press.

RUSSELL F. S. (1956) On the Scyphomedusae *Nausithoë atlantica* Broch and *Nausithoë globifera* Broch. *Journal of the Marine Biological Association of the United Kingdom*, **35**, 363–370.

RUSSELL F. S. (1957) On a new species of Scyphomedusa, *Atolla vanhöffeni* n.sp. *Journal of the Marine Biological Association of the United Kingdom*, **36**, 275–279.

RUSSELL F. S. (1970) *The Medusae of the British Isles*. Vol. II. *Pelagic Scyphozoa, with a supplement to the first volume on Hydromedusae*, Cambridge, Cambridge University Press.

SABINE E. (1821) Invertebrate animals. In: W. E. PARRY, *Journal of a voyage for the discovery of a north-west passage from the Atlantic to the Pacific, performed in the years 1819–20, in His Majesty's Ships Hecla and Griper, under the orders of William Edward Parry, R.N., F.R.S. and commander of the expedition, with an appendix, containing the scientific and other observations*, London, John Murray.

SARS G. O. (1879) Report on the practical and scientific investigations of the cod fisheries near the Loffoden Islands made during the years 1864–1869. *Report of the United States Fisheries Commission*, **1879**(4), 565–611.

SARS G. O. (1890–1895) *Amphipoda. An account of the Crustacea of Norway*, Vol. 1, Copenhagen, Cammermeyers.

SCHELLENBERG A. (1942) Krebstiere oder Crustacea. IV. Flohkrebse oder Amphipoda. *Tierwelt Deutschlands*, **40**, 1–252.

ScheuringL. (1915) Beobachtung über den Parasitismus pelagischer Jungfische. *Biologisches Zentralblatt*, **35**, 181–190.

SHEADER M. (1974) North Sea hyperiid amphipods. *Proceedings of the Challenger Society*, **4**(5), 247.

SHEADER M. and F. EVANS (1974) The taxonomic relationship of *Parathemisto gaudichaudi* (Guérin) and *P. gracilipes* (Norman), with a key to the genus *Parathemisto*. *Journal of the Marine Biological Association of the United Kingdom*, **54**, 915–924.

SHIH C. T. (1969) The systematics and biology of the family Phronimidae (Crustacea: Amphipoda). *Dana Reports*, No. 74, 1–100.

SHOEMAKER C. R. (1914) Amphipods of the South Georgia Expedition. *Science Bulletin of the Museum of the Brooklyn Institute of Arts and Sciences*, **2**(4), 73–77.

SHOEMAKER C. R. (1927) Results of the Hudson Bay Expedition in 1920. V. Report on the marine amphipods collected in Hudson and James Bays, by Frits Johansen in the summer of 1920. *Contributions to Canadian Biology and Fisheries*, New Series, **3**(1), 1–11.

SHOEMAKER C. R. (1945a) The Amphipoda of the Bermuda Oceanographic Expeditions 1929–1931. *Zoologica, New York*, **30**, 185–266.

SHOEMAKER C. R. (1945b) Amphipoda of the United States Antarctic Service Expedition, 1939–1941. *Proceedings of the American Philosophical Society*, **89**(1), 289–293.

SPANDL H. (1927) Die Hyperiiden (exkl. Hyperiidea Gammaroidea und Phronimidae) der Deutschen Südpolar-Expedition 1901–1903. *Deutsche Südpolar-Expedition 1901–1903*, **19**, Zoologie, **11**, 145–287.

STEBBING T. R. R. (1888) Amphipoda. *Report of the Scientific Results of the Voyage of H.M.S. 'Challenger' during the years 1873–1876*, **29**, 1–1737.

STEBBING T. R. R. (1914) Crustacea from the Falkland Islands collected by Mr. Rupert Vallentin, F.L.S.—Part II. *Proceedings of the Zoological Society of London*, **24**, 341–378.

STEEDMAN H. F. (1974) Laboratory methods in the study of marine zooplankton. *Journal du Conseil Permanent International pour l'Exploration de la Mer*, **35**(3), 351–358.

STEP E. (1913) Messmates. *A book of strange companionships in nature*, London, Hutchinson.

STEPHENSEN K. (1923a) Crustacea Malacostraca. V. Amphipoda I. *Danish Ingolf-Expedition*, **3**(8), 1–100.

STEPHENSEN K. (1923b) Revideret Fortegnelse over Danmarks Arter af Amphipoda (1. Del). (Hyperiidea; Gammaridea: Lysianassidae). *Videnskabelige Meddelelser fra Dansk naturhistorisk Foreng i København*, **76**, 5–20.

STEPHENSEN K. (1924) Hyperiidea-Amphipoda 2. Paraphrominidae, Hyperiidae, Dairellidae, Phronimidae, Anchylomeridae. *Report on the Danish Oceanographical Expeditions, 1908–1910, to the Mediterranean and adjacent seas*, **2**, Biology, D4, 71–149.

STEPHENSEN K. (1925) Hyperiidea-Amphipoda 3. Lycaeopsidae, Pronoidae, Lycaeidae, Brachyscelidae, Oxycephalidae, Parascelidae, Platyscelidae. *Report on the Danish Oceanographical Expeditions, 1908–1910, to the Mediterranean and adjacent seas*, **2**, Biology, D5, 151–252.

STEPHENSEN K. (1928) Storkrebs, II. Ringkrebs, 1. Tanglopper (Amfipoder). *Danmarks Fauna*, **32**, 1–399.

STEPHENSEN K. (1929) Amphipoda. *Tierwelt der Nord- und Ostsee*, **14** (X.f), 1–188.

STEPHENSEN K. (1942) The Amphipoda of N. Norway and Spitsbergen with adjacent waters. *Tromsø Museums Skrifter*, **3**(4), 363–526.

STIASNY G. (1934) Scyphomedusae. *Discovery Report*, **8**, 329–396.

STIASNY G. (1940) Die Scyphomedusen. *Dana Reports*, No. 18, 1–40.

STRØM H. (1762) *Physisk og Oeconomisk Beskrivelse over Fogderiet Søndmor, beliggende i Bergens Stift, i Norge*, Første Part, Sorøe.

TATTERSTALL W. M. (1906) The marine fauna of the coast of Ireland. VIII. Pelagic Amphipoda of the Irish Atlantic slope. *Scientific Investigations of the Fisheries Branch, Department of Agriculture and Technical Instruction for Ireland for 1905*, No. 4, 1–39.

TESCH J. J. (1911) Résumé des observations sur le plankton des mers explorées par le Conseil pendant les années 1902–1908: Amphipoda. *Bulletin Trimestriel des Résultats Acquis Pendant les Croisières Périodiques de la Conseil Permanent International pour l'exploration de la Mer*, Part 2, 176–193.

THOMPSON W. (1847) Additions to the fauna of Ireland. *Annals and Magazine of Natural History*, **20**, 237–250.

THORSTEINSON E. D. (1941) New or noteworthy amphipods from the North Pacific coast. *University of Washington Publications in Oceanography*, **4**(2), 50–96.

THURSTON M. H. (1974) Crustacea Amphipoda from Graham Land and the Scotia Arc, collected by Operation Tabarin and the Falkland Islands Dependencies Survey, 1944–59. *Scientific Reports, British Antarctic Survey*, No. 85, 1–89.

THURSTON M. H. (1976a) The vertical distribution and diurnal migration of the Crustacea Amphipoda collected during the SOND Cruise, 1965. I. The Gammaridea. *Journal of the Marine Biological Association of the United Kingdom*, **56**, 359–382.

THURSTON M. H. (1976b) The vertical distribution and diurnal migration of the Crustacea Amphipoda collected during the SOND Cruise, 1965. II. The Hyperiidea and general discussion. *Journal of the Marine Biological Association of the United Kingdom*, **56**, 383–470.

TOULMOND A. and J-P. TRUCHOT (1964) Inventaire de la faune marine de Roscoff. Amphipodes—Cumacés. *Travaux de la Station Biologique de Roscoff*, **1964**, 1–42.

VERRILL A. G. and S. I. SMITH (1874) Report on the invertebrate animals of Vineyard Sound and adjacent waters, with an account of the physical features of the region. In: S. F. BAIRD, *Report on the condition of sea-fisheries of the south coast of New England in 1871 and 1872*. Washington.

VINOGRADOV M. E. (1956) Hyperiids (Amphipoda–Hyperiidea) of the western Behring Sea. (In Russian.) *Zoologicheskii Zhurnal*, **35**, 194–217.

VINOGRADOV M. E. (1976) New species of hyperiid (Amphipoda–Hyperiidea) from the tropical Pacific Ocean. (In Russian.) *Trudy Instituta Okeanologii. Akademia Nauk SSSR*, **105**, 130–134.

VOSSELER J. (1901) Die Amphipoden der Plankton—Expedition. I. Theil. Hyperiidea 1. *Ergebnisse der Plankton-Expedition der Humbolt Siftung*, **2**, G.e., pp. 1–129.

WALKER A. O. (1895) Revision of the Amphipoda of the L.M.B.C. district. *Proceedings and Transactions of the Liverpool Biological Society*, **9**, 287–320.

WERNER B. (1965) Die Nesselkapseln der Cnidaria mit besonderer Berücksichtigung der Hydroida. I. Klassification und Bedeutung für die Systematik und Evolution. *Helgoländer Wissenschaftliche Meeresuntersuchungen*, **12**, 1–39.

WHITE M. G. and D. G. BONE (1972) The interrelationship of *Hyperia galba* (Crustacea, Amphipoda) and *Desmonema gaudichaudi* (Scyphomedusae, Semaeostomae) from the Antarctic. *Bulletin of the British Antarctic Survey*, No. 27, 39–49.

WILTON D. W., J. H. H. PIRIE and R. N. R. BROWN (1908) Zoological log. *Report on the Scientific Results of the Scottish National Antarctic Expedition on the S.Y. 'Scotia' during the years 1902, 1903 and 1904*, **4**, Zoology, (1), 1–103.

YANG W. T. (1960) A study of the subgenus *Parahyperia* from the Florida Current (Genus *Hyperia*: Amphipoda Hyperiidae). *Bulletin of Marine Science of the Gulf and Caribbean*, **10**(1), 11–39.

APPENDIX 1. LITERATURE RECORDS OF ASSOCIATIONS BETWEEN SPECIES OF *HYPERIA* AND OTHER ORGANISMS, PARTICULARLY MEDUSAE

A. *Hyperia galba*

	Chrysaora hysoscella	*Pelagia noctiluca*	*Cyanea capillata*	*Cyanea lamarcki*	*Aurelia aurita*	*Rhizostoma octopus*[1]	'medusae'	*Sarsia tubulosa*	*Haliiholus cirratus*	*?Neoturris pileata*	*Melicertum octocostatum*	*Tima bairdi*	*Beroe* sp.
GOULD (1841)							X						
THOMPSON (1847)						X[2]							
GOSSE (1853)	X[3]												
KINAHAN (1859)							X						
BATE (1862)						X[2]							
BATE and WESTWOOD (1868)						X[2]							
EDWARD (1868a)							X						
EDWARD (1868b)							X						
ROMANES (1876)					X								
ROMANES (1877)					X								
BLANC (1884)			X[4]		X[5]					X[6]			
MEINERT (1890)			X		X								
SARS (1890)					X								
DAHL (1893)			X			X							
BROWNE (1895)						X[7]							
WALKER (1895)						X[8]							
TATTERSALL (1906)	X[9]	X[10]	X		X	X[8]							
EVANS and ASHWORTH (1909)					X							X	
TESCH (1911)	X	X[10]	X		X	X				X[11]	X[12]		
CHILTON (1912)[13]		X[14]											
SCHEURING (1915)			X		X								
STEPHENSEN (1923a)					X								X
CHEVREUX and FAGE (1925)	X	X			X	X[15]							
BIGELOW (1926)			X		X								
SHOEMAKER (1927)					X								
STEPHENSEN (1928)			X		X								
PIRLOT (1929)						X[2]							
STEPHENSEN (1929)			X		X								
PIRLOT (1932)						X[15]							
PIRLOT (1939)		X											
SCHELLENBERG (1942)	X	X[10]	X		X	X					X[12]		X
HOLLOWDAY (1948)						X[16]							
JONES (1948)						X[15]		X					
POULSEN (1950)								X					
BUCHHOLZ (1953)	X	X[10]	X		X	X		X	X		X[12]		X
ALVARADO (1955)	X												
HARDY (1956)	X		X[15]		X	X[15]							
BASSINGDALE and BARRETT (1957)						X							
MARINE BIOLOGICAL ASSOCIATION (1957)	X					X							
OLDEVIG (1959)			X[15]		X	X							X
DAHL (1959a)			X										
DAHL (1959b)			X										
NAGABHUSHANAM (1959)						X							
DAHL (1961)			X										
BOWMAN, MEYERS and HICKS (1963)			X		X								
BRUCE, COLMAN and JONES (1963)						X							
TOULMOND and TRUCHOT (1964)	X		X		X								
CROTHERS (1966)	X					X							
DALES (1966)	X					X							
AGRAWAI (1967)							X						

A. *Hyperia galba*

	Chrysaora hyoscella	*Pelagia noctiluca*	*Cyanea capillata*	*Cyanea lamarcki*	*Aurelia aurita*	*Rhizostoma octopus*[1]	'medusae'	*Sarsia tubulosa*	*Halitholus cirratus*	*?Neoturris pileata*	*Melicertum octocostatum*	*Tima bairdi*	*Beroe* sp.
METZ (1967)			X		X								
HAAHTELA and LASSIG (1967)			X										
RUSSELL (1970)	X	X	X		X	X							
EVANS and SHEADER (1972)			X		X								
LAVAL (1972)	X	X	X		X	X[8]							
BOWMAN (1973)							X						
GOORMAGHTIGH and PARMENTIER (1973)						X							
SHEADER (1974)			X	X	X								

[1] RUSSELL (1970) has maintained *R. pulmo* and *R. octopus* as separate species distinguishable on morphological and geographic grounds.

[2] As *R. cuivieri.*
[3] As *C. cyclonata.*
[4] As *C. capitata.*
[5] As *Medusa.*
[6] As *Stomobranchium.*
[7] As *Pilema.*
[8] As *R. pulmo.*

[9] As *C. isosceles.*
[10] As *P. perla.*
[11] As *Tiara.*
[12] As *Melicertidium.*
[13] Material from Sta. 541 has been redetermined as *H. galba.*
[14] On *Aurelia* according to CHILTON (1912) but *P. noctiluca* (as *P. perla*) according to WILTON, PIRIE and BROWN (1908).
[15] Specific identification not given.
[16] As *R. pulmo* var. *octopus.*

MICHAEL H. THURSTON

B. *Hyperia medusarum*

	Cyanea capillata	*Aurelia aurita*	*Rhizostoma octopus*[1]	'medusae'	*Thaumantias* sp.	*Eucharis multicornis*
STRØM (1762)	X[2]					
SABINE (1821)	X[3]					
THOMPSON (1847)			X[4]			
MEYER and MÖBIUS (1862)[5]		X[6]				
EDWARDS (1868a)[7]				X		
EDWARDS (1868b)[7]				X		
CHUN (1880)						X[8]
SARS (1890)	X[9]	X				
TESCH (1911)	X[9]	X				
NORDGAARD (1912)	X					
CAULLERY (1922)[10]			X[4]	X		
STEPHENSEN (1923a)	X[9]	X			X	
STEPHENSEN (1923b)	X[9]					
BIGELOW (1926)	X	X				
SHOEMAKER (1927)	X	X				
STEPHENSEN (1928)	X	X				
THORSTEINSON (1941)[11]				X		
SCHELLENBERG (1942)	X	X				
OLDEVIG (1959)	X[9]	X				
BOWMAN, MEYERS and HICKS (1963)	X					
DALES (1966)			X[12]			
HAMOND (1967)		X				
BOWMAN (1973)				X		

[1] See footnote 1 under Section A.
[2] As 'Medusae orbiculi margine sedecies emarginato'.
[3] As *C. arctica*.
[4] As *R. cuvieri*.
[5] As *H. latreillei*.
[6] As *Medusa*.
[7] As *Lestrigonus kinahani*.
[8] The ctenophore *E. multicornis* does not occur in the northern waters inhabited by *Hyperia medusarum*. It seems probable that the record of *H. medusarum* given by CHUN (1880) in fact refers to *Hyperoche medusarum*.
[9] Specific identification not given.
[10] As *Hyperina*.
[11] Material redetermined as *H. medusarum*. Was *H. spinigera*.
[12] As *R. pulmo*.

C. *Hyperia gaudichaudi*

	Chrysaora fulgida	*Desmonema chierchiana*	*Beroe* sp.
BROWNE (1908)		X	
CHILTON (1912)		X[1]	
STEBBING (1914)			X
BARNARD (1932)[2]	X[3]		

[1] Identification from BROWNE (1908).
[2] Material redetermined as *H. gaudichaudi*. Was *H. galba*.
[3] *C. fulgida* was the only medusan found in the Discovery collections from Hoetjes Bay (STIASNY, 1934).

D. *Hyperia macrocelphala*

	Desmonema gaudichaudi	'medusae'
DANA (1852)[1]		X
BATE (1862)[1]		X
WALKER (1903)[2]		X
SHOEMAKER (1914)[1]		X
SHOEMAKER (1945b)		X
WHITE and BONE (1972)[3]	X	
BOWMAN (1973)		X
THURSTON (1974)		X

[1] As *Tauria*.
[2] Material redetermined as *H. macrocephala*. Was *H. gaudichaudi*.
[3] Material redetermined as *H. macrocephala*. Was *H. galba*.

E. *Hyperia* sp.

VERRILL and SMITH (1874), SARS (1879), STEP (1913), ORTON (1922), LAMBERT (1936 a,b) and MANSUETI (1963) have reported unidentified *Hyperia* on medusae. VERRILL and SMITH (1874) also reported *Hyperia* on *Salpa*.

APPENDIX 2.   REDETERMINATIONS OF SPECIMENS ASSIGNED TO THE GENUS *HYPERIA*

Those redeterminations pertaining to *H. spinigera* have been dealt with
in the main body of this paper

Author	Original determination	Redetermination
Walker (1903)	*H. gaudichaudii*	*H. macrocephala*
Walker (1907)	*H. gaudichaudii*	*H. macrocephala* (mostly: also specimens of *Hyperiella dilatata*)
Chilton (1912)	*H. gaudichaudii* (Sta. 541)	*H. galba*
Barnard (1930)	*H. galba*	*H. gaudichaudi*
Barnard (1932)	*H. galba* (Sta. WS95)	*Hyperoche* sp.
	(Sta. WS Hoetjes Bay)	*H. gaudichaudi*
White and Bone (1972)	*H. galba*	*H. macrocephala*

Note added in proof—Through the kind offices of Dr. D. E. Hurley, it has been possible
to examine the BANZARE material assigned by him (Hurley, 1960) to *H. spinigera*.
The smaller of the two specimens from Sta. 32 (66°35′S, 61°13′E) is a juvenile male and
belongs to *H. spinigera*, but the larger specimen is an adult male *H. macrocephala*. The
single specimen from Sta. 69 (43°19′S, 93°56′E) is in poor condition, but is tentatively
referred to *H. gaudichaudi*. If this identification is correct, it is of interest in that it is the
first record of the species from a locality remote from land.

The identification of an adult male *H. spinigera* has been checked and found to be correct.

Aspects of the development and biology of post-larval *Valenciennellus tripunctulatus* (Esmark) and *Bonapartia pedaliota* Goode and Bean (Pisces, Stomiatoidei)

Julian Badcock

Institute of Oceanographic Sciences, Wormley, Godalming, Surrey, U.K.

Abstract—The sequences of photophore and pigment development are outlined for *Valenciennellus tripunctulatus* and *Bonapartia pedaliota*. Both species have a protracted metamorphic period in which photophores and pigment are added gradually and sequentially. This contrasts greatly with the mode of development of such genera as *Vinciguerria* and *Cyclothone* in which photophores are pigmented almost simultaneously during a short metamorphic period. The pigmentation of photophores in these two genera is preceded by a post-larval migration to depths occupied diurnally by juveniles and adults. The advantages here would seem that not only are the metamorphic changes achieved rapidly, but also that the newly formed light organs are more likely to be able to match the lower light intensities encountered and thus provide ventral camouflage. Although *Bonapartia* and *Valenciennellus* are not closely related, and structurally their light organs are very different, certain parallels between their respective photophore and pigment developments and their post-larval behaviour can be drawn. As in *Vinciguerria* or *Cyclothone*, these species change during development from an almost transparent post-larva to an adult that relies upon countershading and light organs for camouflage. In a protracted metamorphosis the problem of camouflage is a constantly changing one, but the sequences of photophore and pigment development in *Bonapartia* and *Valenciennelius* appear to be of such design as to overcome this.

INTRODUCTION

Past studies of the family Gonostomatidae (*sensu* Grey, 1960, 1964) have revealed two basic modes of development occurring among its genera (e.g. Sanzo, 1912; Jespersen and Tåning, 1926; Ahlstrom and Counts, 1958; Grey, 1964; Ahlstrom, 1974). In the one, photophores are added gradually and sequentially during the post-larval phase; a process which in some species is relatively prolonged. In the other, photophores are developed during one short phase in life almost simultaneously. Of the twenty genera of gonostomatids (*sensu* Grey, 1960, 1964) currently recognized (Weitzman, 1974), the former mode of development has been noted as occurring in the species of ten (Ahlstrom, 1974). These particular genera may be divided on the basis of their photophore structure and development into two groups, one containing the Maurolicine fishes (*sensu* Ahlstrom, 1974) which includes *Valenciennellus*, and the other the genera *Bonapartia*, *Gonostoma* and *Margrethia* (Ahlstrom, 1974). Considerable variation occurs between these species, however, in the sequences the various photophore series appear, the initial size at metamorphosis, and the size range over which this occurs (Ahlstrom, 1974).

The habits of gonostomatid post-larvae are not particularly well known. Species of *Vinciguerria* and *Cyclothone* are the best documented in terms of development and post-larval behaviour (e.g. Jespersen and Tåning, 1926; Ahlstrom and Counts, 1958; Silas

537

and GEORGE, 1969; GREY, 1964; AHLSTROM, 1974; BOND, 1974). Their post-larvae form light organs almost simultaneously but prior to photophore pigmentation they migrate from the upper 100 m to the depths occupied diurnally by juveniles and adults. During the protracted metamorphosis of *Bonapartia*, *Maurolicus* and *Valenciennellus*, on the other hand, the post-larvae are known to sink gradually with advancing development from depths of about 150 m (JESPERSEN and TÅNING, 1926; KRUEGER, 1972; BOND, 1974; BADCOCK and MERRETT, 1976). The photophores of *Valenciennellus tripunctulatus* (Esmark, 1871) and *Bonapartia pedaliota* Goode and Bean, 1895, are structurally very different. Except for certain photophores, the light organs of *Valenciennellus* are compound in nature, with a common photogenic mass supplying several photophores; whereas those of *Bonapartia* are all separate, individual units. The reviews of GREY (1964) and AHLSTROM (1974) summarize the limited data available on the development of these species but whilst information on the sequential appearance of the various photophore series is provided, the data in general are inadequate to define precisely the sequence of photophore development *within* these series. Moreover, there has been little information on the sequential development of pigment that occurs in these species. In this paper, therefore, the interrelations of photophore and pigment development are shown and these developments are considered in relation to the mode of life.

MATERIALS AND METHODS

This study is based upon collections made by an opening/closing RMT 1 + 8 combination net (BAKER, CLARKE and HARRIS, 1973). One hundred and eighty-eight specimens

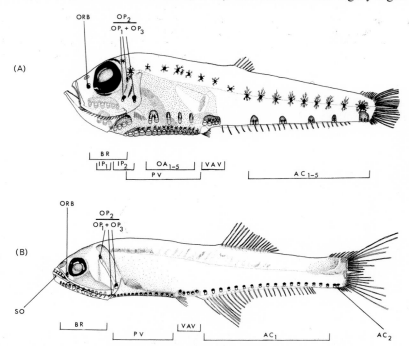

Fig. 1. Photophore nomenclature and disposition in adult (A) *Valenciennellus tripunctulatus*, SL 24.0 mm; (B) *Bonapartia pedaliota*, SL 47.0 mm.

of *Valenciennellus* were examined in detail from a collection taken by RMT 1+8 for vertical distribution studies at 30°N, 23°W (BADCOCK and MERRETT, 1976). *Bonapartia* observations were based upon 45 individuals from 30°N, 23°W and 103 from 32°N, 64°W, caught by RMT 8.

The specimens were fixed in 5–10% saline formol and later transferred to an aqueous storage fluid of 10% propylene glycol, 1% formalin, 0.5% phenoxetol (based on STEEDMAN, 1974). Standard length (SL) of each specimen was measured to the nearest 0.5 mm. Subsequent animal lengths are quoted as standard length.

Photophore counts in *Bonapartia* were made of fully pigmented light organs only since the pre-pigment stage is relatively short. On the other hand, the counts of the largely compound light organs of *Valenciennellus* were considered in relation to three developmental stages: an initial precursor stage where the photophore was present but unpigmented; a second stage during which photophore pigmentation took place; a final phase where development (apart from normal growth in phase with body growth) was completed and the photophore potentially functional. Distinction between the later stages of the second phase and the early stages of the final development was sometimes difficult to maintain.

WEITZMAN (1974) has revised the classification of Stomiatoidei but in this paper, for ease of reference, Gonostomatidae is taken in the sense of GREY (1960; 1964) rather than that of WEITZMAN.

RESULTS

The symbols defining the various photophores are based upon those designated for gonostomatids by GREY (1960, 1964) and AHLSTROM (1974), and their relative positions in the two species summarized in Fig. 1. In all cases except OP_{1-3}, numbering within the respective photophore group has been made in an antero-posterior direction, irrespective of whether that particular photophore is a single unit or part of a compound organ.

Valenciennellus tripunctulatus SL 9.0–30.5 mm

The light organs of *Valenciennellus* are, with the exception of ORB, OP_{1-3} and OA_{3-5}, compound light organs. In 59 individuals with photophores fully developed, counts in the following groups were invariable and as follows: ORB 1; BR, 6; OP, 3; IP_1, 3; IP_2, 4; VAV, 5; OA, 5. The counts of PV varied as 15–18, with frequencies: 15(1), 16(31), 17(17), 18(1). Eight specimens had different counts on each side and are excluded; one count was precluded by damage. With regard to groups AC_{1-5}, counts differed on each side in 26 individuals. Of the remaining 33 specimens counts, with frequencies bracketed, were as follows: AC_1, 2(1), 3(32); AC_2, 3(33); AC_3, 2(1), 3(32); AC_4, 2(31), 3(2); AC_5, 4(33).

Modes of development of compound light organs

The compound light organs contain either a single (BR; IP_1; IP_2; VAV; AC_{1-4}; OA_{1-2}) or double (PV, AC_5) row of photophores. IP_1, IP_2, AC_{1-5} and OA_{1-2} are first discernible as precursor groups bearing the adult complement of unpigmented photophores. Pigmentation, once initiated, is relatively rapid, although consistent in its manner. BR is similarly laid down as a percursor group, but one comprising of only three photophores. These photophores become pigmented, but thereafter further additions are attained by budding in a manner similar to that described for the VAV/AC$_1$ of *Argyripnus atlanticus* Maul, 1952 (BADCOCK and MERRETT, 1972). Photophore additions to the basic light organ are made singly. In many developing specimens a photophore precursor may be seen as a colourless protuberance from the anterior end of the photogenic mass. At the point of protrusion there is a circular white patch in the black pigmentation of the light organ. The photophore precursor becomes pigmented prior to any further additions. VAV, with a group of two photophore precursors, develops in a similar manner, but differs in that photophore protuberances develop from both anterior and posterior ends of the photogenic mass, alternately.

The morphology of PV is relatively complex. In the adult, superficially it normally comprises 16–17 paired photophores in one compound light organ. The anterior end of

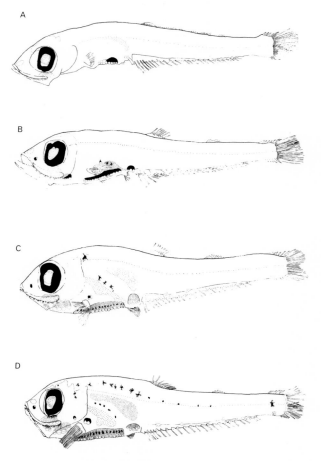

Fig. 2.   Post-larval stages of *Valenciennellus tripunctulatus*. A, early stage 2, SL 10.5 mm; B, early stage 5, SL 12.0 mm; C, stage 7, SL 13.5 mm; D, stage 8, SL 15.0 mm. Precursor units shown unshaded. Day specimens.

the photogenic mass, however, is forked so that the anterior three 'pairs' of photophores arise in two parallel rows of three photophores per branch. The mode of PV development is not clear, but the precursors of the paired photophores are apparently laid down rather than produced by budding (Fig. 2). How the anterior three 'pairs' are developed could not be properly ascertained. Each of the anteriormost 'pairs', however, is derived from a protuberance.

The sequential development of photophores

The sequential development of photophores is summarized in Table 1. For simplicity, animals of like developmental state are grouped in stages, with SL range given in each case. In the following account these stages are utilized as developmental points of reference. The various groups of light organs make an initial appearance in the following order: BR, PV, VAV, ORB, IP_2, IP_1, OA, OP_3, AC_5, AC_1, OP_2, AC_2, OP_1, AC_3, AC_4. Little account, however, of the sequence of photophore formation within the particular groups is presented by Table 1. These sequences are given in Table 2, together with the *earliest* appearance noted of the initial photophores of each group in precursor, partially pigmented, and fully developed form, staged as in Table 1.

No photophores were developed on the smallest individual (9.0 mm). BR was present in the precursor form, but due to abdominal damage the presence or absence of PV precursors could not be ascertained. JESPERSEN (1933), however, recorded a specimen (8.0 mm) in which BR only was present in pigmented form. During PV development usually 1–4 precursors are discernible anterior to the developed photophores (Fig. 2). Photophore development of PV is relatively rapid, all but the ultimate photophore (No. 1) being developed by stage 5 (Table 1). As has been noted above, the ultimate photophore develops by budding. In this context it is perhaps significant that the formation of this photophore is slower than that of the remainder. The initial 15–16 photophores develop during a phase of about 2–3 mm growth; the ultimate develops over a minimum of 2 mm growth (stages 5–10, Table 1).

OA precursors are laid down simultaneously in stage 5, but pigmentation occurs sequentially. OA_1, of the anterior pair, becomes partially pigmented in stage 6; completely so in stage 7 (Fig. 2). The development of OA_2 ensues immediately and is completed during this stage. OA_{3-5} are developed sequentially, being partially pigmented in stage 12 (OA_3, OA_4 then OA_5), and fully developed in stages 12 (OA_3 then OA_4) and 13 (OA_5).

The AC photophore groups do not pigment simultaneously. The precursor groups, however, are all laid down in stages 5–7 (Table 2). AC_{1-4} are laid down either side of the anal fin; AC_5 is a paired compound light organ in the ventral midline of the caudal peduncle (Fig. 2). AC_5 pigmentation starts in stage 8, but only of the anterior three photophore pairs. The ultimate photophore develops in stage 10. AC_{1-4} pigment consecutively being partially pigmented in stages 9 (AC_1), 10 (AC_2), 11 (AC_3) and 13 (AC_4). The process is completed in stages 10 (AC_1), 11 (AC_2), 12 (AC_3) and 14 (AC_4) (Table 2). Thus the development of AC_4 marks the end of photophore appearance (Table 1).

Table 1 shows considerable overlapping of size ranges between various stages, which reflects the natural size variation at any one phase of development. For consistency, observations on pigment development are therefore made with the photophore staging (Table 1) as the reference point.

Table 1.   Valenciennellus tripunctulatus: *photophore development expressed in developmental stages*

Stage	SL range (mm)	BR	PV	VAV	ORB	IP$_2$	IP$_1$	OA	OP	AC$_5$	AC$_1$	AC$_2$	AC$_3$	AC$_4$	No. specimens
0	9.0	.	.	.	.	.	.	.	.	.	.	.	.	.	1
1	8.0*	2	.	.	.	.	.	.	.	.	.	.	.	.	1
2	9.0–11.0	2–4	4–10	.	.	.	.	.	.	.	.	.	.	.	11
3	10.5–12.0	4–5	11–14	2	.	.	.	.	.	.	.	.	.	.	8
4	11.5–12.0	5	14	2–3	1†	.	.	.	.	.	.	.	.	.	2
5	11.0–14.0	4–5	14–16	3–5†	0–1	2–4†	.	.	.	.	.	.	.	.	16
6	14.5	5	(—)	5	1	4	3†	.	.	.	.	.	.	.	1
7	13.5–15.0	5	15–16	5	1	4	3	1–2	.	.	.	.	.	.	15
8	14.5–15.0	5	15	5	1	4	3	2	1	.	.	.	.	.	4
9	15.0–16.0	5	15–16	5	1	4	3	2	0–1	3	.	.	.	.	5
10	15.0–16.0	5	15–17†	5	1	4	3	2	0–2	2–4†	3†	.	.	.	10
11	15.0–17.0	5–6†	15–17	5	1	4	3	2	1–3†	4	3	3†	.	.	15
12	15.0–18.5	5–6	15–17	5	1	4	3	2–4	2–3	4	3	3	3†	.	34
13	17.0–19.0	6	16–17	5	1	4	3	5†	3	4	3	3	3	.	6
14	17.5–30.5	6	15–18	5	1	4	3	5	3	4	3	3	3	2†	59

*From Jespersen (1933).   †Earliest stage of photophore group completion.   (—), No data.

Abdominal and caudal coloration

The abdominal body wall of the adult is thinly developed and relatively transparent. The hypaxial musculature is thickest in the postero-dorsal region and thins as it extends ventrally, such that its minimal development follows the internal diagonal line of the intestine and gonads. Apart from melanophores associated with photophores and a latero-dorsal series (see below), dermal melanophores are absent. The peritoneum of the lateral walls is black and lined externally by a silvering layer (guanine?). The membranes surrounding the swimbladder and gonads are also pigmented, but in the dorsal aspect only. The peritoneal lining of the kidneys is pigmented. The area of the body wall between the gonads and the swimbladder is thus relatively transparent (although the transparency is reduced by the increased development dorsally of hypaxial muscle). The intestine is unpigmented, but both the stomach and oesophagus are black.

The dermal melanophore pattern of the abdominal and caudal regions have been described (BADCOCK, 1969). Single, large (companion) melanophores are associated with each of the photophores of the OA, PV, VAV and AC photophore series; in addition there is a row running the length of the latero-dorsal region (BADCOCK, 1969) (Fig. 1).

The dermal melanophores are developed at a stage later than the internal pigment formation. In the earliest developmental stages at hand (stages 0 and early 2), the animals are colourless apart from the peritoneal section situated postero-dorsally, where the swimbladder lies (Fig. 2). Stomach pigmentation starts later in stage 2—the stage when PV is about half complete—with scattered melanophores spread lightly over its postero-dorsal section. Hereafter, pigmentation spreads anteriorly and ventrally until both stomach and oesophagus are completely pigmented (stage 5). The lateral peritoneal pigmentation does not appear until stages 3–4 when relatively large, individual melanophores are visible, confined to a diagonal line following that of the intestine (Fig. 2). Hereon, lateral pigment is developed more gradually, and coincidentally with the formation of the OA series of

Table 2. *Valenciennellus tripunctulatus: sequence of appearance of pigmented photophores, together with earliest appearance of photophores per series*

Series	Earliest stage for appearance of initial photophore per series			Sequence of appearance of pigmented photophores
	Precursor	Pigmenting	Pigmented	
BR	0	1	1	(6–5) 4, 3, 2, 1
PV	—	—	—	18–15, consecutively to 1
VAV	2	2	3	(3–2) 4, 1, 5
ORB	2?	3	4	1
IP$_2$	3	4	5	3, 2, (4 + 1)
IP$_1$	5	5	6	(3 + 2), 1
OA	5	6	7	1, 2, 3, 4, 5
OP	—	7	8	3, 2, 1
AC$_5$	5	8	9	(1–3), 4
AC$_1$	5	9	10	(1–3)
AC$_2$	5	10	11	(1–3)
AC$_3$	7	11	12	(1–3)
AC$_4$	7	13	14	(1–2)

Dash = no data: parentheses denote simultaneous appearance. Integers in sequence column relate to particular photophores numbered in antero-posterior direction.

photophores. Thus lateral pigmentation is only completed after OA_5 has been developed. The ventro-lateral section of the abdominal cavity posterior to OA_5 is pigmented internally by a dorsal spread of melanophores from the region of the VAV site, also during OA development. Up to the end of OA development (stage 13), the original distinctive, large peritoneal melanophores are still distinguishable. These, however, are absorbed into the black background with the increase in pigment density during the juvenile stages (post-stage 13).

The photophores of adults are superficially obscured by pigment dispersion at night (BADCOCK, 1969), and this power of occlusion is assumed for each photophore as it is developed. Thus, the developed PV photophores are occluded in the smallest night-caught specimens (stage 2), prior to the development of companion melanophores. The companion melanophores of PV are the first to appear, being present during late stage 5 (Fig. 2). Those of VAV are present by stage 10; whilst those associated with OA_{1-2} appear during OA_4 pigmentation.

The first latero-dorsal melanophore lies on the epaxial section above OA_{1-2}. Fourteen to nineteen such melanophores are developed in the adult (BADCOCK, 1969), but 15–16 is more typical. In the typical series, the anterior six or so arch gently, following the dorsal body profile, to a point on a vertical from the posterior margin of the VAV (Fig. 1). Thereafter, the series is straight, with one melanophore usually placed approximately above and one between each AC group. The ultimate melanophore lies dorsal to AC_5. The initial, and anteriormost, dorsal pigment spot appears during late stage 5 (along with the PV companion melanophores). At stages 8–9 the pre-VAV melanophores are more or less all present, so too is the ultimate spot of the series. Post-VAV melanophores develop as minute spots during stages 8–10 (Fig. 2) until the series is completed in the 12th stage.

Valenciennellus has the ability to modify its colour, enabling it to adapt to both diurnal and nocturnal light régimes (BADCOCK, 1969). In the abdominal and caudal regions the principal agents for effecting the change are the dermal melanophores of the dorso-lateral series and those of the photophore series. Their dispersion at night reduces reflectivity, whilst diurnal aggregation enhances the camouflage effectiveness of the silvery peritoneal lining.

Head pigmentation

Melanophores also provide a colour-change mechanism for the head. In the aggregated (daytime) state, the principal sites of pigment occur in adults at the ventral and dorsal margins of the orbit, the inner opercula linings, the brain, and just dorsal to OP_2 where 1–2 large spots exist (Fig. 1) (BADCOCK, 1969). Head pigment develops relatively late. That associated with the ventral margin of the orbit, the brain and OP_2, appear in stages 8–10 (Fig. 2). Pigmentation of the dorsal margin of the eye normally follows immediately (stage 10), and during stage 11 an external streak of opercular pigment between OP_2 and OP_3 is formed. A further superficial spot is added antero-ventrally to OP_3 in stage 12. The pigment of the opercular internal lining starts developing as a diffuse layer providing a backing for the silvery reflective layers of the operculum during stage 12. It becomes fully developed during the juvenile stages (post-stage 13) where it incorporates both opercular and cheek linings. The external head pigment, on the other hand, remains relatively discrete. It is these latter pigment spots that effect the colour change of the head.

Vertical distribution

Valenciennellus tripunctulatus occurs at differing depths and behaves differently in various areas (BADCOCK, 1970; KRUEGER, 1972; CLARKE, 1974; BADCOCK and MERRETT, 1976) but its greatest depth of occurrence is about 550 m. Where depth of capture and animal size have been compared (KRUEGER, 1972; CLARKE, 1974; BADCOCK and MERRETT, 1976) results agree in that animal size increased with increased depth. At 30°N, 23°W the depth distribution spans 100–500 m, although the species principally occupies 200–400 m (BADCOCK and MERRETT, 1976). Post-larvae inhabit the upper reaches of the distribution and with growth progress to the deeper realms where mature adults occur. BADCOCK (1970) and CLARKE (1974) have reported *Valenciennellus* as undertaking short nocturnal migrations into shallower waters, off Fuerteventura (Canary Is.) and Oahu (Hawaiian Is.) respectively. KRUEGER (1972) found no indication of such migrations, however, among the Bermudan population. In addition he found that size/depth relations were maintained at all times. These findings led KRUEGER to suggest that the migration noted by BADCOCK (1970) may be an artefact, due to the presence of post-larvae and juveniles in the shallower night catches. This was not the case. The SOND data were insufficient to interpret distributions in terms of population structure because of inherent contamination problems (BADCOCK, 1970), and size frequency data were not provided for this species. The unpublished data show, however, that the shallower nocturnal captures were predominated by mature adults, reaffirming an apparent migratory habit. The Bermudan population, on the other hand, is not exceptional in being non-migratory. At 30°N, 23°W the population did not make detectable migrations, although the possibility of the species making vertical migrations of small amplitude could not be eliminated (BADCOCK and MERRETT, 1976).

Bonapartia pedaliota SL 10.0–57.0 mm

Photophore development

The usual adult complement of photophores is as follows: SO, 1: ORB, 1: BR, 12: OP, 3; PV, 14; VAV, 5; AC_1, 17; AC_2, 2 (JESPERSEN and TÅNING, 1919; GREY, 1964; this study). The total variation shown by the study specimens is given in Table 3. A sequence of photophore addition in post-larvae was presented by GREY (1964), based on fewer specimens than used here. The present results are tabulated (Table 3) but inclusion of all specimens examined was precluded by damage to them. The order in which the various photophore groups are initiated is in basic agreement with GREY (1964) and as follows: OP + PV, BR, VAV, ORB + AC_1, AC_2 and SO (Table 3). Whether or not OP_3 or the initial PV appear simultaneously cannot be resolved from the present material but AHLSTROM (1974) states that OP_3 develops first, followed by PV. The later development of the initial BR indicated in Table 3 is confirmed by five damaged specimens, 10.0–10.5 mm, in which they are definitely absent.

With the exception of OP, the sequence in which photophores are added within each group is not shown in Table 3. In each case, however, the sequence is almost invariable and a specific photophore is laid down first. The first formed PV is the posteriormost photophore of the series. Thereafter, photophores are added consecutively in an anterior direction along the abdomen (Figs. 3 and 4). The first formed BR lies approximately in the centre of the BR series. The following three photophores are laid down posterior

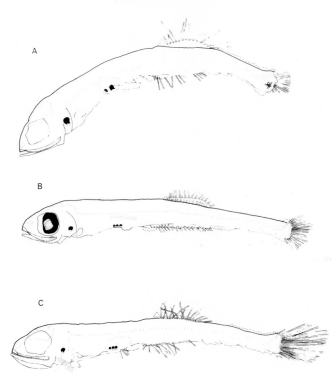

Fig. 3.   Post-larval stages of *Bonapartia pedaliota*. A, stage 2, SL 10.0 mm; B and C, stage 3, SL 12.0 mm.
Precursor units shown unshaded. Night specimens.

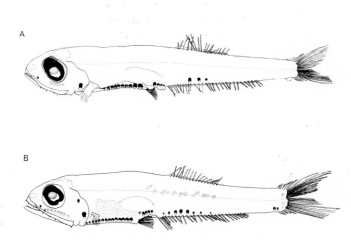

Fig. 4.   Post-larval stages of *Bonapartia pedaliota*. A, stage 5, SL 15.5 mm; B, stage 8, 19.5 mm. Night specimens.

Table 3. Bonapartia pedaliota *photophore stagings*

Stage	SL range (mm)	OP	PV	BR	VAV	ORB	AC$_1$	AC$_2$	SO	No. specimens
0	—	.	.	.	.	.	.	.	.	
1	—	$\frac{0}{0+1}$	.	.	.	.	.	.	.	
2	10.0	$\frac{0}{0+1}$	1	.	.	.	.	.	.	1
3	12.0–12.5	$\frac{0}{0+1}$	2–3	2	.	.	.	.	.	4
4	13.5	$\frac{0}{0+1}$	6	4	2	—	.	.	.	2
5	12.5–15.5	$\frac{0}{0+1}$	6–10	(3)–(6)	2–4	1*	1–4	.	.	14
6	15.0–16.5	$\frac{1}{0+1}$	9–10	5–6	4–(5)	1	3–4	1	.	5
7	15.5–18.5	$\frac{1}{0+1}$	11–12	5–8	(4)–6*	1	4–5	(1)2*	.	18
8	18.0–21.5	$\frac{1}{0+1}$	(12)–14*	7–10	5–6	1	6–13	2(3)	.	22
9	20.5–23.0	3*	13–15	9–10	5	1	12–14	(1)2	.	7
10	24.5–38.5	3	13–15	(11)–13*	5–6	1	16–18*	2	.	22
11	34.5–57.0	3	13–15	11–13	5	1	16–18	2(3)	~1	16

* Adult complement generally attained: parentheses indicate infrequent occurrence.

to this, and the remainder of the series form more or less alternately at anterior and posterior ends. AC$_1$ is initiated prior to AC$_2$ (Table 3). The 6th AC$_1$ photophore, normally lying dorsal to the interspace of the 11th and 12th anal rays, forms first. Subsequent photophore formation follows the number sequence 5, 7, 8, (9), 4, 9, 3, 2, 1 + 10. Thereafter, numbers 11–(16) 17 (18) form consecutively. AC$_2$ develop in the order 2, 1. The initial VAV photophore can be detected as unpigmented and partially pigmented in stage 3 (Table 3, Fig. 3). The development sequence is 2, 3, 4, 5, 1.

Pigment development

In addition to eye and swimbladder peritoneal pigment there is also a caudal spot present in pre-photophore post-larvae (JESPERSEN and TÅNING, 1919). Further pigmentation occurs in stage 5 when the first lateral peritoneal spots are distinguishable. The major developments in internal pigment occur during stages 7 and 8. In the more advanced stage 7 specimens the stomach has begun pigmenting at the dorsal posterior end. Completion of this is attained by an anteriorward progression during stage 8 (Fig. 4). The peritoneal pigment remains relatively lightly developed. During stage 8 the first meningeal spots occur. The first appearance of the latero-dorsal dermal pigment is less easy to pinpoint since it is susceptible to abrasion. The least developed specimens in which its presence was noted were in stage 7.

Throughout the remaining stages of development pigment already laid down is gradually intensified. Head pigmentation (meningeal pigment apart) does not occur until stage 10.

Table 4.   Bonapartia pedaliota: *vertical
distribution and size of animal by night off Bermuda*

Depth (m)	SL range (mm)	Average SL (mm)	No. specimens
102–200	12.0–21.5	18.0	28
205–300	10.0–46.5	20.7	42
295–400	22.0–39.5	31.4	14
405–505	42.0–43.5	42.7	2
505–600	47.0–57.0	52.5	5
605–700	16.0	16.0	1

Opercular pigment is laid down first, followed by that of the mouth lining, the isthmus, the gill arches and finally the branchiostegal membranes. The latter occurs only in individuals greater than 32 mm.

KRUEGER and BOND (1972) have reported diel colour changes in *Bonapartia*, with the dorsum being especially darkened in night specimens. Abrasion of the individuals of this study precludes accurate assessment of the colour change, but certainly melanophores appear more dispersed along the dorsum and in the caudal region in nocturnally caught specimens.

Vertical distribution

At 30°N, 23°W *Bonapartia* occurred at 100–500 m depth (BADCOCK and MERRETT, 1976) whilst off Bermuda (32°N, 64°W) it occurs in 100–700 m depth (KRUEGER and BOND, 1972; BOND, 1974; IOS unpublished data). The Bermudan specimens taken by RMT 8 were considerably less in number, and represented by a smaller size range, by day than by night, despite day and night fishing being equal in effort. Only slight, if any, diel vertical migration seems likely. The night samples taken off Bermuda provide the best insight into the intra-specific depth relations, and the trends indicated by them are also implied in the small day collection and in the 30°N, 23°W collection. Table 4 shows the relation of size of animal to depth of capture. Whilst size of animal does not accurately reflect developmental state, it is a reasonable indicator. Thus a gradual, rather than sudden, downwards migration accompanies the developmental advancement of the individual. In general these observations concur with BOND's (1974) Bermudan results.

DISCUSSION

The adult characters of *Valenciennellus* and *Bonapartia* are acquired gradually by individuals, without a rapid metamorphosis. The structural and morphological changes that occur over this protracted metamorphic period, however, are great and the adults and post-larvae are adapted in different ways to their respective environments. The two species are typical upper mesopelagic forms and show a marked relationship between size of animal and the depth normally inhabited. In general, mesopelagic fishes are camouflaged in several ways, for example, by the utilization of such devices as transparency, countershading, lateral compression (see review, DENTON, 1970) and possibly photophores (e.g. CLARKE, 1963; DENTON, GILPIN-BROWN and WRIGHT, 1972). Larval and post-larval forms tend to transparency, whilst adults are often silvery. In adults, the platelets constituting the reflective layers are arranged and orientated such that light is differentially reflected to best advantage (e.g. DENTON, 1970). Nevertheless a system of this nature can only function satisfactorily under conditions where the ambient light orientation is constant. In the upper mesopelagic region, during daylight, such conditions are satisfied. The juveniles and adults of *Valenciennellus* and *Bonapartia* are appropriately equipped for this environment, and furthermore have the capacity to reduce their bodily reflectivity by pigment dispersion (BADCOCK, 1969; KRUEGER and BOND, 1972) thus providing for adaptation to a nocturnal light régime predominated by bioluminescence (point sources of light). The pre-photophore post-larva in each of these species is also well adapted, being laterally compressed, relatively transparent, with body pigment particularly restricted. The eye, with a silvery iris, is more elongate than in the adult, which reduces the area of ventral shadow (Figs. 2–4). The pigmented peritoneal section of the swimbladder is characteristic of post-larvae of the direct development type gonostomatids (*sensu* GREY, 1964) but it also occurs in some gonostomatids that ultimately develop photophores simultaneously. Thus it is found in *Cyclothone* spp. and *Vinciguerria attenuata* (Cocco, 1838), although oddly not in other *Vinciguerria* spp. (JESPERSEN and TÅNING, 1926; AHLSTROM and COUNTS, 1958; SILAS and GEORGE, 1969). Similar swimbladder peritoneal pigment is found in certain myctophid larvae (TÅNING, 1918). In the natural environment of the post-larva during daytime this pigment would be visible from the ventral aspect. On the other hand, were the swimbladder not pigmented so, its refractive and reflective properties might render its owner even more conspicuous.

With advancing development from the pre-photophore form the body becomes more substantial and less transparent, increasing the problem of camouflage. Although *Valenciennellus* and *Bonapartia* are not closely related [WEITZMAN (1974) places them in separate families] the development of their respective post-larvae through the protracted period of gradual, yet extensive, change is remarkably similar, suggesting that such parallels have been maintained by the action of common selective pressures. The most striking feature common to both species is the close relation apparent between the development of photophores and that of pigment. Among the earliest formed photophores are those of the BR, PV and VAV series. Furthermore, the first formed photophores of these series are located directly ventral to the eye (BR) and swimbladder peritoneal pigment (PV and VAV) (Figs. 2–4). Later additions to the basic pigment in the preanal region are made in conjunction with the development of the photophores, such that there is always a photophore lying ventral to newly formed body pigment. It is also worth noting that pigmentation of the stomach starts at the posterior end, dorsal to the best developed PV, and develops anteriorwards.

Parallels with other species with protracted metamorphic periods may be drawn. In *Gonstoma elongatum* Günther, 1878, *Margrethia obtusirostra* JESPERSEN and TÅNING, 1919, and *Maurolicus muelleri* (GMELIN, 1788) photophore development similarly first occurs at relatively small animal sizes and the data available (JESPERSEN and TÅNING, 1926; GREY, 1964; AHLSTROM, 1974) indicate that, as in *Bonapartia* and *Valenciennellus*, the sequences of photophore and pigment developments are complementary.

The feeding habits of *Bonapartia* are unknown (GREY, 1964), but a recent study (MERRETT and ROE, 1974) showed that *Valenciennellus* of 11–19 mm feed selectively, primarily upon luminescent calanoid copepods of the genus *Pleuromamma* Giesbrecht, 1898, and to a lesser extent upon the luminescent ostracod *Conchoecia curta* Lubbock, 1860. Furthermore, their evidence showed that feeding occurred only by day; the only time, in fact, when the distributions of predator and prey coincided. Although prey and predator size were not compared in this study (MERRETT and ROE, 1974) the data provided are sufficient to corroborate the interpretation of others (e.g. McALLISTER, 1967) of regarding stomach pigmentation as an adaptation to prevent the unwanted transmission of prey luminescence through the stomach wall. It is also pertinent, in this context, to observe that this pigment develops first in the dorsal aspect of the stomach and progresses anteriorly and ventrally (p. 543) and that the early formed lateral peritoneal pigment follows the line of the intestine (Fig. 2). The pigmentation of the *Bonapartia* stomach conforms with the pattern above.

Nocturnal pigment dispersion is more easily seen in *Valenciennellus* since the dermal melanophores, the prime effectors of the colour change system, are more discrete and readily identifiable. During post-larval development not only are these added sequentially, but the sequence is such that they provide a partial (diel) colour change which improves in effectiveness with advancing development. As a result, dorsal and lateral camouflage is maintained during the post-larval period of change. It is in the ventral aspect that the major problem of camouflage arises.

DENTON, GILPIN-BROWN and WRIGHT (1962) have shown the photophores of *Argyropelecus* and *Chauliodus* to be well adapted to fulfil the function of maintaining ventral camouflage as postulated by CLARKE (1963). The functions attributed to the light organs of deep-sea fishes are many (see reviews, McALLISTER, 1967; NICOL, 1967) but it is an attractive proposition that under suitable conditions of ambient illumination *Valenciennellus* and *Bonapartia* should utilize their photophores primarily as camouflage agents, especially in view of their respective photophore dispositions and the habitat they normally occupy. In the post-larvae the juxtaposition of pigment and photophores restricts the ventral shadowing inherently caused to the areas of photophore development. Since the photophores and pigment are added gradually during the protracted metamorphic stage it seems imperative that each photophore should become functional as it develops, as has been speculated to be the case in *Argyripnus* (BADCOCK and MERRETT, 1972). Some evidence for this has been obtained with *Bonapartia*. Using hydrogen peroxide, luminescence has been induced in the photophores of the incomplete BR series in post-larvae (P. J. HERRING, personal communication). The photophores of other series could not, however, be made to luminesce and similar treatment of adults gave identical results (P. J. HERRING, personal communication). In view of the partial success in eliciting luminescence in both adults and post-larvae, one must conclude that either the peroxide method is not altogether suitable in this case or else the animals used were not in good enough condition. However, it is likely that all the photophores become operational as they are developed.

With luminescence observations lacking for *Valenciennellus*, the question remains as to whether developed photophores luminesce before the completion of the whole group. Since the ability of occlusion is developed immediately after each photophore is pigmented (p. 544) and, in the case of BR, PV, VAV and AC_5, prior to the completion of the ensuing photophore, it seems likely that this is indeed the case.

It is now worth considering the daytime light régime encountered by these species. KAMPA (1970) made detailed underwater daylight measurements at a position (28°07′N, 16°22′W) near to where the eastern North Atlantic specimens of this study were collected

(30°N, 23°W). Her measurements were made under conditions of a calm sea and 0.25 cumulus cloud cover (KAMPA, 1970). Between 190 m and 615 m depth the wavelength of maximum transmission shifted from about 470 nm to 474 nm; whilst irradiance values at these wavelengths decreased from $6 \times 10^{-1}\,\mu W/cm^2/nm$ at 190 nm depth to $4.7 \times 10^{-4}\,\mu W/cm^2/nm$ at 480 m (KAMPA, 1970). The post-larvae, then, inhabit very considerably more illuminated waters than do juveniles and adults. What intensity measurements of fish photophores are available are probably conservative estimates (NICOL, 1967; DENTON, GILPIN-BROWN and WRIGHT, 1972), but even so it is unlikely that the post-larvae can fully match the daylight intensities of downwelling light in their environment. Even if the photophores do become functional upon development, as seems likely, of what value would they be to the post-larva, in the context of ventral camouflaging? *Valenciennellus* and *Bonapartia* are at the most very limited vertical migrators. It is feasible, then, that the post-larval photophores could realize a camouflaging capacity over a limited range during the rapid changes in light intensity of dusk and dawn periods—times when large numbers of potential predators migrate through their ranks. In addition, since the post-larvae sink slowly with advancing development, the difference between the ambient and the photophore light intensities should decrease with increased animal size and the ventral shadowing, therefore, become less conspicuous.

The species of *Vinciguerria* and *Cyclothone* have been fairly comprehensively documented in terms of development and post-larval vertical distribution (e.g. JESPERSEN and TÅNING, 1926; AHLSTROM and COUNTS, 1958; SILAS and GEORGE, 1969; AHLSTROM, 1974; BOND, 1974). The metamorphic changes occurring in these genera and, for example, *Bonapartia* are just as great, yet in the former group metamorphosis is relatively sudden and short: a near simultaneous pigmentation of photophores is preceded by a post-larval migration to much greater depths. This seems advantageous since, in the realms of lower ambient light intensities, the photophores are more likely to be able to match the downwelling light. The problems of transitional adaptation associated with a protracted metamorphosis, then, are of a lesser consequence. Nevertheless, the developmental sequences occurring in *Bonapartia* and *Valenciennellus, Gonostoma elongatum, Margrethia* and *Maurolicus* appear highly adaptive, affording to these species an effective camouflage throughout the post-larval metamorphic period.

Acknowledgements—I would like to thank the following all of whom read the manuscript and provided useful comment and discussion: Dr. E. H. AHLSTROM (National Marine Fisheries Service, La Jolla), P. M. DAVID, Dr. P. J. HERRING (I.O.S. Wormley), Professor N. B. MARSHALL (Queen Mary College, London) and N. R. MERRETT (I.O.S.). Thanks also go to Miss R. LARCOMBE who drew the figures.

REFERENCES

AHLSTROM E. H. (1974) The diverse patterns of metamorphosis in gonostomatid fishes—an aid to classification. In: *The Early Life of Fishes*, J. H. S. BLAXTER, editor, Springer-Verlag, New York, pp. 659–674.

AHLSTROM E. H. and R. C. COUNTS (1958) Development and distribution of *Vinciguerria lucetia* and related species in the Eastern Pacific. *U.S. Fish and Wildlife Service, Fishery Bulletin*, **58**, 363–416.

BADCOCK J. (1969) Colour variation in two mesopelagic fishes and its correlation with ambient light conditions. *Nature*, **221**, 283–285.

BADCOCK J. (1970) The vertical distribution of mesopelagic fishes collected on the SOND Cruise. *Journal of the Marine Biological Association of the United Kingdom*, **50**, 1001–1044.

BADCOCK J. and N. R. MERRETT (1972) On *Argyripnus atlanticus*, Maul 1952 (Pisces, Stomiatoidei) with a description of post-larval forms. *Journal of Fish Biology*, **4**, 277–287.

BADCOCK J. and N. R. MERRETT (1976) The vertical distribution and associated biology of midwater fishes in the eastern North Atlantic. I. In 30°N: 23°W, with developmental notes on certain myctophids. *Progress in Oceanography*, **7**, 3–58.

BAKER A. DE C., M. R. CLARKE and M. J. HARRIS (1973) The NIO combination net (RMT 1 + 8) and further developments of rectangular midwater trawls. *Journal of the. Marine Biological Association of the United Kingdom*, **53**, 167–184.

BOND G. W. (1974) Vertical distribution and life histories of the gonostomatid fishes (Pisces, Gonostomatidae) off Bermuda. *Report to the U.S. Navy Underwater Systems Center*, contract No. NOO140-73-C-6304: 1-276.

CLARKE T. A. (1974) Some aspects of the ecology of stomiatoid fishes in the Pacific Ocean near Hawaii. *Fishery Bulletin, United States*, **72**, 337–351.

CLARKE W. D. (1963) Functions of bioluminescence in mesopelagic organisms. *Nature*, **198**, 1244–1246.

DENTON E. J. (1970) On the organisation of reflecting surfaces in some marine animals. *Philosophical Transactions of the Royal Society of London*, Series B, **258**, 285–313.

DENTON E. J., J. B. GILPIN-BROWN and P. G. WRIGHT (1972) The angular distribution of the light produced by some mesopelagic fish in relation to their camouflage. *Proceedings of the Royal Society of London*, Series B, **182**, 145–158.

GREY M. (1960) A preliminary review of the family Gonostomatidae, with a key to the genera and a description of a new species from the tropical Pacific. *Bulletin of the Museum of Comparative Zoology, Harvard College*, **122**, 57–125.

GREY M. (1964) Family Gonostomatidae. In: *Fishes of the western North Atlantic*, Y. H. OLSEN, editor, Memoir. Sears Foundation for Marine Research, Number 1, part 4, pp. 78–240.

JESPERSEN P. (1933) *Valenciennellus tripunctulatus. Faune Ichthyologique de l'Atlantique Nord*, Number **14**, 80–81.

JESPERSEN P. and A. V. TÅNING (1919) Some Mediterranean and Atlantic Sternoptychidae. Preliminary note. *Videnskabelige Meddelelser fra Dansk naturhistorik Forening i Kjobenhavn*, **70**, 215–226.

JESPERSEN P. and A. V. TÅNING (1926) Mediterranean Sternoptychidae. *Report on the Danish Oceanographic Expeditions, 1908–1910 to the Mediterranean and adjacent seas*, **2**, Biology (A 12), pp. 1–59.

KAMPA E. M. (1970) Underwater daylight and moonlight measurements in the eastern North Atlantic. *Journal of the Marine Biological Association of the United Kingdom*, **50**, 397–420.

KRUEGER W. H. (1972) Biological studies of the Bermuda Ocean Acre. IV. Life History, vertical distribution and sound scattering in the gonostomatid fish *Valenciennellus tripunctulatus* (Esmark). *Report to the U.S. Navy Underwater Systems Center, Washington*, Contract No. NOO140-72-0315: 1-37.

KRUEGER W. H. and G. W. BOND (1972) Biological studies of the Bermuda Ocean Acre. III. Vertical distribution and ecology of the bristlemouth fishes (family Gonostomatidae). *Report to the U.S. Navy Underwater Systems Center, Washington*, Contract No. NOO140-72-0315: 1–50.

MCALLISTER D. E. (1967) The significance of ventral bioluminescence in fishes. *Journal of the Fisheries Research Board of Canada*, **24**, 537–554.

MERRETT N. R. and H. S. J. ROE (1974) Patterns and selectivity in the feeding of certain mesopelagic fishes. Marine Biology, **28**, 115–126.

NICOL J. A. C. (1967) The luminescence of fishes. *Symposium of the Zoological Society of London* No. 19, 27–55.

SANZO L. (1912) Comparsa degli organi luminosa in una serie di larve di *Gonostoma denudatum* Raf. *Report: Comitato Talassographico Italiana*, Memoria 9: 23 pp.

SILAS E. G. and K. C. GEORGE (1969) On the larval and post larval development and distribution of the mesopelagic fish *Vinciguerria nimbaria* (Jordan and Williams) (family Gonostomatidae) off the west coast of India and the Laccadive Sea. *Journal of the Marine Biological Association of India*, **11** (1 & 2), 218–250.

STEEDMAN H. F. (1974) Laboratory methods in the study of marine zooplankton. A summary report on the results of Joint Working Group 23 of SCOR and UNESCO, 1968–1972. *Journal du Conseil. Conseil permanent international pour l'Exploration de la mer*, **35**, 351–358.

TÅNING A. V. (1918) Mediterranean Scopelidae (*Sauras, Aulopus, Chlorophthalmus* and *Myctophum*). *Report on the Danish Oceanographical Expeditions, 1908–10, to the Mediterranean and adjacent seas*, Biology 2(A7), 1–154.

WEITZMAN S. H. (1974) Osteology and evolutionary relationships of the Sternoptychidae, with a new classification of stomiatoid families. *Bulletin of the American Museum of Natural History*, **153**(3), 329–478.

Oral light organs in *Sternoptyx*, with some observations on the bioluminescence of hatchet-fishes

Peter J. Herring

Institute of Oceanographic Sciences, Wormley, Godalming, Surrey, England

INTRODUCTION

THE light organs in the hatchet-fish genera *Argyropelecus*, *Polyipnus* and *Sternoptyx* are similarly distributed, and predominantly ventrally located and directed (BAIRD, 1971). Their function is generally regarded as one of ventral camouflage, and in this respect their efficiency in *Argyropelecus olfersi* (Cuvier) has been elegantly demonstrated by the anatomical and physiological studies of DENTON and his colleagues (DENTON, 1970; DENTON, GILPIN-BROWN and ROBERTS 1969; DENTON, GILPIN-BROWN and WRIGHT, 1970, 1972). The only light organs not ventrally directed are the pre-orbital organs of *Argyropelecus* and *Polyipnus*, which shine into the eyes. Among other possible functions it has been suggested that these organs may act as reference standards enabling the fish to monitor its own output in relation to the down-dwelling daylight (NICOL, 1962; cf. LAWRY, 1974). There is some ambiguity in the literature concerning the presence of a pre-orbital organ in *Sternoptyx*. GARMAN (1899) did not include a pre-orbital organ in his description of *S. obscura* but noted that "the silver area below the orbit contains indications of a lantern". BRAUER (1908) described an orbital organ in *S. diaphana*, but this is posterior in position, and, as he noted, probably equivalent to the dorsal pre-opercular organ of *Argyropelecus*. SCHULTZ (1961) and BAIRD (1971) both figure an anterior pre-orbital organ, though the latter author states that the pre-orbital is "ventrally located in *Sternoptyx*".

This paper describes a pair of light organs present in all species of *Sternoptyx*, which may be analogous to the pre-orbital organs in *Argyropelecus* and *Polyipnus*. It also reports some observations on the luminescence of hatchet fishes which relate to the hypotheses noted above.

MATERIALS AND METHODS

Animals were obtained mainly during cruises of R.R.S. *Discovery* in the eastern North Atlantic in 1974. Additional specimens of *Sternoptyx* were examined on board R/V *Alpha Helix* during the 1975 South East Asia Bioluminescence Expedition in the Banda and Halmahera Seas. Material for microscopy was fixed in 5% glutaraldehyde in 0.1 M phosphate buffer pH 7.4, post-fixed in 1% OsO_4 in the same buffer, dehydrated in acetone and embedded in Araldite. Sections were cut on a Huxley microtome using glass knives.

Bioluminescence was monitored with an EMI 6097B photomultiplier (S11 response) and recorded on an S.E. Laboratories S.E. 3006 ultra-violet recorder. Angular distribution measurements were made as described by HERRING (1976), with an acceptance angle of 21°.

RESULTS

Sternoptyx

Observations on freshly caught specimens of *S. diaphana* Hermann, *S. obscura* Garman and *S. pseudobscura* Baird have shown that this genus is alone among the hatchet-fishes in having a greatly enlarged light organ at the antero-ventral margin of each orbit. These organs are most easily seen not externally but from within the mouth. They appear as two elliptical patches, covering quite a wide area of the roof of the mouth (Fig. 1). The organs have no pigment surrounding them, no reflectors, and no coloured 'filters', each of which are found in other light organs. In fresh animals the organs are sometimes spontaneously luminous, whether or not the ventral series of organs are also luminescing, or they may luminesce if mechanically stimulated. In either event each organ glows steadily for periods of up to 30 minutes, fading gradually. In non-luminescing specimens the organ can always be induced to luminesce by local application of 1% hydrogen peroxide in seawater. The luminescence is readily visible in the open mouth, and some feeble light is occasionally visible through the cheek. Although in adults the oral light organs (thus designated in order to describe their effective site) are not easily seen from the exterior, they are much more obvious in the less pigmented juveniles.

Histology of the oral organ of *S. diaphana* shows it to be made up largely of a closely packed mass of cells, polygonal in outline, each containing a large number of granules or spherules of variable sizes and staining densities (Fig. 2). This appearance is almost identical to that of one type of cell which has been described and illustrated by BRAUER (1908) and NUSBAUM-HILAROWICZ (1920) in other light organs of the same species. Sections of the photogenic tissue of the composite abdominal organs of the same specimen of *S. diaphana* showed a rather different general appearance, with much more of the tissue being composed of secretory vesicles or droplets, a feature also noted by the earlier workers. Such extracellular secretory droplets were rare in sections of the oral organs. Another difference between the photogenic tissue of the oral and abdominal organs, readily observed in fresh specimens, is their fluorescence. The tissues of the former are not fluorescent in long wavelength ultraviolet light, whereas those of the latter have a bright blue-green fluorescence. A similar blue-green fluorescence has been observed in the abdominal organs of *Argyropelecus* species and *Polyipnus polli* Schultz, though it is not usually as intense as that seen in *Sternoptyx*.

The photogenic tissue of the abdominal photophores of both *Sternoptyx* and *Argyropelecus* luminesces when homogenized in seawater. Subsequent addition of 1% H_2O_2 to the homogenate intensifies the luminescence and a further addition of ferrous iron to 10^{-3} M induces a brief flash, after which no further luminescence can be elicited. This chemiluminescent response also occurs in searsiid fishes (HERRING, 1972) and should not be regarded as indicative of any particular biochemical pathway for normal luminescence. No data are available on the detailed biochemistry of hatchet-fish luminescence, though TSUJI and HANEDA (1971) have reported that cell-free extracts of the ventral

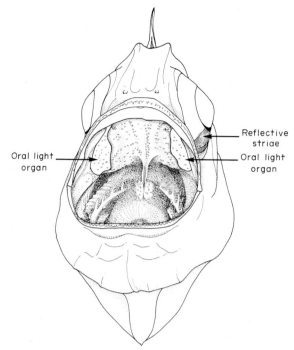

Fig. 1. (A)   Anterior view of the head of *Sternoptyx diaphana* showing the positions of the oral organs and the reflective striae on the iris.

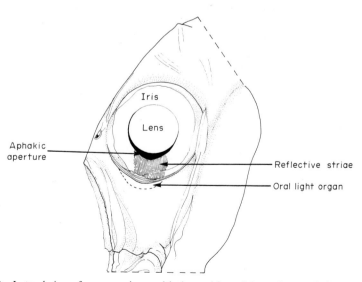

Fig. 1. (B)   Lateral view of same specimen with the position of the oral organ indicated in outline.

25μm

(A)

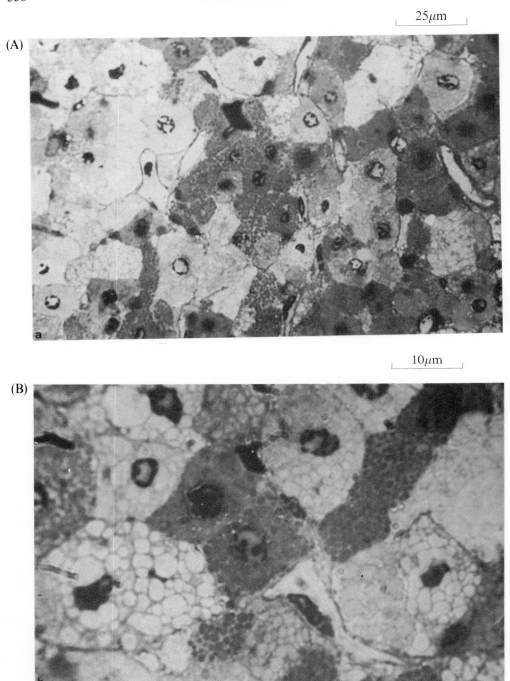

10μm

(B)

Fig. 2. Transverse sections of the oral organ of *S. diaphana*. (A) General view showing the great variety of staining densities in different cells. (B) Higher magnification of a group of cells, some with dense vesicles and others with apparently 'empty' vesicles. Araldite sections, 0.5 μm stained with toluidine blue.

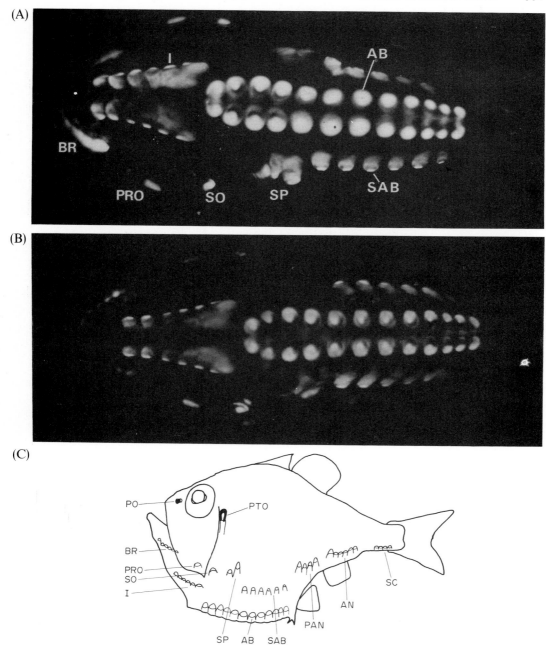

Fig. 3.   Photograph of two specimens of *Argyropelecus olfersi* in ventral view by their own luminescence. (A) Photographed with GAF 500 film and (B) with Ektachrome X (both approximately 15 minutes at f2.8). In neither case are the pre-anal (PAN), anal (AN) or sub-caudal (SC) organs visible. The supra-abdominal organs (SAB) in particular show the bright dorsal spot at the aperture of the photogenic tissue and the ventral diffusion of the light over the reflector surface. Also visible are the abdominal (AB) supra-pectoral (SP) sub-opercular (SO) pre-opercular (PRO) isthmus (I) and branchiostegal (BR) organs. (C) Diagrammatic view of the light organ arrangement in *A. olfersi* (nomenclature following BAIRD, 1971).

photophores of both *Polyipnus spinosus* Günther and *Sternoptyx diaphana* cross-react with the luciferin and luciferase of the ostracod *Cypridina*. They further report that extracts of the myctophid *Diaphus elucens* (Brauer) have similar properties.

Argyropelecus olfersi

Non-luminescing moribund specimens freshly removed from the trawl, either by day or by night, on a number of occasions subsequently showed spontaneous luminescence. In these specimens there was typically a rapid rise to maximum intensity over a period of 30–60 seconds, followed by a steady luminescence for several minutes and then a slow decline in intensity, often extending over at least 30 minutes (at approximately 21°C). All the light organs in a particular specimen responded together, including the pre-orbital organ, though the various groups were not always in precise synchrony. Differences in synchrony became most noticeable in the time of final extinction. Isolated organs, or groups of organs, continued to luminesce when removed from the animal.

Specimens frozen at −60°C while luminescing continued to do so when thawed several days later, though at a much lower intensity. Two specimens similarly thawed after a period of 2 months at −60° failed to luminesce.

A luminescing organ of, for example, the supra-abdominal series emits a diffuse blue light over the whole outer surface of the wedge-shaped reflector unit (DENTON, 1970). This appearance changes little over a wide range of angles of observation, though the intensity of emission changes substantially at different angles (DENTON, GILPIN-BROWN and WRIGHT, 1972). A brighter point of light at the dorsal margin of the reflector unit is visible particularly in ventral view (Fig. 3) and corresponds to the aperture of the photogenic tissue capsule.

The luminescence of an intact animal is much brighter in ventral than in lateral view, but nevertheless still has a significant lateral component. A weak glow remains visible from many of the ventrally directed organs when observed at right angles to their axes. This was particularly apparent in those with the largest reflector surfaces, e.g. the supra-abdominal and supra-pectoral organs. BERTELSEN and GRONTVED (1949) succeeded in photographing the lateral emission from the abdominal series in *A. olfersi*. In many, but not all, specimens the brightest organ in lateral view was the post-orbital organ, whose lateral luminescence was often very bright in comparison to the feeble lateral emission of the ventral organs. BERTELSEN and GRONTVED (1949) also noted the brightness of this organ. Its lateral emission was visible not so much from the reflector region as from the keyhole-like aperture in the dark pigment surrounding the photogenic tissue itself. There is some silvering over this region but light is readily emitted through it (Fig. 4f-i).

DENTON (1970) has postulated that the half-silvering on the outer surface of the ventral photophores of *Argyropelecus* determines, by multiple reflection, the angular distribution of emitted light. This angular distribution matches that of downwelling daylight (DENTON, GILPIN-BROWN and WRIGHT, 1972). An experimental test of the effect of the half-silvered surface was possible utilizing the continued luminescence of isolated organs. The external surface of the supra-abdominal and supra-pectoral organs can be readily peeled from the organs, without causing any other apparent damage. These groups of organs were therefore cut from the flank of a fish within a rectangle of body wall. Although any adhering droplets of water could act as lenses and might affect the angular distribution of emitted

(a)

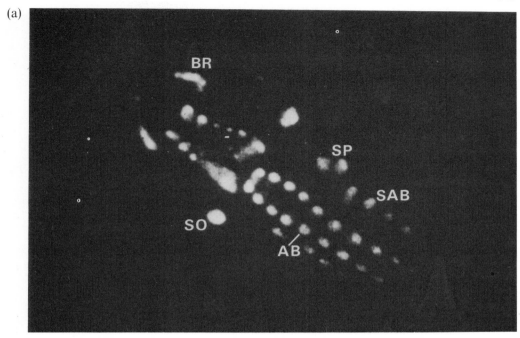

(b)

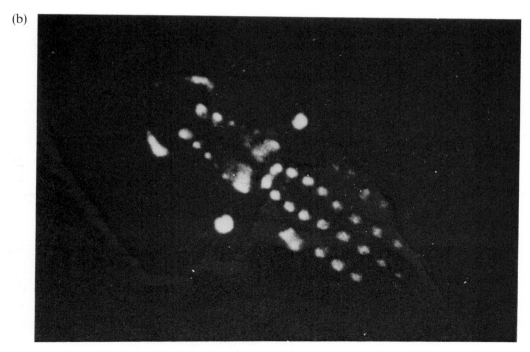

Fig. 4.   Series of image intensified pictures of luminescing *A. olfersi*. The animal was rotated about its antero-posterior axis from (a) in ventral view to (i) in lateral view. Compare (a) and (b) with Fig. 3 and note how in lateral view only the post-orbital (PTO) and sub-opercular organs (SO) are visible.

(c)

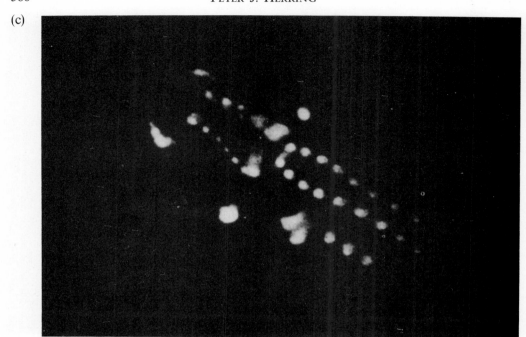

(d)

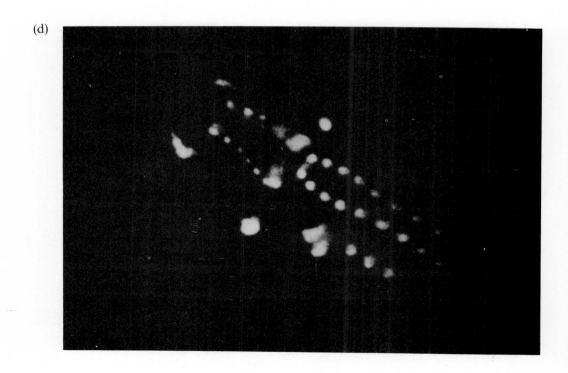

(e)

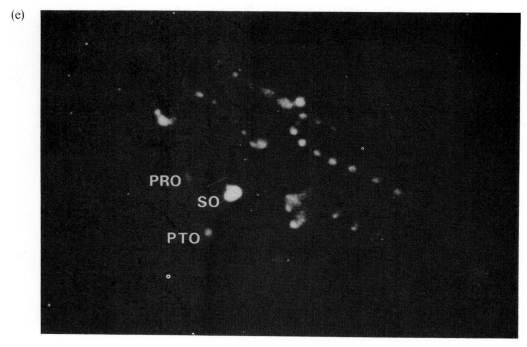

(f)

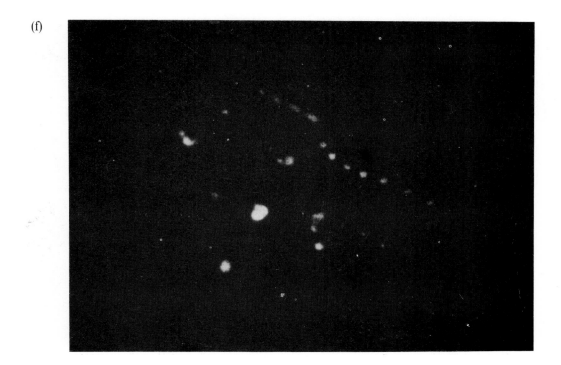

(g)

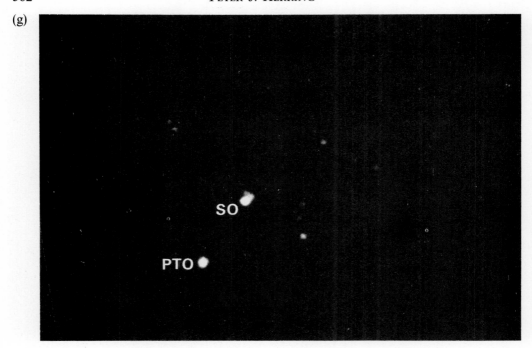

(h)

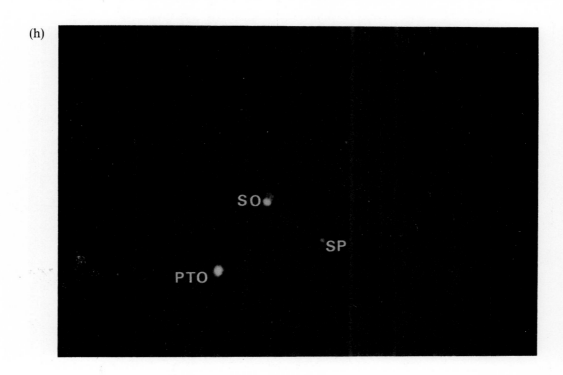

(i)

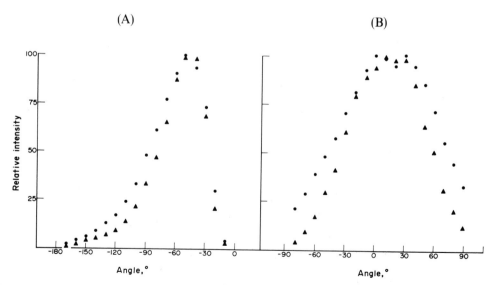

Fig. 5. Angular distribution of light from (A) isolated and (B) *in situ* supra-pectoral and supra-abdominal organs before (▲) and after (●) removal of the surface reflector layer. Rotation was from ventral view (0°) to one or other side. In (A) the peak emission occurs at some 45° from the ventral direction. This is probably caused both by the isolated preparation no longer conforming to the slight curve of the fish's flank and by the arrangement of the preparation itself partially masking the photomultiplier in near ventral view. The double-peaked nature of the curves in (B) reflects the absence of the abdominal organs and the consequent separation of the contributions of the supra-abdominal organs of each side.

light, no such droplets were observed on the preparation. The angular distribution of light from this preparation was then measured in air before and after the surface reflector was removed. The visible effects are quite dramatic, in that the emitted light from the peeled organ no longer appears as a diffuse source spread over the reflector region, but rather as a bright line of light concentrated at the level of the 'filter' pigment in the aperture of the photogenic region. A larger illuminated area is seen only when the organ is viewed from the ventral region. At this angle some light is reflected glancingly off the internal reflector surface. The measured change in angular distribution (Fig. 5A) demonstrates how the half-silvered outer surface serves to reduce the lateral component of emitted light. Similar angular measurements were made with the organs *in situ*: light emission from whole animals

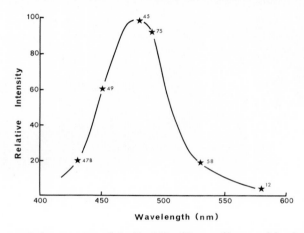

Fig. 6. Bioluminescence emission spectrum of *A. olfersi*. The Wratten filters used for the determination of each point are indicated on the curve.

was restricted to the supra-pectoral and supra-abdominal groups by masking all other light organs, except for the abdominal ones which were excised. These measurements also demonstrate an increased lateral component in the absence of the surface reflectors (Fig. 5B) with the effect being greatest at right angles to the axes of the organs.

Emission spectra

The emitted light of *Sternoptyx, Argyropelecus* and *Polyipnus* species is clear blue to the dark-adapted eye, and attempts were made to determine the emission spectrum of the light from *A. olfersi* using such filters as were available on board ship. The percentage transmission of the bioluminescence through Wratten filters 12, 45, 47B, 49, 58 and 75 was determined on six specimens and gave mean values of 2.5, 2.6, 7.5, 11.5, 7.5 and 12.5%, respectively. These results, corrected for the transmission of each filter and the quantum efficiency of the photomultiplier tube, were used to construct the emission curve in Fig. 6. The maximum emission at 470–475 nm is similar to that observed in many other luminous marine animals. More accurate measurements of the spectral emission from the oral organs of two specimens of *Sternoptyx diaphana* were obtained with an online spectrofluorimeter system WAMPLER and DeSA, 1971) and yielded corrected emission maxima at 473 and 475 nm respectively (J. E. WAMPLER, personal communication).

DISCUSSION

The spectral emission of the bioluminescence of *A. olfersi* is that which would be anticipated from the work of DENTON, GILPIN-BROWN and WRIGHT (1970) who showed that the magenta 'filter' pigment through which the light must pass has a 'window' in the range 460–480 nm. This filter pigment is present in all three hatchet-fish genera, though its distribution in the photophores of *Sternoptyx* differs from that in *Argyropelecus* and *Polyipnus*. In *Sternoptyx* it is present not only in the aperture of the photogenic tissue but may also extend in a narrow thread down the long axis of the reflector region of the organ (described as a 'Strang' by BRAUER, 1908). In the other two genera it is present only as a small block of cells in the aperture of the organ (the B cells of BASSOT, 1966). The filter pigment is similar in appearance and position to that in many other fishes. It is, however, not fluorescent. In this respect it differs from that in *Valenciennellus* (Sternoptychidae), *Ichthyococcus* (Photichthyidae *sensu* WEITZMANN, 1974) and many, if not all, Stomiatoidea, in all of which it has a deep red fluorescence (HERRING, unpublished).

The pre-orbital organs of *Argyropelecus* and *Polyipnus* are similar to the other light organs in construction, but lack reflectors. In *Sternoptyx* the oral organ lacks not only a reflector but all other accessory structures, and has a proportionally larger volume of photogenic tissue. Its homology with the pre-orbital organs of *Argyropelecus* and *Polyipnus* must be doubtful.

The depth distribution of *Sternoptyx* species is very different from that of *Argyropelecus* and *Polyipnus*. *Polyipnus* is a relatively shallow-living species (BAIRD, 1971) probably most abundant in the upper 500 m, and sometimes reaching the surface at night. Recently acquired closing net data on the vertical distribution of *Argyropelecus* and *Sternoptyx* at selected stations in the eastern North Atlantic have shown that the genera are virtually mutually exclusive. *Argyropelecus* is most abundant at 500 m and above, while the species of *Sternoptyx* have maximum abundance below 600 m, and extend to 1500–2000 m (BADCOCK and MERRETT, 1976; BADCOCK, personal communication). If the normal function of the pre-orbital organ is to act as a reference standard for comparison with ambient daylight such a function is less likely to be of significance in *Sternoptyx*, living at these greater depths, in many cases below the significant penetration of daylight. Nevertheless could the oral organ of *Sternoptyx* serve a similar function? It is possible that some light might be reflected off the antero-ventral edge of the iris into the lens. The reflective characteristics of the iris in this region change abruptly (GARMAN, 1899; ZUGMAYER, 1911, plate II, fig. 5). The change is due to the presence of a large number of radially arranged striae, superficial to the blood vessels of the iris (Fig. 1). They are visible only with difficulty in preserved specimens, but in fresh animals the high reflectivity which they impart to this area of the iris is very striking. Their position coincides not only with the oral light organ, but also with the small aphakic aperture. Light from the oral organ might be reflected off them, or perhaps even transmitted along them if they have the properties of light guides. However, when the oral organ of moribund specimens is luminescing light is *not* normally visible at this ventral junction of the eye-ball and the orbit. The oral organs seem unnecessarily large and ill-sited to serve this purpose alone. If, on the other hand, their main (or sole) function is to illuminate the mouth then they are more akin to the oral organs of, for example, *Neoscopelus*, *Chauliodus* and many stomiatids, whose suggested purpose is prey attraction (MARSHALL, 1954).

The light organs of hatchet fishes seem ideally suited to fulfil a camouflage role, as

DENTON and his co-workers have shown. The relative reduction in the size of the light organs of *Sternoptyx* compared with those of *Argyropelecus* and *Polyipnus* is probably another reflection of the different camouflage requirements at deeper levels. It is significant in this context that the most obvious difference in the light organs of *Sternoptyx* is the great reduction of the reflector region. It may even be that *Sternoptyx* uses its ventral light organs only when in the upper part of its depth range, where ambient daylight is still of significance, though of lower intensity than that normally experienced by *Argyropelecus*. WEITZMAN (1974) regards *Sternoptyx* as highly advanced relative to *Argyropelecus* and *Polyipnus*, and much of its physiological specialisation must be related to its deeper habitat.

The lateral emission observed from the post-opercular organ in *A. olfersi* might also act as a means of maintaining contact between individuals in a school. Although BAIRD (1971) considers that neither *Argyropelecus* nor *Sternoptyx* school in the classical sense, groups of individuals have been observed *in situ* (BEEBE, 1935; PERES, 1958; TREGOUBOFF, 1958). Beebe's observations on hatchet-fishes confirm the normality of the appearance of luminescence in trawled specimens. He notes that "I could plainly see the lights of their downwardly-directed photophores when the fish swam above me or turned partly over" and also remarked on how the light was diffused over the area of the reflector: "In this species (*Argyropelecus*) these burned steadily, and each showed a colourful swath directed downwards—the little iridescent channels of glowing reflections beneath the source of the actual light." Clearly the organization of the reflector systems serves not only to alter the angular distribution of light, but also to spread its emission over a considerably greater area. No doubt this greatly enhances its efficiency in uniformly camouflaging the ventral regions of the fish, particularly against predators of limited visual acuity. The ease with which BEEBE was able to observe luminescing hatchet-fish *in situ* and to distinguish between *Sternoptyx* and *Argyropelecus* simply by their luminescent patterns, does raise some questions about the efficiency of the presumed camouflage. The apparent anomaly of their ready visibility yet seemingly elaborate camouflage development must relate to such factors as their normal orientation in the water, the rapidity of control of the luminescence and the visual capabilities of their predators. The intimate relationships involved are unlikely to be finally resolved without further observations of the animals in their normal environment.

Acknowledgements—I am most grateful to Professor E. J. DENTON and to J. BADCOCK for suggesting some improvements to the original manuscript. N. R. MERRETT and J. BADCOCK also gave unstintingly of their time in many discussions of some of the points raised in the paper. P. PARKES, of Oxford Scientific Films, took the film from which the stills in Fig. 4 have been selected, using a Rank 'Owl Eye' image-intensifier (courtesy of Anglia Survival). The Alpha Helix East Asian Bioluminescence Expedition was supported by NSF grants OFS 74 01830 and OFS 74 0283 to the Scripps Institution of Oceanography, and BMS 74 23242 and ONR contract N00014-75-C0242 to the University of California at Santa Barbara.

REFERENCES

BADCOCK J. and N. R. MERRETT (1977) Aspects of the distribution of midwater fishes in the eastern North Atlantic. In: *Oceanic sound-scattering prediction*, N. R. ANDERSEN and B. J. ZAHURANEL, editors, *Marine Science*, Series 3, Plenum Press, 209–242.

BAIRD R. C. (1971) The systematics, distribution and zoogeography of the marine hatchet-fishes (Family Sternoptychidae). *Bulletin of the Museum of Comparative Zoology, Harvard*, **142** (1), 128 pp.

BASSOT J. M. (1966) On the comparative morphology of some luminous organs. In: *Bioluminescence in Progress*, F. H. JOHNSON and Y. HANEDA, editors, Princeton University Press, pp. 557–610.

BEEBE W, (1935) *Half mile down*, The Bodley Head, London, 344 pp.

BERTELSEN E. and J. GRONTVED (1949). The light organs of a bathypelagic fish *A. olfersi* (Cuvier) photographed by its own light. *Videnskabelige Meddelelser fra Dansk naturhistorisk Forening i Kjøbenhavn*, Vol. III, pp. 163–167.

BRAUER A. (1908) Die Tiefsee-Fische. II. Anatomischer Teil. *Wissenschaftliche Ergebnis der Deutsche Tiefsee Expedition* Valdivia, Vol. 15, pp. 1–175.

DENTON E. J. (1970) On the organisation of reflecting surfaces in some marine animals. *Philosophical Transactions of the Royal Society of London*, B, **258**, 285–313.

DENTON E. J., J. B. GILPIN-BROWN and B. L. ROBERTS (1969). On the organization and function of the photophores of *Argyropelecus*. *Journal of Physiology*, **204**, 38.

DENTON E. J., J. B. GILPIN-BROWN and P. G. WRIGHT (1970). On the 'filters' in the photophores of mesopelagic fish and on a fish emitting red light and especially sensitive to red light. *Journal of Physiology*, **208**, 72–73 P.

DENTON E. J., J. B. GILPIN-BROWN and P. G. WRIGHT (1972) The angular distribution of the light produced by some mesopelagic fish in relation to their camouflage. *Proceedings of the Royal Society of London*, **B**, **182**, 145–158.

GARMAN S. (1899) Reports on an exploration off the west coast of Mexico, Central and South America, and off the Galapagos Islands. *Memoirs of the Museum of Comparative Zoology, Harvard*, **24**, 1–431.

HERRING P. J. (1972) Bioluminescence in searsid fishes. *Journal of the Marine Biological Association of the United Kingdom*, **52**, 879–887.

HERRING P. J. (1976) Bioluminescence in decapod Crustacea. *Journal of the Marine Biological Association of the United Kingdom* **56**, 1029–1047.

LAWRY J. V. (1974) Lantern fish compare downwelling light and bioluminescence. *Nature, London*, **247**, 155–157.

MARSHALL N. B. (1954) *Aspects of deep-sea biology*, Hutchinson, 380 pp.

NICOL J. A. C. (1962) The luminescence of fishes. In: *Aspects of Marine Zoology*, N. B. MARSHALL, editor, *Symposium of the Zoological Society of London*, No. 19, 27–55.

NUSBAUM-HILAROWICZ J. (1920) Études d'anatomie comparée sur les poissons provenant des campagnes scientifiques de S.A.S. Le Prince de Monaco. *Résultats des Campagnes Scientifiques, Monaco*, **58**, 1–115.

PERES J. M. (1958) Trois plongées dans le canyon du Cap Sicie, effectuées avec le bathyscaphe F.N.R.S. III de la Marine Nationale. *Bulletin de l'Institut Océanographique, Monaco*, **55**, No. 1115.

SCHULTZ L. P. (1961) Revision of the marine silver hatchetfishes (Family Sternoptychidae). *Proceedings of the United States National Museum*, **112**, 587–649.

TREGOUBOFF G. (1958) Prospection biologique sous-marine dans la region de Villefranche-sur Mer au cours de l'année 1957. I. Plongées en bathyscaphe. *Bulletin de l'Institut Océanographique, Monaco*, **55**, No. 1117.

TSUJI F. I. and Y. HANEDA (1971) Luminescent system in a myctophid fish *Diaphus elucens*. *Nature, London*, **233**, 623–624.

WAMPLER J. E. and R. J. DESA (1971) An on-line spectrofluorimeter system for rapid collection of absolute luminescence spectra. *Applied Spectroscopy*, **25**, 623–627.

WEITZMAN S. H. (1974) Osteology and evolutionary relationships of the Sternoptychidae, with a new classification of stomiatoid families. *Bulletin of the American Museum of Natural History*, **153**, 327–478.

ZUGMAYER E. (1911) Poissons provenant des campagnes du yacht Princesse Alice (1901–1910). *Résultats des Campagnes Scientifiques, Monaco*, **35**, 1–174.

Whale marking: a short review

SIDNEY G. BROWN

Whale Research Unit, Institute of Oceanographic Sciences, c/o British Museum (Natural History), Cromwell Road, London SW7 5BD.*

INTRODUCTION

WHALERS in the days of open-boat whaling occasionally found old harpoons in the bodies of the whales they killed, evidence of the animal's escape from a previous hunt. Records of these finds can sometimes provide information on the whale's migrations or longevity. It was from similar finds of American bomb-lances in blue whales (*Balaenoptera musculus*) killed in the Barents Sea in 1888 and 1898 (MURRAY and HJORT, 1912), and from his knowledge of the results of fish marking experiments, that Hjort in Norway about the year 1914 conceived the idea of marking whales by firing a mark harpoon into them in order to obtain information on their migrations, and on the size of the stock and the rate of exploitation by the whaling industry (ANON., 1920, p. 105). A 'Hjort mark harpoon' (No. 171) was fired into a blue whale from Shackleton's *Endurance* on 18 December 1914 near 62°S, 18°W (SHACKLETON, 1919).

A number of other people also apparently had the idea of marking whales at about this time. TURNER (1924) claims to have thought of it around 1913 and, according to Harmer (ANON., 1915, p. 3), Barrett-Hamilton mentioned the idea prior to leaving for South Georgia in October 1913. Larsen and Sörlle refer in 1918 to metal cases or cylinders being fired into whales to mark them (ANON., 1920, p. 92). However, the first actual experiment in marking whales at sea appears to have been carried out by Captain Tasuke Amano, a Japanese whaling captain and gunner, who fired a 'marking rod' into a blue whale while whaling off Oshima, Japan, in February 1910. He captured the same whale further north off Kinkazan in June 1912. No details of this marking rod are available but he continued his experiments on sperm whales (*Physeter catodon*) using a brass rod 39 cm long, engraved with the date and position of marking (OMURA, 1955; OMURA and OHSUMI, 1964). Apart from Captain Amano's first marking rod, there do not appear to have been recoveries from any of these early marking experiments.

In 1924 Hjort in Norway and Harmer with the Discovery Committee in England began new experiments in whale marking. The Discovery drawing-pin type mark was successfully fired from the Norwegian research vessel *Michael Sars* during a joint cruise in the waters around Iceland and the Faroes (HARDY, 1940). This mark was designed to penetrate just below the surface of the blubber, leaving the numbered disc visible flush with skin. Many whales were marked with this mark around South Georgia and a few elsewhere but no marks were ever recovered and there is no doubt that the mark, which penetrated no more than 6.5 cm into the blubber, was quickly rejected. Realizing that this mark was a

* Present address: Whale Research Unit, c/o British Antarctic Survey, Madingley Road, Cambridge CB3 0ET.

failure, the Discovery Committee considered several other designs before the present form of Discovery mark was adopted. It consists of a metal tube approximately 23 cm long fitted with a leaden ballistic head. It is fired from a modified 12-bore gun and is designed to bury itself completely in the body of the whale and to be found when the carcass is flensed at the whaling factory. The mark bears a serial number and an address for return. A reward is paid for each mark returned with information on the date and position of capture, together with the species, sex, length and other data relating to the animal in which the mark was found.

In the 1932/33 Antarctic whaling season during an experimental cruise around South Georgia some 200 whales were marked with this new mark. Several of the marks were recovered later in the same season and in the following one, and an extensive marking programme using this mark was then undertaken by the Discovery Investigations (RAYNER, 1940).

WHALE MARKING BY THE DISCOVERY INVESTIGATIONS, 1932 TO 1939

The Discovery Committee whale-marking programme in the Antarctic covered five whaling seasons (1932/33 and 1934/45 to 1937/38 inclusive). There were eight marking expeditions, four on the South Georgia whaling grounds using hired whale-catcher boats, and four on the pelagic whaling grounds by R.R.S. *William Scoresby* (RAYNER, 1940, 1948). In addition a few whales were marked in the Antarctic from R.R.S. *Discovery II* and in the 1938/39 season certain German whaling expeditions co-operated with the Discovery Committee in marking whales on the pelagic grounds. A few whales were also marked in more northern waters of the Atlantic and eastern Pacific during the voyages of R.R.S. *William Scoresby* and of the German expeditions to and from the Antarctic.

In all a total of 4988 of the commercially important species of whales were estimated to have been effectively marked* within the six Antarctic whaling Areas (Table 1). Fin whales (*Balaenoptera Physalus*) comprised 74% of the total and, as a result of the four marking expeditions based on South Georgia, 51% of all the whales marked were in Area II. Marking was concentrated on fin, blue and humpback (*Megaptera novaeangliae*) whales, the species from which large catches were being taken and from which recoveries could be expected. Negligible numbers of sei (*Balaenoptera borealis*), sperm, minke (*B. acutorostrata*) and right (*Eubalaena australis*) whales were marked and there have been no recoveries from these species.

At the close of the 1938/39 whaling season marks had been recovered from 190 blue, fin and humpback whales (3.8%) during processing on pelagic factory ships in the Antarctic, or from factory ships or land stations in South Georgia, South Africa and Australia.

WHALE MARKING IN THE ANTARCTIC, 1945–1975

Several organizations and whaling companies in Japan, the Netherlands, Norway, South Africa and the United Kingdom co-operated in whale-marking programmes in the

---

* Effective marks are those which completely penetrate the blubber and pass into the underlying musculature. Marks which protrude from the blubber are not counted as effective hits (unless later recovered from the whale) since they are usually lost within a short time.

Table 1. *Antarctic whale marking—1932/33 to 1938/39 (Discovery Investigations)*

Species	Whaling areas (south of 40°S)							Recoveries			
	I 120°W –60°W	II 60°W –0°	III 0° –70°E	IV 70°E –130°E	V 130°E –170°W	VI 170°W –120°W	All areas	To 1938/39		To 1974/75	
								No.	%	No.	%
Blue	6	298	281	110	—	—	695	37	5.3	46	6.6
Fin	159	2210	1172	125	—	7	3673	118	3.2	313	8.5
Humpback	39	38	100	389	—	—	566	35	6.2	37	6.5
Sei	—	11	—	—	1	—	12	—	—	—	—
Minke	—	2	1	—	—	—	3	—	—	—	—
Right	—	6	—	—	—	—	6	—	—	—	—
Sperm	—	1	28	4	—	—	33	—	—	—	—
All species	204	2566	1582	628	1	7	4988	190	3.8	396	7.9

Table 2.   *Antarctic whale marking—*1945/46 *to* 1974/75 (*International scheme*)

Species	Whaling areas						All areas	Recoveries to 1974/75	
	I	II	III	IV	V	VI		No.	%
Blue	64	25	108*	22	43	10	272	49	18.0
Fin	121	272	817	215	174	82	1681	377	22.4
Humpback	40	13	45	113	335	27	573	14	2.4
Sei	2	61	36	91	67	14	271	37	13.7
Minke	—	—	—	6	—	2	8	—	—
Right	—	1	—	19	—	1	21	—	—
Sperm	20	21	131	150	79	19	420	5	1.2
All species	247	393	1137	616	698	155	3246	482	14.8

* Includes three pygmy blue whales.

Antarctic from the 1945/46 whaling season onwards. In 1953 all the British, Dutch and Norwegian companies shared the expenses of a special marking cruise in the whale catcher *Enern* to the pelagic grounds before the beginning of the 1953/54 baleen whale season (CLARKE and RUUD, 1954; RUUD and ØYNES, 1954). This international co-operation continued in the following year with a second cruise by the same vessel (RUUD and RAVNINGER, 1955). In 1955 an international marking scheme was launched under the auspices and with the support of the International Whaling Commission. This scheme, which is co-ordinated by the Institute of Oceanographic Sciences in England, developed to embrace all whale marking in the southern hemisphere with the exception of that carried out by the Soviet Union. Progress reports showing the distribution of marked whales and giving details of recovered marks are published at regular intervals. On the pelagic whaling grounds in recent years marking under the scheme has been carried out entirely by Japanese expeditions with the invaluable co-operation of the Japanese whaling companies and research laboratories.

In the twenty-five seasons during which marking has been carried out up to the close of the 1974/75 season, a further 3246 large whales were marked in the Antarctic under the international scheme (Table 2). Fin whales again comprise more than half (52%) of the total whales marked but in addition to fin, blue and humpback whales, 271 sei and 420 sperm whales have been marked. The marking of all species has been more widely distributed in the six whaling areas than in the pre-war seasons. The percentage of recoveries from all species combined (14.8%) is substantially higher than for the pre-war marking and, apart from humpback whales, the percentage recoveries from the individual species are also higher.

The extensive marking programme carried out by expeditions from the Soviet Union is part of an entirely separate scheme, although there is close liaison with the international scheme as far as the exchange of data from recovered marks is concerned. In the twenty seasons from 1952/53, when the USSR Antarctic marking programme began, to 1971/72 inclusive, a total of 1224 large whales have been marked (Table 3). Humpback whales account for 42% of the total but numbers of fin, sei and sperm whales have also been marked and the marking is widely distributed in the whaling Areas. The percentage of recoveries for all species (11.0%) is rather less than for post-war marking in the international scheme.

Table 3. *Antarctic whale marking—1952/53 to 1971/72 (U.S.S.R. scheme)* (compiled from IVASHIN, 1973)

Species	Whaling areas						All areas	Recoveries to 1971/72	
	I	II	III	IV	V	VI		No.	%
Blue							18	2	11.1
Fin	37	156	129	16	15	6	359	43	12.0
Humpback	24	8	17	207	228	33	517	63	12.2
Sei	10	25	20	14	46	14	129	14	10.9
Minke							23	—	—
Right							47	—	—
Sperm	4	30	36	17	44		131	13	9.9
All species							1224	135	11.0

WHALE MARKING IN OTHER REGIONS

1. *Southern Hemisphere outside the Antarctic*

North of the six Antarctic whaling areas, the international scheme includes marking off the coasts of Africa, Australia and South America, and among the island groups of the southwest Pacific (Table 4). In South African waters marking began in 1962 and has been concentrated on sperm whales (BEST, 1969). Among the few baleen whales marked here are eleven Bryde's whales (*Balaenoptera edeni*), and this species has also been marked in New Zealand waters. The marking of humpback whales off both east and west coasts of Australia began in 1949 and continued in most years until 1963 when attention was turned to sperm whales, especially in Western Australian waters (CHITTLEBOROUGH, 1965). Around New Zealand and at Tonga, Fiji, New Caledonia, New Hebrides and Norfolk Island marking between 1952 and 1961 organized by DAWBIN (1956, 1959, 1964) concentrated on humpback whales, but some sperm whales were marked in New Zealand waters in 1963

Table 4. *Marking in the southern hemisphere outside the Antarctic—1949 to 1975 (International scheme)*

	Blue	Fin	Hump-back	Sei	Bryde's	Minke	Right	Sperm	Total
South Africa (1962–1975)	—	2	—	6	11	2	—	416	437
Australia (1949–1972)	1	2	1333	5	—	—	1	422	1764
New Zealand and S.W. Pacific Islands (1952–1963)	—	—	915	8	6	—	—	54	983
South America (1958–1966)	2	12	—	12	—	—	2	121	149
Elsewhere	2	8	3	58	—	2	4	261	338
Total	5	24	2251	89	17	4	7	1274	3671

Table 5.   *Marking in the North Pacific*

	Blue	Fin	Hump-back	Sei/ Bryde's	Right	Gray	Sperm	Total
Canada (1955–1967)	—	19	4	5	—	5	83	116
Japan (1949–1972)	64	866	394	515	14	1	2678	4532
U.S.A. (1962–1970)	76	56	44	29	—	5	176	386
U.S.S.R. (1954–1966)	8	51	72	43	20	29	1223	1446
Total	148	992	514	592	34	40	4160	6480
Recoveries	16	209	25	101	—	—	226	577
%	10.8	21.1	4.9	17.1	—	—	5.4	8.9

(GASKIN, 1965). Much of the marking in New Zealand waters was carried out south of 40°S and is therefore, strictly speaking, within the northern boundary of Antarctic Area V. Similarly, a few whales marked in Chilean waters fall within Area I. However, all this New Zealand and Chilean marking has been included in Table 4, together with the results of the other marking programmes carried out north of Antarctic waters, rather than in Table 2 which relates to Antarctic programmes only. In South America marking concentrated on sperm whales off the coasts of Chile, Peru and Ecuador and around the Galapagos Islands (CLARKE, 1962; MEJÍA G., 1964).

Elsewhere in the southern hemisphere north of 40°S, some whales, mostly sei and sperm whales, have been marked during Antarctic cruises. Also included here are 167 whales (including 157 sperm whales) marked during an international cruise sponsored by government agencies in Australia, South Africa, the UK and the USA, in the Indian Ocean in 1973/74 (GAMBELL, BEST and RICE, 1975).

2. *North Pacific*

Programmes of marking large whales in the North Pacific have been carried out by Canada, Japan, the USA and the USSR. Table 5 has been compiled from published reports and Japan and the USSR have continued their programmes in more recent seasons. Since it began in 1949 Japanese marking has been widely distributed and has included seven species, though sperm whales comprise 59% and fin whales 19% of the total number of whales marked (OMURA and OHSUMI, 1964; OHSUMI and MASAKI, 1975). Sperm whales also predominate and are widely distributed in the USSR marking (IVASHIN and ROVNIN, 1967). The Canadian and American marking programmes (RICE, 1974) have been mainly in coastal waters. In the North Pacific it has not always been possible to positively identify individual sei and Bryde's whales and they have been grouped together in Table 5. Marks have been returned from 577 (8.9%) of the 6480 whales of all species. The largest numbers of returns from individual species are from fin whales (21.1%) and sei/Bryde's whales (17.1%). No returns can be expected from the protected right whales and gray whales (*Eschrichtius robustus*). Marking in the North Pacific is organized separately from that in the southern hemisphere and co-ordination of the records is undertaken by the Whales Research Institute in Tokyo.

Table 6.   Marking in the North Atlantic—1950 to 1975

	Blue	Fin	Hump-back	Sei/Bryde's	Minke	Right	Sperm	Total
Canada (1960–1973)	20	287	190	30	12	8	109	656
France (1965–1969)	—	2*	—	—	—	–	4	6
Iceland (1965–1972)	—	17	—	—	—	–	6	23
Norway (1954–1975)	—	27	20	—	207	–	2	256
U.K. (1950–1955)	—	9	—	—	—	–	14	23
Total	20	342	210	30	219	8	135	964
Recoveries	—	46	1	3	4	–	1	55
%	—	13.5	0.5	10.0	1.8	–	0.7	5.7

* Marked in the Mediterranean Sea.

3. *North Atlantic*

Smaller numbers of whales have been marked in the North Atlantic than elsewhere. The Canadian marking programme (MITCHELL, 1974) has been the largest, including seven species and extending over a wide area of the North Atlantic. The Norwegian marking JONSGÅRD and CHRISTENSEN, 1968) has been mainly in Arctic waters and in recent years has been concentrated on minke whales. Icelandic marking (JÓNSSON, 1965) in the Denmark Strait, and French and British marking elsewhere (including the Mediterranean Sea) accounts for a few additional fin and sperm whales. Marks have been returned from 55 (5.7%) of the 964 whales of all species marked, the largest number from fin whales (13.5%), see Table 6.

SUMMARY FOR ALL REGIONS

Complete figures for the total numbers of large whales marked with Discovery-type marks in all regions are not yet available. Table 7 summarizes the earlier tables but does not include whales marked in the USSR programme in the Antarctic since the 1971/72 season, nor any marked in this programme in the southern hemisphere outside the Antarctic. Also not included are whales marked in the North Pacific by the USSR after 1966 and by Japan after 1972. With these qualifications, Table 7 is believed to include all large whales marked to the end of 1975. The largest numbers have been marked in the Antarctic in the pre-war and post-war years, and in the North Pacific. Fin (34%), humpback (22%) and sperm whales (30%) account for 86% of the total of 20,573 whales marked.

The percentage of marks recovered varies widely in the different species and in different regions, reflecting the numbers of whales marked and the intensity of whaling on the species concerned. Figures are not available for all regions but they range for all species combined from 5.7% in the North Atlantic to 14.8% in the Antarctic in the post-war period.

MIGRATIONS AND MOVEMENTS OF WHALES DEMONSTRATED BY RECOVERED MARKS

With information from the world-wide development of modern whaling for rorquals added to the records from earlier open-boat whaling for right, humpback and gray whales, much evidence accumulated from the seasonal catches of whalers in different latitudes and from sightings in coastal waters, for the existence of regular seasonal migrations in some species of the large baleen whales (Kellogg, 1929).

For humpback whales in the southern hemisphere Matthews (1937) assembled evidence for a migration southwards to Antarctic waters for feeding in the southern summer and a return migration northwards to sub-tropical and tropical waters for breeding in the southern winter. One of the immediate results of the Discovery Investigations programme of marking in the Antarctic was direct confirmation of these migrations by the recovery from humpback whales killed off Western Australia and off Madagascar of marks which had been fired into them in the Antarctic (Rayner, 1940). The marking of humpback whales off Australia, New Zealand and among the southwest Pacific islands since 1949 has provided further evidence of their annual migrations southwards and northwards between these regions and the Antarctic (Chittleborough, 1965; Dawbin, 1964, 1966). Examples of some of these migrations as shown by recovered marks are illustrated in Fig. 1. In the North Pacific the migration of humpbacks between the Aleutian Islands area and the Ryukyu Islands, and between Bonin Islands waters and the Aleutian Islands area has been demonstrated (Omura and Ohsumi, 1964; Ohsumi and Masaki, 1975).

For fin whales, one mark illustrating northward migration from the Antarctic feeding grounds in summer to South African waters in winter, and one showing southward migration from the coast of Brazil to the Antarctic, were returned from the Discovery Investigations marking (Rayner, 1940; Brown, 1954). More recent recoveries from the international scheme programmes have added examples of the southward migration of

Table 7. Numbers of the large whale species, including the minke whale, marked throughout the world*

	Blue	Fin	Hump-back	Sei/Bryde's	Minke	Right	Gray	Sperm	Total
Antarctic (1932–1939)	695	3673	566	12	3	6	—	33	4988
Antarctic (1945–1975 International and U.S.S.R.)	290	2040	1090	400	31	68	—	551	4470
Southern hemisphere outside Antarctic (1949–1975)	5	24	2251	106	4	7	—	1274	3671
North Pacific (1949–1972)	148	992	514	592	—	34	40	4160	6480
North Atlantic (1950–1975)	20	342	210	30	219	8	—	135	964
All regions	1158	7071	4631	1140	257	123	40	6153	20573

* Antarctic–U.S.S.R. scheme to 1972 only; North Pacific–U.S.S.R. to 1966 and Japan to 1972 only.

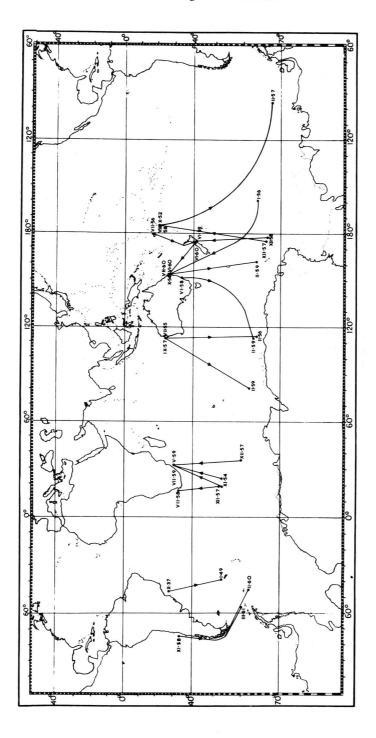

Fig. 1. Examples of seasonal migrations of humpback and fin whales in the southern hemisphere demonstrated by recovered marks. Humpback whales in Australian, New Zealand and Pacific waters; fin whales in South American and South African waters.

fin whales marked off the Chilean coast into the Atlantic sector of the Antarctic (Area II) and northwards from Antarctic Area III to both Atlantic and Indian Ocean coasts of South Africa (Fig. 1 and BROWN, 1962).

In the southern hemisphere returns from sei whales have demonstrated migrations from South African waters off Durban in winter southwards to Antarctic Areas III and IV in summer, and from Western Australia to Antarctic Area IV (BROWN, 1971, 1976). IVASHIN (1971) records a mark showing northward migration from the Antarctic to the coast of Brazil and in the North Pacific similar migrations of sei whales between the breeding grounds and feeding grounds are recorded by OHSUMI and MASAKI (1975).

Seasonal migrations in blue whales have not yet been demonstrated by marking, no doubt because only small numbers of this species have been caught at land stations in the southern hemisphere or on the pelagic whaling grounds in the North Pacific.

In the Antarctic and the North Pacific, marking has shown that some blue, fin and sei whales return year after year to the same region of the whaling grounds, while others disperse to a greater or lesser extent from the place where they were marked (BROWN, 1954, 1968; OHSUMI and MASAKI, 1975; RAYNER, 1940). The picture of migrations and movements of blue, fin and humpback whales in the southern hemisphere revealed by marking led BROWN (1959) to suggest that there is in these species a pattern of segregation and concentration of the populations in the breeding areas and of their dispersal and association on the feeding grounds. It may be that this applies in all baleen whale species which have a well-defined breeding–feeding rhythm necessitating migration from breeding grounds in lower latitudes to feeding grounds in higher latitudes.

Recovered marks illustrating long-distance movements of sperm whales in the North Pacific are reported on by IVASHIN and ROVNIN (1967) and OHSUMI and MASAKI (1975). Similar movements in the North and South Atlantic are noted by MITCHELL (1975) and BEST (1969). One return from a sperm whale marked in the North Atlantic (21°33′N, 17°55′W) in October 1961 and killed off South Africa (33°20′S, 16°52′E) in March 1966, provides the first record of the movement of any species of whale from one hemisphere to the other (IVASHIN, 1967). BEST (1969) and IVASHIN (1971) both report the record of a male sperm whale (10.7 m length) marked in Antarctic Area III (62°22′S, 26°25′E) in December 1967 and shot off Durban, South Africa, in the following May, as the first record of migration from the Antarctic to warmer waters in this species and confirming evidence for such movements from the presence of Antarctic diatoms on males in temperate waters. A southern movement of a male whale from Chilean coastal waters to Antarctic Area I (66°01′S, 83°03′W) is noted by BROWN (1976). It is not yet certain that these movements represent regular seasonal migrations in male sperm whales comparable to those occurring in baleen whales.

INFORMATION ON THE AGES OF WHALES FROM RECOVERED MARKS

Marks returned from whales many years after they were fired into the animals provide some information on the longevity of the different species. The number of years elapsing between marking and recovery of the mark is usually less than the age of the animal concerned since few calves are marked, most animals being marked as adults of unknown age. In Table 8 the oldest marks returned from the different species in the southern

Table 8. *The oldest marks returned from different species to the end of* 1975

Species	Southern hemisphere		North Pacific	
	Oldest return in years	Maximum possible years	Oldest return in years	Maximum possible years
Blue	14	32	11	16
Fin	37	42	18	26
Humpback	17	30	11	12
Sei	11	42	11	26
Bryde's	—	—	19	26
Sperm	22	40	19	26

hemisphere and North Pacific are listed, together with the maximum possible number of years which could have been recorded by returns from the first year in which the species were marked, up to the end of 1975. The maximum possible periods vary because not all species were first marked in the same year and because of the different dates on which blue and humpback whales were protected and therefore ceased to be a source of returns.

More important in relation to age, however, are marks returned after a period of years from whales for which biological collections are also available. These marks can provide an important check on the methods of age determination by means of counts of corpora albicantia in the ovaries of females, or layers in ear plugs from baleen whales, or in teeth from sperm whales (OHSUMI, 1962; MACKINTOSH, 1965; BEST, 1970). An important development in this connection is the production by KOZICKI and MITCHELL (1974) of a prototype mark containing a drug designed to produce permanent marking of ear plug laminations to enable their rate of formation to be accurately calculated.

NEW DEVELOPMENTS IN MARKING LARGE WHALES

The Discovery mark has provided much information on the movements and migrations of some species; some information on longevity and important checks on methods of age determination. It has also been used in estimating population sizes, exploitation rates and fishing mortality coefficients (CHAPMAN, 1974). It is, however, an internal mark and its return depends on the death of the whale. The reduction in catches of some species and the total protection of others has led to a decline in the numbers of marks returned and therefore of the amount of information available. Attention has recently been turned to the development of an externally visible mark and MITCHELL and KOZICKI (1975), and others, are experimenting with modified Discovery marks incorporating a visible plastic 'streamer' to enable marked animals to be recognized in the wild. A completely different approach to tracking large whales was attempted by SCHEVILL and WATKINS (1966) who attached radio beacons to right whales but without success. Further developments along these lines are forecast by NORRIS, EVANS and RAY (1974).

REFERENCES

ANON. (1915) Minutes of evidence. *Inter-departmental Committee on Whaling and the Protection of Whales*, Colonial Office, Miscellaneous No. 298, 209 pp.

ANON. (1920) *Report of the Inter-departmental Committee on research and development in the Dependencies of the Falkland Islands*, H.M.S.O., iii, 164 pp.

BEST P. B. (1969) The sperm whale (*Physeter catodon*) off the west coast of South Africa 4. Distribution and movements. *Investigational Report Division of Sea Fisheries South Africa*, No. 78, 1–12.

BEST P. B. (1970) The sperm whale (*Physeter catodon*) off the west coast of South Africa 5. Age, growth and mortality. *Investigational Report Division of Sea Fisheries South Africa*, No. 79, 1–27.

BROWN S. G. (1954) Dispersal in blue and fin whales. *Discovery Reports*, **26**, 355–384.

BROWN S. G. (1959) Whale marks recovered in the Antarctic seasons 1955/56, 1958/59, and in South Africa 1958 and 1959. *Norsk Hvalfangst-Tidende*, **48**, 609–616.

BROWN S. G. (1962) A note on migration in fin whales. *Norsk Hvalfangst-Tidende*, **51**, 13–16.

BROWN S. G. (1968) The results of sei whale marking in the Southern Ocean to 1967. *Norsk Hvalfangst-Tidende*, **57**, 77–83.

BROWN S. G. (1971) Whale marking—progress report, 1970. *Twenty-first Report of the International Whaling Commission*, pp. 51–55.

BROWN S. G. (1976) Whale marking—progress report, 1975. *Report and papers of the Scientific Committee of the International Whaling Commission*, 1975, pp. 31–38.

CHAPMAN D. G. (1974) Estimation of population parameters of Antarctic baleen whales. In: *The Whale Problem. A Status Report*, W. E. SCHEVILL, editor, Harvard University Press, pp. 336–351.

CHITTLEBOROUGH R. G. (1965) Dynamics of two populations of the humpback whale, *Megaptera novaeangliae* (Borowski). *Australian Journal of Marine and Freshwater Research*, **16**, 33–128.

CLARKE R. (1962) Whale observation and whale marking off the coast of Chile in 1958 and from Ecuador towards and beyond the Galápagos Islands in 1959. *Norsk Hvalfangst-Tidende*, **51**, 265–287.

CLARKE R. and J. T. RUUD (1954) International co-operation in whale marking: The voyage of the *Enern* to the Antarctic 1953. *Norsk Hvalfangst-Tidende*, **43**, 128–146.

DAWBIN W. H. (1956) Whale marking in South Pacific Waters. *Norsk Hvalfangst-Tidende*, **45**, 485–508.

DAWBIN W. H. (1959) New Zealand and South Pacific whale marking and recoveries to the end of 1958. *Norsk Hvalfangst-Tidende*, **48**, 213–238.

DAWBIN W. H. (1964) Movements of humpback whales marked in the south west Pacific Ocean 1952 to 1962. *Norsk Hvalfangst-Tidende*, **53**, 68–78.

DAWBIN W. H. (1966) The seasonal migratory cycle of humpback whales. In: *Whales, Dolphins, and Porpoises*, K. S. NORRIS, editor, University of California Press, pp. 145–170.

GAMBELL R., P. B. BEST and D. W. RICE (1975) Report on the international Indian Ocean whale marking cruise 24 November 1973–3 February 1974. *Twenty-fifth Report of the International Whaling Commission*, pp. 240–252.

GASKIN D. E. (1965) New Zealand whaling and whale research 1962–4. *New Zealand Science Review*, **23**, 19–22.

HARDY A. C. (1940) Whale-marking in the Southern Ocean. *Geographical Journal*, **96**, 345–350.

IVASHIN M. V. (1967) The whale-traveller. *Priroda*, No. 8, 105–107.

IVASHIN M. V. (1971) Some results of whale-marking carried out from Soviet ships in the southern hemisphere. *Zoologicheskii Zhurnal*, **50**, 1063–1078. (In Russian).

IVASHIN M. V. (1973) Marking of whales in the southern hemisphere (Soviet materials). *Twenty-third Report of the International Whaling Commission*, pp. 174–191.

IVASHIN M. V. and A. A. ROVNIN (1967) Some results of the Soviet whale marking in the waters of the North Pacific. *Norsk Hvalfangst-Tidende*, **56**, 123–135.

JONSGÅRD Å. and I. CHRISTENSEN (1968) A preliminary report on the *Harøybuen* cruise in 1968. *Norsk Hvalfangst-Tidende*, **57**, 174–175.

JÓNSSON J. (1965) Whale marking in Icelandic waters in 1965. *Norsk Hvalfangst-Tidende*, **54**, 254–255.

KELLOGG R. (1929) What is known of the migrations of some of the whalebone whales. *Smithsonian Report for 1928*, pp. 467–494.

KOZICKI V. M. and E. MITCHELL (1974) Permanent and selective chemical marking of mysticete ear plug laminations with Quinacrine. *Twenty-fourth Report of the International Whaling Commission*, pp. 142–149.

MACKINTOSH N. A. (1965) *The stocks of whales*, Fishing News (Books) Ltd., 232 pp.

MATTHEWS L. H. (1937) The humpback whale, *Megaptera nodosa*. *Discovery Reports*, **17**, 7–92.

MEJÍA G. J. G. (1964) Marcación de cacholotes frente al Perú. *Informes Instituto de Investigacion de los Recursos Marinos*, No. 24.

MITCHELL E. (1974) Progress report on whale research, Canada. *Twenty-fourth Report of the International Whaling Commission*, pp. 196–213.

MITCHELL E. (1975) Progress report on whale research, Canada. *Twenty-fifth Report of the International Whaling Commission*, pp. 270–282.

MITCHELL E. and V. M. KOZICKI (1975) Prototype visual mark for large whales modified from 'Discovery' tag. *Twenty-fifth Report of the International Whaling Commission*, pp. 236–239.

MURRAY J. and J. HJORT (1912) *The Depths of the Ocean*, Macmillan, xx, 821 pp.

NORRIS K. S., W. E. EVANS and G. C. RAY (1974) New tagging and tracking methods for the study of marine mammal biology and migration. In: *The Whale Problem. A Status Report*, W. E. SCHEVILL, editor, Harvard University Press, pp. 395–408.

OHSUMI S. (1962) Biological material obtained by Japanese expeditions from marked fin whales. *Norsk Hvalfangst-Tidende*, **51**, 192–198.

OHSUMI S. and Y. MASAKI (1975) Japanese whale marking in the North Pacific, 1963–1972. *Far Seas Fisheries Research Laboratory Bulletin*, No. 12, pp. 171–219.

OMURA H. (1955) Age of whales. *Geiken Tsushin*, No. 45, pp. 1–6. (In Japanese.)

OMURA H. and S. OHSUMI (1964) A review of Japanese whale marking in the North Pacific to the end of 1962, with some information on marking in the Antarctic. *Norsk Hvalfangst-Tidende*, **53**, 90–112.

RAYNER G. W. (1940) Whale marking. Progress and results to December 1939. *Discovery Reports*, **19**, 245–284.

RAYNER G. W. (1948) Whale marking II. Distribution of blue, fin and humpback whales marked from 1932 to 1938. *Discovery Reports*, **25**, 31–38.

RICE D. W. (1974) Whales and whale research in the eastern North Pacific. In: *The Whale Problem. A Status Report*, W. E. SCHEVILL, editor, Harvard University Press, pp. 170–195.

RUUD J. T. and P. ØYNES (1954) International co-operation in whale marking: The second period of the voyage of the *Enern* to the Antarctic 1953. *Norsk Hvalfangst-Tidende*, **43**, 383–393.

RUUD J. T. and R. RAVNINGER (1955) Report on the whale marking voyage of the *Enern* to the Antarctic 1954. *Norsk Hvalfangst-Tidende*, **44**, 309–315.

SCHEVILL W. E. and W. A. WATKINS (1966) Radio-tagging of whales. *Woods Hole Oceanographic Institution Reference 66–17*, 1–15.

SHACKLETON E. H. (1919) *South*, Heinemann, p. 12.

TURNER H. M. S. (1924) Whaling in the South Atlantic. *Cairo Scientific Journal*, **12**, 83–93.

Dentinal layer formation in sperm whale teeth

Ray Gambell

Whale Research Unit, Institute of Oceanographic Sciences, U.K.*

Abstract—Sperm whale teeth are made up largely of dentine, which is laid down in alternating translucent and opaque laminae on the pulp cavity edge of the tooth. Earlier studies on the nature of this edge showed that the translucent lamina is deposited in late winter and spring in sperm whales of both sexes on the east and west coasts of South Africa, and in the summer months in male sperm whales in the Antarctic.

New investigations into the timing of lamina formation in sperm whales from Western Australia using acid digested tooth samples showed that in this area the translucent lamina was laid down at different times in the two sexes. The female whales were similar to the South African animals, forming this lamina in late winter and spring. The incidence of the newly formed translucent lamina reaches a peak in the males in summer, as in the Antarctic male sperm whales.

The Antarctic and Western Australian male sperm whales are similar to each other but have a generally older age distribution than the males off South Africa. It appears that different population groups may have different times of lamina formation in the dentine of their teeth. However, only one pair of translucent and opaque laminae, constituting a growth layer, is deposited each year in the teeth of sperm whales in the southern hemisphere.

INTRODUCTION

SPERM whale teeth comprise a central core of dentine and a relatively thin covering layer of cement. The enamel layer present in the foetus is soon worn away. The tooth has a simple root, and the substance of the dentine is formed from the germinal layer lining the central pulp cavity in a series of cone-shaped zones of alternating translucent and opaque tissue. When a tooth is bisected longitudinally these zones appear as dark and light laminae respectively by reflected light, each pair constituting a growth layer (INTERNATIONAL WHALING COMMISSION, 1969).

The dentinal growth layers are now commonly used for age determination in the sperm whale, although there is still a lack of direct evidence on their rate of deposition. Tooth layer counts from nine long-term whale mark recoveries in the North Pacific indicate a rate of accumulation of less than two layers a year. In fact, four of the whales had values for the tooth layer counts divided by the number of years that had elapsed between marking and recovery of less than 1.5 (OHSUMI, KASUYA and NISHIWAKI, 1963; BEST, 1970).

Sperm whale teeth from the North Pacific have been examined by Japanese and Soviet workers to determine the seasonal changes in the thickness of the most recently formed layer in the dentine. OHSUMI, KASUYA and NISHIWAKI (1963) found that the width of the opaque lamina increased in their samples from May to November, suggesting that a single growth layer is formed each year. BERZIN (1971), with apparently similar material collected

---

*Present address: International Whaling Commission, The Red House, Station Road, Histon, Cambridge, CB4 4NP.

from March to October, considered that two growth layers are deposited each year. In this case it is clear that the two groups of workers are interpreting the same structures within the dentine in different ways. In particular, fine translucent bands in the opaque laminae are counted by Berzin, but are not included in the Japanese analysis (KLEVESAL and KLEINENBERG, 1967).

The more generally accepted standard used by other workers throughout the world corresponds most closely to the Japanese interpretation. In the following analysis and discussion the rather irregularly appearing and fine translucent lines are neglected, and the alternating translucent and opaque laminae for counting purposes are considered to be only the major zones discernible in the dentine. Even this approach does not entirely remove the problems of interpretation, for some teeth have a very pronounced tendency to form double laminae (GAMBELL, 1972). A certain subjective skill is then still involved in the layer counts.

SOUTH AFRICAN AND ANTARCTIC SPERM WHALES

In an attempt to settle the problem of the rate of deposition of dentine in the teeth, GAMBELL and GRZEGORZEWSKA (1967) assessed the distance of the most recently formed translucent lamina from the pulp cavity edge in samples of acid-etched teeth collected in different months of the year. The material studied was obtained from two geographically separate areas; 731 whales from the sub-tropical fishery at Durban on the east coast of South Africa, covering the months from February to October, and 269 whales from the Antarctic fishery from the land stations on South Georgia and on floating factory ships operating between 10°E and 91°W longitudes, in the period October to March. The results showed that the last formed translucent lamina reached a peak of occurrence 'on', 'near' and 'far' from the pulp cavity edge in each of these two sets of samples. Thus it was concluded that two pairs of translucent and opaque laminae are formed each year in the teeth of the sperm whales in the southern hemisphere, *if it is assumed that the Durban and Antarctic samples can be considered in combination.*

The Antarctic samples contained only male sperm whales, while the Durban material was made up of both males and females. The pattern of layer formation was essentially similar in the two sexes in the Durban samples. In addition, it was noted that similar results were obtained from both maxillary and mandibular teeth in both areas. The results are summarized in Fig. 1, where the monthly mean percentages of maxillary and mandibular teeth combined having each of the three edge-type classifications are plotted.

BEST (1970) carried out an independent study on sperm whale teeth collected at Durban from March to September, as well as from March to October at Donkergat on the west coast of South Africa. He used a very similar method of approach in classifying the nature of the most recently formed lamina in thin sagittal sections of maxillary teeth. He concluded that the general trends in the monthly proportions of teeth forming translucent, thin-opaque and thick-opaque laminae are the same in both sexes. The results for 390 whales of both sexes combined at Donkergat are illustrated in Fig. 2. They show that the proportion of whales forming a translucent lamina increases rapidly in spring, as in the Durban data of GAMBELL and GRZEGORZEWSKA (1967). The early season incidence of the translucent category is much lower in the west coast samples, but the evidence from both areas is similar in showing that the opaque lamina is most common in winter. This leads to the conclusion that only one lamina of each kind, and thus only one growth layer, is formed each year off Donkergat.

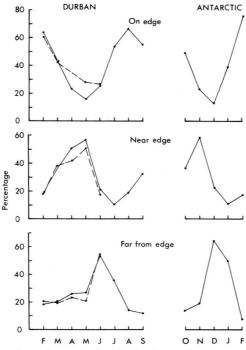

Fig. 1.   Position of the last-formed translucent dentinal lamina in relation to the pulp cavity edge of teeth from male (solid line) and female (broken line) sperm whales sampled at Durban and in the Antarctic.

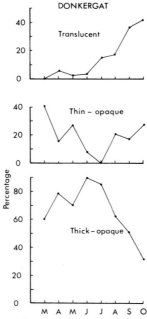

Fig. 2.   Nature of the most recently formed dentinal lamina on the pulp cavity edge of teeth from sperm whales sampled at Donkergat (from data kindly supplied by Dr. P. B. Best).

WESTERN AUSTRALIAN SPERM WHALE TEETH

More recently, samples of mandibular teeth collected from March to December of 1973 and 1974 have kindly been made available to us for analysis by Mr. J. L. Bannister from the fishery conducted off Western Australia. These provide welcome additional material from the southern hemisphere to extend the geographical range of the sperm whales studied, and for comparison with the earlier samples.

Method

The teeth collected were mainly first mandibulars, as these are generally the most suitable for age-determination purposes. They are usually the straightest and least worn of the tooth row as well as the most easily removed (BERZIN, 1961; INTERNATIONAL WHALING COMMISSION, 1967). After removal from the carcases at the Albany whaling station, the teeth were cleaned and stored dry before being shipped to London for study.

In the laboratory the straight teeth were bisected longitudinally with a diamond-impregnated circular saw, and a few curved teeth with a band saw. The cut surface of one-half of each tooth was then supported in a trough of 10% formic acid for 30 hours at room temperature. This modification of the acid etching technique developed by BOW and PURDAY (1966) causes differential digestion of the translucent and opaque laminae, and converts them into ridges and grooves respectively when the tooth has been rinsed and dried. The appearance of the ridges can be enhanced by lightly shading the surface of each tooth with a pencil.

As in the earlier work of GAMBELL and GRZEGORZEWSKA (1967), the distance of the most recently formed ridge, corresponding to the translucent layer, from the pulp cavity edge was classified into one of three categories. These are on-the-edge, near-the-edge and far-from-the-edge, and were judged by the naked eye. The last two groups were distinguished as less or more than 50% of the width of adjacent growth layers.

Results

No significant differences were evident between the results from the two seasons, so the numbers of whales with teeth falling in each of the categories are shown in Table 1 for each month of both seasons combined. These data are plotted as percentage values in Fig. 3 for comparison with the other southern hemisphere analyses shown in Figs. 1 and 2.

The most surprising feature of the Australian results is the difference evident in the pattern of lamina formation between the two sexes. The translucent lamina is most frequently seen on the edge of the teeth of the females in September, and there is a general increase during most of the season to this peak, followed by a decline. The far-from-edge maximum occurs a little earlier in August, while the near-edge category shows a fall from the start of the season to a minimum in September followed by an increase in spring. These timings for the peaks and troughs in the patterns of lamina formation in the Western Australian female sperm whales are therefore essentially similar to those demonstrated for the males and females off both coasts of South Africa.

The teeth from male sperm whales off Western Australia have the highest occurrence

WESTERN AUSTRALIA

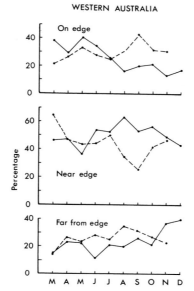

Fig. 3. Position of the last-formed translucent dentinal lamina in relation to the pulp cavity edge of teeth from male (solid line) and female (broken line) sperm whales sampled in Western Australia.

of the translucent lamina on-the-edge in the early part of the year, and the incidence falls during the winter and spring. This pattern is the reverse of that observed for the far-from-edge category, which tends to increase in frequency throughout the sampling period. The near-edge type reaches a maximum in August. Thus the patterns in lamina formation in the teeth from these male sperm whales are quite different from those in the females in the same area as well as those in the South African whales.

Table 1. *Numbers of sperm whale mandibular teeth from Western Australia, showing the position of the latest formed translucent dentinal lamina in relation to the pulp cavity edge*

	♂♂				♀♀			
	On	Near	Far	Total	On	Near	Far	Total
March	5	6	2	13	3	9	2	14
April	21	34	17	72	4	7	4	15
May	34	31	19	84	10	13	7	30
June	21	33	7	61	7	11	7	25
July	5	10	4	19	7	14	7	28
August	14	54	17	85	7	8	8	23
September	14	37	18	69	15	9	11	35
October	25	66	25	116	22	29	19	70
November	14	51	38	103	8	12	6	26
December	4	10	9	23	2	1	0	3
Total				645				269

DISCUSSION

From the evidence summarized in Fig. 1, GAMBELL and GRZEGORZEWSKA (1967) concluded that a translucent lamina is laid down on the growing edge of the average sperm
whale tooth between June and August off Durban. Similarly, from material including that
illustrated in Fig. 2, BEST (1970) considered that a translucent dentine lamina is deposited
in spring in the South African whales. He further suggested that as his data gave no
evidence for a second translucent lamina being formed about 6 months earlier (March/
April), only one such lamina is formed each year. Similar conclusions on the timing can
now be made for the Western Australian females (Fig. 3).

In the Antarctic, GAMBELL and GRZEGORZEWSKA (1967) suggested that a translucent
lamina is formed between December and February according to the data for male sperm
whales shown in Fig. 1. The information from male sperm whales off Western Australia
(Fig. 3) also indicates that this lamina must be deposited during this same period. Although
these months are the close season for the Western Australian whale fishery, the high
levels of the far-from-edge and on-edge categories in November/December and March/
May respectively point firmly to this conclusion.

Thus it seems that there are variations in the time at which the translucent laminae,
and conversely the opaque laminae, are formed in sperm whales in different parts of the
southern hemisphere. It is instructive now to consider if there are any identifiable features
linking the male and female sperm whale populations off South Africa with the females
off Western Australia, but separating them from the Antarctic and Western Australian
males.

An obvious possibility lies in the known latitudinal segregation and seasonal movements
which occur in this species. Females are seldom found further south than 45–50°S, but
the males and particularly the bigger and older bulls segregate and penetrate into higher
latitudes.

A mathematical model of the segregation of the male sperm whales in the North Pacific
constructed by OHSUMI (1966) also seems applicable to the southern hemisphere, on the
evidence of cyamid infestation and seasonal densities (BEST, 1974). The males appear to
segregate at an increasing rate from 12 years of age onwards. By about 18 years, half of
them have segregated to higher latitudes, and the rate remains constant after age 25 years
when it varies between 75% and 90%.

BANNISTER (1974) has pointed out that the modal length and age of the Western
Australian catches are higher than those of catches in similar latitudes at Donkergat and
by pelagic fleets. The age composition of the male sperm whales caught at Donkergat in
1963 and 1964 as given by BEST (1970) is very similar to that at Durban in 1962, 1963,
1965 and 1967, illustrated in Fig. 4. The modal age found at Donkergat is 18 years and
at Durban is 17 years. This can be compared with the Western Australian age data for
1973 and 1974 plotted in Fig. 4. The age composition in these recent seasons is close to
that found by BANNISTER (1974) in 1962–1966, when full recruitment occurred between
21 and 25 years. The present modal age is 24 years.

The Western Australian male sample is much more like the Antarctic whales shown
in Fig. 4 rather than the South African material considered here in terms of its age
structure. The size and age compositions of the South African and Western Australian
catches of female sperm whales are also rather different (Fig. 4), but no age-related differences of migration or segregation are known in this sex. It seems reasonable on these

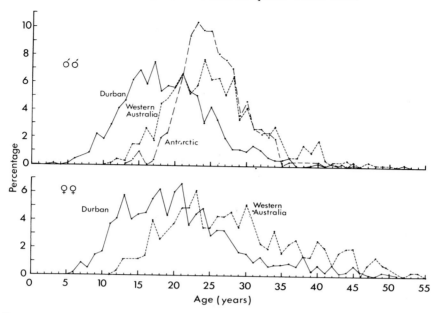

Fig. 4. Percentage age compositions of the male (upper) and female (lower) sperm whales sampled at Durban, in the Antarctic and Western Australia.

grounds to suggest that such observed differences and similarities reflect the degree of affinity of the whales concerned. This may extend into other characters bearing on such factors as the social structure, organization, behaviour and basic metabolism of the sperm whales in various populations in the southern hemisphere.

CONCLUSION

The alternating translucent and opaque laminae in the dentine of sperm whale teeth are composed of mineral and organic material. The reason for the different appearances of the laminae is not known for certain, and neither is the causal mechanism. It has been suggested that there may be a connection with variations in the perfection of mineralization of the dentine, due to fluctuations in vitamin D intake and synthesis, perhaps resulting from variations in direct ultraviolet dosage. Both of these factors would be very variable from year to year as well as from place to place. However, the laminae are deposited regularly throughout the life of the whale so long as the pulp cavity remains open. It now appears that the time of year at which the laminae are formed varies between certain groups of sperm whales in the Southern Hemisphere. In males and females off South Africa, and females off Western Australia, the translucent and presumably more highly mineralized lamina is laid down in spring. In the Antarctic males and those off Western Australia, this lamina is formed in summer. It is suggested that these differences may reflect fundamental variations between these populations of whales, which could be the result of nutritional and/or genetic variations between populations.

Although the bigger bulls are known to migrate between the Antarctic and more temperate waters (BANNISTER, 1974; BEST, 1974), it appears that in all sperm whales only one translucent and one opaque lamina is formed each year. The number of dentinal growth layers is thus a good index of the absolute age of the sperm whale.

Acknowledgements—Grateful thanks are due to Mr. J. L. BANNISTER (Western Australian Museum, Perth) for providing the sperm whale teeth which formed the new material for this study. Thanks are also due to Dr. P. B. BEST (Sea Fisheries Branch, Cape Town) for copies of his original data quoted in this paper.

REFERENCES

BANNISTER J. L. (1974) Whale populations and current research off Western Australia. In: *The whale problem: a status report*, W. E. SCHEVILL, editor, Harvard University Press, Cambridge, Mass., pp. 239–54.

BERZIN A. A. (1961) Materialy po razvitiyu zubov i opredeleniyu vozrasta kashalota. *Trudy Soveshchanni. Ikhtiologickeskaya Kommisya*, Moscow, **12**, 94–103.

BERZIN A. A. (1971) *Kashalot*, Pishchevaya Promyshlennost, Moskva, pp. 1–367.

BEST P. B. (1970) The sperm whale (*Physeter catodon*) off the west coast of South Africa. 5. Age, growth and mortality. *Investigational Report. Division of Sea Fisheries, Republic of South Africa*, No. 79, pp. 1–27.

BEST P. B. (1974) The biology of the sperm whale as it relates to stock management. In: *The whale problem: a status report*, W. E. SCHEVILL, editor, Harvard University Press, Cambridge, Mass., pp. 257–93.

BOW J. M. and C. PURDAY (1966) A method of preparing sperm whale teeth for age determination. *Nature, London*, **210**(5034), 437–438.

GAMBELL R. (1972) Sperm whales off Durban. *Discovery Reports*, **35**, 199–358.

GAMBELL R. and C. GRZEGORZEWSKA (1967) The rate of lamina formation in sperm whale teeth. *Norsk Hvalfangsttidende*, **56**(6), 117–121.

INTERNATIONAL WHALING COMMISSION (1967) Sperm whale sub-committee meeting report. *Report of the International Commission on Whaling*, **17**, 120–127.

INTERNATIONAL WHALING COMMISSION (1969) Report of the meeting on age determination in whales. *Report of the International Commission on Whaling*, **19**, 131–137.

KLEVESAL G. A. and S. E. KLEINENBERG (1967) Opredelenie vozrasta mlekopitayushchikh po sloistym strukturam zubov i kosti. *Akademiya Nauk SSSR, Moscow*, pp. 1–144.

OHSUMI S. (1966) Sexual segregation of the sperm whale in the North Pacific. *Scientific Reports of the Whales Research Institute, Tokyo*, No. 20, 1–16.

OHSUMI S., T. KASUYA and M. NISHIWAKI (1963) Accumulation rate of dentinal layers in the maxillary tooth of the sperm whale. *Scientific Reports of the Whales Research Institute, Tokyo*, No. 17, pp. 15–35.

Observations on diving behaviour of the sperm whale
Physeter catodon

C. LOCKYER

Whale Research Unit, Institute of Oceanographic Sciences,
c/o British Museum (Natural History), Cromwell Road, London SW7 5BD*

Abstract—Observations, made from ships, on diving behaviour of sperm whales are analysed for different sizes of animals. Depth, duration, descent and ascent rates, respiratory rhythm and social groupings are all considered. During chasing, most sperm whales dive for periods less than 10 minutes to less than 400 m, although large whales can dive to 1100 m and for periods of up to an hour. Average ascent and descent rates from and to different depths for various durations range between 70 and 158 m/min and 87 and 168 m/min, respectively. The respiratory rate between dives generally varies between 4 and 7 blows/min. There appears to be a similar pattern of social groupings during diving as that seen at the surface by various authors. The smaller whales tend to be more social than the large bulls which are frequently solitary on dives.

METHODS

SPERM whales are usually abundant off the east coast of South Africa in February when they are generally observed to be migrating northwards up the coast for the winter (GAMBELL, 1972).

Observations on sperm whale diving behaviour were made during a series of cruises operating off South Africa for the prime purpose of marking whales as part of a fishery monitoring programme. The cruises took place at the following times and within the geographical limits indicated:

1. February 1–21, 1972, between latitudes 29°S and 32°S and longitudes 29°E and 33°E.
2. February 1–21, 1973, between 30°S and 35°S and 25°E and 33°E.
3. November 24, 1973–February 3, 1974, between 20°S and 40°S and 30°E and 70°E.
4. January 18–31, 1975, between 30°S and 38°S and 21°E and 32°E.

Surface-water temperatures varied between 17.7°C and 27.1°C in latitudes near 40°S and 20°S respectively. Noon air temperatures ranged between 18°C and 34°C in the same latitudes respectively.

The cruises were operated by chartering a whale catcher of about 610 gross tons, with a top cruising speed of about 14 knots.† The vessels were equipped with Simrad 24-kHz sonar sets capable of emitting a ping in an effective range up to 2500 m at angles of tilt down to 90° to the horizontal.

The depth of the whale tracked on sonar was calculated from the range and angle of tilt to the horizontal. The angle of tilt was always rounded to the nearest 5° and the range was usually recorded to the nearest 10 m. The sonar was inadequate for accurately

---

* Present address: Whale Research Unit, c/o British Antarctic Survey, Madingley Road, Cambridge CB3 0ET.
† 1 knot = 0.51 m s⁻¹.

tracking whales close to the surface when the angle of tilt was 0–5°. These limitations reduced the accuracy to which the depths could be measured. Most observations were made at 35–70° angle of tilt and at a range much less than 1000 m.

Irregular ups and downs of about 25 m depth change recorded on dives were treated as insignificant, especially on shallow dives when the results of the sonar were inaccurate because of the shallow angle of tilt. During the dive period, the ship usually drifted at the surface, or moved very slowly (perhaps 3 knots) just to keep the whales within 1000 m range. Immediately prior to their surfacing, the ship would steam quickly towards the whales in order to close within 10 m of their position. Only rarely did the vessel have to steam to stay with whales during their dives. However, both the movements of the vessel at the surface and of the whales whilst tracking, particularly at long range, introduced Doppler shift errors into the estimation of the depth of dive. This emphasizes the probable insignificance of the irregular ups and downs.

Whales could be easily observed from the bridge and gun platform at the bow of the vessel while at the surface, and could be tracked underwater by means of sonar.

In March 1972 a few observations on diving behaviour were made from a Cessna 310P aeroplane flying at 125–155 m above sea level at 130–135 knots. These aerial observations, whilst limited by the inability to track whales underwater, had the advantage of being of animals undisturbed by a ship's machinery and gear. This disturbance by the ship's presence almost certainly influenced the behaviour of the whales, so that this factor must be borne in mind when relating observed behaviour to natural behaviour.

All observations on diving behaviour were aimed at covering the entire dive sequence including pre-dive period at the surface, blowing intervals, dive depth and duration, resurfacing time and recovery period, until either the whale was lost or left after marking. All time intervals were recorded using a 1-hour stopwatch with a sweep second hand. The number of whales diving in a group was defined as the number of animals submerging simultaneously in the same area, remaining together during the dive, and resurfacing simultaneously. Occasionally, a whale joined the group underwater, or one of the original group separated, but the number in the group was always determined as the least number remaining together. The body lengths of the whales followed were recorded to the nearest foot (and later converted to metres) in whaling tradition. The gunners aboard the whaling vessels were sufficiently experienced to estimate the size of whales chased to this level of accuracy. Any errors in size estimation were mostly within ±0.3 m.

RESULTS AND DISCUSSION

A. *Depth and duration of dives*

1. *Dive depth/duration correlation and frequency*

The observed frequencies of dives, by maximum depth and total dive duration, by individual whales, regardless of size or inclusion in a group dive are shown in Table 1. There is a positive correlation between depth and duration ($r = 0.755 \pm 0.014$). The greatest proportion of all dives was to depths less than 400 m for periods of less than 10 minutes. Dives below 800 m and for more than 30 minutes formed less than 5% of all dives observed. KOOYMAN (1968) and KOOYMAN and ANDERSEN (1969) have shown that short shallow dives to less than 100 m for under 5 minutes are most common for the Weddell seal,

Table 1. *Depth and duration correlation of individual sperm whale dives; all sizes of individuals at all times of day are included*

Depth (m)	Time (min)												Total
	0–5	–10	–15	–20	–25	–30	–35	–40	–45	–50	–55	–60	
0–99	112	37	1										150
100–199	118	36		1									155
200–299	45	79	12										136
300–399	7	80	30	14	2	5							138
400–499	5	44	66	20	2		2						139
500–599		34	31	10	21	8							104
600–699		8	13	5	14	2	1	2					45
700–799		3	6	9	8		9		2				37
800–899					1	8	2					5	16
900–999				2					8				10
1000–1099													–
1100–1199						1							1
Total	287	321	159	61	48	24	14	2	10	–	–	5	931

Leptonychotes weddelli, which is known to be able to dive to 600 m depth and for periods up to 44 minutes. These authors also found a correlation between dive depth and duration for the Weddell seal. The circumstances surrounding these observations differ, however, in that the sperm whales were being harassed by the pursuing vessel and might not have been performing dives true to their natural behavioural pattern, whereas the seal was virtually free and unharassed in its behaviour.

2. Variation of diving habits with group number and body size

The sperm whale data have been further analysed by examining the dive depth and duration achieved by different size classes of animals. The animals have been grouped roughly into five categories ranging from calves to harem master size adult bulls. The whales observed have been grouped roughly by development and sexual maturity, according to size, based on BEST (1974). The numbers of observations in each category are not always large, but both Table 2, illustrating depth with size, and Table 3, showing duration with size, indicate a trend for greater potential diving ability with an increase in body size, and consequently with age, and sexual and social development. KOOYMAN (1968) found that young Weddell seal pups were not as able in diving performance as the adults.

Nevertheless, sperm whales are able to dive very soon after birth. Two observations suggest this. Firstly, in 1972 a 4.3-m calf accompanying a 10.6-m whale was recorded as diving for $16\frac{1}{2}$ minutes, to 670 m. The depth/time profile of this dive is shown in Fig. 9. Average birth size is estimated to be about 4.0 m (BEST, 1968; GAMBELL, 1972). Secondly, in 1973, the events surrounding the birth of a sperm whale were witnessed (GAMBELL, LOCKYER and ROSS, 1973), and the 3.7-m calf, separated from its mother, remained submerged for 7 minutes about 1 hour after birth.

The diving ability of adults and larger sperm whales appears very great. HEEZEN (1957) reviewed incidences of sperm whales becoming entangled in bottom cables. The indication was that these deep excursions down to depths of as much as 620 fathoms (1138 m) were

Table 2. *Cumulative percentage of individuals diving to different depths, throughout the day*

Size of whale in metres	Depth of dive in metres												Sample size
	0–99	100–199	200–299	300–399	400–499	500–599	600–699	700–799	800–899	900–999	1000–1099	1100–1199	
4.0–7.9 calves and adolescents	100.0	80.0	55.0	50.0	45.0	35.0	10.0						20
8.0–9.4 sexually immature ♂♂ and ♀♀	100.0	80.9	71.0	65.6	57.2	24.5							131
9.5–10.9 sexually mature ♀♀ and pubertal ♂♂	100.0	86.4	68.2	50.9	33.2	21.6	12.3	6.5	3.6	1.3			633
11.0–11.9 maturing ♂♂	100.0	66.1	53.8	50.8	35.4	23.1	20.0	16.9	1.5				65
12.0–16.4 sexually mature ♂♂	100.0	84.1	67.1	47.6	42.7	21.9	19.5	14.6		3.7		1.2	82

Table 3. Cumulative percentage of individuals diving for different periods of time, throughout the day

Size of whale in metres	Duration of dive in minutes												Sample size
	<5	<10	<15	<20	<25	<30	<35	<40	<45	<50	<55	<60	
4.0–7.9 calves and adolescents	100.0	80.0	50.0	15.0	10.0								20
8.0–9.4 sexually immature ♂♂ and ♀♀	100.0	70.2	32.8	6.1	3.0	0.8							131
9.5–10.9 sexually mature ♀♀ and pubertal ♂♂	100.0	68.1	32.1	17.7	10.9	5.8	3.0		2.0			0.8	633
11.0–11.9 Maturing ♂♂	100.0	70.8	47.7	27.7	20.0	6.1		1.5					65
12.0–16.4 sexually mature ♂♂	100.0	71.9	43.9	28.0	18.3	15.8	13.4	3.7	2.4				82

made by large bulls. The deepest dive observed on these marking cruises was to 1140 m by a 13.8 m bull. NEMOTO and NASU (1963) suggested that sperm whales dive to the sea bottom to depths exceeding 600 m, in quest of food, since the stones, sand and deep bottom-dwelling organisms are often found in their stomachs. There is a record by CLARKE in WOOD (1972) of two bull sperm whales diving for periods of 53 minutes and 1 hour 52 minutes. On capture, the stomach of one animal yielded two specimens of *Scymnodon*, a small shark, thought only to live on the sea bottom; the sounding in the area was about 3200 m.

The average dive depth and duration for different size classes was about the same for all sizes of whale. The mean dive depth for all sizes of whales fell between 315 and 360 m, and the mean dive duration centred around 10½ minutes. This is perhaps not unexpected since the over-riding factor influencing many dives was alarm which probably affected all sizes and ages of whales equally. The depth and duration range have been shown to increase with body size, but deeper dives were less usual, consequently the depths and durations of most dives were similar for all sizes of sperm whales. This may be a reflection of the social behaviour and whale size groupings. The data in Table 4 show in the first column the average number of whales of similar size on any group dive. The second column indicates the average total number of whales of mixed sizes including the particular size considered on a group dive. The third column shows for a particular size class, the percentage participation in group dive, including other size categories as an indication of the frequency of size mingling.

In summary, the results in Table 4 show that the smallest whales usually dived (94% of the dives) with several other larger whales, although frequently they were the sole representatives of their size class in the group. The middle-size whales also usually dived

Table 4. *Average number of individuals together on a group dive, and average depth and duration of the dive, throughout the day*

	1	2	3
Size of whales in metres	Average number of whales of same size on group dive	Average number of whales of mixed sizes on group dive	Mixed size dives as % of all group dives
4.0–7.9 calves and adolescents	1.2	5.9	94
8.0–9.4 sexually immature ♂♂ and ♀♀	3.6	5.6	47
9.5–10.9 sexually mature ♀♀ and pubertal ♂♂	4.1	4.6	17
11.0–11.9 maturing ♂♂	2.2	2.4	10
12.0–16.4 sexually mature ♂♂	1.7	1.9*	2*

* Some doubt about the mixed size association because the bull surfaced separately.

with a few other whales, although less frequently with whales of other sizes. The large bulls generally dived solo or in pairs, and hardly ever associated with other sized animals on a dive.

The numbers of whales seen in groups whilst diving were always less than the numbers first observed when the vessel initially approaches a school of sperm whales. As the vessel approaches a school, the whales generally dived at a distance of 300 m from the vessel, and then dispersed, breaking into smaller groups (cf. GAMBELL, 1968). The social grouping of the sperm whale, well documented by OHSUMI (1971) and GAMBELL (1972), appears to be reflected in the size composition of the diving groups, so that diving behaviour in this respect is probably governed by social organization.

B. *Diving pattern*

1. *Dive depth/time profile*

A number of dive profile types have been recognized during chasing. The majority of dives were relatively shallow and of short duration. Many dives in this category were quick down and up manoeuvres, and some assumed a 'V'-shaped profile to a depth somewhere between 50 m and 200 m (Fig. 1). Others were very shallow, some being little more than prolonged pauses between blows (Fig. 2).

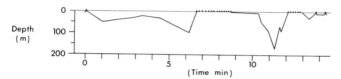

Fig. 1.   An example of a series of short and shallow dives by a bull sperm whale ($\sim$ 12.8 m) which was one of a pair (5 February 1973).

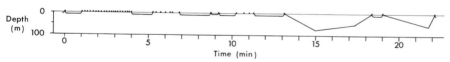

Fig. 2.   An example of series of shallow dives interspersed with prolonged pauses between blows by a lone bull sperm whale ($\sim$ 13.7 m, 20 December 1973).

Other dives of intermediate duration and depth were generally 'V' or 'U'-shaped, and were probably prolongations of the type described in Fig. 1. Two intermediate depth and duration dives profiles are shown in Fig. 3. These, like the shallow dives, were possibly precipitated by alarm to chasing.

Very deep dives generally commenced with a rapid descent to the maximum depth, and then, either an excursion at this depth, or a rapid ascent. An example of the deep 'V'-shape profile is shown in Fig. 4. An example of a 'U'-shape profile of an excursion dive at depth is shown in Fig. 5.

Variations often occurred as a combination dive which, for example, may have involved a rapid descent and then a slow partial ascent, a shallow excursion, then resurfacing (Fig. 6). Groups of whales sometimes split up or joined during long excursions (Fig. 7),

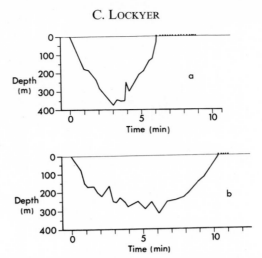

Fig. 3.   Two examples of dives of intermediate depth and duration by lone sperm whales: (a) ~9.7 m, (b) ~10.6 m
(both 15 February 1973).

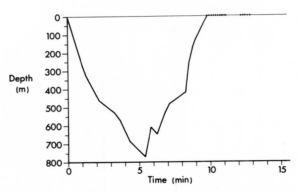

Fig. 4.   An example of a deep dive by a sperm whale (~9.7 m) in a group of three (5 February 1973).

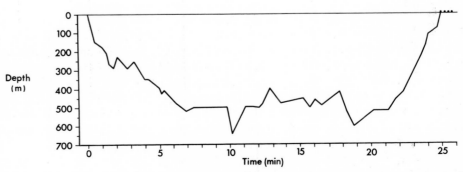

Fig. 5.   An example of a prolonged excursion at depth by a sperm whale (~9.1 m) in a group of three
(8 February 1973).

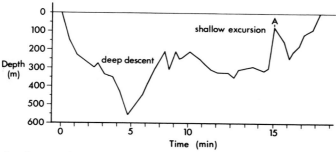

Fig. 6. An example of a complex dive by a lone sperm whale (~9.7 m) during which disturbance by the approaching catcher may have halted ascent at point A (8 February 1973).

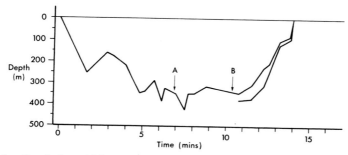

Fig. 7. Profile of a dive during which regrouping occurred. Initially two whales (~9.5–10.0 m) and a calf (~4.6 m) dived. Fast swimming underwater occurred from point A to point B. Close to B a third whale joined the group and all four whales surfaced together (31 January 1975).

and frequently the whales were 'talkative' during deep dives—that is there was much clicking (as described by WORTHINGTON and SCHEVILL, 1957; PERKINS, FISH and MOWBRAY, 1966; and WATKINS and SCHEVILL, 1975) and creaking, audible on the sonar receiver, especially when groups of animals reconvened underwater. These deep dives may have been not just for the purpose of avoiding the vessel, but for some social communication or feeding. Deep dives may be performed to get out of the sonar range. WATKINS and SCHEVILL (1975) found that pingers apparently upset the usual communicatory behaviour of sperm whales, at short range. The whales probably need to echo-range and produce sounds to communicate in order to locate each other at depths where light intensity is very low, or negligible below about 200 m.

KOOYMAN and ANDERSEN (1969) and KOOYMAN (1968, 1975) described different dive profiles for the Weddell seal. These profiles were generally similar to those of the sperm whale, but also included many shallow horizontal excursions between breathing holes.

In the sperm whale different types of dive may follow each other without set sequence (e.g. Fig. 8). The most significant pattern here appears to be the change from a few leisurely blows after the initial dive to prolonged and frequent blowing after about half an hour of chasing. This increase in respiratory activity is discussed in more detail later.

In all the dive profiles shown, there are frequently irregular ups and downs, in some cases of about 100 m amplitude. Vagaries of about 25 m depth change are in some instances due to the limitations of the method of recording, and sudden depth changes of even 100 m may be insignificant at 1000-m range. The profiles are portrayed as jerky

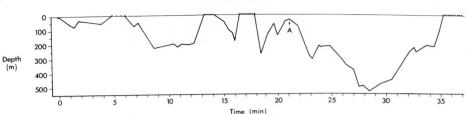

Fig. 8. A series of dives performed by a lone bull sperm whale (~ 12.8 m) which was possibly alarmed by the catcher's approach at A (5 February 1973).

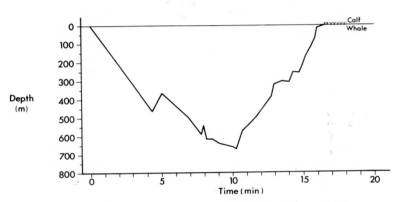

Fig. 9. An example of a dive profile for a whale (~ 10.6 m) accompanied by a calf (~ 4.3 m). Note the two initially resurface together, but the calf has a subsequently higher blow frequency (9 February 1972).

movements, due to the method of recording, and the true profiles probably would have been quite smooth. The ups and downs are particularly apparent during long deep excursions such as in Fig. 5, when the errors introduced by Doppler shift during tracking at long range were most apparent. However, similar ups and downs occurred during long excursions of the Weddell seal (KOOYMAN, 1968), when depth change was frequently quite sudden, although a different tracking method was employed.

Very often, when sperm whales remained at a particular depth, especially at shallow and intermediate depths, they rapidly swam horizontally at about 11–12 knots for a considerable period, and they appeared to be trying to escape from the following vessel. This speed was not generally maintained for more than a few minutes without avoidance tactics coming into play. These tactics included altering course sharply at the same time as surfacing and blowing. The pursuing vessel was at times forced to complete a tight circuit, while the whales again altered direction and escaped. Such an incident is shown in Fig. 10. The track of the ship is seen in the background, with a circle of smooth water inside it. The two whales swimming away in the centre were at this point abeam of the vessel, having outmanoeuvred it.

2. Descent and ascent rates

In Table 5 the mean descent rates have been calculated for dives of different duration to different maximum depths. Table 6 gives similar data on mean ascent rates. For all

Fig. 10.   Photograph showing the result of tactical movements by sperm whales used to avoid a pursuing vessel. The track of the ship describes a tight circle enclosing a patch of water into which the pursued animals veer off abeam of the vessel, and out of reach.

dive/duration categories, the rates were normally distributed except on shallow dives of short duration which were showed by occasional high rates. For these categories, the logarithmic mean is also given. The ranges of rates were far wider in the shallow depth/ short time categories. It is possible that very rapid ascent/descent rates are only possible to and from shallow depths after short periods because the physiological condition of the whale becomes more complex during deeper and longer dives.

The means of each category where valid standard deviations could be calculated (excepting the category 0–199 m, < 10 min, where the arithmetic mean is inappropriate for the non-Gaussian distribution) have been compared with each other using Student's t-test (BAILEY, 1959) for small samples.

Table 5. . *Observed ranges and mean (arithmetic) descent rates for sperm whales in m/min (100 m/min = 3.24 knots) with the sample size (N)*

Maximum dive depth (m)		Dive duration in minutes		
		< 10	10–20	⩾ 20
0–199	range	21–600 m/min	29	
	mean	127 ± 102		
	N	58	1	
		(log. mean = 102)		
200–399	range	39–375	26–360	80
	mean	132 ± 67	106 ± 80	
	N	41	14	1
		(log. mean = 118)		
400–599	range	81–250	57–180	61–373
	mean	133 ± 52	110 ± 38	132 ± 99
	N	10	16	8
600–1199	range	114–142	91–308	57–228
	mean	128	168 ± 72	87 ± 45
	N	2	7	13

Table 6. *Observed ranges and mean (arithmetic) ascent rates for sperm whales in m/min (100 m/min = 3.24 knots) with the sample size (N)*

Maximum dive depth (m)		Dive duration in minutes		
		< 10	10–20	⩾ 20
0–199	range	23–420	68	
	mean	114 ± 80		
	N	66		
		(log. mean = 94)		
200–399	range	60–333	45–164	38–102
	mean	124 ± 64	91 ± 32	
	N	47	12	2
		(log. mean = 112)		
400–599	range	101–222	66–296	45–154
	mean	141 ± 35	126 ± 25	89 ± 38
	N	11	25	6
600–1199	range	142–174	99–168	75–143
	mean		133 ± 25	93 ± 21
	N	2	7	11

Results in Table 5 indicate that only the mean in the category 600–1199 m 10–20 min, is significantly different from the other means at the 95% level. There is no consistent trend for descent rates to fall off or increase either with depth or duration, so that descent rates, except sometimes on shallow dives of short duration, are similar regardless of the depth/time category. The reason for the significantly different rate on the start of medium duration deep dives is unclear. The descent rate averaged 122.4 m/min for all dives observed, equivalent to 3.96 knots; an easy swimming pace for the sperm whale. GASKIN (1964) observed sperm whales usually at a rate of 3–4 knots.

Mean ascent rates, in contrast, show many significant differences. Faster ascent rates occurred from greater depths yet the ascent rate decreased with prolongation of a dive, although larger sample sizes are needed to confirm these trends, they probably reflect the internal physiological and chemical changes which occur at pressure, plus perhaps fatigue effects after a long dive.

KOOYMAN (1968) observed that the ascent rate of the Weddell seal decreased with greater dive time, and ascribed this to the chemical changes in the seal, specifically to the need to remove the excessive build up of dissolved nitrogen in the blood and tissues. A slower ascent rate after longer dives is perhaps understandable considering two factors:

1. Increased solution of nitrogen in the blood and tissues with increased duration on a dive necessitates slower ascents to maintain the equilibrium of the gas.
2. Fatigue resulting from a long dive would not permit as much swimming effort during surfacing as would be possible after a short dive.

The faster ascent rates from greater depths are not so clearly understandable. It is possible that shallow dives do not present the whale with the same urgency in surfacing, in that the surface is more quickly attainable. Possibly physical factors of the water, such as salinity changes, temperature gradients and pressure alterations with depth may affect the buoyancy and swimming control of the whale so affecting the surfacing speed. In addition CLARKE (1970) proposed an automatic control system in the spermaceti organ of the head of the whale, which he theorized could trim the ballast of the whale through thermal-density changes. SCHENKKAN and PURVES (1973) also ascribed various functions connected with diving to the spermaceti organ, but whilst not disputing CLARKE's theory showed convincingly from anatomical studies that the method of cooling CLARKE had postulated was in error. Clearly there is much to learn about the diving physiology of the sperm whale, and its control of descent and ascent.

KOOYMAN and ANDERSEN (1969) observed that the Weddell seal had a descent rate of 104 m/min to 360 m on a 6 min 50 sec dive, and ascended from 200 m at 120 m/min. They also commented that the porpoise *Steno bredanensis* can descend at the rate of 100 m/min, and *Tursiops truncatus* can ascend at 143 m/min from 100 m. These observations were for single events, but none the less tend to confirm that rapid and constant rates of ascent and descent in various marine mammals are both possible and normal.

The two most rapid ascent and descent rates observed for the sperm whale were an ascent of 675 m/min (21.9 knots) by a single 7.3 m whale on a dive to a maximum depth of 225 m for 2 min 8 sec; and a descent of 760 m/min (24.6 knots) by a group of eight 10–11 m whales on a dive to a maximum depth of 190 m for 5 min 50 sec. These rates seem exceptionally high. The whales swimming at these speeds must be moving at about their maximum speed. BERZIN (1972) in his review of speed of movement of sperm whales concluded that a speed of the order of 20 knots could be developed only when a sperm whale was seriously frightened or wounded, and could be maintained only for a short while. The descent speed of 24.6 knots could be higher than the horizontal maximum swimming speed because gravity and buoyancy control by the whales may have added to the speed. As these speeds seem unusual, they have been omitted from Tables 5 and 6.

3. *Influence of the environment on diving pattern*

(a) *Presence of vessels.* GAMBELL (1968) observed that a school of sperm whales readily disperses when a vessel approaches, and forms smaller groups. This was observed regularly during marking cruises, implying that the mean numbers of whales diving together in a group during chasing probably constitute the minimum group sizes usually found.

On numerous occasions after sighting a school and giving chase the whales would simultaneously dive when the ship approached within 300 m, scattering in different directions underwater before resurfacing in smaller groups. It was frequently clear that diving to shallow depths was an escape response to the vessel. The usual practice was for the ship to idle waiting in the area until the whale's return to the surface was imminent and then to steer full ahead to meet the whale. As a result whales often were alarmed by the sudden surface engine noise, and plunged straight down again from a relatively shallow depth (Figs. 6 and 8). The turning point where the ascent halted and the whale descended again was usually between 50 m and 100 m. It is possible at this depth for the whale to see the course and movement of the ship as well as hear it.

Very occasionally, a whale was observed to defaecate near the surface prior to a dive whilst being pursued. A parallel observation was when a newly born calf, separated from its mother, became confused and swam headlong into the side of the vessel. The calf defaecated (perhaps meconium) and appeared stunned (GAMBELL, LOCKYER and ROSS, 1973).

(b) *Time of day*. In Table 7 the relationships, for the 9.5–10.9 m sized whales between the dive depth and duration with time of day are examined.

This was the only size category for which sufficient data were available to examine the effects of time of day. The number of whales in a group diving together tended to increase throughout the day, groups in the evening being almost double the size of groups in the morning. However, comparing the 95% confidence limits, there appears to be some overlap of mean ranges of average group size, so this trend needs confirmation with more data. The whales are probably naturally very flexible in their social groupings, but the chasing may have disrupted the normal pattern of grouping.

Table 7.   *Analysis of group diving behaviour at different times of the day, for whales of 9.5–10.9 m length*

Dive details for the group	Time of day		
	04.30–10.59 hr morning (light)	11.00–15.59 hr day (light)	16.00–20.30 hr evening (light to dark)
Average group number of whales of 9.5–10.9 m only, and also with other sizes on a dive, with standard error of mean and 95% range; n = sample size	3.07 ± 0.33 $2.43 - 3.72$ $n = 41$ groups	4.21 ± 0.29 $3.64 - 4.78$ $n = 100$ groups	6.62 ± 1.24 $4.18 - 9.06$ $n = 13$ groups
Average depth of dive in m, with standard error	284.2 ± 2.6	315.0 ± 1.8	465.4 ± 6.0
Average duration of dive in min, with standard error	9.2 ± 0.5	8.1 ± 0.3	17.9 ± 1.2

The average depth and duration of dives showed a definite trend to increase in the evening hours, compared with those in the morning and middle of the day hours, which were fairly similar. It would be interesting to know if this trend is continued into the night hours.

The frequency distributions of the depth and duration data closely followed a Poisson distribution, with the greatest number of dives being performed in shallower depths for short periods. The right-hand tails of the distributions, however, extended as the day progresses.

KOOYMAN (1975) found that the depth and duration of dives by the Weddell seal decrease at night-time, and the diurnal pattern may be influenced greatly by visibility. However, visibility is not likely to contribute greatly to diving potential in the sperm whale which frequents depths exceeding 200 m where darkness virtually prevails anyway. None the less, it is possible that the sperm whale may prefer to remain in shallower depths in daylight because of the added visibility. However, the vertical migrations of the whale's prey may also have an important but unknown influence.

(c) *Wind and sea conditions.* The depth and duration of group dives were analysed according to the wind speed prevailing. Wind speeds varied from 5 to 35 knots although just over 50% of all observations were in wind speeds of less than 15 knots.

The observed data were compared with an expected distribution of dive depth or duration with wind speed, assuming no correlation between these factors. Calculation of χ^2 values indicated that there was no significant correlation between dive duration and wind speed. However, results of χ^2 calculation for dive depth with wind-speed data (Table 8) indicate a highly significant correlation ($p < 1\%$). The percentage values, given in parentheses in Table 8, show that there was a tendency for whales to leave the top 100 m of the sea surface in higher wind speeds, perhaps in response to the rougher water conditions here.

Table 8. *Observed (O) and expected (E) dive depth distribution with wind speed for sperm whales, and observed values as percentages of total sample in each wind-speed category*

Wind speed (knots)	Dive depth in metres				Total
	< 100	100–200	> 200–400	> 400	
< 15	$O = 25$ $E = 18.8$ $\%O = 19$	$O = 27$ $E = 25.5$ $\%O = 20$	$O = 34$ $E = 40.8$ $\%O = 26$	$O = 46$ $E = 46.9$ $\%O = 35$	132
15–24	$O = 11$ $E = 14.3$ $\%O = 11$	$O = 16$ $E = 19.3$ $\%O = 16$	$O = 31$ $E = 30.9$ $\%O = 31$	$O = 42$ $E = 35.5$ $\%O = 42$	100
$\geqslant 25$	$O = 1$ $E = 3.9$ $\%O = 4$	$O = 7$ $E = 5.2$ $\%O = 26$	$O = 15$ $E = 8.3$ $\%O = 55$	$O = 4$ $E = 9.6$ $\%O = 15$	27
Total	37	50	80	92	259

$\chi^2 = \dfrac{\sum (O - E)^2}{E} = 17.26$:   degrees of freedom = 6.

There were two occasions during chasing when single animals headed into the sea at the surface, and appeared to experience interruption of the breathing rhythm. Frequently blowing or breathing did not occur when the whales broke surface, perhaps because of a wave washing over the head. However, the surfacing rhythm remained constant, only some surfacings permitting successful blowing. A whale chased under these conditions did not keep this course for long before diving. It is noteworthy that both observations were in fine weather, with a wind speed less than 15 knots and a sea state 2. Weather and sea conditions have little affect on sperm whale diving and blowing behaviour during chasing. However, it would be worthwhile investigating if wind and wave direction affect the course taken by unharassed sperm whales, particularly in rough weather.

GAMBELL (1968) mentioned that the majority of whales off Durban are found in the current lines where the waters are notably food-rich. During marking cruises off South Africa, the general impression was that sperm whales favoured the current lines from a check on surface temperatures. Aerial observations, when the demarcation of current lines is usually obvious from water colour changes, also tended to confirm Gambell's findings. Most coastal dives were observed in depths charted near or over the 1000-fathoms (2835 m) line.

C. Surface behaviour of individuals between dives

1. Respiratory rate

Observations during March 1972, from the air, gave a range of 3 to 8 blows/minute, smaller whales under 9.5 m tending to blow more frequently than bigger ones. On several occasions 9.5–10.9 m whales were seen to be resting at the surface in small groups, and blowing at the rate of 3 to 4 blows/minute, the higher rate again being for smaller animals. Dives observed were for periods less than 6 minutes. Other observations showed eight sperm whales of 8.0–9.4 m blowing 6 to 8 blows/minute after actively diving for 18 minutes. GAMBELL (1968) observed undisturbed bulls over 12 m length sounding for 35–60 minutes yet blowing 'very little' and lolling after diving.

For whales observed from shipboard, blow times were recorded between dives during chasing. Initially an analysis was done whereby size of whale, duration of previous dive and position of the dive in a dive series were all taken into consideration. However, unlike the observations made from the air, no persistent correlation between these factors was observed. The usual rate appears to be between 4 and 7 blows/minute for all sizes of whales during chasing.

In Figure 9 the blowing frequencies for a 10.6 m whale and a 4.3 m calf are shown after a $16\frac{1}{2}$ minute dive. The greater blowing frequency of the calf compared with the larger whale is clear. Extremes of the blowing range were by a 16.1 m bull which blew at a rate of about 2 blows/minute during the recovery period after two short dives of < 5 minutes immediately after a 29 minute dive to depths exceeding 600 m, and by a 4.3 m calf (Fig. 9) which blew at an average rate of 10 blows/minute during the 10 min 50 sec period following the dive to 670 m; only the initial recovery rate of 8 blows/minute is shown in Fig. 9, but this increased irregularly to 12 blows/minute in the last 3 minutes before the next dive to 200 m lasting $7\frac{1}{2}$ minutes.

Timings of blows from both shipboard and aerial observations fall within a similar range, although undisturbed resting whales seem to blow less frequently than actively swimming and diving ones. This is probably the result of a fall in metabolic rate when the whales are merely lolling at the surface. BERZIN (1972) quoted sperm whales as blowing 4–6 blows/minute. GASKIN (1964), who made observations both from shipboard and the air, found that undisturbed whales blow 2–3/minute.

2. Length of recovery interval at the surface between dives

Table 9 shows the average recovery period at the surface between dives. Some sample sizes are small, and as there were no consistent differences between whales of various body lengths, all data have been combined. The overall average recovery periods indicate that duration increases with prolongation of a dive, although blowing frequency does not necessarily increase. Thus the oxygen debt may be paid off by spending a longer time breathing after a prolonged dive. However, such is the variability in surface behaviour, that there is no way of accurately predicting the time spent at the surface between dives.

Table 9.   *Average time spent by sperm whales at the surface between dives, during chasing. Sample sizes in parentheses*

Body length (m)	Previous dive duration (minutes)		
	< 10 min	10–20 min	> 20 min
4.0–5.4	2.17 (3)		
5.5–7.9	2.26 (4)	3.00 (2)	
8.0–9.4	1.02 (16)	2.23 (7)	
9.5–10.9	1.40 (75)	3.21 (13)	3.04 (11)
11.0–11.9	1.49 (15)	2.18 (10)	4.00 (4)
12.0–14.9	1.55 (33)	2.40 (9)	3.28 (6)
15.0–16.4	1.27 (3)		
Total mean with standard error	1.42 ± 0.01 (149)	2.46 ± 0.04 (41)	3.22 ± 0.05 (21)

Often a whale may have only partially recovered from a deep dive when alarmed by chasing, so that a series of dives was performed before the fatigued whale was forced to remain at the surface for a recovery period of more than 3 minutes. Under these circumstances the whale remained lolling even at the approach of a vessel. One apparently fatigued whale showed reluctance to dive even after two marks had been fired into its back blubber. The average intervals observed between dives during chasing, in Table 9, almost certainly represent only a proportion of the total recovery period needed. This would explain some whales' exhaustion after multiple dives, because the whales had reached the limit of their physiological endurance and needed to recover not just from the preceding dive but the whole succession of dives.

KOOYMAN (1968) had very similar observations on dive recovery of Weddell seals. Usually the seals spent 2–3 minutes at the surface ventilating between deep dives. However, after a long dive of more than 20 minutes duration, the seal appeared reluctant to perform another deep dive and tended to remain near the surface. KOOYMAN explained that this behaviour might possibly aid maintenance of the nitrogen equilibrium in the tissues.

SUMMARY

Observations on diving behaviour in sperm whales were made during four whale-marking cruises aboard a whaling vessel, between 1972 and 1975 inclusive, off South Africa. Sperm whales were mostly observed to dive to less than 400 m for periods under 10 minutes. Dives below 800 m and for over 30 minutes formed less than 5% of all observed dives. Calves and juvenile whales appeared less able to achieve long deep dives than older and mature whales, although the average depth and duration of all dives appeared to be about the same for different size classes.

The younger animals seemed more sociable in their diving habits, and the number of whales in a group decreased in the larger body length classes. However, calves and juveniles were mostly observed in groups of whales of mixed sizes, whereas groups of larger-sized whales were generally of the same size. These size associations during diving seemed to be merely a reflection of the overall social groupings.

A number of dive profiles were observed, although basically a dive was either 'V'-shaped, with a rapid descent followed by a rapid ascent, or involved a horizontal excursion between descent and ascent. This pattern was consistent for all depths of dives. Occasionally a complex dive was observed, where a shallow excursion preceded or followed a deep stage. Descent and ascent rates were found to be comparable with those of other marine mammals, means to and from various depths ranging from 87–168 m/min and 70–158 m/min respectively.

Diving behaviour was found generally not to be affected greatly by weather conditions. The grouping behaviour of whales on dives may be affected by time of day, but this is probably a reflection of social behaviour and may not be directly associated with diving. However, indications are that whales probably remain at shallower depths for shorter durations during the morning and middle of the day, compared with afternoon and evening.

Ships and sonar almost certainly irritated or upset sperm whales which performed escape dives. However, actual responses to particular stimuli are difficult to define with certainty.

Sperm whales of all sizes tended to blow regularly at about 4–7 blows/minute, regardless of diving activities, perhaps less when lolling and undisturbed by a ship's presence. However, more time was needed at the surface after longer dives, for recovery.

Acknowledgements—I would like to thank the crews of the chartered whale catcher vessels, *C.G. Hovelmeier* and *Pieter Molenaar*, owned by The Union Whaling Co. Ltd., Durban, in particular sonar operators Mr. S. IRELAND, Mr. R. CROMBIE and Mr. W. CHRISTENSEN for their help and co-operation. Also, I am grateful to Dr. R. GAMBELL for his careful records of sperm whale dives observed during cruise 3 when I was not present. Thanks are also due to the flying crew of the Aircraft Operating Co., Durban, who enabled me to make aerial observations of sperm whale behaviour in 1972.

REFERENCES

BAILEY N. T. J. (1959) *Statistical methods in biology*, English Universities Press Ltd., London, 200 pp.
BERZIN A. A. (1972) *The sperm whale* (translated from Russian), A. V. YABLOKOV, editor, Israel Program for Scientific Translations, Jerusalem, 394 pp.
BEST P. B. (1968) The sperm whale (*Physeter catodon*) off the west coast of South Africa. 3. Reproduction in the female. *Investigational Report Division of Sea Fisheries South Africa*, No. 66, pp. 1–32.
BEST P. B. (1974) The biology of the sperm whale as it relates to stock management. In: *The whale problem*, W. E. SCHEVILL, editor, Harvard University Press, Cambridge, Massachusetts, pp. 257–93, 419 pp.
CLARKE M. R. (1970) Function of the spermaceti organ of the sperm whale. *Nature London*, **228**(5274), 873–874.
GAMBELL R. (1968) Aerial observations of sperm whale behaviour. *Norsk Hvalfangsttidende*, **57**(6), 126–138.

GAMBELL R. (1972) Sperm whales off Durban. '*Discovery*' *Reports*, **35**, 199–358.
GAMBELL R., C. LOCKYER and G. J. B. ROSS (1973) Observations on the birth of a sperm whale calf. *South African Journal of Science*, **69**(5), 147–148.
GASKIN D. E. (1964) Recent observations in New Zealand waters on some aspects of behaviour of the sperm whale (*Physeter macrocephalus*). *Tuatara*, **12**(2), 106–114.
HEEZEN B. C. (1957) Whales entangled in deep sea cables. *Deep-Sea Research*, **4**(2), 105–115.
KOOYMAN G. L. (1968) An analysis of some behavioural and physiological characteristics related to diving in the Weddell Seal. In: *Biology of the Antarctic Seas*, 3, G. A. LLANO and W. L. SCHMITT, editors, Antarctic Research Series (Washington), **11**, 227–261.
KOOYMAN G. L. (1975) A comparison between day and night diving in the Weddell Seal. *Journal of Mammalogy*, **56**(3), 563–574.
KOOYMAN G. L. and H. T. ANDERSEN (1969) Deep diving. In: *The biology of marine mammals*, H. T. ANDERSEN, editor, Academic Press, New York and London, pp. 65–94, 511 pp.
NEMOTO T. and K. NASU (1963) Stones and other aliens in the stomachs of sperm whales and specific soluble substances. *Scientific Reports of the Whales Research Institute*, Tokyo, **17**, 83–91.
OHSUMI S. (1971) Some investigations on the school structure of sperm whale. *Scientific Reports of the Whales Research Institute*, Tokyo, **23**, 1–25.
PERKINS P. J., M. P. FISH and W. H. MOWBRAY (1966) Underwater communication sounds of the sperm whale, *Physeter catodon*. *Norsk Hvalfangsttidende*, **55**(12), 225–229.
SCHENKKAN E. J. and P. E. PURVES (1973) The comparative anatomy of the nasal tract and the function of the spermaceti organ in the Physeteridae (Mammalia, Odontoceti). *Bijdragen tot de Dierkunde*, **43**(1), 93–112.
WATKINS W. A. and W. E. SCHEVILL (1975) Sperm whales (*Physeter catodon*) react to pingers. *Deep-Sea Research*, **22**, 123–129.
WOOD G. L. (1972) *The Guinness Book of animal facts and feats*, Guinness Superlatives Ltd., Enfield, 384 pp.
WORTHINGTON L. V. and W. E. SCHEVILL (1957) Underwater sounds from sperm whales. *Norsk Hvalfangsttidende*, **46**(10), 573–575.

The development of the 'Gloria' sonar system from 1970 to 1975

J. S. M. RUSBY and M. L. SOMERS

Institute of Oceanographic Sciences, Wormley, Godalming, Surrey, U.K.

Abstract—This paper summarizes the main developments which have taken place in the GLORIA long-range sonar system during the past 5 years of use, and describes how these have led to the design of a Mark 2 sonar.

The most significant factors in this progression have been the successful introduction of analogue correlation processing techniques which gave the original design an additional 20 dB signal-to-noise gain: and the use of this excess gain to demonstrate that the sonar would provide good-quality records to a range of 27 km in the deep sea using one-third of the original array height. During this period the sonar was also successfully demonstrated on the continental shelf, making geological and fisheries surveys out to a range of 13 km.

The paper concludes with a brief description of the Mark 2 design, which will use a neutrally buoyant nose-towed vehicle with a diameter approximately one-third that of the original vehicle, with port and starboard arrays covering a swathe 50 km wide in the deep sea. It will be handled by a portable cradle and winch module for use on different ships.

1. INTRODUCTION

THE purpose of this paper is to summarize the lessons learnt and the developments which have resulted in the 5 years since the GLORIA (Geological Long Range Inclined Asdic) system was first demonstrated.

In 1960 the Institute of Oceanographic Sciences, then called the National Institute of Oceanography, developed the first side-scan sonar (TUCKER and STUBBS, 1961) based on earlier work first reported by CHESTERMAN, CLYNICK and STRIDE (1958). As a result of the success of this type of instrument for surveying outcrops and sediment distribution on the continental shelf, Sir George Deacon, F.R.S. initiated a study in 1965 to determine whether the same acoustic principle could be used in the deep ocean to obtain sound pictures, or sonographs, of the ocean floor. The study showed that it should be possible to do this, that is to 'scale-up' the geometry and performance required by a factor of 25 accommodating oceanic depths of 8 km and ranges of 22 km, provided that

1. a high-power, low-frequency array with the necessary directivity could be made, and
2. towed at a depth of at least 100 m in order to
   (a) ensure sufficient radiated power without cavitation,
   (b) reduce the initial downward refraction of the sound rays,
   (c) lower the received noise level from the towing ship.

By 1969 such a sonar had been constructed and demonstrated at sea (RUSBY, 1970). The array was housed in a streamlined glass-fibre body, 10 m long with a diameter of 1.75 m (see Figs. 1 and 2) towed by armoured electric cable from R.R.S. *Discovery* at a depth of 130 m and a speed of 6 knots. 30 msec sound pulses at a frequency of 6.5 kHz and an

611

Fig. 1. The GLORIA sonar vehicle being serviced in its constant tension davit onboard R.R.S. *Discovery*. The external shell of the vehicle is in three parts, made of glass fibre strengthened by aluminium frames and bulkheads. The centre section is a high strength filament wound tube with bulkheads at each end to support the rotatable 3-tonne rectangular array in bearings. The forward bulkhead also carries the array training gear and the high pressure air vessels for deballasting the free-flooding vehicle. At the after end the rear bulkhead carries the array distribution boxes, including fuses, directivity change switches, and resolvers to steer the received beam. Two gyros are fitted which give the roll/pitch attitude of the vehicle and yaw information used for the resolvers and by the servo control loop for the trim rudders. The vehicle is 10 m long and weighs over 6 tonnes in air and 3 tonnes in water, the load is supported by a ball and socket joint in the centrally mounted towing head.

acoustic power of 50 kW were transmitted every 30 seconds to give a maximum range of 22 km. The dimensions of the sonar array were such that it provided a narrow acoustic beam in the horizontal plane, $2\frac{1}{2}°$ wide, with a greater angular width in the vertical plane of $10°$. The array was mounted on bearings in the towed vehicle so that it could be adjusted remotely to radiate the sound beam at the optimum angle to the sea floor, either to port or to starboard. Sonographs were built up in the usual way by displaying the backscattered signals from successive transmissions to the sea floor contiguously on a line recorder as the array moved through the water (see Fig. 3).

2. MAIN DEVELOPMENTS

(a) *Towing system*

From the first trials carried out, initially in the Clyde estuary and then in the western Mediterranean, it was clear that the towing system would be likely to prove unsatisfactory in heavy weather.

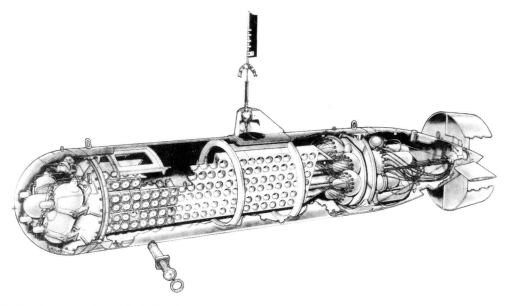

Fig. 2.   A cutaway view of the inside of the GLORIA vehicle. The array contains 6×24 lead zirconate titanate transducers with an efficiency of 90% capable of radiating 500 acoustic watts per element. The array frame can be remotely rotated from the towing ship to launch the sound at the correct angle to the sea floor, either to port or to starboard. It produces a beam which is narrow in the horizontal plane, $2\frac{1}{2}°$ wide, and $10°$ in the vertical plane. The received signals from different sections of the array are phase controlled to ensure that the beam stays 'locked' to the bearing of the last transmission. The vehicle is suspended from the ball and socket termination seen in the towing head, and is guided by servo-controlled rudders to reduce yaw oscillations when surveying.

The vehicle was launched from a specially constructed constant tension davit, and then towed round to a position about 100 m astern of *Discovery*. The ship then got slowly up to towing speed as the flooding valves on the vehicle were remotely operated. The vehicle then sank to a depth dependent on the forward speed of the ship, the hydrodynamics of the armoured towing cable and its own negative buoyancy and drag. Model trials had shown that the main problem of such a simple towing system was the relatively high coupling between the heave motion of the stern towing point and the vehicle. This had been reduced in the earlier third scale design study experiments by introducing a rubber accumulator system on the poop of the towing vessel. In the full-scale trials 60 m nylon ropes were used running over sheaves mounted outboard along each side of *Discovery* and connected by wire rope bridles to the armoured cable (see EDGE, 1974). The heave isolation provided by this system proved inadequate in heavy weather, and the high inertia and vertical drag of the body caused large snatch loadings to occur on the vehicle cable termination.

To overcome this problem, a suggestion made by Dr. J. C. Swallow, F.R.S. was developed, to place a subsurface float in the armoured cable catenary and so decouple the vehicle from the ship. Following tank trials a streamlined glass fibre float filled with polyurethane foam was constructed which had a net positive buoyancy equal to the flooded weight of the sonar vehicle and armoured cable beneath it (see Fig. 4 and EDGE, 1974). The buoyancy and trim of the float could be adjusted by fitting balance weights

Fig. 3.   Sonar beam configuration in the deep sea, showing how the back-scattered information from an inclined beam to the sea floor is built up line by line into a sonograph as the sonar is towed through the water. The first return will be from beneath the sonar vehicle, giving an echo-sounding profile of the sea bed; as the range increases so the ray angles to the sea floor decrease until at the furthest ranges the information appears as a near plan view on the record. When making a mosaic the information can either be replotted to correct for the ray angles involved, or sufficient overlap can be maintained between adjacent runs so that only the un-distorted, near plan view, part of each record need be used. The diagram shows reflectors or scatterers as light-toned, shadows as dark. The same 'convention' is used in all the sonographs reproduced in this paper.

fore and aft. With a nose-down trim of about 15° the towing system shown in Fig. 4 was found to be stable, and could be towed in force 8–9 conditions with a typical suspended load variation of ±200 kg on a standing load of 3500 kg with the float running at a depth of 15–20 m. The float depth was found to vary little with towing speed in the range 3–7 knots, any significant variation due to speed could be compensated by remotely deballasting or flooding the foreward and aft compartments of the sonar vehicle.

    This system was subsequently used in the North Atlantic and Mediterranean for all the surveys carried out from 1970 to 1975. Although it allowed the sonar to run through gale conditions the vehicle still had to be deployed and recovered in good weather, either of these operations typically took 3–4 hours to complete. If possible, the vehicle was usually left submerged for 10–14 days to survey about 10^5 km^2 of deep sea floor at a speed of 7 knots.

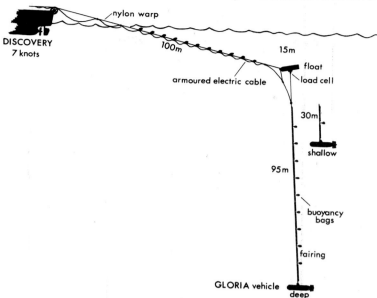

Fig. 4.   Modified towing configuration using a streamlined sub-surface float to decouple the heave of the towing ship from the flooded sonar vehicle containing some 20 tonnes of water.

(b) *Signal processing*

The next development undertaken was concerned with the way the returned acoustic signals were processed.

Initially the sonar system was designed to transmit a simple 12 or 30 millisecond 50-kW constant frequency pulse, detected in the usual minimum bandwidth of $1/\tau$ and with a range resolution of $\bar{c}\tau/2$, where τ is the pulse length and $\bar{c}$ is the mean sound velocity. As a reserve the equipment also included a digital correlator which enabled 100-Hz-wide linear swept frequency pulses of 4 seconds duration to be processed. The potential advantage of the correlator system could be considerable as the sonar was power limited by the onset of cavitation at the radiating face of the array and amplitude limited by the maximum strain permitted in the sound transducers. Since the system was able, if required, to transmit these long pulses at a power level of 50 kW the energy gain in the water would be typically $4/(30 \times 10^{-3}) = 133$ over the short pulse system. A correlation system, or matched filter, with this type of pulse coding provided a range resolution of $c/2B$, where B is the pulse bandwidth, so there would also be an improvement in range resolution by a factor of 3. With an effective matched filter bandwidth of $1/B$, the theoretical signal/noise power gain of the long pulse correlator system over a short pulse CW system using the same radiated power would be $B \times$ long pulse length = 400 or 26 dB (STEWART and WESTERFIELD, 1959; ALLEN and WESTERFIELD, 1964; SOMERS, 1970). However, it was estimated that due to the finite sampling and clipping of the waveform at the input to the digital correlator, plus certain phase discrepancies introduced by the circuits, the theoretical gain would be about 23 dB. The question was whether this considerable gain would be realized in practice, particularly when viewing a reverberation field of sound scatterers as well as individual reflectors.

In practice there was found to be a very useful gain of about 20 dB for single targets against a noise background, but the tonal range of the received sonographs from reverberation fields typified by sea bed back-scattering was poorer due to the loss of amplitude information by hard clipping at the input to the correlator.

In 1971 an analogue drum correlator was introduced into the system with excellent results. Not only was the high gain of the original digital correlator maintained but the sonograph tonal quality was enhanced since the original amplitude information was now

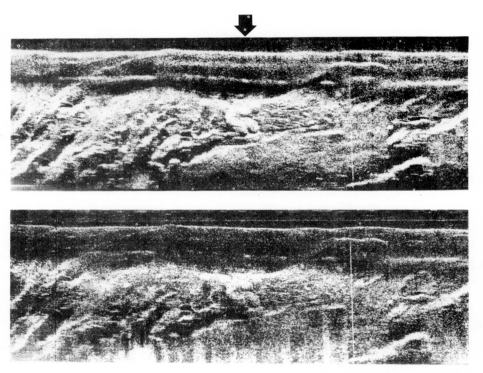

Fig. 5.   Isometric sonographs of the same piece of ground on the Mediterranean Ridge sough of Crete. The top one is recorded from an analogue drum correlator, and the bottom from a digital correlator with hard clipping at the input. The tonal range is less in the sonograph from the digital correlator and some of the detail has been lost. The records portray fractured sediments at a depth of 2000 m, and cover an area of 13 × 40 km, typical relief is from 20–200 m high. Direction of radiation is shown by the broad arrow.

conserved. Figure 5 compares two identical sonographs, where the returned signals have been processed by the two different correlators (SOMERS, 1973). After 1971 the analogue correlator was used exclusively, and the 'spare' 20 dB of gain was to lead eventually to a proposed change in design for a new, Mark 2, sonar.

(c) *Display*

For the first surveys a wet paper recorder was used for display purposes. It suffered from a limited dynamic range of 15–18 dB, and there was always the possibility of the paper drying out, particularly when the 30 seconds sweep speed was in operation for a maximum range of 22 km. It acted as an acceptable means of directly monitoring the sonar performance, but there was an obvious need for an off-line quality display accommodating more closely the full dynamic range of the signals from the detected correlator output.

The solution we found was to modify a standard photographic picture receiver used by newspaper offices. This type of recorder uses a crater tube as a modulated light source to expose photographic paper which is loaded in sheets on a revolving drum. The crater tube carriage is mounted on a lead screw and moves parallel to the drum axis; the paper loading, unloading and processing are quite automatic. The main modifications were concerned with the gearing of the lead screw and drum, so that the machine could be synchronized with a fast playback of the previously recorded signals from the correlator. After 'time varied gain' had been applied these signals had a maximum dynamic range of about 35–40 dB, which could be accommodated by a good-quality FM tape recorder, and by the picture receiver.

Unfortunately the records produced by the machine suffered from a distortion which exaggerated the down range dimension by 3 or 4 to 1 compared with the orthogonal dimension along the ship's track. It was not possible to increase the line width produced by the crater tube to compensate. So a further photographic process was introduced which removed this distortion by the use of an anamorphic camera. The prints from the picture receiver are carefully placed in sequence on a mounting board, and this board is moved past a stationary slit through which it is photographed on to moving 35-mm negative film. This isometric negative can then be printed to any desired enlargement, and is a particularly valuable source material for making mosaics of a large survey. The records still contain distortion at the closer ranges produced by the steep ray angles involved (see Fig. 3), but the effect of this can be minimized by ensuring that adjacent widths of insonified ground are made to overlap by, say, 20% in a survey. When only the bearing of geological 'trends' are to be plotted from the records this '$\cos \theta$' distortion can be manually corrected.

During intensive, large-scale surveys it was found possible to carry out the above display processing at sea so that isometric sonographs were available for mounting in a mosaic within 24 hours of recording. Figure 6 is an example of such a mosaic, showing ridge trends and fracture zones in the neighbourhood of the FAMOUS area of the Mid-Atlantic Ridge southwest of the Azores [WHITMARSH and LAUGHTON, 1975; WHITMARSH and LAUGHTON 1976].

(d) Continental shelf trials

As explained in the Introduction, the GLORIA design was intended to produce side-scan sonar records of the deep sea floor for geological purposes. But in 1971 it was recommended that it should be given a trial on the continental shelf, mainly to see whether pelagic fish shoals could be detected at long ranges. This recommendation was based on the results of WESTON and REVIE (1971) in detecting pilchard shoals off the north Cornish coast with a sonar employing a comparable frequency.

A trial was arranged in September 1971 in the Minch off western Scotland when a herring stock would be present. In spite of the fact that a summer thermocline existed it was found possible in this experiment to detect large herring shoals out to a range of 13 km, as long as the sonar array was towed in an isothermal part of the water column. By tracking backwards and forwards along a 13-km base line *Discovery* was able to search

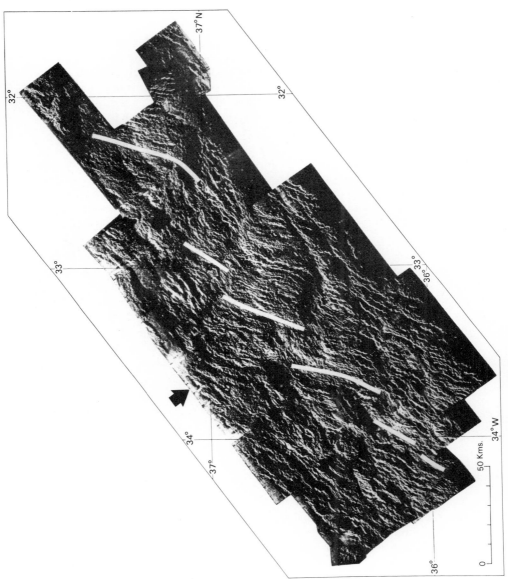

Fig. 6. A mosaic of sonographs on the crest of the Mid-Atlantic Ridge southwest of the Azores covering an area of 31,000 km². The area was viewed towards the southeast (see broad arrow) from eleven parallel tracks separated by about 11 km, each covering a strip up to 27 km wide. The area is dominated by linear features, believed to be fault scarps, which lie parallel to the spreading axes (marked by white lines) and which are often continuous from one fracture zone to the next. The dark areas of the fracture zones indicate sediment infilling.

the same 13×13-km² square every hour. In this way large aggregations of herring could be plotted and their movements relative to the tidal stream deduced (see Fig. 7, and RUSBY, SOMERS, REVIE, McCARTNEY and STUBBS, 1973). A purse seine vessel was under the direction of *Discovery* during this experiment and was able to make good catches by remote guidance from the GLORIA sonar plots on *Discovery*.

Fig. 7.   An isometric sonograph from the shallow waters of the Minches near the Hebrides off western Scotland, covering an area of 13×16 km². It records a 6-km-long aggregation of herring (arrowed) off the edge of a rock bank (outlined in white). The movement of such fish aggregations could be plotted relative to readily distinguishable rock features (such as A, B and C) as *Discovery* scanned the same area at intervals of an hour. The herring are in mid-water during daylight in a water depth of 140 m off the edge of the bank, the water depth over the rocks on the bank is typically 60 m. The broad arrow gives the direction of radiation.

In 1973 a second trial in the same fishery was organized, this time in November when the water was isothermal. It had been hoped that larger ranges might be obtained under the winter conditions, but again the maximum useful range was 13 km. A more extensive area was covered on this occasion with the vehicle deployed and it was found that useful geological information could be recorded in these typical shelf depths of about 50–100 m. Figure 8 shows an example of such a sonograph, running northwest from a point south of Barra Head to the edge of the continental shelf (RUSBY and REVIE, 1975), in which a variety of features and textures were displayed.

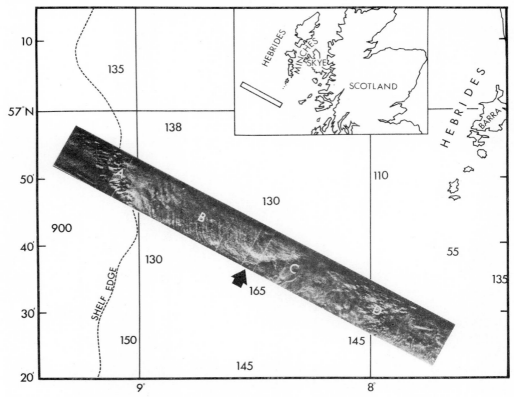

Fig. 8.   A 110 × 13-km² isometric sonograph giving a plan view of part of the shelf south of the Hebrides off western Scotland. Water depths are in metres. A = gullies on the continental slope. B = region of curved longitudinal sand patches overlying gravel and stones. C = region of mud. D = extension of the Lewisian block south of Barra Head.

In 1975 the sonar was used again in shallow water, to view the shelf edge and slope at the entrance to the English Channel and in the Bay of Biscay, and again geologically useful records were obtained, particularly of the canyon formations on this continental slope (see Fig. 9 [BELDERSON and KENYON (1976)]).

On all these surveys it was necessary to shorten the length of armoured cable beneath the subsurface float from about 100 to 30 m (see Fig. 4), in order to give adequate clearance beneath the suspended vehicle. It was felt that this severe reduction in cable length would lead to an increase in yaw coupling between the towing ship and the vehicle. In fact any increase was not noticeable, possibly because the effects of yaw were not so pronounced at the shorter ranges used, 13 km, compared with the maximum range of 22 km employed in the deep Atlantic water. Certainly yaw is a problem at the longer range, leading to the elongation of point targets in a direction parallel to the ship's track, which can blur the record and lead to confusion in the interpretation of the sonograph (see COOPER, 1974). The vehicle was provided with both a gyro-controlled active rudder as well as a means of electronically locking the received beam to the bearing of each previous transmission, but neither of these systems answered the problem completely.

Fig. 9.   A 13 × 50-km isometric sonograph of canyons on the continental slope west of Brittany near 45°40′N. The sonar was towed along the shelf edge and viewed downslope (see broad arrow). Depths varied from 200–1000 m directly under the vehicle, to about 2500 m at the maximum range of 13 km at the bottom of the slope. The reasonably even 'illumination' achieved over this 10° slope was due largely to the use of the wider vertical beam (30°) when only the centre two rows of the six-row array were activated. The small arrow indicates a 20° course change.

Both the ability to detect fish and also to define geological features and sea bed forms in the shallow and intermediate depths of the continental shelf and continental slope are important additional attributes of the sonar. These demonstrations of its capabilities in shallower water were to play a part in the decision to design and construct a Mark 2 sonar.

(e) *Reduction in sonar array height*

From 1971 onwards it was clear that, with the power-handling capability of the sonar allied to the correlation processing gain, there was excess signal to noise available in the design. With this knowledge, one important and simple experiment which had to be carried out was to discover if the sonar could function satisfactorily using only two rows of the available six rows of transducers in the vehicle. A remotely operated switch had originally been installed in the vehicle to vary the transducer row sensitivity for beam shading, the only modifications required were to rewire some connections so that in one position this switch isolated the top and bottom two rows of transducers, and to rematch the power amplifier to the new array impedance.

First trials with this facility were conducted south of the Azores in 1973, where a comparison was made between the performance of the sonar using the whole array and only the centre two rows. The effect of this was to broaden the width of the beam in the vertical plane from 10° to 30°, and to decrease the total directivity gain by about 12–15 dB, hence the need to have excess signal-to-noise gain available. The results showed that the sonographs obtained with only the centre two rows activated were superior in quality, due to the more even insonification provided by the wider vertical beam. Later in 1973 the same mode was used in the eastern Mediterranean, and it was also employed in 1975 in the survey of the Mid-Atlantic Ridge near 45°N and the Biscay continental margin survey (see Fig. 9).

3. GLORIA MARK 2

From the start it was clear that the main weakness of the system lay with the problem of handling the bulky vehicle under tow as well as in deployment and recovery. Certainly the introduction of the sub-surface float allowed the sonar to be operated in gale-force conditions, but launching and recovery still required calm sea for a guaranteed period of 3–4 hours or more if there were complications. This usually implied working in the lee of land. And swimmers were required to connect the various wire bridles and ancillary gear which circumscribed the geographical areas in which the sonar could be operated. Also the system was completely 'tied' to one ship, by virtue of the installation of the constant-tension davit and sonar power amplifier.

These factors, combined with the positive results of the correlation processing, continental shelf and array height-reduction experiments, led to a proposal for a new design which would have the following improved characteristics:

1. A reduction in sonar vehicle diameter by a factor of 3.
2. A modular handling and electronic system capable of being fitted on different vessels.
3. Simultaneous scanning to port and starboard to a maximum range of 27 km (15 miles) in the deep sea.
4. A towing speed of 8 knots.

It was decided that since any advantage in changing the sonar frequency would be marginal, the transducers designed for the original sonar should be used in the new version. With 6 rows operating, the Mark I array contained 6×24 transducers, more than enough for the projected new design containing port and starboard arrays of 2×30 transducers each.

Another important development conspired to 'shape' our thinking about the new vehicle. The Institute had recently developed high-power transducers with titanium diaphragms working in filled-nylon housings. These had been proved in a new deep-sea echo-sounder for the past 3 years, with no sign of deterioration or corrosion. So the proposal arose for a neutrally buoyant Mark 2 vehicle built up for ease of construction and servicing from a number of filled-nylon blocks, each containing two port- and two starboard-facing transducers. Such a neutrally buoyant vehicle could be nose-towed, with good heave and yaw isolation on a long unfaired cable, and furthermore could be handled by a hydraulically operated cradle and winch mounted with a power pack on a portable base fitted to the poop of the towing vessel (see Fig. 10).

The necessary modifications to the existing transducers have been successfully tested and model towing trials of the proposed configuration have been carried out at sea with good yaw and heave isolation achieved.

The design of the vehicle, electronics and handling gear is complete and construction has started. Each array will be divided into six sections for transmission and reception, and each section will be fed from its own phase-controlled power amplifier module. Signal phase on transmission and reception will be controlled by resolvers directed by a yaw gyro mounted in the vehicle so that the beam will be electronically steered and 'locked' to a given bearing independent of the yaw of the vehicle. It is hoped in this way to overcome the yaw problems of the earlier design in which only the received signals could be phase-controlled, with successive transmissions simply being radiated at right angles to the array axis. Good sonar isolation between the arrays has been assisted by coding the FM transmissions to port and starboard in a different form.

It is believed that the lightness and slimness of the new vehicle together with its simple configuration will enable its deployment and recovery to be carried out in moderately rough conditions, without the use of swimmers, so it should be possible to operate the sonar anywhere in the ocean. In addition its relative simplicity and portability should allow a number of users to make both geological and biological surveys both in the deep sea and on the continental shelf.

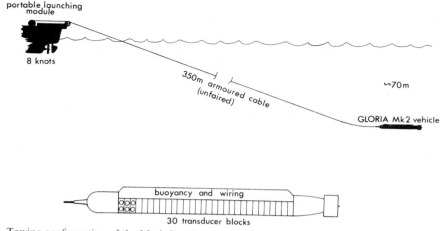

Fig. 10. Towing configuration of the Mark 2 vehicle. The vehicle will be 8 m long and will weigh 2 tonnes in air. In water it is designed to be neutrally buoyant so that it can be nose-towed on a long cable to reduce heave and yaw coupling. Thirty filled-nylon transducer blocks will carry a port and starboard pair of transducers capable of covering a swathe 50 km wide on the deep sea floor. The arrays will be split into six sections, and these sections will be electronically steered by resolvers, both on transmission and reception, so that the beam can remain 'locked' on a given bearing during a survey. The armoured electric cable and vehicle will be handled by a portable unit containing the hydraulic power pack, winch and launching cradle. The modular power amplifiers for each section, and the signal processing and display equipment, will be housed in a portable cabin.

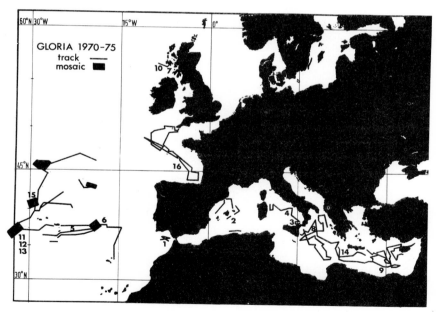

Fig. 11. GLORIA track chart for 1970–1975. The numbers refer to the following references (see reference list): 1. KENYON and BELDERSON (1973); 2. BELDERSON, KENYON and STRIDE (1970); 3. BELDERSON, KENYON and STRIDE (1974a); 4. BELDERSON, KENYON and STRIDE (1972); 5. LAUGHTON, WHITMARSH, RUSBY, SOMERS, REVIE, McCARTNEY and NAFE (1972); 6. LAUGHTON and WHITMARSH (1974); 7. RUSBY, SOMERS, REVIE, McCARTNEY and STUBBS (1973); 8. BELDERSON, KENYON and STRIDE (1974b); 9. KENYON, STRIDE and BELDERSON (1975); 10. RUSBY and REVIE (1975); 11. LAUGHTON and RUSBY (1975); 12. WHITMARSH and LAUGHTON (1975); 13. WHITMARSH and LAUGHTON (1976); 14. STRIDE, BELDERSON and KENYON (1977); 15. SEARLE and LAUGHTON (1976); 16. BELDERSON and KENYON (1976).

5. CONCLUSIONS

Five years of operation and development have answered the basic acoustic question of whether useful long-range side-scan records could be obtained in the deep sea. Discoveries resulting from a number of GLORIA surveys, made in conjunction with other more standard geophysical techniques, are witness to this (see Fig. 11 and list of references). And as an additional bonus, it has been found that such a technique has a useful geological and biological role to play in the shallow waters of the continental shelf and margin.

The resolution of such a system is limited by the horizontal angular beam width employed, $2\frac{1}{2}°$, and the effect of this is most noticeable at the longest ranges. In the deep sea, where it is economically and acoustically desirable to use the maximum range of 22–27 km, this broadening of the beam, coupled with vehicle yaw and the mechanism of backscattering, tends to highlight those targets with trends which face or lie parallel to the ship's course. So it has been found important, particularly when the maximum range is employed, to include control tracks on other bearings from the main courses used in a survey. When this control is exercised the sonar has proved to be a powerful tool for elucidating the trends of tectonic, erosional and sedimentary features such as ridges, faults, canyons and depositional channels found in the deep sea and on the continental margin. This ability to view the 'grain' of the sea floor over large areas and distances has been its main contribution to knowledge, particularly in the neighbourhood of the Mid-Atlantic Ridge and in the eastern Mediterranean, especially since its ability to resolve texture changes is limited where there is no relief on the sea floor.

Acknowledgements—On the occasion of the publication of this birthday volume to Sir GEORGE DEACON the authors welcome the opportunity to thank him publicly for the enthusiastic support he has given to the team over a period of some 10 years, during the difficult times as well as the good. Without that leadership, and the willingness to bear risk, this project would have died at an early stage.

In addition they would like to thank all their colleagues at the Institute who have contributed so much from their fund of specialized knowledge to GLORIA. And finally to Captain Howe, O.B.E., whose fine seamanship enabled her to be towed thousands of miles behind *Discovery* without injury.

REFERENCES

ALLEN W. B. and E. C. WESTERFIELD (1964) Digital compressed time correlators and matched filters for active sonar. *The Journal of the Acoustical Society of America*, **36**, 121–139.

BELDERSON R. H. and N. H. KENYON (1976) Long range sonar views of submarine canyons. *Marine Geology*, **22**, M64–M74.

BELDERSON R. H., N. H. KENYON and A. H. STRIDE (1970) 10-km wide views of the Mediterranean deep sea floor. *Deep-Sea Research*, **17**, 267–270.

BELDERSON R. H., N. H. KENYON and A. H. STRIDE (1972) *Sonographs of the sea floor*, Elsevier, Amsterdam, pp. 185.

BELDERSON R. H., N. H. KENYON and A. H. STRIDE (1974a) Features of submarine volcanoes shown on long range sonographs. *Journal of the Geological Society*, **130**, 403–410.

BELDERSON R. H., N. H. KENYON and A. H. STRIDE (1974b) Calabrian Ridge, a newly discovered branch of the Mediterranean Ridge. *Nature*, **247**, 453–454.

CHESTERMAN W. D., P. R. CLYNICK and A. H. STRIDE (1958) An acoustic aid to sea-bed survey. *Acustica*, **8**, 285–290.

COOPER R. (1974) *An analysis of the effect of vehicle yaw on GLORIA sonographs*, Institute of Oceanographic Sciences, Report No. 6.

EDGE R. H. (1974) Handling and towing the long range side scan sonar vehicle GLORIA on R.R.S. *Discovery*. In: *OCEAN 74*, Vol. 1 (*Proceedings of the Institute of Electrical and Electronics Engineers, International Conference on Engineering in the Ocean Environment, Halifax,* (1974), Institute of Electrical and Electronics Engineers, New York, pp. 307–315.

KENYON N. H. and R. H. BELDERSON (1973) Bed forms of the Mediterranean undercurrent observed with side-scan sonar. *Sedimentary Geology*, **9**, 77–99.

KENYON N. H., A. H. STRIDE and R. H. BELDERSON (1975) Plan views of active faults and other features on the lower Nile cone. *Geological Society of America Bulletin*, **86**, 1733–1739.

LAUGHTON A. S. and J. S. M. RUSBY (1975) Long range sonar and photographic studies of the median valley in the FAMOUS area of the Mid-Atlantic Ridge near 37°N. *Deep-Sea Research*, **22**, 279–298.

LAUGHTON A. S. and R. B. WHITMARSH (1974) The Azores-Gibraltar plate boundary. In: *Geodynamics of Iceland and the the North Atlantic (Proceedings of the NATO Advanced Study Institute, Reykjavik, 1974)*. Reidel, Dordrecht, pp. 63–81.

LAUGHTON A. S., R. B. WHITMARSH, J. S. M. RUSBY, M. L. SOMERS, J. REVIE, B. S. McCARTNEY and J. E. NAFE (1972) A continuous east–west fault on the Azores–Gibraltar Ridge. *Nature*, **237**, 217–220.

RUSBY J. S. M. (1970) A long range side-scan sonar for use in the deep sea (GLORIA project). *International Hydrographic Review*, **47**, 25–39.

RUSBY J. S. M. and J. REVIE (1975) Long-range sonar mapping of the continental shelf. *Marine Geology*, **19**, M41–48.

RUSBY J. S. M., M. L. SOMERS, J. REVIE, B. S. McCARTNEY and A. R. STUBBS (1973) An experimental survey of a herring fishery by long-range sonar. *Marine Biology*, **22**, 271–292.

SEARLE R. C. and A. S. LAUGHTON (1976) Preliminary tectonic results from a sonar study of the Mid-Atlantic Ridge crest near Kurchatov Fracture Zone. *Bulletin de la Société Géologique de France*, **18**, 797–799.

SOMERS M. L. (1970) Signal processing in project GLORIA, a long range side-scan sonar. In: *Electronic engineering in ocean technology*, D. G. TUCKER, editor (*Proceedings of the Institution of Electronic and Radio Engineers, Conference, Swansea, 1970*), Institution of Electronic and Radio Engineers, London, pp. 109–120.

SOMERS M. L. (1973) Some recent results with a long range side-scan sonar. In: *Signal processing (Proceedings of the NATO Advanced Study Institute, Loughborough, 1972)* Academic Press, London, pp. 757–767.

STEWART J. L. and E. C. WESTERFIELD (1959) A theory of active sonar detection. *Proceedings of the Institute of Radio Engineers*, **47**, 872–881.

STRIDE A. H., R. H. BELDERSON and N. H. KENYON (1976) Evolving miogeanticlines of the East Mediterranean. *Philisophical Transactions of the Royal Society of London*, A, **284**, 255–285.

TUCKER M. J. and A. R. STUBBS (1961) A narrow beam echo-ranger for fishery and geological investigations. *British Journal of Applied Physics*, **12**, 103–110.

WESTON D. E. and J. REVIE (1971) Fish echoes on a long range sonar display. *Journal of Sound and Vibration*, **17**, 105–112.

WHITMARSH R. B. and A. S. LAUGHTON (1975) The fault pattern of a slow-spreading ridge near a fracture zone. *Nature*, **258**, 509–510.

WHITMARSH R. B. and A. S. LAUGHTON (1976) A long-range sonar study of the Mid-Atlantic Ridge crest near 37°N (FAMOUS area) and its tectonic implications. *Deep-Sea Research*, **23**, 1005–1023.

Outer Ridges of Orogenic Arc Systems

ROBERT H. BELDERSON and NEIL H. KENYON

Abstract—A broad (several hundred kilometres wide) archlike structure known as the 'outer basement high' has been described on the sea floor external to the deep trench of many of those orogenic arc systems that front onto oceanic crust. This paper proposes that three deformed outer ridges in the eastern Mediterranean (Calabrian, Hellenic and Cyprus Outer Ridges) are also fundamentally the same feature, but that their morphological expression differs from the oceanic outer ridges in accordance with their much thicker sedimentary cover and thicker crustal setting. A search has been made for further examples of such outer ridges comprised of strongly folded and faulted sediments. One of these is the newly recognized Barbados Outer Ridge which is developed in thick continental rise sediments external to the Barbados Ridge of the Lesser Antilles Arc System. A similar feature is also suggested at the eastern end of the Indonesian Arc System. Intra-continental arc systems may include an external, outer ridge-like structure (such as the Jura fold belt), but these are generally much narrower in their surface expression (perhaps only tens of kilometres or less wide). The sediment-covered outer ridges are considered to be modern examples of well-developed miogeanticlines, while the 'outer basement high' is thought to be a more rudimentary development of such a feature.

INTRODUCTION

THE convex sides of many modern arc systems face outwards towards a deep ocean floor which bears a relatively thin sediment cover ($\frac{1}{2}$ km or so) above oceanic crust. External to the deep trench there is usually a broad but low archlike structure which has been variously noted (such as a "well-developed regional rise in topography" (WATTS and TALWANI, 1974), "an outer swell or high" (COLEMAN, 1975), or a "low submarine ridge" (SEELY, VAIL and WALTON, 1974)). A general impression of the scale of this outer ridge can be gained from the thirty-five projected topographic profiles of Pacific trenches shown by HAYES and EWING (1970, Fig. 19). Attention has perhaps not been focused on this feature because it is usually only minor in amplitude (less than 1 km in height, but several hundred kilometres wide) relative to the more massive internal components of the arc system.

The outer ridge shows a close correlation with an outer gravity high which was considered by WATTS and TALWANI (1974) to be an important part of the regional gravity field of island arcs. It has generally been explained by the advocates of plate-tectonics as an upward flexure or bulge in the lithosphere oceanward of where it bends down into a subduction zone. For instance, HANKS (1971) proposed that the Hokkaido Rise, a 730-m high swell oceanward of the Kurile Trench, is produced by a horizontal compressive stress acting in a direction normal to the trench axis. More recently PARSONS and MOLNAR (1976) have proposed that the outer ridge can be explained solely by a bending moment applied to oceanic lithosphere (though this does not, of course, preclude horizontal compression).

The present paper draws attention to further examples of outer ridges which, although varying in their morphological expression, are considered to be fundamentally the same type of feature. It is important, for the purposes of this paper, that these tectonic outer ridges should not be confused with *depositional* sedimentary ridges, such as the Blake–Bahama Outer Ridge, nor with the 'accretionary prism' of Karig and Sharman (1975) which, when thick enough, is seen as a ridge sometimes termed 'the sedimentary arc'. Such an accretionary prism is situated *internal* to the deep trench of the arc system.

OUTER RIDGES WITH DEFORMED SEDIMENTARY BLANKET

East Mediterranean Outer Ridges

The Calabrian, Hellenic and Cyprus Arc Systems of the East Mediterranean region each include an outer ridge, known respectively as the Calabrian, Hellenic and Cyprus Outer Ridges (Figs. 1 and 2). The morphology of these outer ridges has been described in detail by Stride, Belderson and Kenyon (1977). They consist essentially of long and broad (up to about 150 km wide) but low (up to about 1 km) ridges situated external to their respective arc systems. The outer ridges are composed of tectonically disturbed sediments, with a low magnetic relief, in marked contrast to the more internal Calabrian, Hellenic and Cyprus Arcs with their exposures of metamorphic and igneous rock. Superimposed on each outer ridge is a smaller scale folded and faulted topography which has been observed with long-range (14 km) side-scan sonar (Fig. 3), sub-bottom profiler and narrow-beam echo-sounder. This relief shows a gradation from the outside of the arc inwards, from simple folds, to folds heavily faulted and slumped along the strike, to an additional series of cross-faults (probably strike-slip). The fundamental tectonic distinction

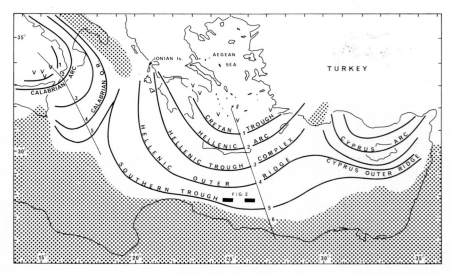

Fig. 1. The active arc systems of the East Mediterranean, showing the positions of the ridges and troughs (as located by Stride, Belderson and Kenyon, 1977): (1) tensional basin backed by a volcanic constructional arc, (2) thrust arc, (3) trench, (4) outer ridge, (5) outer trough, (6) the foreland (shown stippled). The orogenic polarity of the arc systems is directed radially outwards. V in Figs. 1, 2, 4 and 5 marks the position of Late Tertiary and Quaternary volcanoes of mainly andesitic composition.

between the outer ridges and the internal parts of their orogenic arcs is that the former are folded and faulted, but not strongly thrust, while the latter are associated with major thrusting and imbrication.

Barbados Outer Ridge

A good analogy with the three East Mediterranean Outer Ridges is found towards the southern end of the Lesser Antilles Arc System (Figs. 2 and 4). A thick pile of continental rise sediment derived from South America here lies oceanward from the Arc. This allows the 'sedimentary' arc to be fully developed in the form of the Barbados Ridge, capped by the Island of Barbados. However, the available contoured relief (KEAREY, PETER and WESTBROOK, 1975, Fig. 2) shows that the feature hitherto known as the Barbados Ridge consists in reality of two parallel ridges, the Barbados Ridge (on which the island of Barbados stands) and an Outer Ridge, separated by a trough. The structure of the feature has been described by WEEKS, LATTIMORE, HARBISON, BASSINGER and MERRILL (1971) as a "greatly fractured anticlinorium consisting of two parallel arches with a central syncline". They were able to trace this 'anticlinorium', as well as the Lesser Antilles volcanic arc, and the intervening trough, into and across the South American continental shelf.

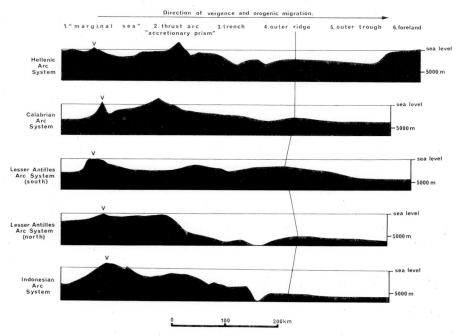

Fig. 2. Correlation of the main ridges and troughs of the Hellenic, Calabrian, Lesser Antilles and Indonesian arc systems (lines of profiles are located in Figs. 1, 4 and 5). The upper three profiles cross an Outer Ridge of deformed sediments and the lower two cross an 'Outer Basement High' in an open ocean setting. Vertical exaggeration is X8 which is not great enough to show the areas of small-scale topographic roughness. (These profiles were drawn using U.S. Defense Mapping Agency Hydrographic Center Chart No. 310; KEAREY, PETER and WESTBROOK, 1975; and *Geological-Geophysical Atlas of the Indian Ocean*, Moscow, 1975).

Reflection profiles (CHASE and BUNCE, 1969; BUNCE, PHILLIPS, CHASE and BOWIN, 1970; LOWRIE and ESCOWITZ, 1969; WESTBROOK, BOTT and PEACOCK, 1973) show strong tectonic deformation within the Barbados Outer Ridge, and echo-sounder records over the small-scale tectonic relief superimposed upon the ridge show the same massed interlocking hyperbolae so distinctive of the Calabrian, Hellenic and Cyprus Outer Ridges. Because of its comparable tectonic deformation and location in relation to the arc system as a whole, we equate the Barbados Outer Ridge with the outer ridges of the East Mediterranean. If this is correct, then it follows that the Barbados Trough is equivalent to the Hellenic Trough complex, and the Barbados Ridge to the Hellenic Arc. Both systems are backed by a marginal sea (eastern Caribbean, or Aegean) within which is built a mainly andesitic volcanic-constructional arc.

The main elements of the southern part of the Lesser Antilles Arc system can be extended into features on the northern part of the arc (Figs. 2 and 4), which is more typical of arcs facing the open ocean. WESTBROOK, BOTT and PEACOCK (1973) suggested that Barbados Island and its associated negative gravity anomaly is situated *above* a buried trench. However, although the Barbados Ridge cannot be traced as a continuous bathymetric feature much further north than Barbados Island (where it is probably interrupted by a series of arc-transverse structures), the negative gravity anomaly continues northwards

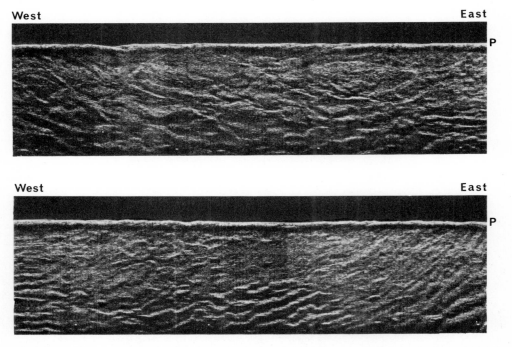

Fig. 3. Sonographs, each showing a true plan view of a 14-km-wide and 50-km-long strip of characteristic topography on the outer part of the Hellenic Outer Ridge. Simple, branching folds, with a wavelength of about 1 km and crest length traceable up to 16 km, are best developed towards the bottom part of the lower sonograph. They have a general trend parallel to the Outer Ridge. Over the area of the upper sonograph and parts of the lower one, the folds are disrupted by faulting and slumping largely parallel to their strike. There are also indications of cross-faulting in the western part of the upper sonograph. Strong echoes appear white and shadows black. P represents the profile of the sea floor beneath the ship's track. Sonographs are located in Fig. 1 (the upper sonograph is the westernmost).

along the *inner* wall of the Puerto Rico Trench (BUNCE, PHILLIPS, CHASE and BOWIN, 1970). The Barbados Ridge is therefore probably equivalent to the break of slope outside the volcanic arc (the trench-slope break of DICKINSON, 1973, and KARIG and SHARMAN, 1975), while the Barbados Trough is a largely sediment-filled extension of the much deeper Puerto Rico Trench. The sequence across the arc system in this northern region is complicated by the presence of the early Tertiary volcanic arc of the 'Limestone Caribbees' external to the Late Tertiary volcanic arc. The Barbados Outer Ridge also has its equivalent outside the Puerto Rico Trench, in the form of the Puerto Rico Outer Ridge (BUNCE, PHILLIPS and CHASE, 1974). This is underlain by both a mantle and basement high bearing a relatively thin (400 m or so) sediment cover described by these authors as 'convoluted', and is also (as with many other outer ridges) associated with a free-air gravity anomaly high. Indeed, a study of gravity and seismic models by WESTBROOK (1975) has already suggested the presence of an outer gravity high beneath the Barbados Outer Ridge which is not obviously expressed because of the load of sediments covering the supposed basement high.

Indonesian Outer Ridge

As suggested by CHASE and BUNCE (1969), there is a similarity in setting between Barbados island and the Island of Timor, which is situated on the 'sedimentary arc', external to the volcanic Indonesian Arc but internal to the Java Trench. Outside the Java Trench there is an outer basement high (Figs. 2 and 5) such as those mentioned above. Any eastward structural continuation of this Indonesian Outer Ridge might be expected to trend along the outer edge of the Sahul Shelf off northern Australia, but the thicker

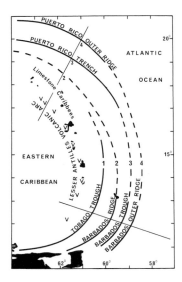

Fig. 4.   The Lesser Antilles arc system, showing a correlation between the ridges and troughs (numbered as in Fig. 1 caption) in the region of thick sediments in the south and the region of thin sediments overlying oceanic basement in the north. The proposed Barbados Outer Ridge is a newly recognized feature. The deep oceanic floor east of the outer ridge is considered equivalent to the 'foreland' zone of Fig. 1. The lines are dashed where zones cannot be recognized directly from the topography alone.

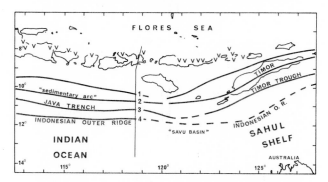

Fig. 5.   The ridges and troughs of the Indonesian arc system (numbered as in Fig. 1 caption). The Indonesian Outer Ridge in the open ocean setting in the west can possibly be traced through the 'Savu Basin' on to the upper continental slope north of Australia. Both the deep oceanic floor and the Sahul Shelf south of the outer ridge are considered equivalent to the 'foreland' zone of Fig. 1.

crustal setting and different lithological composition would be expected to influence the nature of any outer ridge in this region. Nevertheless there are significant morphological and structural indications of such a feature. For example, a series of shallow carbonate banks (rising almost to sea level) are found along the upper continental slope to the north of the Sahul Shelf. These are situated along the crest of an anticlinal arch (VEEVERS, 1971, Fig. 10) separated from the continental shelf by the 'Wyrallah Syncline'. Magnetic anomalies suggest that this arch is a deep-seated structure. Between this and the outer basement high in the deep sea to the west there is an apparent gap called the 'Savu Basin' (Fig. 4). A *Glomar Challenger* bathymetric and seismic profile through this area (VEEVERS and HEIRTZLER, 1974, Figs. 11 and 12, from 1900 hrs 22 November 1972 to 0300 hrs 23 November 1972 centred on about 121°30′E, 11°50′S) reveals a small-scale topography which is very similar in appearance to that typical of the East Mediterranean and Barbados Outer Ridges. This is described by VEEVERS and HEIRTZLER as "characteristically shaped in overlapping rounded hillocks, 2 to 3 km across and 150 m high, in the style of abyssal hills". These features are unlikely to be volcanic abyssal hills, on account of the sedimentary setting and the quiet magnetics. We therefore suggest, in the absence of a more detailed survey of topographic trends, that this short section of the outer Indonesian Arc most closely resembles the outer ridges of the East Mediterranean. The lateral transition from a deep sea to a continental setting would then be analogous to the proposed structural continuation of the Hellenic Outer Ridge located to the west of the Ionian Islands, Greece (STRIDE, BELDERSON and KENYON, 1977).

INTRA-CONTINENTAL ANALOGUES

Some arc systems front wholly onto an oceanic-type crust and some onto continental-type crust at one or both extremities, while others front wholly onto continental crust (e.g. the Alps and Carpathians). An analogy between the Hellenic Outer Ridge and the Jura fold belt of the western Alps was first suggested by GIERMANN (1966) and has been supported by STRIDE, BELDERSON and KENYON (1977). Other possible examples of such intra-continental outer ridges are found in the Appenine Foredeep (JACOBACCI, 1962,

Fig. 1); in the Fore-Balkan zone of Bulgaria (FOOSE and MANHEIM, 1975, Fig. 3) which consists essentially of an anticlinorium, and is backed by a zone of thrusts, nappes and metamorphic rocks (the Stara Planina zone); in the Gorgan Spur structural zone of the Alborz Mountains, northern Iran (STÖCKLIN, 1974, Figs. 5 and 6, Zone 1); and in the buried Kahta-Kastel geanticline of south-east Turkey (RIGO DE RIGHI and CORTESINI, 1964, Fig. 8). Further occurrences of such a feature may often lie wholly or partially buried beneath post-orogenic molasse deposits.

DISCUSSION

KARIG and SHARMAN (1975) have noted the wide variation in structural style and lithologic content of the strongly thrust 'accretionary prism'. They suggest that the morphologic variations from a mere structural high or terrace on the outer slope of the volcanic arc to a more or less continuous ridge seem to depend more upon relative rates of sediment supply into the trench above the 'subduction zone' than on the rate or duration of subduction. The best developed of these accretionary prisms are found where an arc fronts onto ocean floor bearing a relatively thick sedimentary cover, such as occurs adjacent to an important source of supply (e.g. the Indonesian Arc, particularly towards its eastern termination opposite Australia and its western end opposite the Bengal Fan; the Antilles Arc, in the vicinity of South America; the Aleutian Arc, towards its Alaskan end). Likewise, the morphology of the outer ridge depends upon the thickness of sediment cover on the external oceanic floor. When this is sufficiently thick, a deformed sedimentary outer ridge develops atop the outer basement high. Hence there is sometimes a lateral transition from outer basement high to a folded and faulted sedimentary outer ridge, and from this to an intra-continental analogue. The outer ridge should be included in any 'ideal' arc system, the full development of which COLEMAN (1975) related to age, but which should also depend upon the thickness of sediment cover on the external oceanic floor.

The two modes of origin of the outer ridge mentioned in the introduction were either a horizontal compressive stress or a bending moment acting on a strong, though elastic oceanic lithospheric plate as it approaches an island arc. Tensional stretching within the upper part of such flexures seems to account for normal faulting noted in the basement of some Pacific outer ridges. What then is the cause of the compressional deformation observed in the sediments over the East Mediterranean and Barbados outer ridges? Perhaps the analogy with the Jura helps to provide an answer. There the folds of the sedimentary blanket, which overlies a faulted basement swell, are considered by many geologists to result from a process of décollement.

STRIDE, BELDERSON and KENYON (1977) consider that the three outer ridges of the East Mediterranean and their associated arc systems fit well into a eugeanticlinal-miogeanticlinal scheme such as AUBOUIN (1965) convincingly proposed (using the Tertiary Hellenides as a type example). It would be surprising if there were found to be no modern equivalents of AUBOUIN's miogeanticlines, and these authors have suggested that the three East Mediterranean outer ridges are, indeed, examples of modern miogeanticlines. These sediment blanketed geanticlines are located in the correct, external situation relative to the heavily thrust eugeanticline, with its included crystalline rocks, and the ubiquitous faults and folds of the outer ridges are developed even in Quaternary sediments. This geosynclinal scheme can be applied to all the arc systems described here (Figs. 1, 4 and 5). The miogeanticlines can be of the Gavrovo (Hellenides) type, sometimes with massive reef limestones, (AUBOUIN, 1965), situated at shelf depths (and from time to time rising above sea level) as in the case of the Hellenic Outer Ridge near the Ionian Islands, or possibly the Indonesian Outer Ridge north of the Sahul Shelf (which is capped by carbonate banks). The East Mediterranean and southern Lesser Antilles miogeanticlines are of AUBOUIN's Briançonnais (Western Alps) type, situated at greater depths and the site of present-day relatively reduced (pelagic) sedimentation. Elsewhere, the 'outer basement high' is thought to be a more rudimentary development of the latter type of miogeanticline. There are differences between the geosynclinal model of AUBOUIN (1965, Fig. 16) in that the basement of the external part (or miogeosynclinal realm) need not be continental crust, nor need a continental foreland necessarily be present.

Contrary to the now ruling orthodoxy of plate tectonic theory, the driving force of the demonstrably outward migrating orogens of the East Mediterranean is thought to be related to the outward spreading of mantle diapirs that are located beneath the internal basins (VAN BEMMELEN, 1972). Thus a succession of eugeanticlines and miogeanticlines are developed progressively outwards in time, in the manner envisaged by Aubouin.

In the case of the Pacific outer ridges, if this feature is due to horizontal compression, it could equally be formed by compression from the inside of the arc, as from the outside. Thus KREBS (1975) has explained the origin of the southwest Pacific Island Arc-Trench and mountain systems in terms of diapir-like upwelling from the asthenosphere (see particularly his Fig. 8). If this hypothesis is correct, the need for a 'subduction zone' disappears. In which case plate accretion at sea-floor spreading axes is perhaps adequately explained by the expanding earth hypothesis.

Acknowledgements—We thank ARTHUR STRIDE for critical review of the manuscript, COLIN PELTON for assistance in preparing diagrams, and many colleagues for their work both at sea and ashore in the maintenance and operation of the long-range side-scan sonar system. Scripps Institute of Oceanography, as operator for the Deep-sea Drilling Project, is thanked for the supply of copies of Leg 27 seismic and echo-sounder profiles.

REFERENCES

AUBOUIN J. (1965) *Geosynclines*, Elsevier Publishing Company, 335 pp.

BUNCE E. T., J. D. PHILLIPS and R. L. CHASE (1974) Geophysical study of Antilles Outer Ridge, Puerto Rico Trench, and northeast margin of Caribbean Sea. *American Association of Petroleum Geologists Bulletin*, **58**, 106–123.

BUNCE E. T., J. D. PHILLIPS, R. L. CHASE and C. O. BOWIN (1970) The Lesser Antilles Arc and the eastern margin of the Caribbean Sea. In: *The sea*, Vol. 4, Part 2, A. E. MAXWELL, editor, John Wiley & Sons, pp. 359–385.

CHASE R. L. and E. T. BUNCE (1969) Underthrusting of the eastern margin of the Antilles by the floor of the Western North Atlantic Ocean, and origin of the Barbados Ridge. *Journal of Geophysical Research*, **74**, 1413–1420.

COLEMAN P. J. (1975) On Island Arcs. *Earth Science Reviews*, **11**, 47–80.

DICKINSON W. R. (1973) Widths of modern arc-trench gaps proportional to past duration of igneous activity in associated magmatic arcs. *Journal of Geophysical Research*, **78**, 3376–3389.

FOOSE R. M. and F. MANHEIM (1975) Geology of Bulgaria: a review. *American Association of Petroleum Geologists Bulletin*, **59**, 303–335.

GIERMANN G. (1966) Gedanken zur Ostmediterranen Schwelle. *Bulletin de l'Institut Océanographique, Monaco*, **66**, 1–16.

HANKS T. C. (1971) The Kuril Trench–Hokkaido Rise system: large shallow earthquakes and simple models of deformation. *Geophysical Journal of the Royal Astronomical Society*, **23**, 173–189.

HAYES D. E. and M. EWING (1970) Pacific boundary structure. In: *The sea*, Vol. 4, Part 2, A. E. MAXWELL, editor, John Wiley & Sons, pp. 29–72.

JACOBACCI A. (1962) Evolution de la fosse mio-pliocène de l'Apennin apulo-campanien (Italie méridionale). *Bulletin de la Société Géologique de France*, **4**, 691–694.

KARIG D. E. and G. F. SHARMAN (1975) Subduction and accretion in trenches. *Geological Society of America Bulletin*, **86**, 377–389.

KEAREY P., G. PETER and G. K. WESTBROOK (1975) Geophysical maps of the eastern Caribbean. *Journal of the Geological Society*, **131**, 311–321.

KREBS W. (1975) Formation of Southwest Pacific Island Arc-trench and Mountain systems: plate or global-vertical tectonics? *American Association of Petroleum Geologists Bulletin*, **59**, 1639–1666.

LOWRIE A. and E. ESCOWITZ, editors (1969) *Kane 9*, Global ocean floor analysis and research data series, 1, U.S. Naval Oceanographic Office, 971 pp.

PARSONS B. and P. MOLNAR (1976) The origin of outer topographic rises associated with trenches. *Geophysical Journal of the Royal Astronomical Society*, **45**, 707–712.

RIGO DE RIGHI M. and A. CORTESINI (1964) Gravity tectonics in foothills structure belt of south-east Turkey. *American Association of Petroleum Geologists Bulletin*, **48**, 1911–1937.

SEELY D. R., P. R. VAIL and G. G. WALTON (1974) Trench Slope model. In: *The geology of continental margins*, C. A. BURK and C. L. DRAKE, editors, Springer Verlag, Berlin, pp. 249–260.

STÖCKLIN J. (1974) Northern Iran: Alborz Mountains. In: *Mesozoic–Cenozoic orogenic belts*, A. M. SPENCER, editor, Geological Society Special Publication No. 4, pp. 213–234.

STRIDE A. H., R. H. BELDERSON and N. H. KENYON (1977) Evolving miogeanticlines of the East Mediterranean. *Philosophical Transactions of the Royal Society of London* (A), **284**, 255–285.

VAN BEMMELEN R. W. (1972) *Geodynamic Models*, Elsevier Publishing Company, 267 pp.

VEEVERS J. J. (1971) Shallow stratigraphy and structure of the Australian continental margin beneath the Timor Sea. *Marine Geology*, **11**, 209–249.

VEEVERS J. J. and J. R. HEIRTZLER (1974) Bathymetry, seismic profiles, and magnetic-anomaly profiles. In: *Initial Reports of the deep sea drilling project*, Vol. 27, J. J. VEEVERS, J. R. HEIRTZLER *et al.*, U.S. Government Printing Office, pp. 339–381.

WATTS A. B. and M. TALWANI (1974) Gravity anomalies seaward of deep-sea trenches and their tectonic implications. *Geophysical Journal of the Royal Astronomical Society*, **36**, 57–90.

WEEKS L. A., R. K. LATTIMORE, R. N. HARBISON, B. G. BASSINGER and G. F. MERRILL (1971) Structural relations among Lesser Antilles, Venezuela and Trinidad-Tobago. *American Association of Petroleum Geologists Bulletin*, **55**, 1741–1752.

WESTBROOK G. K. (1975) The structure of the crust and upper mantle in the region of Barbados and the Lesser Antilles. *Geophysical Journal of the Royal Astronomical Society*, **43**, 201–242.

WESTBROOK G. K., M. H. P. BOTT and J. H. PEACOCK (1973) Lesser Antilles subduction zone in the vicinity of Barbados. *Nature, Physical Science*, **244**, 118–120.

A bathymetric survey of the eastern end
of the St. Paul's Fracture Zone

T. J. G. FRANCIS*

Cooperative Institute for Research in Environmental Sciences,
University of Colorado/NOAA, Boulder, Colorado 80309

Summary—A detailed bathymetric survey has been conducted at the eastern end of the St. Paul's Fracture Zone in the vicinity of its junction with the median valley to the south. The transform fault part of the fracture zone appears to be offset about 6 km to the south of the inactive section of the east. There is evidence that the deep hole at the junction of rift and fracture zone is a caldera of recent origin.

INTRODUCTION

IN December 1974 four ocean-bottom seismographs (FRANCIS, PORTER, LANE, OSBORNE, POOLEY and TOMKINS, 1975) were deployed from R.R.S. *Shackleton* at the eastern end of the St. Paul's Fracture Zone, close to its junction with the short section of spreading axis connecting it to the Romanche Fracture Zone to the south. The purpose of the deployment was to observe microearthquake activity within a range of about 30 km of the OBS array, whether it occurred on the spreading axis of the ridge or along the fracture zone itself. In support of this operation a detailed bathymetric survey was carried out in the vicinity of the OBS. The purpose of this paper is to describe the results of that survey and its geological interpretation. The microearthquake observations, which have not yet been fully investigated, will be the subject of a separate paper.

METHOD

During the course of the cruise *Shackleton* paid two visits to the deployment area, on the first occasion to launch the OBS and on the second to recover them. Between these visits, when the OBS were recording on the seabed, the ship operated up to 400 nautical miles† (740 km) away carrying out underway geophysical measurements over the Sierra Leone Rise. Each time the ship arrived in the area operations began with the mooring of a radar transponder buoy, approximately at the center of the OBS array. Thus navigation was based on a combination of radar fixes on the transponder buoy and satellite navigation. The ship's radar could detect the transponder out to ranges of 12–17 nautical miles (22–23 km). On the passages into and out of the area and on those tracks beyond

---

* On leave from Institute of Oceanographic Sciences, Blacknest, Brimpton, Reading, England.
† 1 nautical mile = 1,853 km.

radar range, navigation was based solely on satellite fixes. The complete track chart is shown in Fig. 1.

The ship was fitted with a Kelvin Hughes MS38 precision depth recorder which was operated continuously along the tracks shown in Fig. 1. This instrument operates at a frequency of 10 kHz and with the transducer fitted to R.R.S. *Shackleton* has a cone angle (2θ) of 40° to the half-power points. In addition to the PDR a magnetometer was towed along the longer track lines shown in Fig. 1, but, because of its relative sparsity and the nearness of the magnetic equator the magnetic data added little to the bathymetric and has not been included in this study. Seismic reflection profiling equipment was operated only on the passage tracks into and out of the area.

The bathymetric chart obtained for the area is shown in Fig. 2 and a simplified version of the same, together with OBS locations and earthquake epicentres, is shown in Fig. 3.

ECHO CHARACTER

In an area of rugged topography with little sedimentary cover, seismic reflection profiles do not yield much useful information. The ideal acoustic tool in such terrain would have

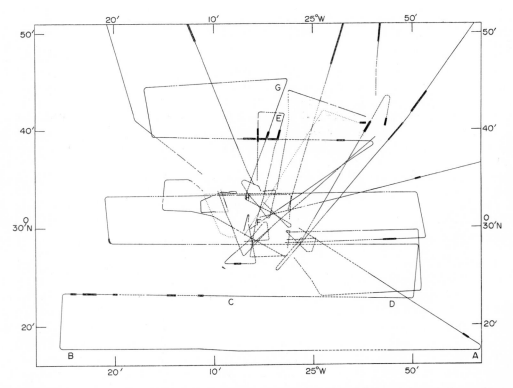

Fig. 1.   Tracks of R.R.S. *Shackleton* in the survey area. Examination of the PDR records made it possible to distinguish three types of bottom: unsedimented (dashed lines); partially sedimented (thin solid lines); sediment ponds (thick solid lines) (see Figs. 4, 5, 6 and 7). Along dotted lines the PDR records were of insufficient quality to make a distinction.

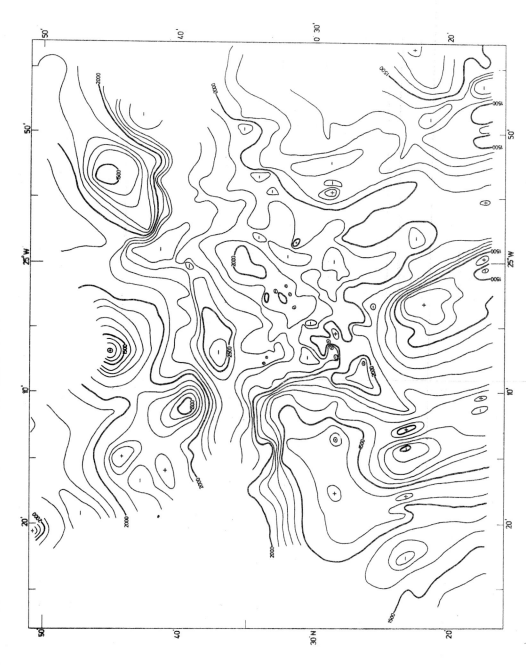

Fig. 2.   Bathymetric map based on the tracks shown in Fig. 1. Depths in corrected fathoms (MATTHEWS, 1939). Contour interval 100 fathoms.

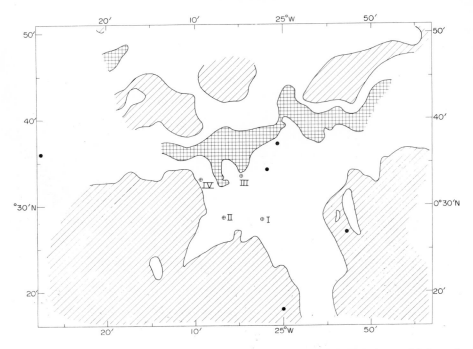

Fig. 3.   Simplified bathymetric map derived from Fig. 2. Depths shallower than 1900 corrected fathoms hatched diagonally. Depths greater than 2300 corrected fathoms cross hatched. Earthquake epicenters reported by the I.S.C. for 1963–1972 are shown by solid dots. Circled crosses show sites of OBS deployment.

a narrow beam. It was found in the case of the St. Paul's Fracture Zone survey, however, that the wide beam echo-sounding records contained much information which could be interpreted in relation to the sedimentary cover. In general, little attention has been given to echo-sounding records beyond measuring depths, but by a combination of echo-sounding and bottom photography LAUGHTON, HILL and ALLAN (1960) were able to correlate echo-character with the type of bottom.

Essentially three different types of echo could be identified on our PDR records:

1. Fuzzy echoes with little variation in signal strength recorded over a depth range of 100–200 fathoms.* These were observed over the youngest area of sea floor and some steep slopes (Figs. 4, 5, 6 and 7). Submersible and photographic observations of the floor of the median valley of the Mid-Atlantic Ridge further north (BALLARD, BRYAN, HEIRTZLER, KELLER, MOORE and VAN ANDEL, 1975) show it to be composed of a rough pile of pillow basalts with negligible sedimentary cover. There is no reason for suspecting the youngest areas of sea floor in our survey area to be different. Such a seabed is extremely rough at wavelengths comparable to that used by the PDR ($\lambda = 15$ cm). In this situation specular reflection does not occur, but sound energy can be scattered back to the ship's echo-sounder from the whole area of sea floor illuminated. LAUGHTON, HILL and ALLAN (1960) have shown that the length of the echo obtained from a flat rough bottom is

---

* 1 fathom = 1.8288 m.

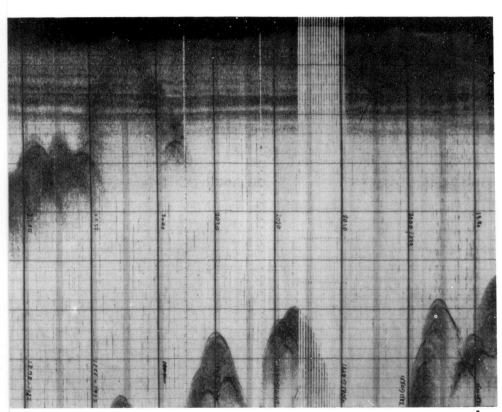

AN
ALLEY

A

sec).
alley
high

$d(\sec \theta - 1)$ where d is the water depth and 2θ the effective beam width of the echo sounder.

Setting $d = 2000$ fm and $\theta = 20°$,

$$d(\sec \theta - 1) = 128 \text{ fm}.$$

This is in reasonable agreement with the echo length obtained from the median valley floor.

2. Sharp echoes, comparable in length to the outgoing pulse, forming hyperbolic traces on the PDR record. These were observed over the mountainous terrain flanking the median valley (Figs. 4, 5) and over the older seafloor to the north of the fracture zone (Figs. 6, 7). Most of the hyperbolae differed little from those calculated for point reflectors at the appropriate depth, thus the scale of the reflecting surfaces must have been small. But the sharpness of the echoes indicates that specular reflection was taking place. Since the ocean floor acquires thicker sedimentary cover as it moves away from the spreading axis, these hyperbolic echoes are thought to indicate that the bottom is sufficiently sedimented to remove its short wavelength roughness ($\lambda \sim 15$ cm) but not, of course, so much as to alter the hilly nature of the terrain.

The sharp hyperbolic echoes become distinct at a distance of from 5–10 km from the axis of the median valley (Figs. 4, 5). The sedimentation over these mountainous regions but since the level of biological productivity is high in this region of the equatorial Atlantic the sedimentation rate is quite high, of the order of 10 mm/1000 yr (KOBLENTZ-MISHKE, VULKOVINSKY and KABANOVA, 1970; BÉ and TOLDERLUND, 1971). Taking a half-spreading rate of 1.5 cm/yr for the vicinity of the St. Paul's Fracture Zone (MORGAN, 1968), the sediment thickness to be expected at a distance of 5–10 km from the spreading axis lies in the range 3–7 m. This is clearly enough to obliterate the short wavelength roughness of the basalt floor.

3. Strong echoes indicating a smooth and flat-lying seabed (Figs. 6, 7). These are clearly associated with sediment ponds and were more common to the north of the fracture zone than on the younger seafloor to the south.

The type of echo observed is indicated along the tracks shown in Fig. 1.

DISCUSSION

Both the median valley and the fracture zone are clearly expressed in the bathymetry (Figs. 2, 3). The south-southeast trend of the median valley away from the fracture zone is also evident from the lack of sediment (Fig. 1). As the median valley approaches the fracture zone it widens rather abruptly, possibly because a subsidiary fracture zone at 0°28′N offsets it to the west. The deepest water in the whole survey area occurs at the junction of the median valley with the main fracture zone. This is a situation commonly observed elsewhere on the Mid-Atlantic Ridge. The existence of such deep holes at the junctions of fracture zones with the central rift of mid-oceanic ridges has been explained by SLEEP and BIEHLER (1970) in terms of the increased loss of hydraulic head of the viscous material upwelling along the ridge axis which occurs there.

An interesting feature of the fracture zone is that the active part to the west of the junction with the rift appears to be offset about 6 km to the south of the inactive part to the east. If the present pole of opening for the Atlantic Ocean is at 58°N, 36°W

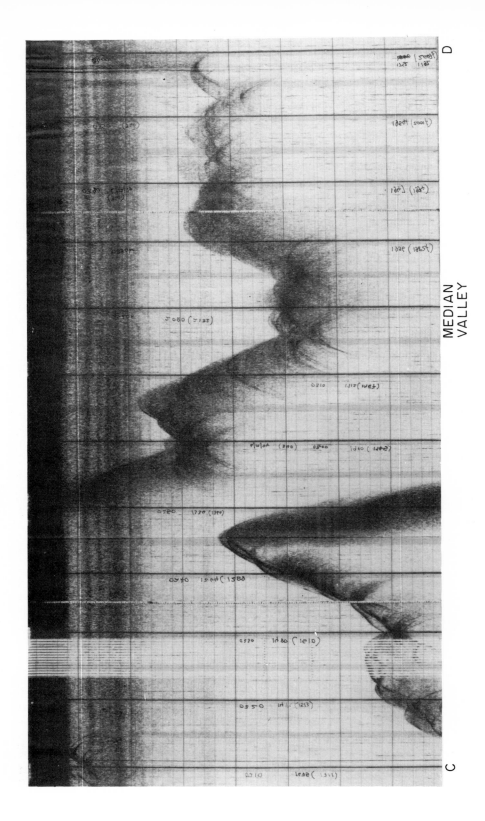

MEDIAN
VALLEY

Fig. 5.   PDR record along east–west profile CD in Fig. 1. See caption of Fig. 4.

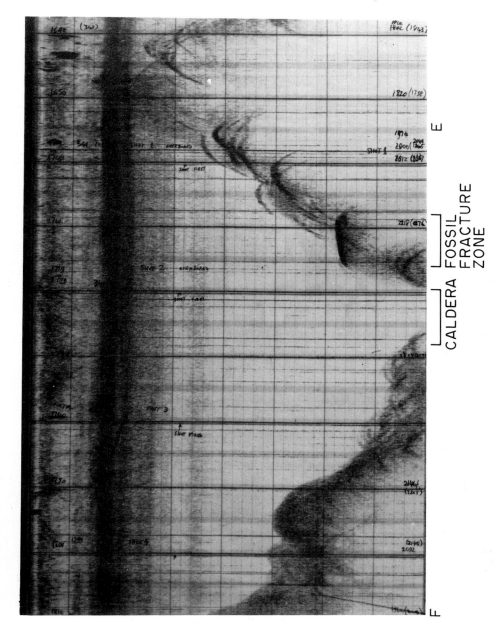

Fig. 6.   PDR record along north–south profile EF in Fig. 1. The older crust to the north of the fracture zone is more sedimented than the younger to the south. Although echoes from the caldera floor are lost in the scattering layers, they would not be if this floor were appreciably sedimented. The thickest sediment lies up the north wall and is interpreted as the fossil axis of the fracture zone.

(MORGAN, 1968), the trend of the St. Paul's Fracture Zone at 25°W should be 083–263°. Earthquake epicenter data (BARAZANGI and DORMAN, 1969) and the generalized bathymetry (HEEZEN, BUNCE, HERSEY and THARP, 1964) for the whole of the St. Paul's Fracture Zone suggest a nearly east–west trend. The trend of the fracture zone as defined by the 2300-fathom contour (Fig. 3) is about 078–258°. Rather than accept this as its true trend, an alternative interpretation is that the active and inactive sections each trend roughly east–west but are offset from each other. This latter interpretation is reinforced by a consideration of the sediment distribution. The active section of the fracture zone to the west of and including the deep hole has little sedimentary fill. North of the deep hole, however, a prominent pond of sediment lies some 200 fathoms higher up the north wall (Figs. 6, 7). This pond lines up exactly with the trend of the inactive part of the fracture zone to the east (Figs. 1, 3). It is implausible that such a pond could form without much sediment reaching a deeper area to the south. The deep hole, therefore, must be a collapse feature—

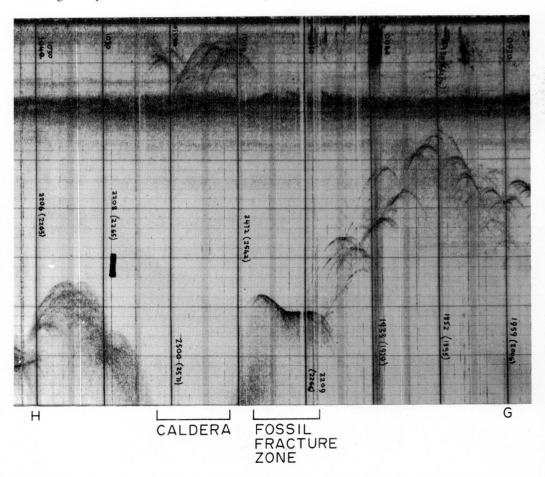

Fig. 7.   PDR record along north–south profile GH in Fig. 1. As on profile EF the older crust to the north of the fracture zone is more sedimented than the younger to the south. Note that there is little sediment in the caldera floor compared to that in the pond on the north wall. The latter is interpreted as the fossil axis of the fracture zone.

a caldera—of relatively recent origin, whilst the sediment pond to the north marks the fossil axis of the fracture zone. The occurrence of caldera collapse along the spreading axes of slow spreading ridges has been proposed by FRANCIS (1974). It is clear from the echo-sounder profiles crossing the deep hole that it is, at least on its northern side, fault bounded. The eastward extent of the present transform fault trend suggests that the caldera collapse and the shift of the fracture zone axis occurred within the last 700,000 years.

Acknowledgements—I thank Captain G. H. SELBY SMITH, the officers and men of the R.R.S. *Shackleton* for their assistance at sea, Dr. E. J. W. JONES who was chief scientist, and Mrs. P. A. COURTNEY who did most of the data reduction. The paper was written at the Cooperative Institute for Research in Environmental Sciences, University of Colorado/NOAA, Boulder, Colorado, to whom I am most grateful for a fellowship. The work was supported by the Natural Environment Research Council, U.K.

REFERENCES

BALLARD R. D., W. B. BRYAN, J. R. HEIRTZLER, G. KELLER, J. G. MOORE and VAN ANDEL (1975) Manned submersible operations in the FAMOUS area: Mid-Atlantic Ridge. *Science*, **190**, 103–108.

BARAZANGI M. and J. DORMAN (1969) World seismicity maps compiled from ESSA, Coast and Geodetic Survey, Epicenter data 1961–1967. *Bulletin of the Seismological Society of America*, **59**, 369–380.

BÉ A. and D. S. TOLDERLUND (1971) Distribution and ecology of living planktonic Foraminifera in surface waters of the Atlantic and Indian Oceans. In: *Micropalaeontology of oceans*, B. M. FUNNEL and W. R. REIDEL, editors, Cambridge University Press, London.

EWING M., G. CARPENTER, C. WINDISCH and J. EWING (1973) Sediment distribution in the oceans: the Atlantic. *Geological Society of America Bulletin*, **84**, 71–88.

FRANCIS T. J. G. (1974) A new interpretation of the 1968 Fernandina Caldera collapse and its implications for the Mid-Oceanic Ridges. *Geophysical Journal of the Royal Astronomical Society*, **39**, 301–318.

FRANCIS T. J. G., I. T. PORTER, R. D. LANE, P. J. OSBORNE, J. E. POOLEY and P. K. TOMKINS (1975) Ocean bottom seismograph. *Marine Geophysical Research*, **2**, 195–213.

HEEZEN B. C., E. T. BUNCE, J. B. HERSEY and M. THARP (1964) Chain and Romanche fracture zones. *Deep-Sea Research*, **11**, 11–33.

KOBLENTZ-MISHKE O. J., V. V. VOLKOVINSKY and J. G. KABANOVA (1970) Plankton primary production of the world ocean. In: *Scientific exploration of the South Pacific*, National Academy of Sciences, Washington, D.C.

LAUGHTON A. S., M. N. HILL and T. D. ALLAN (1960) Geophysical investigations of a seamount 150 miles north of Madeira. *Deep-Sea Research*, **7**, 117–141.

MATTHEWS D. J. (1939) *Tables of the velocity of sound in pure water and sea water for use in echo-sounding and sound-ranging*, 2nd edition, Hydrographic Department, Admiralty, London.

MORGAN W. J. (1968) Rises, trenches, great faults and crustal blocks. *Journal of Geophysical Research*, **73**, 1959–1982.

SLEEP N. H. and S. BIEHLER (1970) Topography and tectonics at the intersection of fracture zones and central rifts. *Journal of Geophysical Research*, **75**, 2748–2752.

Geochemical studies of organic-rich sediments from the Namibian Shelf
I. The organic fractions

R. J. MORRIS and S. E. CALVERT

Institute of Oceanographic Sciences,
Wormley, Godalming, Surrey GU8 5UB, U.K.

Abstract—An organic-rich, diatomaceous sediment core from the shelf off Walvis Bay has been examined for the composition and variation in some of the major organic fractions, together with analyses of the lipid and amino acid content. Many intact and partially altered planktonic pigments and lipids are present, indicating good preservation over the 1000-year period of accumulation of the sediment.

The fatty acid and fatty alcohol compositions were similar to that in marine diatoms and planktonic crustaceans although the longer chain unsaturated acids and alcohols (C_{20}–C_{22}) were in lower concentration and odd-numbered normal and branched-chain C_{15}–C_{17} acids were present. Some influence of bacterial activity may be indicated by the presence of odd-chain acids.

The n-alkanes have distribution maxima at C_{17} at the sediment surface and C_{25-27} and C_{17-19} at depth. CPI values are all low. No polyunsaturated hydrocarbons were found. Aromatic hydrocarbons were present in all core samples.

A number of sterols were present at the sediment surface, with cholesterol being the most abundant. These compounds are identical to those found in diatom and crustacean lipids with the exception of methylene cholesterol.

The amino acid composition changes slightly with depth in the core, acidic acids falling and valine, arginine and aromatic acids increasing with depth.

Fulvic acid levels are high at all depths in the core, whereas humic acid levels are significantly lower. Conversion of the planktonic organic material to higher molecular weight compounds appears to be slow.

INTRODUCTION

THE organic fraction of marine sediments may be derived from plant and animal sources and from terrestrial and marine environments. This source material undergoes varying degrees of alteration in the water column before final deposition, and a series of poorly understood diagenetic changes after deposition. The resultant complexity of the composition of the organic fraction of marine sediments produced by the interplay between these processes has recently been amply illustrated by, for example, the work of COOPER and BLUMER (1968), BROWN, BAEDECKER and KAPLAN (1972), FARRINGTON and QUINN (1973), and GASKELL, MORRIS, EGLINTON and CALVERT (1975) on Recent sediments and of AIZENSHTAT, BAEDECKER and KAPLAN (1973) and SIMONEIT and BURGLINGAME (1974) on pre-Recent sediments collected by the Deep-Sea Drilling Project. Information on the general and specific composition of recently sedimented organic material is vital to a better understanding of the cycling of organic carbon in the marine environment and for studies of the early stages of alteration of source material towards deeply buried carbon-rich residues.

We report here the composition of some of the major fractions of the organic material in sediments accumulating on the shelf off Namibia (South West Africa). Since this area receives organic material overwhelmingly from planktonic sources, we are able to study the composition and alteration of organic material from a single source.

The Namibian shelf lies beneath the Benguela Current, a site of intensive seasonal upwelling (Hart and Currie, 1960) and periods of exceptionally high primary production (Steeman Nielsen and Jensen, 1957). Hydrographic descriptions of the current by Hart and Currie (1960), Stander (1964), Shannon (1966), Vissler (1969), Jones (1971) and Calvert and Price (1971a) show that upwelled water is derived from depths of about 200 m at the shelf edge, and that strong vertical and horizontal gradients of nutrient concentrations are present. Due to the upwelling of subsurface water, and the high production and consequent decomposition of organic matter in the shelf waters, oxygen concentrations in the water are quite low and decrease toward the bottom (Calvert and Price, 1971a). The dense plankton blooms found off Walvis Bay consist mainly of diatoms (Hart and Currie, 1960; Kollmer, 1963).

The Namibian shelf is one of the deepest in the world (Simpson and Du Plessis, 1968) and, considering the lack of runoff from the adjacent Namib Desert (Logan, 1960), the depositional terrace is surprisingly wide (Van Andel and Calvert, 1971). Modern sediments on the shelf range from calcareous sands on the outer shelf to organic-rich, diatomaceous oozes, containing abundant fish debris, on the inner shelf (Senin, 1968; Eisma, 1969; Avilov and Gershanovich, 1970; Calvert and Price, 1971b). The nearshore oozes contain high concentrations of H_2S and benthos is rare or absent; they correspond to the *azoic zone* of Copenhagen (1934, 1953).

Some aspects of the organic geochemistry of the Recent sediments off Namibia have recently been reported. Romankevich and Baturin (1974) have determined the major classes of organic matter present in a range of sediment samples from the area off Walvis Bay. Boon, Leeuw and Schenck (1975) and Boon, Rypstra, Leeuw and Schenck (1975) have identified the major fatty acids in the sediments which were released upon alkaline hydrolysis.

In this paper, we have investigated the composition of the organic matter in a single sediment core sample collected off Walvis Bay from the point of view of the relative proportions of some of the major compound classes present and in terms of the variation in the composition of the lipid and protein amino acid fractions in some more detail.

MATERIALS AND METHODS

Sampling

The sediment core was collected in October 1968 on R.V. *Argo* of the Scripps Institution of Oceanography during Expedition CIRCE at 22°56.4′S, 13°59.6′E in 143 m water depth (Fig. 1). A corer with a 20-cm diameter fibre-glass barrel and with a sphincter valve and core catcher (Burke, 1968) was used. The core, 81 cm in length, was sealed and frozen in a vertical position immediately after collection and has been stored at $-10°$ to $-20°C$ until sampled.

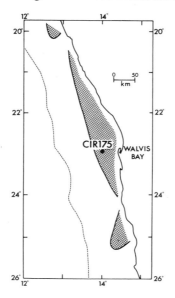

Fig. 1.   Position of sediment core (CIRCE 175) used in this study. Shaded areas show the distribution of diatomaceous ooze on the inner shelf (see CALVERT and PRICE, 1971b). Dashed line marks the edge of the continental shelf.

Sample preparation

Samples (5 g) for mineralogical and chemical analyses were air-dried at 100°C and ground to a fine powder in a tungsten carbide mill. For organic analyses, samples having a wet weight of 150–250 g were taken from the centre of the core (well away from the core barrel) at the intervals indicated in Table 1. These samples were not dried: they were checked for the absence of any large benthic organisms which would have affected the organic composition and were then divided into two parts: one part for amino acid, fulvic acid and humic acid analyses and the other part for lipid analysis.

Analyses

(a) *Carbon and carbonate* were determined gravimetrically using a Leco carbon analyser following methods given in GASKELL, MORRIS, EGLINTON and CALVERT (1975).

(b) *Opal* (*diatomaceous silica*) (see CALVERT, 1966) was determined by X-ray diffraction techniques following methods described by EISMA and VAN DER GAAST (1971).

(c) *Quartz* was determined by X-ray diffraction techniques following methods given in CALVERT (1966).

The carbon, carbonate, opal and quartz data have been corrected for the diluting effect of sea salt in the dried samples by determining the Cl contents by X-ray emission methods (see CALVERT and MORRIS, 1977).

(d) *Organic fractions.* All chemical solvents were Aristar grade (British Drug Houses); other reagents were analytical grade (B.D.H.). All-glass extraction equipment was used and blanks of the extraction and analytical schemes showed no significant contamination.

TABLE 1

MAJOR ORGANIC FRACTIONS (EXPRESSED AS % DRY WT.) OF THE SEDIMENT SAMPLES

DEPTH OF SAMPLE (CM)	ORGANIC EXTRACT I (unbound lipids)	ORGANIC EXTRACT II (readily hydroly-able lipids)	ORGANIC EXTRACT III (strongly bound lipids)	AMINO ACIDS	HUMIC ACIDS	FULVIC ACIDS	FINAL * RESIDUE
0 – 5	0.90	0.76	0.14	4.88	0.90	9.67	7.60
5 – 10				3.60	0.99	15.30	5.40
10 – 15	–	–	–	3.78	1.04	19.26	5.70
20 – 30	0.61	0.53	0.20	2.96	0.81	21.94	4.25
40 – 50	0.34	0.31	0.17	2.32	0.78	14.44	5.17
60 – 70	0.23	0.38	0.13	2.03	0.51	16.67	2.56

*   Residue from amino acid and humic extractions.
     May include varying amounts of salt.

The analytical schemes applied are shown in Fig. 2. The methods used for thin-layer chromatography (TLC) and amino acid analysis have been described previously (MORRIS, 1973a, 1975). Wherever possible operations were carried out under oxygen-free nitrogen. Three lipid extracts of the sediment samples were made. Firstly, the unbound lipids were extracted, unaltered, with $CHCl_3$/MeOH (2:1) (Organic Extract I); secondly, the weakly bound lipids were released by mild acid hydrolysis (Organic Extract II); finally, strong acid hydrolysis was used to release any strongly bound lipids (Organic Extract III). FARRINGTON and QUINN (1971) have suggested that some of the fatty acids may be 're-bound' to the sediment following hydrolysis, but in this work repeating both the 2 N HCl and 6 N HCl hydrolysis/extraction steps resulted in < 2% additional extractable fatty acid.

The total amino acids (protein amino acids + amino acids bound to fulvic and humic acids—see, for example, HARE (1972), RASHID (1972) and MULLER (1975)) were liberated by 24-hour, 6 N HCl hydrolysis of the fresh sediment. This acid hydrolysis resulted in a yellow-brown aqueous extract and it is thought that some of the lower molecular weight fulvic acids were lost during this treatment. Therefore the amounts of fulvic acid, liberated by subsequent alkaline hydrolysis of the sediment residue, may be less than the total amounts present.

Gas liquid chromatographic (GLC) analyses of the lipid fractions were performed using a Perkin–Elmer model F-11 or a Pye model 104 chromatograph. Combined gas chromatographic-mass spectrometric (GC-MS) analyses were carried out with a Pye 104 gas chromatograph coupled to an AEI Model MS9 mass spectrometer (BALLANTINE, ROBERTS and MORRIS, 1975). Methylation (BF_3/methanol) of free fatty acids, acetylation (pyridine/acetic anhydride) of free fatty alcohols and silylation (bis-trimethylsilylacetamide) of sterols, together with a TLC clean up in each case, preceded GLC and GC-MS analyses.

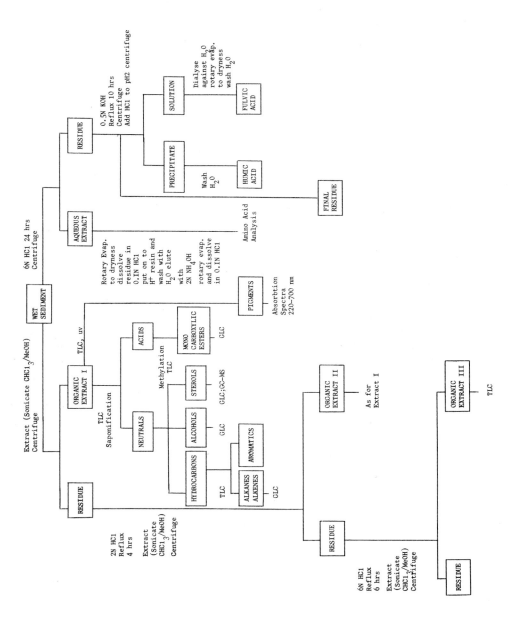

Fig. 2.   Extraction and separation scheme for the organic analyses. For details, see text.

Where necessary catalytic hydrogenation of the methyl esters (Ackman and Burgher, 1964; Morris, 1973a) was used as an additional means of structural identification. The saturated and unsaturated long-chain hydrocarbons were separated from the other hydrocarbon classes by TLC (Morris, 1973b) prior to GLC. Insufficient material was available for quantitative separation of the alkane/alkene fraction.

Column conditions were as follows:

(a) Fatty acids and fatty alcohols: 2-m × 5-mm glass column packed with 100–120 mesh Chromosorb G (AW-DMCS) coated ($3\frac{1}{2}\%$) with high performance DEGS either held isothermally at 180°C or temperature programmed from 120°C to 200°C at 10°C/min; $2\frac{1}{2}$-m × 5-mm glass column packed with 80–100 mesh Chromosorb W (AW-DMCS) coated (4%) with SILAR 5CP held isothermally at 210°C.

(b) Hydrocarbons: 4-m × 5-mm glass column packed with 80–100 mesh Chromosorb W (AW-DMCS) coated (3%) with Apiezon L, temperature programmed from 80° to 270°C at 10°C/min.

(c) Sterols: $2\frac{1}{2}$-m × 4-mm glass column packed with 100–120 mesh Diatomite CQ coated (1%) with Dexsil 300 GC held isothermally at 260°C; $2\frac{1}{2}$-m × 4-mm glass column packed with 100–120 mesh Diatomite CQ coated (1%) with SILAR 5CP held isothermally at 220°C.

(e) *Radiocarbon dating and stable carbon isotope analyses* were carried out on 100-g samples (wet weight) by D. D. Harkness of the Scottish Universities Research Reactor Centre at East Kilbride. Organic carbon only, rather than the total carbon, was used for these analyses. ^{13}C values are expressed per mil relative to the P.D.B. standard.

RESULTS

The sediment consisted of a fluid, dark olive-green diatomaceous ooze, containing high concentrations of H_2S. There was no evidence of lithological variation over the full length of the core.

The radiocarbon ages of the core sections (Fig. 3) suggest a rapid sedimentation rate,

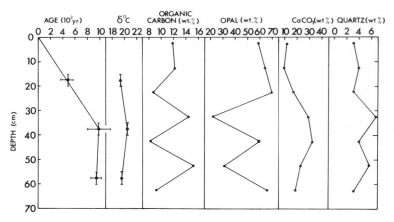

Fig. 3. Compositional and age data for the core subsamples. Samples for age dating and for carbon isotopic compositions were from relatively large intervals, whereas other analyses were carried out on smaller samples taken at the points indicated.

of the order of 0.35 mm yr^{-1} for the upper part, assuming a zero age for the sediment surface. The similar ages at 37- and 57-cm depth may be due to slumping or reworking of the sediment. The age of a near-surface sediment from a core collected in the same position, as reported by VEEH, CALVERT and PRICE (1974), is significantly higher than the age of the uppermost sample reported here. On the other hand, the age of the $CaCO_3$ fraction in the same sample analysed by VEEH, CALVERT and PRICE (1974) is modern. Where ages of the organic carbon and the carbonate fraction in the same sediment sample are available, the organic carbon age is almost invariably younger (EMERY and BRAY, 1962).

The age data given in Fig. 3 is based on the organic carbon and appears to be consistent with the extrapolated age/depth curve as drawn.

The stable carbon isotopic composition of the organic material in the sediment is fairly constant in the three samples examined, ranging from -19.5 to $-20.5\%_{oo}$. $\delta^{13}C$ values less than $-23.0\%_{oo}$ are generally accepted as indicating a marine planktonic source (see discussion in GASKELL, EGLINTON, MORRIS and CALVERT, 1975); a totally marine input of organic material over the last 1000 years is therefore indicated for the sediment examined.

Organic carbon and opal contents, which, as emphasized previously, are derived entirely from the plankton, together make up the major fraction of the sediment (Fig. 3). They vary antipathetically over fairly wide limits down the core, a situation produced to a large measure by the constraint imposed by the constant sum of all components. However, the concentration of $CaCO_3$, composed of comminuted mollusc and brachiopod fragments, varies independently of carbon and opal. The inverse correlation between carbon and opal may therefore also reflect variable compositions, in terms of the ratio of dispersed organic matter to skeletal silica, of the planktonic input to the sediment.

The concentrations of quartz, as a measure of terrigenous material, are very low in all samples examined (Fig. 3). The sediment is essentially composed of skeletal debris and dispersed organic matter.

The distribution of the major organic fractions in the core samples are shown in Table 1. The results (expressed as % dry wt) are similar to those reported by ROMANKEVICH and BATURIN (1974) for several surface samples from this area, although they found rather higher levels of extractable lipids (free bitumens) and lower levels of humics (fulvic + humic acids). The major trend with depth in the core is a decrease in the amounts of amino acids and unbound/weakly bound lipids and an increase in the fulvic acid content. This suggests that the more reactive natural product compounds are being incorporated into higher molecular weight compounds with time in the sediment, although the high level of fulvic acid at even the surface horizon indicates considerable compositional change to the natural product input within the water column prior to sedimentation.

Organic extracts

Solvent extracts of marine sediments often contain 'elemental' sulphur (GASKELL, MORRIS, EGLINTON and CALVERT, 1975 and references therein) and in this work sulphur was found in all of the lipid extracts. This identification was confirmed by X-ray emission spectrometry. The H_2S present in marine sediments is produced by sulphate-reducing bacteria (BUTLIN, 1949; THODE, KLEEREKOPER and McELCHERAN, 1951) and is the largest reservoir of S in such sediments (KAPLAN, EMERY and RITTENBERG, 1963). Although we have no data on the actual sulphide levels in these sediments, a deposit of S is formed

TABLE 2

MAJOR LIPID CLASSES IN ORGANIC EXTRACTS I AND II (μg/g dry sediment)

Depth of Sample (cm)	EXTRACT I						EXTRACT II			
	Fatty Acids	Hydrocarbons (Alkanes)	Alcohols	Sterols	Phytol		Fatty Acids	Hydrocarbons	Sterols	Phytol
0 – 10	350	170	30	11	330		730	200	T	1
20 – 30	230	120	–	6	–		270	53	ND	–
40 – 50	110	150	–	0.6	–		140	35	ND	–
60 – 70	30	90	2	ND	20		170	40	ND	–

–	=	NOT ANALYSED
ND	=	NOT DETECTED
T	=	< 0.5 μg

on the surface of the dried sediment samples (Hart and Currie, 1960). The oxidation of pore water sulphide during the work-up procedures is presumably giving rise to the high concentrations of S in the lipid extracts (see Chen, Moussavi and Sycip, 1973). TLC and GLC separations were not, however, affected by the presence of S.

Approximate quantification of the major lipid classes identified from organic extracts I and II (Table 2) is based on summed GLC responses. The levels are high compared with other marine sediments for which data are available, including highly productive areas such as the Northwest African margin (Gaskell, Morris, Eglinton and Calvert, 1975). They are substantially lower, however, than levels reported for Recent Black Sea sediments where Peake, Casagrande and Hodgson (1974) report up to 4200 μg fatty acids/g dry sediment and up to 2500 μg sterols/g dry sediment. It is not clear from the data given by these authors whether they are dealing with sapropels, which are known to contain up to 20% organic carbon in some Black Sea cores (Ross, Degens and MacIlvaine, 1970), rather than 'normal' Recent Black Sea sediments which contain considerably less organic matter.

(a) Pigments

TLC of the unbound lipids (Organic Extract I) showed the presence of many pigment bands ranging from strongly polar to very non-polar. A mixture of green to grey-green bands which fluoresced strongly under UV light (porphyrins) and yellow, orange and red bands which did not fluoresce (carotenoids) were seen. Seven pigment fractions of different polarity were separated. Tentative identification of the pigments was based on spectral analysis in both CHCl$_3$ and 80% acetone. The major non-polar pigment was a yellow carotenoid identified as a mixture of α and β carotenes (454, 466, 485, 497 nm maxima). A grey-green porphyrin was the next least polar major pigment (410, 667 nm maxima), identified as phaeophytin. This was followed by a complex mixture of green to grey/green porphyrins (broad soret-band 412–420 nm and 665 nm maxima) thought to include chlorophylls, phaeophorphyrins and chlorophyllides. Small peaks at 365 and 608 nm suggested the presence of bacteriochlorophyll and medium polar, yellow and red caroteroids (peaks and shoulders around 445 and 475 nm) were seen. The polar pigment fractions (40–60% of total extract) (green/brown) gave very strong soret bands (410–418 nm) suggesting a complex mixture of porphyrins.

Organic extract II (weakly bound lipids) contained much lower levels of medium polar pigments with almost no carotene, whilst organic extract III (strongly bound lipids) contained only strongly polar pigment material.

A measure of the total chlorophyll/phaeophytin levels in the lipid extracts was obtained by determining the phytol levels (SCHULTZ and QUINN, 1974). High levels were found in organic extract I from the surface horizon (Table 2) (330 µg phytol/g dry sediment–1000 µg chlorophyll/phaeophytin/g dry sediment), decreasing with depth in the core (60–70 cm horizon, Extract I: 20 µg Phytol/g dry sediment–60 µg chlorophyll/g dry sediment). The surface horizon value is very similar to that previously reported by ALDERSHOFF (750 µg/g dry sediment, in BRONGERSMA-SANDERS, 1951) who determined the chlorophyll content of a surface sample from the Walvis Bay area using the absorption at 660 nm of an alcoholic extract.

(b) *Lipids*

Visualization of the TLC plate in I_2 vapour showed that a number of non-polar and medium polar lipid compounds were present in Organic Extracts I, in addition to the pigments. Their R_f values corresponded to hydrocarbons, wax esters/sterol esters, glycerides, free fatty acids/free fatty alcohols and sterols. Organic Extracts II contained only hydrocarbons and free fatty acid/free fatty alcohol and Extracts III contained only polar compounds. The polar fractions of all the lipid extracts were found by X-ray emission to contain P, but whether this is due to the presence of intact phospholipids is unknown.

(i) *Fatty acids.* The relative distributions of the major fatty acids in Extracts I and II are given in Fig. 4; 14:0, 16:0, 16:1, 18:0 and 18:1 fatty acids are the major components

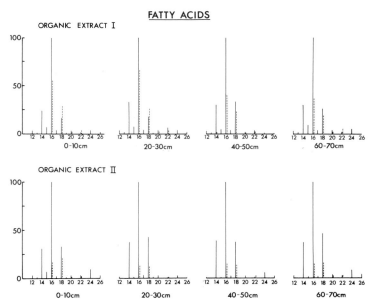

Fig. 4. Distribution of fatty acids in organic Extracts I and II determined from gas chromatogram peak areas. Amounts expressed relative to the most abundant component as 100%. Solid lines correspond to alkanoic acids; dashed lines to alkenoic acids.

with far more unsaturation (16:1/18:1) in Extracts I than in Extracts II. Traces of what appeared to be polyunsaturated C_{20} and C_{22} acids were also seen in Extract I from the surface horizon. A number of branched-chain fatty acids were found, anteiso/iso 15:0 (2.5–5.0% total fatty acids) and anteiso/iso 17:0 (1.2–2.3% total fatty acids) being the major types. C_{19} and C_{20} isoprenoid acids were tentatively identified but at low levels (<1.5% total fatty acids). Direct comparison with the results of BOON, LEEUW and SCHENCK (1975) is difficult, as they analysed the total fatty acids liberated after alkaline hydrolysis. The general distribution of fatty acids is similar, but these workers report rather lower levels of C_{18} acids and higher levels of C_{14} acids.

BOON, LEEUW and SCHENCK (1975), assuming the major fatty acid input to these sediments came from diatoms, compared some of the data available on fatty acid compositions of marine diatoms with their sediment compositions and concluded that major compositional changes had occurred. From their review of diatom compositions, they concluded that unsaturated C_{14}, C_{16} and C_{20} fatty acids were the major diatom acids, whilst C_{14}, C_{16} and C_{18} saturated acids were more minor components. In the sediment they found low levels of unsaturated acids, high levels of 14:0 and 16:0 acids and many odd-numbered, branched chain acids. They suggested that the algal fatty acids were being rapidly resynthesized into more saturated straight chain and branched chain bacterial fatty acids before their incorporation into the sedimentary organic matter.

Many workers have suggested that odd-numbered normal, iso and anteiso acids have a bacterial origin; indeed there are reports of these acids being major components of some bacterial lipids (KATES, 1964; KANEDA, 1967). Their presence (particularly C_{15} and C_{17} acids) in marine sediments has been taken as an indicator of mirco-organism activity (COOPER and BLUMER, 1968). However, insufficient data are available on the lipid composition of either marine diatoms or marine bacteria to be able to support the idea of total conversion from diatom lipid to bacterial lipid.

The majority of studies on marine diatom lipids have used material cultured in the laboratory and have revealed very few examples of a characteristic diatom fatty acid composition. Most notable was the finding that C_{18} polyunsaturated acids were either very minor components (DEMORT, LOWRY, TINSLY and PHINNEY, 1972) or completely absent (CHUECAS and RILEY, 1969). Comparison of fatty acid analyses of naturally occurring diatoms with those of the same species cultured in the laboratory reveals significant differences. LEWIS (1969) found 16:0, 16:1 and 20:5 to be the major acids in Arctic diatoms whereas BOTTINO (1974) reported 14:0, 16:0 and 18:1 to be the main ones in Antarctic diatoms. These may be compared with 16:0 and 16:1 (ACKMAN, TOCKER and MCLACHLAN, 1968) and 16:1, polyunsaturated C_{16} acids and 20:5 (CHUECAS and RILEY, 1969) which were found to be the major acids of diatoms grown under different culture conditions. It seems likely that environmental and culture conditions are responsible for some of these differences; REITZ and MOORE (1972) have found, for example, that different forms of dietary carbon have an influence of the fatty acid composition of *Euglena*. The fatty acid composition of diatoms has also been found to vary with the stage of maturity (ACKMAN, JANGAARD, HOYLE and BROCKERHOFF, 1964; PUGH, 1971). Thus, the analysis of mature cultures might not reflect the true fatty acid composition of a normal pelagic population during a period of high growth rate.

Attempts to characterize the fatty acid composition of marine bacteria are equally confusing. BLUMER, CHASE and WATSON (1969) found that NH_3-oxidizing strains had a simple lipid composition, 96–100% consisting of 16:0 and 18:1 acids, but that nitrate-oxidizing bacteria had a more complicated composition which included C_{14}–C_{19} acids. OLIVER and COLWELL (1973a, b) found that the major fatty acids in twenty strains of gram-negative marine and estuarine bacteria were 16:0, 16:1 and 18:1. One species, however, exhibited a very different fatty acid pattern with branched-chain acids forming the major class (see also KATES 1964; KANEDA, 1967). Eight species of marine bacteria have been analysed by KUNIMOTO and co-workers (KUNIMOTO, 1975; KUNIMOTO, ZAMA and IGARASHI, 1975). They report 15:0/15:1, 16:0/16:1 and 17:0/17:1 as being the major acids in seven species, with 18:1 and 16:0 in the other species. The mean ratio of unsaturated/saturated acids in these species was 2.8:1.

The question of the extent of bacterial involvement in the alteration of sedimentary organic material, either by alteration of the natural product input or by replacement by bacterial lipid, is a major one. It must, however, remain unanswered for the moment due to insufficient unequivocal data.

Bearing these considerations in mind, fatty acid analyses of some cultured diatom species which belong to the major groups found in Namibian waters (see HART and CURRIE, 1960) have recently been reported by BOUTRY and BARBIER (1974) who worked on *Chaetoceros* and ORCUTT and PATTERSON (1975) who worked on *Nitzschia*, *Biddulphia* and *Fragilaria*. In nearly all cases, 14:0, 16:0, 16:1 and polyunsaturated C_{20} are the major acids. Planktonic crustaceans, particularly copepods, are likely to be the next important source of sedimentary fatty acids in this area (see later), 16:0/16:1, 18:1, 20:5, 22:6 being their major lipid fatty acids (SARGENT, LEE and NEVENZEL, 1976; MORRIS and CULKIN, 1976). Comparison of these data with the major sediment fatty acids found in the unbound lipid extracts of the sediments investigated here suggests that an almost total loss of poly-unsaturated fatty acids and the appearance of odd-numbered normal and branched-chain C_{15} and C_{17} acids are the major changes to the likely fatty acid input. Comparison with the bound lipid extracts indicates an almost total loss of unsaturation. Bacterial activity is certainly indicated by the presence of the odd chain acids but whether the loss of the polyunsaturated acids is due to auto-oxidation (they are very susceptible to breakdown even in low-O_2 environments), abiological hydrogenation, biological hydrogenation, bacterial replacement or a mixture of these, is not clear.

(ii) *Hydrocarbons*. The distributions of the major *n*-alkanes in Extracts I and II are given in Fig. 5. The identifications are based only on relative retention times; the alkane values quoted may therefore include contributions from more complex branched chain hydrocarbons. Extract I from the surface horizon gives a distribution maximum at C_{17} whilst the deeper samples give a major distribution maximum at C_{25}–C_{27} with a secondary maximum at C_{17}–C_{19}. All have low CPI (Carbon Preference Index) values (1.00–1.05). Extracts II give similar alkane distributions at all depth horizons (maxima at C_{17}–C_{19}) with little high molecular weight alkanes C_{24} and low CPI values (0.81–1.46). Pristane was found in Extracts I and II (0.5–5.0% of the total alkanes) but phytane was not adequately separated from C_{18} alkane. GLC analysis indicated the presence of other branched and unsaturated hydrocarbons but these were not generally further investigated.

From the alkane results there is evidence of a significant contribution from hydrocarbons originating in marine phytoplankton, whose alkane distributions would be expected to maximize at C_{17} (LEE and LOEBLICH, 1971; BLUMER, GUILLARD and CHASE, 1971; CLARK and BLUMER, 1967; WINTERS, PARKER and VAN BAALEN, 1969; YOUNGBLOOD, BLUMER, GUILLARD and FIORE, 1971; YOUNGBLOOD and BLUMER, 1973). The poly-

HYDROCARBONS

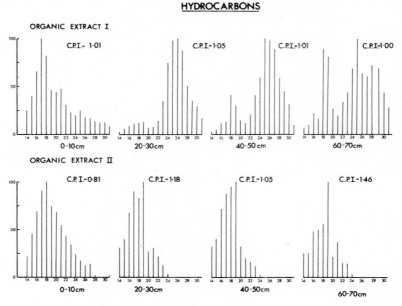

Fig. 5.   Distribution of *n*-alkanes in Organic Extracts I and II determined from gas chromatogram peak areas. Amounts expressed relative to the most abundant component as 100%. Carbon Preference Index (CPI) values are indicated.

unsaturated C_{19} and C_{21} hydrocarbons, known to be major compounds in many diatoms and algae (Lee and Loeblich, 1971; Blumer, Mullin and Guillard, 1970; Youngblood, Blumer, Guillard and Fiore, 1971), were not, however, detected in the sediment extracts, presumably being broken down by autoxidation, reduced to the corresponding C_{19}/C_{21} alkane or replaced by bacterial activity. Welte and Ebhardt (1968) have reported a correlation between the *n*-alkanes and *n*-fatty acids in recent sediments from the Persian Gulf area, indicating that the alkanes may be formed by a reduction, within the sediment, of the corresponding acids. Similarly, Powell, Cook and McKirdy (1975) have suggested that the C_{22} alkane, which is a prominent constituent in some sediment samples, especially phosphorites, has its source in the polyunsaturated $C_{22:6}$ fatty acid common to many marine organisms. In this particular reducing sediment, therefore, C_{14}–C_{22} even-chain alkanes may originate by direct reduction of the corresponding C_{14}–C_{22} fatty acid input, and C_{13}–C_{21} uneven alkanes may originate from either direct phytoplankton input or decarboxylation of the C_{14}–C_{22} fatty acid input. This would be one explanation for the low CPI values of the C_{14}–C_{22} alkanes.

The contribution of bacterial hydrocarbons, as for the fatty acids, is open to question. Davis (1968) has reported a low CPI for the alkanes of sulphate-reducing bacteria, while Johnson and Calder (1973) have postulated microbial activity as being responsible for the observed reduction of odd/even predominance with depth among the *n*-alkanes of the sediment core from a salt-marsh. Iso and anteiso branched hydrocarbons have also been reported in some marine bacteria (Oro, Tornabene, Nooner and Galpi, 1967).

A high content of long-chain alkanes ($> C_{25}$) with a large CPI normally indicates that the parent organic matter was derived from land plants (Powell and McKirdy, 1973;

HUNT, 1974). The major unbound alkanes present in the lower depth horizons of this core are certainly long chain $C_{24}-C_{31}$ but they have a very low CPI (1.00–1.05). The geochemical and ^{13}C data suggest very little terrigenous input to the sediment, so it is suggested that chain extension of the lower molecular weight natural product input may be occurring in this sediment. What is difficult to explain is why the bound (i.e. trapped) and unbound alkane distributions below the surface horizon should be different, unless active resynthesis is restricted to the unbound fraction.

For organic Extracts I and II, TLC of the total hydrocarbons (MORRIS, 1973b) gave a fraction which exhibited considerable UV fluorescence and had an R_f value corresponding to polyaromatic compounds. Ultraviolet spectra of this fraction showed strong absorption between 220–280 nm with shoulders from 280–340 nm suggesting the presence of mono, di- and polyaromatic hydrocarbons. Aromatic hydrocarbons have been reported in a variety of Recent, presumably unpolluted, marine sediments (see, for example, PEAKE, CASAGRANDE and HODGSON, 1974; BORDOVSKIY, 1974; HUNT, 1974; YOUNGBLOOD and BLUMER, 1975; BLUMER and YOUNGBLOOD, 1975; HITES and BIEMANN, 1974). The evidence suggests natural forest fires as their origin (BLUMER and YOUNGBLOOD, 1975; YOUNGBLOOD and BLUMER, 1975) followed by aeolian transport and eventual deposition. The presence of aromatic compounds in these sediments may therefore be an indication of at least one terrigenous input, albeit a minor one, to this sedimentary environment.

(iii) *Alcohols*. Significant levels of fatty alcohols were found in the unbound lipid fractions from the surface and 60–70 cm horizon (see Table 2). The major alcohol was 16:0 (> 60% total alcohols) with 14:0, 18:0 and 20:0 (7–16%) and traces of 22:0.

Wax esters have not been reported in phytoplankton lipids but have been reported in marine bacteria (KUNIMOTO, 1975; KUNIMOTO, ZAMA and IGORASHI, 1975) although no fatty alcohol compositions were given. Wax esters are known to be major lipids in many pelagic marine crustaceans and fish (SARGENT, LEE and NEVENZEL, 1976 and references therein; MORRIS and CULKIN, 1976 and references therein), normally with 16:0 as the major fatty alcohol together with 14:0, 18:0/18:1, 20:1, 22:1 although 20:1 and 22:1 are the principal alcohols of the upper-water, wax-ester-rich, calanoid copepods.

HART and CURRIE (1960) report a significant, though generally patchy spasmodic, distribution of zooplankton in the waters off Walvis Bay. The copepods were the most important planktonic crustacean with the calanoid copepods as the single largest group (KOLLMER, 1963; UNTERUBERBACKER, 1964).

As with the fatty acids, a comparison of the likely natural product fatty alcohol input with the sediment alcohol composition suggests that the major change is a loss of the unsaturated alcohols, especially 20:1 and 22:1 whose input to the sediment might be expected to be significant in view of the abundance of the calanoid copepods. Once again the relative importance of autoxidation, biohydrogenation or bacterial replacement is unknown.

(iv) *Sterols*. Extract I from the surface horizon contained a fairly complex mixture of sterols, but the levels of sterols dropped quickly with depth in the core (Table 2). Only very low levels of sterols were found in any of the Extracts II. Cholesterol was the major sterol found with brassicasterol (or 24S isomer-crinosterol) and β-sitosterol (or 24S isomer-clionasterol). The surface horizon also contained stigmasterol, campesterol, cholestanol, sitostanol, 22 (trans)-24-norcholesta-5,22-dien-3β-ol and 22 (trans) cholesta-5,22-dien-3β-ol.

In each case identification was by comparison of GLC retention times and mass spectra with those of an authentic sample run under identical GC-MS conditions (Ballantine, Roberts and Morris, 1975). The sterol composition of marine diatoms has been analysed by several workers. In two species, Kanazawa, Yoshioka and Teshima (1971) found brassicasterol as the major sterol, whilst Rubinstein and Goad (1974) reported crinosterol in a single species and Tornabene, Kates and Volcani (1974) reported both brassicasterol and clionasterol in another species. Boutry and Barbier (1974) found cholesterol and methylene cholesterol to be the major sterols in a *Chaetoceros* species with campesterol, stigmasterol (or fucosterol) and β-sitosterol. Orcutt and Patterson (1975) examined eleven species and reported crinosterol, cholesterol, campesterol and stigmasterol as the most common sterols. In marine pelagic crustaceans, cholesterol is by far the major sterol (normally > 75%) (Roberts, Ballantine and Morris, unpublished data).

The sterols found in this sediment are similar to those which have been reported as being major components of diatom and crustacean lipids, with the exception of methylene cholesterol. Possibly this sterol is less stable than the others; its concentration in seawater is certainly reported to decrease with increasing depth in the water column (Saliot and Barbier, 1973).

(c) *Amino acids*

The amino acid content of the sediment decreases with depth in the core (4.88–2.03% dry wt) (Table 1). The amino acid composition (Table 3—relative % composition) changes little with depth, the relative levels of the acidic acids falling slightly while that of valine, the aromatic acids and argenine increase. Morris (1975) has suggested that from the similarity of the many published protein amino acid results for phytoplankton and zooplankton, there is a fairly qualitatively predictable input of amino acids from plankton into a sediment. Comparison of this predicted input with the amino acid composition of the Namibian sediment shows some differences (see Table 3). In the sediment there has been a relative decrease in glutamic acid, arginine and lysine levels, all major components of planktonic protein, and a corresponding increase in serine, proline, glycine and alanine levels. Possible explanations of these changes are that amino acids like glutamic, arginine and lysine are more susceptible to breakdown in this reducing environment, or are more strongly held within the sediment by metal complexation (Mopper and Degens, 1972), or that bacterial activity is responsible.

CONCLUSIONS AND SUMMARY

The sediment accumulating in the nearshore area off Namibia is extremely organic-rich and contains many intact and nearly intact planktonic pigments and lipids which have obviously undergone little diagenetic change even after 1000 years. Such good preservation of lipid compounds has only been found in one other marine environment, namely the Black Sea (see, for example, Drozdova and Gorskiy, 1972).

It is apparent from the levels of fulvic acid and polar lipid material that, in quantitative terms, considerable diagenesis of the total natural product input has taken place during the period of accumulation of the sediment. However, the much lower levels of humic acid suggest that conversion to the higher molecular weight compounds in this environment is fairly slow (see Nissenbaum and Kaplan, 1972).

TABLE 3

THE TOTAL AMINO ACID COMPOSITION OF THE SEDIMENT
SAMPLES (RELATIVE WT. % COMPOSITION)

AMINO ACID	SEDIMENT SAMPLES						PREDICTED PLANKTON INPUT (MORRIS, 1975)
	0–5 cm	5–10 cm	10–15 cm	20–30 cm	40–50 cm	60–70 cm	
Aspartic Acid	10.2	10.3	9.7	8.5	8.7	7.3	9.2
Threonine	5.8	6.0	5.8	5.9	5.5	5.4	4.5
Serine	9.2	10.1	9.4	9.0	8.7	8.9	5.0
Glumatic Acid	7.4	7.4	6.8	6.7	7.1	6.7	13.0
Proline	9.2	7.7	7.2	7.2	6.9	7.1	2.7
Glycine	19.0	19.7	18.9	16.7	19.1	17.0	7.0
Alanine	9.8	10.2	9.7	9.2	9.7	9.8	5.8
Valine	4.2	4.4	7.0	6.9	7.3	7.2	5.0
Cystine	0.2	0.2	0.1	0.1	0.1	T	1.5
Methionine	1.2	1.1	1.2	1.0	0.9	1.3	3.0
Isoleucine	2.9	2.8	2.4	3.1	2.8	3.5	4.5
Leucine	6.1	5.4	5.5	6.6	5.9	7.9	7.2
Tyrosine	2.5	2.4	2.5	3.4	2.5	3.4	4.5
Phenylalanine	3.4	3.0	2.9	4.5	4.0	4.6	4.5
Lysine	5.8	5.9	5.5	5.3	5.5	5.6	9.2
Histidine	1.4	1.6	1.5	1.5	1.4	1.6	2.6
Arginine	0.9	0.9	1.6	3.8	2.3	1.9	8.1
Ornithine	0.6	0.7	1.0	0.5	1.4	0.8	1.0

These conclusions are supported by the findings of ROMANKEVICH and BATURIN (1974). These workers felt that the enrichment of these Namibian sediments in bitumens (polar-lipid material) was due to the selective preservation of the lipids during the decomposition of the planktonic input. They further suggested that the abundance of humic acids (fulvic + humic in their work-up) compared to the insoluble organic compounds supported a relatively low degree of transformation of organic matter.

In this work, qualitative changes within the sediment to the probable lipid and protein amino acid input were examined in more detail, the major changes being:

1. a loss of the long chain C_{20}–C_{22} unsaturated fatty acids and fatty alcohols;
2. occurrence of long-chain ($> C_{24}$) alkanes in the unbound lipid fraction;
3. a loss of methylene cholesterol;
4. a significant decrease in the levels of some of the important planktonic protein amino acids.

This preliminary investigation is being extended to a more detailed examination of the pigment/lipid geochemistry of other sediment samples from this area.

Acknowledgements—The authors would like to thank Mr. J. C. ROBERTS and Mrs. M. E. ARMITAGE for much appreciated help in the sterol and amino acid analyses, Mr. P. K. STUDDART for carrying out some of the mineralogical analyses, and D. D. HARKNESS of the Scottish Research Reactor Centre for generously providing $\delta^{13}C$ data and radiocarbon dates.

Expedition CIRCE was supported by contracts of the U.S. Office of Naval Research and grants of the National Science Foundation to the Scripps Institution of Oceanography and Oregon State University. We thank T. H. VAN ANDEL for providing ship time for sample collection.

REFERENCES

Ackman R. G. and R. D. Burgher (1964) Employment of ethanol as a solvent in small-scale catalytic hydrogenation of methylesters. *Journal of Lipid Research*, **5**, 130–132.

Ackman R. G., P. M. Jangaard, R. J. Hoyle and H. Brockerhoff (1964) Origin of marine fatty acids. I. Analysis of the fatty acids produced by the diatom *Skeletonema costatum*. *Journal of the Fisheries Research Board of Canada*, **21**, 747–756.

Ackman R. G., C. S. Tocher and J. McLachlan (1968) Marine phytoplankter fatty acids. *Journal of the Fisheries Research Board of Canada*, **25**, 1603–1620.

Aizenshtat Z., M. J. Baedecker and I. R. Kaplan (1973) Distribution and diagenesis of organic compounds in JOIDES sediments from Gulf of Mexico and Western Atlantic. *Geochimica et cosmochimica acta*, **37**, 1881–1898.

Van Andel T. H. and S. E. Calvert (1971) Evolution of sediment wedge, Walvis Shelf, South-West Africa. *Journal of Geology*, **79**, 585–602.

Avilov I. K. and D. Y. Gershanovich (1970) Investigation of the relief and bottom deposits on the south-west African Shelf. *Okeanologiya*, **10**, 301–306 (Engl. translation pp. 229–232).

Ballantine J. A., J. C. Roberts and R. J. Morris (1975) Sterols of the cockle *Cerastoderma edule*. Evaluation of thermostable liquid phases for the gas–liquid chromatographic-mass spectrometric analysis of the trimethylsilyl ethers of marine sterols. *Journal of Chromatography*, **103**, 289–304.

Blumer M., T. Chase and S. W. Watson (1969) Fatty acids in the lipids of marine and terrestrial nitrifying bacteria. *Journal of Bacteriology*, **99**, 366–370.

Blumer M., R. R. L. Guillard and T. Chase (1971) Hydrocarbons of marine phytoplankton. *Marine Biology*, **8**, 183–189.

Blumer M., M. M. Mullin and R. R. L. Guillard (1970) A polyunsaturated hydrocarbon (3,6,9,12,15,18-heneicsoahexaene) in the marine food web. *Marine Biology*, **6**, 226–235.

Blumer M. and W. W. Youngblood (1975) Polycylic aromatic hydrocarbons in soils and recent sediments. *Science*, **188**, 53–55.

Boon J. J., J. W. de Leeuw and P. A. Shenck (1975) Organic geochemistry of Walvis Bay diatomaceous ooze. I. Occurrence and significance of the fatty acids. *Geochimica et cosmochimica acta*, **39**, 1559–1565.

Boon J. J., W. I. C. Rijpstra, J. W. de Leeuw and P. A. Shenck (1975) Phytenic acid in sediments. *Nature*, **258**, 414–416.

Bordovskiy O. K. (1974) Principal features of the chemical composition of organic matter in sea and ocean-basin sediments. *Okeanologiya*, **14**, 448–456 (Engl. translation pp. 369–375).

Bottino N. R. (1974) The fatty acids of Antarctic phytoplankton and euphausiids. Fatty acid exchange among trophic levels of the Ross Sea. *Marine Biology*, **27**, 197–204.

Boutry J. L. and M. Barbier (1974) La diatomee marine *Chaetoceros simplex calcitrans* Paulsen et son environment. I. Relations avec le milieu de culture; etude de la fraction insaponifiable, des sterols et des acides gras. *Marine Chemistry*, **2**, 217–227.

Brongersma-Sanders M. (1951) On conditions favouring the preservation of chlorophyll in marine sediments. *Proceedings of the 3rd World Petroleum Congress*, Section I, pp. 401–413.

Brown F. S., M. J. Baedecker and I. R. Kaplan (1972) Early diagenesis in a reducing fjord, Saanich Inlet, British Columbia. III. Changes in organic constituents of sediment. *Geochimica et cosmochimica acta*, **36**, 1185–1203.

Burke J. C. (1968) A sediment coring device of 21-cm diameter with sphincter core retainer. *Limnology and Oceanography* **13**, 714–718.

Butlin K. R. (1949) Some malodorous activities of sulphate-reducing bacteria. *Proceedings Society of Applied Bacteriology*, **12**, 39–42.

Calvert S. E. (1966) Accumulation of diatomaceous silica in the sediments of the Gulf of California. *Bulletin Geological Society of America*, **77**, 569–596.

Calvert S. E. and R. J. Morris (1977) Geochemical studies of organic-rich sediments from the Namibian shelf. II. Metal-organic associations. In *Voyage of Discovery*, M. V. Angel, editor, *Deep-Sea Research*, supplement to Vol. **24**, 667–680.

Calvert S. E. and N. B. Price (1971a) Upwelling and nutrient regeneration in the Benguela Current, October 1968. *Deep-Sea Research*, **18**, 505–523.

Calvert S. E. and N. B. Price (1971b) Recent sediments of the South West African Shelf. ICSU/SCOR. Working party 31 Symposium. Cambridge, 1970: The Geology of the East Atlantic Continental Margin, F. M. Delany, editor, *Institute of Geological Sciences Report No. 70/16*, pp. 171–185.

Chen K. Y., M. Moussavi and A. Sycip (1973) Solvent extraction of sulfur from marine sediment and its determination by gas chromatography. *Environmental Science and Technology*, **7**, 948–951.

Chuecas L. and J. P. Riley (1969) Component fatty acids of the total lipids of some marine phytoplankton. *Journal of the Marine Biological Association of the United Kingdom*, **49**, 97–116.

CLARK R. C. and M. BLUMER (1967) Distribution of n-paraffins in marine organisms and sediment. *Limnology and Oceanography*, **12**, 79–87.

COOPER R. L. and M. BLUMER (1968) Linear, iso and anteiso fatty acids in Recent sediments of the North Atlantic. *Deep-Sea Research*, **15**, 535–540.

COPENHAGEN W. J. (1953) The periodic morality of fish in the Walvis Region. *Investigational Report, Division coast. Investigational Report, Fisheries and Marine Biological Survey Division, Union of South Africa*, **3**, 1–18.

COPENHAGEN W. J. (1953) The periodic mortality of fish in tne Walvis Region. *Investigational Report, Division of Fisheries, Union of South Africa*, **14**, 1–34.

DAVIS J. B. (1968) Paraffinic hydrocarbons in the sulphate-reducing bacterium *Desulphovibrio desulfuricans*. *Chemical Geology*, **3**, 155–160.

DEMORT C. L., R. LOWRY, I. TINSLY and H. K. PHINNEY (1972) Biochemical analysis of some estuarine phytoplankton species. I. Fatty acid composition. *Journal of Phycology*, **8**, 211–216.

DROZDOVA T. W. and Y. N. GORSKIY (1972) Conditions of preservation of chlorophyll, pheophytin and humic matter in Black Sea sediments. *Geochemistry International*, **9**, 208–218.

EISMA D. (1969) *Sediment sampling and hydrographic observations of Walvis Bay, S.W. Africa, Dec. 1968–Jan. 1969*, NIOZ Internal Publication 1969–1. Texel, The Netherlands.

EISMA D. and S. J. VAN DER GAAST (1971) Determination of opal in marine sediments by X-ray diffraction. *Netherlands Journal Sea Research*, **5**, 382–389.

EMERY K. O. and E. E. BRAY (1962) Radiocarbon dating of California basin sediments. *Bulletin of the American Association of Petroleum Geologists*, **46**, 1839–1856.

FARRINGTON J. W. and J. G. QUINN (1971) Comparison of sampling and extraction techniques for fatty acids in recent sediments. *Geochimica et cosmochimica acta*, **35**, 735–741.

FARRINGTON J. W. and J. G. QUINN (1973) Biogeochemistry of fatty acids in Recent sediments from Narragansett Bay, Rhode Island. *Geochimica et cosmochimica acta*, **37**, 259–268.

GASKELL S. J., R. J. MORRIS, G. EGLINTON and S. E. CALVERT (1975) The geochemistry of a recent marine sediment off north-west Africa. An assessment of source of input and early diagenesis. *Deep-Sea Research*, **22**, 777–789.

HARE P. E. (1972) Amino acid geochemistry of a sediment core from the Cariaco Trench. *Carregie Institution of Washington, Yearbook No. 71*, pp. 592–596.

HART T. J. and R. T. CURRIE (1960) The Benguela Current. *Discovery Reports*, **31**, 123–298.

HITES R. A. and W. G. BIEMANN (1974) Identification of specific organic compounds in a highly anoxic sediment by GC/MS and HRMS. *Abstracts 168th National American Chemical Society Meeting 'ANAL 40 (1974)'*.

HUNT J. M. (1974) Hydrocarbon geochemistry of Black Sea. In: *The Black Sea—Geology, Chemistry and Biology*, D. A. ROSS and E. T. DEGENS, editors, *American Association of Petroleum Geologists Memoir*, **20**, 499–504.

JOHNSON R. W. and J. A. CALDER (1973) Early diagenesis of fatty acids and hydrocarbons in a salt march environment. *Geochimica acta*, **37**, 1943–1955.

JONES P. G. W. (1971) The southern Benguela Current region in February 1966: Part I. Chemical observations with particular reference to upwelling. *Deep-Sea Research*, **18**, 193–208.

KANAZAWA A., M. YOSHIOKA and S. TESHIMA (1971) The occurrence of brassicasterol in the diatoms *Cyclotella nana* and *Nitzschia closterium*. *Bulletin of the Japanese Society for Scientific Fisheries*, **37**, 899–903.

KANEDA T. (1967) Fatty acids in the genus *Bacillus*–I. Iso and anteiso fatty acids as characteristic constituents of lipids in 10 species. *Journal of Bacteriology*, **93**, 894–903.

KAPLAN I. R., K. O. EMERY and S. C. RITTENBERG (1963) The distribution and isotopic abundance of sulphur in recent marine sediments off southern California. *Geochimica et cosmochimica acta*, **27**, 297–331.

KATES M. (1964) Bacterial lipids. In: *Advances in lipid research*, R. PAOLETTI and D. KRITCHEVSKY, editors, Academic Press, New York, **2**, pp. 17–90.

KOLLMER W. E. (1963) The pilchard of South West Africa (*Sardinops ocellata Pappe*): Notes on zooplankton and phytoplankton collections made off Walvis Bay. *Investigational Report Marine Research Laboratory, South West Africa*, No. 8, pp. 1–78.

KUNIMOTO M. (1975) Lipids of marine bacteria. II. Lipid composition of bacteria of marine type. *Bulletin of the Faculty of Fisheries, Hokkaido University*, **25**, 342–350.

KUNIMOTO M., K. ZAMA and H. IGARASHI (1975) Lipids of marine bacteria. I. Lipid composition of marine *Achromobacter* species. *Bulletin of the Faculty of Fisheries, Hokkaido University*, **25**, 332–341.

LEE R. F. and A. R. LOEBLICH (1971) Distribution of 21:6 hydrocarbon and its relationship to 22:6 fatty acid in algae. *Phytochemistry*, **10**, 593–602.

LEWIS R. W. (1969) The fatty acid composition of arctic marine phytoplankton and zooplankton with special reference to minor acids. *Limnology and Oceanography*, **14**, 35–40.

LOGAN R. F. (1960) The Central Namib Desert, South-West Africa. *National Academy of Science, Natural Research Council*, Washington, Publication No. **758**, 162 pp.

MOPPER K. and E. T. DEGENS (1972) Aspects of the biogeochemistry of carbohydrates and proteins in aquatic environments. *Technical Report Woods Hole Oceanographic Institution*, No. 72–68, pp. 1–118.

MORRIS R. J. (1973a) The lipids of marine zooplankton. Ph.D. Thesis, University of Southampton, pp. 11–27.

Morris R. J. (1973b) Uptake and discharge of petroleum hydrocarbons by barnacles. *Marine Pollution Bulletin*, **4**, 107–109.

Morris R. J. (1975) The amino acid composition of a deep-water marine sediment from the upwelling region north-west of Africa. *Geochimica et cosmochimica acta*, **39**, 381–388.

Morris R. J. and F. Culkin (1976) Marine lipids. I. Analytical techniques and fatty acid ester analyses. In: *Oceanography and Marine Biology: an Annual Review*, H. Barnes, editor, George Allan & Unwin, **14**, 391–433.

Muller P. (1975) Diagenese stickstoffhaltiger organischer Subatanzen in oxischen und anoxischen marinen Sedimenten. *Meteor Forschungsergebnisse, Reihe, C*, No. **22**, 1–60.

Nissenbaum A. and I. R. Kaplan (1972) Chemical and isotopic evidence for the *in situ* origin of marine humic substances. *Limnology and Oceanography*, **17**, 570–582.

Oliver J. D. and R. R. Colwell (1973a) Extractable lipids of gram-negative marine bacteria: fatty acid composition. *International Journal of Systematic Bacteriology*, **23**, 442–458.

Oliver J. D. and R. R. Colwell (1973b) Extractable lipids of gram-negative marine bacteria: phospholipid composition. *Journal of Bacteriology*, **114**, 879–908.

Orcutt D. M. and G. W. Patterson (1975) Sterol, fatty acid and elemental composition of diatoms grown in chemically defined media. *Comparative Biochemistry and Physiology*, **50B**, 579–583.

Oro J., Tornabene, D. W. Nooner and E. Gelpi (1967) Aliphatic hydrocarbons and fatty acids of some marine and fresh-water micro-organisms. *Journal of Bacteriology*, **93**, 1181–1818.

Peake E., D. J. Casagrande and G. W. Hodgson (1974) Fatty acids, chlorins, hydrocarbons, sterols and carotenoids from a Black Sea core. In: *The Black Sea—Geology, Chemistry and Biology*, D. A. Ross and E. T. Degens, editors, American Association of Petroleum Geologists Memoir 20, pp. 505–523.

Powell T. G., P. J. Cook and D. M. McKiroy (1975) Organic geochemistry of phosphorites: Relevance to petroleum genesis. *American Association of Petroleum Geologists Bulletin*, **59**, 618–632.

Powell T. G. and D. M. McKirdy (1973) The effect of source material, rock type and diagenesis on the *n*-alkane content of sediments. *Geochimica et cosmochimica acta*, **37**, 623–633.

Pugh P. R. (1971) Changes in the fatty acid composition of *Coscinodiscus eccentricus* with culture-age and salinity. *Marine Biology*, **11**, 118–124.

Rashid M. A. (1972) Amino acids associated with marine sediments and humic compounds and their role in solubility and complexing of metals. *Proceedings 24th International Geological Congress, Montreal*, **10**, 346–353.

Reitz R. C. and G. S. Moore (1972) Effects of changes in the major carbon source on the fatty acids of *Euglena gracilis*. *Lipids*, **7**, 217–220.

Romankevich Y. A. and G. N. Baturin (1974) The biogeochemical composition of the sediments on the West African Shelf (5–23°S. lat.) *Okeanologiya*, **14**, 660–664 (Engl. translation, pp. 529–533).

Ross D. A., E. T. Degens and J. Macilvaine (1970) Black Sea: recent sedimentary history. *Science*, **170**, 163–165.

Rubinstein I. and L. J. Goad (1974) The occurrence of (24S)-24-methylcholesta-5,22E-dien-3β-ol in the diatom *Phaeodactylum tricornatum*. *Phytochemistry*, **13**, 485–487.

Saliot A. and M. Barbier (1973) Sterols from sea water. *Deep-Sea Research*, **20**, 1077–1082.

Sargent J. R., R. F. Lee and J. C. Nevenzel (1976) Marine waxes. In: *Chemistry and Biochemistry of Natural Waxes*, P. Kollatukudy, editor, N. Holland (in press).

Schultz D. M. and J. G. Quinn (1974) Measurements of phytol in estuarine suspended organic matter. *Marine Biology*, **27**, 143–146.

Senin Y. M. (1968) Characteristics of sedimentation on the shelf of South-West Africa. *Litologiya: Poleznye Iskopayemye*, No. 4, pp. 108–111 (English translation, pp. 476–479).

Shannon L. V. (1966) Hydrology of the south and west coasts of South Africa. *Investigational Report, Division Sea Fisheries, Republic of South Africa*, No. **58**, 1–62.

Simoneit B. R. and A. L. Burlingame (1974) Study of organic matter in DSDP (JOIDES) cores, degs. 10–15. In: *Advances in Organic Geochemistry, 1973*, B. Tissot and F. Bienner, editors, Editions Techniq, pp. 629–645.

Simpson E. S. W. and A. du Pleiss (1968) Bathymetric, magnetic and gravity data from the continental margin of south-western Africa. *Canadian Journal of Earth Sciences*, **5**, 1119–1123.

Stander G. H. (1964) The Benguela Current off South West Africa. *Investigational Report, Marine Research Laboratory, South West Africa*, No. 12, pp. 1–43.

Steeman-Nielsen E. and A. E. Jensen (1957) Primary oceanic production. *Galathea Report*, **1**, 49–136.

Thode H. G., H. Kleerekoper and D. McElcheran (1951) Isotopic fractionation in the bacterial reduction of sulphate. *Research*, **4**, 581–582.

Tornabene T. G., M. Kates and B. E. Volcani (1974) Sterols, aliphatic hydrocarbons and fatty acids of a non photosynthetic diatom, *Nitzschia alba*. *Lipids*, **9**, 279–284.

Unteruberbacher H. K. (1964) The pilchard of South-West Africa (*Sardinops ocellata*): Zooplankton studies in the waters off Walvis Bay with special reference to the copepods. *Investigational Report, Marine Research Laboratory, South West Africa*, No. 11, pp. 1–42.

VEEH H. H., S. E. CALVERT and N. B. PRICE (1974) Accumulation of uranium in sediments and phosphorites on the South West African shelf. *Marine Chemistry*, **2**, 189–202.

VISSER G. A. (1969) Analysis of Atlantic waters off the West Coast of Southern Africa. *Investigational Report, Division Sea Fisheries, Republic of South Africa*, No. 75, pp. 1–26.

WELTE D. H. and G. EBHARDT (1968) Distribution of long chain *n*-paraffins and *n*-fatty acids in sediments from the Persian Gulf. *Geochimica et Cosmochimica acta*, **32**, 465–466.

WINTERS K., P. L. PARKER and C. VAN BAALEN (1969) Hydrocarbons of blue green algae: geochemical significance. *Science*, **193**, 467–468.

YOUNGBLOOD W. W. and M. BLUMER (1973) Alkanes and alkenes in marine benthic algae. *Marine Biology*, **21**, 163–172.

YOUNGBLOOD W. W. and M. BLUMER (1975) Polycylic aromatic hydrocarbons in the environment: homologous series in soils and recent marine sediments. *Geochimica et cosmochimica acta*, **39**, 1303–1314.

YOUNGBLOOD W. W., M. BLUMER, R. L. GUILLARD and F. FIORE (1971) Saturated and unsaturated hydrocarbons in marine benthic algae. *Marine Biology*, **8**, 190–201.

Geochemical studies of organic-rich sediments from the Namibian Shelf II. Metal–organic associations

S. E. CALVERT and R. J. MORRIS

Institute of Oceanographic Sciences,
Wormley, Godalming, Surrey GU8 5UB, U.K.

Abstract—The concentrations of Cu, Mo, Ni, Pb and Zn in several lipid, humic acid, fulvic acid and final residue ("kerogen") fractions of diatomaceous oozes collected off Walvis Bay have been determined by X-ray emission methods. Highest concentrations of all five metals occur in the humic acids, Ni and Pb are in very low concentration in the fulvic acids, Ni is moderately enriched in the final residue, and all five metals are present in the lipid fractions in moderate amounts but with no clear fractionation. The metal concentrations in the fractions examined do not account for the total metal contents of the bulk sediments.

Information on the reported metal–organic matter associations in soils, peat and coal and in some marine sediments, and the metal contents of marine plankton, together with experimental work on metal reactions with fulvic and humic acids, has been used to suggest that low molecular weight degradation products of marine organisms, and also perhaps the intact lipid residues, are responsible for the complexing or chelation of metals in sediments during diagenesis and that continual polymerization of such material transfers the metals to the insoluble, relatively unreactive high molecular weight fraction (humic acid).

INTRODUCTION

THE possible association of trace metals with organic matter in marine sediments has been frequently reported. Such an association has been taken to imply that the metals were either transported to the sediment by sedimenting organic material or have been absorbed by the organic fraction of the sediment after deposition. Thus, BRONGERSMA-SANDERS (1965; 1969) has proposed that the metal enrichment of many bituminous shales is due to the supplies of metals by plankton in areas of upwelling. On the other hand, VOLKOV and FOMINA (1971) have argued that the concentrations of trace metals in some Black Sea sediments cannot have been contributed solely by plankton, based on the present organic content of the sediments, but were more likely extracted from seawater by settling seston. A more important process may be the binding of metals present in the pore solutions by organic material during diagenetic reactions within the Recent sediments.

The evidence for metal–organic associations in Recent sediments has commonly been a statistical correlation between the metal and organic carbon contents of a sediment (CURTIS, 1966). Such correlations are displayed, for example, by Cu, Ni, Mo and V with carbon in Black Sea sapropels (VOLKOV and FOMINA, 1971), by Mo with carbon in Mediterranean sediments (PILIPCHUK, 1972) and by U with organic carbon in Mediterranean and Black Sea sediments (KOCHENOV, BATURIN, KOVALEVA, YEMEL'YANOV

667

and SHIMKUS, 1965; BATURIN, KOCHENOV and SHIMKUS, 1967). LITTLE-GADOW and SCHAFER (1974) have demonstrated statistically significant positive correlations between Cu, Fe, Hg, Pb and Zn and organic carbon in sediments from the German Bight, which have suggested preferential absorption of the metals by organic matter.

In only a few cases has evidence for the presence of metals in organic phases been demonstrated. NISSENBAUM, BAEDECKER and KAPLAN (1972) reported a trace element analysis of the dissolved organic matter in the pore water of anoxic sediments from a fjord which showed high concentrations of Cu, Ni, Pb and Zn. VOLKOV and FOMINA (1971) have reported trace metal analyses of some extracted organic fractions of Black Sea sapropels which showed only moderate enrichments compared to the metal contents of the total sediments while PILLAI, DESAI, MATTHEW, GANAPATHY and GANGULY (1971) have found high concentrations of metals in the humic and fulvic acid fractions of several shallow marine sediments. Likewise, COOPER and HARRIS (1974) have shown that the trace metal contents of a larger series of organic extracts of sediments from a polluted river were generally quite high and that the distribution of the various metals in the different fractions was quite variable.

The association of trace metals with organic material in soils, coal and peat has been recorded and intensively investigated for a long period (GOLDSCHMIDT, 1954; KRAUSKOPF, 1955; MANSKAYA and DROZDOVA, 1968; SZALAY and SZILAGYI, 1969; SAXBY, 1969). The important part played by the humic fraction, the most abundant class of compounds in these materials, in complexing trace metals has been frequently stressed and its role in controlling soil fertility has led to a voluminous literature (see SCHNITZER and KHAN, 1972).

The presence, and possible biochemical significance, of trace metals in marine organisms has been reviewed by VINOGRADOV (1953), GOLDBERG (1957) and BOWEN (1966). Much of the data available refer to metal concentrations of whole organisms rather than the metal content of specific organic compounds or classes of compounds which could be used to study biochemical pathways. Studies by LUNDE (1971; 1973 a, b) of the metal contents of the lipid phase of marine organisms have shown the likely levels of some trace metals in such extracts and have indicated the complexing ability of lipid components, especially the phospholipids. Work on the occurrence of metal–organic complexes in seawater has been reported by SLOWEY, JEFFREY and HOOD (1976) and WILLIAMS (1969) and the probable importance of organic chelators in controlling oceanic production has been stressed by JOHNSTON (1964) and BARBER and RYTHER (1969). However, STUMM and MORGAN (1970) have drawn attention to the paucity of data on such complexes and to the difficulty of isolating and characterising soluble chelates in natural waters.

The composition of the sediments accumulating on the continental shelf off Namibia have been reported by CALVERT and PRICE (1970, 1971). On the inner part of the shelf, especially in the area off Walvis Bay, the sediments contain high concentrations of organic carbon and diatomaceous silica (opal). Several trace elements are enriched in these sediments, including Cu, Ni, Pb and Zn (CALVERT and PRICE, 1970), U (VEEH, CALVERT and PRICE, 1974) and Mo (CALVERT and PRICE, unpublished). Significant correlations between Cu, Mo, Ni, Pb, Zn and organic carbon contents are found, and in view of the unusual bulk composition of the sediments, being composed almost entirely of dispersed organic material and skeletal debris, some association between the metals and organic material has been inferred although the mechanism of metal enrichment is unknown (CALVERT and PRICE, 1970).

An investigation of the distribution and the composition of some of the organic components of a sediment core from this area by MORRIS and CALVERT (1977) has provided an opportunity to examine the presence and the distribution of some of the enriched trace metals in some organic fractions.

MATERIALS AND METHODS

Details of the core collection, subsampling scheme, sample preparation and methods of extraction of the organic fractions are given in MORRIS and CALVERT (1977). Concentrations of Cu, Mo, Ni, Pb and Zn were determined in the fractions listed in Table 1, where some brief details of the extraction procedures are also given.

TABLE 1

ORGANIC FRACTIONS USED FOR TRACE METAL ANALYSIS*

FRACTION	EXTRACTION METHOD	COMPOSITION
Organic Extract I	$CHCl_3$/MeOH, sonicate wet sediment. Centrifuge.	Unbound lipids
Organic Extract II	2N HCl, reflux 4 hours; $CHCl_3$/MeOH, sonicate residue from I. Centrifuge.	Bound lipids
Organic Extract III	6N HCl, reflux 6 hours; $CHCl_3$/MeOH, sonicate residue from II. Centrifuge.	Strongly bound lipids
Humic Acid	6N HCl, reflux 24 hours; residue + 0.5N KOH, reflux 10 hours; precipitate with HCl at pH2. Wash with H_2O. Redissolve in 0.5N KOH.	Higher molecular weight humic compounds
Fulvic Acid	Supernatant from humic acid extraction; dialysed against H_2O, rotary evaporated to dryness.	Lower molecular weight humic compounds
Final Residue	From humic acid extraction.	Refractory, insoluble organic material + resistant minerals (quartz, feldspar, mica).

* See Morris and Calvert (1977, Fig. 2)

Attention was restricted in this study to the location of organically bound metals rather than the determination of the metal contents of all sediment components. We did not attempt, for example, to determine the metal content of the amino acid fraction because the extraction procedure used for this fraction (MORRIS and CALVERT, 1977) resulted in the presence of those metals originally contained in the carbonate fraction and in the less resistant aluminosilicates.

Concentrations of Cu, Mo, Ni, Pb and Zn were determined in some untreated bulk sediments, dried at 100°C and ground in tungsten carbide, by X-ray emission spectrometry, using the method of ANDERMAN and KEMP (1958) to correct for matrix difference between samples and standards. A series of international rock standards (U.S. Geological Survey, G-2, BCR-1, AGV-1, GSP-1, PCC-1, DTS-1; Centre de Recherchés Pétrographiques et Géochimiques BR, GA, GH, DRN; Zentales Geologisches Institut TS, TB, BM; and National Institute of Metallurgy, South Africa, NIM-G, -N, -P, -D, -L and -S) were used together with a supplementary series for Mo made by spiking a Mo-free silicate base with known amounts of specpure MoO_3. Tungsten and molybdenum anodes were used at 60 kV and 20 mA, together with a scintillation counter. Precision was better than $\pm 5\%$ for all elements.

Analyses of the Cl and total S contents of the same samples were also carried out by X-ray emission methods. The samples were mixed with $LiBO_2$ and La_2O_3 and pressed into 31-mm diameter discs with a chromatographic cellulose backing at 10 tons pressure. Standards were prepared by spiking a silicate-rock base with NaCl and K_2SO_4. Samples and standards were irradiated under vacuum with a Cr anode at 50 kV and 30 mA and a flow-proportional counter with appropriate pulse height selection was used. The chlorine contents were used to correct the metal, carbon, carbonate, opal and quartz contents of the samples for the diluting effect of sea salt in the dried samples, and for the contribution of sea-salt sulphate to the total S values.

The chemical analyses of the organic fractions were also carried out by X-ray emission spectrometry. The organic extracts I, II and III were made up to a standard volume in Aristar grade (B.D.H.) chloroform and the humic acid was made up to volume in 0.5 N Analar KOH. Small volumes (usually 0.25 ml) of the extracts were deposited as thin films on 6-µm-thick mylar film supported on polypropylene sample cups. Standards consisted of both aqueous and chloroform-based solutions (except for Mo) of metals using the respective cyclohexylbutyrate salts. The films were mounted in nylon holders and irradiated using a W anode at 60 kV and 20 mA for Mo, Ni and Zn and with a Mo anode at the same power settings for Cu and Pb. Linear calibration curves were obtained over the range of interest. Sensitivities (3σ above background) were 0.04 µg Cu, 0.42 µg Mo, 0.05 µg Ni, 0.32 µg Pb and 0.03 µg Zn and precision was $\pm 5\%$ for all metals.

Accurately weighed samples of the fulvic acid and final residues were used in solid form and were mounted as thin layers in 31-mm-diameter chromatographic cellulose discs under 10 tons pressure. Standards consisted of NBS Bovine Liver and Orchard leaves, Bowen's Kale (BOWEN, 1969) and cyclohexylbutyrate salts mounted in the same fashion as the samples. For Mo, a separate series of standards were made up by adding known amounts of specpure MoO_3 to Bowen's Kale. The discs were mounted in nylon cups and irradiated using the conditions given above for the analyses of the evaporated solutions.

RESULTS

Concentrations of Cu, Mo, Ni, Pb and Zn, together with the S and organic carbon values, of the total sediment are given in Table 2. It should be stressed that these analyses were carried out on small sediment subsamples from the core which do not coincide with the larger samples used for the organic extractions. The metal concentrations are similar to those given by CALVERT and PRICE (1970, 1971) for the surface organic-rich sediments

TABLE 2

TRACE METAL, ORGANIC CARBON AND SULPHUR
CONTENTS OF UNTREATED SEDIMENTS*

DEPTH (cm)	Cu (ppm)	Mo (ppm)	Ni (ppm)	Pb (ppm)	Zn (ppm)	C (wt.%)	S (wt.%)
0- 5	69	59	115	4	61	11.9	1.19
10-15	79	42	107	16	64	12.2	1.19
20-25	83	36	86	10	136	8.6	0.68
30-35	73	37	145	8	112	14.3	1.02
40-45	72	42	107	5	64	8.1	0.94
50-55	75	51	167	5	38	15.1	1.11
60-65	74	38	85	5	91	9.2	0.65

* Salt-corrected basis.

accumulating off Walvis Bay. In this particular case, the metal concentrations vary erratically down the core, a result also found for the carbon, carbonate, opal and quartz contents (MORRIS and CALVERT, 1977).

Results of the analyses of the organic fractions are given in Table 3 and Fig. 1. All three lipid extracts contain significant concentrations of the five metals. Nickel and Zn appear to be more concentrated in Extract I, Mo is more concentrated in Extract III, and

TABLE 3

TRACE METAL CONTENTS (PPM) OF EXTRACTED ORGANIC FRACTIONS

DEPTH (cm)	ORGANIC EXTRACT I					ORGANIC EXTRACT II					ORGANIC EXTRACT III				
	Cu	Mo	Ni	Pb	Zn	Cu	Mo	Ni	Pb	Zn	Cu	Mo	Ni	Pb	Zn
0-10	51.6	64.0	141.6	113.6	243.2	46.0	48.0	48.0	91.6	48.4	61.6	58.8	60.0	88.0	98.4
20-30	45.6	60.0	141.6	114.0	116.4	70.0	38.0	66.8	103.6	27.6	59.6	84.0	64.4	82.8	61.6
40-50	39.6	66.0	121.6	94.8	79.2	48.0	48.0	64.0	132.8	29.2	43.4	72.3	49.5	172.9	42.5
60-70	43.2	83.2	180.4	127.6	78.4	51.6	44.8	59.2	93.2	27.6	40.3	57.7	40.7	159.3	46.0

DEPTH (cm)	FULVIC ACID					HUMIC ACID					FINAL RESIDUE				
	Cu	Mo	Ni	Pb	Zn	Cu	Mo	Ni	Pb	Zn	Cu	Mo	Ni	Pb	Zn
0- 5	26	102	ND*	ND	160	877	2368	333	491	894	45	35	–	ND	67
5-10	24	51	ND	4	52	707	914	621	414	689	23	45	275	12	62
10-15	23	41	ND	ND	58	686	1168	409	671	817	37	48	184	ND	57
20-30	25	36	ND	ND	69	683	1000	500	283	817	48	41	207	26	70
40-50	25	41	5	ND	62	533	437	533	267	933	75	63	204	4	98
60-70	20	56	ND	4	50	797	1525	491	306	712	33	35	123	64	58

* ND = not detected

S. E. CALVERT and R. J. MORRIS

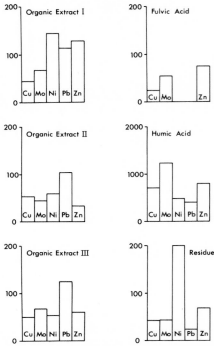

Fig. 1. Distribution of mean trace metal contents of the organic fractions. Vertical scales in ppm. Note different scale for the Humic Acid Fraction.

Cu and Pb appear to be equally concentrated in all three extracts. Apart from Zn, which decreases in composition down the core in Extract I, and Cu, which also decreases in the same sense in Extract III, there appears to be no clear variation in the concentrations of the metals in the various core subsamples.

In the fulvic acid fraction, Ni and Pb could not be detected in most of the subsamples and was close to the limit of sensitivity in the remaining subsamples. Cu was present in rather constant amount in all samples but in lower concentration than in the organic extracts. Mo and Zn were both present in similar concentrations to those in the organic extracts, the surface sample being much more concentrated than other samples.

The concentrations of all metals analysed were highest in the humic and fraction. This was most noticeable for Mo, which reached 2368 ppm in the surface sample. No clear systematic variation in concentrations down the core could be detected.

In the final residue, Ni showed the highest concentrations, reaching 275 ppm in the 5-10-cm horizon. Cu, Mo and Zn concentrations were similar to those in the fulvic acid fraction, and again showed no systematic change with depth. Pb concentrations were markedly erratic, being below the limit of detection in two samples and reaching 64 ppm in the lowermost sample.

Table 4 shows the contribution in ppm of each of the extracted fractions to the metal content of the total sediment, based on the content of each organic fraction as given by MORRIS and CALVERT (1977, Table 1). The metal contents of the lipid extracts make an insignificant contribution to the metal content of the total sediment. On the other hand, the fulvic and humic acids and the final residue do make a contribution to the total

TABLE 4

METAL CONCENTRATIONS (PPM) OF THE TOTAL SEDIMENT
CONTRIBUTED BY THE ORGANIC FRACTIONS

DEPTH (cm)	ORGANIC EXTRACT I					ORGANIC EXTRACT II					ORGANIC EXTRACT III				
	Cu	Mo	Ni	Pb	Zn	Cu	Mo	Ni	Pb	Zn	Cu	Mo	Ni	Pb	Zn
0–10	0.5	0.6	1.3	1.0	2.2	0.3	0.4	0.4	0.7	0.4	0.1	0.1	0.1	0.1	0.1
20–30	0.3	0.4	0.9	0.7	0.7	0.4	0.2	0.3	0.6	0.1	0.1	0.2	0.1	0.2	0.1
40–50	0.1	0.2	0.4	0.3	0.3	0.1	0.1	0.2	0.4	0.1	0.1	0.1	0.1	0.3	0.1
60–70	0.1	0.2	0.4	0.3	0.2	0.2	0.2	0.2	0.3	0.1	0.1	0.1	0.1	0.2	0.1

DEPTH (cm)	FULVIC ACID					HUMIC ACID					FINAL RESIDUE				
	Cu	Mo	Ni	Pb	Zn	Cu	Mo	Ni	Pb	Zn	Cu	Mo	Ni	Pb	Zn
0– 5	2.5	9.9	0	0	15.5	7.9	21.3	3.0	4.4	8.0	3.4	2.7	–	0	5.1
5–10	3.7	7.8	0	0.6	7.9	7.0	9.0	6.1	4.1	6.8	1.2	2.4	14.8	0.6	3.3
20–25	4.4	7.9	0	0	11.2	7.1	12.1	4.2	7.0	8.5	2.1	2.7	10.5	0	3.2
25–30	5.5	7.9	0	0	15.1	5.5	8.1	4.0	2.3	0.6	2.0	1.7	8.8	1.1	3.0
45–50	3.6	5.9	0.7	0	8.9	4.1	3.4	4.1	2.1	7.3	2.9	3.2	10.5	0.2	5.1
65–70	3.3	9.3	0	0.7	8.3	4.1	7.8	2.5	1.6	3.6	0.8	0.9	3.1	1.6	1.5

amounts of metals in the sediment, and this is most marked for Mo and Zn in the fulvic acids, Cu, Mo, Pb and Zn in the humic acids and Ni in the residue.

In spite of the high concentrations of metals in some of the organic fractions (Table 2), the total amounts of the metals contributed by the organic material examined, apart from Pb, do not account for the total metal concentrations of the bulk sediment. The remaining amounts of metals are present in the skeletal debris, both calcareous and siliceous, the aluminosilicates, authigenic phases, such as pyrite, and in other organic fractions not separated by the methods used here.

DISCUSSION

The analyses reported here demonstrate that significant concentrations of trace metals are present in several classes of organic material from a relatively shallow-water sediment containing high concentrations of organic material derived overwhelmingly from planktonic sources (see MORRIS and CALVERT, 1977). The relative amounts of the metals appear to vary from one fraction to another, and the concentrations of all five metals examined are highest in the humic acid fractions.

VOLKOV and FOMINA (1971) have reported some trace metal analyses of higher molecular weight polar lipids (benzene/alcohol extractable), and humic and fulvic acid fractions of Recent sapropels from the Black Sea. They found mean concentrations of 5.6 ppm Cu and 2.1 ppm Ni in the lipid fraction, 1.1 ppm Cu and no detectable Ni in the humic acid fraction and 11.3 ppm Cu and 4.6 ppm Ni in the fulvic acid fraction, results which contrast markedly with those reported here. These same authors also stated that most of the Mo in sapropelic Black Sea sediments was associated with the fulvic acid fraction, although no data were given.

In distinct contrast, PILLAI, DESAI, MATHEW, GANAPATHY and GANGULY (1971) found that the humic acid fractions of three shallow-water marine sediments contained between 815 and 1430 ppm Cu and between 1061 and 2520 ppm Zn, while the fulvic acid fractions contained only 223 ppm Cu and no detectable Zn. However, in experiments designed to examine the complexing ability of the humic and fulvic acids from these same sediments, these authors found that the fulvic acid was much more effective than the humic acid. These results are consistent with those of RASHID (1971) who showed experimentally that lower molecular weight humic compounds from sediments from the Scotian shelf complexed larger quantities of both di- and tri-valent metals than the higher molecular weight humic fraction.

In sediments from a polluted river, COOPER and HARRIS (1974) showed that the less polar lipids, more polar lipids, asphalt (higher molecular weight polar lipids) and humic acid fractions contained moderately high concentrations of Cd, Cr, Cu, Fe, Mn, Ni, Pb and Zn, and that the relative proportions of the metals varied in the different fractions from one sampling site to another. The greatest variation appeared to be in the asphalt fraction, where Cu, for example, varied between 9 and 680 ppm and Fe varied between 73 and 10,000 ppm.

Interpretation of the wide variations in the concentration of trace metals in various organic fractions from marine and freshwater sediments reported by several groups of workers is made difficult by the fragmentary information on the general composition of much of the organic material in such sediments and on the relative importance of different classes of organic compounds in complexing and chelating metals in the marine environment. Some progress on this problem has been made, however, by studies of soil organic matter and to some extent by recent investigations of the trace metal composition of marine organisms.

The association between trace metals and organic matter in soils, as well as plant material, peat and coal, is well known (GOLDSCHMIDT, 1937, 1954; VINOGRADOV, 1959; MANSKAYA and DROZDOVA, 1968). More particularly, it is clear that the humic and fulvic acid fractions of soils play the most important role in the trace element balance in the soil. These two fractions differ in molecular weight, ultimate analysis and functional group content, the fulvic acids having a lower molecular weight but a higher content of oxygen-containing functional groups per unit weight than humic acids (SCHNITZER and KHAN, 1972). The high exchange capacity of soil organic matter is ascribed almost entirely to these fractions (VAN DIJK, 1971) and RASHID (1969) has shown that humic acids in marine sediments also account for a significant fraction of the total exchange capacity. The high chelating or complexing ability of humus is ascribed to the carboxyl and phenolic hydroxyl groups (MORTENSEN, 1963).

GRIFFITH and SCHNITZER (1975) have recently reported the composition of metal–organic complexes isolated from some tropical soils which had number-average molecular weights (see HANSEN and SCHNITZER, 1969) of 250 to 350 and which contained 6.8 to 8.6% by weight metal, mainly Al and Fe with minor amounts of Si, Mg, Ca and Cu. They showed that the Fe and Al were bonded with the negatively charged carboxyl and phenolic hydroxyl groups of the fulvic acids.

A great deal of experimental work on the reaction between metals and humic and fulvic acids from soils and marine and freshwater sediments (for summaries and reviews see MORTENSEN, 1963; MANSKAYA and DROZDOVA, 1968; SCHNITZER, 1971; SIEGEL, 1971; SCHNITZER and KHAN, 1972; PAULI, 1975) has shown that the affinities of humic com-

pounds for metals follow the Irving–Williams order (IRVING and WILLIAMS, 1948), that is
Cu > Ni > Pb > Co > Zn > Fe > Mn. Experiments with humic acids extracted from
marine sediments by RASHID (1974) show the same sequence. Although the humic acid
fraction is present in higher concentration than the fulvic acid faction in soils, the fulvic
acids nevertheless show a greater chelating ability for trace metals (RASHID, 1971; DESAI,
MATHEW and GANGULY, 1972). Complexation and chelation of metals may therefore take
place predominantly during the early stages of the formation of humic compounds in
soils and marine sediments. As suggested by NISSENBAUM and KAPLAN (1972), soluble
organic complexes consisting of cellular degradation products, including amino acids,
carbohydrates and possibly lipids and polycarboxylic acids, polymerize to form insoluble
fulvic acids. A soluble organic complex from the pore water of an anoxic marine sediment
studied by NISSENBAUM, BAEDECKER and KAPLAN (1972) showed some high concentrations
of several trace metals, including Ag, Cu, Ni, Mn, Mo, Pb and Zn. Polymerization of
such material to fulvic acid, which is also chemically reactive, and further diagenesis,
which increases the molecular weight of this material together with some loss of functional
groups, produces humic acids which may represent a relatively stable and less reactive
reservoir of organically bound metals, both in soils and in marine sediments.

A possible supply of metals to sediments in an organically bound form is clearly
represented by the sedimentation of whole or fragmented planktonic organisms, and in
the area investigated here, where primary and secondary production is very high, such a
source may be important. It is well known that many marine organisms concentrate some
trace elements to very large extents (GOLDBERG, 1957) and a large body of information
on the general levels of metals in such organisms is available (VINOGRADOV, 1953;
BOWEN, 1966). While it is clear that much of the older data are suspect, because of the
problem of contamination (BOWEN, 1966) and because of the use of questionable analytical
techniques (BROOKS and RUMSBY, 1965), more recent analyses of bulk plankton and
individual plankters can be used to estimate a probable contribution of several trace
elements to sediments from planktonic sources. Some recently reported representative
contents of the five trace metals considered here in phytoplankton and zooplankton are
shown in Table 5. The most recent determinations appear to yield substantially lower
values and are clearly significantly lower than the concentrations of metals in the bulk
sediments and in the organic fractions of the sediments studied here.

MARTIN (1970) has argued that the concentrations of some trace metals in mixed zooplankton samples collected from 100 m depth in the Caribbean area are significantly higher
than the concentrations of the same metals in surface samples. Although such differences
could have been caused by the presence of different taxa in the samples, which could
have different compositions and/or different migration patterns, MARTIN explained the
results by the adsorption of metals from seawater onto copepod exoskeletons which would
take place to a greater extent during food-dependent lowered moulting rates in the deeper
populations. The enhancement of metal concentrations in crustacean exoskeletons by
adsorption-exchange is supported by the work of PAQUEGNAT, FOWLER and SMALL (1969)
who showed that the Zn contents of the moulted exoskeletons of *Euphausia pacifica* were
14 times higher than in the muscle tissue of the same individuals and that the total Zn
content of marine organisms is considerably more than the actual maximum requirement
of this element.

MARTIN (1970) suggested that the adsorption of metals by exoskeleton material could
play a significant part in biogeochemical recycling in the sea, and if the process were

substantiated, a supply of metal-enriched skeletal debris to sediments accumulating in shallow water could be important. However, more recent analyses reported by LEATHERLAND, BURTON, CULKIN, MCCARTNEY and MORRIS (1973) and MARTIN and KNAUER (1973) show that the metal concentrations of bulk plankton samples and individual zooplankters collected from a wide range of water epths are similar, and fall at the lower end of the range of values given by MARTIN (1970) for surface samples.

TABLE 5

SOME RECENT DETERMINATIONS OF THE TRACE
METAL CONTENTS (PPM DRY WEIGHT BASIS) OF
MARINE PLANKTON

METAL	1	2	3	4	5	6	7	8	9
Cu	238	27	60		5	11	16	10	6
Mo	2				2	2	2	2	2
Ni	29	27	68		2	8	4	2	4
Pb	101	30	47		5	2	2	3	2
Zn		188	417	28–105	24	180	69	113	110

1.   NICHOLLS, CURL and BOWEN (1959). Calamus finmarchicus,
        surface, NW Atlantic. Single analysis.

2.   MARTIN (1970). Mixed zooplankton, mainly Undinula vulgaris,
        surface, Puerto Rico area. Medians of 10 samples.

3.   MARTIN (1970). Mixed zooplankton, mainly Pleuromamma xiphias,
        100 m. depth, Puerto Rico area. Medians of 9 samples.

4.   LEATHERLAND, BURTON, CULKIN, MC CARTNEY and MORRIS (1973).
        Range of 7 analyses of various individual zooplankters
        from 0-1255m depth, NE Atlantic.

5.   MARTIN and KNAUER (1973). Phytoplankton. Median of 28 surface
        samples, NE Pacific.

6.   MARTIN and KNAUER (1973). Mixed zooplankton, 0-100 m. depth,
        NE Pacific. Medians of 9 samples for Ni, Pb and Zn and
        of 14 samples for Cu.

7.   MARTIN and KNAUER (1973). Euphausiids, 0-100m. depth, Monterey
        Bay. Medians of 9 samples.

8.   MARTIN and KNAUER (1973). Copepoda, 0-100m. depth, Monterey
        Bay. Medians of 10 samples.

9.   MARTIN and KNAUER (1973). Radiolaria, 0-100m. depth, Monterey
        Bay. Medians of 6 samples.

Although the metals considered here, apart from Pb, are required for metabolic activities in many organisms (BOWEN, 1966; UNDERWOOD, 1971), the metal content of specific classes of organic compounds in marine organisms is poorly known. LUNDE (1973a) has shown that the lipid fraction of laboratory prepared fish meal contains a small amount of Zn (range 0.54–3.97 ppm; number of samples 6), as well as some other metals, and that the phospholipid component is capable of further complexing contaminant metals during the factory production of fish meal (LUNDE, 1971). This is confirmed by metal contents of the lipid phases of tunny and halibut which are positively correlated with the phosphorus contents, present as phospholipid, in the same samples (LUNDE, 1973b).

On the basis of a simple comparison between the metal concentrations in oceanic plankton, as reported, for example, by MARTIN and KNAUER (1973), and in the Namibian shelf diatomaceous oozes (Tables 2 and 3), it is clearly not possible to account for the metal concentrations in the sediments by direct supplies from bulk planktonic organic material. An enrichment mechanism must therefore be involved. One possibility is by adsorption of metals from seawater by settling seston, as suggested by VOLKOV and FOMINA (1971), a process which may also be included in that discussed by MARTIN (1970). Alternatively, and somewhat more likely, an enrichment mechanism within the sediment, during the diagenetic alteration of organic matter and other sediment components, may operate.

The organic fraction of the sediments on the Namibian shelf contains many intact and only partially degraded components considered to be derived directly from the plankton (MORRIS and CALVERT, 1977). Some of these fractions contain relatively high concentrations of trace metals, notably the lipid fractions (Table 3). While these fractions are not present in sufficient abundance to contribute a significant proportion of the metals to the total sediment, they may nevertheless represent a sediment component with a high complexing or chelating affinity for trace metals (cf. LUNDE, 1971, 1973 a,b), which are probably present in reasonably high concentration in the anoxic pore waters of these sediments. Also present are the high molecular weight soluble and insoluble condensation and polymerization products (NISSENBAUM and KAPLAN, 1972), which also have high complexing affinities (RASHID, 1971). During the progressive enrichment of the sediment organic matter in humic acids during diagenesis, at the expense of the other fractions, a large concentration of trace metals may therefore build up in the insoluble high molecular weight material. This fraction presumably undergoes further diagenesis to form carbon-rich kerogen (NISSENBAUM and KAPLAN, 1972). It is interesting to note that a high concentration of Ni was found in the final residue ('kerogen') of the Nambian shelf sediment studied here, recalling the well-known presence of Ni in crude oil and petroleum (see SAXBY, 1969).

VOLKOV and FOMINA (1971) have recently drawn attention to the possible importance of sulphides, more particularly pyrite, in controlling the trace metal geochemistry of organic-rich marine sediments. They have argued that this effect is important particularly for Mo, which is a trace metal showing the most marked enrichments in such sediments (see CALVERT, 1976). Although the fulvic acid extracts of Black Sea sapropels contain high concentrations of Mo, pyrite in these same sediments also contains up to 915 ppm Mo which can be extracted with $0.1 \, n$ NaOH, the reagent used to extract the humic compounds (VOLKOV and FOMINA, 1971). The ease of extraction of the Mo from the pyrite was explained by VOLKOV and FOMINA by the presence of a thinly dispersed layer of MoS_3 on the surface of the pyrite. Other metals present in the pyrite included Cu (up to 1562 ppm) and Ni (up to 3133 ppm), but these were not extracted by NaOH (see VOLKOV and FOMINA, 1971, table 3). Although not discussed by VOLKOV and FOMINA, this would imply that the Cu and Ni were present in the pyrite lattice, a puzzling implication in view of the similarity of Ni and Fe, but the dissimilarity of Cu and Fe, in ionic sizes (MOHR, 1959). This enrichment mechanism may also apply to some degree to the Namibian shelf sediments, because pyrite is present, as indicated by the salt-corrected S values in Table 2, and in view of the enrichment of Mo in the fulvic acid fractions (Table 3). However, the humic acid fractions contain considerably higher concentrations of Mo, and indeed of the other metals determined, so that these metals are more likely bonded to the humic compounds and do not remain in solution when the humic acid fraction is precipitated (see Table 1).

The intriguing possibility that pyrite contributes significantly to the trace metal content of organic-rich sediments in general is therefore unresolved, and must await further data on the composition of mineral phases in sediments which are formed during diagenetic reactions within sediments, reactions which we have also suggested lead to the enrichment of some organic fractions in these same metals.

CONCLUSIONS

The data reported here have shown significant trace metal contents of the extracted organic fractions which do not, however, account for the total levels of metals in the bulk sediments. The particularly high concentrations of metals in the humic acid fractions are consistent with a substantial body of information on metal contents of this fraction in soils. This confirms earlier suggestions that the organic fraction is responsible to some extent for the minor metal composition of this particular sediment.

The metal concentrations of the organic fractions of the sediment are generally much higher than the most recent analyses of marine planktonic organisms. It is therefore suggested that a secondary concentration mechanism involving the diagenetic alteration of organic and inorganic sediment components and the recycling of metals within the deposited sediments is responsible for the concentrations of metals in the organic fractions. In particular, chemically reactive planktonic components plus low molecular weight humic material is capable of complexing metals in the sediment. Subsequently, during condensation and polymerization reactions, the metals are transferred to the higher molecular weight (humic acid) fraction.

The possibility that pyrite (FeS_2) contributes a substantial proportion of the Mo to some Recent sediments is considered unlikely in the present case, although the problem is unresolved. The data obtained here show that Mo, and other metals, are more highly concentrated in the humic acid fraction compared with the fulvic acid fraction; it is in the latter that one would expect to find high metal concentrations if they are released from pyrite during alkali extraction.

Acknowledgements — We thank P. K. Studdart and C. H. Batchelor for help with the analytical work. The sediment samples were collected during Expedition CIRCE of the Scripps Institution of Oceanography which was supported by grants from the National Science Foundation and contracts of the Office of Naval Research with the University of California and Oregon State University. T. H. van Andel kindly provided ship-time for the collections.

REFERENCES

Anderman G. and J. W. Kemp (1958) Scattered X-rays as internal standards in X-ray emission spectroscopy. *Analytical Chemistry*, **30**, 1306–1309.

Barber R. T. and J. H. Ryther (1969) Organic chelators: factors affecting primary production in the Cromwell Current upwelling. *Journal of Experimental Marine Biology and Ecology*, **3**, 191–199.

Baturin G. N., A. V. Kochenov and K. M. Shimkus (1969) Uranium and rare metals in the sediments of the Black and Mediterranean Seas. *Geokhimiya*, No. 1, pp. 41–50.

Bowen H. J. M. (1966) *Trace elements in biochemistry*, Academic Press, London, 241 pp.

Bowen H. J. M. (1969) Standard materials and intercomparisons. In: *Advances in activation analysis*, J. M. A. Leuiham and S. J. Thomson, editors, Academic Press, London, pp. 101–113.

Brongersma-Sanders M. (1965) Metals of Kupferschiefer supplied by normal seawater. *Geologische Rundschau*, **55**, 365–375.

Brongersma-Sanders M. (1969) Permian wind and the occurrence of fish and metals in the Kupferschiefer and marl slate. In: *Sedimentary ores*, C. J. James, editor, Univ. of Leicester Press, pp. 61–68.

Brooks R. R. and M. G. Rumsby (1965) The biogeochemistry of trace element uptake by some New Zealand bivalves. *Limnology and Oceanography*, **10**, 521–527.

Calvert S. E. (1976) Mineralogy and geochemistry of nearshore sediments. In: *Chemical oceanography*, 2nd edition, Vol. 6, J. P. Riley and R. Chester, editors, Academic Press pp. 187–280.

Calvert S. E. and N. B. Price (1970) Minor metal contents of recent organic-rich sediments off South West Africa. *Nature*, **227**, 593–595.

Calvert S. E. and N. B. Price (1971) Recent sediments of the South West African Shelf. In: *Geology of the East Atlantic Continental Margin*, E. M. Delany, editor, Institute of Geological Sciences Rept. 70/16, pp. 171–185.

COOPER B. S. and R. C. HARRIS (1974) Heavy metals in organic phases of river and estuarine sediment. *Marine Pollution Bulletin* **5**, 24–26.

CURTIS C. D. (1966) The incorporation of soluble organic matter into sediments and its effect on trace element assemblages. In: *Advances in organic geochemistry*, G. D. HOBSON and M. C. LOUIS, editors. Pergamon Press, pp. 1–13.

DESAI M. V. M., E. MATHEW and A. K. GANGULY (1972) Interaction of some metal ions with fulvic acid isolated from marine environment. *Journal of the Marine Biological Association of India*, **14**, 391–394.

HANSEN E. H. and M. SCHNITZER (1969) Molecular weight measurements of polycarboxylic acids in water by vapour pressure osmometry. *Analytica Chimica Acta*, **46**, 247–254.

IRVING H. and R. J. P. WILLIAMS (1948) Order of stability of metal complexes. *Nature*, **162**, 746–747.

GOLDBERG E. D. (1957) Biogeochemistry of trace metals. In: *Treatise on marine ecology and paleoecology*, Vol. 1, J. W. HEDGPETH, editor, Geological Society of America Memoir 67, pp. 345–357.

GOLDSCHMIDT V. M. (1937) The principles of distribution of chemical elements in minerals and rocks. *Journal of the Chemical Society*, pp. 655–673.

GOLDSCHMIDT V. D. (1954) *Geochemistry*, Clarendon Press, Oxford, 730 pp.

GRIFFITH S. M. and M. SCHNITZER (1975) The isolation and characterisation of stable metal–organic complexes from tropical volcanic soils. *Soil Science*, **120**, 126–131.

JOHNSON R. (1964) Seawater, the natural medium of phytoplankton. II. Trace metals and chelation, and general discussion. *Journal of the Marine Biological Association, U.K.*, **44**, 87–109.

KOCHENOV A. V., G. N. BATURIN, S. A. KOVALEVA, X. M. YEMELYANOV and K. M. SHIMKUS (1965) Uranium and organic matter in the sediments of the Black and Mediterranean Seas. *Geokhimiya*, No. 3, pp. 302–313.

KRAUSKOPF K. B. (1955) Sedimenatry deposits of rare metals. *Economic Geology*, 50th Anniv. Vol., Part I, pp. 411–463.

LEATHERLAND T. M., J. D. BURTON, F. CULKIN, M. J. McCARTNEY and R. J. MORRIS (1973) Concentrations of some trace metals in pelagic organisms and of mercury in North-east Atlantic Ocean water. *Deep-Sea Research*, **20**, 679–685.

LITTLE-GADOW S. and A. SCHAFER (1974) Schwermetalle in der Sedimenten der Jade. *Senkenbergiana Maritima*, **6**, 161–174.

LUNDE G. (1971) Activation analysis of trace elements in lipids with emphasis on marine oils. *Journal of the American Oil Chemist's Society*, **48**, 517–522.

LUNDE G. (1973a) Trace metal contents of fish meal and of the lipid phase extracted from fish meal. *Journal of the Science of Food and Agriculture*, **24**, 413–419.

LUNDE G. (1973b) Analysis of trace elements, phosphorus and sulphur in the lipid and non-lipid phase of halibut (*Hippoglossus hippoglossus*) and tunny (*Thunnus thynnus*). *Journal of the Science of Food and Agriculture*, **24**, 1029–1038.

MANSKAYA S. M. and T. V. DROZDOVA (1968) *Geochemistry of organic substances*, Pergamon Press, 345 pp.

MARTIN J. H. (1970) The possible transport of trace metals via moulted copepod exoskeletons. *Limnology and Oceanography*, **15**, 756–761.

MARTIN J. H. and G. A. KNAUER (1973) The elemental composition of plankton. *Geochimica et Cosmochimica Acta*, **37**, 1639–1653.

MOHR P. A. (1959) The distribution of some minor elements between sulphide and silicate phases of sediments. *Contributions Geophysical Observatory University College of Addis Ababa*, A-2, 18 pp.

MORRIS R. J. and S. E. CALVERT (1977) Geochemical studies of organic-rich sediments from the Namibian Shelf. I. The organic fractions. In: *Voyage of Discovery*, M. V. ANGEL, editor, *Deep-Sea Research* supplement to Vol. **24**, 647–665.

MORTENSEN J. L. (1963) Complexing of metals by soil organic matter. *Proceedings of the Soil Science Society of America*, **27**, 179–186.

NICHOLLS G. D., H. CURL and V. T. BOWEN (1959) Spectrographic analyses of marine plankton. *Limnology and Oceanography*, **4**, 472–478.

NISSEMBAUM A. and I. R. KAPLAN (1972) Chemical and isotopic evidence for the *in situ* origin of marine humic substances. *Limnology and Oceanography*, **17**, 570–582.

NISSENBAUM A., M. J. BAEDECKER and I. R. KAPLAN (1972) Studies on dissolved organic matter from interstitial water of a reducing marine fjord. In: *Advances in organic geochemistry 1971*, H. R. VON GAFTNER and H. WEHNER, editors, Pergamon Press, pp. 427–440.

PAQUEGNAT J. E., S. W. FOWLER and L. F. SMALL (1969) Estimates of the zinc requirements of marine organisms. *Journal of the Fisheries Research Board of Canada*, **26**, 145–150.

PAULI F. W. (1975) Heavy metal humates and their behaviour against hydrogen sulfide. *Soil Science*, **119**, 98–105.

PILIPCHUK M. F. (1972) Some problems of the geochemistry of molybdenum in the Mediterranean Sea. *Litologiya i Poleznye Iskopaemye*, No. 2, pp. 25–31 (English translation, pp. 167–173).

PILLAI T. N. V., M. V. M. DESAI, E. MATHEW, S. GANAPATHY and A. K. GANGULY (1971) Organic materials in the marine environment and the associated metallic elements. *Current Science*, **40**, 75–81.

Rashid M. A. (1969) Contribution of humic substances to the cation exchange capacity of different marine sediments. *Maritime Sediments*, **5**, 44–50.

Rashid M. A. (1971) Role of humic acids of marine origin and their different molecular weight fractions in complexing di- and tri-valent metals. *Soil Science*, **111**, 298–305.

Rashid M. A. (1974) Absorption of metals on sedimentary and peat humic acids. *Chemical Geology*, **13**, 115–123.

Saxby J. D. (1969) Metal–organic chemistry of the geochemical cycle. *Reviews of Pure and Applied Chemistry*, **19**, 131–150.

Schnitzer M. (1971) Metal–organic matter interactions in soils and water. In: *Organic compounds in aquatic environments*, S. J. Faust and J. V. Hunter, editors, Marcel Dekker, New York, pp. 297–315.

Schnitzer M. and S. U. Khan (1972) *Humic substances in the environment*, Marcel Dekker, New York, 327 pp.

Siegel A. (1971) Metal–organic interactions in the marine environment. In: *Organic compounds in aquatic environments*, S. J. Faust and J. V. Hunter, editors, Marcel Dekker, New York, pp. 265–295.

Slowey J. F., L. M. Jeffrey and D. W. Hood (1967) Evidence for organic-complexed copper in seawater. *Nature*, **214**, 377–378.

Stumm W. and J. J. Morgan (1970) *Aquatic chemistry*, Wiley—Interscience, New York, 583 pp.

Szalay A. and M. Szilagyi (1969) Accumulation of microelements in peat humic acids and coal. In: *Advances in organic geochemistry*, P. A. Schenck and I. Navenaar, editors, Pergamon Press, London, pp. 567–577.

Underwood E. J. (1971) *Trace elements in human and animal nutrition*, Academic Press, London, 543 pp.

Van Dijk H. (1971) Cation binding of humic acids. *Geoderma*, **5**, 53–67.

Veeh H. H., S. E. Calvert and N. B. Price (1974) Accumulation of uranium in sediments and phosphorites on the South West African Shelf. *Marine Chemistry*, **2**, 189–202.

Vinogradov A. P. (1953) *The elementary chemical composition of marine organisms*, Sears Foundation for Marine Research, Memoir 2, 647 pp.

Vinogradov A. P. (1959) *Geochemistry of rare and dispersed elements in soils*, Chapman & Hall, London.

Volkov I. I. and L. S. Fomina (1971) Dispersed elements in sapropel of the Black Sea and their inter-relationship with organic matter. *Litoligiya i Poleznye Iskopaemye*, No. 6, pp. 3–15 (English translation, pp. 647–656).

Volkov I. I. and L. S. Fomina (1972) The role of iron sulfides in the accumulation of minor elements in Black Sea sediments. *Litologiya i Poleznye Iskopaemye*, No. 2, 18–24 (English translation, pp. 161–166).

Williams P. M. (1969) The association of copper with dissolved organic matter in seawater. *Limnology and Oceanography*, **14**, 156–158.

Studies by the American CLIMAP group considered as a foundation for understanding Quaternary events on the Continental Shelves peripheral to Great Britain and Ireland

L. H. N. COOPER

Associate at the Marine Biological Laboratory, Plymouth

Abstract—Hypotheses for the cold periods of the Quaternary concerned with stationary permanent firnfields in the English Channel and Celtic Sea needed more supporting evidence. This has now been provided by the members of the United States CLIMAP group. However, their method of Q-mode factor analysis was read at first with severe reservations. These have now vanished but only after a critical study of the literature of pelagic foraminiferans and publication of a study of the foraminiferans recovered from the Greenland–Iceland–Norwegian Sea.

It is suggested that these methods developed by the CLIMAP group and still in process of rapid development provide the means for studies in the lands bordering the eastern North Atlantic of first the Devensian (Weichselian, Midlandian) and then of the seven major cold cycles recognized in the Atlantic as having occurred in the last 600,000 years.

INTRODUCTION

GEORGE DEACON is one to whom scientific research has to be fun to the man who does it, otherwise it is unlikely to achieve a quality worth spending money upon. Moreover, to him there could be no worth-while scientific discovery without adventure, but equally he has rigorously curbed speculation unbridled by well-established knowledge. He has attracted to his laboratory a remarkable staff who have shared his love for scientific adventure wisely disciplined. His influence on me came largely during invigorating walks over Dartmoor during which he encouraged publication of ideas in which only he and one other colleague of similar vision could see sense. In this volume to commemorate his seventieth birthday, I am presenting a bridge-building exercise over the shelves peripheral to Great Britain and Ireland to connect deep-sea and land-based Quaternary geology.

A HYPOTHETICAL FIRNFIELD IN THE ENGLISH CHANNEL AND CELTIC SEA

During more than 40 years spent in systematic study of the English Channel and Celtic Sea as they are today, the hypothesis evolved that at times during the Quaternary, when oceanic sea level was lower by 100 m or more, the area was filled with long-standing compacted locally fallen snow or firnfield. Initially one may conceive the bed of the Channel developing as tundra. At times the tundra would have become morass similar to the Vasuigan Swamps of Siberia (58°N, 76°E), well to the south of the contemporary zone of permafrost. In winter these are completely icebound; in spring the ice is first replaced by an inland sea and as this drains dense thickets of cedars, larches and pines may flourish. Winter snowfall upon the Vasuigan Swamps is today exceeded by summer ablation under a continental climate so that no firnfield builds up. But whenever snowfall exceeded ablation in the more oceanic climate of the English Channel, then locally fallen snow should have accumulated to harden into a firnfield. Since this firnfield accumulating on what is now sea-bed had negligible potential energy, it left no geological record. This concept would seem to be identical with the ideas of KELLAWAY, REDDING, SHEPHARD-THORN and DESTOMBES (1975).

Drainage not only from the English Channel firnfield but from southern England, northern France and the watersheds of the Rhine, Meuse and Scheldt river systems had to reach the ocean as seasonal floods, often catastrophic. It would seem to be an axiom that due to the greater altitude of the sun over France than 170 km further north over England, local melt water would have tended to drain south towards the coast of Britanny and to have there collected in the then existing bays. When drainage from the firnfield-marginal lakes was augmented by spates of melt water from northern France or from the Rhine, conditions were well suited for the rapid construction of 'windgaps' or spillways across the necks of headlands separating the embayments.

When later all the ice melted and the sea level rose, these headlands remained as islands and as evidence that very powerful flows of water had cut them off. Cabioch's illustrations of large boulders on the line of this lateral drainage (BOUHOT, VALÉRIAN, PAILLÉ, BOILLOT and CABIOCH, 1965, Fig. 4; CABIOCH, 1972, planche I, 1) provide a measure of the power of these rivers along the southern edge of the firnfield.

The powerful ballmill conditions prevailing during torrential floods should have triturated much rock to sand or even finer material. Except for the special case where 'wind gaps' were cut across limestone promontories, this sand or fine material should contain no appreciable amount of calcium carbonate.

There were probably many variants on the route of flood waters but the sills shown in Table 1 are likely to have been part of a main channel.

Table 1

Sill	Depth in metres
Alderney Race	40
South of Alderney	42
East of Sark	46
Plateau de Barnouic—Cotes du Nord	49
Les Sept Iles—Ile Tome	44
Plateau des Triagos—cliffs near Lannion	59
Passage de Fromveur (Ushant)	51

These are the features believed to have been created during the most catastrophic periods of ablation. Tentatively one may infer that the surface of the adjacent firnfield relative to present sea level then lay considerably above −40 m O.D. abreast of Alderney and considerably above −51 m O.D. at Ushant, also that the slope downwards towards the southwest was of the order 40 mm/km. More detailed study may well reveal that some of the useful coastal navigation channels such as the Chenal de la Bigne (48°41′N, 2°00′W) and Chenal d'Erquy (48°39′N, 2°29′W) were also overspill channels.

Along the southern coast of Devon and Cornwall today there are channels between the present land and the several Mewstones and other offlying islands as far west as the Runnelstone. It is suggested that these mark the course of summer drainage from the watershed north of the English Channel firnfield. On this hypothesis the similar channel at Kettles Bottom, between Lands End and the Longships, belongs to the drainage system not of the English Channel but of lateral drainage from the Bristol Channel.

Since many who are most knowledgeable have seemed unwilling to accept the concept that the English Channel ever contained a considerable firnfield, further evidence was sought from the bed of the Celtic Sea and its continental slope and from the developing literature on cores from the bed of the eastern North Atlantic. At the time of my most active work in those areas (1950–1963), appropriate techniques had not been developed. Difficulties in assessing recent developments are compounded by the need for students of the Quaternary, trained in divers basic disciplines, fully to comprehend minutiae in the taxonomy, physiology, and ecology of the pelagic Foraminifera.

The part played during the Quaternary by the Greenland–Iceland–Shetland Ridge

ARRHENIUS (1952) first recognized the cyclic nature of the dissolution of abyssal calcareous sediments. The greatest rate of dissolution occurs a few thousand years after the termination of a glacial stage. SHACKLETON and OPDYKE (1976) have now shown that this lag has followed every glacial stage throughout the last 1.5 million years. This finding leads to an assured understanding of the changing vigour of vertical oceanic circulations, much better than has been possible by theoretical manipulation of the equilibrium equations relating pressure and temperature, bicarbonate, carbonate and calcium ions to concentrations of $^{12}CO_2$ and $^{14}CO_2$. Subtly small changes in basic assumptions had allowed the writer to come to a spectrum of facile but widely differing conclusions. Selection may now be more surely made.

Forty years ago search for an origin of the North Atlantic Deep Water had become abortive but by the mid-fifties an answer had started to appear (COOPER, 1955, 1956a), viz. that it is not at the surface but at a depth of 500–1000 m where cold water from the Norwegian–Greenland Sea overspills the ridge to become heavily admixed with North Atlantic Central Water. During the 'Little Ice Age' (A.D. 1550–1700 and especially in the last decade of that period; MANLEY, 1952; MASON, 1976) conditions were much more favourable for the formation of North Atlantic Deep Water than they have been since. But during truly glacial periods, when oceanic permanent ice cover extended south of the Ridge, conditions were different again. Though the temperature over a completely ice-bound Arctic Ocean was lower than over the largely open ocean of today or of the Little Ice Age, the rate of exchange of heat through the insulating shield of ice was almost certainly reduced (cf. HUNKINS, 1975). Rapid air–sea exchange of heat and of kinetic energy most favourable for bulk formation of North Atlantic Deep Water was most likely then to occur over ice-free saline water south of the ice edge. Indeed, since summer melting of seasonal ice was likely to produce a seal of brackish and therefore lighter water, the zone for optimum production of Deep Water may well have been mostly south of the limit of seasonal ice. Arguments *ex hypothesi*, such as these, rapidly fragment and require control by observations such as those the CLIMAP group are making.

When using our knowledge of the present to interpret the past the geological youth of the Greenland–Iceland–Scotland ridge must always be in mind. It is essential to assess the sill depths of this ridge during the prolonged cooling down during the late Caenozoic and the changes that these sill depths have undergone during the Quaternary. Such changes would have controlled variations in the nature of the overspill from the Arctic into the North Atlantic which then affected the structure of the other oceans. Today the thermodynamic initiation and control of formation of North Atlantic Deep Water occurs north of the ridge in an area east of Jan Mayen. In some years, but not in all, much colder water with density exceeding σ_t 28.00 is formed there. It is sufficient to build a pond of sufficiently heavy water above the sill levels to maintain the overspill to the North Atlantic (COOPER, 1955, 1956a; WORTHINGTON, 1970). Though water below this depth cannot overspill, it is not inert since it provides a buffer stock of water able to hold the overspill at a constant salinity level which in the Holocene has been around 34.92‰. The essential increase in density by bulk cooling is much assisted by stormy weather at the time of minimum air temperature and is impeded by the presence of floating ice. Conditions which favour equilibration with atmospheric oxygen should also favour equilibration of the ocean with $^{12}CO_2$ and $^{14}CO_2$. The creation of new heavy water in the Greenland–Iceland–Norwegian Sea requires that an equal volume of water must overspill the ridge into the North Atlantic where it mixes vigorously with North Atlantic Central Water having oxygen less than 70% saturated (assessed under atmospheric pressure) and correspondingly enriched with carbon dioxide. The water in the Central Water and some of the carbon dioxide has arrived mainly from below by upward displacement. The carbon dioxide is supplemented by multi-stage decay of faeces and biological detritus raining down from the productive surface waters. These complex processes result in the North Atlantic Deep Water acquiring a considerable apparent radio-carbon age shortly after its synthesis as a recognizably new Atlantic watermass (COOPER, 1956b).

In the early days of the water-supply industry lead pipes were much used and were liable to be attacked and dissolved by the waters which flowed through them. Agents such as humic and fulvic acids strongly supplemented the attack due to carbonic acid. The comprehensive adjective 'aggressive' was then most useful. Similarly at the interface with the ocean bed agents such as 'gelbstoffe' and chelators in minute amounts but supplementing the carbon dioxide may justify the same broadly descriptive adjective 'aggressive'.

During a period of climatic amelioration, a succession of relatively mild winters will not produce a supply of new aggressive water to sink to the sea bed. Given sufficient lapse of time, the abyssal water will become saturated with calcium carbonate at the ambient hydrostatic pressure. Dissolution of calcareous deposits at all depths in the North Atlantic will cease until a period of climatic deterioration reinvigorates the vertical circulation first in the Greenland–Iceland–Norwegian Sea and then, after overspill and vigorous mixing, in the North Atlantic.

The Mediterranean outflow over the sill of the Strait of Gibraltar also contributes by a mixing process to abyssal water but today the process is much less efficient than the Arctic overspill. When the Mediterranean region was more arid than it is today, salinity could have become high enough to carry the warm but dense overspill to abyssal depths in the Atlantic. By contrast, when the Mediterranean climate was more pluvial, the directions of the surface and sill-depth exchanges through the Strait of Gibraltar would have become reversed. Thus the timing of contributions from the Arctic and European Mediterranean Seas may have been opposed. Arguments from the Southern Hemisphere need to take account of the completely different history and topography of land and ocean and of the absence of essential information from the inner Weddell Sea before the very recent arrival there of ice-breaking research ships.

With the disappearance of icefloes and icebergs from the Greenland–Iceland–Norwegian Sea, a few cold and stormy winters could have given rise to much cold, heavy, well-oxygenated and chemically aggressive water able to flush out the ocean abyss. A supply of aggressive water introduced in a few cold winters could then continue to dissolve calcareous deposits for centuries.

Though the flow of geothermal heat upwards through the bed of the ocean is small, it is not trivial. Ultimately on a geological time scale conditions must have evolved some-where in the world for free interchange of abyssal water with the heaviest surface water then existing on the planet. This is a statement for the ocean of the Principle of Le Chatelier that when the vertical circulation is submitted to a stress, the circulation responds in such a way as to neutralize that stress. SHACKLETON and OPDYKE (1976) have demonstrated that the Le Chatelier principle has operated after every glacial stage during the past 1.5 million years. This statement is over-simple since, as with so many applications of the Principle of Le Chatelier, there was considerable hysteresis presumably due to the insu-lation provided by unbroken ice cover; vertical circulation was therefore much damped. There is, however, an inconvenient rider which needs to be checked since during each glacial cycle there should have been two opportunities for efficient ventilation of the ocean abyss with aggressive oxygenated water. This would also have occurred during initial cooling down as cold conditions transgressed towards the equator. However, when this was rapid, stagnation of abyssal water beneath the ice may have come about equally rapidly. A thin layer of stagnating, 'saturated', unaggressive water separated the abyssal bed from a large volume of aggressive water. Only molecular diffusion and geothermal heat were available to create limited vertical circulation close to the abyssal bed. There is no evidence that this ever happened anywhere in the North Atlantic.

The CLIMAP Group

Rapid developments in the quantitative assessment of deep sea cores from the eastern North Atlantic has been due almost entirely to United States laboratories which in 1971 joined forces as the CLIMAP (Climate Long Range Investigation, Mapping and Prediction) project. Fundamental to the advances of the CLIMAP group have been methodological publications by KLOVAN and IMBRIE (1971) and IMBRIE and KIPP (1971). My first hyper-critical reaction to these papers was adverse—that the evidence would not support the deductions drawn. To check this, an assessment was made of the literature on the foraminiferans upon which the Q-mode factor analysis of IMBRIE and KIPP is largely based. By the spring of 1975 my most persistent doubts had vanished. It seemed that others may have had similar difficulty in accepting this revolutionary new method of handling taxonomic evidence quantitatively and so would benefit from presentation of the arguments which have led to this Pauline conversion.

After completion of the draft submission for this volume, it became clear that publication of the CLIMAP report (CLINE and HAYES, 1976) would immediately make this paper obsolescent. I am indebted to R. M. Cline, the co-editor, and Dorothy Merrifield, book publisher for the Society, for making uncorrected page proofs available to me prior to publication. Progress by the CLIMAP group had been even greater than anticipated, necessitating a complete redrafting and withdrawal of some parts of my original manuscript.

There are many cross-references between the authors contributing to the Memoir as undocumented oral communications. Exceptionally speedy and fundamental as has clearly been the informal coordination of the CLIMAP authors, nevertheless, when applying their work in the European environment, this 3-year lag in cross-referencing needs to be always in mind.

My early criticisms of the methodological paper by IMBRIE and KIPP have been completely answered so that the Q-mode factor analysis is seen now as a powerful means for developing late Quaternary studies in the British Isles and Western Europe. There remains no need to present in detail the evidence and arguments which have led to this change of view.

A mathematical method for determining oceanographic parameters from faunal data

When the new micropalaeontological method for quantitative palaeoclimatology was published by IMBRIE and KIPP (1971), it became clear that much critical laboratory experience had already been gained. IMBRIE, VAN DONK and KIPP (1973) and KIPP (1976) have later elucidated a number of obscure points. These arose from uncertainty as to how problems of continuous taxonomic variation and intergrading could be handled by a mathematical method which seemed to require sharp taxonomic boundaries to be drawn. The procedure is best stated by verbatim citation from KIPP (1976) who combined three different starting-points, as follows.

(1) The raw palaeontological data from core tops are factor-analysed into varimax assemblages. Start 1 represents the acquisition of taxonomic data on the sea bed (matrix X_{ct}). These data are presented graphically in accompanying figures as species distribution maps. The line of calculation beginning at Start 1 produces a varimax matrix **B** and an assemblage description matrix **F'**. The columns of the **B** matrix are presented as distribution maps of the assemblages.

(2) A least-squares technique is used to write a set of palaeoecological equations relating the varimax assemblages to observed oceanographic parameters. Start 2 represents the acquisition of oceanographic data. Regression techniques combine these data with the faunal information (matrix **B**) to produce a set of equations **K**.

(3) The fossil data from a core are described in terms of the core-top varimax assemblages. These assemblages are used with the palaeoecological equations to obtain estimates of palaeoenvironments. Start 3 represents the application of the matrices **F'** and **K** to any set of samples (X_c). The final result is a set of environmental estimates (**Y**) for the ocean as it is today.

No paraphrase can get round the need to understand precisely the meanings of each of the technical terms here used. Individuals are rare who combine adequate mastery of the advanced mathematical handling techniques, of the taxonomy, physiology and ecology of the Foraminifera and of the Quaternary sciences to which the results have to be applied. Progress in Europe in applying the methods so successfully used by the CLIMAP group in North America will be rapid only if division of labour is achieved between specialists of similar high quality and ability to work together.

Biological indicators of very cold Quaternary oceanic climates

On the basis of a long core from the Caribbean, PHLEGER (1948) built a prescient working hypothesis for relating foraminiferan faunas to alternation of Pleistocene glacial and interglacial stages and speculated on the possibility of correlating such faunas with North American and European Pleistocene sequences.

Two species of foraminiferans, *Globigerina pachyderma* (Ehrenberg) and *G. quinqueloba* Natland, have proved excellent indicators of cold climates. Specimens of *G. pachyderma* may coil either to the left or to the right (BANDY, 1959; ERICSON, 1959); in chemical terms they are enantiomorphs consequent upon asymmetric biochemical synthesis. PARKER (1971) has stated that "the reason for these coiling-direction changes are not known. In fact we do not know whether we are dealing with races, subspecies, or even different species."

Other palaeobiologists (CIFELLI and SMITH, 1969) have well expressed the problem facing the Quaternary student: "There is not sufficient evidence to date, to prove that the five-chambered form presently found in the high-latitude northern plankton is, in fact, *G. pachyderma*. It is also conceivable that *G. pachyderma* so commonly found in surface sediments at high latitudes, is a recently extinct or almost extinct species. In plankton tows we have found a few specimens closely resembling and possibly being the living representative of *G. pachyderma*. The problem of the taxonomy of the plexus of specimens found in cold waters of the high latitudes exists, but the 'central' form should perhaps be separately designated taxonomically while at the same time recognising that it intergrades with the previously designated and elsewhere distinct taxa."

KIPP (1976) resolved a closely allied taxonomic problem, earlier stated by ERICSON and WOLLIN (1956), by introducing a separate category for 'intergrades' between *Globoquadrina dutertrei* (d'Orbigny) and right-coiling *Globigerina pachyderma* (Ehrenberg) to reduce inconsistencies in species counts made by a large number of workers with varied micropalaeontological experience.

In the northern hemisphere interpretation of the evidence provided by left-coiling *G. pachyderma* as a climatic indicator is made difficult by the complicated oceanic circulation patterns imposed by the irregular distribution of land. Since the circumpolar circulation around Antarctica is much less impeded by land the patterns of distribution of *G. pachyderma* there during ice ages should have been simpler and more symmetrical. KENNETT (1968) and MALGREN and KENNETT (1972) have used this in a diagnostic study on sinistral and dextral *G. pachyderma*. They concluded that in an apparently continuous cline, it grades from a southerly sinistrally coiled, dominantly 4-chambered (umbilical view) form with a greatly thickened test, through an intermediate stage which is dominantly sinistral and $4\frac{1}{2}$-chambered or 5-chambered, to a dominantly 4-chambered dextral form to the north.

The Lyellian principle of studying a presently glaciated Arctic ocean as a key to the past in the North Atlantic applies. Present-day polar circulations have been described by COACHMAN and AAGAARD (1974). The records of planktonic foraminiferans in the water and on the abyssal sea bed in the Norwegian–Iceland–Greenland Sea in the last few hundred years have been examined using Q-mode factor analysis by a member of the CLIMAP group (KELLOGG, 1975, 1976). In six cores selected for detailed study, he found that percentages of 95% or more of the left-coiling form of *G. pachyderma* appear to characterize conditions near or under winter ice, i.e. that left-coiling *G. pachyderma* thrives in water which may be ice-covered in winter but open in summer. Historical records show that Arctic ice-cover may quite recently have been much more extensive than today so that it is possible that this species may also flourish under more permanent ice. Organisms which grow free or permanently attached beneath floe ice may include stages in the life-cycle of left-coiling *G. pachyderma*.

KELLOGG handles the uncertainties in the factor analysis with vigour, not only mathematically but biologically. He (1975, p. 31) writes: "ideally the IMBRIE and KIPP (1971) technique should be used where one is able to extract information from a larga data matrix consisting of numerous samples containing many species. This limitation might be partially overcome in the Norwegian Sea if a way could be found to split the admittedly highly variable morphology of sinistral G. pachyderma in some ecologically significant manner." "This study pushes the temperature scale to the freezing point." "For much of the lower portion of this range ($-1.7°C$ to $4.0°C$) only one assemblage (G. pachyderma sinistral) dominates. Thus the [Q-mode factor] equations are incapable of distinguishing between these temperatures and no real meaning should be assigned to variations estimated within this range."

KELLOGG's discussion goes far to reconciling the temperature of $6.4°C$ in the North Atlantic at $52°35'N$, $21°56'W$ reported by SANCETTA, IMBRIE and KIPP (1973) at a time when the CLIMAP group believe the oceanic ice had reached its southernmost extension. This report had made it hard to accept that the polar front ever reached $43°N$.

If studies of Quaternary history recorded in the sea bed west of the British Isles and France are not to bog down in profitless argument, it is desirable better to understand the complexities surrounding the taxa which include left-coiling G. pachyderma and to extend this understanding to historic (Holocene) time north of the Greenland–Iceland Ridge and especially north of $77°N$.

Observations on the Moving Polar Front in the eastern North Atlantic

Icebergs first appeared in the Labrador Sea during the Pliocene 3 million years ago. The Labrador current (characterized by the appearance of Globigerina pachyderma and Globorotalia inflata) probably first developed at this time. Atlantic cores are likely therefore to provide a basis for a European glacial chronology not only for the whole of the Pleistocene or Quaternary but for the last million years of the Pliocene as well (BERGGREN, 1972). It is essential therefore appropriately to sample not only the bed of the eastern North Atlantic but the continental shelf bordering the British Isles. Many areas are likely to prove unrewarding, since the record may have later been expunged by moving ice in ice shelves or by fans of detritus deposited by turbidity currents. One may hope for survival of late Pliocene and early Pleistocene evidence only in favoured places. The ocean bed should be best sampled well away from submarine canyons, whereas the shelf needs sampling where it is deepest and beneath the range of scouring shelf ice. La Chapelle Bank and the Meriadzek Terrace are amongst some of the more promising sites.

The ash bands

Just as thermometric fixed points such as the freezing-point of sulphur have much helped in developing and applying practically the precision thermodynamic scale of temperature, so will 'ash bands' and similar extensive markers help to establish a precision scale of duration for events during the Devensian (Würm, Weichselian, or Wisconsin). As the work of the CLIMAP group has evolved, its members have been grappling with means of defining the sea–ice margin in terms of evidence deduced from sea-bed cores. There are three regimes: (1) permanent sea–ice cover, (2) semi-permanent sea–ice cover of varying seasonal extents, and (3) no cover. Many critics are unwilling to accept that there was ever much permanent sea–ice cover in the North Atlantic south of the Greenland–Iceland–Shetland Ridge; they would seem to think primarily in terms of drifting icebergs. The arguments pertinent not only to the Late Devensian but to all ice ages are convincingly discussed by MCINTYRE, KIPP, BÉ, CROWLEY, KELLOGG, GARDNER, PRELL and RUDDIMAN (1976, p. 56 *inter alia*).

RUDDIMAN and MCINTYRE (1973, 1976) mapped the distribution of bands of volcanic ash in the eastern North Atlantic. One band closely dated at around 9300 B.P. and another (RUDDIMAN and GLOVER, 1972) less precisely cross-dated at around 67000 B.P. were strongly developed at the coring stations nearest to Ireland. Clearly these ash bands extend towards the European shelf and, hopefully, even on to it. They hold great promise of bridging the gaps in the evidence between the deep seas, the shallow shelves and the several land areas. The 9300 B.P. ash band helps to define the boundary of the polar front which, though locally abrupt in both space and time, may vary regionally in timing by as much as 7000 years along 50°N between Great Britain and Newfoundland or more than 9000 years in a northwest–southwest direction. It provides Time's Arrow as the Late Devensian polar front transgressed the North Atlantic and is matched by good radiocarbon dating, whereas the ash band at around 67000 B.P. cannot be so matched. The intrinsic value of the earlier ash band for correlating events in the Early Devensian is that much the greater.

Regions to the southwest of the British Isles would have first warmed up past a critical temperature. There the early deglacial warming of eastern North Atlantic waters between 13700 and 12500 yr B.P. marked the first northward penetration of sub-polar and temperate waters of the North Atlantic Drift along the glaciated coasts of Europe. There was a very rapid readvance at about 10000 (or 10220?) years B.P. Clearly it is desirable to check this oceanic date with the dates assigned to readvances of the ice in Scotland (e.g. the Loch Lomond?) or Ireland (e.g. the Carlingford?).

On the ocean-floor as on land, the record (RUDDIMAN and MCINTYRE, 1976) becomes more difficult to read as we go back in time. Nevertheless marker bands of ash, turbidite and once-living organisms are enabling many cross correlations to be developed. Seven complete climatic cycles are recognized as having occurred since 600000 B.P. Correlations of each of these with Great Britain and the Continent may need to be brought ashore across Ireland, i.e. information derived from core RC9-225 (54°58.6'N, 15°23.5'W) and others need to be connected by cores from the continental shelf west of Ireland with the Irish sites listed by MITCHELL, COLHOUN, STEPHENS and SYNGE (1973).

The maximum glaciation around 18000 B.P.

RUDDIMAN and MCINTYRE (1973) and MCINTYRE, KIPP, BÉ, CROWLEY, KELLOGG, GARDNER, PRELL and RUDDIMAN (1976) established by Q-mode factor analyses the extent of the glaciation of the North Atlantic about 18000 B.P., a time more or less synonymous with the terms Late Weichselian, Late Devensian or Late Wisconsin. They demonstrated that the polar assemblage of foraminiferans had an extension sweeping southward from a small area off Greenland and Labrador to a dominant position throughout the North Atlantic north of 42°N at this time. Thus for a short time during the Late Devensian the polar front lay more than 700 km south of the edge of the Celtic Sea and continuously from 30000 to 15000 years B.P. it was more than 250 km to the south. This evidence is compatible with the bed of the Celtic Sea having been for 15,000 years covered with either firnfield or with seawater at the freezing-point and carrying a semipermanent cover of ice.

690 L. H. N. COOPER

McINTYRE, KIPP, BÉ, CROWLEY, KELLOGG, GARDNER, PRELL and RUDDIMAN (1976, figs. 16 and 17) have presented two maps which illustrate as anomalies the extent by which sea surface temperatures in August and in February in years near to 18000 B.P. fell below those experienced today. Whereas in some areas the winter and summer anomalies differed from each other by as much as 10°C, in the eastern North Atlantic in latitudes abreast of the Celtic Sea and Bay of Biscay the anomalies in the two months were the same, viz. 10°C. The lands bordering these seas are today considered as textbook examples of equable maritime climate with considerable, dependable year-round precipitation; 18,000 years ago, near to open ocean, the climate should have been similarly equable but much more of the precipitation would have been as snow. But by contrast with today, the polar front then lay between the Celtic Sea and the zone of intense evaporation in the trade wind zone only about 2°C colder than now. A simple thermodynamic heat engine operating between a heat source in 30°N and a heat sink in 50°N should have been capable of maximum work about 80% greater than today. Nevertheless due to the thermostatic control exerted by ice and water at around the freezing-point, this higher capacity for work will have resulted not in wide fluctuations of temperature, but in rapid short-term alternations of firnfield, meltwater and seawater. Fluctuations in the invasions of seawater beneath a floating ice shelf of rapidly varying thickness open up many interesting possibilities of the controls over the periglacial and glacial climates of southern Britain and the peripheral shallow seas.

The CLIMAP investigators have also been able to add a Q-factor analysis for data from 100 m depth and clearly it is only a matter of time before deeper oceanic water, inhabited by some foraminiferans such as *Globorotalia scitula*, becomes similarly treated. This information will be highly relevant for studies of the Quaternary history of the British shelf seas.

Globorotalia scitula

KIPP's (1976, Fig. 12) illustration of the relative abundance of *Globorotalia scitula* (Brady) on the North Atlantic sea bed calls to mind the present-day distribution of Mediterranean water in the Atlantic. Even in the Strait of Gibraltar this is never at the surface and in the eastern North Atlantic it is mostly deeper than 1000 m. KIPP (1976, p. 22) described the species as deep-living and in the Atlantic not found in surface tows. If today it is endemic in any part of the western or eastern Mediterranean or the northern Adriatic Sea, variations in its distribution during the Quaternary could increase our understanding of reversals of flow through the Strait of Gibraltar. To the writer it has always seemed likely that the northern Adriatic was a sea area well suited for ice-age 'over-wintering' of zooplankton now adjusted for success in the English Channel and Celtic Sea. This hypothesis has been difficult to sustain unless one also assumes that there have been several reversals of the direction of flow of the exchange currents in the Strait of Gibraltar and that there has been a considerable seismic uplift of what is now the bed of the northern Adriatic. A study on the lines of the CLIMAP study might provide an assured answer.

SUMMARY

A hypothesis is stated that during periods of low sea level during the Quaternary, the English Channel and Celtic sea carried a substantial, almost stationary firnfield built from locally fallen snow. This firnfield held trifling potential energy and so left no geological record. The islands which fringe the north coast of France may owe their origin to over-spilling spates of summer meltwater from the English Channel firnfield and from the snowfields of northwestern Europe. These eroded 'windgaps' short-circuited headlands separating marginal seasonal freshwater lakes or embayments.

An associated hypothesis concerns the effect on the oceanography of the North Atlantic and the adjacent shallow seas of the formation of the ridge connecting Greenland through Iceland and the Faeroes to Scotland during the later Cainozoic and especially during the Quaternary.

The necessary basic oceanic substructure for these and other hypotheses is now being provided by the American CLIMAP project, initiated in 1971 by a group of North American laboratories. My assessment first based on publications by individuals now in the group has been overtaken by the publication in August 1976 of a Geological Society of America Memoir (CLINE and HAYES, 1976) of which an uncorrected proof copy has been most kindly provided by the Society. The paper first submitted for this volume has been drastically rewritten in the light of this uncorrected proof. The hypothesis of an extensive firnfield on the sea bed south of Great Britain and Ireland receives strong support.

The mathematical equations used for the varimax analysis are set out by verbatim citation. While some temporary limitations for North Atlantic studies may be assessed from work undertaken in the Greenland–Iceland–Norwegian Sea, volcanic ash bands are being developed as precision time markers for the ocean bed and hold promise beneath shallow seas.

The CLIMAP deduction that about 18000 B.P. the polar front reached as far south as 42°N lat., the latitude of northwestern Spain and of Massachusetts, supports the hypothesis that at that time more or less permanent but more or less stationary firnfields could have developed in the English Channel and northern Celtic Sea and that a floating ice shelf was possible in the southern Celtic Sea.

The distribution of *Globorotalia scitula* (Brady) on the bed of the North Atlantic suggests that this species may have value for assessing variations in the exchanges of water which may have occurred in the Strait of Gibraltar.

REFERENCES

ARRHENIUS G. (1952) Sediment cores from the East Pacific. *Report of the Swedish Deep Sea Expedition*, **5**, Göteborg.

BANDY C. L. (1959) Geologic significance of coiling ratios in the foraminifer *Globigerina pachyderma* (Ehrenberg). *Geological Society of America Bulletin*, **70**, No. 12, pt. 2, p. 1708 (abstract).

BERGGREN W. A. (1972) Cenozoic biostratigraphy and paleobiogeography of the North Atlantic, in Laughton, Berggren *et al.*, 1972. *Initial Reports on a Deep Sea Drilling Project*, **12**, 965–1001.

BOUHOT G., J. VALÉRIEN, A. PAILLÉ, G. BOILLOT and L. CABIOCH (1965) Essai d'une caméra legère de télévision sous-marine dans la Manche Occidentale. *Research Film*, **5**, No. 4, 320–330.

CABIOCH L. (1972) La station biologique de Roscoff et son rôle dans l'exploration des fonds de la Manche et de l'Atlantique. *Cahiers de Biologie marine*, **13**, 589–595.

CIFELLI R. and R. K. SMITH (1969) Problems in the distribution of recent planktonic Foraminifera and their relationships with water mass boundaries in the North Atlantic. In: *Proceedings of the First International Conference of Planktonic Microfossils, Geneva 1967*, **2**, 68–81.

CLINE R. M. and J. D. HAYS, editors (1976) Investigation of late Quaternary Paleoceanography and Paleo-climatology. *Geological Society of America, Memoir*, **145**, pp. 1–464. (Seen only as uncorrected proof and without appendices.)

COACHMAN L. K. and K. AAGAARD (1974) Physical oceanography of Arctic and subarctic Seas. In: *Marine geology and oceanography of the Arctic Seas*, Y. HERMAN, editor, Springer-Verlag, pp. 1–72.

COOPER L. H. N. (1955) Deep water movements in the North Atlantic as a link between climatic changes around Iceland and the biological productivity of the English Channel and Celtic Sea. *Journal of Marine Research*, **14**, 347–362.

COOPER L. H. N. (1956a) Hypotheses connecting fluctuations in Arctic climate with biological productivity in the English Channel. *Papers on Marine Biology and Oceanography, Deep Sea Research*, Suppl. to Vol. **3**, pp. 212–223.

COOPER L. H. N. (1956b) On assessing the age of deep oceanic water by carbon-14. *Journal of the Marine Biological Association of the United Kingdom*, **35**, 341–354.

ERICSON D. B. (1959) Coiling direction of *Globigerina pachyderma* as a climatic index. *Science, New York*, **130**, 219–220.

ERICSON D. B. and G. WOLLIN (1959) Correlation of six cores from the equatorial Atlantic and the Caribbean. *Deep-Sea Research*, **3**, 104–125.

HUNKINS K. (1975) The oceanic boundary layer and stress beneath a drifting ice floe. *Journal of Geophysical Research*, **80**, 3425–3433.

IMBRIE J. and N. G. KIPP (1971) A new micropaleontological method for quantitative paleoclimatology: application to a late Pleistocene Caribbean core. In: *The late Cenozoic glacial ages*, K. K. TUREKIAN, editor, Yale Univ. Press, pp. 71–181.

IMBRIE J., J. VAN DONK and N. G. KIPP (1973) Paleoclimatic investigation of a late Pleistocene Caribbean deep-sea core: Comparison of isotonic and faunal methods. *Quaternary Research*, **3**, 10–38.

KELLAWAY G. A., J. H. REDDING, E. R. SHEPHARD-THORN and J.-P. DESTOMBES (1975) The Quaternary history of the English Channel. *Philosophical Transactions of the Royal Society, London*, A, **279**, 189–218.

KELLOG T. B. (1975) Late Quaternary climatic changes in the Norwegian and Greenland Seas. In: *Climate of the Arctic. 24th Alaska Science Conference, Fairbanks, Alaska. August 15 to 17, 1973*, G. WELLER and S. A. BOWLING, editors, Fairbanks, Alaska, University of Alaska, Geophysical Institute, pp. 3–36.

KELLOGG T. B. (1976) Late Quaternary climatic changes: evidence from deep-sea cores of Norwegian and Greenland Seas. *Geological Society of America, Memoir*, **145**, 77–110.

KENNETT J. P. (1968) Latitudinal variation in *Globigerina pachyderma* (Ehrenberg) in surface sediments of the southwest Pacific Ocean. *Micropaleontology*, **14**(3), 305–318, pl. 1.

KIPP N. G. (1976) New transfer function for estimating past sea-surface conditions from sea-bed distribution of planktonic foraminiferal assemblages in the North Atlantic. *Geological Society of America, Memoir*, **145**, 3–41.

KLOVAN J. E. and J. IMBRIE (1971) An algorithm and Fortran IV program for large scale Q-mode factor analysis. *Journal of the International Association of Mathematical Geology*, **3**(1).

McINTYRE A., N. G. KIPP, A. W. H. BÉ, T. CROWLEY, T. KELLOGG, J. V. GARDNER, W. PRELL and W. F. RUDDIMAN (1976) Glacial North Atlantic 18,000 years ago: a CLIMAP reconstruction. *Geological Society of America, Memoir*, **145**, pp. 43–76.

MALGREN B. and J. P. KENNETT (1972) Biometric analysis of phenotypic variation: *Globigerina pachyderma* (Ehrenberg) in the South Pacific Ocean. *Micropaleontology*, **18**(2), 241–248.

MANLEY G. (1952) *Climate and the British scene*, Fontana Library, 382 pp.

MASON B. J. (1976) Towards the understanding and prediction of climatic variations. *Quarterly Journal of the Royal Meteorological Society*, **102**, 473–498.

MITCHELL G. F., E. A. COLHOUN, N. STEPHENS and F. M. SYNGE (1973) Ireland. In: *A Correlation of Quaternary deposits in the British Isles*, G. F. MITCHELL, L. F. PENNY, F. W. SHOTTON and R. G. WEST, editors, Geological Society of London, Special Report, **4**, 99 pp.

PARKER F. L. (1971) Distribution of planktonic Foraminifera in recent deep-sea sediments. In: *The micropalaeontology of oceans*, B. M. FUNNELL and W. R. RIEDEL, editors, Cambridge University Press, pp. 289–307.

PHLEGER F. B. JR. (1948) Foraminifera of a submarine core from the Caribbean Sea. *Göteborgs K. Vetenskap-och Vitterhetssamhälles Handlingar*, VI, Ser. B, **5**, No. 14, 1–9.

RUDDIMAN W. F. and L. K. GLOVER (1972) Vertical mixing of ice-rafted volcanic ash in North Atlantic sediments. *Geological Society of America Bulletin*, **83**, 2817–2836.

RUDDIMAN W. F. and L. K. GLOVER (1975) Subpolar North Atlantic circulation at 9300 yr B.P.: faunal evidence. *Quaternary Research*, **5**, 361–389.

RUDDIMAN W. F. and A. McINTYRE (1973) Time-transgressive deglacial retreat of polar waters from the North Atlantic. *Quaternary Research*, **3**, 117–130.

RUDDIMAN W. F. and A. McINTYRE (1976) Northeast Atlantic paleoclimatic changes over the past 600,000 years. *Geological Society of America, Memoir*, **145**, 111–146.

SANCETTA C., J. IMBRIE and N. G. KIPP (1973) Climatic record of the past 130,000 years in North Atlantic deep-sea core V23-82: correlation with the terrestrial record. *Quaternary Research*, **3**, 110–116.

SHACKLETON N. J. and N. D. OPDYKE (1976) Oxygen-isotope and paleomagnetic stratigraphy of Pacific Core V28-239: Late Pliocene to Latest Pleistocene. *Geological Society of America, Memoir*, **145**, pp. 449–464.

WORTHINGTON L. V. (1970) The Norwegian Sea as a Mediterranean basin. *Deep-Sea Research*, **17**, 77–84.

Index